BELL, PAUL

Environmental Psychology

5th

Hove: Lawrence Erlbaum   0805860886

ENVIRONMENTAL PSYCHOLOGY

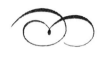

FIFTH EDITION

Environmental Psychology

FIFTH EDITION

Paul A. Bell
Thomas C. Greene
Jeffery D. Fisher
Andrew Baum

LAWRENCE ERLBAUM ASSOCIATES, PUBLISHERS
Mahwah, New Jersey   London

Publisher	Earl McPeek
Acquisitions Editor	Brad Potthoff
Market Strategist	Katie Matthews
Developmental Editor	Peggy Howell
Project Editor	Michele Tomiak
Art Director	Susan Journey
Production Manager	James McDonald

Cover photo: Copyright © 2001 PhotoDisc

ISBN: 0-15-508064-4
Library of Congress Catalog Card Number: 00-105961

Address for Domestic Orders
Harcourt College Publishers, 6277 Sea Harbor Drive, Orlando, FL 32887-6777
800-782-4479

Address for International Orders
International Customer Service
Harcourt, Inc., 6277 Sea Harbor Drive, Orlando, FL 32887-6777
407-345-3800
(fax) 407-345-4060
(e-mail) hbintl@harcourt.com

Address for Editorial Correspondence
Harcourt College Publishers, 301 Commerce Street, Suite 3700, Fort Worth, TX 76102

Web Site Address
http://www.harcourtcollege.com

Harcourt College Publishers will provide complimentary supplements or supplement packages to those adopters qualified under our adoption policy. Please contact your sales representative to learn how you qualify. If as an adopter or potential user you receive supplements you do not need, please return them to your sales representative or send them to: Attn: Returns Department, Troy Warehouse, 465 South Lincoln Drive, Troy, MO 63379.

Printed in the United States of America

3 4 5 6 7 8 9   039   9 8 7 6 5

Harcourt College Publishers

PREFACE

Environmental psychology has grown and expanded over the past four decades. We are a product of our environment, our biology, and the interaction of the two. Individual components of the physical environment impact us continually, such as with noise, weather, and confined spaces. More interesting for our field is the impact of larger-scale environments—workplaces, neighborhoods, cities. At the same time, we modify our environment more than any other creature. We can modify it for our comfort through careful implementation of design principles (including aesthetics, engineering, and accommodation of cultural needs), or we can modify it with disregard for the impact we have on other people and on the entire ecological system upon which we are dependent. This book is about all of the above. It is about how our biological nature and our proclivity to process information continually lead us to interact with the environment in ways that can be harmonious with or destructive to our greater ecological system.

The way we approach the discipline is grounded in empirical science. We have a growing database of laboratory and field research, with enormous samples and individual case studies, and interdisciplinary and international cooperation; this enables us to integrate theory, research, and application through a wealth of scientific information. The discipline is increasingly applied, dealing with everyday environmental problems and practical solutions, yet it remains fundamentally grounded in scientific research and theory.

FEATURES

This fifth edition of our textbook maintains the focus on application of science and theory to the solution of problems involving the natural and built environments. It is updated with very recent research yet grounded in historical per-spectives. We emphasize the relevance of issues for the past, present, and future. We tie the diverse material together with an eclectic environment–behavior model. In every chapter we deal with micro-environmental influence and macro-environmental influence, including short-term and long-term effects.

The illustration program, so important to this book, has been carefully redone, and we have selected new photographs that help students better understand their environment. Throughout the text, topics are well integrated with multiple cross-references to other chapters. The key terms appear at the end of the chapter, and boxes focus on interesting studies or events related to each chapter.

CHAPTER TOPICS

Our **introductory chapter** views the characteristics that distinguish environmental psychology from other fields, provides the relevant historical context, and summarizes the research methods central to the discipline. **Chapter Two** lays the groundwork of the biological human being and our relationship with nature, including biophilia and biophobia. There we introduce environmental attitudes, values, and ethics, and the principles that apply to environmental assessment. Here we also introduce the concept of placemaking and attachment to place. **Chapter Three** introduces models of environmental perception and cognition. How we process information about natural and built environments, how we store environmental information, and how we negotiate our way through environments involves some very basic as well as applied principles. **Chapter Four** reviews the major theories of human–environment interaction, such as overload, behavior constraint, stress, and Barker's ecological psychology. **Chapter Five** explores sound and noise and the ways noise can impact us in our schools, our homes, and our workplaces. Though it is alarming to note the effects noise has on children's classrooms, it is reassuring to find that abatement procedures can be effective when properly implemented, some even in the design phase of classroom creation. In **Chapter Six** we explore weather variables and their impact on us, including El Niño and global warming, heat and violence effects, and seasonal influences on mood as well as the question of lunar effects on mental health. In **Chapter Seven** we examine disturbed environments on a large scale, including natural disasters, technological catastrophes, and pollution effects. An added feature of this chapter is the framing of pollution issues in terms of environmental justice. Use of space is explored in **Chapters Eight** and **Nine** on personal space, territoriality, and crowding. These chapters cover classic and more recent work on nonhuman animal crowding, privacy regulation, and effects of design alternatives on spatial effects. **Chapter Ten** covers the large-scale environment of the city, including material on commuter stress and road rage, designs for crime prevention, amelioration of city problems through environmental interventions, and restoring natural areas in cities. In **Chapter Eleven** we elaborate on the design process and specific architectural features, with a major section on how the design of college campuses illustrates the relevant principles. **Chapter Twelve** examines environmental psychology applied to residences (homes, neighborhoods) and institutional settings (hospitals, prisons, nursing homes). **Chapter Thirteen** applies environmental psychology to work environments, learning environments (classrooms, museums), and leisure settings. This approach provides a comparison and contrast of implementing designs in built and natural environments. In **Chapter Fourteen** we conclude by reviewing how psychological principles can be used to preserve and restore the environments upon which we all depend.

As with previous editions, we conceived *Environmental Psychology* for use in environment and behavior courses (environmental psychology, social ecology, architectural psychology, ecological psychology, environmental design, and the like). It can also be used as a supplement in more specialized courses.

We continue to be excited about this field and are pleased to hear others express similar sentiment. As in the past, we invite your feedback on ways to improve our product.

New to This Edition

We have opened a **Web site** to refer readers to other relevant Web sites. New developments happen so fast that it is impossible to update the textbook itself in as timely a manner as we would like, but the Web site gives us an opportunity to present contrasting perspectives on new and old developments. We invite readers to send us other Web sites they believe should be listed.

The fifth edition of *Environmental Psychology* has an international perspective, and environmental occurrences and their effects are screened through that view of a larger world. This edition also has more on technological interventions. Many references are new or updated. Nearly 100 new photographs have been added, replacing most of those seen in previous editions.

We have retained the **glossary** of important terms, and now these terms are page referenced in order to help students whose background is only tangential to the field. We have tried to restrict the list of terms to those that are introduced for the first time in a specific chapter. We have used italics rather than boldface for terms that are used only once in the book, reserving the glossary for terms that occur multiple times.

New to the fifth edition are **computerized test banks** for Mac and Windows. The traditional **Test Bank/Instructor's Manual** in print is also available.

Acknowledgments

As with the previous edition, the fifth edition is a joint effort of four authors whose time commitments shift with career demands. We again thank Ross Loomis for his material from the first edition. No undertaking of this magnitude happens through the efforts of only the authors, so we would like to acknowledge the assistance of those who worked diligently to make it happen.

Our individual assistants have included Christopher Kulik, Heather Ferguson, Jessica Manchester, Crystal Martin, Sara Sam, Anna Warden, Stephen Arnold, Tracy Rudney, Beth Visnich, Lori McBurney, and Michele Hayward.

Special thanks for assistance with photograph goes to Helen Cooney, Dawn Nannini, Steve Elias, Yvonne Myers, Delores and Lew Scarbrough, Bob Swerer, Frank Vattano, Harleen Alexander, John Buffington, Louise McDonald, Stan and Keri Jurgens, and Anna Warden. Some of the figures were drawn in collaboration with David Abraham, Christopher Kulik, Jessica Manchester, and Anna Warden.

Special thanks also go to the editorial and production staff at Harcourt, including Peggy Howell, Developmental Editor, Michele Tomiak, Project Editor, Susan Journey, Art Director, and James McDonald, Production Manager.

Our thoughtful reviewers Jim Rotton, Florida International University, Julia Gover Hall, Drexel University, Lee A. Jackson, University of North Carolina–Wilmington, Robert Bechtel, University of Arizona, Richard Ryckman, University of Maine, Gary Evans, Cornell University, Lloyd K. Stires, Indiana State University, George I. Whitehead, Salisbury State University, and Anna Warden have provided most helpful direction, and we thank you.

We also wish to acknowledge our deep debt to our students over the lifetime of this book. They are too numerous to name individually, but their support, critiques, questions, and enthusiasm have challenged and enriched both the authors and the text.

And as every author knows, spouses tolerate a writing schedule that is not easy to accommodate. Patty, Mela, Allison, and Carrie have stood by us yet one more time as we put things aside to work on "the manuscript." Thanks also to our children, Breanna, Benjamin, Andy, Aaron, Molly, Jesse, and Callie, who have been (mostly) patient through the project, and to Guy and Doll (Figure 12-2), who provided diversionary entertainment each day!

CONTENTS

CHAPTER 4   THEORIES OF ENVIRONMENT–BEHAVIOR RELATIONSHIPS   97

CHAPTER 7   DISASTERS, TOXIC HAZARDS, AND POLLUTION   205

CHAPTER 10 THE CITY   333

ENVIRONMENTAL PSYCHOLOGY

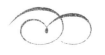

FIFTH EDITION

CHAPTER 1

THE WHY, WHAT, AND HOW OF ENVIRONMENTAL PSYCHOLOGY

Why Study Environmental Psychology?

In a play, the stage and scenery provide the context of what is going on: the kind of room the characters are in, the way it is decorated, and the amount and nature of its furnishings help us to interpret what is happening. They provide meaning for the actors' and actresses' actions and determine where they can walk, lean, or otherwise interact with props. For the play, the stage and scenery are the **environment** in which the story unfolds. The meaning of behavior on the stage and what can and cannot be done are enabled and limited by this environment. The theater would be far less entertaining or educational without the context provided by its environment.

In real life, our behavior also occurs in the context of an environment, one that is constantly changing and rich in information. Unlike the setting on a stage, however, it provides more than meaning. Our environment also provides us with basic needs for life, including food, water, and air to breathe. It is also modified by our actions, and is irrevocably altered whenever one of us changes it. Our environment includes all of our natural and built surroundings, and is a delicately balanced system that can easily be bruised or damaged. Whenever we change some part of it, other parts also change, and these other changes may be unintended or even dangerous. Concerns about what we were doing to our environment reached unprecedented prominence in the 1960s and continue to be an issue of serious concern into the 21st century as consequences of years of neglect have become apparent. The depletion of the ozone layer of the atmosphere, global warming and increasingly destructive storms, massive oil spills in coastal waters, choked transportation corridors and other urban ills, and endangering of other species because of human exploitation of resources reflect these problems. In truth, we have done much to remedy the damage we have done to the environment and have taken steps to pre-

vent some new problems. Yet, the pollution of our air and water, increasing energy use, crowding, noise, toxic accidents, and other environmental problems continue (see Figures 1-1 through 1-5). What more can be done to deal with the situation?

In a sense, environmental psychology deals with "environment" at two different levels. On the one hand, environmental psychology is concerned with environments as the context of behavior: Our moods and behaviors are meaningful only if they can be understood in terms of their context. In this sense, the environment determines which behaviors are possible, how difficult or successful they may be, and so on. You cannot sit unless there is a chair, and cannot walk where a wall blocks your access. These environmental features make some actions possible and others difficult or impossible. Such **affordances** are possibilities allowed or provided by an environment and are strong determinants of behavior. In the classic *Gestalt* insight studies by Kohler (1970), the solution to an ape's dilemma was provided by the environment: The ape found himself in a cage, with bananas hanging overhead and out of reach, and few if any resources available. The environment did not provide any way to jump high enough to get them, but provided tools (e.g., pieces of sticks that could be put together and used to knock the bananas down). The insight of which Kohler wrote was the recognition of a solution to the problem (how to get the bananas) that was afforded by the environment. Similar affordances are reflected in the building materials for homes (usually what is plentiful in the area, such as brick or wood) or in the pathways worn across lawns where people choose to walk.

Environmental psychology, then, is very concerned with the environment as a determinant or influence on behavior and mood. As we will discuss later in this chapter, environmental influences on mood and behavior are pervasive

Figure 1-1 Increasing evidence indicates that the civilization of the Anasazi ancestors of today's Navajo and Hopi peoples disappeared 800 years ago because of a drastic climate change and overuse of natural resources. Other civilizations may have met a similar fate for the same reasons. Since the industrial revolution, Earth's surface has warmed considerably, and many scientists attribute this global warming to atmospheric accumulation of waste gases from burning fossil fuels, as well as to human modification of vegetation over large regions. Could we be headed for the same fate as the Anasazi, where the ruins of our civilization will become an intriguing tourist attraction?

Figure 1-2 Researchers have established that the availability of informal spaces for socializing with neighbors helps build a sense of community and is associated with lower crime rates. After decades of emphasizing the privacy of a fenced back yard or patio, designers are now reintroducing front porches with sitting areas in order to encourage residents to socialize informally with neighbors. Some communities are even adding front porches to building codes.

Figure 1-3A & B These two offices are almost the same size and have similar furnishings. The major noticeable difference is that one is on an interior corridor with no window and the other has a window with a pleasant view of green vegetation. Research indicates that sunlight entering an office is associated with higher job satisfaction and lower intentions to resign from the job.

Figure 1-4 Road rage is a growing concern, especially in urban areas where traffic congestion is high and stress from commuting on freeways is well documented. Recent research suggests that driving a route through nature-dominated scenery both reduces stress and helps prevent heightened reactions to future stressful events.

Figure 1-5 Programs to encourage alternative modes of transportation can be effective if carefully planned. Mass transit and bicycling help prevent traffic congestion and reduce air pollution. Research shows that college students and others place a very high value on bike paths, clean air and water, and efforts to preserve wildlife and other natural resources.

and important. In this way, our field of study is similar to social psychology or developmental psychology. The environment provides meaning and affects behavior just as do social settings or age or developmental stages. One can act silly in certain settings (at home in your room) or at specific ages without fear of embarrassment, but there are combinations of developmental, social, and environmental conditions (such as during an oral examination or at a funeral or wedding) in which silliness is inappropriate or out of place.

At a second level, environmental psychology is also concerned with the *consequences* of behavior on the environment, and more broadly, with larger scale environmental problems such as pollution, recycling, and ecosystem issues. This is a very different focus, though it follows from the basic premise that behavior and the environment mutually affect each other. So how does behavior affect the environment?

Environmental psychology has evolved, in part, to provide some answers to this question. Some environmental problems are heavily influenced by human action. Air pollution, a common problem in large cities, is one of these

problems, and because human behavior causes it, it seems plausible that modification of behavior will offer one of the best ways to curb or eliminate it. Principles of learning, motivation, perception, attitude formation, and social interaction help explain why we ever engaged in and accepted polluting behavior in the first place. Principles of developmental psychology, social psychology, abnormal psychology, and physiological psychology help explain the deleterious effects of pollution on humans. Furthermore, research on attitude change, behavior modification, social behavior, and personality can suggest some steps that will be necessary to change behavior in order to reduce pollution.

Environmental psychologists incorporate both of these levels of environmental influence—how environments influence people and how people influence environments—in studying how we interact with specific environments. In the past, building design was primarily concerned with structural integrity and appearance. Designers now also consider how buildings affect the people using them. Thus, we want to know both how the design influences people and how we can modify the design to facilitate the function for which the setting is intended. Principles of crowding, privacy, personal space, and environmental perception, as well as noise, temperature, air circulation, and cost may all be factors in how a building is designed and how well it serves its intended function. College residence halls may now be designed to accommodate the social needs of students as well as more traditional concerns such as cheap housing for large numbers of residents and control of their noise. Housing projects, racked by major failures years ago, may now include a range of behavioral criteria in design and construction (e.g., promotion of a sense of security and community, accommodation of disabilities).

In the process of suggesting possible solutions for environmental problems, psychologists are gaining considerable practical knowledge about relationships between behavior and environment as well as gaining invaluable information about conceptual or theoretical models of human behavior. For this reason, environmental psychology not only is practical but also provides a meaningful focus of traditional psychological disciplines. The tendency to depict environmental psychology as an applied field is partly due to the fact that much of what the field studies involves opportunities to improve management of our surroundings.

However, research in a number of areas, as well as the development of theories to describe behavior across different situations (see Chapter 4), reflects the fact that environmental psychology is also concerned with building basic knowledge of human behavior and how it interacts with the environment. In this chapter, we will talk about many aspects of environmental psychology. First, we must define our field and consider its characteristic approach and basic assumptions. Environmental psychology is distinctive in many ways, and these factors shape the methods that can be used to study environment–behavior relationships. Hence, we will also discuss methodological issues: How does one go about studying the processes and problems encompassed by environmental psychology? Many obstacles to reliable study of these phenomena are difficult to overcome, and designing and adapting procedures and measures to do so is one of the field's evolving challenges. Finally, we will discuss some ethical issues associated with the field and preview the content areas to be discussed in the rest of the book.

What Is Environmental Psychology?

DEFINITIONS OF ENVIRONMENTAL PSYCHOLOGY

The preceding discussion of environmental psychology should convince you that the field offers present-day relevance for the discipline of psychology as well as the exciting possibility of a unique perspective on environmental problems. Yet, this does not really *define* "environmental psychology." Like most areas of psychology, it is easier to list what environmental psychologists do than to define the field, and we certainly could define the field as "what environmental psychologists do" (Proshansky, Ittelson, & Rivlin, 1970).

Over the past few decades, experts defining the field have attempted to emphasize the influence of the environment on people as well as the influence of people on the environment. They have also emphasized these influences in built and, more recently, natural environments, with a focus on the larger scale rather than on specific individual components of environments (for further discussion, see Sime, 1999; Stokols & Altman, 1987). Altogether, we think this definition is appropriately brief and encompassing:

Environmental psychology is the study of the molar relationships between behavior and experience and the built and natural environments. To elaborate on the implications of this definition, we will next describe certain characteristics of environmental psychology that make the field unique and further delimit its scope.

CHARACTERISTICS OF ENVIRONMENTAL PSYCHOLOGY

The two primary distinctions between environmental psychology and other fields of psychology are (1) the perspective it takes in studying its subject matter and (2) the kinds of problems or settings that are selected for study. The perspective that environmental psychology takes is that since environmental effects on behavior are important, much of our research should involve naturalistic studies of behavior in the built and natural environments. That means the settings chosen for study are likely to be outside the laboratory (though not always), and the problems studied are likely to be how we do or do not adjust to the normal and disturbed features of those settings. We shall describe some characteristics of this perspective, drawing on several historic sources (e.g., Proshansky, 1976; Wohlwill, 1970). Our list of characteristics of environmental psychology is by no means exhaustive but simply reflects the unique perspective of the field.

First and foremost among these characteristics is an emphasis on studying environment–behavior relationships as a unit, rather than separating them into supposedly distinct and self-contained components. Traditional approaches to the study of sensation and perception assume that environmental stimuli are distinct from one another and that the perception (or response to) the stimulus, being distinct from the stimulus itself, can be studied somewhat independently of it. Environmental psychology looks upon the stimulus and its perception as a unit that contains more than just a stimulus and a response. The stimulus–response perceptual relationship between an urban landscape and an urban inhabitant, for example, depends not just on the individual

stimuli in the landscape. It also depends on the patterning, complexity, novelty, and movement of the contents of the landscape and on the past experience of the perceiver (e.g., whether he or she is a long-time resident or a newcomer); his or her ability to impose structure on the landscape; his or her auditory (perceived through the ear) and olfactory (perceived through the nose) associations with the landscape; and his or her personality characteristics. In environmental psychology, all these things make up one holistic environmental-perceptual behavior unit. Like Gestalt psychology, which influenced American psychology during the mid-20th century, the whole is greater than the simple sum of its parts. This is what "molar" refers to in our definition of the field and helps to draw boundaries between environmental psychology and many other areas of psychology.

To the environmental psychologist, a shopping mall consists of not just separate episodes of people engaging in individual activities. It is considered a physical setting containing a high density of people who interact with one another and with the physical setting in very predictable ways, and who experience certain pleasant and unpleasant consequences of these conditions (Figure 1-6). Thus, the environmental setting both facilitates and constrains or limits the behavior that occurs in it. Furthermore, as the occupants of this setting move about, they change some aspects of the environment, such as the concentration of people and the ability of others to move about unimpeded. If the behavior is studied in isolation, separate from these particular environmental conditions, the conclusions derived from the studying process will inevitably be limited. The environment cannot be studied separately from the behavior, and the behavior cannot be studied separately from the environment, without losing valuable information. This does not mean that environmental psychologists never take a close look at a particular environment–behavior relationship in a field or laboratory setting, but it does mean they assume from the beginning that such dissection of an integral unit cannot tell the whole story.

A second and related assumption underlying environmental psychology is that envi-

Figure 1-6   To the environmental psychologist, this shopping mall consists not just of separate episodes of people engaging in individual activities. It is a physical setting containing a high density of people who interact with one another and with the environment in very predictable ways, and who experience certain pleasant and unpleasant consequences of these conditions.

ronment–behavior relationships are really *inter*relationships: The environment influences and constrains behavior, but behavior also leads to changes in the environment. Consider the issue of energy resources and pollution. The availability of certain energy sources in the environment determines whether certain types of energy-consuming behavior will occur—but that behavior, in turn, determines the type of pollution that will result. Oil price increases may motivate conservation efforts and affect many aspects of one's lifestyle. This may lead to a reliance on other forms of energy, to new forms of pollution, and so on. With continued consumption, energy resources are differentially affected, and this, in turn, can shift consumption patterns. Note that this example also demonstrates that environment–behavior relationships need to be studied as units in order to see the whole picture.

A third characteristic is that environmental psychology is less likely to draw sharp distinctions between applied and basic research than are other areas in psychology. Some fields of psychology engage in theoretical or basic research as the primary means of understanding behavior. The major goal of such research is to gain knowledge about the subject matter through discovering cause–effect relationships and building theories. If such research also leads to the solution of a practical problem—which it often does—that is well and good, but a practical application is not necessarily a goal of that research. Applied research, on the other hand, is intended from the start to solve a practical problem, and it is valued not for its theoretical relevance but for its specific utility. Theory building may result from applied research, but is not its primary focus.

In contrast, environmental psychology usually undertakes a given piece of research for both applied and theoretical purposes at the same time. That is, almost all research in environmental psychology is *problem-oriented* or intended to be relevant to the solution of some practical issue, and thus the cause–effect relationships and theoretical material evolve from this focus. Research areas, such as the effects of pollution on behavior, changing environmentally destructive behavior, and the design of environments for efficient human use, are concerned with applications and practical matters, yet much of the factual content and theoretical underpinnings of environmental psychology derive directly from this type of research. Once again, this does not mean that environmental psychologists cannot take a practical problem into the laboratory for controlled study, but it does mean that the laboratory research of an environmental psychologist is oriented toward solving real-world problems (Figure 1-7).

If environmental psychology studies molar relationships among environment, mood, and behavior, and it is characterized by a focus on behavior *in* the environment, by the assumption that environment and behavior mutually affect each other, and by an interest in environmental problems, how does this translate into research and theory? An example here may help. If one is interested in the effects of crowded living conditions, options of ways to proceed may be limited. Because residential crowding is a *chronic* condition lasting for months or years, it is unrealistic to assume that one could reproduce it in the laboratory with human research participants. Use of nonhuman animals could help here, though relationships between

Figure 1-7 The laboratory research of environmental psychologists is oriented toward solving real-world problems. In this photograph, a researcher is evaluating the impact of tourist aircraft noise on ratings of tranquility, solitude, annoyance, and scenic beauty as participants view slides of national parks in the presence of simulated helicopter noise.

environment and behavior are likely to differ across species. More importantly, the fact that residential crowding occurs in people's home territories makes it different from crowding in any other setting. As a result, a study of crowded housing would of necessity be of a residential environment—and researchers have studied a variety of them, ranging from crowded prisons to college residence halls, public housing, and high-rise apartment buildings. The belief that behavior and environment mutually affect each other could lead researchers to include measures of what residents do to reduce crowding (such as personalizing their space, altering their space design) and how it affects experience. The problem-focused nature of environmental psychology could lead to development of an intervention to reduce crowding or distress. In Chapter 9, we discuss the development of re-

search on crowding, consistent with the previously mentioned assumptions of environmental psychology.

A fourth characteristic is that environmental psychology is part of an interdisciplinary and international field of study of environment and behavior. Environmental perception, with its emphasis on the perception of a whole scene, is relevant to the work of landscape architects, urban planners, builders, and others in related fields. The study of the effects of the physical environment (noise, heat, and space) on behavior is relevant to the interests of industrialists, lawyers, and architects, as well as prison, hospital, and school officials. The design of environments is of concern not only to architects and designers but also to anthropologists, museum curators, traffic controllers, and office managers, to name but a few. Moreover, changing environmentally destructive behavior is of concern to everyone who is aware of the dangers of pollution, urban blight, and limited natural resources in all parts of the world. Perhaps the need for this type of interdisciplinary perspective is reflected in the growth of related fields, such as urban sociology, behavioral geography, urban anthropology, and recreation and leisure planning. Throughout this textbook we will draw on these and other disciplines in order to explain environmental–psychological phenomena. You will also find as we progress through the book that environmental psychology has a strong international emphasis. A brief look at any year's issues of journals such as the *Journal of Environmental Psychology* or *Environment and Behavior* will show multiple disciplines and multiple countries represented.

In summary, environmental psychology is characterized by the following: (1) study of environment–behavior relationships as a unit; (2) study of the interrelationships of environment and behavior; (3) a relative lack of distinction between applied and theoretical research; (4) an interdisciplinary and international appeal; and (5), as we will shortly see, an eclectic methodology (i.e., a rich mixture of methods). Let us turn now to a description of the methodology of environmental psychology.

WHERE DID ENVIRONMENTAL PSYCHOLOGY COME FROM?

The scientific study of the relationships between environment and behavior can be traced back to studies from 100 years ago (e.g., Gulliver, 1908; Trowbridge, 1913). Nineteenth-century psychologists had studied human perception of environmental stimuli such as light, sound, weight, pressure, and so on, and emphasis on learning and the advent of behaviorism led to intensive study of such environmental events as reinforcement schedules and early childhood experience. By the 1940s, a modest amount of research on environment–behavior links had been reported, including early work in behavioral geography, the psychology of cognitive maps of environments, and urban sociology (Moore, 1987). However, these studies did not systematically approach the interaction of environment and behavior in its fullest sense. The studies of how design factors affect the development of social relationships among students reported by Festinger, Schachter, and Back (1950) represent a turning point in the development of systematic study of environment and behavior.

During the 1950s, work in this area slowly increased. Lewin (1951) had conceptualized the environment as a key determinant of behavior, and even though his emphasis was primarily on the social environment, the importance of his theory for environmental psychology is often discussed. Barker and his colleagues compiled extensive systematic research on environment and behavior relationships during this period, examining effects of environments on the behavior of children, comparing behavior in small towns and in schools (see Barker, 1990 for a historical review). Research on spatial behavior, psychiatric ward design, and other aspects of environment–behavior relationships also developed during this period (e.g., Hall, 1959; Osmond, 1957). Architects and behavioral scientists began what has become a long-standing collaboration in an effort to achieve another objective: designing buildings to facilitate behavioral functions.

Other lines of work have also fed into the present field of environmental psychology. Already noted work by Barker on ecological psychology (see the box on page 14) emphasized the ways in which the entire environment influences the types of behavior that will occur within it, and work by E. T. Hall (1959, 1966) in *proxemics*, or how we use space, as well as the work of researchers interested in the effects of crowding (Calhoun, 1962, 1964) have stimulated volumes of research on these areas of human–environment interaction. Research in environmental psychology began to flourish around the world, and other work in perception and cognition played a significant role in environmental psychology as well. With the advent of concerns over energy use and preservation of the natural environment, more and more researchers are looking into ways of changing our wasteful and destructive practices of interacting with the environment.

By the mid-1970s, these developments led a few psychology departments to offer formal programs of study in environmental psychology, and many more departments began to offer courses with that title. Textbooks on the subject emerged, journals devoted to the field (such as *Environment and Behavior* and the *Journal of Environmental Psychology*) were started, and organizations such as the Environmental Design Research Association were formed. The American Psychological Association has officially recognized environmental psychology (in conjunction with population psychology) as one of its divisions, and international societies, such as the International Association for the Study of People and Their Surroundings, have become active. Interest in the issues addressed by environmental psychology remains strong in the 21st century.

How Is Research in Environmental Psychology Done?

Are environmental psychologists and other psychologists similar or different in the way they view research? As mentioned earlier, two unique qualities of environmental psychology are that it studies environment–behavior relationships as whole units and that it takes a more applied focus than other areas of psychology. These qualities affect environmental psychologists' approaches to research in several ways. Most important is the fact that they tend to conduct research in the actual setting that concerns them and thereby preserve the integrity of that setting (e.g., Proshansky, 1972; Winkel, 1987). Thus, they are especially inclined to use techniques that take them to field settings, although they may also opt for the careful controls of laboratory research when appropriate.

RESEARCH METHODS IN ENVIRONMENTAL PSYCHOLOGY

Basically, environmental psychologists have the same "arsenal" of research methods as other psychologists—they just use it somewhat differently. It includes experimental methods, correlational methods, and descriptive methods. We will describe each of these techniques, first noting general strengths and weaknesses and then evaluating their appropriateness for research in environmental psychology. It will become apparent that, because of the different research values held by environmental psychologists, their choice of methods frequently differs from that of other psychologists. The description of their methods is brief and introductory and should give you enough background to understand the methodological issues in the rest of the text.

Experimental Research

Only one methodology allows researchers to identify with certainty the variable that is caus-

ing the effects they observe in a study. It is called the **experimental method,** in which the researcher systematically varies an **independent variable** (e.g., heat) and measures the effect on a **dependent variable** (e.g., performance). Usually two or more levels of an independent variable are used (e.g., 70, 90, or 100° F for levels of heat), and often multiple dependent measures are used (e.g., measures of mood and performance). Two forms of control are necessary in experimental research. First, only the independent variable is allowed to differ between experimental conditions, so that all other aspects of the situation are the same for all experimental conditions. When variables other than the ones being studied also vary across different conditions, they are considered **confounds.** Second, participants are randomly assigned to experimental treatments. This **random assignment** makes it improbable (with a sufficient number of participants) that differences between treatment conditions are caused by factors other than the independent manipulation (e.g., different personality types). In other words, experiments should be high in **internal validity:** They should be conducted in such a way that the effect on the dependent variable is due to differences in the independent variable and not due to any other factors. Experimental methodologies may be used in both laboratory and field settings, although it is clearly more difficult to manipulate variables and establish controls in the field.

While experimental methodologies have predominated in many traditional areas of psychology, they have not dominated research in environmental psychology to the same extent. Although the fact that they permit causal inference is an advantage, for environmental psychologists the liabilities of experimental methods frequently outweigh their benefits. One problem is that the degree of control required often creates an artificial situation, which destroys the integrity of the setting. This makes findings from these studies less generalizable to the real world; that is, it reduces **external validity.** Further, it frequently is possible to maintain the control necessary for an experiment only over a brief period, which makes most experimental studies short term. Since many environmentally caused effects do not manifest themselves over a short term, this is a problem (e.g., depletion of a natural resource may take decades or longer).

However, experimental studies in the laboratory can be useful for studying environmental issues. For example, as we will see in Chapter 5, Glass and Singer (1972) used artificial laboratory conditions to specify some of the psychological aspects of exposure to noise and were able to discover relationships that would have been difficult, if not impossible, to find in field studies or nonexperimental investigations. In these studies, participants in a laboratory were exposed to predictable and unpredictable noise, and some were provided with a sense of control over the noise by virtue of having a way to shut off the noise if they wished. Predictable noise had few negative effects on participants, while unpredictable noise had several effects. More important, the sense of control attenuated the negative consequences of unpredictable noise. The nature of this phenomenon and the need to isolate individual causes made laboratory experimentation the only feasible way to study these relationships.

An alternate approach to experimental laboratory techniques is to conduct field experiments. By transferring many aspects of experimental science to a field setting, we can increase realism and generalizability and still have enough control over the variables we are studying to be able to derive causal relationships. Participants are still randomly assigned to conditions and the experimenter manipulates the independent variables. Field experiments are difficult to set up, however, and the manipulations may appear somewhat artificial. Any artificiality reduces **experiential realism,** defined as the extent to which the research experience resembles that of the real world and impacts the participants as intended.

An example of the value of field experimentation is provided by a renowned study of territoriality conducted by Edney (1975). In general, research on territoriality has been difficult to carry out in the laboratory because it requires the experimenter to induce feelings of ownership in research participants. Since territoriality already exists in one's home environment, Edney decided to run a field experiment using students' dormitory rooms as the laboratory. He randomly assigned half the participants to their own room (the "resident" condition) and half to the rooms of other students as "visitors." Participants performed a variety of tasks within this context. The results, reviewed in more detail in Chapter 8, showed that people experience more control when on their home ground than when visiting the territory of another, and perceive their own territory as more pleasant and private. More important for our present purposes is that Edney successfully used a naturalistic setting to observe an environmental phenomenon and to study its effects in a systematic, causal manner. Because participants were randomly assigned (i.e., to resident and visitor conditions), a degree of control was established over extraneous variables. Experiential realism and external validity were enhanced by the field setting, so this study represents the best of both experimental and field research.

For a variety of reasons, researchers are often unable to do research in the field. The appropriate settings may not be available, the logistics of doing a field study may be too great, or sufficient control may not be attainable. Some researchers have responded by using **simulation methods**—introducing components of a real environment into an artificial setting. By simulating the essential elements of a naturalistic setting in a laboratory, experiential realism and external validity are increased, and some experimental rigor is retained.

Simulation techniques are useful for studying numerous aspects of human–environment behavior. We noted in Figure 1-4 that road rage is of increasing concern. Yet how can we practically or ethically manipulate road rage on our freeways in order to study it? An alternative approach is to use a simulated drive down a roadway, employing a computerized driving simulator. Ellison-Potter, Bell, and Deffenbacher (2000) did just this in showing that situational factors such as anonymity and hostile bumper stickers increased simulated aggressive driving. Interestingly, as we observed in Figure 1-4, a simulated drive through nature-dominated scenery reduces driving stress (Parsons, Tassinary, Ulrich, Hebl, & Grossman-Alexander, 1998). With the increased sophistication of computers and computer-aided design systems, sophisticated computer-graphic simulations can aid research on various environments (e.g., homes, offices, and parks).

A more traditional means to view the natural environment experimentally is by showing participants photographic slides of a wide range of settings. In such a simulation, researchers might vary the complexity of urban and rural scenes (e.g., Herzog & Smith, 1988; R. Kaplan, 1987) that participants evaluate. This would provide information about how complexity affects preference in urban and rural contexts. Overall, slides or computer-displayed photos offer several advantages as a simulation of the real environment: They are easy to present to a small or large group, they are inexpensive to produce and obtain, and they allow a wide variety of scenes to be shown at one time. Figure 1-8A shows how experimenters displayed pictures of vistas at the Grand Canyon in order to study how participants weighted different attributes of a hypothetical trip, including driving time and how much air pollution obscured the scenery (Bell, Malm, Loomis, & McGlothin, 1985).

Correlational Research

In **correlational research** the researcher does not or cannot manipulate aspects of the situation and cannot randomly assign participants to various conditions. The relationship between *naturally occurring* situational variations and some other variable can be assessed through careful observation of both, using this method. Assume that a researcher wants to see how much time visitors at the Grand Canyon spend looking at different vistas depending on the visibility of distant landmarks as air pollution varies periodically (Figure 1-8B). By observing the naturally occurring variations in visibility and viewing time, the researcher can make a statement about whether changes in one are related to changes in the other. However, visibility is not under the researcher's control but is instead manipulated by time or other factors, participants are not randomly assigned to time, and the type of control characteristic of experimental studies is not exercised. As a result, a causal inference cannot be made. Unable to randomly assign people to times when visibility is high or low, one cannot rule out the possibility that the observed relationship between visibility and viewing behavior may be caused by a third variable. For example, some people may believe that a haze is natural moisture and makes the scene mystical. Further, without an experiment, we know nothing about the *direction* of a relationship between two variables, because we are unsure which variable is the antecedent and which is the consequent. Thus, correlational methods are relatively low in *internal* validity.

Although correlational methods are clearly inferior to experimental methods in terms of ability to explain the "why" of a reaction to environmental conditions, they offer certain important advantages for the environmental psychologist. First, it is impossible or unethical to manipulate many environmental conditions that are studied, making experimental research out of the question. When this is the case, such as in studies of disasters, correlational methods permit the researcher to use the natural, everyday environment as a laboratory. In such research, artificiality is not a problem, and generalizability—or external validity—is greater. What types of correlational research are conducted by environmental psychologists? Two groups of studies can be identified. One group determines the association between naturally

Figure 1-8A & 8B   The photo on the top shows a display used in an experiment in which hypothetical trips to the Grand Canyon are displayed. The researchers manipulated driving time and different levels of vista visibility associated with varying amounts of air pollution (Bell, Malm, Loomis, & McGlothin, 1985). The experiment allows high internal validity but perhaps low external validity. In the photo on the bottom, a researcher records the time visitors to the Grand Canyon spend viewing vistas and correlates viewing time with pollution-related visibility (Ross, Haas, Loomis, & Malm, 1984). Such correlational research is high in external validity but lacks the controls of high internal validity laboratory experimentation.

occurring environmental change (e.g., natural disasters) and the behavior of those in the setting. Another group assesses relationships between environmental conditions and archival data (e.g., the relationship between housing density and crime rate). **Archival data** means data that can be found in historical records such as police reports or meteorological records.

Descriptive Research

Experimental studies provide causal information, and correlational research tells us if relationships exist between variables. **Descriptive research** reports characteristics or reactions that occur in a particular situation. Since such research is not constrained by a need to infer causality or association and often need not generalize to other settings, it can be quite flexible. The main requirement is that measurements be **valid** (i.e., they should measure what they profess to measure) and **reliable** (i.e., they should occur again if repeated). Under these conditions, we can assume the results are an accurate representation of reality.

In general, descriptive techniques are used more frequently in environmental psychology than in other areas of psychology. Their use is prompted partly by the phenomena being studied. We must often answer such basic questions as "What are the patterns of space utilization?" before using more sophisticated methodologies to test for underlying causes. In other words, descriptive research may be needed to identify behaviors that occur in a particular setting, so that they can then be studied in other ways. Descriptive research conducted by environmental psychologists includes studies of people's movements in physical settings, studies of the ways people perceive cities, and studies of how people spend their time in various settings. (For an example of this type of descriptive study, see the box on page 14.) Two types of descriptive research that are becoming increasingly important are environmental quality assessment and user satisfaction studies, in which environments are evaluated in terms of satisfaction or other

BARKER'S BEHAVIOR SETTINGS:
One Example of Descriptive Research

Probably the most extensive program of descriptive research ever conducted by an environmental psychologist was performed by Roger Barker. Barker's research centered on the concept of **behavior settings,** which he described as public places (e.g., churches) or occasions (e.g., auctions) that evoke their own typical patterns of behavior. Barker believed that the behavior setting is the basic "environmental unit" and that research that describes behavior settings in detail "identifies discriminable phenomena external to any individual's behavior" (Barker, 1968, p. 13) and that these settings have an important bearing on that behavior.

Fourteen years of such descriptive research were summarized in the book *The Qualities of Community Life* (Barker & Schoggen, 1973). Here, the behavior settings of two towns, "Midwest" (located in the midwestern United States, with a population of 830), and "Yoredale" (located in England, with a population of 1,310), were detailed. The descriptions were based on the reports of trained observers. Some of their findings are quite interesting and certainly tell us something about the character of the two towns. For example, Midwest had twice as many behavior settings involving public expression of emotions, and the structure of the settings provided children in Midwest with 14 times as much public attention as Yoredale children. Religious behavior settings also were more prominent in Midwest than in Yoredale, as were educational–government settings. However, in Yoredale, more time was spent in behavior settings related to physical health and art. We will describe Barker's behavior setting approach in more detail in Chapter 4.

characteristics by people who use them. For the most part, these studies rely on asking people about their needs, quality of life, and satisfaction. However, quite a number of different measurement techniques are used in research employing experimental, correlational, and descriptive methods.

DATA COLLECTION METHODS

Many of the ways in which environmental psychologists measure variables that they are studying are common in all areas of psychology. Other methods are more eclectic, borrowing from several fields, and a few measurement strategies are more or less specific to environmental psychology. The important thing to keep in mind when evaluating and choosing different data collection methods is that the assessment of behavior, mood, or response to environmental conditions should be as unobtrusive as possible. Measuring responses to a situation should not change the way the setting is perceived.

Ideally, research participants should not be aware of what you are measuring or when you are measuring it; we refer to techniques that allow this as **unobtrusive measures.** This is not always possible, and many measurement strategies have evolved.

Self-Report Measures

The most obvious way to measure moods, thoughts, attitudes, and behavior is to ask participants how they feel, what they are thinking, or what they do or have done; we call this approach the use of **self-report measures.** By interviewing participants, having them answer questionnaires, and using projective techniques, a great deal of important information can be obtained. Thus, if you are interested in the effects of noise on mood, you might ask people living in noisy and quiet areas how they feel during noisy periods, all of the time, or in whatever frame of reference you are investigating. The directness of measurement inherent in this technique is a clear advantage, but several problems characterize self-report as well.

First, self-report measures require that what you are assessing is something of which people are aware. These measures are also influenced by participants' interpretations, and, therefore, a number of sources of bias have to be taken into account. In the event that you are studying controversial issues, such as the impact of building high-level nuclear waste repositories near communities, self-reports may reflect more than just how people feel or what they think. If you ask people whether construction of such a repository would cause them to feel anxious or stressed, their stated preferences for construction might not be their true preferences. People who are opposed to the project might believe that if they say they would feel very anxious and stressed, the construction might not occur. Conversely, people in favor of the project could minimize negative replies to bolster the likelihood of construction. For such a case, the responses collected might not reflect mood as much as people's "votes" for or against the project.

Another problem is that people may not interpret questions or response options in the same ways. The ways in which concepts are understood or defined may vary, resulting in misleading answers to questions that the researcher thinks are clear. In crowding studies, for instance, researchers frequently ask participants whether they feel crowded or to rate how crowded they feel. The value of doing this is dependent on all people having similar definitions of crowding. However, Mandel, Baron, and Fisher (1980) found that this is not the case, and that men and women differ in their notions of crowding. When given a choice between two definitions of crowding, one dealing with there being too many people in a setting and the other dealing with there not being enough space, men chose evenly, while women were more likely to choose the definition emphasizing numbers of people. Thus, participants responding to self-report measures may have different ideas from the experimenter about what questions and answers mean and may differ from one another in these interpretations as well.

Regardless of these problems, self-report measures are often the only way to collect certain types of data, and as a result, effort has been directed toward minimizing these and other sources of bias. One way to do this is to develop measures that are standardized, or for which norms are available. Standardization of questionnaires or surveys is achieved by testing them on several different samples to estimate how people respond to them; these norms or estimates of "normal" responding can then be used for comparison with unique samples of respondents to which these instruments are administered. Thus, symptom checklists such as the Symptom Checklist 90 (SCL-90; Derogatis, 1977) are given to several different samples and norms developed so that responses of those in a specific study can be compared with how people in different types of groups typically respond. Finding that symptom reports of people living in crowded urban areas approximate those of psychiatric inpatients, while those of uncrowded residents are more like "normal" nonpatient responses, tells us more than just that crowded people report more symptoms than do people who are not crowded. It also gives us an idea of how intense their discomfort may be and whether it is enough of a problem to require some action.

The most common ways of collecting self-report data are by constructing and administering questionnaires and by interviewing people. Questionnaires are easy to administer and relatively inexpensive to produce and distribute, require little skill to administer, can be given to large numbers of people at a time, and can accommodate participants' desire for anonymity by not requiring them to give their names. However, it requires a great deal of experience and many validation studies to construct a good questionnaire, so many researchers opt to use questionnaires constructed by others. One advantage to this is obvious: The questionnaire has already been used in other studies so we have an idea of how good it is. However, scales such as the Perceived Stress Scale (Cohen, Kamarck, & Mermelstein, 1983) or Moos and Gerst's (1974) University Residence Environment Scale will be useful only if they measure concepts that you are also trying to study.

Interviews are not used as often as are questionnaires, partly because they are more costly and time consuming. Ordinarily, it will take longer for people to participate in interviews than to complete questionnaires, and usually only one person can be interviewed at a time. As with questionnaires, skill and experience are needed to construct questions and code responses in interviews. However, when using interviews, participants can be asked to explain inconsistencies in responses or to expand on their answers. People may also be more likely to voice honest opinions orally than when asked to put them in writing.

Another form of self-report measure is cognitive mapping, which is used to create "maps of the mind." Through a variety of procedures, described in Chapter 3, such images are transposed to paper. Cognitive maps are extremely valuable to researchers as a means of understanding how people code spatial information about their everyday environment. In addition to examining the mapping of city environments, studies have looked at how college campuses, local neighborhoods, and even nations are perceived. Through the use of these techniques, perceptions of various demographic groups can be measured and compared, and factors that afford qualitatively different perceptions can be identified.

Observational Techniques

A major measurement technique in environmental psychology, probably second in use only to questionnaires, is direct **observation.** In this method, people watch others and report their behavior and interactions in a given setting. These techniques can take many forms, ranging from informal observation of an environment, to a recorded narrative of what is seen, or to structured observation in which areas of the setting are preselected and particular behaviors

are recorded on special coding forms (e.g., Harvey et al., 1999). The advantage of observational methods over other techniques is the opportunity to gain first-hand knowledge of the way people behave in natural settings, as described in the box on page 14.

Unlike self-report measures, which assume that participants are able to express themselves, observational methods measure actions people may not even be aware they are performing. Since they can also be used without the participant's knowledge (i.e., can be unobtrusive), they minimize responses that are the result of people knowing they are being watched.

Observational methods have a number of disadvantages as well. One is that human error may be made in coding behavior. Examples include misidentifying one behavior for another, or being unable to code all the activity because it is occurring too quickly. The researcher using observational methods must also interpret the behaviors that are seen, and his or her interpretation may not be the same as the purpose of the behavior in the mind of the people being observed. Observational methods are also time dependent, which means that the investigator must be present when the behavior under study is taking place. This can often be inconvenient and time consuming, especially for behaviors that are infrequent. Some of these problems can be alleviated through the use of instrumentation (e.g., photographic equipment), but errors in coding and inferences based on them pose problems. However, use of these methods can yield valuable information as seen in the box on page 17.

If you are interested in how people react when the distance between them and others is small, you could ask individuals how they would react or how much space they would want. A better way to study this is to observe people under varying conditions in which they are close to others, recording whether they move away, how much they look at, talk with, or touch other people, and how far they stand or sit from others. While it is possible that participants could estimate how far away they might sit or whether they would leave, it is less likely that they could

BEHAVIOR MAPPING:
Observing People in Places

Few techniques are available to observe and record information about a large number of people in a given area. From such a mass of activity, an interpretable measure of behavior must be constructed. One specialized means of accomplishing this task is **behavior mapping,** which is concerned with accurately recording people's actions in a particular space at specific times. In this technique, observers record the behaviors occurring in one or more settings with the use of a preconstructed coding form developed through a series of steps (Bell & Smith, 1997; Ittelson, Rivlin, & Proshansky, 1976). First, the area to be investigated is defined. It may be a large hospital ward, a series of classrooms, or even a single room. The observers initially make narrative observations of the behavior occurring in the setting, either by taking notes or by tape-recording their impressions. From this information, categories of behavior and interactions are organized and listed on a coding form. Using such forms, the observers code actions that occur in each area of the setting during the period of research.

Behavior mapping can serve a variety of purposes (Ittelson et al., 1976). It may be used to describe behaviors in the setting. In this context, schemes can be developed to code interactions among specified individuals and also to index the type of interaction and where it is taking place. Mapping may also be used to compare behaviors occurring in different situations and settings or behaviors in the same setting at different times of day. It is also a means of learning about the utilization of equipment and facilities (e.g., whether areas are used as intended). Finally, behavior mapping can be employed to predict the use of new facilities.

report how much eye contact or touching they might exhibit. Since these behaviors are important aspects of how people use space, an observational study is probably better in this case. Sometimes these techniques can be combined with instrumentation in which mats or benches with switches hidden beneath them detect distances people maintain from one another under a variety of environmental conditions (Barnard & Bell, 1982; Kline & Bell, 1983).

Instrumentation is helpful at times when direct human observation is not the most productive, economical, or feasible way to collect data. Behavioral events may be sporadic, taxing the attention of the observer and wasting time in long waits with little opportunity to collect data. The area being observed may be too large for one or even several individuals to cover. In these cases, the researcher must either create a device that will do the job or choose from available instruments.

One type of instrumentation that functions quite well as a surrogate observer is photo-graphic equipment. With increasing availability of photographic supplies at reasonable cost, photographs, videotapes, and computer storage of images are being widely used by researchers. These media preserve records of the environment and events in it for future reference. They may be viewed repeatedly, even for different purposes and different studies.

Finally, engineers, architects, and designers have developed techniques to measure the full range of ambient conditions, such as the amount of light, noise, temperature, humidity, and air motion (see Fraser, 1989 for a description of these measures). We will describe many of these measures in Chapters 5 and 6.

Task Performance

In some studies the effects of environmental conditions on people's abilities to perform is of interest. Some occupational settings may be characterized by high-volume intermittent noise, confinement in isolation, high levels of density, and thermal extremes. It is important to determine how these conditions affect performance. Tasks used to assess environmental effects on performance may deal with manual dexterity and eye–hand coordination, performance on cognitive tasks, or with virtually any other aspects of performance. We will discuss for the moment only one task used in environmental research, although a brief search through the literature would reveal many others.

One of the most important aspects of using task performance as a measure of some environmental condition or change is to select a task that requires the kind of skills or effort you want to study. For instance, if you are interested in how some independent variable affects tolerance for frustration, you might want to measure persistence on difficult tasks. This has often been done by using a frustration tolerance task developed by Feather (1961). In this measure, which is also discussed in relation to Glass and Singer's (1972) noise research in Chapter 5, four line drawings are presented on separate pieces of paper. Participants are given each type of drawing and are told to trace each line without going over any line twice and without lifting the pencil from the sheet. If they make an error, they are to start on a new form. Participants are also told that if they complete a particular form or give up on it, they should go on to the others. Unknown to them, however, two puzzles are not solvable, and constitute the measure for frustration tolerance. All the experimenter has to do to measure persistence is to count the number of discarded forms, the amount of time spent on the unsolvable drawings, or both.

Trace Measures

Physical traces, that is, evidence of specific activities (e.g., cigarette butts in an ashtray as a means of measuring cigarette smoking or wear patterns on a lawn as a measure of traffic patterns), can be used to assess the effects of different settings as well. These *trace measures* are called **erosion measures** if they signify something taken away or worn down (e.g., wear patterns on carpet) or **accretion measures** if they signify something left behind (e.g., fingerprints

on a display case). For example, researchers might measure the volume of litter left in different types of recycling bins or might count the number of "No Trespassing" signs or barriers such as iron bars over windows to measure territorial defense in different communities.

Choosing Measures

With this array of possible measurement strategies, how does one go about selecting the measures for a particular study? Obviously, many factors are involved, including cost, whether we have certain types of instruments, and so on. The most important determinant is the question you want to study: If you are interested in arousal, physiological measures might be used with self-reported mood measures and, perhaps, performance measures. If behavior is the key variable, observation and self-report might be used. There are, however, some issues that are relevant in all studies of environment–behavior relationships.

Many measures are obtrusive—their use means that participants are *aware* they are being studied. Obtrusive measures are easier to use, but when people are aware of being measured (as well as of *what* is being measured), their responses may be different than they would have been if the measures had been disguised. With unobtrusive measures, the observer is not in sight and people are not told about being watched, though this may give rise to ethical concerns. Similarly, use of instruments such as cameras can be unobtrusive. Many of these measures, however, have been created for specific study purposes. For example, Bickman and his colleagues (1973) dropped stamped, addressed envelopes in high- and low-density college residence halls and studied helping behavior as a function of density by comparing rates at which letters were found and mailed. Students were not aware of being in a study when they found and mailed the letter. Similarly, Webb and others (1981) proposed assessing the popularity of various environmental settings in museums by measuring the number of nose and handprints on the display case. For an extensive discussion of other clever and useful unobtrusive measures see Webb et al. (1981).

The bottom line in doing research in environmental psychology is to apply measurement techniques that address the questions you are asking, that disturb the setting as little as possible, and that allow you to study real people in real environments. Field studies combining self-report, observation, and task performance, such as a study reported by Fleming, Baum, and Weiss (1987), are one way to achieve this. By observing people's behavior in their neighborhoods, gathering extensive self-report data as well as physiological data bearing on arousal, and measuring tolerance for frustration on a challenging task, it was possible to document several aspects of living in crowded urban neighborhoods. Integrated studies of laboratory, field, and archival data are also useful in developing a comprehensive picture of the problem under investigation.

ETHICAL CONSIDERATIONS IN ENVIRONMENTAL PSYCHOLOGY RESEARCH

Before concluding our discussion of what environmental psychologists study and how they do so, we should describe some of the ethical problems and considerations that arise in all environmental research. As you may have noticed, many design and measurement techniques require that the participant be unaware that an investigation is taking place. This frequently improves the validity of research in a number of important ways. Unfortunately, it also raises a number of ethical questions. The American Psychological Association (APA) and government agencies have issued standards concerning protection of human research participants. Most colleges and universities have review boards to advise the investigator on difficult and ethical issues in research design.

Many ethical considerations appear relevant to environmental research. Especially important are lack of full and **informed consent** by the participant and **invasion of privacy.** We will limit our discussion to these two topics,

but the interested reader is encouraged to read Sieber (1998) or any recent guidebook on behavioral research methods.

Informed Consent

Whenever possible, participants should be informed of all aspects of a research project so that they can decide whether or not they wish to participate. The assumption is that a lack of such information restricts freedom of choice. However, careful consideration suggests that informed consent is not always possible or desirable. For example, the researcher working with the developmentally disabled may find it impossible to fully explain a highly technical study. Further, many field studies must be performed unobtrusively, or participants' knowledge would bias the results to the extent that they are misleading. Before unobtrusive field research is undertaken, an assessment has to be made concerning the extent to which human welfare and dignity are in jeopardy, and these concerns must be weighed against the value of the experiment. In effect, the researcher should assure himself or herself that the major issues to be illuminated by the study justify the slight discomfort to people who are not offered an opportunity to give consent.

Another issue related to informed consent concerns whether people who participate in experiments without their knowledge (such as observations of pedestrians at a crosswalk) should be told about the study later. Is it better to leave people unaware so that they will not be upset by the realization of having been in an experiment? Or is it the right of all participants to receive a full explanation of the purpose and intent of the study? Informing participants after the experiment has taken place may oversensitize them to the possibility of future research or observation taking place in everyday settings. For some people, the fear of being unwitting participants in research at other times might be quite distressing. On the other hand, there are strong ethical concerns (e.g., the participant's right to know) that the researcher must weigh before withholding such information.

Invasion of Privacy

What is the rationale for assuming it is permissible under some conditions to observe people without their knowledge? Obviously, an invasion of privacy is involved in such situations. Since people in public settings realize they are under informal observation by others, most researchers believe formal observation should be no more threatening. However, if research participants in a public setting become aware of being observed and choose not to participate, the research should provide them with an alternative route or area that is not being monitored. While potentially this leads to selection bias in people involved in the study, it may importantly protect their right to privacy.

The assumption that under some conditions researchers have the "right" to observe people requires us to judge when behavior falls in the public domain and when it should be considered private. In Chapter 8 we describe a famous and controversial study conducted in a men's lavatory. The right to privacy is clearly a question that must be considered in this and other settings. Alternative methods of observation when the research involves privacy issues should always be considered. It is the responsibility of every researcher to weigh ethical questions along with methodological validity questions when designing behavioral studies.

Preview of the Content Areas of Environmental Psychology

Thus far we have described the characteristics of environmental psychology and reviewed briefly the methodological perspectives of its practitioners. The remainder of this textbook is devoted to an examination of the contents of the field, including empirical findings and theoretical perspectives. As indicated in Figure 1-9, we will begin with a discussion of nature and human nature; humans are biological creatures who manipulate their environment. Our interactions with the built and natural environments have biological underpinnings but are heavily influenced by experience—experience that leads us to form attitudes about environments and to assess those environments. Next we will discuss environmental perception and cognition, examining the ways in which environments are perceived, how these perceptions are retained and altered by situational factors, and how we negotiate our way through environments. We will then look at ways in which the environment influences behavior, beginning with theoretical perspectives on environment–behavior relationships. We will also see how stress and other reactions to the environment are influenced by such factors as noise, weather, disasters, air pollution, personal space, and crowding. Then we will examine the behavioral relationships involved in defined settings such as cities, residential settings, hospitals, prisons, work environments, learning environments, and recreation areas. In doing so, we will see how knowledge of environment–behavior relationships can be used to design environments for maximum human utility. Finally, we will conclude with intervention strategies for modifying environmentally destructive behavior and improving our relationship with the environment. As we proceed through these topics, we invite you to visit our Web site to obtain information on other useful and current Web sites relevant to each topic: http://www.colostate.edu/Depts/Psychology/envment.htm.

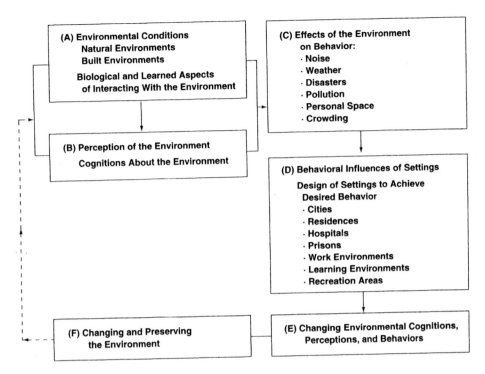

Figure 1-9 Organization of the book.

SUMMARY

Environmental psychology is concerned with studying environmental issues by drawing on the knowledge and techniques of many areas within psychology, and as such it serves as a meaningful focus for these areas. We define environmental psychology as the study of the molar relationships between behavior and experience and the built and natural environments. The distinguishing characteristics of environmental psychology include the following: (1) environment–behavior relationships are integral units; (2) environment–behavior relationships are reciprocal or two way; (3) the contents and theory of the field are derived primarily from applied research; (4) the field is interdisciplinary and international in nature; and (5) environmental psychology employs an eclectic methodology.

Methods employed by environmental psychologists include experimentation, from which cause and effect can be inferred; correlation, which is suitable for certain field settings but ambiguous in inferring cause and effect; and description, which is often a necessary first step in new areas of research. Specific research techniques are either obtrusive, in which individuals know they are being studied, or unobtrusive, in which they do not know they are part of research.

Specific measures employed in environmental psychological research include self-report measures, observation of actual behavior or of trace or accretion measures, instrumentation, and task performance. Each method and measurement technique has advantages and disadvantages, and ethical considerations such as informed consent and privacy rights must be weighed in selecting alternative procedures.

KEY TERMS

accretion measures	dependent variable	experimental method	random assignment
affordances	descriptive research	external validity	reliable
archival data	environment	independent variable	self-report measures
behavior mapping	environmental	informed consent	simulation methods
behavior settings	psychology	internal validity	unobtrusive measures
confounds	erosion measures	invasion of privacy	valid
correlational research	experiential realism	observation	

SUGGESTED PROJECTS

1. Examine our Web site at http://www.colostate.edu/Depts/Psychology/envment.htm. Select three other Web sites referred to there, and compare the methods used in the studies and findings reported. How do they compare in internal and external validity?

2. Try an experiment that involves the importance of context. Tape-record a passage from a play or movie, and play it back for people after describing either the actual setting for the scene or a completely different setting. Do people tell you the same things about the two settings?

3. Look up a year's worth of articles in *Environment and Behavior* or *Journal of Environmental Psychology*. What disciplines are represented? In what countries are the researchers based?

4. Make a list of environmental problems you would like to see psychology try to solve. As this course progresses, annotate your list to include the psychological principles and research you think would be applicable to solving the problems you named.

NATURE AND HUMAN NATURE

INTRODUCTION

VALUES AND ATTITUDES

THE CHANGING MEANING OF NATURE
CONTEMPORARY VALUES: THE ROLE OF HUMANS
  IN NATURE
ENVIRONMENTAL ATTITUDES

ENVIRONMENTAL ASSESSMENT

QUALITY ASSESSMENTS
  Indices of Environmental Quality
  Affective Appraisals
THE SCENIC ENVIRONMENT: LANDSCAPE AESTHETICS
  AND PREFERENCE
  The Descriptive Approach: Experience and Artistic
    Judgment of Landscapes
  Physical-Perceptual Approaches to Scenic Evaluation
  Psychological Variables in Landscape Assessment
    Biological Influences: Biophilia and Biophobia
    Berlyne's Aesthetics: Formalizing Beauty
    The Kaplan and Kaplan Preference Model
    Evaluation of the Psychological Approach

RESTORATIVE EFFECTS OF NATURE

STRESS REDUCTION
ATTENTION RESTORATION THEORY
RESTORATIVE NATURE RECONSIDERED

NATURAL LANDSCAPES AS PLACES
SUMMARY
KEY TERMS
SUGGESTED PROJECTS

 # Introduction

Road trip! School is out, and before summer jobs begin, Cheryl and Anna are beginning a two-week vacation in the California Sierra Nevada: a sojourn to a land of calendar pictures — snow-capped peaks, dramatic glacial valleys, and roaring streams.

Cheryl: "This is going to be so much fun, it makes me smile just to think about it."

Anna: "You bet, hanging out in the mountains with my best friend. What could be better?"

Cheryl: "Yep, back to nature with Anni; I'm psyched."

Upon their return two weeks later, much had changed. It began with their first night in the mountains. Anna wanted to stay in an old lodge with a great view of the lake and a continental breakfast. Cheryl had planned on a "leisurely" 6-mile hike to a remote location as far as possible from any signs of human activity. Their compromise was to stay in an established U.S. Forest Service campground with a fire grate, a picnic table, and a nearby shower facility. When they arrived in Yosemite National Park, Anna was awestruck with the beauty of the glacial valley. Cheryl, on the other hand, could not recover from her disgust with the intrusions of humanity and human constructions into the landscape that was so dear to the wilderness advocate John Muir.

A sad ending would be the dissolution of their friendship, but this is not a sad story. Yes, they quickly realized how different they were in their expectations for the trip and in their definitions of nature and the natural. But after a week of hikes, sunrises, campfires, and waterfalls, the women each realized a profound

change. As Cheryl put it, "This place has become a part of me, and you have become a part of this place." Friends they remain.

To frame our examination of humans and their relationships with nature and natural landscapes we might begin by evoking the title of a book, *Humanscape: Environments for People* (S. Kaplan & R. Kaplan, 1978). *Humanscape* was a book of readings assembled in the 1970s by Stephen and Rachel Kaplan. Although the book itself may be somewhat dated, consider the name again: *Humanscape*. Ponder the word . . . allow yourself to daydream a bit. If you were skilled enough to paint a picture of such an imaginary place, what would it look like? Perhaps the title implies a *humane* place, a place where humans live and prosper. Would it be a wilderness landscape like the nature depicted on calendars gracing office walls? If so, (somewhat ironically) a humanscape would seem to be a place that is nearly unmodified by human activity. On the other hand, history convinces us that, at least in contemporary Europe and North America, wild landscapes are almost immediately modified when humans arrive. Perhaps the humanscape of your imagination is a place that reflects all sorts of human activity, like a farm, or even a city.

Where *do* we belong? Our attitudes toward built, modified, or natural environments seem to reflect conflicts between positive and negative characteristics of each (see Figure 2-1). The city is a source of stimulation and opportunity, but also a place of noise, crime, and pollution. Wilderness may provide an escape from urban ills, but the price is often a loss of convenience and comfort, and a different set of dangers. Even pastoral landscapes of cultivated fields and small villages are difficult to distinguish reliably from suburbia—a place that is increasingly regarded with ambiguity.

Many authorities report that natural landscapes are generally preferred over urban scenes (e.g., R. Kaplan & S. Kaplan, 1989; Ulrich, 1993). Built settings with at least some natural elements are also preferred over those without (Herzog, 1989; Sheets & Manzer, 1991). But what is the definition of "naturalness"? Often "natural environment" is never explicitly defined. When it is, the definition usually equates natural with the absence of humans or contrasts *natural* with *built* environments. Is nature found only in a wilderness? If so, what are the characteristics of a true wilderness? Many North Americans would agree with the view, codified by wilderness legislation, that wilderness exists only in places that show the absence of human activity.

Of course, many people never have an opportunity to visit a wilderness. Persons from an urban background may have a much different definition of nature than those from more rural areas, but elements of nature may be important for them, too (Figure 2-2). Some researchers define nature in terms of the presence of discrete natural elements such as vegetation or water, acknowledging the potential for small natural elements such as street trees or artificial ponds to temper the city (Mausner, 1999).

Perhaps there can even be too much naturalness, or, at least, too much wildness. Is a campground without fire grates and outhouses poorer than a remote site with little evidence of humans and no "improvements"? For some, natural areas may be associated with fear (as might be the case if one fears bears or snakes), disgust (with being dirty, for instance), or a lack of modern comforts (Bixler & Floyd, 1997). It may also be that experience with some kinds of natural environments may undercut one's appreciation for others. For example, viewing commercial presentations of spectacular scenery may increase support for preservation of national parks but decrease appreciation for more subtle, local natural areas (Levi & Kocher, 1999).

Even if humans are part of nature, what are the characteristics of our species? So-called primitive societies use fire, make clothing, and build shelter. Confirmed backpackers rely on tents, well-made boots, and other forms of technology. Is it our *human nature* to modify our surroundings? In recent centuries, our numbers have multiplied and our ability to use (or misuse) technology has grown. Perhaps *our* nature has gotten us into trouble.

Figure 2-1 Attitudes toward the environment reflect conflicts between positive and negative features.

Figure 2-2 Elements of nature may be important to residents of urban areas.

A theme that runs through much of this chapter concerns a somewhat different issue regarding the nature of humans. We will find ourselves confronting one of the oldest controversies in psychology. Simply, there is substantial disagreement about whether human behavior comes to us fairly automatically (na-tivism) or whether it is highly dependent on either cognitive processes or learning (empiri-cism). The controversy extends to explanations for our feelings for natural places. Are our reactions to nature automatic and common to nearly all humans, or are they the result of our individual learning history and culture? ▣

Values and Attitudes

Chapter 14 will discuss methods of modifying attitudes and behavior to preserve the environment. For now, we will examine more broadly how our attitudes and value systems color our relationships with the natural world. Let us begin with a question: What is your attitude toward air pollution, wilderness landscapes, or litter? If you found our question easy to answer, you must have an intuitive understanding of the term *attitude,* yet a formal definition has sometimes been elusive. Attitudes cannot be directly observed but must be inferred from behavior, including self-reflections and self-reports. Most theorists would agree that **attitudes** represent a tendency to *evaluate* an entity such as an object or an idea in a positive or negative way (Eagly & Chaiken, 1993). Most also agree that attitudes are based on affective (emotions), behavioral (actions), and cognitive (thoughts) components (e.g., Baron & Byrne, 2000; Taylor, Peplau, & Sears, 2000).

Although there is considerable overlap, an **ethic** or **value** is usually thought of as a broader construct than attitude and represents standards held by a person, culture, or religion. Informally we might think of these values as based on a collection of specific, related attitudes underlain by some abstract principle that gives them both a generality and a moral tone. Although this may be a chicken-versus-egg dilemma, we will assume that specific attitudes develop within a normative or value-based context. Thus, we begin by examining some historic and contemporary nature-related values.

THE CHANGING MEANING OF NATURE

In seeking to understand the way modern North Americans value nature, perhaps we should begin by examining the historic attitudes held by Europeans (see Nash, 1982 or Oelschlaeger, 1991 for reviews). Do the terms "gorgeous," "inspiring," "relaxing," or "refreshing" sound reasonable when describing the Alps? If so, you may be surprised to learn that in medieval Europe "terrible," "horrible," or even "disgusting" would be much more likely descriptors. In fact, Europeans so abhorred the wilderness that travelers sometimes insisted on being blindfolded so that they would not be confronted with the terror of untamed mountains and forests! Furthermore, European Christians inherited a Biblical prejudice: The Garden of Eden was a paradise from which humanity was ejected, and the desert wilderness was the land of hardship to which we were banished. Saint Francis of Assisi was a notable exception who believed that wild creatures had souls and preached to them as equals. However, the Church branded his views as heretical, perpetuating the dominant view of wilderness as profane (Nash, 1982).

During the period of Enlightenment, European attitudes moderated. Fueled partly by scientific discoveries, natural phenomena were seen by some as complex and marvelous manifestations of God's will. By the end of the 1600s, European intellectuals were increasingly fascinated, rather than repulsed, by nature. Nevertheless, this attitude was primarily a luxury enjoyed by privileged city dwellers rather than those who were forced to contend more intimately with the dangers of untamed wild lands.

Eventually, some Europeans sailed for North America seeking a land they had been told was a paradise. Most found anything but the "easy life." Eastern North America was, of course, a wild forest before European settlers began clearing it for farming. Whatever their original attitude, early European settlers in North America found that the necessities of food, shelter, and safety depended on overcoming the new American wilderness:

> The pioneer, in short, lived too close to the wilderness for appreciation. Understandably his attitude was hostile and his dominant criterion utilitarian. The *conquest* of wilderness was his major concern. (Nash, 1982, p. 24)

Nash compared the environmental values of the Puritan settlers of New England with those held by the Colonialists of the Middle Atlantic and southern states. The Puritans found themselves in a threatening environment of harsh winters and poor soil. The combination of this harsh environment and their conservative religious tradition led the Puritans to view the wilderness around them as a hostile, threatening landscape inhabited by servants of the devil (referring to the natives). Thus, the Puritans saw themselves as envoys from God whose mission was to pacify the wilderness and to break the power of evil. As their already poor farmland was exhausted by ill-advised farming practices, the descendants (both genetic and intellectual) of the Puritans moved westward, clearing forests and fencing prairies in an effort to conquer the vast American wilderness.

Although their principal attitude toward nature was also utilitarian, the settlers of the Middle Atlantic colonies benefited from a more hospitable environment and expressed somewhat different attitudes (Nash, 1982). Many were of the Anglican faith, and most were better educated, wealthier, and more likely to study

and appreciate natural phenomena than were the Puritans of New England. Virginian Thomas Jefferson may have epitomized the attitudes of the late 18th-century gentleman/naturalist (Figure 2-3). He believed that nature could be better managed through understanding rather than conquest, and as U.S. President, Jefferson charged the Lewis and Clark expedition of 1803 with providing detailed reports of natural phenomena. In spite of his more benign attitude toward nature, the vistas he most appreciated were still pastoral, rural landscapes of farms and country lanes, not the true wilderness. Yet in Jefferson, and those like him, we can see the beginnings of an attitude of conservation and curiosity rather than exploitation and loathing for nature and wild things.

It may have taken the development of Romanticism in Europe in the 18th and early 19th centuries to persuade Americans to look at the true wilderness with pleasure rather than disdain. The Romantic tradition grew largely from an urban literary elite who found themselves attracted to the contrasting rugged vastness of wilderness. Wilderness was the inspiration for the evolving concept of the sublime: a sense of awe and reverence, sometimes mixed with elements of fear (e.g., Burke, 1757; Kant, 1790). America did not have the cultural traditions, material wealth, or power of Europe, but size and diversity of wilderness lands was one domain in which the New World could compare favorably with the older cultures (Nash, 1982). In the decades following the American Revolution, the wilderness became a source of national pride with a growing minority of Americans. Soon American writers like James Fenimore Cooper and painters like Thomas Cole and Albert Bierstadt (see Figure 2-4) began to celebrate and romanticize the vistas of the untamed lands of North America. Fear or hostility still dominated most people's feelings about wilderness, but a new minority began to celebrate the wild lands as places of beauty.

Clearly, North American attitudes toward wilderness landscapes have changed since the colonial period (Merchant, 1992; Nash, 1982; Oelschlaeger, 1991). Still, for Americans in the

Figure 2-3 Thomas Jefferson's landscape at Monticello. As a designer, Jefferson believed that nature could be better managed through understanding rather than conquest.

21st century, reactions to the wilderness are often ambivalent. On the one hand there is a literary, artistic, and philosophical tradition that associates the wilderness with beauty and even religious experiences. An appreciation of wild lands is one legacy of leaders like Thomas Jefferson and artists like Thomas Cole. Late 19th-century writers like Henry David Thoreau and John Muir established a literary and philosophical tradition more recently articulated by modern environmentalists like Edward Abbey, Annie Dillard, Wendall Berry, Aldo Leopold, Bill McKibben, and Wallace Stegner. These authors are by no means the first to appreciate nature, of course. Although their influences on North American attitudes are less apparent, the importance of both nature and culture are at least as well demonstrated by Chinese and Japanese painters who preceded Western artists in celebrating wilderness landscapes by more than a thousand years.

CONTEMPORARY VALUES: THE ROLE OF HUMANS IN NATURE

Perhaps wild lands will always inspire both fear and appreciation. Even restricting ourselves to 20th-century European and North American history, it seems clear that the appreciation of nature is heavily influenced by culture and fashion. If attitudes are so transitory, is there any reason to make the ethical leap to suggest that any particular relationship with nature is "best"? Although many psychologists might prefer to avoid such value-laden questions, the applied nature of environmental psychology makes it difficult for us to maintain such reserve. Briefly,

Figure 2-4 *Looking Up the Yosemite Valley.* Albert Bierstadt's painting from the Hudson River School of American landscape painters.

one reason (which we will elaborate shortly) is because there is biological evidence that contact with certain natural landscapes can have restorative effects on modern humans. More in line with our present discussion, however, is the conviction that our planet is facing a global ecological crisis (see Chapters 6 and 14) and that our survival depends on our ability to change the way we do things! In figuring out how to turn things around, it would help to better understand differing ethical views of our relationship with nature and the wilderness.

We might begin at the end of the 19th century when there was a convergence between the appreciation of nature sparked by Romanticism and the new realization that humans were part of an interconnected web of life, a view fostered by Darwin's *Origin of Species* (1859). Americans began to realize that the supply of natural resources was finite. Under President Theodore Roosevelt's leadership, the federal government began efforts to manage natural resources to conserve them for human use. According to Oelschlaeger (1991), this became entrenched as **resourcism** (resource conservation), which remains the dominant American perspective on natural lands. Notice assumptions of **homocentrism** or **anthropocentrism** in this view. Anthropocentric (or homocentric) values emphasize the utility of nature for the use of humans as a species. Natural landscapes are seen as stockpiles of raw material to be transformed into the wants and needs of *humans* (Merchant, 1992; Oelschlaeger, 1991). On public lands, for instance, a manager's job is to use rational means such as scientific discoveries to maximize the output of natural resources for human use.

Preservationism is a less common, but still influential view that differs from resource conservation by emphasizing a holistic view of nature that assumes that an intact ecosystem is greater than the sum of its parts. An ecosystem has evolved into a complex system of interdependent parts, and changes to any one of these may have devastating effects on any of the others. Thus, preservationists value programs that maintain intact ecosystems such as wilderness areas (see Figure 2-5). According to Oelschlaeger, preservationism rejects a strictly economic approach to valuing nature in favor of species diversity, rarity, or beauty (see the box on page 31). Nevertheless, critics charge that preservationists often retain an essentially anthropocentric worldview when they advocate the preservation of intact ecosystems. You may have heard individuals argue that we should preserve tropical rain forests to avoid unwittingly destroying some plant or animal that could provide a cure for cancer. Without belittling the goal, we must point out that this argument still assumes the anthropocentric goal of managing nature for the benefit of humans.

Perhaps you have already anticipated another ethic. **Ecocentrism,** also referred to as *biocentrism,* maintains that natural ecosystems possess value in their own right, independent of their value to humans. Humans have no special standing, and ethical human actions will be those that promote all life on Earth. Aldo Leopold's (1949) **land ethic** is a well-known proposal for an ecocentric worldview.

> In short, a land ethic changes the role of *Homo sapiens* from conqueror of the land-community to plain member and citizen of it. (Leopold, 1949, p. 204)

Leopold's land ethic is remarkable in its simplicity. Unlike a piecemeal aggregation of specific attitudes that may or may not result in consistently pro-environmental behaviors, adoption of a land ethic (if successful) could result in a complete restructuring of a person's or a culture's values in an ecocentric direction.

For about 20 years the best-known measure of environmental values has been a scale called the **New Environmental Paradigm (NEP).** The NEP is based on the proposition that a new understanding of the relationship between humans and the natural world is emerging, one in which humans are an integral part of nature rather than alienated from it (Dunlap et al., 1992, 2000). As a measuring tool, the NEP may measure general environmental concern rather

Figure 2-5 Can natural landscapes be "managed" as measurable (economic) commodities?

⌒ THE PRICE OF PRICELESS NATURAL ASSETS

We have noted that preservationism rejects a strictly economic valuation of natural assets such as a forest or a wildlife preserve. As much as we would like to think that a rain forest or an endangered species or a view of the Grand Canyon is priceless, sometimes circumstances force an economic valuation on these natural assets. For example, sometimes policy makers have to calculate the logging value of a forest in terms of its board feet of lumber and compare this with the economic value of managing the forest for recreational benefits (e.g., entry fees for hiking or skiing). At other times the natural asset is damaged, such as when a seashore is devastated by an oil spill. While we might calculate the economic value of lost fishing in such a case, how do we assess deeply held feelings of aesthetic value (e.g., the lost scenic beauty), bequest value (wanting to leave the asset for the enjoyment of future generations), or existence value (an ecocentric approach that says the asset has a right to exist independent of human intervention)? In lawsuits following catastrophes such as oil spills, the courts typically use a method called *contingent valuation* to determine the financial liability of those responsible for the damage. Contingent valuation involves asking a panel to judge how much more they would pay for an improved asset, such as how much more they would pay in entry fees to use a forest that has improved hiking trails. An alternative approach is the *paired-comparison method,* in which a panel selects the preferred item in each of many pairs of items, such as $10,000 versus a wildlife preserve, or a wildlife preserve versus a bike trail, or a bike trail versus $10,000. Over enough pairs and a large enough panel, it is possible to derive a rank ordering of the items, from lowest value to highest. Evidence suggests that people can readily form this rank order of values, and that we do indeed place very high values on natural assets (e.g., Clarke et al., 1999; Peterson et al., 1996).

than more specific clusters of values (Schultz & Zelezny, 1999). In line with our review of the evolution of environmental thought, other researchers have proposed that a person's values are better characterized as matching categories that roughly approximate Merchant's (1992) egocentric, anthropocentric, and ecocentric value systems (Stern & Dietz, 1994).

Deep ecology is a related—and increasingly popular—term. Deep ecology is a form of ecocentrism based on a critique of modern technology, science, and political structures. Many deep ecologists would assert that we have arrived at a global crisis because our culture is dominated by a mechanistic worldview that is perpetuated by science and adopted to serve the domination of capitalism (e.g., Merchant, 1992). If these critics are correct, humankind's relations to the natural world endanger it, so we must promote sociocultural change.

A recent movement that attempts to rectify damage humans have caused to natural environments is termed **green justice** or **environ-mental justice,** which we cover in more detail in Chapter 7. Adherents to this movement seek a balance of the interests of nature and ecology over mere anthropocentric interests. The interested reader is referred to the fall 1994 issue of the *Journal of Social Issues* and the winter 1996 issue of *Social Justice,* which are devoted entirely to the topic.

In reviewing the historic development of environmental values we have introduced a number of terms, some of which are nearly synonymous. Figure 2-6 provides what may be a useful summary. We have really been discussing two sets of values. Nature-based values such as Merchant's (1992) egocentrism, anthropocentrism, and ecocentrism represent the way one conceives of nature or the environment. Resource values (exploitation, conservation, preservation), on the other hand, reflect a person's opinion about how individuals or the society of which the person is a part should behave toward the environment. It seems unlikely that a person with genuinely ecocentric values could advocate

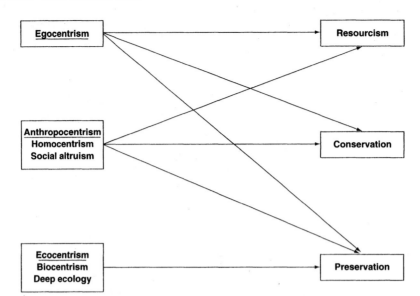

Figure 2-6   Modern values toward nature.

exploitation. However, it is entirely possible that a person could advocate wilderness preservation for his or her own pleasure, for the benefit of humankind, or because the person believes that plants and animals have a moral right to exist and prosper independent of human concerns. Thus, the same apparent behavior could result from quite different values.

ENVIRONMENTAL ATTITUDES

Values are important life goals or standards and are generally viewed as influencing attitudes and behaviors (Olson & Zanna, 1994; Schultz & Zelezny, 1999). In general, values are seen as determinants of attitudes, which, in turn, shape behavior. Values might be distinguished from attitudes in that values are broader and less clearly integrated into an interconnected system. A student who is interested in attitudes might wish to consult a recent social psychology textbook (e.g., Baron & Byrne, 2000; Meyers, 1999; Taylor et al., 2000) or review (e.g., Eagly & Chaiken, 1993). Briefly, although there is evidence that some attitudes may arise at least partly from genetic sources (e.g., Arvey et al., 1989; Keller et al., 1992) or may be triggered directly by sensory input (e.g., Zajonc, 1984), most theo-

rists believe that attitudes are primarily learned (Baron & Byrne, 2000). Thus, attitude formation probably involves many of the principles of classical conditioning, instrumental conditioning, and social learning familiar to introductory psychology students. We should note that although we concentrate on natural environments in this chapter, our discussion of attitudes is also applicable to built environments.

What do you suppose would be the result if you were to ask 50 of your friends whether unspoiled nature is (1) beautiful and (2) important? We will predict that most would say that nature is both. Why, then, is so much litter removed from almost every North American recreation area? As we have said, we often assume that values influence attitudes and that attitudes influence behavior. For example, if someone thinks that wilderness is desirable, wilderness landscapes are more likely to be perceived as inviting. In turn, a person who is attracted to wild landscapes should be more likely to engage in activities in wild settings. But how strong are the links between attitudes and behavior? For years, social psychologists were frustrated by findings that on the surface, at least, attitudes were often not consistent with behaviors. With additional research, psychologists have begun to understand the attitude-behavior link, and (with some relief) can demonstrate that attitudes do predict a variety of behaviors (Grob, 1995; Meyers, 1999; Schultz & Zelezny, 1999).

Kaiser, Wolfing, and Fuhrer (1999) propose that the initial failure to demonstrate a strong relationship between attitudes and behavior results from three sources: (1) a lack of a unified concept of attitude among researchers; (2) because of differences in the way attitudes and behaviors are measured; and (3) because some research fails to recognize that constraints on people's behavior encourage them to act in counter-attitudinal ways. Perhaps the most influential theory of attitudes is the model proposed in Fishbein and Ajzen's **theory of reasoned action** or **theory of planned behavior** (Ajzen & Fishbein, 1980; Kaiser, Wölfing, & Fuhrer, 1999), that a person's *intentions* to perform a particular behavior are the immediate antecedents to the actual *behavior.* These intentions themselves result from both one's *attitude* and the subjective, value-based assessment of the *norms* of one's society or group. So, norms, together with values and attitudes, determine behavioral intentions, which in turn predict overt behaviors (Figure 2-7).

Initially, Fishbein and Ajzen expected that a person's actual behavior and his or her reports of intentions to behave would be highly correlated. It is now clear that a number of variables affect behavior directly without operating on behav-ioral intentions (Chaiken & Stangor, 1987). For example, Ajzen (e.g., Ajzen & Madden, 1986) adds a dimension of *perceived control* reflecting the degree to which an individual is affected by real or perceived obstacles that would constrain his or her intended actions (see Chapter 4 for more on perceived control). According to Fishbein and Ajzen (1975), a general attitude may not predict a specific behavior; but a multiple-item scale measuring components of an attitude is more likely to predict a class of behaviors. A pro-environmentalist may not always keep the thermostat low in the winter to save energy, but someone who adheres to several pro-environmental concepts probably does engage in more pro-environmental behaviors (recycling, car pooling, water conservation) than someone who is not concerned with the environment.

Another research approach assumes that some sort of attitude activation is necessary before an attitude can direct behavior in a particular situation (Fazio, 1990; Fazio & Zanna, 1981; Fazio et al., 1986). According to this view, the strength of the association between an attitude and a particular object or situation will determine the degree to which that attitude is activated and, thus, exert influence on behavior. This strength will vary depending upon such factors as direct experience with the attitude object and the number of times the attitude has been expressed (Baron & Byrne, 2000; Chaiken & Stangor, 1987). At the extreme, an attitude

Figure 2-7   Fishbein and Ajzen's model of attitude change.

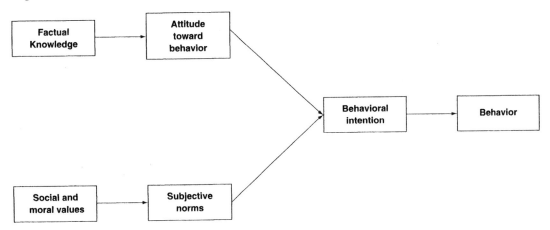

might be inaccessible or unformed in memory. It could even be that a dedicated environmentalist has never even considered the idea that trash left after a hockey game is a form of litter.

Finally, several researchers believe that attitudes can actually *follow* behavior (Bem, 1971; Festinger, 1957). That is, it may be that if we first change behaviors, attitudes consistent with those behaviors will develop in order to maintain consistency between our behavior and our attitudes as we perceive them or wish them to be perceived by others. Although there is evidence that attitudes do sometimes become more similar to actual behavior, this observation does not always hold true. Just because we are paying for pollution-control devices on our cars does not mean that our attitudes toward air pollution are changing (O'Riordan, 1976). It could be, of course, that attitudes both precede behaviors and follow them. We will revisit the issue of attitude–behavior linkage in Chapter 14.

Environmental Assessment

QUALITY ASSESSMENTS

Environmental assessment broadly encompasses efforts to describe environments or their components (Craik & Feimer, 1987). In the United States, for example, the National Environmental Policy Act of 1969 (NEPA) has been one factor stimulating the development of programs to assess environmental dimensions such as air and water quality. Monitoring these and other characteristics of environments can assist in documenting the effects of historic environmental changes and in predicting the future impacts of proposed projects.

Indices of Environmental Quality

Using modern technology, it is possible to assess pollution levels, noise levels, property deterioration, and other directly measurable aspects of the environment. Such measures can be incorporated into an objective indicator or **Environmental Quality Index (EQI).** Although these indices themselves are presumably objective physical measures, the term *quality* implies a subjective evaluation. For example, the concentrations of a known chemical toxin considered acceptable by one person or organization may be quite different from that acceptable to another. These differences of opinion reflect contrasting attitudes, the beliefs or feelings influenced by our individual learning and background.

In some instances the goal of assessment is not to determine the presence or level of some physical constituent of environmental quality, but rather, the perceived environmental quality as estimated by a human observer. This assessment method may not require sophisticated technology (although it does require careful attention to psychological measurement techniques). Typically, some sort of self-report scale asking for subjective assessment of environmental quality is employed, which results in a **Perceived Environmental Quality Index (PEQI)** (Craik & Zube, 1976). The PEQI is designed to serve a number of assessment purposes. As a measure of average responses of an affected population it may be one component of environmental impact statements or provide baseline data for evaluating environmental intervention programs. It also facilitates comparison of trends in the same environment over time, comparison of different environments at the same time, and detection of aspects of the environment that observers use in assessing quality. A somewhat different analysis may demonstrate individual or group differences in environmental perception. Currently, PEQIs exist for assessing air, water, and noise pollution, residential quality, landscapes, scenic resources,

AN EXAMPLE OF ENVIRONMENTAL ASSESSMENT: Visibility and the Perception of Air Pollution

As we will see in Chapter 7, air pollution has a number of negative effects on human health. One additional concern that has received increased attention (e.g., Stewart, 1987; Stewart et al., 1983) is the need to protect visual air quality. In the United States, the National Park Service, the U.S. Forest Service, and others are concerned about the impact of air pollution on the scenic vistas in parks and wilderness areas (Figure 2-8). As part of the amended Clean Air Act of 1977, the U.S. Congress sought to protect and even enhance the visual air quality (defined as the absence of discoloration or human-caused haze) of many pristine areas. The federal land manager is charged with the complex problem of determining whether a given change in visual air quality will have an impact on visitor enjoyment. Since visual air quality is based upon human perceptions and emotional reactions, measures of this phenomenon must be based upon or validated against human responses

Figure 2-8   Assessing the impact of air pollution on scenic vistas is a complex process.

(Craik, 1983; Stewart et al., 1983). Two critical issues parallel the distinction we have drawn between PEQIs and EERIs. First, it is necessary to determine how much of an increase in haze is required to cause a perceptible change in the environment. In addition to the concentration and composition of pollution, the detectability of haze is dependent upon factors such as color, whether it is layered in a band (layered haze does not occur naturally), and the angle of the sun. According to the Clean Air Act legislation, demonstrating that haze is detectable is not enough. It is also necessary to determine whether haze, even if it is detectable, significantly changes a visitor's experience. As you might expect, different individuals and different organizations disagree on the definition of "significant."

outdoor recreation facilities, transportation systems, and institutional or work environments (e.g., Craik & Feimer, 1987; Craik & Zube, 1976).

PEQIs provide an estimate of the perceived presence of environmental qualities, but not our feelings or emotional reactions to them (Craik & Feimer, 1987; Ward & Russell, 1981). Instead, an **Environmental Emotional Reaction Index (EERI)** assesses emotional responses such as annoyance or pleasure (e.g., Russell & Lanius, 1984; Russell & Pratt, 1980; Russell et al., 1981). Thus, the absolute measured level of sound might be reflected in an EQI, the human perception of this sound in the environment would result in a PEQI, and the emotional reactions engendered by these perceptions would be best characterized by an

EERI. These indices may yield very different results. For example, a moderate level of sound might prompt an extremely negative emotional reaction if the respondent wished quiet surroundings for study, but high levels might enliven a party. (The box on this page discusses some of the issues in assessing air quality, just one example of the distinction among different types of environmental assessment.)

Affective Appraisals

Just what are emotional reactions to environments? Russell and Snodgrass (1987) observe that definitions of emotions (often referred to by psychologists as **affect**) are ambiguous. Emotional reactions may be relatively long-term tendencies to feel love toward some individual, or short-term affective states. In the present discussion, we will focus upon affective appraisals, which are emotions directed toward something in the environment. How many terms could be used to create an EEQI describing the affective quality of a place? We can think of dozens, perhaps hundreds, but Russell and his colleagues (e.g., Russell & Lanius, 1984) have developed a circular ordering of 40 descriptors of places (see Figure 2-9) that include many commonly used emotional terms. Notice that these adjectives can be represented as a circular array in a space defined by two underlying bipolar dimensions. The horizontal axis ranges from unpleasant to pleasant, and the vertical axis ranges from not arousing to arousing. To pick two examples, the model implies that a serene environment should be pleasant but not arousing, whereas a frenzied environment is both arousing and unpleasant.

How might we account for the fact that in using the same psychological dimensions for evaluating identical environments, individuals often differ in their preferences? One answer

Figure 2-9   According to the Russell and Lanius model of the affective quality of places, emotional reactions to environments can be described by their relative position on unpleasant–pleasant and not arousing–arousing continua. Note that we have few words for emotional neutrality.

Adapted from Russell, J. A., and Lanius, U. F., 1984. Adaptation level and the affective appraisal of environments. Journal of Environmental Psychology, 4, 119–135.

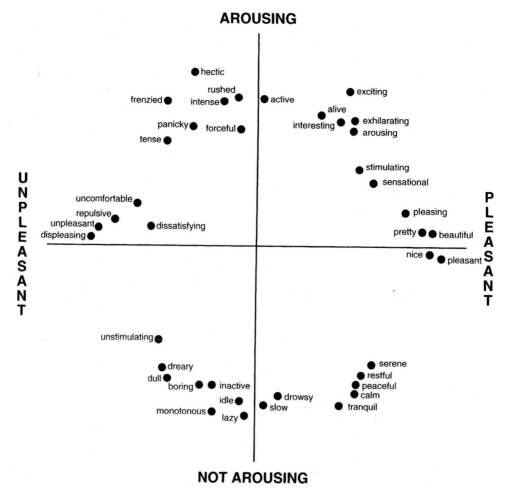

lies in the concept of **adaptation level** (Helson, 1964; Wohlwill, 1974). Individuals may have different levels of preference for complexity, causing the objectively measurable level of complexity in one scene to be too low for one individual but too high for another. In other words, experience may lead different individuals to prefer different levels of complexity. Wohlwill refers to an individual's optimum level on any one dimension as his or her adaptation level, and deviations from that optimum require adaptive measures (e.g., arousal reduction or sensation seeking). Russell and Lanius (1984) provide an interesting example of the effects of adaptation on emotional appraisals of landscape scenes. Recall that their model of affective appraisal of environments posits that emotional reactions can be described by two independent dimensions, pleasure and arousal. Russell and Lanius (1984) found that exposure to a slide of known emotional appraisal (say, gloomy and unarousing) would be associated with a tendency to evaluate a subsequent target scene in a direction emotionally away from the first stimulus (in our example, toward exciting and less gloomy). Stated simply, adaptation to one landscape is likely to bias affective evaluations of subsequent scenes in a predictable fashion. In Chapter 4 we will see how adaptation level can be used to explain not only individual differences in environmental evaluation, but also individual differences in responses to environmental stimulation.

THE SCENIC ENVIRONMENT: LANDSCAPE AESTHETICS AND PREFERENCE

Picture what you consider to be a beautiful landscape. Is your imaginary scene one of snow-capped peaks? A rocky seashore? Perhaps you imagine a pastoral scene of rolling hills, covered wooden bridges, and rustic fences? Do you think the scene you are imagining is much like that imagined by people the world over when responding to the same question, or are there differences among individuals and cultures? What can we learn about humans as a species from their landscape preferences? What elements make certain scenes almost universally viewed as beautiful and inspiring? Theoretical questions like these have attracted the attention of a number of environmental psychologists and other behavioral scientists (see Kaplan & Kaplan, 1989; Ulrich, 1986; Zube, Sell, & Taylor, 1982 for reviews).

A primary impetus for investigations of landscape aesthetics was provided by governmental legislation of the 1960s and 1970s that required the inventory and preservation of scenic resources (Zube et al., 1982). As an example, suppose we are building a new road to a remote forest recreation area. We want the new roadway to provide access and attractive vistas; but we do not want it to become an unpleasant intrusion for either the motorists or for hikers and campers in the recreation area. How do we provide access and maintain scenic quality?

The Descriptive Approach: Experience and Artistic Judgment of Landscapes

Not surprisingly, many of the most widely applied principles for landscape assessment and management evolved from the design tradition of landscape architecture. This approach emphasizes design principles derived from experience and artistic judgment. Vast areas of public lands such as national parks, national forests, and national wilderness areas have been assessed using a descriptive approach. The **descriptive approach** emphasizes the importance of contrasts in lines, forms, colors, or textures that are likely to draw attention (Figure 2-10). This description is not inconsistent with more empirically derived perceptual data. The human visual processing system is specialized for the detection of contrasts and is particularly "hard-wired" to detect certain simple lines or shapes (e.g., Goldstein, 1999; Heft, 1983; Hubel & Wiesel, 1979) and to seek a focal point or other source of organization (Ulrich, 1979).

Given that these principles help to determine what will receive attention, what determines whether the scene is evaluated as pleasant or unpleasant? In general, experts surmise that

Figure 2-10 Contrasts in lines, forms, colors, and textures are likely to draw attention.

natural landscape components are preferred to those that are the result of human activity. For example, natural scenes in which contrast is high often receive positive evaluations. The contrast of snow-capped mountain peaks with the green valleys at their feet probably heightens their visual appeal. Similarly, many of the scenic areas of the American Southwest are particularly striking because they showcase the brilliant hues of desert sandstone. On the other hand, one would probably wish to minimize contrasts that draw attention to utility lines, mines, and commercial establishments.

It seems safe to assume that landscape architects and other natural resource specialists are among the most sensitive and knowledgeable observers of landscapes. On the other hand, their design training may have also led them to perceive landscapes in ways that differ from the general public (Kaplan & Kaplan, 1989). In relying on the artistic (as opposed to empirical or data-based) tradition, the descriptive approach may be deficient in terms of reliability and validity (Daniel & Vining, 1983; Kaplan & Kaplan, 1989; Ulrich, 1986). Thus, there has been in-

creasing interest in supplementing expert opinions with preference models based on the responses of recreationists and other users. In the United States, for instance, some of the most consistent support for behavioral research has been provided by the U.S. Department of Agriculture's Forest Service (Kaplan & Kaplan, 1989; Ulrich, 1993). How might behavioral scientists differ from design professionals addressing issues of landscape aesthetics? The term *scientist* suggests that method is the source of one difference. Although we have made it clear that environmental psychologists are quite eclectic in their methodology, most would gravitate to an empirical approach emphasizing objective observations of individual users rather than design professionals. In addition, we can assume that psychologists are likely to be most interested in investigating psychological variables: behaviors or mental events that reveal the environment as it is filtered and focused by perception (see Chapter 3). As you might have predicted, psychologists have contributed both methods and psychologically based variables to the literature of landscape assessment.

Physical-Perceptual Approaches to Scenic Evaluation

One of the most direct extensions of psychological methodology is represented by what we will call the **physical-perceptual approach.** These strategies emphasize characteristics of the physical environment that can be related statistically to judgments of preference or landscape quality. Naturalness and the presence of water or vegetation are examples of physical landscape characteristics that might be used to predict negative or positive evaluations of scenic quality (Figure 2-11). In an early study, Shafer, Hamilton, and Schmidt (1969) assessed the preferences of individuals for landscapes in the Adirondack Mountains of New York state and found that preferences were associated with such factors as the area of immediate vegetation multiplied by the area of distant vegetation, vegetation multiplied by the area of water, and so on. The preponderance of evidence finds that scenic quality is related to such characteristics as the compatibility of adjacent land uses, ab-solute relative relief (i.e., differences in height, such as from valley to mountaintop or canyon rim to the valley floor), ruggedness, naturalness, and forest management practices.

Vining, Daniel, and Schroeder (1984) have extended the same basic model to forested residential landscapes. Presumably, identification of manageable characteristics that are likely to be perceived as unsightly (as may be the case in our new roadway example) may help to avoid conflicts in areas of high visibility or quality. Similarly, Im (1984) applied the physical-perceptual approach to study the relationship between landscape characteristics and visual preferences in the enclosed environment of a college campus. In this instance, visual preferences were most positively affected by the slope of the ground and tree canopy or vegetation coverage, whereas the height ratio (described as the height of the landscaped "walls" in a scene) was negatively related to preference.

The physical-perceptual approach is empirically based and more in keeping with the traditions of behavioral science than art or

Figure 2-11   Physical landscape characteristics may predict evaluation of scenic quality.

professional practice. In general, this research shows both cross-cultural agreement and consistency within a culture (e.g., Wellman & Buhyoff, 1980; Zube et al., 1982). Part of the consistency may involve the nature of the judgments made about a scene. In general, judgments of *preferences* for a scene show variability and individual differences, whereas judgments of *quality* or value seem to be more consistent and have less individual variation. That is, several individuals may agree that a group of scenes is striking, untarnished, and of high scenic quality; nevertheless, one person may prefer desert landscapes, another mountains, and a third, seascapes. Overall, these statistical models do a very respectable job of predicting assessments of scenes and have been frequently applied by resource managers, though rarely by designers (Im, 1984).

A shortcoming of the physical-perceptual approach is that the predictors it generates do not always make intuitive or theoretical sense (Ulrich, 1986; Weinstein, 1976). This criticism is not terminal (who says reality has to be easily understood?), but the predictive equations developed in one setting may be appropriate for only a specific type of landscape. Although the model is attractive because it emphasizes objective characteristics of the environment, psychologists are also intrigued by the possibility that more purely *psychological* variables might arrive at constructs that are more easily applied to human experience.

Psychological Variables in Landscape Assessment

Our emphasis will now move from quantification of perceptions of physical features of the environment to an examination of psychological or cognitive processes that underlie aesthetic judgments. Predictors such as complexity and coherence are typical examples of those found in models derived from this approach, and these variables are primarily located in human perception or cognition rather than inherent in the objective landscape. Physical measures of complexity or similar psychological predictors in a scene are difficult to obtain, so measures of these

factors must usually be obtained from subjective judgments. In a typical procedure, a panel of judges evaluates scenes on dimensions such as complexity, ambiguity, spaciousness, or uniqueness, and then the same or another panel judges the quality or beauty of the scene. First, let us examine a potential biological explanation for the influence of these psychological properties.

Biological Influences: Biophilia and Biophobia   In 1984, Edward O. Wilson used the term **biophilia** to describe what he believed to be a human need for contact with nature. According to Wilson, this need is a modern manifestation of a genetic predisposition to be attracted to other living organisms. Steeped in evolutionary theory, this view emphasizes that human history did not begin in the relatively short period of the last 10,000 years for which we have evidence of settlements and agriculture. If the history of our species is short, the history of civilization is much shorter still—perhaps only the most recent 1 percent of human history (Altman & Chemers, 1980; Wilson, 1993). Thus, Wilson asserts that humans are a species whose bodies, and especially whose brains, evolved in an environment dominated by the need to survive in nature.

It would be surprising if the influence of these earlier environments has already vanished in the short time since the advent of urban environments. Some psychologists believe that humans behave in accordance with functional-evolutionary principles (e.g., Kaplan & Kaplan, 1989). The function of much of human behavior is to further our chances of survival and is guided by inherited behavioral tendencies acquired by our species through evolution. What causes us to behave in a functional manner? Such behaviors in humans, like other animals, are presumably based not on some rational evaluation of a situation, but instead on a predisposition to *like* environments in which we are prepared to function well.

Gibson's (1979) concept of affordance, on which we will elaborate in Chapter 3, is central to many modern explanations of human affinity for nature and landscape aesthetics (Summit & Sommer, 1999). Briefly, certain features of the environment are said to "afford" shelter, food gathering, or an opportunity to survey the landscape. For example, humans may have evolved a preference for certain tree shapes (Sommer & Summit, 1995; Summit & Sommer, 1999) or landscape characteristics (Heerwagen & Orians, 1993) because these objects or settings were more likely to support human survival during the evolution of our species.

One of the strongest arguments for biophilia is presented by its converse, **biophobia** (Ulrich, 1993). Biophobia might be understood as an example of prepared learning (Seligman, 1970). Prepared learning refers to a propensity to learn quickly and to retain aversions to certain objects and situations that have threatened humans throughout evolution. Although modern technology has minimized the dangers of encounters with spiders or snakes, a propensity to quickly learn or to retain learned fears might persist in the gene pool. Research generally supports this proposition for humans and other primates (Cook & Mineka, 1989, 1990; Kendler et al., 1992; Ulrich, 1993), especially the proposal that learned fears of certain biophobic objects or situations will be resistant to extinction. Perhaps, then, our preferences for certain types of landscapes have some connection to innate fears and loves of features within them.

Not all humans fear spiders or snakes, and relatively few individuals who do have themselves been bitten. According to Ulrich (1993), whether particular individuals develop these fears will depend on their own experiences or those of people around them. Fear may never be learned if the object is never encountered. On the other hand, there is ample evidence that humans can learn vicariously, that is, by observing the reactions of others. Thus, one of the important functions of human culture and communication may be to allow individuals to learn of natural dangers without themselves being endangered.

The positive effects of biophilia are not as well documented as their phobic converses, and the research most commonly targets reactions to natural physical environments rather than to animals (Ulrich, 1993). The general argument for biophilia is similar to its phobic counter-part: Because our species evolved in a natural environment, we may have a biologically prepared readiness to learn and to retain positive responses to certain aspects of nature. Ulrich proposes three potential responses to biophilic nature: attention/approach/liking; physical and psychological restoration; and enhanced cognitive performance (see Figure 2-12).

Berlyne's Aesthetics: Formalizing Beauty

Berlyne (e.g., 1974) was among the first to develop a general model of aesthetics—a model that has more recently been applied to questions of environmental aesthetics. Berlyne's proposed important roles for *complexity*, or the extent to which a variety of components make up an environment; *novelty*, or the extent to which an environment contains new or previously unnoticed characteristics; *incongruity*, or the extent to which there is a mismatch between

Figure 2-12 Our awe-filled reaction to the splendor of nature may have a biological basis, according to the biophilia hypothesis.

our environmental factor and its context; and *surprisingness*, defined as the extent to which our expectations about an environment are disconfirmed.

Berlyne also distinguished between two types of exploration. *Diversive exploration* occurs when one is understimulated and seeks arousing stimuli in the environment, as when one is "trying to find something to do." *Specific exploration* occurs when one is aroused by a particular stimulus and investigates it to reduce the uncertainty or to satisfy the curiosity associated with the arousal. Berlyne (1974) suggested that aesthetic judgments can be characterized along two dimensions. The first dimension is called *uncertainty-arousal*. As uncertainty or conflict increases, arousal associated with specific exploration increases. The second factor is called *hedonic tone*, which is similar to pleasantness. This factor is related in a curvilinear (inverted-U) fashion to uncertainty. As uncertainty increases, hedonic tone (degree of pleasantness) first increases, then decreases. Apparently, we are happiest with intermediate levels of stimulation or uncertainty and do not care for excessive stimulation or excessive arousal. Consequently, those environments that are intermediate in complexity and novelty and surprisingness should be judged as the most beautiful, whereas environments that are extremely high or low in terms of these properties should be judged as less beautiful or even ugly. These properties are sometimes called *collative stimulus properties,* meaning they create perceptual conflict, and how we resolve the perceptual conflict leads to aesthetic inference.

Berlyne's theory emphasizes the importance of information in aesthetic judgments. In this sense, his approach is quite consistent with the notions of information overload and adaptation level we will discuss in Chapter 4. This also means that his model has elements of what we term a *constructivist* model of perception (see Chapter 3). **Constructivism** posits that perception is a fairly active process in which sensory information is analyzed, compared with past experience, and manipulated before perceptual judgments are made. Constructivist theories might be contrasted with *nativistic* approaches

such as biophilia or Gibson's direct perception of affordances (see Chapter 3). Affordances, as we saw earlier, are qualities of an environment that support the functioning and survival of an animal. For humans, for instance, trees may afford shelter. Perhaps it is surprising that one of the best-known modern models of landscape aesthetics manages to combine these two historically conflicting views.

The Kaplan and Kaplan Preference Model

Berlyne assumed that identifiable properties in an objective array of stimuli (complexity, for instance) result in predictable judgments of beauty or ugliness. On the other hand, our examination of environmental perception suggested that there are considerable individual differences in perceptions of environments and that people react quite differently to scenes based on their content. For example, it seems plausible to generate two scenes, one urban and one depicting wilderness, that are about the same on all of Berlyne's dimensions. Yet we know that there is a large body of research literature attesting to the importance (perhaps based on our genetic history) of nature. Culture probably supplies other content categories. Some people, if given a choice, would live in upstate New York, others in tropical Florida, and others in the desert of Arizona.

Steven Kaplan (1975, 1987) and Rachel Kaplan (1975) describe the procedures they used in constructing their model of environmental preference. Basically, these researchers collected a large number of slides of various landscapes and asked respondents to classify them according to certain schemes (similar–dissimilar, like–dislike, and so on). Next, the researchers statistically identified the elements in the scenes that led to this classification and evaluation. In this way, they derived several factors that can be used to predict preferences for various types of environments.

The Kaplan model combines nativistic and constructivist elements. Basically, they say that we humans will like or prefer those landscapes in which the traits of our particular species are most useful. We are neither strong nor fast. We have a poor sense of smell. We are awkward in water, and (without technology) we fly very poorly. What we do seem to be good at is processing and remembering information and being opportunistic at making use of it. Perhaps it is reasonable to expect that people will generally be attracted to scenes in which they are likely to function most effectively. Kaplan suggests that we will be attracted to environments that are (or were, prehistorically) survivable.

What does a scene that supports humans look like? For one thing, it should contain certain resources or contents such as water or abundant food that humans can exploit. Certainly humans, like other animals, have a pressing need for food, water, and shelter. But how do humans differ from other species? Again, the Kaplans' answer is that humans are good at, and even like, processing information (on occasion our students question the proposal that they find processing information pleasant, but the popularity of games of knowledge and skill indicates otherwise).

Information processing is a cognitive perspective about which we will learn more in Chapters 3 and 4. In this context the Kaplans conclude that humans have a fondness for environments that provide rapid, comprehensible information. Scenes that exhibit both may offer prospect and refuge (Appleton, 1975; Greenbie, 1982). *Prospect* is the ability to gain an open, unobstructed view of the environment, whereas *refuge* is provided by safe, sheltered places where a person might hide. Prospect and refuge are simultaneously high in parklike scenes that show open but bounded space. Some researchers (e.g., Balling & Falk, 1982; Heerwagen & Orians, 1993) note the resemblance between these landscapes and the African savanna, where many believe our species evolved. Perhaps our parks, cemeteries, and campuses are con-

structed approximations of the ancient environment that shaped the evolution of our species.

Apparently, two general dimensions underlie the categories: content and spatial configuration (Kaplan & Kaplan, 1989; R. Kaplan, S. Kaplan, & Ryan, 1998). One of the most striking content dimensions is the presence of nature. For instance, in groups of photographs depicting natural scenes, those with any sign of human activity are usually singled out in the classification process. In scenes of urban environments, those with even modest natural elements are identified. The second major dimension is spatial configuration, characterized by the bipolar qualities of openness versus closeness and defined versus undefined space. We seem to prefer scenes that facilitate travel and wayfinding by not being either too open and without definition or so closed that they obstruct our vision and travel.

If the Kaplans and others are correct, people will be attracted to scenes in which human abilities to process information are stimulated and in which this processing will be successful. In more psychological or information-based terms, people will like scenes that are understandable and make sense. In addition, however, people will also prefer scenes that are not too simple or dull. We like scenes that are engaging and involving—scenes that contain some mystery, for example (see the box on pages 46–47).

The Kaplans have organized information dimensions into a preference matrix with four main components (see Table 2-1):

1. **Coherence,** or the degree to which a scene "hangs together" or has organization—the more coherence, the greater the preference for the scene. (Figure 2-13A)

Table 2-1	ORGANIZATION OF THE KAPLAN AND KAPLAN MODEL OF ENVIRONMENTAL PREFERENCE[*]	
Characteristics of Information	**Understanding**	**Exploration**
Immediate	Coherence	Complexity
Inferred or Predicted	Legibility	Mystery

[*]*Adapted from S. Kaplan, 1987.*

Figure 2-13A Coherence.

Figure 2-13B Legibility.

Figure 2-14A Complexity.

Figure 2-14B Mystery.

2. **Legibility,** or the degree of distinctiveness that enables the viewer to understand or categorize the contents of a scene—the greater the legibility, the greater the preference. (Figure 2-13B)
3. **Complexity,** or the number and variety of elements in a scene—the greater the complexity (at least for natural scenes), the greater the preference. (Figure 2-14B)
4. **Mystery,** or the degree to which a scene contains hidden information so that one is drawn into the scene to try to find out this information (e.g., a roadway bending out of sight on the horizon)—the more mystery, the greater the preference. (Figure 2-14B)

At least two of these dimensions (complexity, coherence) are very similar to Berlyne's collative properties. A distinction between the Kaplan model and the Berlyne perspective, however, is that the Kaplans emphasize the informational content of a scene in a functional or ecological sense as one basis of preference judgments. For example, coherence and legibility relate to understanding or "making sense" out of the environment (being able to comprehend it and what is going on in it). Complexity and mystery can be considered aspects of "involvement" with the environment, or the degree to which

one is stimulated or motivated to explore and comprehend it. Table 2-1 represents these components in one version of the Kaplan model (S. Kaplan, 1987). As can be seen, one dimension of this matrix is the "Understanding" versus "Exploration" distinction. The other dimension revolves around the degree of effort required to process environmental information or the immediacy in time of the components of the environment. That is, coherence and complexity are thought to require less inference or analysis, whereas legibility and mystery require more cognitive processing. Although the relative importance of each element is not clear, coherence and complexity may require only moderate levels in order to facilitate information processing, whereas the more legibility and mystery in a scene, the better—in terms of preference judgments.

Even the most biologically oriented researchers do not suppose that we all have identical landscape preferences. For example, there may be age-related variation in landscape preferences (Balling & Falk, 1982; Bernaldez et al., 1987; Lyons, 1983; Zube et al., 1983). Balling and Falk (1982) report that children prefer savanna-like environments but that these preferences can be modified and become less and less powerful over a lifetime. Perhaps eventually, familiarity with other types of environments, especially those of "home," supersedes childhood

MYSTERY:
Intrigue or Fear?

According to the Kaplans' model of landscape preference (Kaplan & Kaplan, 1982; S. Kaplan, 1987), mystery is an element that increases interest and involvement in a scene by providing the promise of further comprehensible information. Typical examples of scenes with high mystery are those featuring paths curving out of sight or in which part of the environment is obscured or shadowed (R. Kaplan, S. Kaplan, & Ryan, 1998). But perhaps you are wondering whether high mystery is always a positive predictor of preference? Imagine yourself walking alone at night past a dark, curving alley. Would the scene possess mystery? Would the dark, unknown quality of the scene enhance your preference (see Figure 2-15A)?

You may not be surprised to learn that researchers have found that mystery can contribute either to *positive* environmental preference or to *fear,* depending on the situation (Herzog & Miller, 1998). Particularly in urban settings, vegetation or architectural features that allow concealment may enhance feelings of danger and fear (Fisher & Nasar, 1992; Nasar & Fisher, 1993; Nasar & Upton, 1997).

There are several ways to deal with this apparent ambiguity. For example, S. and R. Kaplan (1982) essentially refine their definition of mystery. They suggest that the term is properly applied in instances in which new information is not forced upon the perceiver, but is only suggested or implied (see Figure 2-15B). Thus, the viewer must have the ability to control the incoming information by choosing whether or not to remain in or move into a scene. Having control should reduce or eliminate fear (see Chapters 4 and 5 for a more

Figure 2-15A Although mystery is heightened by hidden information, possibly dangerous scenes are not preferred.

preferences for savanna. In her critique of Balling and Falk, Lyons (1983) agrees that landscape preferences diverge with age (as well as sex and place of residence) but suggests that the functional-evolutionary perspective underestimates the importance of culture in determining preferences.

Kaplan and Kaplan emphasize the role of familiarity in assessing scenic value. In general, the familiar, especially the "old and genuine" aspects of a scene, make it more desirable. Furthermore, those who are more familiar with a landscape may include locals as opposed to tourists and (in a different sense) experts as opposed to laypersons. Ultimately, we await a theory of landscape aesthetics that successfully accounts for both culture and biology (e.g., Bourassa, 1990).

Evaluation of the Psychological Approach

The Berlyne conceptualization of aesthetics and the Kaplan and Kaplan preference model are but two specific examples of what we have termed the *psychological approach* to assessment. Daniel and Vining (1983), Heerwagen and Orians (1993), Ulrich (1986), and Zube et al. (1982) review other psychological models. It is encouraging to note that in most cases, dimensions such as complexity, coherence, and ambiguity or mystery are found to predict scenic value by researchers using different methodologies. Unfortunately, we do not yet know how many of these dimensions we need in order to assess a scene adequately, and the way we combine the dimensions in judging one scene is not always universal. For example, complexity may complete discussion of the importance of perceived control in a variety of environmental situations). Ulrich (1977) took an alternate perspective when he suggested that mystery will be positively related to preference in situations with little risk, but inversely related in threatening situations.

Although fear may be incompatible with pleasantness, Herzog and Miller (1998) are among those who point out that some element of danger or controlled risk can actually be attractive (perhaps you remember playing scary nighttime games in your neighborhood as a child). Herzog and Miller (1998) investigated the paradoxical role of mystery and concluded that mystery can be a positive predictor of both preference and danger, but that the two responses are generally incompatible with each other. Their findings indicate that mystery's role is to contribute to whichever reaction (fear or preference) dominates.

Finally, Bernaldez et al. (1987) report some interesting differences in the way mysterious elements are evaluated by people of different ages. According to these researchers, whether a scene exhibiting darkness and shadow is perceived primarily as mysterious or risky and dangerous differs with age. It appears that a childhood fear of darkness and the unknown changes, and by young adulthood such environments take on a stimulating or artistic quality. We might conclude that the psychological character of the observer (including maturity) acts with the setting to determine the perceived character of an environment, and the presence of mystery will heighten the existent affective tone.

Figure 2-15B In this instance, the trail may invite you to move into the landscape to acquire more information.

best predict quality in one scene, and mystery may best predict quality in another.

Although the Kaplan and Kaplan model hypothesizes that content and spatial organization have their underlying roots in human evolution, another criticism of their theory comes from Roger Ulrich's research on biophilia. Like the Kaplans, Ulrich (e.g., 1993) emphasizes the importance of nature as a component in landscape judgments. For Ulrich, however, these judgments seem to be more purely based on biology and classically conditioned learned associations. Affective reactions would occur almost instantaneously, without the need for the more cognitive processing implied by dimensions such as legibility or mystery.

Restorative Effects of Nature

Whatever the reason for our affinity for natural elements, there is evidence that visiting natural places, or even viewing photographs of natural places, may have a restorative effect. Interestingly, although authorities agree that natural scenes or experiences have this positive potential, there is some controversy regarding the mechanism of this effect (Hartig & Evans, 1993; Kaplan, 1995; Ulrich et al., 1991). Two prominent explanations are stress reduction and attention restoration theory, and each grew from one of the approaches to landscape assessment we have just discussed. We will describe each of these explanations in much more detail and in a broader context in Chapter 4, but for now let us see how they can be applied to our reactions to natural scenes.

STRESS REDUCTION

Some of the most direct support for biophilia comes from studies showing that contact with certain types of nature creates what are called *restorative responses;* settings that foster these responses are termed **restorative environments** (see Figure 2-16). Perhaps life has included high levels of stress in all eras of human existence. Whether the source is a dangerous predator or the pressure of a deadline, there is ample evidence that humans pay a price for their stressful existence (see Chapter 4). Countering this stress, restorative responses may include reduced physiological stress, reduced aggression, and a restoration of energy and health. According to the functional-evolutionary perspective, humans should have a biologically prepared affiliation for certain restorative natural settings, but no such prepared response to urban environments since these have generally affected only a few generations of human experience.

Roger Ulrich (1979) demonstrated that viewing a series of nature scenes could lessen the effects of the stress induced by a college course examination. A subsequent study (Ulrich, 1984) compared the postsurgical recovery rates for hospitalized patients whose rooms overlooked either a small stand of trees or a brown brick wall. Those with the more natural view had fewer postsurgical complications, enjoyed faster recovery times, and required fewer painkillers. Other studies suggest that exposure to natural scenes can reduce presurgical tension and anxiety (Ulrich, 1986). According to Ulrich, the results are consistent with the hypothesis that attention-holding properties of scenes can work

Figure 2-16   Natural environments may restore psychological resources sacrificed to the demands of modern life.

two ways. As a component of dangerous encounters with nature, attention may be associated with stress, whereas attention to peaceful natural environments may result in calmative, restorative physiological effects.

A 1991 study by Ulrich is particularly instructive. The study investigated the effects of viewing videotapes of natural or urban scenes during a short recovery period following a stressful video. The stress-inducing video, a 10-minute black-and-white film intended to reduce industrial accidents, depicted simulated blood and mutilation. After viewing the film, participants watched one of six 10-minute color videos of everyday nature or urban scenes. Viewing scenes of water or a parklike setting not only resulted in more positive feelings, but was also associated with lower levels of several measures of stressful arousal (including blood pressure, skin conductance, and muscle tension). Unlike nature-dominated videos, urban scenes failed to show stress recovery effects. Perhaps most interestingly, both the stress-inducing movie and the nature video were associated with heart rate slowing (a response that is characteristic of heightened attention), whereas the urban scenes were not. In another study we mentioned in the previous chapter (Parsons et al., 1998), participants who had previously experienced different types of mild stress showed lower subsequent levels of physiological stress after viewing a videotaped trip through nature-dominated scenes. In fact, the researchers found some limited support for the notion that viewing nature-dominated simulations could produce a kind of inoculation, which could moderate the consequences of a *future* stressor.

ATTENTION RESTORATION THEORY

An alternative explanation (R. Kaplan & S. Kaplan, 1989; S. Kaplan, 1995) for the restorative psychological benefits of nature is provided by **Attention Restoration Theory (ART).** As outlined by Steven Kaplan (1995), tasks that require mental effort draw upon directed attention (see Figure 2-17). In order to accomplish these tasks, an individual must expend effort to achieve focus, to delay expression of inappropriate emotions or actions, and to inhibit intrusive distractions. Of course, problem solving also

Figure 2-17　Attention Restoration Theory (ART) proposes that the fascination of restorative environments restores our ability to direct attention to life's challenges.

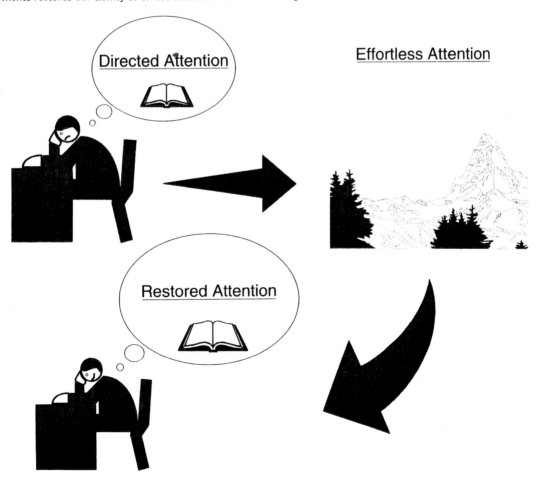

requires other resources—a store of knowledge that may be consulted, for instance—but Kaplan proposes that directed attention may be particularly fragile, and thus, important to recapture.

Any project, if sufficiently intense and sufficiently prolonged (even if the task itself is pleasant), is eventually likely to lead to **directed attention fatigue.** Kaplan cites the mental exhaustion that follows the examination period at the end of a college semester as a familiar example. We need a way to "recharge." Sleep helps, but sleep is insufficient. According to ART, it is necessary to rest directed attention by finding a different, involuntary attention that requires little effort. Fascination provides this effortless attention.

Natural settings are seen as an important source of fascinating things to draw one's attention. In fact, nature has an abundance of fascinating objects. Some are *soft fascinations* such as clouds, sunsets, or leaves flickering in the sunlight. These soft fascinations require little effort to capture our attention. Importantly, they are also compatible with human wants and needs (unlike other fascinating things like snakes and spiders). Restorative natural elements provide for fascination, but with the additional opportunity for reflection in an environment that is both removed from day-to-day tasks and compatible with human wants and needs.

RESTORATIVE NATURE RECONSIDERED

We have discussed two explanations for the apparent restorative effects of natural scenes. Attention Restoration Theory proposes a somewhat more complex explanation for the restorative effects of nature or natural elements than the mechanisms of stress reduction. Ulrich and his colleagues (Ulrich, Simons, Losito, Fiorito, Miles, & Zelson, 1991) point out that some (biophobic) elements of nature, such as snakes, are fascinating, but probably don't lead to restoration. Although Ulrich et al. assert that the speed of biophilic or biophobic responses seems to make the cognitive mediating processes implied by ART difficult, much work needs to be done to understand both the ART and stress-reduction models.

Note that both theories grew from research programs that were initially developed as explanations of landscape preference. In the case of ART at least, some (e.g., Herzog & Barnes, 1999; Herzog & Bosley, 1992) have wondered whether the aesthetic variable *preference* is synonymous with the restorative concept of *soft focus* (renamed "tranquility"). Although preference and tranquility are, in fact, highly correlated, Herzog and his colleagues have found that the information variables of *mystery* and *coherence* are more positively related to preference than to tranquility, and that the relationship varies with setting category (field/forest versus desert landscapes, for instance).

Although stress reduction and ART currently can be characterized as competing theories, it is quite possible that a combination of these two explanations will ultimately evolve (Hartig & Evans, 1993; Kaplan, 1995). There is little doubt that many people believe that experiences with nature have restorative effects, supported in part by the growing literature in wilderness recreation that we will review in Chapter 13.

Natural Landscapes as Places

Sometimes affective evaluations are attached to specific geographical locations or settings that have acquired special meaning (e.g., Fredrickson & Anderson, 1999; Steel, 1981; Stokols, 1990; Tuan, 1974). The term **place attachment** refers to the sense of rootedness people feel toward certain places, a phenomenon sometimes called *a sense of place.* To illustrate, consider the difference between "house" and "home." Do these two labels bring different thoughts to mind? Place-centered attitudes are personal, are highly valued, and may even be perceived as spiritual or religious (Fredrickson & Anderson, 1999; Mazumdar & Mazumdar, 1993; Roberts, 1996). Many researchers emphasize that these components of place are complex, with dynamic interrelationships that defy simplification into cause–effect relationships between discrete components (e.g., Low & Altman, 1992; Steele, 1981).

The experience of place is likely to be private and different from one person to the next. Many examinations of place are **phenomenological,** that is, based on people's subjective description of their experiences, and thus, at odds with the empiricism that dominates behavioral science. Of course, important phenomena are not only those that are conveniently studied. It would be hard to deny the importance of home (see Chapter 12), the effects of forced relocation of the poor or elderly, or the special relationship some people feel for a particular landscape (Chapter 13). Indeed, in later chapters we will see that we readily develop attachments to built environments as well as natural ones.

Methodologically, place assessments are likely to be more integrative and holistic than traditional landscape assessments. Most of the studies of landscape aesthetics that we have reviewed are based upon evaluations of slides or photographs of natural scenes. Although there is evidence that this method is adequate for assessing the *content* depicted by a photo, reactions may differ when people are asked to evaluate the *places* the photograph represents (Scott & Canter, 1997).

What do we know about natural places? They are permeated by affect, partly because of their aesthetic character but also because of their association in memory with events, persons, or feelings. Think of a place that has special meaning for you: a place, perhaps, where you would like to take your closest friend. Is part of what makes the place special its aesthetic appeal? Is part of your affection based on the accumulated memories that you associate with it? As you see in Figure 2-18, the meaning of a place results from the accumulated interactions between an individual's life history and a setting (Steele, 1981). Place experiences often include some feeling of ownership. Ownership in this case is a psychological phenomenon that does

Figure 2-18   Meaning of a place involves interactions between life history, the physical and cultural setting, and managerial actions.

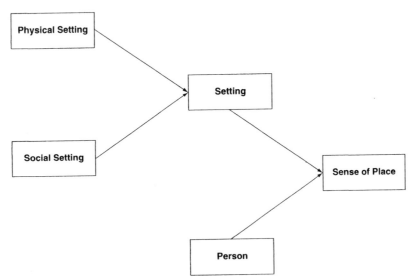

not require legal title to a piece of land or a building, but rather, a sense that the person has some uncommon, special relationship with that particular setting. This quality of psychological ownership has taken many names. It may be termed "territory" (Altman, 1975) or "kinship" (Mitchell et al., 1991), or it may be drawn as a distinction between the landscapes of "insiders" or "outsiders" (Relph, 1976; Riley, 1992).

Although much of what characterizes a place may be very personal, groups and even entire communities can develop place attachments. In fact, Newell's (1997) study of residents of the United States, Ireland, and Senegal revealed that there were more similarities in what people named as favorite places than differences across the three cultures. People from all three cultures were likely to name their own homes, places, or belongings. In support of the biophilia hypothesis, natural places were heavily represented, although the particular type of natural environment was likely to reflect local landscape experiences (pastoral countryside in Ireland, for instance).

Furthermore, the importance of place may vary from individual to individual. In fact, an attachment to place is often based on the history of social interactions at a particular location in addition to distinctive landscape or architectural character (Fredrickson & Anderson, 1999).

Whether or not a person develops a deep affective relationship with a particular place is based partly on the whims of history, but some individuals seem more likely to develop place attachments. As research continues, it may be that we will identify "place people" as a personality type (e.g., Mitchell et al., 1991; Steele, 1981; Williams et al., 1992). In one investigation of forest landscapes that we will revisit in Chapter 13, Mitchell et al. distinguished between use-oriented visitors and attachment-oriented users. Whereas use orientation resulted in a focus on activities and activity-based experiences, attachment orientation indicated an emotional bond with the setting.

What seems clear is that places are both the *objects* of people's interest and the *causes* of moods, feelings, and other reactions. Thus, places might best be understood in terms of transactions between physical settings and people acting in them. Place may seem remote from the examination of visual sensation that began this chapter. Like Barker's ecological psychology (see Chapter 4), however, place represents a synthesis of many of the trends in environmental psychology: dedication to holistic analysis, suspicion of strict cause–effect relationships, and an emphasis on real-world environments. We will have more to say about place attachment in Chapters 11 and 12.

SUMMARY

We have covered much ground in our examination of the relationship between humans and nature. For each of us this relationship may be based on both learning and biological predisposition. Much of the learned component may be understood in terms of the affects and values that comprise attitudes. We also briefly reviewed several contemporary attitudes or ethical positions with regard to the relationship between humans and nature. Increasingly there is evidence for a fairly direct effect of certain forms of nature on our physiological function,

and a companion biological influence on landscape preferences.

Although the link between attitude and behavior is imperfect, we believe that the evidence now shows that attitudes predict behavior. We also briefly reviewed several contemporary attitudes or ethical positions with regard to the relationship between humans and nature.

In addition to these theoretical issues, we reviewed contributions from environmental psychology to the assessment of natural environments. Appraisals may include the assessment of physical qualities (EQIs), perceived quality (PEQIs), or emotional reactions prompted by a particular setting (EERIs). In particular, we examined contemporary approaches to the assessment of landscape preferences and visual quality. Increasingly there is evidence for a fairly direct effect of certain forms of nature on our physiological function, and a companion biological influence on landscape preferences.

Other theorists have emphasized a more cognitive explanation for landscape preferences.

Biologically oriented and cognitive theorists agree that experiences in natural environments can restore the concentration and positive affect many of us seem to sacrifice to the demands of modern lifestyles.

Finally, we introduced a less empirical discussion of place. The concept of place attempts to integrate the character of a setting with the personal, often powerful emotions and memories an individual associates with it

KEY TERMS

adaptation level	directed attention	green justice	physical-perceptual
affect	fatigue	homocentrism	approach
anthropocentrism	ecocentrism	land ethic	place attachment
Attention Restoration	empiricism	legibility	preservationism
Theory (ART)	environmental	mystery	resourcism
attitudes	assessment	nativism	restorative
biophilia	Environmental Emo-	New Environmental	environments
biophobia	tional Reaction In-	Paradigm (NEP)	theory of reasoned
coherence	dex (EERI)	Perceived Environ-	action or theory of
complexity	environmental justice	mental Quality	planned behavior
constructivism	Environmental Qual-	Index (PEQI)	values
deep ecology	ity Index (EQI)	phenomenological	
descriptive approach	ethic		

SUGGESTED PROJECTS

1. Review the discussion of environmentally related values. Which of these views is closest to your own? For example, do you think your views are more anthropocentric or ecocentric? With respect to natural environments, are your views closest to resourcism or ecocentrism? Hypothetically, how would your attitudes and behavior change if you were to adopt one of the other ethical perspectives?

2. Ask a few friends to provide adjectives to describe three quite different campus or local environments. Can you place each of the adjectives generated by your friends into the bipolar model of affective reactions suggested by Rus-

sell and his colleagues? (Refer to Figure 2-9.) Is there any difference between the assessments of friends who grew up in fast-paced city environments and those from more rural homes?

3. List five outdoor locations in the local environment that you find visually pleasant. Using the text's discussion of landscape aesthetics and preference, list those features of each location that may have led to your positive evaluations. Does a location have plants and other "green" features? Water? Is it legible or mysterious?

4. Ask five friends or classmates to describe a place that is special to them. Try to pick out the most important themes in each description. Are their descriptions based on physical characteristics of the setting? Are other people (friends or family, for instance) mentioned as significant contributors to the meaning of the place? Do they recount episodes of personal experiences or emotional associations? Finally, after completing this exercise, do you feel you know the people you interviewed better?

CHAPTER 3

ENVIRONMENTAL PERCEPTION AND COGNITION

Introduction

Consider for a moment two imaginary individuals, a man from the metropolitan area of Atlanta, Georgia, and a woman from the rural region around Fishtail, Montana. What might their reactions be if we brought them together at her home in Fishtail on a March day with the temperature in the low 60s? The individual from Atlanta might find the day a bit chilly compared with his warmer and more humid climate, while the person from Fishtail might consider it quite warm and pleasant after a long winter in the foothills of Montana's mountains. The city resi-dent might appreciate Fishtail's clean, crisp air, but might be unnerved by the remoteness of the little hamlet. How would the woman from Fishtail in our fictional story differ in her perceptions of this, her long-time home?

Perception presents us with a "picture" or best guess as to the present state of the environment around us. Of course, the specific form of the perceptual input varies. It may be the visual image of a landscape vista, the smells and sounds of a city street, or even the text of this book.

Often this information *seems* to be stored and recalled as images or maps. The physiological processes that allow sensation are grounded in observable biological processes and events. However, the world as we "see" it also includes elements retrieved from experience and memory. One of the hottest debates in psychology concerns whether or not the subjective experience of "seeing" a mental image or map corresponds in any way to the actual mental *representation* or storage system. In this chapter we will also try to understand how this information is stored, retrieved, and referenced in our daily interactions with the world around us. Humans are not "stuck" in the environment of the present or even conventional reality. Think of the room in which you are reading right now.

Can you imagine how it would look if you were somehow able to view it from a window in the ceiling? Even when we are not actively viewing, hearing, or smelling an environment, we can experience it mentally. We acquire facts and opinions about the world around us, and we remember emotional reactions to environments from experience. Presumably, we can use this mental representation of the physical environment to make plans, to understand the terrain around us, or to solve problems involving an environmental context—finding a dry cleaning establishment, for example. In general, researchers refer to this ability to imagine and think about the spatial world as **environmental cognition.**

Characterizing Environmental Perception

Perhaps when you opened this book you expected to read about ways of using psychology to address environmental problems such as pollution, crowding, noise, or global warming. Perhaps you expected to learn about psychology applied to the human-dominated processes of designing buildings and cities. You may even have expected to read about psychology applied to environments such as wilderness areas in which humans are dwarfed by their surroundings and our intrusion is minimized. We consider all of these topics, or content domains, to be within the scope of this book. Across the variety of issues and environments, a common principle applies: Humans change the environment in both intentional and unintentional ways and are, in turn, changed by the places we inhabit. In almost every case, our actions are informed by clues about the state of this interaction—clues gathered from the world by our senses and reconciled with information from our prior experiences. Although we will focus on vision, the modality through which most humans acquire most of their knowledge about the envi-

ronment, we should recognize that other senses may exert powerful influences.

Of course, a recognition of the fundamental role of information is not unique to environmental psychology. Historically, psychologists have made a distinction between two processes that gather and interpret environmental stimulation. The term **sensation** has been applied to the relatively straightforward activity of human sensory systems in reacting to simple stimuli such as an individual sound or a flash of light (e.g., Goldstein, 1999). Your awareness and evaluation of striking architecture, sublime landscapes, or distasteful dumps are probably founded on the sensations created by an array of photons of light stimulating individual receptor cells in your eyes. Although sensation is clearly a critical first step, our focus will be on **perception,** a term that is applied to the more complicated processing, integration, and interpretation of complex, often meaningful stimuli like those we encounter in everyday life (see Figure 3-1).

Differing views of the process of perception underlie some of the most enduring theoretical

Figure 3-1 Perception: processing the sensory information encountered in everyday life.

debates in all of science and philosophy. As we emphasized in Chapter 1, the most important distinction between environmental psychology and other fields is not the particular *setting* for behavior, nor even the content areas such as crowding, personal space, or perception, but rather the perspective the field takes on studying its subject matter. In this chapter we will use the processes of perception to illustrate the distinctive approach of environmental psychology to some of these enduring theoretical issues.

Perceptual processes were among the earliest topics for 19th-century psychologists as they tried to establish their infant field. Initially, many researchers attempted to understand how individuals differed in sensing lights, colors, or other phenomena. These researchers hoped that carefully trained individuals could learn to report their sensory experiences accurately. For example, a researcher might ask, "In looking at this orange, is the sensation of color more or less important than the sensation of shape?" Soon, however, psychologists began to doubt the veracity of such *introspective* reports. Can a person really inspect and accurately report his or her own sensory experiences? Researchers began to insist that psychology was a science, and that science could investigate only *observable* (physical, behavioral) phenomena, not unobservable mental events. Another way of saying this is that psychology was moving away from subjective reports of personal experiences (known as *phenomenology*) toward a preoccupation with externally observable events as the only legitimate source of data (a position known as *empiricism*). New studies focused on biological events that led to sensations. The structures of the eyes or ears, for example, were seen as the neurological basis for the simple sensations generated by points of light or sound.

Using the empirical approach, scientists believe that they have now demonstrated that at least part of what we perceive is based on a simple, mechanical transmission of a sensory message from one part of the nervous system to another. For example, a point of light striking a small area on the eye's retina apparently begins a neural signal that moves through individual cells in several layers of the retina, to the thalamus (a structure located deep in the brain), and, eventually, to the occipital lobe at the back of the brain. Two researchers, David Hubel and Torsten Wiesel, eventually won a Nobel prize for their demonstration that certain small and identifiable areas of the brain become electrically stimulated by very specific patterns of light. For instance, one group of cells is stimulated when a *horizontal* line of light falls across an area of the retina, a different group of cells become stimulated when the light is *vertical,* and other cells seem to respond to *movement* going only in a certain direction. From this so-called *primary visual receiving area* in the back of the brain's cortex, messages spread to adjacent parts of the parietal and temporal lobes of the brain. So we think your "mind's eye," as you read the words on this page, begins with activity in the very back of your head that then spreads to a number of other areas of your cerebral cortex (Goldstein, 1999).

Imagine a place that is very special to you. Perhaps this is a place where you would like to take a close friend if he or she hasn't seen it. Perhaps this is a place known only to yourself, or perhaps the location is famous. It may be wild or urban, but what is most important is that it is special to you. Now that you have pondered for a few moments, consider a question. Do you think the methods of science and empiricism can capture all of the perceptual experiences that make the place meaningful? Even if they could adequately describe all of the physical characteristics, would they be able to understand the influence of this place on you? We think that you may be convinced that understanding environmental perception is quite a complex task and one that may require a variety of modes of inquiry.

When we say that perception involves experience and memory, we imply that cognitive processes are involved. Later in this chapter we will examine our memory for environments, especially as a basis for finding our way. In addition to cognitive processes, our feelings about the environment influence our perception of it, and our perceptions influence our feelings. To pick an extreme example, one person viewing a machine-cut swath in a forest might view it as an ugly scar, whereas another might see it as an attractive sign of jobs and prosperity. Thus, environmental perception includes both an assessment of what is in a scene and an evaluation of the good and bad elements. These affective or evaluative components and the beliefs that underlie them are the roots of the attitudes we hold toward an environment. In this chapter we focus on perception as a process for gathering information about the world and as a source of affective responses and associations.

Environments are rich in stimuli; in fact, the environment contains more information than we can comprehend at once, so we must selectively process it. Right now, make a conscious effort to process all of the stimulation coming from the environment around you. You may hear the sounds of others, a cough, for example, the turning of pages, or someone shifting positions. Can you feel the pressure of your chair, the temperature of the room, perhaps a draft from a nearby door? You may detect the odor of someone's perfume or the printer's ink on the pages of this book, and you may find yourself distracted by activity outside a window. On reflec-

tion, it is quite an accomplishment to make sense out of all this confusion! Again, we foresee an important role for cognitive processes, specifically, *information processing*. As we will see in Chapter 4, an inability to process important information because of an information overload is one explanation for some of the detrimental effects environments have on us. On the other hand, we actually seek certain levels of comprehensible information. Perhaps you agree that this inclination may underlie our attraction to such apparently different environments as exciting amusement parks and informative museums.

Finally, the perceptual process involves actions by us. We bring expectations, experiences, values, and goals to an environment. The environment provides us with information, and we perceive it through activity. Part of this activity is simple exploration to orient ourselves in an environment; part of it is designed to find strategies for using the environment to meet needs and goals; and part of it is related to establishing confidence and feelings of security within the environment. Since social and cultural factors, such as sex roles, socioeconomic status, and exposure to modern architecture, influence what one learns or has the opportunity to experience, it stands to reason that a factor such as culture influences perception.

PERSPECTIVES ON ENVIRONMENTAL PERCEPTION

Traditional Approaches to the Perception of Size, Depth, and Distance

Traditional laboratory investigations have focused on *object perception* (Ittelson, 1970, 1973, 1978; Kaplan & Kaplan, 1982), that is, the patterns of sensation that allow us to scan our memories and to recognize distinct objects with which we have had some prior experience. But the day-to-day challenges of life in the complex environments of the real world are not so simple. We must not only recognize objects, but also locate them in the context of three-dimensional space, to know how far away they are, how fast they are moving, and their importance to us (Kaplan, 1982). Thus, although environmental psychologists recognize that laboratory studies of simple stimuli offer a useful—and, in fact, necessary—foundation, they find their special challenge in the almost overwhelming complexity that characterizes real-world stimuli such as landscapes, buildings, and cities.

Although we have highlighted the differences between environmental perception and more traditional approaches, much of what we know of environmental perception is, nevertheless, based on conventional perceptual theory. Consider the interacting experiences of size, distance, and depth. Among the most powerful cues (at least for those of us brought up in North America or Europe) is **linear perspective.** Not until the 1400s did artists discover the depth-producing convention that lines that are parallel in a landscape will converge as they grow farther away. Receding railroad tracks are perhaps the most familiar example (Figure 3-2).

Forced perspective is an interesting three-dimensional application of the same principle (Figure 3-3). For example, Cinderella's Castle in the "Magic Kingdom" of Disney World, like many of the buildings in all of the Disney theme parks, uses forced perspective to make scaled-down buildings appear as tall as their grander inspirations (Marling, 1997). Of course, there are a number of other examples of the use of

linear perspective and other cues to create architectural effects. Occasionally, visual illusions or other sources of misinterpretation lead to less happy, even life-threatening consequences. Chapter 13 provides examples of such misinterpretations in aviation and other workplaces.

Holistic Analysis

One traditional goal of experimental psychology (and science in general) is to carefully control all possible causes of a phenomenon in an effort to simplify understanding. Much of what we know about human perception is based on laboratory investigations that control the variety and complexity of stimuli in order to determine their cause more easily. These procedures maximize what we call *internal validity* (see Chapter 1). For example, science has learned much from inspection of individual neurons in the human visual system. Perhaps someday we will fully understand the steps of sensation as a cascade of simple, interacting, almost mechanical events. But isn't some of the richness of behavior lost in our attempts to simplify it? As we said in Chapter 1, the most important characteristic of environmental psychology is a desire to study environment–behavior relationships as *holistic units* rather than separating them into smaller component stimuli and responses. Consider a study of landscape perception as an example. A fairly simple study might ask participants to rate the attractiveness of slide photographs; we might think of attractiveness as a dependent measure. But what are the independent or predictor variables? Colors in the photographs (there might be thousands)? Objects like trees, waterfalls, or animals? Weather conditions? Should we include the topographic relief created by mountains, plateaus, and river canyons as well as human intrusions like litter or pollution's haze? Add other modalities like smell and sound, and traditional experimental control over the individual stimuli seems absurd. But our desire to know about human reactions to the environment in which we participate does not seem absurd at all. We would really like to know what makes one landscape in all of its richness and complexity

Figure 3-2 The edges of this path in the snow appear to converge in the distance, a depth cue known as linear perspective

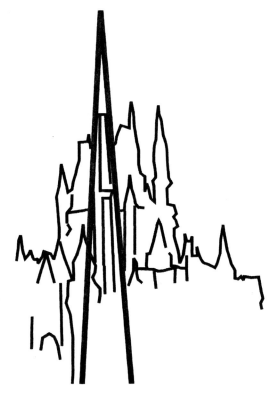

Figure 3-3A & 3B   Converging lines of forced perspective make Walt Disney World's "Cinderella's Castle" appear larger than it really is.

more attractive than another. Some more molar (larger) analysis seems necessary.

Whereas conventional approaches to perception often discuss how a sensory mechanism detects a single aspect of an object in the environment, in environmental perception we are concerned with a more holistic, encompassing process. The systems approach is one perspective that emphasizes this complex interaction of environmental stimuli and the personality of the perceiver that ultimately forms the experienced environmental unit. Although the patterns of mutual influence are complex, presumably the total system is a construction of these interacting but still separable parts (Altman & Rogoff, 1987). A **transactional approach** goes one step further and proposes that the system cannot be divided into separate elements or discrete relationships. Rather, the experienced environment is an event in time whose components are so in-

termeshed that no part is understandable without the simultaneous inclusion of the complex texture of all aspects of the instant, and it is the interaction among the components that is of interest for study. In Chapter 4 we will examine one of the best examples of a transactional approach, the ecological psychology of the late Roger Barker and his colleagues that we briefly encountered in the box on page 14.

Gestalt Psychology   You may recognize that this emphasis on holistic, global responses is similar to the position taken in the first half of the 20th century by the founders of **Gestalt perception.** As you may know, the Gestalt psychologists rejected the notion that an understanding of human perceptual processes could be furthered by reducing these processes into smaller and smaller basic units. Instead, the founders of Gestalt psychology (Max Wertheimer, Wolfgang

Kohler, and Kurt Koffka) concluded that the whole is different from solely a simple sum of its component parts. Some of the earliest demonstrations of this principle involved the apparent movement effect. Although both a child's flip book and motion pictures are made up of dozens of stationary scenes, flashing pictures one after another creates perceived movement (Goldstein, 1999; Rock & Palmer, 1990). The animated sequence may be built of discrete scenes, but its impact can best be understood as a moving whole. Just as a melody is different from a collection of its component notes and a dance is different from its steps, a landscape is more than just an array of light particles. Shape and melody are examples of what Gestalt psychologists called *emergent properties*.

Gestalt psychologists attempted to specify rules by which we organize small parts into cohesive wholes and why some of these objects become the focus (*figures*) of our attention and others, the *ground*. These so-called laws of organization outlined by the Gestalt psychologists are discussed in most introductory psychology textbooks and in courses devoted to sensation and perception (Figure 3-4). One overriding principle of organization is what the Gestalt psychologists called *Pragnänz*. Basically, this principle states that when there is some ambiguity in the visual array, the viewer will perceive the simplest shape consistent with the information available (Figure 3-5). Beyond a wealth of striking examples and illusions, however, the place of Gestalt theory in modern psychology's understanding of perception is unclear (Goldstein, 1999). In general, the Gestalt principles have withstood the test of time as descriptions of repeatable phenomena, but the original explanations the Gestalt psychologists offered for these phenomena have not (Goldstein, 1999).

Experimental evaluations of the Gestalt principles continue (see Baylis & Driver, 1995; Rock & Palmer, 1990). Even if our understanding of the mechanisms is incomplete, the

Figure 3-4   Examples of Gestalt psychology's laws of organization. (A) Proximity: Do you see six lines or three pairs of lines? (B) Closure: Do you see a circle and square even though the figures are not complete? (C) Similarity: Do you see 25 letters or a pattern of X's and O's?

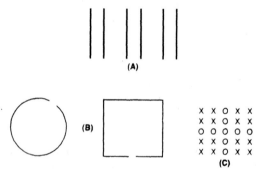

Figure 3-5A & 5B   Although these buildings are of very different ages, the use of similar columns creates coherence on the Davidson College campus.

importance of Gestalt psychology in environmental perception seems certain because Gestalt psychology has had a disproportionate influence on architects and other design professionals. In fact, Gestalt psychology is probably *the* most influential theory of perception on designers in this century (Lang, 1987). This prominence reveals a good deal about the needs of architects and about the relationship between psychology and the design professions (see Chapter 11). An architect needs to understand the visual effect of his or her design. What elements of a building's shape will be perceived as dominant? Can a new building facade be harmoniously joined to an old one? Gestalt psychology recognizes the importance of holistic analysis and offers workable answers to some of these pragmatic questions.

NATIVISM VERSUS LEARNING

As we saw in the previous chapter, another of the oldest controversies in psychology concerns the degree to which human perception comes to us fairly automatically (*nativism*) versus the view that perception is highly dependent on learning through direct observation (another manifesta-

tion of *empiricism*). In psychology, this controversy often focuses on nature (e.g., unchanging genetic influences) and nurture (learning) in human behavior. Perception of depth (and the sometimes attendant fear of heights) might illustrate both sides of the issue. We have already observed that the depth and size cues received from linear perspective appear to be at least partially constructed from our processing of learned cues. On the other hand, the Gestalt psychologists believed that perception was quite automatic, requiring little learning. As we will see, more recent theorists (e.g., Gibson, 1979; Ulrich, 1993) also propose that many aspects of perception are unlearned or very quickly learned and automatic to all humans with normally functioning visual systems.

Brunswik's Probabilism

One theory that seems particularly applicable to environmental perception and our understanding of individual learning differences is the probabilistic model of Egon Brunswik (1956, 1959). His approach, also known as the **lens model** (Figure 3-6), envisions the perceptual process as analogous to a lens wherein stimuli from the

Figure 3-6 Brunswik's lens model. Environmental stimuli become focused through our perceptual efforts. Distal cues are based on objective features of the environment and are of different importance in accurate perception (ecological validity). These cues are, in turn, weighted and processed differently by individuals (cue utilization) in making perceptual judgments.

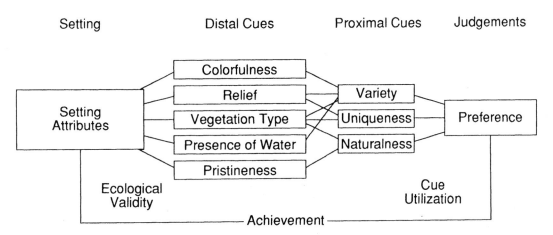

environment become focused and perceived through our perceptual efforts. Brunswik differentiated between *distal* stimulus variables that are the source of incoming sensory patterns (a distant mountain, for instance) and *proximal* stimuli (the actual pattern of light on the retina of an observer's eye). The proximal stimulus includes a great deal of complexity. Embedded in this complexity is both redundant information and ambiguity about the nature of the sensory world. Consider the world that lies behind this page as you read. Some things are completely covered by the book and are unknowable except in memory. Other objects may be partly occluded (perhaps by part of your own hand if you are holding the edge of the book) and provide incomplete information. In noncontrived situations, your knowledge of the sensory world is almost always incomplete. The actual lens in Brunswik's model represents the mental processes that search for relevant cues and weight those that experience has demonstrated to be most important in drawing perceptual conclusions.

Perhaps an example will illustrate the fundamentals of Brunswik's model. Suppose that you and a friend are hiking a mountain trail in the wilderness of Montana. Suddenly you notice movement in the bushes to your right! Your perceptual processes become focused on gathering information from the environment so that you can identify the stimulus and decide on the appropriate behavior. In little more than an instant, you decide to carefully back off from what you perceive to be a foraging grizzly. Seconds later, your companion laughs and points to a very small, albeit very loud, chipmunk. How did you make such a mistake? We know that not all stimuli presented by the environment are equally useful in accurate perception. Some of the information presented by the environment may be insufficient, superfluous, or even misleading. The noises coming from the bushes beside the trail provided useful but insufficient information to make an accurate perceptual decision. Noise is one characteristic of animal movements, whether from a chipmunk or a grizzly bear, so it is one useful cue, but perhaps not as useful in this instance as knowing the ani-

mal's size. Each of the stimuli emanating from the environment might be assigned a probability weight based upon its usefulness in supporting accurate perception. In Brunswik's terms, these stimuli vary in their *ecological validity*.

However, two observers might differ in their interpretation of the situation, even when each receives the very same stimulus array. You and your companion might weight the environmental information differently in making a best guess or probabilistic judgment. Perhaps you were just thinking of reports of grizzly bear attacks in nearby Glacier National Park, or perhaps your friend was daydreaming and did not even hear the noises. Thus, not only do certain environmental stimuli differ in their objective usefulness (**ecological validity**), but individuals may weight them more or less appropriately (**cue utilization**) because of past experiences, personality, or other differences. Rather than simply determining the judgments of the average observer, Brunswik's model fosters *policy capturing,* a procedure that determines the idiosyncratic patterns of weights assigned by individual judges. Subsequent analysis may reveal that certain individuals (perhaps possessing common background or personality traits) share similar weighting profiles or policies, a valuable insight that might have been overlooked by traditional averaging techniques. (For a recent application of the lens model to architecture, see Gifford et al., 2000.)

Ecological Perception of the Environment

The perceptual experience consists of many "significances," or meaningful stimuli or events that reach our awareness. That is, we are most likely to notice those things of significance to us as members of the human species, perhaps because they help us to survive. A name for this perspective from the traditional study of perception is **functionalism.** According to this view (e.g., Kaplan & Kaplan, 1982; Ulrich, 1993), our perceptions are molded by the necessity to "get along" with the environment. For example, we compare present sensations with past ones in order to see whether the present stimuli signal danger or serve as cues for food or shelter. Often theorists suggest that these functional processes have evolved biologically as part of our species' adaptation to environmental demands. Of course, evolution implies change, but by emphasizing the importance of biological or genetic influences, these functional theories are quite nativistic with respect to the perceptions of an individual human.

Our discussion of perception of the environment (particularly from the functional perspective) would not be complete without considering J. J. Gibson's **ecological perception.** Gibson was critical of standard perceptual theory's emphasis on the processing of individual cues in static, "snapshot" visual images. According to Gibson (1979), rather than perceiving individual features or cues that we organize into recognizable patterns, we respond to (detect or tune in) meaning that already exists in an ecologically structured environment. We may overlook some of this embedded meaning, but it is readily available to an appropriately attuned organism mobile enough to experience it (see also Heft, 1981, 1989).

We have noted that the conventional approach to perception considers perception of the external environment as a function of a variety of interpretive psychological processes, that is, a stimulus activates a specific nervous system receptor, and the pattern of receptor stimulation is interpreted with the memory of past experiences to get information about the environment. From the conventional perspective, we have to interpret disconnected stimuli in order to construct something meaningful about the environment. Gibson considered these to be special, rather artificial circumstances. For him, perception of the environment is more direct and less interpretive than this. That is, perceptual patterns convey much information quite directly—without elaborate processing by higher brain centers. Furthermore, Gibson believed that perception is much more holistic, so that properties of the environment are perceived not as distinct points but rather as meaningful entities. Let us develop the Gibsonian approach to perception a bit further by exploring the concept of affordances.

Perception of Affordances According to Gibson, we receive much valuable information directly through our perception of the environment. Gibson viewed organisms as actively exploring their environment, encountering objects in a variety of ways. Through this process, we experience the surface of an object, its texture, and angles from different perspectives. This allows us to perceive an object's *invariant functional properties* (i.e., "useful" properties of an object that do not change, such as "hardness"). The invariant functional properties of objects as they are encountered in the course of an organism's active exploration are termed *affordances* (Figure 3-7), a term we encountered in the previous two chapters. The notion of affordances will become clearer as we look at a few examples. If an object is solid rather than liquid or gaseous, if it is inclined toward the ground at an angle other than 90°, and if at least part of it is higher than the organism, then that object affords shelter. If an object is solid and rigid, if it is raised off the ground, and if its top surface is fairly parallel to the ground, then the object affords sitting or "sittability." If an object is malleable, can be placed in the mouth whole or in pieces, and is of such biochemical substance as to provide nourishment, then it affords "eatability."

Obviously, what affords shelter, sittability, and eatability for a fish does not necessarily do

Figure 3-7 This arrangement of stones affords "sittability."

so for a human; what affords these things for a human does not necessarily do so for an elephant. In this sense, affordances are species-specific (although there is, of course, overlap across species). Whereas a tree affords shelter to a bird and food to certain insects, it affords fuel (among other things) to humans. For this reason, affordances must be viewed from an ecological perspective.

From this perspective, we can see that affordances involve perceptions of the ecologically relevant functions of the environment. To perceive affordances of the environment is to perceive how one can interact with the environment. It is through perception of affordances that an organism can find its niche in the environment. An **ecological niche,** according to Gibson, is simply a set of affordances that are utilized. In this regard, humans possess a remarkable talent: We can alter an environment so that it affords anything we want, such as more expensive shelters, more beautiful scenery, and so on. In doing so, we may change the affordances of that environment with respect to other humans and other organisms. When we dam a river to create a lake, which affords us water, recreation, and flood protection, we may also change the immediate environment so that it affords life support for fish and waterfowl but does not afford life support for groundhogs and bats or for the farmer who lost his or her home or cropland to the lake. Perception of this changed environment, then, depends on the gain and loss of affordances for each organism. Certainly, many of the changes we humans have imposed upon our environment for our short-term benefit have had severe long-term consequences for both our species and others. Our skill in manipulating the affordances of our environments is both wonderful and dangerous.

Some Implications for Environmental Perception

In one sense, Brunswik's and Gibson's theories could hardly have been more different. Brunswik emphasized perception as a process of probabilistic calculations influenced by individual differences whereas Gibson insisted that perceptual "truth" lies in the environment and can be perceived directly with little or no complex interpretation. The controversy continues (Kaplan, 1995; Ulrich et al., 1991). What Gibson and Brunswik shared, however, was a belief that perception can best be understood by examining the complexities and challenges of perceivers in the real world. Like the Gestalt theorists, their emphasis on holistic, molar analysis anticipated the approach to perception that is most characteristic of modern environmental psychology.

HABITUATION AND THE PERCEPTION OF CHANGE

It is apparent that environmental perception is a very complex and involving process. Because all we know of our world is filtered through perception, perceptual processes underlie much of the balance of the content of this textbook. For the moment, let us briefly examine habituation and the perception of change.

Thus far we have talked about perception without regard to time. That is, we have noted some of the principles and properties of environmental perception as if perception is constant from one moment to the next. Once we consider time as a variable in environmental perception, three important phenomena emerge: perception of movement, habituation (adaptation), and perception of change. The latter two are especially important in environmental psychology.

Habituation or Adaptation

What happens if a perceivable stimulus does not change across time? The answer involves what is known as **habituation** or **adaptation:** If a stimulus is constant, the response to it typically becomes weaker over time. Many who live near freeways, for example, at first find it difficult to sleep, but after a few nights they become habituated to the noise and have little trouble sleeping. Should they have guests some night, however, the guests are likely to be bothered by the noise.

Explanations for adaptation or habituation tend to be either cognitive or physiological (Evans et al., 1982; Glass & Singer, 1972). Sometimes the distinction is made that "habituation" refers to a physiological process and "adaptation" to a cognitive process. Often, however, the two terms are used interchangeably.

Physiological explanations of habituation emphasize the notion that the receptors themselves fire less frequently upon repeated presentation of a stimulus. Cognitive explanations of the phenomenon propose a cognitive reappraisal of the stimulus as less deserving of attention after repeated presentation. The first time you hear a loud noise, you allocate considerable attention to it to find out what it is and to determine whether it is a potential source of threat. Once you know that it is a train, a trash truck, or your neighbor's car, you probably evaluate it as nonthreatening to your well-being and thus attend to it less. However, from a cognitive perspective, our example may reflect more of a response bias than a perceptual shift. That is, rather than actually perceiving the noise as less noxious, nearby residents may simply learn to respond to it less intensely or less frequently (e.g., Evans et al., 1982).

Adaptation is not always successful in eliminating unpleasant environmental stimuli, of course. If the stimulus is too unpleasant, it may well continue to be perceived as annoying (e.g., Loo & Ong, 1984). Furthermore (as we will discuss in more detail in Chapters 4 and 5), even adaptation that appears successful may require the mobilization of the body's physical or cognitive resources and eventually contribute to a general breakdown that may be manifested in stress disorders.

An important factor in adaptation (again, refer to Chapters 4 and 5) is the predictability or regularity of the stimulus. We are more likely to adapt to a constant hum in the background than to the irregular noise of a jackhammer. Bursts of noise that come at regular or predictable intervals are easier to adapt to than unpredictable stimuli but more difficult to adapt to than constant stimuli. Once we adapt to a stimulus and the stimulus ceases (as in the interval between bursts of noise), our adaptation to the stimulus also dissipates somewhat. When the stimulus recurs, we must adapt again. Furthermore, unpredictable stimuli require that more attention be allocated for evaluation of the stimuli as threatening or nonthreatening. Thus, predictability is an important variable in the adaptation process.

Perception of Change

If we readily adapt to environmental stimulation, will we perceive change in such things as air pollution and urban blight? If we live in an area where air pollution is high, and we adapt to it, how can we perceive changes in the level of pollution? Sommer (1972) suggests that the answer lies in the **Weber–Fechner function** of psychophysics. This function, derived from the research of the late 19th century, is based on the amount of increment (increase or decrease) in intensity of a stimulus that is required before a difference is detected between the new and old intensities. Stated simply, this law says that the intensity of a new stimulus required for it to be perceived as different from the present stimulus is proportionate to the present stimulus. To use an economic example, there seems to be more of a difference between $1.00 and $2.00 than between $1,000,001 and $1,000,002. It takes only a small increment to detect a difference in very low-intensity stimuli, but a much larger increment is needed for high-intensity stimuli. This function (though not as mathematically accurate as more modern psychophysical functions) generally applies to all forms of stimulation, including light, sound, pressure, and smell. Sommer suggests that the law applies not only to individual stimuli in a laboratory, but to urban pollution as well. That is, a community with little pollution might become alarmed when clouds of brown smog suddenly appear, but large urban areas with heavy smog should require extremely high levels of additional pollution before becoming alarmed. Similarly, we might expect strip zoning in small communities where careful neighborhood planning exists to be noticeable enough to spur the community to action against such blight. Larger communities where strip zoning is commonplace, however, would probably not care as much when one more fast-food chain appears on the strip.

Sommer proposes that we take advantage of the Weber–Fechner phenomenon in changing detrimental environmental behaviors. Any time we are asked to change our lifestyles to preserve the environment, there is resistance. But what if the change in lifestyle is so small as to go unnoticed? We might be able to make subtle changes that have a great impact on the environment. Requiring that beverages be sold in returnable containers, for example, is not as drastic a measure as banning beverages in all containers. Requiring that recyclable containers be sep-

arated from other trash is even a smaller step than banning nonreturnable containers, and so on. In other words, if the perceivable change is small, we will be less resistant to it than if it is large.

Furthermore, change that is rapid (such as movement or burning) is more easily detected than change that is slow (such as growth). There is ecological survival value in knowing that one's environment is changing rapidly. Imagine, for example, the importance of prompt reaction if a forest fire endangers your home. Unfortunately, comparable damage that occurs slowly (as when pollution from cities kills trees) is less noticeable.

Overview of Environmental Cognition

For much of the first half of the 20th century, psychologists shied away from discussing such "cognitive" matters as memory and images. Their reluctance reflected the objection of behaviorists and others to the study of the "unobservable and unmeasurable" events that occur as we process mental information. With the so-called cognitive revolution of the latter part of that century, some psychologists "rediscovered" these complex operations of memory, thinking, problem solving, and imagery (see Evans, 1980; Gärling & Evans, 1991; Golledge, 1987; Golledge & Stimson, 1997). Directly observable or not, information from memory gives us important clues to those aspects of the environment that are most salient or important to us. Have you ever been lost? Particularly in remote regions, people report that being lost is a threat to survival and a profoundly troubling challenge to our self-confidence. When applied to large-scale environments, memorable features may be useful for finding our way from one place to another and back again. Easily remembered environments are easier to travel through, and the opportunities they provide are more apparent. Simply being more comprehensible may make environments more aesthetically pleasing (see

Chapter 2). Finally, inspection of environmental memories may offer a way to "get inside our heads" for insights into the way humans store, process, and retrieve information. Many find this to be one of the most fundamental, fascinating, and controversial areas in all of psychology.

AN INFORMAL MODEL OF SPATIAL COGNITION

The ability to capitalize on a rich and varied environment is at least partially dependent upon the human propensity to store geographical information. Useful as maps and charts may be, however, people more typically travel through a familiar environment without these aids. How is this accomplished? Many psychologists suggest that all humans carry with them an organized mental representation of their environment, commonly referred to as a *cognitive map*. Simply stated, a **cognitive map** is a mental framework that holds some representation for the spatial arrangement of the physical environment. The term *cognitive map* may be unfortunate, however (e.g., Kitchin, 1994; Kuipers, 1982). Is the cognitive map something a research participant draws to represent his or her spatial memory? Is it a stored image that roughly corresponds to the actual spatial environment? Is it simply a metaphor: a mental structure that is used as if it were a physical map? Is it simply an unfortunate, but convenient, fiction? More accurately, perhaps we should refer to a "cognitive collage" consisting of many different bits of information, some pictorial, some episodic, some maplike (Hirtle & Sorrows, 1998; Tversky, 1993). Given the long history of the term, we will cautiously adopt "cognitive map," but we caution you to think of it as an inexact, perhaps even inaccurate, description of the spatial environment.

No matter what the status of these "maps," it does seem clear that they are not the same as a cartographer's in either physical form or in content. They are sketchy, incomplete, distorted, simplified, and idiosyncratic. We might think of them as composed of three elements: places, the spatial relations between places, and travel plans (Gärling et al., 1984). In this instance, *place* refers to the basic spatial unit to which we attach information such as name and function and perceptual characteristics such as affective quality or affordances (in subsequent chapters we develop a somewhat more complex understanding of place). Depending upon the scale of the particular cognitive map we are consulting, a place may be a room, a building, a town, a nation, or a planet. In addition, cognitive maps reflect *spatial characteristics*, such as the distance and direction between places and the inclusion of one place within another (your room is inside a building, which is itself within the boundaries of a town, and so on). Finally, Gärling et al. (1984) propose the concept of *action plans* as important bridges between the mental world of cognitive maps and the navigation and other behaviors that they support.

Whether maps are stored in the mind or on paper, we might ask, "What do maps do for us?" We have already suggested that a primary use is to facilitate *wayfinding*. **Wayfinding** is the adaptive function that allows us to move through an environment efficiently to locate valuable items like food, shelter, or meeting places within the environment (Downs & Stea, 1977; Evans, 1980). This leads us to propose an informal model (see Figure 3-8) that emphasizes travel from one place to another as a primary goal of spatial cognition. Before we begin a journey, we

Figure 3-8   An informal model of spatial cognition. Instructions from other humans, printed maps, and memories of past travels help an individual form an action plan for a proposed journey. The success or failure of this plan as it is carried out becomes stored in memory and leads to future travel plans and to an evolving cognitive map for future reference.

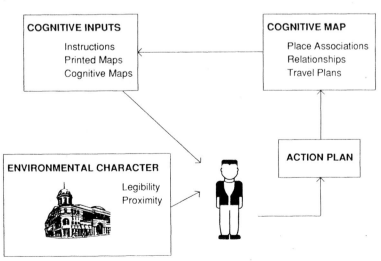

construct an action plan, that is, a strategy or itinerary for our movements (Gärling et al., 1986). Our plan will need to include some sort of information about the relative locations of places. Without this information we would have to search haphazardly, hitting or missing desired locations in a very inefficient way. In a new environment we may need to depend upon a physical reference such as an atlas or a friendly passerby in order to formulate our travel plan. (Of course, we may later use our own cognitive maps to communicate locations to others and to understand others' communications about location to us.) Being able to "visualize" the directions someone gives us and associating the directions with familiar landmarks and paths enhances our wayfinding ability. In the absence of such physical aids or in well-known environments, we consult the spatial representation in our memory, our cognitive map.

Notice that although our figure recognizes a cognitive map as a source of information for the construction of plans, acquisition of the map is itself the result of previous experience in the environment. We begin our subsequent discussion with the stored information residing in cognitive maps and then turn to their application in wayfinding. This is really a cyclical system, and our starting point is arbitrary.

Cognitive Maps

The topic of cognitive mapping has fascinated not only environmental and cognitive psychologists, but also researchers in geography, anthropology, and environmental planning and design. Cognitive maps are a very personal representation of the familiar environment that we all experience. Take a few moments to think about the layout of your campus. Try to imagine several vistas and the paths you most frequently take. Now, on a clean sheet of paper try to draw a map of the campus showing important features so that a stranger could use your sketch to find his or her way around. This is, of course, your personal cognitive map; we will refer to this type of drawing as a **sketch map.** You will probably want to refer to it often as you continue reading this chapter.

HISTORY OF COGNITIVE MAPPING

Investigation of cognitive maps is not really a new idea (cf. Trowbridge, 1913). Modern study of these maps has its most direct roots in the work of E. C. Tolman (1948), who described the way in which rats learn to "map" the environment of an experimental maze. Tolman's basic strategy over a number of experiments was to first train rats to take a particular path in a maze in order to reach a food reward. When the path was later blocked, the rats seemed able to switch to another previously unused path that led toward the goal. Some rats would choose a path never before used and pass up one that had been reinforced if the new path was a more direct route to the goal box (Tolman, 1948; Tolman, Ritchie, & Kalish, 1946). Thus, the rats seemed to have learned not just a series of turns or responses, but also a general idea of the location of the reward relative to the starting position. In order to describe this place information that his rats had apparently learned, Tolman coined the term *cognitive map.*

An Image of the City: Kevin Lynch

At first, few investigators were interested in pursuing the study of cognitive maps. Although it would be an exaggeration to say that Tolman's work was forgotten, it was not until the publication of *The Image of the City* by the urban planner Kevin Lynch (1960) that there was widespread interest in understanding the formation and use of humans' cognitive maps.

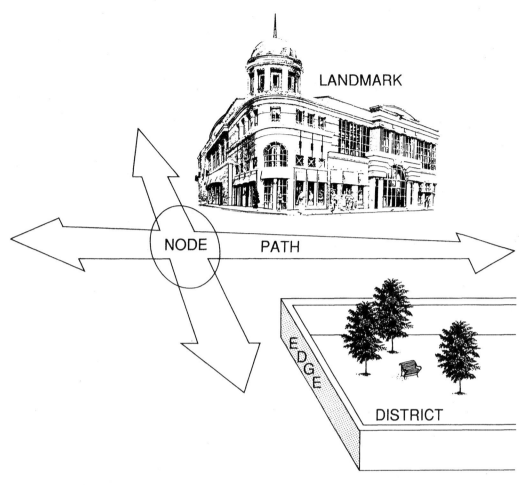

Figure 3-9 A diagram illustrating all five of Lynch's elements in a cognitive map.

As a planner, Lynch was among the first in his field to try to understand such subjective concerns as people's feelings about the quality of their environment and how their perceptions could be used in environmental design. *The Image of the City* remains a classic reference in cognitive mapping. In it, Lynch simultaneously established a field of inquiry, a methodological approach to data collection, and a vocabulary for describing features of cognitive maps—a vocabulary that is still used widely. For both historical and pragmatic reasons, then, Lynch's approach seems to merit a detailed discussion.

Lynch based his initial ideas about cognitive maps on data gathered in Boston, Jersey City, and Los Angeles. Lynch asked participants in

his studies to draw sketch maps of their city, to give detailed descriptions of certain routes such as the path from home to work, and to list the most distinctive and vivid elements of their respective cities. Upon comparing these data, he identified different elements that seemed common across the three different cities (see Figure 3-9).

Elements of Cognitive Maps

Return to your sketch map of campus. Are there obvious streets and buildings designated on it? Are there broad areas you could designate as "fraternity row" or "dormitory area" or "athletic complex"? Lynch found that five categories of features could be used to describe and analyze cognitive maps: paths, edges, districts, nodes, and landmarks. **Paths** are shared travel corridors such as streets, walkways, or riverways. **Edges** are limiting or enclosing features that tend to be linear but are not functioning as paths, such as a seashore or wall. Notice that in some instances one person's path (the rail line of a commuter train) may be another person's edge (if the rail line divides a town). **Districts** are larger spaces of cognitive maps that have some common character such as "Fraternity Row," or the "Chinatown" found in many cities. **Nodes** are major points where behavior is focused, typically associated with the intersections of major paths or places where paths are terminated or broken, such as a downtown square, a traffic circle, or the interchange of two freeways. Finally, **landmarks** are distinctive features that people use for reference points. Usually landmarks are visible from some distance, as in the case of the Washington Monument or a tall building in a city. Can you identify examples of these five categories on your campus map?

Additional Early Observations

The basic elements (paths, landmarks, nodes, edges, districts) outlined by Lynch seem well established (Aragones & Arredondo, 1985; Evans, 1980), although some have suggested that these elements are most applicable to environments on the scale of cities (for which Lynch developed them) rather than smaller or larger units of analysis. Lynch was part of the team that helped to maintain the delightful pedestrian corridor stretching from Boston's Commonwealth Avenue, through the Public Garden and Boston Common, and (with minor breaks) all the way to a revitalized Boston Harbor. Central in the plan is the Quincy Market, a successful downtown redevelopment that attracts both tourists and residents (see Chapter 10). Other early researchers who were inspired by Lynch noticed stylistic differences in people's cognitive maps. Lynch's associate Donald Appleyard (1970), for example, used sketch maps to evaluate the images of residents in a city in eastern Venezuela. The maps seem to fit into one of two categories: those predominantly made up of elements that one might encounter sequentially in traveling from one place to another, such as paths (**sequential maps**), or those that instead emphasize spatial organization (a common term for this type of bird's-eye view is **survey knowledge**) such as landmarks or districts (**spatial maps**) (see Figure 3-10). At least for these city dwellers, Appleyard reported that most maps were sequential, that is, rich in paths and nodes. This interest in the acquisition of cognitive spatial knowledge and the distinction between sequential and survey knowledge remains very current.

Figure 3-10 Idealized examples of sequential (left) versus spatial cognitive maps.

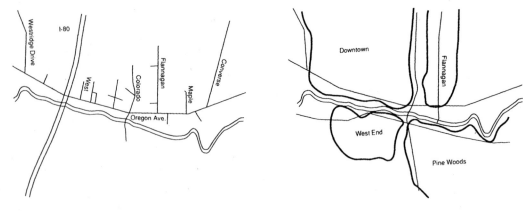

CURRENT PERSPECTIVES

More and more researchers began to discover cognitive mapping in the following three decades, leading to the present variety of research goals and methods all described loosely by the terms *cognitive map* or *spatial cognition.* As we said, cognitive mapping has evolved from several disciplines, especially psychology, planning, and geography. This reflects the excitement of interdisciplinary research so characteristic of environmental psychology. At least partly because of differences among these disciplines, however, the general topic of cognitive mapping is a loosely organized literature based upon a variety of methods and research goals.

One might think of an individual's finished cognitive map as representing a personal understanding of his or her environment. Certainly some features of the physical environment possess characteristics that are likely to cause them to be perceived as more important or distinctive, and thus, more likely to be stored in memory. Planners and geographers have been especially concerned with these physical characteristics (Gärling & Evans, 1991; Kitchin, 1994; Kitchin, Blades, & Golledge, 1997). Often the features of concern to these researchers are those that make an environment *legible* (see Figure 3-11). *Legibility,* another term popularized by Kevin Lynch and that we encountered in the previous chapter, reflects the degree to which an environment is easily learned and remembered. Clearly, legibility is important in facilitating travel within a building or outdoor environment, but it may also influence our affective reactions to the environment. Indeed, in Chapter 2 we discussed legibility as one important predictor of landscape preference.

In addition to the physical characteristics of a city or rural environment itself, you might expect that individual or cultural differences

Figure 3-11   Scenes that allow aerial or long-range perspectives are often rated as highly legible.

would cause people to place varying weights on certain environmental features. For example, you are likely to know more about the area of a college campus nearest your dorm or along your most frequent path to class. We might expect your map to be somewhat different from that of a person living in a different location or from that belonging to a faculty member. Of course, almost all researchers agree on the importance of these individual differences in experience. This has been a particularly important area of investigation for psychologists, especially those with an interest in social or developmental psychology.

METHODS OF STUDYING COGNITIVE MAPS

Given the variety of disciplines and specialties that have found interest in the general topic of cognitive mapping, you may not be surprised to learn that there are almost as many methodological techniques for gathering and analyzing cognitive maps as there are researchers. This diversity is exciting, but often the source of serious difficulties because data gathered using one method cannot be easily compared with data produced by another. In fact, these methodological problems may be among the most serious faced by researchers in the area (Evans, 1980; Kitchin, 1996; Milgram & Jodelet, 1976). Let us begin by describing some of the most common approaches.

Sketch Maps

As you recall, Kevin Lynch (1960) employed several methods in his seminal investigations of people's responses to the spatial environments of Los Angeles, Jersey City, and Boston. His primary method, however, was to ask individuals to draw a sketch map of their city (Figure 3-12). This approach has remained among the most popular, and it was responsible for establishing the vocabulary of cognitive mapping terms such as *paths, landmarks, districts, nodes,* and *edges.* Sketch maps provide a rich source of data. They have several liabilities, however, and these seem

to become more and more serious as researchers become more sophisticated in their research questions.

Sketch maps are flawed if observed differences in responses aren't primarily the result of differences in the mental maps that they are meant to measure. We have already observed that sketch maps require participants to take a perspective that places them in the air above the terrain being mapped—a perspective they are unlikely to have experienced. It is likely that participants vary in drawing ability or their ability to take hypothetical perspectives. Not surprisingly, different tests of cognitive mapping ability arrive at different results (Kitchin, 1996). Consider the fairly straightforward problem of accounting for the scale of the field upon which individuals draw their maps. Classically, they receive a sheet of blank paper and are asked to reproduce a particular city, neighborhood, campus, or other spatial environment. This method seems unbiased in tapping the resident mental image of a spatial environment, but the scale and orientation the person chooses for his or her maps will be idiosyncratic, making comparison across individuals problematic. For example, how can one reasonably compare two maps of a city, one that includes suburbs and outlying areas, another confined to the city proper? Providing people with several existing landmarks or an outline of the area reduces this variability— but also introduces artificial constraints on the map. Consider the task of a researcher faced with the job of analyzing the sketch map you created in response to our challenge earlier in this chapter. How could a researcher combine your map with those drawn by others? Should the analysis include a list of named buildings or other landmarks and paths? Probably, but can the researcher correctly identify unlabeled (but drawn) buildings and streets? What locations are labeled on a map but in the wrong place? The researcher may notice **distortions:** streets that intersect at the wrong angle, for instance, or missing curves in rivers, but how can these distortions be quantified?

In spite of these liabilities, sketch maps are still common. They generate extremely rich data

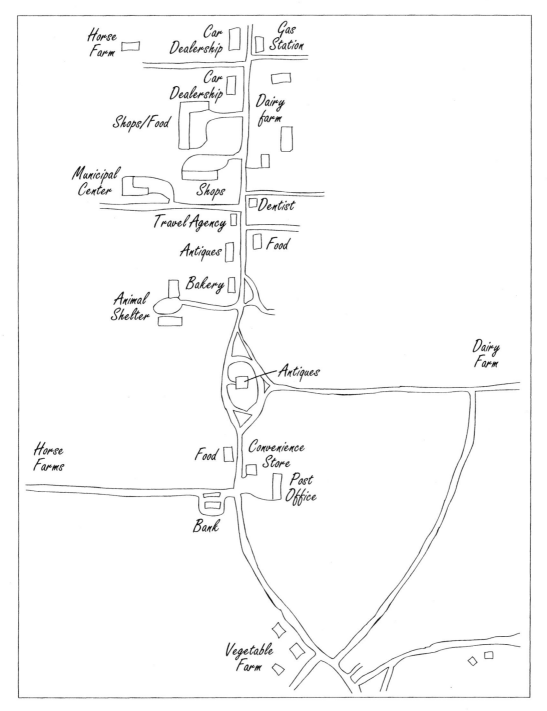

Figure 3-12   A student's sketch map of her hometown.

in a manner that usually seems to be ecologically valid (realistic and reasonable). In fact, sketch maps are probably about as valid and reliable as other methods of assessing spatial knowledge (Blades, 1990; Newcombe, 1985; Pinheiro, 1998).

Mapping Reactions to Remembered Environments

Geographers like Peter Gould and Rodney White (1982) present a different approach to mental maps. Whereas the primary focus of the methods presented so far is to reproduce the person's mentally stored image of an environment, Gould's approach recovers not a person's cognitive map, but rather, characteristics or qualities assigned to places within a person's environment that are compiled and subsequently represented graphically on a map. We call such a representation a *Gouldian map*. Various statistical approaches have been employed, but all depend on asking people to rate or rank a number of different points according to some evaluative dimension (preference as a place of residence, for example). The final results can be superimposed as shaded regions (these may correspond loosely to Lynch's districts) on an accurate basemap. These shaded regions represent the collective assessments of subjective qualities such as attractiveness (Gould & White, 1982; Lloyd & Steinke, 1986; Mace & Greene, 1997) or familiarity (e.g., Gale et al., 1990), and can be on virtually any scale. For example, Figure 3-13 shows Gould and White's desirability ratings for areas of the United States based upon data collected from California residents. Notice that areas of high preference include the West Coast and New England. On the other hand, the Deep South and South Dakota receive lower ratings. How would residents of a different location, say, the Deep South, respond? As you can see from the figure, people tend to like their regions, even if others in the nation are less favorably impressed. This seems reasonable, if for no other reason than self-selection. In Chapter 11 we will explore the use of cognitive mapping techniques as part of the planning process on a small college campus.

Figure 3-13　A preference map of the United States as reported by California residents.

Recognition Tasks

In his early investigations of residents' images of Boston, Kevin Lynch also asked participants to report whether they recognized photos of landmarks that were interspersed in a collection of pictures of unfamiliar locations. Lynch seems to have included this *recognition task* as a reliability check for his more familiar sketch map procedure. Milgram and Jodelet (1976) revived this approach because it avoids many of the problems inherent in having people with varying abilities draw sketch maps. Unfortunately, the procedure limits our ability to compare the orientations and geographical distances between spatial elements that are often evident both in various mapping techniques and in direct-distance estimates as discussed in the next section. In addition, this technique emphasizes *recognition* (the ability to recognize a place you have seen before) over *recall,* which asks you to remember and reproduce as much as you can without the assistance of photos to jog your memory. To illustrate, would you typically draw and label the location of your favorite dry cleaner on a sketch map of your hometown? Probably not. Would you recognize the same dry cleaning establishment from a picture of it? Probably. An emphasis on recognition over re-call is not necessarily a liability. Some (e.g., Passini, 1984) believe that recognition tasks more closely approximate the way most of us deal with movement within familiar environments (we will return to this issue in our discussion of wayfinding later in this chapter). Still, it should be clear that these recognition tasks are quite different from the standard sketch map technique, and thus, are not directly comparable.

Distance Estimates and Statistical Map Building

A number of researchers have also employed an approach that avoids sketch mapping by asking people to simply estimate the distances between locations in a large-scale environment. Of course, these distances probably represent some of the information included in a person's sketch map, and being able to estimate distance is an

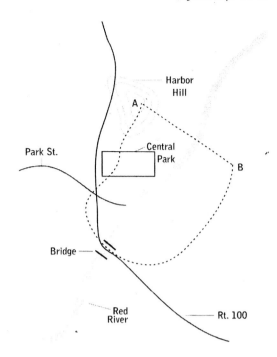

Figure 3-14   The *functional* distance from point A to point B is considerably longer than the *euclidian* (crow fly) path because of the need to cross a river at the bridge. The *cognitive* distance may be longer still because of the increased information created by the "clutter" of street crossings and other environmental changes.

important tool for someone wanting to travel in his or her environment. We can distinguish between at least three measures of distance (see Figure 3-14). First, there is a physical measurement of the distance between two points "as the crow flies," often referred to as *euclidian* distance. Many years ago, Festinger, Schachter, and Back (1950) used the term *functional distance* to denote the actual travel distance for a person walking or driving between locations in the target environment. Finally, *cognitive distance* (Golledge, 1987; Hanyu & Itsukushima, 1995) can refer to either euclidian or functional distance, but in either case, it is based upon a cognitive or perceptual estimate rather than an actual measurement.

One well-known statistical approach to assessing cognitive estimates of euclidian distance is **multidimensional scaling (MDS).** MDS is a statistical procedure in which participants estimate the distances between a number of buildings or other locations in the environment. Given the distance between each point and each of a number of other points, a computer can generate something resembling a map by optimally placing each location on a two-dimensional map so as to minimize errors in distance estimates. You might be able to test an informal form of this scaling technique without knowledge of the underlying mathematics. Ask a group of your friends to estimate the distance between each pairing of 10 buildings. Now calculate the mean (average) estimated distance between each pair of locations. Cut a piece of yarn to a scaled length representing each mean distance, and stretch out all of the yarn pieces so that they connect to tacks (representing the buildings). If all of your friends were absolutely accurate for each distance estimate, the procedure should generate an accurate map. Of course, consistent distortions are interesting, too. For instance, your friends might consistently exaggerate the distance of unpleasant or uncommon travel paths.

Although multidimensional scaling eliminates some of the problems of other methods (e.g., differences in drawing skills), it, too, has problems. Later in this chapter we will encounter one of the most serious: There is increasing evidence that a person's estimate of the distance from point A to point B is *not* necessarily the same as his or her estimate of the distance from point B to point A!

Our description might have prompted you to consider other problems with MDS. Unlike an unbounded sketch map, the procedure focuses not on *what* paths or landmarks a person remembers but on the estimated distances between locations presented by the researcher. In general, distance estimates may be more easily quantified than sketch maps, but they also lack some of the intuitive ecological or face validities.

ERRORS IN COGNITIVE MAPS

Types of Errors

Cognitive maps are rough approximations rather than perfect representations of the physical environment. In fact, we can identify several sources of error that frequently occur in them (e.g., Appleyard, 1970; Byrne, 1979; Evans, 1980; Downs & Stea, 1973; Pinheiro, 1998). First, cognitive maps tend to be *incomplete*. We often leave out minor paths and details, but sometimes we even omit districts and landmarks. Second, we often *distort* our representation of the environment by placing things too close together, too far apart, or aligning them improperly. Most errors in cognitive maps of cities occur at street intersections, where people have a tendency to misestimate the size of intersection angles. Acute intersection angles are often overestimated, and obtuse angles are underestimated. Figure 3-15 shows an intersection in a small town. As you might expect from our discussion, the intersection is depicted as a right angle in almost all cognitive maps of the area. Perhaps the error was compounded in this example by the designers of two of the buildings adjacent to the intersection. The building on the left (a bowling alley) actually matches the acute angle of its side of the intersection, whereas the building on the right matches the obtuse angle of its street boundaries. The irregularities of the buildings come as a complete surprise to even long-time residents of the village. We also tend to represent nonparallel paths as being parallel, nonperpendicular paths as being perpendicular, and curved paths as being straight.

People also have a tendency to overestimate the size of familiar or liked areas in their cognitive maps . For example, Milgram and Jodelet (1976) found that Parisians seem to increase the size of their home neighborhood out of proportion with the rest of Paris. These errors may be quite telling. For instance, Pinheiro (1998) asked Brazilian students to draw sketch maps of the entire world. In addition to geographical size, the size of the countries in the sketch maps (and whether they were even included) was heavily influenced by indicators of geopolitical, military, and economic power. Affective qualities assigned to environments also affect sketch maps. Seibert and Anooshian (1993) report that participants across several age groups did not typically include disliked items or areas on sketch maps.

Figure 3-15 A nonperpendicular intersection in a small town that is remembered as perpendicular by even long-time residents.

A third type of error involves *augmentation,* or the addition of features to a map that are not there. Appleyard's Venezuelan study (1970) provides a classic example of these augmentations. A European engineer included a nonexistent railroad line in his sketch map because experience led him to predict a rail connection between a steel mill and a mining port. In this instance, the engineer's experience led him to infer a logical, but nonexistent, map component. Notice that this same phenomenon (sometimes referred to as *inferential structuring*) may often lead individuals to make correct assumptions, but these may properly be called augmentations if the person's cognitive map represents features that have never actually been experienced. In these instances, the researcher is unlikely to recognize the augmentation and will miscode an interesting error as an accurate response.

Altogether, then, our cognitive maps are clearly not always very accurate representations of the physical environment. Understanding the sources of these errors may well give us insights into the effect of individual differences in such factors as experience, age, skill, or personality.

Furthermore, some researchers are using insights gained from either the errors people make in spatial cognition tasks or differences in the speeds at which features can be recalled to learn more about the basic processes underlying human memory.

Familiarity and Socioeconomic Class

The types of errors in our maps, as well as the degree of detail in them, vary according to several factors. As you might expect, a number of studies have shown that the more familiar you are with an environment, the more accurate and detailed are your cognitive maps of it (e.g., Appleyard, 1970, 1976; Evans, 1980; Gärling et al., 1982).

Several authors report that familiarity probably explains the frequent observation that people from higher socioeconomic status groups draw more thorough maps than the poor (e.g., Appleyard, 1976; Ramadier & Moser, 1998). That is, upper- and middle-class individuals probably have more experience with broader areas of a city than lower-class individuals whose mobility is restricted, primarily by the lack of easy access to transportation. In the classic study, Donald Appleyard reported that motorists (generally from the upper class) had the most sophisticated maps of a Venezuelan town, whereas those forced to walk produced less sophisticated sketches. In sum, those with more travel experience make better sketch maps (Beck & Wood, 1976). Even more important than just being mobile, those who must attend to the passing environment (drivers, for instance) are more likely to process street names, directions, addresses, and distances. Thus, public transportation users who may well travel great distances but attend only to the passing sequence of stops do not produce the richness or accuracy of the cognitive maps of drivers. In sum, the longer we have experience with an area and the more movable we are within it, the more thorough our cognitive maps.

Ramadier and Moser (1998) broadened our understanding by introducing the term *social legibility* to characterize the cultural distance between an individual and the surroundings. In a study conducted in Paris, sketch maps and other measures for students from sub-Saharan Africa differed from those of European students (Spain, Italy, and Portugal). Apparently, for the Europeans, culturally derived expectations assisted the student's understanding of the city, whereas the African students were less able to understand culturally based physical cues. (See Figure 3-16.)

Perhaps it is only "common sense" that the quantity of information stored in memory increases with exposure, and that the opportunity to experience an environment differs by socioeconomic class, age, and perhaps, gender. It is important not to underestimate the value of these observations, however, nor their usefulness for planners and geographers. Nevertheless, perhaps a more interesting area for psychological research centers not just on the importance of familiarity and experience in adding to the quantity of stored information, but also on the qualitative changes in cognitive maps (and the memory storage and retrieval processes underlying them). Some early suggestions of such qualitative changes can be found in Appleyard's (1970) study in Venezuela. You will recall that Appleyard distinguished between *sequential* sketch maps emphasizing paths and nodes and *spatial* maps featuring a high proportion of landmarks and districts. Appleyard noted that maps were more spatial for long-term residents than for newcomers, and that spatial elements were more prominent in familiar areas of the city. Later research systematically investigated this phenomenon. Evans et al. (1981) reported that the basic path and node structure appears to be learned first, and then as an individual spends more time in the environment, he or she fills in other details such as landmarks. Thus, as an individual becomes more familiar with an environment, his or her cognitive map of it becomes more spatial.

On the other hand, Heft (1979) reported that adults rely more on landmarks to learn a route through a novel path network the first time they traverse it as compared with later occasions. This would seem to be the reverse of

Figure 3-16 The role of cultural expectations in large-scale maps. A far away planet? The so-called "Continental Drip" view of a well-known land mass (*Hint:* Turn the book upside down).

the path-primacy effect. Perhaps elements such as landmarks are used for wayfinding, but are not always represented in sketch maps. In support of this view are several comparisons between adults and children that suggest that one important difference between the maps drawn by people of different ages is that adults are more likely to attend to landmarks that lie at critical points on a route, such as the point at which one has to make a turn, than are children.

Gender Differences

Do males and females differ in their cognitive mapping abilities? Several researchers (e.g., Maccoby & Jacklin, 1974) have reported that males may possess superior visual and spatial skills, at least on paper-and-pencil tasks. If this generalization is true, one might expect males to be superior in their ability to draw complete and accurate cognitive maps. Perhaps having "a good sense of direction" is more important to the self-esteem of males than it is for females (Bryant, 1982), consistent with the popular notion that "men don't like to ask for directions." If so, one might expect males to be superior to females for motivational reasons, even if they possess no native superiority in mapping.

Some earlier researchers found evidence for gender differences in the final product of cognitive mapping exercises, although they concluded that these sex differences are most likely due to differences in familiarity with an area (e.g., Evans, 1980). Appleyard (1976), for example, found men's maps to be slightly more accurate and extensive than women's but attributed this difference to the higher exposure of men to the city. Abu-Obeid (1998) reported that newcomer males learned new environments faster, but that with more experience in the environment, females and males performed similarly on wayfinding tasks. Even some researchers who have given individuals both cognitive maps and paper-and-pencil tasks have found sex differences on the paper-and-pencil tasks, but not for spatial memory (McNamara, 1986).

More theoretically interesting than simple measures of overall competence in drawing cog-

nitive maps are a limited number of studies suggesting that the cognitive maps of men and women are about equally accurate, but stylistically different. Again, some hint of differences between the maps of males and females appeared in Appleyard's early investigations; females seemed to be somewhat more spatially oriented than males. Other researchers have concluded that females are as accurate overall as males in their maps, but that women emphasize districts and landmarks, whereas males are more likely to emphasize the path structure (Galea & Kimura, 1993; McGuinness & Sparks, 1979; Pearce, 1977). In addition, Ward, Newcombe, and Overton (1986) have reported that males are more likely to voluntarily give compass directions or distance estimates phrased in measurements such as mileage than are females when asked to give directions based upon a map. Nevertheless, when instructed to phrase their directions using these dimensions, females were as successful as males.

In a pair of related experiments, McGuinness and Sparks (1979) found that women included fewer paths between landmarks, included more landmarks, were less accurate in placing buildings with respect to the underlying spatial terrain, but were more accurate than males in the placement of buildings with respect to their distance from one another. Interestingly, the second experiment of the pair demonstrated that females actually did know the locations of many roads and paths that they had not voluntarily included in sketch maps. As was the case with Ward et al. (1986), it seems that women often remember the location of these features, but do not always include them in their maps unless specifically asked to do so. McGuinness and Sparks concluded that whereas females seem to approach the organization of topographical space by grouping landmarks and establishing their distance from one another, males are more likely to begin with a network of roads and paths, which may provide a somewhat more accurate framework.

Affect may be one source of male–female differences. Males seem more confident in wayfinding (Devlin & Bernstein, 1995), and females express higher levels of anxiety when wayfinding (Schmitz, 1997). Thus, many male–female differences may reflect that anxious people generally do not do as well on mental rotation tasks and are likely to have less complex knowledge of the orientation of the environment and underlying distances (Lawton, 1996; Schmitz, 1997).

Holding (1992) examined the prediction that buildings in the same hierarchical cluster are closer together than equidistant buildings belonging to different clusters. For this task, distance estimates of females were more affected by cluster membership than were those of males. In general, males may begin by setting up an organizational framework of paths and nodes for their sketch maps and then superimpose features such as landmarks and districts on this established framework. On the other hand, women may be more likely to try to establish individual relationships between landscape elements or clusters of elements without this organizing framework provided by path networks.

In conclusion, males and females are probably equally capable of mapping their surroundings, but some stylistic differences await further investigation. Even if these differences are valid, the source of them is unclear. They may be explained by differences in experience, familiarity, or the socialization process, but a biological component cannot yet be entirely eliminated.

ACQUISITION OF COGNITIVE MAPS

We have noted several instances in which spatial cognition does not match cartographic maps. In general, cognitive maps become more similar to cartographic maps as an environment becomes more familiar (e.g., Evans et al., 1981). The two most common situations in which to observe this process are with children (for whom many environments will be unfamiliar) and with adult newcomers.

Children's Maps

Much of the interest among developmental psychologists and others investigating children's

spatial cognition is based upon the implication that the changes that occur in these maps reflect not only a change in the amount of information in memory, but also a change in the type of information and the way it is used (see Heft & Wohlwill, 1987). For example, differences between children and adults may reflect not just less experience, but a very different approach to problem solving than that employed by adults.

The most influential theory of cognitive development as applied to spatial cognition may still be the one proposed by Jean Piaget and his colleagues (e.g., Piaget & Inhelder, 1967). In one classic study, Piaget asked children to sit in a chair and to view a table on which were placed three model mountains (see Figure 3-17). Three other chairs were placed around the table, upon one of which was seated a doll. From a set of drawings the child was asked to select a view of the scene as it would appear to the doll. Children younger than 7 or 8 typically chose not the view from the doll's perspective, but the view they themselves saw. Piaget termed this *egocentrism*.

According to Piaget, during the egocentric phase, the child's frame of reference is centered on his or her own activities. Environmental features in the child's spatial image are disconnected and the environment is fragmented. Later, the child's map is oriented around fixed places in the environment that the child has explored, but not necessarily the place he or she now occupies. These known areas are, however,

Figure 3-17 Piaget's three-mountain problem.

disjointed. Finally, the child's frame of reference assumes the characteristics of a spatial survey map with a more objective representation of the environment.

Ironically, one effect of Piaget's conclusions may have been to reduce interest in children's spatial abilities because his research led to the conclusion that children could not understand and use maps until about the age of 7. Recent research leads us to temper Piaget's conclusions from the three-mountain experiment (Matthews, 1992). For example, although there are some changes in children's ability to interpret aerial photographs between kindergarten and grade two (e.g., Blades & Spencer, 1987; Stea & Blaut, 1973), children seem better able to make use of aerial photographs and maps than Piaget would have predicted (e.g., Blades & Spencer, 1987; Rutland et al, 1993). Even 3-year-olds have at least some ability to form spatial representations (DeLoache, 1987), although this ability improves significantly with age (Heth, Cornell, & Alberts, 1997; Rutland et al., 1993).

Many studies support Piaget's observation of spatial egocentrism. Somewhat more controversy surrounds whether these findings reflect a truly different way of thinking (as Piaget would imply) or a slow increase in the quantity of environmental information and cognitive skills. Much of the research on children's cognitive maps is based upon studies that have employed models to simulate environments. It may be that the relatively poor mapping abilities demonstrated by children participating in studies that employ this method may have resulted at least partly from the artificial methodology of the research rather than actual mapping deficits (e.g., Cornell & Hay, 1984; Evans, 1980). One fairly reliable difference between older children and adults and younger children seems to be the way they make use of landmarks. Although even 8-year-olds notice and can remember distinctive landmarks, older children and adults seem more likely to place the landmarks into a reference system that integrates distinctive landmarks with other environmental attributes (Cornell, Heth, & Skoczylas, 1999; Heth et al., 1997).

Adult Map Acquisition

Although Piaget believed that children's strategies for spatial problem solving differed from those of adults, many researchers conclude that the cognitive maps of adults in new environments also develop from route to survey knowledge in a manner much like the age-related changes observed in children (Golledge et al., 1985; McDonald & Pellegrino, 1993). Individuals knit together a cognitive map from the accumulation of information acquired by traveling different routes in a new environment. Eventually the person is able to take "shortcuts" like the one demonstrated in Figure 3-18. McDonald and Pellegrino conclude that the processes may generally follow a sequence from landmarks to routes to survey knowledge.

Adults are likely to have at least one advantage over children; they are more likely to understand and have access to published printed maps. Of course, the purpose of most maps is to provide an accessible and permanent record of spatial information, so maps should be valuable aids for spatial learning. Is information learned from a map different from that acquired from experience? People may learn from maps quite differently than from actually moving through an environment (Thorndike & Hayes-Roth, 1982). Map learners are privy to a bird's-eye view of the environment, and thus, may acquire survey knowledge because a map provides direct access to global relationships of distance and location. On the other hand, although spatial learning based on actual navigation in the environment may be more difficult to obtain, it benefits from the advantages of ecological context and, perhaps, more accurate representation of the travel distances for each leg of a journey.

In their review, McDonald and Pellegrino (1993) differentiate between *primary* and *secondary* spatial learning. Primary learning involves direct experience moving through the environment, whereas secondary learning comes from studying maps or other environmental descriptions. Over time, the spatial representations acquired through actual navigation

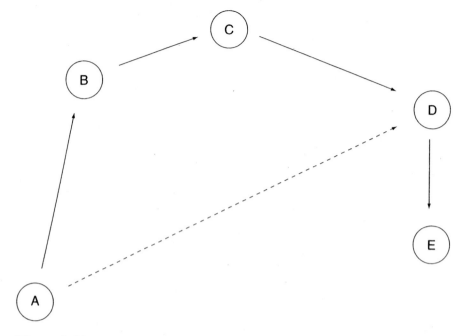

Figure 3-18 The solid line indicates the routes a person has traveled between five landmarks. If he or she has developed a survey map, it will also be possible to take a previously untraveled shortcut indicated by the dashed line.

become more like that of survey knowledge. In instances in which the environment is relatively simple with streets laid out in rectangular grids, navigation may quickly lead to more accurate survey knowledge than that gained from maps (Thorndyke & Hayes-Roth, 1982). Most maps present an aerial view, and this perspective presumably facilitates the development of survey knowledge.

Despite these advantages, learning about an environment from maps sometimes results in certain distortions that seem to appear in maps generated from primary experiences. For instance, the orientation of a cartographic map you use to learn a new environment may affect the orientation of your subsequent memory (MacEachren, 1992; Warren, 1994; Warren & Scott, 1993). If you study a map that is drawn and labeled according to the convention that north is "up," you may always assume that locations east of the center of the map are "right" and those west are "left." This presents no prob-

lem as long as you travel northward, but perhaps you have experienced the common confusion of reorienting your mental map when you travel in a southerly direction. Cognitive maps constructed from actual experience seem not to suffer from this orientation specificity. In our discussion of wayfinding at the close of this chapter we will revisit orientation problems as we consider "you-are-here" maps in places like airports and shopping centers.

Although there are important individual differences, spatial skills eventually decline for older adults (e.g., Arbuckle, Cooney, Milne, & Melchior, 1994). For both younger and older adults, prior experience in similar environments can facilitate cognitive mapping, but violations of typical expectations seem to have a more detrimental effect on older persons. Although there are several possible explanations for this effect, Arbuckle et al. conclude that this deficit primarily reflects an age-related decline in ability to ignore irrelevant prior knowledge.

MEMORY AND COGNITIVE MAPS

We have seen some characteristics of sketch maps and other physical representations of human cognition. Notice, however, that cognitive maps themselves have no external physical existence; they reside only in our minds. Let us turn now to a fundamental question: Exactly how *is* a cognitive map represented in memory? Psychologists have differing opinions on the matter (Evans, 1980; Searleman & Hermann, 1994), and investigations of this representational question have sparked sophisticated studies by both environmental and cognitive psychologists. Some of the most interesting questions concern the exact form of the mental representation and the organization and structure of a memory or retrieval process (McNamara, 1986).

The Form of the Representation

One of the fundamental issues is the form of the mental representation of spatial knowledge. One view is that we have an image or mental "picture" of the environment in our memory. This view, termed the *analogical* or **analog representation** (meaning the mental map is an analogy of the real world), says that the cognitive map roughly corresponds point for point to the physical environment, almost as if we have a file of slide photographs of the environment stored in the brain (e.g., Cornoldi & McDaniel, 1991; Kosslyn, 1980, 1983).

Another view, the *propositional* approach, advocates more of a meaning-based or **propositional storage** of material. The environment is represented as a number of concepts or ideas, each of which is connected to other concepts by testable associations such as color, name, sounds, and height. When we call on this propositional map we search our memory for various associations, and these are reconstructed and represented as a mental "image" or in a sketch we draw (Johnson-Laird, 1996; Pylyshyn, 1973, 1981).

Current thinking combines these two approaches, concluding that cognitive maps contain both propositional and analogical elements (e.g., Evans, 1980; Gärling et al., 1984; Kosslyn, 1980, 1983; Tye, 1991). For example, most information about the environment may be stored in memory through propositions, but we can use this propositional network to very quickly mentally construct an analogical image that has many of the qualities of a photograph. We may then use this image, rather than the propositional network some researchers suppose to underlie it, to solve spatial cognition problems.

Distance

Some understanding of the distances between locations is necessary if we are to use a cognitive map for wayfinding. If maps are analogs of the real world, distance may also be represented in the stored memory itself. For example, when people are asked to judge whether a pair of states (e.g., Georgia and Mississippi) are closer together than another pair (e.g., Michigan and Iowa), the more similar the distances within the two pairs, the longer it takes to make a decision (Evans & Pezdek, 1980). Moreover, recall of distance between two points takes longer the greater the distance on a map (Kosslyn, Ball, & Reiser, 1978). From such evidence some have concluded that cognitive representations of the environment require scanning in order for judgments to be made about them. The more information we must scan, the longer it takes to make judgments about spatial relationships (e.g., Kosslyn, 1983).

An analogical storage of spatial information is not the only explanation for many of the observed distance effects, however. Perhaps a longer pathway also provides more opportunities to acquire the bits of knowledge that make up propositions. Generally, the more information we must scan in our memory while making a "mental journey" through an environment, the farther the distance we assume we have traversed. For example, judgment of traversed distance, that is, the distance we have traveled over a given period of time, is in part dependent upon the number of pieces of information we encounter as we travel. Investigations of this so-called clutter effect reveal that students walking a path designated by a line of tape placed on a floor judge a path to be longer the more right-angle turns it contains (Sadalla & Magel, 1980; Thorndyke, 1981). In general, the more intersections or objects a path crosses, the longer the path is judged to be (Sadalla & Staplin, 1980a, 1980b).

As you may remember from our brief discussion of multidimensional scaling, recent findings indicate that the estimated distance between point A and point B on a cognitive map is not necessarily the same as the distance from point B to point A (McNamara & Diwadkar, 1997; Newcombe, Huttenlocher, Sandberg, Lie, & Johnson, 1999). This irreversibility casts doubt on the regularity assumed by so-called euclidean models, but has sparked a growing interest in other models of cognitive structure.

Structure

One line of reasoning begins with the assumption that humans are limited in their ability to process incoming information. Too much information may tax our perceptual and cognitive abilities, resulting in cognitive overload (see

Figure 3-19 Relationships between clusters and reference points may affect distance estimates and knowledge of even well-known areas.

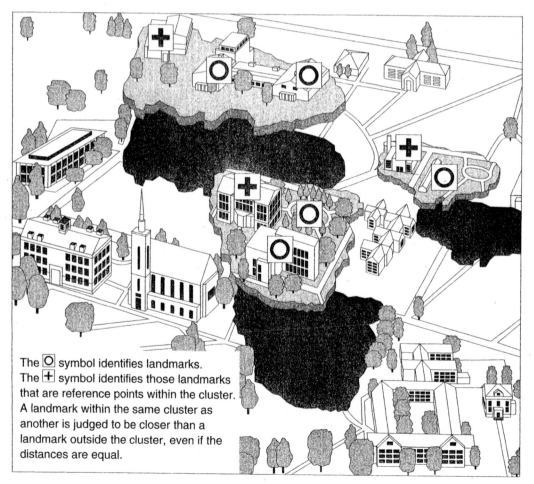

The ⊙ symbol identifies landmarks.
The ✚ symbol identifies those landmarks that are reference points within the cluster. A landmark within the same cluster as another is judged to be closer than a landmark outside the cluster, even if the distances are equal.

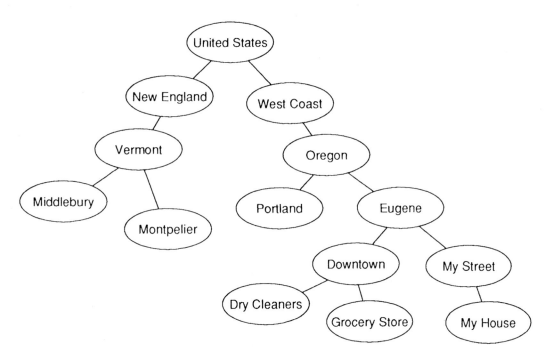

Figure 3-20   An example of a semantic network as it might underlie a person's cognitive map.

Chapter 4 for an additional discussion of environmental load as a theoretical explanation for a variety of behaviors). Perhaps you have read elsewhere that humans seem to benefit from strategies that organize the complex information they wish to remember (lists of letters or numbers, for instance) into a smaller number of meaningful "chunks." There is evidence that people also divide spatial information in a manner similar to chunking (Allen, 1981; Allen & Kirasic, 1985). Although the criteria for inclusion in a chunk or cluster may differ, good candidates are landmarks that are both near to each other and similar in architecture and use. Perhaps this amounts to a "rediscovery" of what Lynch termed "districts." Let us resurrect our discussion of cognitive distance. There is considerable evidence that landmarks within the same cluster are judged to be closer to each other than to a third, equidistant point outside their cluster (Hirtle & Jonides, 1985; Holding, 1992; McNamara et al., 1989). Moreover, each cluster may itself be represented by a *reference point,* a sort of "best example" that symbolizes all the lo-

cations within the cluster (Couclelis et al., 1987; Ferguson & Hegarty, 1994; Sadalla et al., 1980). As illustrated in Figure 3-19, we might imagine a world of well-known clusters or districts within which distance estimates are fairly accurate, but between which knowledge is less precise.

Perhaps clusters are organized in some orderly fashion in memory. Some time ago, Collins and Quillian (1969) demonstrated that the retrieval of information from semantic memory (memory for concepts) sometimes acts as if it is based upon a hierarchical memory network. That is, information may be stored according to some organizational system that is based on ordered categories. This is typically presented as a tree diagram illustrating the relationships between concepts as branches like those in Figure 3-20. Presumably, some sort of sequential search of levels in these categories occurs when one is asked to determine relationships between concepts. The exact form of these *semantic networks* is controversial and the subject of a great deal of research in cognitive psychology (see Best, 1998, for a readable review).

For our present purposes, what is most interesting is the idea that some form of a network might also describe the way in which spatial information is represented in memory. There is some evidence that for spatial memory at least, there may be an upper limit to how much a person can remember (Byrne, 1979; Tversky, 1981). A networked storage process would be a rather economical system in the memory space it requires because information common to all members of a spatial cluster or category (perhaps symbolized by the reference point) could be stored only once. All points in Nebraska are west of all points in Ohio, for example. Although theoretically efficient, this storage system might be subject to certain types of errors that would make some memories more difficult or time consuming to retrieve than others.

An important study reported by Stevens and Coupe (1978) provides an opportunity to experiment with an interesting example. First, draw a rough map of the United States. Now, indicate the locations of San Diego, California, and Reno, Nevada. Do not read further until you have done so. Finished? Except for those living near the West Coast, most people place San Diego west of Reno apparently because they think of California, the superordinate category to which San Diego belongs, as being west of Nevada. As you will see upon consulting a U.S. map, San Diego is actually east of Reno! As is often the case, things probably are not so simple. A simple tree diagram or *hierarchical network* cannot explain all of the phenomena we have observed in cognitive maps. Some propose refinements of the network model; others propose quite different memory structures that do not depend upon hierarchical storage. We will have to leave the complexities of these arguments to cognitive psychologists (see Best, 1998). At the present time, however, versions of the network model (e.g., Gärling et al., 1984; Kaplan & Kaplan, 1982) remain popular within the cognitive mapping literature. Perhaps McNamara's (1986) "partially hierarchical" structure represents the data as well as any. This means that memories may be stored according to hierarchical principles, but that there remain some interconnections between areas that cut across this hierarchical structure. However the exact process occurs, most studies, whether they are lab or field studies, indicate that spatial information is acquired quickly and that forgetting is minimal.

Wayfinding

Most of the research we have presented to this point has focused on a rather static, plain-view map of the environment residing in memory. (For now, we will lay aside the argument concerning the specific form of this representation.) Other authors (e.g., Cornell & Hay, 1984; Gärling, Böök, & Lindberg, 1986; Passini, 1984) are interested in wayfinding, the process by which people actually navigate in their environments.

One of the most profoundly troubling experiences we can face is being lost. In such an instance, our human capabilities of information processing and storage have deserted us, and because most of us are dependent upon others and technology, our very survival may be threatened. Being truly lost may be a relatively rare phenomenon, but newcomers commonly experience the stress and anxiety that accompany disorientation in both buildings and natural environments (e.g., Cohen et al., 1986; Hunt, 1984). For some groups, this stress may be particularly serious, even life threatening (Heth & Cornell, 1998; Hunt, 1984).

ACTION PLANS AND WAYFINDING

Gärling et al. (1986) propose one model of wayfinding that may prove useful in organizing our

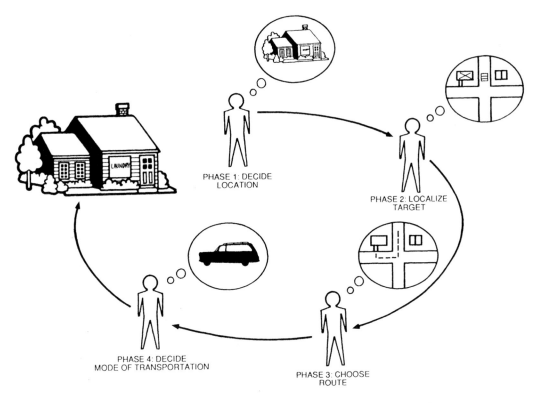

PHASE 1: DECIDE
LOCATION

PHASE 2: LOCALIZE
TARGET

PHASE 3: CHOOSE
ROUTE

PHASE 4: DECIDE
MODE OF TRANSPORTATION

Figure 3-21   Wayfinding: A hypothetical example of a trip to retrieve one's laundry.

discussion (see Figure 3-21). We will provide a hypothetical example to illustrate the different steps of the model. Imagine that a friend has asked you to drop off some clothes at a dry cleaning establishment. First, we determine a destination. Should you take your friend's dry cleaning to your favorite establishment, or to one closer to your friend's home for his or her convenience? Assuming that you decided to choose a dry cleaner near your friend's home, the second step requires the new target destination to be localized; that is, you must determine the general location of the target environment. If you are unfamiliar with your friend's neighborhood, you may need to use a telephone book or some other source to pinpoint the target. Third, you choose a route between your present location and the dry cleaner, again requiring you to ask directions or to refer to a map if you are unfamiliar with the neighborhood. Finally, you must make a choice of travel mode, depending

upon factors such as the distance to the destination and the availability of transportation.

Notice that the model emphasizes an internal psychological process that lets us anticipate or rehearse what will eventually be our actual behavior in moving through the environment. Thus, Gärling et al. have adopted the concept of *action plans* (Russell & Ward, 1982) as links between stored environmental information and wayfinding behavior.

Like cognitive maps, action plans for wayfinding often reveal spatial asymmetries. For example, many people use one route traveling to or from work, school, or home, and a different route for the return journey. These differences may be quite practical. One of your authors avoids a steep hill that imposes itself on the bicycle route into his office by taking a more gradual path. On the way home, however, he delights in a route that ends in a swooping, more direct ride down the hill. Not all route asymmetries are so obvious, however. Bailenson, Shum, and Uttal (1998) propose that cognitive "road climbing" explains many asymmetries. Road climbing is a cognitive heuristic (mental shortcut) by which people select a long and straight route from the point of origin instead of alternate paths. In fact, Bailenson et al.'s participants preferred a path that was initially long and straight over alternatives that were up to 50% shorter!

A good cognitive map would be one excellent wayfinding aid (Abu-Obeid, 1998), but some authorities doubt whether a person actually needs a detailed map, either mental or on paper, to find a travel goal. For example, Passini (1984) suggests that wayfinding might best be viewed as a sequence of problem-solving tasks that require a certain amount of stored environmental information. This may be an easier task than drawing your route on a sketch map for at least two reasons. First (assuming you have at least some experience in the environment in question), you are facing a task of recognition. Instead of recalling a cognitive map, you may need only to recognize a particular environmental feature such as a landmark as you encounter it, and to make a correct decision (such as to turn left) when in its presence. Second, wayfinding is in some way self-correcting. If you find yourself suddenly moving into unfamiliar terrain, you may retrace your steps to the point where you erred and try again. Thus, errors in wayfinding need not be cumulative. You may have slightly misjudged the distance or direction from one building to another. Once you do manage to find your way to this key decision point, minor errors earlier in the journey are no longer of any consequence. Traditional cognitive maps are less forgiving.

Thus, the process of wayfinding may be based on a series of unfolding expectancies (Cornell et al., 1994, 1999). Incidentally, the process of route planning and encoding may not have to be purposeful. Familiarity and incidental route learning may occur even when one's attention is directed elsewhere.

You may recall that Byrne (1979) demonstrated that intersections between roads are typically remembered as right angles, even when the actual angle varies from 60° to 90°. Byrne suggested, in fact, that precise information concerning the shape of intersections may be missing from memory entirely. Perhaps, for wayfinding purposes at least, it is sufficient to have a network map that preserves only the connections between steps along a route, but which requires neither knowledge of the distances between choice points nor the precise angle at which routes join. On the other hand, it may be that although one can travel successfully along a predetermined route by recognizing a succession of choice points, a more sophisticated navigation system would allow a person to arrive at the same location via a number of different (perhaps shorter) routes and to find a new location based upon its location with reference to some known landmarks. This more sophisticated type of wayfinding may require the richer spatial understanding characteristic of cognitive maps.

SETTING CHARACTERISTICS THAT FACILITATE WAYFINDING

Earlier in this chapter we noted Kevin Lynch's (1960) emphasis on legibility, which largely determines the degree to which an environment facilitates cognitive mapping. Although legibility has long had a place in the literature of both behavioral science and planning, relatively little sustained effort has addressed the need to create more legible buildings (Abu-Ghazzeh, 1996). Gärling et al. (1986) and Abu-Ghazzeh describe three characteristics of physical settings that are likely to affect wayfinding: the degree of differentiation, the degree of visual access, and the complexity of the spatial layout.

Differentiation refers to the degree to which parts of the environment look the same or are distinctive. In general, buildings that are distinctive in shape, easily visible, well maintained, and free-standing are better remembered. In the context of interior environments, for example, Evans et al. (1980) demonstrated the effectiveness of color coding in improving wayfinding in a building's interior.

In addition to differentiation, the ability to learn a new environment may depend upon the **degree of visual access.** This is the extent to which different parts of the setting can be seen from other vantage points. Of course, Lynch (1960) recognized the importance of visual access in what he termed landmarks. Evans et al. (1982) also speak of **transition,** or direct access from a building to the street. Finally, **complexity of spatial layout** refers to the amount and difficulty of information that must be processed in order to move around in an environment. Too much complexity undermines both navigation and learning. For example, Weisman (1981) found that simple floor plans facilitated wayfinding in campus buildings. Simplicity was even more important than familiarity with the setting in predicting wayfinding difficulties. Taken to an extreme, no amount of familiarity may be able to compensate for extreme architectural complexity (Moeser, 1988). We hasten to distinguish between the complexity of a route or route network, as the term is used here, and the complexity of a particular facade, which should contribute to differentiation as discussed above.

MAPS

Map learners are privy to a bird's-eye view of the environment, and thus, may acquire what we have called survey knowledge. Thus, a map provides direct access to global relationships of distance and location. On the other hand, spatial learning based upon direct learning (actually navigating in the environment) may be more difficult to obtain, but it does benefit from the advantages of ecological context and perhaps yields a more accurate representation of the travel distances for each leg of a journey. Complexity of the environment is one important variable, so if the environment is relatively simple, with streets or hallways laid out in rectangular grids, navigation may quickly lead to more accurate survey knowledge than that gained from maps (Thorndyke & Hayes-Roth, 1982). On the other hand, for complex environments with many nonperpendicular paths, maps may remain the most efficient method of route learning (Moeser, 1988). As a rule, people who learn from maps acquire better survey knowledge, whereas direct learning facilitates route knowledge (Rossano, West, Robertson, Wayner, & Chase, 1999; Taylor & Tversky, 1996).

You-Are-Here Maps

One problem with maps is that people sometimes have difficulty translating maps into usable navigation tools (Butler, Acquino, Hissong, & Scott, 1993; Levine, 1982; Thorndyke & Hayes-Roth, 1982). For example, have you ever consulted a **you-are-here map** in a shopping center, museum, or subway terminal? Was the map easy to read and understand, or did you find yourself nearly as confused after reading the map as when you began? As we will see in Chapter 11, legibility is an important attribute of architectural design, and we might expect maps or signs to be important wayfinding aids. At least one study shows maps to be inferior to hallway signs (Butler et al., 1993), apparently because maps require more study time and require more memory store as opposed to a simple directional sign.

Given their advantages in providing a more flexible "bird's-eye view," can posted maps be made more usable? Marvin Levine and his associates (Levine, 1982; Levine, Marchon, & Hanley, 1984) have explored the design and placement of you-are-here maps and have outlined several simple principles that dramatically improve the usefulness of these orientation aides.

Structure Matching The first problem faced by a you-are-here map user is **structure matching,** that is, the need to pair known points in the environment with their corresponding map coordinates. If a person reading the map is unable to accomplish this task, even an accurately drawn map will not be very useful. Technically, Levine argues, two known points on both the map and in the terrain provide the minimum amount of information necessary for a person

Figure 3-22   Structure matching in you-are-here maps. In this map, labels and caricature map symbols allow the user to match the map with the surrounding terrain.

to relate any object in the environment with its map symbol. A viewer must know not only where he or she is (as would a person viewing Figure 3-22), but also the location of a second pair of points. For example, the figure shows buildings that can be easily identified both on the map and in the environment visible to the visitor. Although in the future this is accomplished by attaching a sign to Building L, we note that the same end might be achieved by using a caricature map symbol that resembles the building as it would be seen from the position of the person reading the map (rather than an aerial or blueprint perspective).

A second way of providing two-point correspondence is to carefully place the map near an asymmetrical feature. This allows the visitor to pinpoint his or her location and that of nearby features. In addition, Levine encourages the use of a bipart you-are-here symbol (also in Figure 3-22). Here both the map the viewer is reading and the position of the visitor are indicated, technically fulfilling the need for two points and allowing the viewer to correctly bring the map and the environment into correspondence.

Orientation   As former Boy or Girl Scouts may know, a map is most easily used if it is placed parallel to the ground and turned so that it is oriented with the terrain. Thus, a goal that is ahead of you on the map is ahead in the terrain, and something to the right on the map is to your right in the environment. In some instances, you-are-here maps in a building can be displayed horizontally so that the map is properly oriented. In most cases, however, practical reasons require the map to be hung vertically on the wall. Although it may not be obvious, correct alignment of these maps may be critical to ensuring that they are easily understood and used by visitors. Levine, Marchon, and Hanley (1984) propose that wayfinding maps are best when what is forward on the ground is up on the map (Figure 3-23). This **forward-up equivalence** also ensures that what is to the right in the terrain is to the right on the map, and so forth. Levine et al.'s experimental data (1984) show that misalignment of you-are-here maps by 90° or more seriously misleads people, even those who have been alerted to the misalignment!

Unfortunately, the Levine et al. (1984) study also showed that this principle is regularly violated in airports, offices, and other buildings. The severity of the violation may range from being a small inconvenience to shoppers, to potentially life-threatening in the case of fire evacuation maps in an office complex.

SIMULATIONS FACILITATING SPATIAL LEARNING

Although some have expressed concern that even carefully prepared photographic simulations of routes may be inferior to actual walks as wayfinding training aids (Cornell & Hay, 1984), other studies have successfully employed slide photographs as environmental simulations (e.g., Cohen et al., 1986; Hunt, 1984). In fact, casually acquired familiarity may sometimes never achieve the level of spatial understanding achieved by individuals given planned instruction (Moeser, 1988). It also follows that if some method could be found to accelerate spatial familiarity, some of the distress associated with relocation could be reduced. Some have focused on the need to assist children in adjusting to new spatial environments. For example, Cohen and his associates (Cohen et al., 1986) investigated the effect of two spatial familiarization experiences on the attitudes of 5- and 6-year-old kindergarten boys. Two weeks before the start of school, some of the boys were given either an on-site tour or a simulated tour accompanied by

Figure 3-23 Forward-up equivalence aids in orienting you-are-here maps.

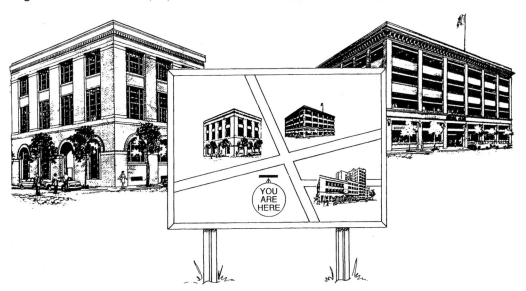

a scale model of their school. Boys who received either familiarization treatment felt more secure and comfortable several weeks after the start of school than a control group that received no training.

In another study with a quite different population, Hunt (1984) investigated procedures for improving the wayfinding abilities of senior citizen volunteers in an unfamiliar nursing home. One group was given a site visit in which they individually received a guided tour through the experimental building. Members of the second experimental group were individually shown photographs of the building ordered in the same sequence as experienced by those on the guided tour. As they viewed the slide photographs, this group could also inspect three-dimensional models of the building's floor plan and exterior. Participants in both treatment conditions were generally superior to a control group on a variety of on-site wayfinding tasks. This result was not surprising; it confirms the usefulness of some prior exposure to an environment, whether simulated or in the form of a tour. More interesting, however, are the differences between the simulation- and site-visit groups. The groups were similar in their ability to find their way to places along a previously learned route, but members of the simulation group were superior in their ability to find new locations, in their ability to identify photographs of building landmarks, and in their understanding of the exterior shape and the spatial configuration of the building. In sum, both groups could learn sequential routes, but the simulation group apparently had a richer and more flexible mental image (presumably because of their exposure to the bird's-eye views provided by the scale models).

The advent of inexpensive and powerful computers promises to offer new ways to learn unfamiliar environments. Some researchers have demonstrated that virtual reality "trips" through an environment can be effective, even more than maps (Rossano et al., 1999). Hirtle and Sorrows (1998) reported an ambitious campus wayfinding project that may soon be widely adopted. They created a highly interactive World Wide Web site that provided redundant map-like, verbal, and pictorial information that users could tailor to their own needs and styles.

Of course, much of the wayfinding information we acquire comes directly from other people—our friends, acquaintances, or a helpful stranger. Information may include oral instructions, simple sketch maps, or more complex drawings. Interestingly, verbatim instructions (either written or oral) seem superior to more complex or graphic maps that emphasize the overall geography or survey knowledge (Kovach, Surrette, & Aamodt, 1988). Just as not everyone is equally successful in wayfinding, not everyone is equally skilled in giving spatial information. Vanetti and Allen (1988) suggest that the ability to give useful route instructions depends upon both spatial skills and verbal ability. Unless a person knows a spatial layout, he or she is not likely to give useful instructions. On the other hand, if that person is unable to express those instructions clearly, pure spatial knowledge will not be of much use. To examine these ideas, Vanetti and Allen divided participants into high and low spatial ability and high and low verbal ability groups. Interestingly, there was little difference between the groups in the ability to follow route instructions, but those with high spatial ability were more likely to suggest a more efficient route to others.

WHEN WAYFINDING FAILS: LOST AND FOUND

Sometimes, thankfully rarely, a person becomes disoriented in an environment that is so remote that authorities must form a search-and-rescue party. In these life-threatening situations, finding the lost person quickly is critical. A study of lost persons in Canadian wilderness areas (Heth & Cornell, 1998) revealed that a knowledge of the underlying geography and the recreational activity the lost person was engaged in (mountain bikers versus hikers, for instance) were important in defining the size of the search area. Another factor was the psychological state of the lost person because those who are despondent or who walk away to avoid others behave differently from most recreationists.

CONCLUSION

We will conclude our discussion of wayfinding by noting that people are generally more successful at wayfinding than in cognitive mapping. This observation may be most clearly seen at the extremes of the age spectrum. In spite of the data we reviewed regarding possible deficiencies in children's cognitive maps, particularly their tendency for environmental egocentrism, children beyond kindergarten age seem quite competent at wayfinding. We have already characterized wayfinding as primarily a recognition task and distinguished it from sketch maps that emphasize recall. In addition, many measures of cognitive mapping ability such as sketch maps depend upon skills such as drawing ability that are not so clearly or so often demanded as wayfinding skills in the real world. Finally, perhaps some individuals, particularly children, are intimidated or overwhelmed by the complexity of the task requested by many cognitive map studies, but perform well when faced with an ecologically valid situation.

SUMMARY

Whereas the conventional approach to perception examines the way the brain interprets messages from the sensory organs concerning specific elements in the environment, environmental perception views the perceptual experience as more encompassing, including cognitive, affective, interpretive, and evaluative responses. Moreover, environmental perception is likely to consider the person–environment relationship from a holistic systems or transactional perspective. Environmental perception involves activity on our part, especially in terms of exploring the environment to determine what needs it meets. In addition, exposure to a particular environment may result in adaptation or habituation—the weakening of a response following repeated exposure to a stimulus.

The line between perception and cognition is a hazy one. Cognition integrates memory and experience with a judgment of the present derived from perception to help us think about, recognize, and organize the layout of an environment. Cognitive maps are our mental representations of this layout and can be analyzed through a variety of methods. The best-known approach to cognitive mapping is that of Kevin Lynch, who emphasized the major elements: paths, landmarks, nodes, edges, and districts.

Cognitive maps are not perfectly accurate representations of the environment; they contain distortions, omissions, and other errors. These errors often reflect the importance of familiarity with an environment. Current thinking suggests that cognitive maps may be stored as images, propositions, or both. Propositions, in particular, are often thought of as organized into networks, but the specific form of storage remains controversial.

Action plans serve as the bridge between stored mental images or facts and actual behavior in the environment. The process of using stored spatial information along with maps and other aids is called wayfinding. It seems that wayfinding may involve both recognition of landmarks and other features at choice points, and the recall of a more sophisticated survey or spatial map. Architectural features that make an environment more distinctive or simpler to understand may improve wayfinding. Other attempts to convey spatial information, such as signs and training programs, are likely to improve wayfinding abilities and to reduce the stress of disorientation.

KEY TERMS

adaptation	ecological perception	lens model	spatial maps
analog representation	ecological validity	linear perspective	structure matching
cognitive map	edges	multidimensional	survey knowledge
complexity of spatial	environmental	scaling (MDS)	transactional approach
layout	cognition	nodes	transition
cue utilization	forward-up	paths	wayfinding
degree of visual access	equivalence	perception	Weber–Fechner
differentiation	functionalism	propositional storage	function
distortions	Gestalt perception	sensation	you-are-here maps
districts	habituation	sequential maps	
ecological niche	landmarks	sketch map	

SUGGESTED PROJECTS

1. Ask several friends to draw cognitive maps of your campus. Are the major components similar to one another and to the map we asked you to draw while reading this chapter? Do the maps differ by academic major or year in school? Are there any instances of distortions that are consistent with our examination of cognitive maps?

2. Re-read the section on multiple dimensional scaling. Now create a chart showing each combination of pairs for 10 campus locations. Pick one pair that is intermediate in distance and likely to be very familiar to everyone who spends time on campus. Use this as your measurement standard (because many people don't "think" in feet, yards, or meters). Ask your class or a few friends to estimate the distance between each location and every other location in terms of the unit of measurement you constructed, and calculate means for each pairing. Finally, cut yarn or string in lengths proportional to the mean distance estimates between buildings, and connect these to tacks (representing the buildings themselves). With a little stretching here and there, you should find that your procedure has captured a fairly accurate map of campus.

3. Inventory the you-are-here maps in your town or campus. Do any of these maps violate any of the principles outlined by Levine (1982)? What could be done to improve these maps as wayfinding aids?

4. Make up a short questionnaire to administer to your friends. Ask them to indicate the direction they would travel to get from one to the other of 10 pairs of cities. We suggest that you include in your list Reno, Nevada, to San Diego, California; Oklahoma City, Oklahoma, to Lexington, Kentucky; Windsor, Ontario, to Albany, New York; the Atlantic entrance of the Panama Canal to the Pacific entrance; and London, England, to Minneapolis, Minnesota. Inspect an atlas and construct your own answer key. What kinds of errors did your friends make? Were the pairs of cities we chose particularly difficult? Did you find support for the Stevens and Coupe (1978) position that errors reflect the organizational hierarchy of cognitive maps? Why or why not?

CHAPTER 4

THEORIES OF ENVIRONMENT–BEHAVIOR RELATIONSHIPS

Introduction

You awake one morning anticipating an important job interview with a lab on campus. Unfortunately, your roommate had different ideas about when the alarm should be set, and a blaring radio awakens you an hour earlier than anticipated. But with your adrenalin pumping, you cannot filter out the noise of the morning and thus cannot go back to sleep. Grudgingly, you shower and get dressed and prepare for breakfast. But another roommate used the last of the coffee to stay up studying the night before, and you are now in a grumpy mood. Heading to your early morning class, you encounter a construction project that requires a 5-minute delay; but the delay seems like half an hour, as irritable as you have become. So concerned about getting to class on time, you don't see a stop sign, and the ever-present local law enforcement issues you a ticket. As you finally arrive for class, the only seats available are in the front row, where you are under intense scrutiny from the instructor, or in the back, where it is difficult to see. You
take a seat in the back and discover that the first part of the hour is devoted to a half-hour video you have seen in three other classes. Your mind wanders. Then the instructor announces that the final exam must be changed to a day when you are already scheduled to be on a trip with your family; on that trip, you had been planning to visit your uncle who had recently been forced out of his home because a nearby waste dump had been discovered leaching toxic chemicals into the neighborhood. You feel as if you are losing control of the events in your life. You retreat to a nearby scenic park where you can collect your thoughts. It is a refreshing experience, and at last you feel restored. Unfortunately (the same term used when your day began), the time passes quickly and before you realize it, you have missed the job interview. Rushing to the lab, you discover that the manager has had so few applications, your lack of promptness is forgiven and you receive an interview anyway. You retreat to the comforts of your room and receive a call from

a friend you have not seen for several months. Sharing the experiences of the day seems to help. Eventually, you fall asleep exhausted.

Although it is unlikely that all of the above events would happen to you in one day, they are typical of the things we encounter in a life's journey as we interact with our physical and social environments. Although the events and the reaction to them seem disjointed, environmental psychologists are interested in developing uni-

fied theories that would explain such disparate person–environment interactions. In this chapter we will examine the use and development of some of the relevant theories that are employed in environmental psychology today. We will begin with a general discussion of the concept and function of theory, then examine some specific psychological theories that have evolved to explain the nature of environment–behavior relationships, and conclude with our own synthesis of these various orientations. ▦

The Nature and Function of Theory in Environmental Psychology

The scientific method is really little more than a specific way of gaining knowledge. Scientists, whether devotees of environmental psychology or any other field, assume there is a great deal of order in the universe that can be discovered with appropriate methodology. Before the application of scientific inquiry, however, this universal order is perceived more as chaos or uncertainty than as something systematic. Science (or more specifically, the scientific method) is simply a set of procedures for reducing the uncertainty, thereby gaining knowledge of the universal order. It is to these procedures that we owe our progress thus far in environmental psychology. Other approaches to gaining knowledge do exist, of course, such as the methods of religion. In religion, the basis of reducing uncertainty is tradition, faith, revelation, authority, and in many cases, experience. The basis of reducing uncertainty in the scientific approach, on the other hand, is a mathematical prediction of observable events. Once we can predict with a degree of certainty what will happen to phenomenon "A" (e.g., crime or violence) when a change occurs in phenomenon "B" (e.g., population density), we have taken a giant step toward a scientific understanding of these phenomena.

Suppose, for example, that we want to apply scientific methods to discover the principles involved in getting people to reduce air pollution. We assume that such principles exist (e.g., appealing to conscience, implementing government regulations, administering punishment), and that through scientific inquiry we can not only discover them, we can also predict how much air pollution will be reduced by applying the principles in varying amounts and combinations. Moreover, the principles should also predict the positive and negative consequences of their application (e.g., cleaner air, better health, potentially reduced profits and productivity of an industry, higher utility bills). Such predictions, however, are rarely perfect. To the extent that our predictions of the phenomena are not perfect, uncertainty remains about the portion of the ordered universe under study, but we continue our quest for knowledge to reduce this uncertainty further.

Thus, scientists assume that events in the universe are related to other events in the universe, and that through scientific inquiry these relationships can be discovered and their consequences predicted. Using scientific methods, environmental psychologists observe fluctuations in some phenomena (e.g., climate changes, inadequate space in an office) and predict their subsequent impacts (e.g., violence, reduced productivity, efforts to modify interior design of space). Research in psychology, then, is the search for the antecedents of our various behaviors and thoughts—antecedents including environmental factors, biological influences, and intrapsychic (cognitively or emotionally generated) events (cf. Franck, 1984).

We should note that the assumption of causation in science involves a philosophical notion of **determinism.** In an absolute sense, a deterministic system implies the opposite of "free will." In a softer interpretation, we can study many interrelated determinants of an outcome, including personal choice (cf. James, 1979; Rotton, 1986). Environmental psychologists are often more concerned with analyzing and conceptualizing patterns or shapes of relationships than with a narrower focus on antecedent–consequent dependencies in an environment–behavior system. Altman and Rogoff (1987), for example, note the value of a *transactional approach* (a view we encountered in Chapter 3), which concentrates on the patterns of relationships rather than on specific causes, although R. Kaplan (1987) cautions that there can be problems of inference in some research that takes a transactional perspective. We should point out that from the perspective of a purist, once we decide that a phenomenon, psychological or otherwise, is indeterminate or is unpredictable, we are really saying that this phenomenon is not within the realm of scientific inquiry and that we cannot gain knowledge about it through scientific methods. Yet by refining our methods and changing our conceptualizations as new information becomes available, we are able to improve our understanding of (i.e., reduce our uncertainty about) environment–behavior relationships (Clitheroe, Stokols, & Zmuidzinas, 1998).

Hypotheses, Laws, and Theories

How do we proceed with scientific research in environmental psychology? Most likely, we start with simple observations. We might observe,

for example, that as the concentration of inmates in prisons increases, violence in prisons goes up. We have observed two phenomena, prison population density and prison violence, and we have noted a relationship between the two: As one increases, the other increases (i.e., they are *positively correlated*). We might then hypothesize, or formulate a hunch, that high population density in prisons leads to increased aggression and violence. (We should caution that actual studies show the effects of prison crowding to be complex, as we discuss in Chapter 9.) The next step in scientific methodology is critical—testing the hypothesis. A **hypothesis** is a proposition; in science it is empirically testable. All methods of gaining knowledge generate hypotheses. What makes science unique is the method of verifying the hypotheses. Whereas religion may rely on faith, tradition, or individual experience to verify hypotheses, science insists that hypotheses be verified by publicly observable (**empirical**) data. Recall that we described in Chapter 1 several means of acquiring such data. When these observable data do not support the hypothesis, the scientist must either modify the hypothesis or generate an entirely new hypothesis and test it again. If we took the question into the laboratory and found, for example, that putting several individuals into a small room did not increase their level of aggressiveness, we would have to reject the idea that crowding causes aggression and come up with a more complicated hypothesis. For example, maybe it is only under conditions of deprivation, such as boredom, that crowding leads to increased aggression. This is a testable proposition. Indeed, as we shall discover in Chapter 9, several investigators have looked at the combined effects of crowding and other variables, such as poverty, on aggression.

Once we have gathered a number of empirical facts, we can proceed to a more abstract and theoretical level to explain these facts. A word on the distinction between empirical laws, theories, and models may be helpful here. **Empirical laws** are statements of simple observable relationships between phenomena (often expressed in mathematical terms) that can be demonstrated time and time again. Such things as the law of gravity, the law that magnetic opposites attract, and the law of effect in psychology (i.e., behaviors that lead to pleasurable consequences are likely to be repeated) are easy to demonstrate at an empirical level. Theories usually involve more abstract concepts and relationships than empirical laws and consequently are broader in scope. Theories are not, as a rule, demonstrable in one empirical setting but are inferred from many empirical relationships. Examples include the theory of evolution, the theory of relativity, and equity theory in psychology (dissatisfaction in a relationship occurs if outcomes are not proportionate to inputs). Finally, a **model** is usually more abstract than an empirical law but is not as complex as a theory. Models are usually based on analogies or metaphors. For example, as was observed in Chapter 3, investigators have assumed that mental maps resemble the ones drawn by cartographers. To take another example, more than one theorist (e.g., Knowles, 1980a) has drawn an analogy between magnetic or gravitational force fields and the distance that strangers maintain between themselves and others. A model is often an intermediate step between the demonstration of an empirical law and the formulation of a theory. The distinction can be made that a model is the application of a previously accepted theoretical notion to a new area, but in practice the terms *model* and *theory* are often used interchangeably.

It may be helpful to think of theory as existing at several levels. **Heuristics** are simple principles that facilitate decision making. For example, the representativeness heuristic says that we decide whether an item fits a category (e.g., whether a sound is meaningless or whether it signals something) based on how representative it is of other items in that category (e.g., whether it resembles other known signals). A model is more elaborate and is based on analogies. A *bivariate theory* simply relates two variables, such as temperature and violence. Well-articulated theories are often quite elaborate and relate multiple concepts to one another.

Basically, a **theory** consists of a set of concepts plus a set of statements relating the concepts to one another. At the theoretical level, we might say that the undesirable effects of high population density in prisons are mediated by the stress associated with high density. That is, high population density leads to stress, and stress in turn may lead to a variety of undesirable consequences, such as increased violence or mental illness. The concept of stress in this example is relatively abstract in that it is not directly observable but rather is inferred from events that are observable. Such inferred phenomena are often termed **intervening constructs** or **mediating variables.** Empirically, we might infer stress from autonomic arousal (e.g., increased blood pressure, heart rate, or skin temperature), from verbal and nonverbal signs of anxiety, or from a disintegrated quality of behavior. The distinction between direct observation and abstract inference is one of the main differences between the empirical and theoretical levels of scientific inquiry (Figure 4-1). We should also mention that a mediating variable is a construct that operates between an antecedent and an outcome, whereas a **moderator variable** modifies or interacts with a relationship between other variables (Baron &

Figure 4-1 Theories usually involve abstract inferences about mediators of empirically observed cause-and-effect relationships.

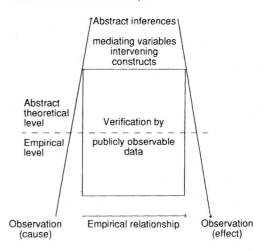

Kenny, 1986; Evans & Lepore, 1997). For example, we might find that noise hurts performance on a reading task because it interferes with attention, but that believing we have control over the noise results in less impaired performance. In this case we would say that attention mediates the relationship between noise and performance and that perceived control moderates (or modifies) the noise–performance relationship.

FUNCTIONS OF THEORIES

We can identify at least three basic functions of theories. First, theories help us to predict relationships between variables, which implies that we can control what happens to one variable by regulating another variable. For example, if we know that certain conditions of prison crowding lead to violence (i.e., cause violence), we can control the violence to some extent by changing the crowded conditions. If our crowding theory says that stress mediates a relationship between crowding and violence, we might also control violence by controlling stress.

A second function of theories is to summarize large amounts of data. Instead of having to know thousands of pieces of data about the levels of stress and violence under thousands of levels of crowding, if we have a good theory we can summarize all this information in a few theoretical statements. Such summaries in turn help us predict events that we may not yet have observed at the empirical level. A third function of theories is the generalization of concepts and relationships to many phenomena, which helps to summarize the knowledge in a particular area. For example, if our theory states that high levels of stress lead to increased levels of violence, this implies that we can generalize the theoretical notion to *any* factor that increases stress, including crowding, noise, poverty, and marital discord. Furthermore, if we can establish that a particular environmental event, such as wild fluctuations in temperature, is stressful, then we can infer from our stress theory that this environmental event will lead to more violence. If

empirical evidence does not suggest that a theory generalizes very well, the theory should be modified or rejected in favor of theories that do offer good **generalizability.** In one sense, the issue of generalizability is troubling for environmental psychologists, since we like to study environment–behavior relationships in the context where they naturally occur. This proclivity may mean that many relationships observed may well not exist in any other context. This is just one reason elaborate theories have been difficult to generate in environmental psychology (cf. Altman & Rogoff, 1987; Clitheroe et al., 1998; R. Kaplan, 1987; Winkel, 1987). Nevertheless, most scientists, including environmental psychologists, would agree that a good theory is high in generalizability.

In addition to these three basic functions, theories are useful in other ways. For one thing, they help to generate additional research by suggesting new relationships between variables. Many scientists assert that the best research is that generated by theories. Another use for theories is in the application of research to practical problems. Solutions to problems are often needed quickly, with little time available for basic research. If theories already exist, they can suggest solutions. In a broader sense, theories can help guide policy decisions. Defensible space theory, for example, can guide the design of public housing or public plazas in order to reduce crime and increase perceptions of safety (Ham-Rowbottom, Gifford, & Shaw, 1999). Theories in environmental psychology, as in any scientific field, must be constantly evaluated, just as hypotheses must be verified. The basic functions of theories suggest the criteria by which theories should be evaluated. First, a theory is valuable to the extent that it predicts. Given two theories about the same environment–behavior relationship, the one that predicts most accurately most of the time is considered more valuable. Second, good theories do a superior job of summarizing many empirical relationships. Third, a valid theory must be very generalizable. Given two theories about the effects of noise on performance, the one that applies to more situations is the more valuable. Fourth, the most useful theories suggest new hypotheses to be tested empirically. In most scientific endeavors, much of the significant research is generated from theories rather than used to construct new theories. Since this research is crucial to our understanding of the phenomena under study, theories that suggest new areas of investigation are highly valued. Let us turn now to some of the theories in common use in environmental psychology.

Environment–Behavior Theories: Conceptualizing Our Interaction With the Environment

As environmental psychology has matured from its infancy over three decades or so, several theories or models have emerged that attempt to give direction to research and bring a degree of unity across a diverse field of inquiry. These theories tend not to be very complex and tend to be restricted to a somewhat limited predictive domain (cf. Clitheroe et al., 1998; Proshansky, 1973; Stokols, 1983). Theories or models restricted to environmental perception or cognition (e.g., the lens model) have been covered in Chapter 3. In this chapter, we will present a set of theories whose predictive domains are largely restricted to the effects of environmental conditions on behavior. The empirical data to which these theories are most applicable will be described in subsequent chapters on such topics as noise, weather, air pollution, personal space, crowding, and urban environments. Specifically, we will examine, for now, the following six theoretical perspectives: (1) arousal; (2) load (overload and underload); (3) adaptation level (AL); (4) behavior constraint; (5) stress; and (6) ecological psychology. We will then see how these various formulations can be integrated into an eclectic model that we will use repeatedly in the remainder of the book. In general, these theoretical approaches suggest that the environment impinges on us and we react to it; how and why this happens varies from theory to theory. All of these approaches, however, imply that we adapt to the stimulation; that is, we change our reaction to it over time, which we describe as *adaptive* or *maladaptive,* depending on the consequences.

Before elaborating on these various approaches, it will be helpful to keep several points in mind. First, theoretical concepts are not always easy to grasp, and the reader may feel overwhelmed with just one reading of this chapter. Full development and application of the material will become clearer in subsequent chapters. Second, we often rely on more than one theory to explain a given phenomenon. As we will see in the discussion of stress, mediators of these various conceptual approaches often occur together, and it is sometimes useful to appeal to more than one approach to explain the data. Finally, different theories are useful at different levels of analysis. For example, ecological psychology is especially applicable to group behavior, whereas the other approaches are often more useful at the individual level of analysis.

THE AROUSAL PERSPECTIVE

One effect of exposure to environmental stimulation is increased arousal, as measured physiologically by heightened autonomic activity, such as increased heart rate, blood pressure, respiration rate, adrenalin secretion, and so on; or behaviorally by increased motor activity; or simply as self-reported arousal. From a neurophysiological perspective, **arousal** is a heightening of brain activity by the arousal center of the brain, known as the **reticular formation** (Hebb, 1972). Berlyne (1960) characterized arousal as lying on a continuum anchored at one end by sleep and at the other end by excitement or heightened wakeful activity. Since arousal is hypothesized to be a mediator or intervening variable in many types of behavior, a number of environmental psychologists have turned to this concept to explain many of the influences of the environment on behavior.

In fact, you may recall that arousal is one of the dimensions along which any environment can be evaluated (Kerr & Tacon, 1999; Russell & Snodgrass, 1987). The arousal model makes distinct predictions about the effects on behavior of *lowered* arousal (i.e., toward the "sleep" end of the continuum) as well as *heightened* arousal, and is quite useful in explaining some behavioral effects of such environmental factors as temperature (Bell & Greene, 1982), crowding (Evans, 1978), and noise (Broadbent, 1971; Klein & Beith, 1985). We should emphasize that pleasant as well as unpleasant stimuli heighten arousal. An exciting date or a thrilling ride at an amusement park can be just as arousing as noxious noise or a crowded elevator.

What happens to behavior when the arousal level of the organism moves from one end of the continuum to the other? As you might expect, several things occur. For one, arousal leads people to seek information about their internal states. That is, we try to interpret the nature of the arousal and the reasons for it. Is the arousal pleasant or unpleasant? Is it due to people around us, to perceived threat, or to some physical aspect of the environment? In part, we interpret the arousal according to the emotions displayed by others around us (Reisenzein, 1983; Schachter & Singer, 1962; Scheier, Carver, & Gibbons, 1979). In addition, the causes to which we attribute the arousal have significant consequences for our behavior. For example, if we attribute the arousal to our own anger, even though it may be due to a factor in the environment, we may become more hostile and aggressive toward others (e.g., Zillmann, 1979). However, attributing the arousal to anger may not be the only reason for increased aggression. According to several theories of aggression (Berkowitz, 1970; Zillmann, 1983), if aggression is the response most likely to occur in a particular situation, then heightened arousal will facilitate aggression. We find, for example, that when

noise increases arousal, it may also increase aggression (Geen & McCown, 1984; Geen & O'Neal, 1969; see also Chapter 5).

Another reaction we have when we become aroused is to seek the opinion of others. We in part compare our actions to those of others to see whether we are acting appropriately and to see whether we are better off or worse off than others (Festinger, 1954; Wills, 1981). This process is known as **social comparison.** We can feel better about our own circumstances if we compare our standing with others who are faring more poorly. Victims of natural disasters, for example, become very aroused by the circumstances and seek to compare their fate with the fate of others (Hansson, Noulles, & Bellovich, 1982).

Arousal also has important consequences for performance, especially as formulated through the **Yerkes–Dodson law.** According to this law, performance is maximal at intermediate levels of arousal and gets progressively worse as arousal either falls below or rises above this optimum point. Moreover, the inverted-U relationship between arousal and performance varies as a function of task complexity. For complex tasks, the optimum level of arousal occurs at a slightly lower level of arousal than for simple tasks, as depicted in Figure 4-2. This **curvilinear relationship** appears consistent with other findings (see page 110) that humans seek an intermediate level of stimulation—too much or too little is undesirable (Berlyne, 1960, 1974). From an environment–behavior perspective, we would expect that as environmental stimulation from crowding, noise, air pollution, or any other source increases arousal, performance will either improve or deteriorate, depending on whether the affected person's response is below, at, or above the optimum arousal level for a particular task (see also Broadbent, 1971; Hebb, 1972; Kahneman, 1973). Apparently, low arousal is not conducive to maximum performance, and extremely high arousal prevents us from concentrating on the task at hand.

The arousal approach fares reasonably well as a theoretical base in environmental psychology, although it does have shortcomings. Perfor-

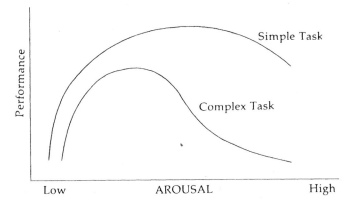

Figure 4-2 The Yerkes–Dodson law predicts an optimal level of performance for simple and complex tasks, with arousal below or above the optimum resulting in performance decrements.

mance and aggression can be predicted from the effects of the environment on arousal, and the arousal notion does generalize to several environmental factors, most notably noise, heat, and crowding. We can also show that viewing scenes of natural environments reduces disturbingly high levels of physiological arousal (Ulrich et al., 1991). Unfortunately, arousal can be difficult to measure with a high degree of confidence and generalizability. Some measures used in research include heart rate, blood pressure, respiration rate, blood vessel constriction, galvanic skin response (or GSR, meaning electrical conductance of skin due to sweating), palmar sweat index (reaction of palm sweat with a chemical), urine secretion, brainwave activity, physical activity level, muscle tension, skin temperature, and self-report scales. Physiological indices of arousal are not always consistent with one another and are often not consistent with self-reports, such as paper-and-pencil measures of arousal (cf. Cacioppo & Petty, 1983; Dienstbier, 1989). Whereas one measure may indicate increases in arousal in a given situation, other measures may show decreased or unchanged arousal. Which measure to choose in predicting behavior thus becomes a serious problem. Nevertheless, the arousal notion is a useful one and will probably continue to be incorporated into those environment–behavior relationships to which it is applicable.

THE ENVIRONMENTAL LOAD PERSPECTIVE

Imagine you are trying to study for three exams you have the following day, but your roommate wants to watch television, there is a loud party next door, and two friends come by to entice you to go out for pizza. How can you possibly study for your exams with all this going on? This situation is similar to the circumstances under which the **environmental load** or **overstimulation** approach explains environment–behavior relationships. Especially useful when describing reactions to novel or unwanted environmental stimuli, the model derives from research on attention and information processing, and it can be described in five parts (Broadbent, 1958, 1963; Cohen, 1978; Easterbrook, 1959; Kaplan, 1995; Milgram, 1970):

1. Humans have a limited capacity to process incoming stimuli and can invest only a limited effort in attending to inputs at any one time.
2. When the amount of information from the environment exceeds the individual's capacity to process all that is relevant, information **overload** occurs. The normal reaction to overload is a type of "tunnel vision" in which we ignore those inputs that are less relevant to the task at

hand and devote more attention to those that are relevant. We often take active steps to prevent less relevant or distracting stimuli from occurring. For example, Ahrentzen and Evans (1984) noted how teachers modify the classroom environment to minimize distractions.

3. When a stimulus occurs that may require some sort of adaptive response (or when an individual thinks such a stimulus will occur), the significance of the stimulus is evaluated by a monitoring process, and a decision is made about which coping response, if any, to employ. Thus, the more intense or unpredictable or uncontrollable an input, the greater its adaptive significance and the more attention paid to it. Furthermore, the more uncertainty generated by an input about the need for an adaptive response, the more attentional capacity allocated to it.

4. The amount of attention available to a person is not constant and may be temporarily depleted after prolonged demands. After attending to prolonged demands, the total capacity for attention may suffer from an overload. For example, after studying hard for several hours, it is difficult to do anything that demands much attention. This state of overload or directed attention fatigue (DAF) can result in increased mental errors, difficulty concentrating, and irritability.

5. Attentional fatigue can be ameliorated through reduced demands on information processing or through restorative environments such as nature trails, parks, museums, or zoos. This latter principle has been labeled *Attention Restoration Theory,* or *ART* (Kaplan & Kaplan, 1989; Kaplan, 1995), which we described in Chapter 2 as a separate theory, but which we now incorporate into the broader load perspective.

What happens to behavior when an overload occurs? The answer depends on which stimuli are given adequate attention and which are ignored. Generally, stimuli most important to the task at hand are allocated as much attention as needed, and less important stimuli are ignored. If these less important stimuli tend to interfere with the central task, ignoring them will enhance performance. If, however, a task requires a wide range of attention, as when we must do two things at once, performance on less important tasks will deteriorate. In an interesting demonstration of this process, Brown and Poulton (1961) required participants driving a car either in a residential area (relatively small number of important inputs) or in the parking lot of a crowded shopping center (relatively large number of important inputs) to listen to a series of taped numbers and determine which numbers changed from one sequence to the next. More errors were made on this secondary numbers task when participants drove in the shopping center, presumably because in the shopping center more attention had to be allocated to important stimuli connected with driving, to the detriment of the less important stimuli of the numbers task. Drawing somewhat on this line of work, some urban planners have argued that successful core city areas should actually make busy streets narrower and slow traffic down, under the assumption that slower traffic will allow less attention to be focused on traffic and more on commercial opportunities along the street (Pacelle, 1996).

According to the overload model, once capacity for attention has been depleted owing to prolonged demands, even small demands for attention may cause overloading. Even *after* exposure to unpleasant or excessive stimulation *has ceased,* behavioral aftereffects, such as decreased tolerance for frustration, errors in mental functioning, and less frequent altruistic behavior, may occur (see Chapters 5 and 6 and the box on page 107 for research examples). The overload model attributes these aftereffects to a reduced capacity to attend to relevant cues.

About a century ago, sociologist Georg Simmel attributed behavioral pathologies in large urban areas to a type of overload (e.g., Simmel, 1957 translation). Milgram (1970) also

⤳ THE AFTERMATH OF OVERLOAD

We have noted that we tend to narrow attention and ignore noncentral information when our processing capacity becomes overloaded, so our performance may deteriorate. Interestingly, for some time after the overload has apparently stopped, we may suffer from the cost of recovering from the overload. Cohen and Spacapan (1978) reported two studies demonstrating this aftereffect of overload.

First, 80 research participants viewed a panel of 12 lights, with each light one of three different colors. When a light came on, the participant had to press one of three keys corresponding to the matching color. In a low-load condition, the time between consecutive illuminations was 0.8 second. In the high-load condition, the time between lights was 0.4 second. Thus, in the high-load condition, participants had to process more information in a given interval of time. Once they had finished the task, participants were asked to work on some paper-and-pencil puzzles that were actually impossible to solve. Those who had been in the high-load condition spent almost 2 minutes less than those in the low-load condition before giving up on the puzzles. Apparently, the higher load reduced tolerance for the frustration of working on the puzzles.

In a second study, 40 students individually walked through a shopping mall and listed and priced various items in the stores according to a prearranged set of instructions for 26 tasks. All participants had 30 minutes to complete the sequence of pricing tasks. High-load participants were given twice as many items as low-load participants to list for each task. In addition, half the participants conducted their mall survey on weekday afternoons (low shopper density) and half on weekend afternoons (high shopper density). The very last pricing task occurred in an isolated hallway. As the participant finished it, an experimenter's assistant (unknown to the participant) standing nearby pretended to lose a contact lens. As the environmental load model would predict, fewer participants helped look for the lens in the high-load than low-load condition (17% versus 57%), and fewer helped in the high- than low-density condition (also 17% versus 57%). Thus, even after an apparent overload, there is a negative impact on task performance and social behavior. Cohen and Spacapan also interpreted their results in terms of aftereffects within the context of the environmental stress model, which we discuss beginning on page 116.

suggested that the deterioration of social life in large urban areas is caused by the ignoring of peripheral social cues and a reduced capacity to attend to them because of the increased demands of everyday functioning. Thus, urban ills, such as bystanders ignoring others in distress, may be due in part to an environmental overload in which the hustle and bustle of everyday life in the city requires so much attention that there is very little left over for "peripheral" social concerns. Some city dwellers may be forced to develop an aloof attitude toward others in order to allocate enough time to everyday functioning (see Chapter 10).

You have probably experienced overload by the end of finals week. What do you do about it? You probably do something to "get away from it all." The notion of overload is particularly germane to the study of leisure environments and other settings we have called restorative environments, as we saw beginning in Chapter 2. Kaplan and Kaplan (1989) suggest that prolonged concentration on a task leads to directed attention fatigue (DAF), which is a state of mental exhaustion similar to overload. Their research suggests that recovery from DAF is most likely in a restorative environment, defined as having four characteristics: (1) *being away,* or something other than your normal environment; (2) *extent,* or providing an experience that is extended in time and space; (3) *fascination,* or being interesting and engaging; and (4) *compatibility,* or the ability of the environment to support what you intend to do. A wilderness experience, for example (Hartig, Mang, & Evans, 1991), or a museum visit (Kaplan, Bardwell, & Slakter, 1993) may meet the requirements of a restorative environment for some people. Those who engage in activity in a restorative environment typically report increased interest and acuity in dealing with the task that led to DAF (e.g., Kaplan et al., 1993). Natural environments are believed by many to be most effective in this regard (Herzog, Black, Fountaine, & Knotts, 1997; Parsons, 1991). We will see more about these benefits of natural environments in Chapter 13.

The environmental load model stands up to theoretical scrutiny about as well as the arousal model. It does predict some of the behavioral consequences of excessive environmental stimulation. However, there are many difficult-to-determine "if's" incorporated into the model, including whether or not in a given situation an overload occurs, whether a specific task is important, whether ignoring less important stimuli facilitates or impairs performance on a particular task, and whether demand has been sufficiently prolonged to deplete attentional capacity.

In terms of generalizability, the model applies to mental and motor performance and to at least some social behaviors. As far as generating research is concerned, the environmental load model does suggest many possibilities, including evaluating whether or not a given environment is likely to produce an overload, and assessing the extent to which attentional depletion contributes to social and environmental problems.

THE UNDERSTIMULATION PERSPECTIVE

The environmental overload perspective suggests that many environment–behavior relationships, especially those leading to undesirable behavioral and affective (emotional) consequences, are a function of too much stimulation from too many sources. A number of theorists have suggested, however, that many environment–behavior problems result from **understimulation,** or *too little* stimulation. **Sensory deprivation** studies (e.g., Zubek, 1969) suggest

that depriving individuals of all sensory stimulation can lead to severe anxiety and other psychological anomalies, although some of these effects may be due to suggestibility from laboratory procedures such as the prominent display of a "panic button" and asking participants to sign liability release forms that disclose the possibility of serious damage (cf. Barabasz & Barabasz, 1985). In a study of students sharing a 10-person living facility, Brown (1992) discovered that solitude was a common experience but could often be felt as aversive. On the other hand, there may well be benefits of reduced stimulation (see the box on page 109). For example, floating in an isolated tank of water may increase creativity and decrease anxiety, hostility, and depression (Forgays & Forgays, 1992; Forgays et al., 1991). Nevertheless, other research has documented the deleterious effects of understimulation on such processes as the maturational development of the young (e.g., Sapolsky, 1997; Schultz, 1965). Drawing on these sources, some theorists suggest that the environment should sometimes be made more complex and stimulating in order to restore excitement and a sense of belonging to individuals' perceptions of their environment.

Even limited sensory deprivation can have predictable effects on us. For example, Antarctic isolation has been shown to modify task performance and psychological outlook (Barabasz & Barabasz, 1986; Suedfeld, 1998; Wood, Lugg, Hysong, & Harm, 1999). Moreover, the isolation of solitary sailing or from being an aircrash survivor in a remote region can generate a "sensed presence" of another individual even when no such person exists (Suedfeld & Mocellin, 1987).

Although cities may have an overstimulating social environment, they may subject inhabitants to an understimulating physical environment. Urbanologist A. E. Parr (1966) contended that fields, forests, and mountains contain an unending variety of changing patterns of visual stimulation, but that urban areas contain the same patterns repeated on every street. In many tract housing developments in particular, the structures all resemble one another. According to Parr, the giant skyscrapers

REST:
The Benefits of Sensory Deprivation

We have noted that reduced stimulation can have deleterious effects on people. In contrast to such findings, many studies suggest that there are actually benefits to boredom. A procedure called **REST**, or Restricted Environmental Stimulation Technique, or Restricted Environmental Stimulation Therapy (Suedfeld, 1980; Suedfeld et al., 1990) involves placing a person in a soundproof, darkened room or into a darkened water tank (Figure 4-3).

The potential benefits of REST seem to cover many areas (Barabasz & Barabasz, 1993). For example, reduced stimulation seems to have favorable outcomes for hyperactive and autistic children (Suedfeld, Schwartz, & Arnold, 1980). Biofeedback seems to have more positive effects when combined with REST (Lloyd & Shurley, 1976; Plotkin, 1978). Hypertension (high blood pressure) can be reduced with REST (Suedfeld, Roy, & Landon, 1982), and REST can help some individuals stop smoking (Suedfeld & Baker-Brown, 1986). REST may enhance some forms of creativity and impair others (Norlander, Bergman, & Archer, 1998).

Figure 4-3 An example of a REST chamber. The participant lies on a fluid-filled mattress in the chamber, and the doors are closed, reducing the noise to minimal levels.

How does REST achieve these results? No one knows for sure, but one proposed avenue is that with reduced extraneous stimulation, participants can better recognize internal states, such as high blood pressure, and thus take more effective self-regulatory steps. Another possibility is that REST disorganizes established mechanisms for maintaining chronic maladaptive patterns, and thus permits new, more adaptive mechanisms to occur. More complex cognitive and physiological avenues have also been explored (Suedfeld et al., 1990).

However it works, REST has become commercially popular. That is, businesses provide REST-type tanks or beds for clients to purchase or rent on a short-term basis. We suspect such ventures will spur more investigations into the value of REST procedures.

lining city streets and the interiors of modern windowless structures instill a sense of enclosure rather than a sense of being drawn to the next horizon. Parr and others assert that this lack of stimulation leads to boredom and is in some way responsible for such urban ills as juvenile delinquency, vandalism, and poor education (cf. Heft, 1979b).

To study some of these problems of understimulation, Wohlwill (1966) advocated scaling environments along a number of dimensions of stimulation, including intensity, novelty, complexity, temporal change or variation, surprisingness, and incongruity. As we will see in Chapters 10 and 12, the desire for these types of stimulation may explain why people leave

cities in great numbers to live in more "natural" environments.

As a theoretical approach by itself, the understimulation angle does help predict some environment–behavior relationships, but it stands in marked contrast to arousal and overload theories that examine the same environments and find too much stimulation. Moreover, some researchers claim that benefits can be derived from deprivation of sensory stimulation (e.g., Suedfeld, 1975). We will reserve further judgment on the understimulation perspective until we have examined a theoretical approach that attempts to consolidate the understimulation and overstimulation views.

ADAPTATION LEVEL THEORY: OPTIMAL STIMULATION

If the research evidence supporting the arousal and overload perspectives suggests that too much environmental stimulation has deleterious effects on behavior and emotions, and if the evidence supporting the understimulation perspective suggests that too little stimulation similarly has undesirable effects, it stands to reason that some intermediate level of stimulation would be ideal. This is the approach taken by Wohlwill (1974) in his **adaptation level (AL)** theory of environmental stimulation. Borrowing from Helson's (1964) adaptation level theory of sensation and perception, Wohlwill began with the assumption, for example, that humans dislike crowds, at least on certain occasions, as when trying to make last-minute Christmas purchases or trying to leave a packed football stadium at the end of the game. On the other hand, most of us do not like total social isolation all day, either. Along these lines, Altman (e.g., 1975) describes environmental mechanisms by which we regulate privacy to achieve the desired level, and Gärling, Biel, and Gustafsson (1998) describe how an optimal level of uncertainty in the environment seems to be preferred. Wohlwill believed that the same applies for all types of stimulation, including temperature, noise, and even the complexity of roadway scenery. What we usually prefer is an optimal level of stimulation (see also Zuckerman, 1979).

Categories and Dimensions of Stimulation

At least three categories of environment–behavior relationships should conform to this optimal level hypothesis, according to Wohlwill. These categories are sensory stimulation, social stimulation, and movement. Too much or too little sensory stimulation is undesirable, too much or too little social contact is undesirable, and too much or too little movement is undesirable. (Do you see the similarity between this notion and the Yerkes–Dodson law described under arousal theory?) These categories in turn vary along at least three dimensions that have

optimal levels. The first dimension is *intensity*. As we have noted, too many or too few people around us can be psychologically disturbing. Too little or too much auditory stimulation has the same unwanted effect. We have all experienced the irritation of neighbors making distracting noise while we were trying to listen to a lecture or concert, of loud stereos playing when we are trying to study, and of children screaming when adults are trying to carry on a conversation. On the other hand, if you have ever been in a soundproof chamber for very long, you know that the absence of external sound becomes very unnerving after only a few minutes.

Another dimension of environmental stimulation is *diversity*, both across time and at any given moment. Too little diversity in our surroundings produces boredom and the desire to seek arousal and excitement. Too much diversity—as in the typical "strip" of fast-food franchises, gas stations, and glaring neon signs common in many towns and cities—is considered an eyesore. As we saw in Chapter 2, considerable research (see Wohlwill, 1974) indicates that the perceived attractiveness and the degree of pleasant feelings associated with a human-built scene are maximized at an intermediate level of diversity.

The third dimension of stimulation is *patterning*, or the degree to which a perception contains both structure and uncertainty. The total absence of structure that can be coded by information-processing mechanisms, such as diffuse light of a constant intensity or a single tone at a constant volume, is disturbing. By the same token, a very complex pattern that contains no predictable structure is also disturbing. To the extent that a modern built environment is so diverse and complex that we have difficulty imposing a perceptual structure on it, we probably experience that environment as stressful. Urban street patterns are a good example of this dimension. Parallel streets in intersecting grid patterns can be monotonous. On the other hand, complex layouts with no predictable numbering or no easy access to major arteries can cause confusion. An intermediate level of patterning, with gently winding streets and cul-de-sacs, is usually comfortable, yet pleasantly stimulating.

Optimizing Stimulation

After assuming the general rule that there are optimal levels of environmental stimulation, Wohlwill introduced a modifier to this rule by further assuming that each person has an optimal level of stimulation, which is based on past experience. Thus, Tibetan people live comfortably at altitudes with so little oxygen that most of us would have difficulty maintaining consciousness; they have adapted to a level of oxygen concentration quite different from what most of us would consider ideal. Similarly, those of us who live in cities probably have a higher level of tolerance for crowds and less tolerance for isolation than do most residents of rural areas. After a person brought up in a rural area has lived in a city for a few months or years, he or she probably acquires a greater tolerance for crowds than the rural resident who never moves to a city. Wohlwill referred to this shift in optimal stimulation level as **adaptation,** defined as a shift in our judgmental or affective responses to a stimulus following continued exposure to it.

Adaptation levels not only differ from person to person as a function of experience, but also change with time following exposure to a different level of stimulation. For example, Bih (1992) described how Chinese students moving to New York City adapted over time, Yamamoto et al. (1992) studied how Japanese adapt to the workplace following university graduation, and Laska (1990) observed how people make long-term adaptations to repeated flooding of their homes. Thus, how one evaluates and reacts to a given environment along a particular dimension is in part determined by how much that environment deviates from one's adaptation level on that dimension. The more an environment deviates from the adaptation level, the more intense the reaction to that environment should be. As you may recall from Chapter 2, there is considerable variation in individual perceptions of what constitutes a "beautiful" or "desirable" environment. Adaptation level theory suggests the reason for this variation involves individual differences in adaptation level along several relevant dimensions. For example, Mocellin et al. (1991) found that anxiety levels did not increase

Figure 4-4 Differences in adaptation level may explain why some people pay heightened attention to environmental risk and others minimize the same risk. Grieshop and Stiles (1989) found that over 25% of their household respondents reported suffering illness from pesticide exposure, yet there was high risk-taking in pesticide use even among those who perceived higher danger. Perhaps adaptation level of perceived risk explains such findings. In Chapter 7 we will review more evidence about perceived risk and adaptation to environmental hazards.

among groups stationed in harsh, isolated Arctic and Antarctic environments; quite likely, adaptation levels were already suitable for these circumstances (cf. Harrison, Clearwater, & McKay, 1991). Why do people differ so much in their assessment of environmental risk, such as perceived danger of toxins? Differences in adaptation level may be the explanation (Vaughan, 1993; see Figure 4-4).

Adaptation Versus Adjustment

Adaptation level theory postulates an interesting environment–behavior relationship in the distinction between adaptation and what Sonnenfeld (1966) termed **adjustment.** Adaptation refers to changing the response to the stimulus, whereas adjustment refers to changing the stimulus itself. Adjustment in this case does not

ENVIRONMENTS AND THE ELDERLY:
Environmental Press and Competence

Adaptation level theory posits that each person has an optimal level of stimulation along several dimensions. A special application of this idea is a model of **environmental press** developed by M. Powell Lawton and Lucille Nahemow to describe environments for the elderly (Lawton, 1975; Lawton & Nahemow, 1973; Nahemow & Lawton, 1973). This model posits that the demands (i.e., press) an environment places on its occupants as well as the competence of the occupants determine the consequences of interacting with the environment. If the impact of the press is within the **environmental competence** of the individual to handle it (i.e., within the adaptation level), positive feelings about the environment occur and the behavior is adaptive. If the press is considerably weaker or stronger than the competence of the individual (i.e., outside the adaptation level), negative feelings and maladaptive behavior occur. Thus, an understimulating nursing-home environment or a fast-paced, crime-ridden neighborhood may be outside the desired adaptation level for press and competence. We will see in Chapter 12 how environments for the elderly can be designed to suit their needs.

refer to the adjustment–maladjustment continuum conceptualized in clinical psychology (i.e., an internal, psychological state), but rather to a mechanism by which we change the environment. For example, adaptation to hot temperatures would involve gradually getting used to the heat so that we sweat more efficiently upon exposure to it. Adjustment would involve either wearing lighter clothes or installing an air-conditioning system so that the temperature stimulus striking our skin is much cooler. For most organisms and for early human societies, adaptation was probably a more realistic option than adjustment. For modern societies with advanced technology, however, adjustment is so clearly a realistic option that we prefer it over adaptation. If we do not like the color of our bedroom walls or the furniture layout of the office, we change these features to something that is more comfortable, that is, to something within our adaptation level. In general, adaptation level theory suggests that when given a choice between adapting and adjusting, people will take the course that causes the least discomfort.

Evaluation of the Optimal Stimulation Perspective: Breadth Versus Specificity

It should be obvious that adaptation level theory incorporates some of the best features of the arousal, overload, and understimulation perspectives. As such, it has rather broad generalizability; it is applicable to physical and social environments as well as to all forms of sensation and perception. Adaptation level theory also suggests that future research might well concentrate on the adaptation process in order to solve many environmental problems. One difficulty that arises with this theory, however, is that since it allows for so much individual variation in adaptation level, it becomes very difficult to make more general predictions about environmental preference and environment–behavior relationships. This problem typically arises in behavioral science theories. The more specific the elements from which predictions are made, the less general the predictions; the more general the predictors, the less precise the predictions.

Another problem with AL theory is that it is often difficult to identify an "optimal" level of stimulation before we make a prediction. We mentioned that in environmental aesthetics it has been proposed that an intermediate level of complexity leads to optimal judgments of beauty. This prediction often proves incorrect, in part because we have difficulty defining what we mean by an intermediate level of complexity. In order for AL theory to work, we would need to see what conditions maximize judgments of beauty, then define that level of complexity as intermediate. What is really needed is more research that quantifies levels of environmental stimulation. Only then can we know how well AL theory predicts environment–behavior relationships.

THE BEHAVIOR CONSTRAINT PERSPECTIVE

According to the theoretical perspectives we have examined thus far, excessive or undesirable environmental stimulation leads to arousal or a strain on our information-processing capacity. Another potential consequence of such stimulation is loss of **perceived control** over the situation. Have you ever been caught in a severe winter storm or summer heat wave and felt there was nothing you could do about it? Or have you ever been forced to live or work in extremely crowded conditions and felt the situation was so out of hand there was nothing you could do to overcome it? This loss of perceived control over the situation is the first step in what is known as the **behavior constraint** model of environmental stimulation (Proshansky, Ittleson, & Rivlin, 1970; Rodin & Baum, 1978; Stokols, 1978, 1979; Zlutnick & Altman, 1972). So important is the feeling of perceived control that some would classify the behavior constraint model as a subunit of a **"control model"** that is more global. Whatever classification one might prefer, the concepts in the model are extremely important to environmental psychologists.

The term *constraint* here means that something about the environment is limiting or interfering with things we wish to do. The constraint can be an actual impairment from the environment or simply our belief that the environment is placing a constraint on us. For example, Tanner (1999) found that both subjectively perceived constraints and objective constraints predicted Swiss adults' reports of ability or inability to reduce automobile usage.

Once you cognitively interpret the situation as being beyond your control, what happens next? When you perceive that environmental events are constraining or restricting your behavior, you first experience discomfort or negative affect (i.e., have unpleasant feelings). You

also probably try to reassert your control over the situation. This phenomenon is known as **psychological reactance,** or simply **reactance** (Brehm, 1966; Brehm & Brehm, 1981; Wortman & Brehm, 1975). Any time we feel that our freedom of action is being constrained, psychological reactance leads us to try to regain that freedom (cf. Strube & Werner, 1984). If crowding is a threat to our freedom, we react by erecting physical or social behaviors to "shut others out." If the weather restricts our freedom, we might stay indoors or use technological devices (e.g., snowplows, air-conditioned cars) to regain control. According to the behavior constraint model, we do not actually have to experience loss of control for reactance to set in; all we need do is *anticipate* that some environmental factor is about to restrict our freedom. Mere anticipation of crowding, for example, is enough to make us start erecting physical or psychological barriers against others.

What happens if our efforts to reassert control are unsuccessful in regaining freedom of action? The ultimate consequence of loss of control, according to the behavior constraint model, is **learned helplessness** (Garber & Seligman, 1981; Seligman, 1975). That is, if repeated efforts at regaining control result in failure, we might begin to think that our actions have no effect on the situation, so we stop trying to gain control even when, from an objective point of view, our control has been restored. In other words, we "learn" that we are helpless. Students who try to change a class schedule but are rebuffed by the registration office numerous times soon "learn" that they are helpless against bureaucracy. Similarly, if efforts to overcome crowding are unsuccessful, we may abandon our attempts to gain privacy and change our lifestyles accordingly. During a very severe winter we sometimes hear reports of individuals "giving up" trying to keep warm when their fuel supplies become depleted, and some people die as a result. While less severe than death, learned helplessness often leads to depression.

The behavior constraint model, then, posits three basic steps: perceived loss of control, reactance, and learned helplessness. Often, components of the model are discussed within the context of the stress or load models. Whether treated within the behavior constraint model or within some other model, it is clear that perceived loss of control has unfortunate consequences for our actions, and that restoring control enhances performance and mental outlook. In a now classic line of research, for example, Glass and Singer (1972) told participants they could reduce the amount of noxious noise in an experiment by pressing a button. Simply telling participants about this option actually reduced or eliminated many of the negative effects of noise, even though participants did not, in fact, press the button. That is, simply perceiving that they could control the noise reduced the adaptive costs of that stress. Perceived control over noise has also been found to reduce its negative effects on aggression (Donnerstein & Wilson, 1976) and helping behavior (Sherrod & Downs, 1974). Moreover, perceived loss of control over air pollution seems to reduce efforts to do anything about the problem (Evans & Jacobs, 1981). Similarly, perceived control over crowding reduces its unpleasant effects (e.g., Langer & Saegert, 1977; Rodin, 1976), and perceived control over crime may motivate us to employ more prevention measures (Miransky & Langer, 1978; Tyler, 1981). Moser and Levy-Leboyer (1985) showed that loss of perceived control over a malfunctioning phone led to acts of aggression, but the availability of information designed to restore control improved the situation. Similarly, Rochford and Blocker (1991) observed that perceived control over the threat of flooding was associated with greater activism to do something about it, and Allen and Ferrand (1999) found that personal control predicted environmentally responsible behavior.

Perceived control also has implications for institutional environments. Topf (1994) asserts that lack of control over noise negatively impacts health of hospital patients and staff alike. Presence of perceived control, on the other hand, appears beneficial to health and well-being. Langer and Rodin (1976) manipulated the amount of control residents of a nursing home had over their daily affairs. For instance,

one group was told the staff would take care of them while another group was told they were responsible for themselves. One group was given plants to raise themselves, while the other group was given plants to be cared for by the staff. After three weeks, residents in the high-control group showed greater well-being and enhanced mood and more activity than those in the low-control condition. Even 18 months later, the high-control group had more positive outcomes (Rodin & Langer, 1977; see also Lemke & Moos, 1986; Rothbaum, Weisz, & Snyder, 1982). Schulz (1976) has also documented positive effects of a perceived-control intervention for the institutionalized elderly. After the intervention was terminated, however, those in the high-control condition showed an especially rapid decline (Schulz & Hanusa, 1978), perhaps because a lessened sense of control produced learned helplessness and depression. Schutte et al. (1992) developed a scale to measure perceived control in the institutional setting and found that it is an efficient indicator of whether or not interventions do indeed increase perceived control.

Types of Control

Several attempts have been made to elaborate on the types of control we can have over our environment. Averill (1973), for example, distinguished among categories of (1) *behavioral control,* in which we have available a behavioral response that can change the threatening environmental event (e.g., turning off a loud noise); (2) *cognitive control,* in which we process information about the threat in such a way that we appraise it as less threatening or we understand it better (e.g., deciding that a contaminant in our water is not toxic); and (3) *decisional control,* in which we have a choice among several options (e.g., choosing to live in a quiet rather than a noisy neighborhood). Behavioral control can be manifested either through regulated administration, in which there is control over who administers the threatening event and when they do it, or stimulus modification, in which the threat can be avoided, terminated, or otherwise

PERCEIVED CONTROL AND RESEARCH ETHICS: A Dilemma

We have noted that perceived control over unpleasant environmental stimulation, such as noise, reduces the negative consequences of exposure to the stimuli. An interesting problem in this regard has arisen in the area of laboratory research on environmental stressors. Gardner (1978) notes that for years it was possible to demonstrate such effects as reduced proofreading speed and accuracy when laboratory participants were exposed to uncontrollable noise. Subsequent research, however, has failed to find these detrimental effects. What went wrong? Gardner provides evidence that the "culprit" is a set of research ethics guidelines established by the federal government and implemented by universities and other research institutions. Among these guidelines is a requirement that participants be informed of potential risks of being in the experiment, even though the risks are minimal. Moreover, participants must be told that they are free to terminate the experiment at any time and must sign an "informed-consent" statement disclosing the risks and the termination provision. Gardner provides evidence that such informed consent procedures amount to giving participants perceived control over the stressor, and thus stress effects are reduced! The situation is then an ethical dilemma: How can one ethically do experimental research on stressful environmental conditions if the ethical procedures in effect eliminate the negative reactions to these conditions? Gardner proposes that where risks are minimal, the need to know about these effects justifies modification of the informed-consent procedures. Such a decision would rest with an Institutional Review Board to ensure the safety of subjects. Alternatively, more emphasis may have to be placed on field research involving observance of naturally occurring instances of the stressor. Ultimately, we suspect the ethical dilemma can never be fully resolved (see also Dill et al., 1982).

modified. Cognitive control can be manifested either through appraisal of the event as less threatening or through information gain about such factors as predictability or consequences. Thompson (1981) has noted that there are some questions about this type of categorization and adds a category of *retrospective control,* in which we perceive present control over a past aversive event. Weisz, Rothbaum, and Blackburn (1984) distinguish between *primary control,* meaning overt control over existing conditions, and *secondary control,* meaning accommodating to existing realities and becoming satisfied with things the way they are (see also Thompson, 1981). These authors note that there may be cultural differences in the emphasis placed on primary versus secondary control (cf. Azuma, 1984; Kojima, 1984). Apparently, the amount of control we have is important: Being able to control both onset and termination of a noise results in better adaptation than control over only onset or only termination of the noise (Sherrod et al., 1977). We may also have

more control over some areas of our lives (e.g., our bedroom) than over others (e.g., our community; see Paulhus, 1983).

It would be an oversimplification to state that the greater the control we perceive over our environment, the better we are able to adapt to it successfully. In fact, under some circumstances, control can lead to increased threat, anxiety, and maladaptive behavior (e.g., Averill, 1973; Folkman, 1984; Thompson, 1981). For example, knowing that you can control a potential flood by building a larger levee may make you worry about the expense and time commitment of the intervention. Or, if your dwelling is built near a toxic-waste dump, perceived control through the option of moving away may heighten concern about losing close neighbors and forsaking the emotional attachment to your dwelling. Interestingly, sometimes we actually prefer less control. Certain evidence even suggests that some elderly people prefer less personal control over health-related decisions and wish that others would make these decisions for them (see Rodin, 1986; Woodward & Wallston, 1987).

Aspects of Helplessness

Just as research has progressed on the perceived-control component of the behavior constraint model, so has research on the reactance and learned-helplessness components. For our present discussion, the work on learned helplessness seems especially important. For example, Hiroto (1974) found that when people were given a chance to terminate an aversive noise, those who had previously been able to control it learned to terminate it. Those who had previously been unable to control the noise, however, responded as if they were helpless and failed to learn the termination procedure. Similarly, a field study with school children found that those who attended noisy schools near Los Angeles International Airport showed more signs of learned helplessness than those from quieter schools (Cohen et al., 1980, 1981).

Learned helplessness effects have been interpreted in terms of attribution theory (e.g., Abramson, Seligman, & Teasdale, 1978; Miller & Norman, 1979; Peterson & Seligman, 1984; Sweeney, Anderson, & Bailey, 1986; Tennen & Eller, 1977). Attributions are inferences about causes of events or about characteristics of people or events. In general, helplessness effects are more likely to occur if we attribute our lack of control over the environment to (1) stable rather than unstable factors (e.g., to our physical or mental inability to do anything about it rather than to our temporary lack of time to act on it); (2) general rather than specific factors (e.g., attributing pollution to all industry rather than to a specific factory); and (3) internal rather than external locus of control (e.g., attributing our discomfort in a crowd to our own preference for open spaces rather than to the behavior of others in the crowd). In a confirmation of this attributional approach, researchers found that those who attributed negative outcomes to global (general) factors showed helplessness deficits in settings both similar and dissimilar to the setting where an initial negative outcome occurred.

Those who attributed the initial negative outcome to specific factors, however, showed helplessness effects only in settings similar to the initial one (Alloy et al., 1984).

Value and Limitations of the Behavior Constraint Perspective

Research on reactance, perceived loss of control, and learned helplessness is certain to continue, whether interpreted from the perspective of the behavior constraint model or from some other perspective. The model itself has considerable, though limited, utility. In instances of perceived loss of control, the model is quite useful in predicting some of the consequences. In cases in which there is no reason to infer perceived loss of control, however, other mediators, such as stress, arousal, and overload, are probably necessary to explain environment–behavior relationships. Moreover, the behavior constraint approach places much emphasis on individual reactions and can minimize the need to look at the entire setting.

THE ENVIRONMENTAL STRESS PERSPECTIVE

Another theoretical perspective, which is widely used in environmental psychology, is to view many elements of the environment, such as noise and crowding, as stressors. Stressors—including job pressures, marital discord, natural disasters, the turmoil of moving to a new location—are considered to be aversive circumstances that threaten the well-being of the person. **Stress** is an intervening or mediating variable, defined as the reaction to these circumstances. This "reaction" is assumed to include emotional, behavioral, and physiological components. The physiological component was initially proposed by Selye (1956) and is often called **systemic stress.** The behavioral and emotional components were proposed by Lazarus (1966, 1998) and are often called **psychological stress.** Since physiological and psychological stress reactions are interrelated and do not occur alone, environmental psychologists usually combine all the components into one theory, or the **environmental stress model** (e.g., Baum, Singer, & Baum, 1981; Evans & Cohen, 1987; Lazarus & Folkman, 1984).

Sometimes the term *stress* is restricted to environmental events, and an additional term, *strain,* is used to describe the consequence within the organism. However, we will use *stress* to refer to the entire stimulus–response situation, *stressor* to refer to the environmental component alone, and *stress response* to refer to the reaction caused by the environmental component.

As we will see, some components of the arousal, environmental load, adaptation level, and behavior constraint approaches fit very well into an environmental stress framework. Overload, for example, can be viewed as one consequence of coping with stress, and heightened arousal is certainly a component of stress. Similarly, an optimal level of stimulation (i.e., stimulation at the adaptation level) should result in little evidence of a stress reaction, but multiple constraints on behavior might be expected to lead to considerable signs of stress.

We will organize our discussion of stress under three subheadings. First, we will consider the *characteristics of stressors,* such as how long they last or how often they occur. Since the degree to which these events actually cause stress is dependent upon how they are interpreted (i.e., whether people notice them and decide that they might be harmful or aversive), we will also discuss the *appraisal of stressors.* Finally, the kinds of *stress responses* that occur (physiological reactions, cognitive factors, coping strategies, aftereffects) will be considered.

Characteristics of Stressors

There are a number of ways we might classify stressors. Lazarus and Cohen (1977) have described three general categories of environmental stressors: cataclysmic events, personal stressors, and background stressors. These vary according to *severity of impact* as well as other dimensions, such as the ease of the **coping** or adaptation process in response to them.

Cataclysmic Events   Natural disasters, war, nuclear accidents, or fires are unpredictable and powerful threats that generally affect all of those touched by them. Such **cataclysmic events** are overwhelming stressors that have several basic characteristics. They are usually sudden, giving little or no warning of their occurrence. They have a powerful impact, elicit a more or less universal response, and usually require a great deal of effort for effective coping. The accidents at Three Mile Island and Chernobyl, the Mount Saint Helens eruption, as well as the more common tornadoes, hurricanes, and other natural disasters (Adams & Adams, 1984; Baum, Fleming, & Davidson, 1983; Hartsough & Savitsky, 1984; Pennebaker & Newtson, 1983; Proulx, 1993; Rotton et al., 1997) can all be considered in this category of stressors.

The powerful onset of sudden cataclysmic events may initially evoke a freezing or dazed response by victims. Coping is difficult and may bring no immediate relief. However, the severely threatening period of such an event usually (but not always) ends quickly, and recovery begins. A tornado may strike for only a brief time, and other cataclysmic events may be over in a few days. When the process is allowed to proceed without a return of the stressor, rebuilding progresses and more or less complete recovery is generally achieved. In the case of Three Mile Island or Love Canal, where rebuilding is not what is needed (nothing was actually destroyed), and the damage already done is less important than the damage that may yet come, recovery may be more difficult (Baum, Fleming, & Davidson, 1983).

One important feature of cataclysmic events, which is in some ways beneficial for the coping process, is that they affect a large number of people. Affiliation with others and comparing feelings and opinions with them have been identified as important styles of coping with such threats, since **social support** can moderate the effects of stressful conditions (e.g., Norris & Kaniasty, 1996). Because people are able to share their distress with others undergoing the same difficulties, some studies have suggested that cohesion results among these individuals (Quarantelli, 1978). Of course, this does not always happen, and residents cannot "band together" to fight a stressor indefinitely. When a stressor persists in an apparently unresolvable manner, problems of a different kind can arise—including learned helplessness.

Personal Stressors A second group of stressors may be termed **personal stressors.** These include such events as illness, death of a loved one, or loss of one's job—events that are powerful enough to challenge adaptive abilities in the same way as cataclysmic events (e.g., Dooley, Rook, & Catalano, 1987; Lehman, Wortman, & Williams, 1987). Personal stressors generally affect fewer people at any one time than do cataclysmic events, and may or may not be expected. Frequently, with personal stressors the point of severest impact occurs early, and coping can progress once the worst is over, although this is not always the case. Often the magnitude, duration, and point of severest impact of cataclysmic events and personal stressors, such as death and loss of a job, are similar. However, the relatively smaller number of people who experience a particular personal stressor at any one time may be significant because there are fewer others to serve as sources of social support. Also, a cataclysmic event such as a flood can result in the loss of a loved one, loss of a job, or other personal stressors.

Background Stressors Less powerful, more gradual, but more chronic and almost routine stressors are termed **background stressors.** Rotton (1990) prefers to divide background stressors into two types. **Daily hassles** (or microstressors) are stable, low-intensity problems encountered as part of one's routine (Lazarus et al., 1985; Zika & Chamberlain, 1987), such as we described in the opening paragraph of this chapter. **Ambient stressors** are "chronic, global conditions of the environment—pollution, noise, residential crowding, traffic congestion—which, in a general sense, represent noxious stimulation, and which, as stressors, place demands upon us to adapt or cope" (Campbell, 1983, p. 360).

Whereas daily hassles (losing things, home maintenance) are unique each day and affect a specific individual, ambient stressors such as pollution impact a larger number of people, are chronic and nonurgent, and are difficult to remove through the efforts of one individual. While many background stressors are mundane and of relatively low intensity, some may not even be noticeable, like certain instances of air pollution (e.g., Evans & Jacobs, 1981). Any one or two background stressors may not be sufficient to cause great adaptive difficulty, but when a number occur together, they can exact a cost over time and may be as serious as cataclysmic events or personal stressors. Regular and prolonged exposure to certain low-level background stressors may even require more adaptive responses in the long run than more intense stressors. For example, long-term exposure to noise (Evans, Hygge, & Bullinger, 1995), neighborhood problems (White et al., 1987), and long-term commuting stress (White & Rotton, 1998) can be quite problematic.

With background stressors, it is often difficult to identify a point at which "the worst is over," and it may not be at all clear that things will get better. In fact, things may go from bad to worse. In addition, the benefits for coping or of having others who "share in the experience" may not be as great as for other types of stressors. This may be because the intensity of background stressors is frequently so low as to never raise the need for affiliation; or, alternatively, social support may not be appropriate in these situations (cf. Campbell, 1983).

Appraisal

A given environmental event may or may not be a stressor in all circumstances, and in the same circumstance it may be a stressor to some individuals and not to others. The probability of an event becoming stressful is determined by a number of factors (Evans & Cohen, 1987), including the characteristics of the event and the way individuals appraise it. Thus, in order for the stress process to begin, there must be cognitive **appraisal** of a stimulus as threatening. To use an environmental example, 90° F (32° C) to a native southerner is not likely to be very stressful in midsummer. To someone living in Barrow, Alaska, however, the mere thought of experiencing 90° F for a few hours a day may well be evaluated as threatening. In other words, the same stimulus that may not be stressful in one situation may be stressful in another—the stimulus has not changed, but the individual's appraisal of it as threatening or nonthreatening has changed. Moreover, cognitive appraisal that an aversive event, such as crowding, is pending is often sufficient to elicit a stress response, even though the physical event itself does not happen (e.g., Baum & Greenberg, 1975).

Lazarus (1966, 1998) suggests that this cognitive appraisal is a function of individual psychological factors (intellectual resources, knowledge of past experience, and motivation) and cognitive aspects of the specific stimulus situation (control over the stimulus, predictability of the stimulus, and immediacy or "time until impact" of the stimulus). The more knowledge one has about the beneficial aspects of a source of noise, or the more control one has over the noise (in terms of terminating or avoiding it), the less one is likely to evaluate that stimulus as threatening, and the less threatening the situation is likely to be.

Types of Appraisal Cognitive appraisal of a situation is more complex than merely assessing its potential threat (see Baum et al., 1982, for a review). Several different types of appraisal are possible. **Harm or loss appraisals** focus on damage that has already been done (Lazarus & Launier, 1978). For example, victims of a natural or technological disaster could be expected to make harm/loss evaluations. In general, rapid loss of resources is associated with traumatic stress (Hobfoll, 1991). In contrast, **threat appraisals** are concerned with future dangers. Environmental toxins such as pesticides may evoke perceived threat to one's health, and threat appraisals may precede exposure to them. Perceived threat from and stress reaction to a chronic toxic hazard is likely to be worse than that associated with a quick-hitting flood (Baum

et al., 1992; Baum & Fleming, 1993). The ability to anticipate potential difficulties allows us to prevent their occurrence but may cause us to experience anticipatory stress. It is hard to say which is worse—seeing one's home destroyed in a hurricane (harm/loss) or not knowing how one will be sheltered from the elements until one can build a new home (threat). As this example suggests, threat and harm/loss appraisals usually go hand in hand (Lazarus & Folkman, 1984). **Challenge appraisals** are different from others because they focus not on the harm or potential harm of an event, but on the possibility of overcoming the stressor. Some stressors may be beyond our coping ability, but we all have a range of events for which we are confident of our ability to cope successfully. Stressors that are evaluated as challenges fall within this hypothetical range (Dienstbier, 1989; Lazarus & Launier, 1978; Tomaka et al., 1997).

Factors Affecting Appraisal A number of factors have been identified that affect our appraisals of environmental stressors. These include the characteristics of the condition in question (e.g., how loud a particular noise is); situational conditions (e.g., whether what we are doing is compatible with or inhibited by the potential stressor); individual differences; and environmental, social, and psychological variables. To cite but one example, the upper-middle-class resident of a large city may be less likely to experience difficulty as a result of urban conditions than a poorer resident of the same city. Or, he or she may be better able to avoid the seamy side of the city and thus less likely to be exposed to aversive urban conditions. Attitudes toward the source of stress will also mediate responses; if we believe that a condition will cause no permanent harm, our response will probably be less extreme than if it carries the threat of lasting harm. If our attitudes are strongly in favor of something that may also harm us, we may reappraise threats and make them seem less dangerous. For example, Elliott et al. (1997) describe how residents near a landfill in Ontario appraised it as less threatening over time.

We described the influence of perceived control when discussing the behavior constraint approach. Perceived control is also an important moderator of stress, providing a sense of being able to cope effectively, to predict events, and to determine what will happen. Giving research participants information about a stressor prior to their exposure to it helps them to plan and predict what will happen. Such information increases perceived control and reduces the threat appraisal made when the stressor is experienced. For example, the stress associated with surgery or aversive medical procedures can be reduced by providing patients with accurate expectations of what they will feel (e.g., Johnson & Leventhal, 1974). Other studies have found that accurate expectations about high levels of density reduce crowding stress (Baum, Fisher, & Solomon, 1981; Langer & Saegert, 1977).

Coping styles or behavior patterns also appear to affect the ways in which events are appraised, as well as which types of coping are invoked. Work on a number of these dimensions, such as **repression-sensitization** (the degree to which people think about a stressor), **screening** (a person's ability to ignore extraneous stimuli or to prioritize demands), and **denial** (the degree to which people ignore or suppress awareness of problems), has indicated that people differing along these dimensions may interpret situations differently (e.g., Bell & Byrne, 1978; Collins, Baum, & Singer, 1983; Mehrabian, 1976–77). A study by Baum et al. (1982), for example, suggested that individuals who cope with overload by screening and prioritizing demands are less susceptible to the effects of crowding than people who do not cope in this way.

Another moderator of stress appraisals may be social support—the feeling that one is cared about and valued by other people, and that he or she belongs to a group. Many have long believed that interpersonal relationships can somehow protect us from many ills (e.g., Cohen & Wills, 1985; Jung, 1984). However, the effects of having or not having social and emotional support have not always been clearly demonstrated (cf.

Ganellen & Blaney, 1984; Hendrick, Wells, & Faletti, 1982). One possible reason is that those from crowded homes or other backgrounds of social distress may respond to others through withdrawal rather than attachment (Evans & Lepore, 1993).

Characteristics of the Stress Response

A distinction is often made between **primary appraisal,** which involves assessment of threat, and **secondary appraisal,** which involves assessment of coping strategies. Appraisals of stressors help determine responses to them. If an appraisal is "negative" and an event is seen as being dangerous, responses that prepare us to cope will ensue. These stress responses involve the whole body. Physiological changes are part of this response, most reflecting increased arousal. At the same time, emotional, psychological, and behavioral changes may also occur as part of the stress response.

Physiological Response   Part of the response to an aversive or stressful stimulus is automatic. Selye's (1956) **general adaptation syndrome (GAS)** consists of three stages: (1) the alarm reaction, (2) the stage of resistance, and (3) the stage of exhaustion. Initially, an **alarm reaction** to a stressor causes autonomic processes (heart rate, adrenalin secretion, and so on) to speed up. The second stage in the stress process, the **stage of resistance,** also begins with some automatic mechanism for coping with the stressor. If heat is the stressor, sweating occurs; if extreme cold is the stressor, shivering may occur. When these homeostatic mechanisms do not restore **equilibrium,** signs of exhaustion or depleted reserves will be observed as an organism enters the last of Selye's three stages, the **stage of exhaustion.** The primary indicants of this stage are ulcers, adrenal enlargement, and shrinkage of lymph and other glands that confer resistance to disease.

Some responses to environmental stress are virtually indistinguishable from those evoked by direct assault on body tissue by pathogens. Recalling Selye's three-stage process, it appears that stress results in heightened secretion of **corticosteroids** during the alarm reaction, followed by a decline in reactivity (as measured by this secretion) through resistance and exhaustion. The **catecholamines**—*dopamine, epinephrine* (adrenalin), and *norepinephrine*—are also active in stress, along with emotional distress (Arnsten, 1998).

Increased catecholamine and corticosteroid secretion is associated with a wide range of other physiological responses, such as changes in heart rate, blood pressure, breathing, muscle potential, inflammation, and other functions. In the brain, stress enhances the activity of an emotional center (the amygdala), making memories of emotional events stronger, and decreases the attention-focusing and organizing and planning operation of the frontal lobes (Arnsten, 1998). All of these effects appear to be stronger in the absence of perceived control.

These findings may also be viewed as consistent with pioneering work by Cannon (1929, 1931), who suggested that epinephrine has a positive effect on adaptation. Epinephrine provides a biological advantage by arousing the organism, thus enabling it to respond more rapidly to danger. When extremely frightened or enraged, we experience an arousal that may be uncomfortable but that readies us to act against the thing that scares or angers us. Thus, stress-related increases in catecholamines may facilitate adaptive behavior.

Some studies have shown superior performance on simple, well-learned tasks following stress reactions. On the other hand, arousal has been associated with impaired performance on complex tasks. Decreases in problem-solving abilities, increases in general negativity, impatience, irritability, feelings of worthlessness, and emotionality may all accompany a stress response, and emotional disturbances such as anxiety or depression may occur (Arnsten, 1998; Evans et al., 1995; Rotton et al., 1997).

Coping Strategies   In the stage of resistance, many coping processes are also cognitive, so

that the individual must decide on a behavioral coping strategy. According to Lazarus (1966, 1998), the coping strategy is a function of individual and situational factors and may consist of flight, physical or verbal attack, or some sort of compromise. Lack of success in the coping process may increase the tendency to evaluate the situation as threatening. For example, Faupel and Styles (1993) found that victims of hurricane Hugo reported more stress if they had engaged in activities to prepare for the disaster; perhaps the experienced devastation in spite of preparation increased perceived threat. Associated with this cognitive coping process are any number of emotions, including anger and fear. To use another example, the stress reaction to a large crowd in a city might consist of evaluation of the crowd as threatening, physiological arousal, fear, and flight to a less crowded area (Figure 4-5).

Many ways of categorizing coping strategies have been developed (see Aldwin & Revenson, 1987). Two useful distinctions employed by Lazarus and his colleagues are (1) *direct action* or *problem-focused*, such as information seeking, flight, or attempts to remove or stop the stressor; or (2) **palliative** or *emotion-focused*, such as employing psychological defense mechanisms (denial, intellectualization, etc.), using drugs, meditating, or reassessing the situation as nonthreatening (see also Roth & Cohen, 1986). To the extent that direct action is not available or practical, palliative strategies become more likely. Interestingly, a sense of humor helps people cope with many types of stress (Martin & Lefcourt, 1983), as does viewing relaxing scenes of nature (e.g., Parsons et al., 1998).

Adaptation   As previously noted, if the coping responses are not adequate for dealing with the stressor, and all coping energies have been expended, the organism will enter the third stage of the GAS, the stage of exhaustion. Fortunately, something else usually happens before exhaustion occurs. In most situations, when an aversive stimulus is presented many times, the stress reaction to it becomes weaker and

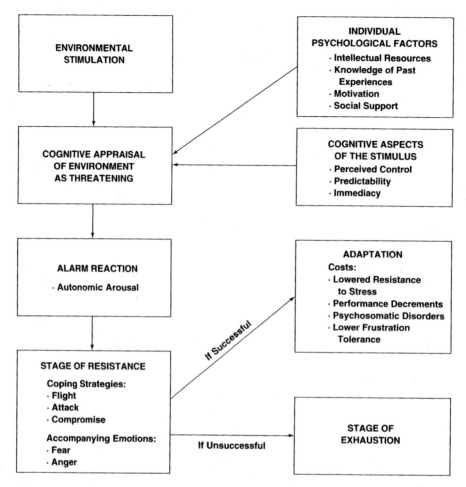

Figure 4-5 The stress model.

weaker. Psychologically, this process is called *adaptation*. Adaptation to a stressor may occur because neurophysiological sensitivity to the stimulus becomes weaker, because uncertainty about the stressor is reduced, or because the stressor is cognitively appraised as less and less threatening.

Visitors to a polluted city, for example, initially may suffer overt physiological symptoms, such as shortness of breath, and may express a great deal of fear about the potential health consequences of exposure to atmospheric pollutants. On successive days in the city, however, these visitors, realizing that they have not died yet, may "lose" the fear of breathing the air (see

Chapter 7). As another example, consider the stress that might build as one moves to a new office; all of the old emotional attachments, productive work habits, and spatial organization could be threatened. But even small improvements in the new office, such as increased lighting or more privacy, may reduce the threat and stress associated with the move (Spreckelmeyer, 1993).

Adaptation to stress is both beneficial and costly. Almost all events in life, from birth, to attending school, to driving on freeways at rush hour, involve some degree of stress. Obviously, the individual who has been exposed to stress and has learned to handle it is better able to deal with the next stressful event in life. In this sense, the "teaching" function of stress is beneficial to the organism as long as the stress can be handled—the experience boosts self-confidence and provides skill development (e.g., Aldwin & Stokols, 1988; Martin et al., 1987). We have seen that environmental stress sometimes improves performance, probably because the arousal associated with stress (if it is not too severe) facilitates performance. Exposure and adaptation to stressful events may also be costly, however. If the total of all stresses at any one time exceeds the capacity of the individual to cope with them, some sort of breakdown, physical or mental, is almost inevitable.

Psychosomatic disorders, performance decrements, and lowering of resistance to other stressors are often the costs of adapting to prolonged or excessive stress. Still another cost, one that we have treated previously as a separate theoretical approach, is the resulting cognitive overload: Our information-processing capacity becomes so overloaded by the stressor that additional processing is difficult and more prone to error. Some costs of adaptation may occur during exposure to the stressor, including performance decrements and physiological wear and tear. Other costs may occur after the stressor is no longer around. For example, as we will see in Chapter 5, even after an aversive noise has stopped, tolerance for frustration, accuracy of mental functioning, and even altruistic behavior (see the box on page 107) may continue to be impaired (see also Cohen, 1980).

Cognitive deficits associated with stress may be caused by behavioral strategies that are used for coping with stress—"tuning out" or narrowing one's field of attention (e.g., Arnsten, 1998; Cohen, 1978). When under stress, we may be unable to concentrate or unwilling to put effort into a task (e.g., Glass & Singer, 1972). In other ways, our coping response may be specific to the stressor being experienced, reflecting the specific causes of our discomfort. People may respond to crowding that is due to too many people by withdrawing and avoiding social contact, whereas their response to crowding that is due to limited space might be aggression

(e.g., Baum & Koman, 1976). A person might respond to job loss actively if the loss was caused by a lack of effort rather than ability, or may become helpless under certain conditions.

Aftereffects Following adaptation to stress, there may be long-term consequences—or **aftereffects**—after the stressor has terminated. These are not specific to certain stressors but appear to reflect more general effects (Cohen, 1980) and fit in with Selye's (1956) notion of limited adaptive energy. As exposure to stress increases, adaptive reserves are depleted, causing aftereffects and reductions of subsequent coping ability. Evidence for the existence of post-stressor effects comes from a number of sources, including research on the effects of noise (e.g., Evans et al., 1995; Glass & Singer, 1972; Rotton et al., 1978; Sherrod & Downs, 1974; Sherrod et al., 1977), crowding (Evans 1979a), electrical shock (Glass et al., 1973), hurricanes (Moore & Moore, 1996; Riad & Norris, 1996), and floods (Tobin & Ollenburger, 1996).

Psychological effects that linger or persist may also reflect consequences of adaptation. Calhoun (1967, 1970) has referred to **refractory periods,** which are periods of time during which an organism recovers from a bout with a stressor. If the refractory period is interrupted by another encounter, increased stress-relevant problems are likely. Recovery from stress, as with recovery from overload, is likely to be facilitated by experiences with natural environments (Ulrich et al., 1991).

Some long-term aftereffects may be physiological. The cascade of neurochemical events in the stress response is thought to have an impact on the immune system. That is, exposure to environmental stressors could lead to impaired functioning of the immune system, and thus in the long run to decreased resistance to infectious diseases and, via immune malfunctioning and other pathways, increased risk of some diseases such as cancer and heart disease (Maier, Watkins, & Fleshner, 1994). Prolonged or sudden elevation of circulating catecholamines may damage body tissue and is suggested as a cause of the development of hardening of the arteries and other diseases of the blood vessels (Schneiderman, 1982). Indeed, the relationship between stress and health is one that is of ever-increasing interest in modern times (e.g., Bernard & Krupat, 1994).

Assessing the Stress Model

When we evaluate the effectiveness of using stress as a mediator for a theoretical approach in environmental psychology, we find that it does an admirable job with the data in its predictive domain. The stress approach does help predict many of the consequences of environmental deterioration as well as the presence or absence of observable effects of such specific stressors as crowding and extremes of heat and cold. In this respect, the stress approach has a great deal of generalizability: It applies to many situations and accounts for the combined effects of many environmental and social stressors that are presented at the same time (e.g., Levine, 1988). Perhaps for this reason the stress approach suggests many directions for new research. If we treat a given environmental event as a stressor, then we should be able to predict its effects, with or without the presence of other stressors, from our knowledge of the effects of other stressors. Furthermore, we should be able to use present knowledge about coping with stress to help control reactions to unwanted environmental stressors. On the other hand, one problem with using only the stress approach as a theoretical inroad in environmental psychology is that the identification of stressors is somewhat ambiguous (e.g., Lazarus et al., 1985). For example, suppose we expose individuals to a particular stimulus and get no stress reaction. Should that stimulus be regarded as something other than a stressor, or did those particular individuals just not evaluate it as threatening under the experimental circumstances? In addition, stress models have some difficulty predicting exactly when individuals will cope with a stressor in different ways; that is, we do not easily predict when someone will use palliative versus direct-action strategies—we know that people use these different strategies, but describing the chosen path

after the fact is easier than predicting it ahead of time.

BARKER'S ECOLOGICAL PSYCHOLOGY

The theoretical perspectives reviewed up to this point have been concerned primarily with the specific effects of the environment on behavior; but, with the exception of the behavior constraint model, they have not been concerned much with the effects of behavior on the environment. Yet, as we have noted many times, behavior inevitably influences the environment. **Ecological psychology** views environment–behavior relationships as two-way streets or, in other words, as ecological interdependencies.

Barker (1968, 1979, 1987, 1990) and his colleagues have been the principal advocates of the ecological psychology perspective. The focus of Barker's model is the influence of the **behavior setting** on the behavior of large numbers of people, which is termed the **extra-individual behavior pattern.** A unique aspect of Barker's approach is that the behavior setting is an entity in itself. It is not an arbitrarily defined social-scientific concept but actually exists and has a physical structure, although it does change over time (Wicker, 1987). In order to understand just how this behavior setting functions, we will first look at some characteristics of the behavior setting, then see how the setting fits into Barker's theory of staffing. For additional reading, you might wish to consult the July 1990 issue of *Environment and Behavior,* which was written by Barker, his students, and colleagues, and is devoted entirely to a commemoration of the functioning of the Midwest Psychological Field Station, which was the setting of much of Barker's early work.

The Nature of the Behavior Setting

A number of behaviors can occur inside a structure with four walls, a ceiling, and a floor; but if we know that the cultural purpose of this structure is to be a classroom, then we know that the behavior of the people in the structure will be quite different than if its purpose is to be a church, a factory, or a hockey arena. The fact that this behavior setting is in a built environment also tells us that the extra-individual behavior will be different from that in the natural environment of a forested wilderness or a desert. This cultural purpose exists because the behavior setting consists of the interdependency between **standing patterns of behavior** and a **physical milieu.** Standing patterns of behavior represent the collective behaviors of the group, rather than just individual behaviors. These behaviors are not unique to the individuals present, but they may be unique to the setting. If the behavior setting is a classroom in a lecture-oriented course, then the standing patterns of behavior would include lecturing, listening, observing, sitting, taking notes, raising hands, and exchanging questions and answers. Since this **en masse behavior** pattern occurs only in an educational behavior setting, ecological psychologists would infer that knowing about the setting helps us predict the behavior that will occur in it. The physical milieu of this behavior setting would include a room, a lectern, chairs, and perhaps a chalkboard and projector and screen.

Once the individuals leave the classroom, the physical milieu still remains, so the standing behavior patterns are independent of the milieu. Yet they are similar in structure, or **synomorphic,** and together create the behavior setting (Figure 4-6). A change in either the standing behavior patterns (as when a club holds a meeting in the classroom) or the physical milieu (such as when the class is held outdoors on the first warm day of spring) changes the behavior setting. Sommer (1998) describes the synomorphic evolution of food co-ops as gradual renovations to the stores in order to make the design more and more consistent with the ideology of the group.

How can we use the behavior setting conceptualization to understand environment–behavior relationships? Perhaps a few examples can best illustrate the utility of this approach (see also Wicker & Kirmeyer, 1976). One very famous application of ecological psychology was described in Chapter 1. In this study, Barker and his colleagues (Barker & Schoggen, 1973; Barker & Wright, 1955) compared a small town in Kansas with one in England. They found, for

Figure 4-6 According to Barker's ecological psychology, knowing about the physical setting tells us much about the behaviors that occur there. In the setting shown, what behaviors can you always expect to see?

example, that in England behavior settings under the control of businesses were more common, and the behavior in them lasted longer. In settings involving voluntary participation, however, Americans spent more time and held more positions of responsibility than did the Britons. (The significance of such findings will be more apparent later in the discussion of staffing.) Wicker (1979, 1987) notes that ecological psychology methods are very useful for such diverse goals as documenting community life, assessing the social impact of change, and analyzing the structure of organizations for such factors as efficiency of operation, handling of responsibility, and indications of status. In addition, as Bechtel (1977) noted, ecological psychology can be useful in assessing environmental design. By carefully examining the behavior setting, one can analyze such design features as pathways, or links between settings, and focal points, or places where behavior tends to concentrate. In the lobby of a building, for example, it is important to separate pathways to various elevators, offices, and shops in order to avoid congestion and confusion. An information center in the lobby, though, would be most useful if placed at a focal point. As another example, open-plan (i.e., no internal walls) designs in schools and offices, although having advantages, often lead to inadequate boundaries between behavior settings, thereby causing interference with the intended functions (e.g., Oldham & Brass, 1979). We will discuss more of these kinds of design implications in Chapters 11, 12, and 13.

Staffing the Setting: How Many Peas Fill a Pod?

What happens if a behavior setting such as a classroom or theater has too few or too many inhabitants for maximum functioning efficiency? Do students at small schools, for example, take on more roles of responsibility than students at larger schools? Studies of these questions from the ecological psychology perspective have led to what is called **staffing theory** (Barker, 1960; Barker & Gump, 1964; Wicker & Kirmeyer,

1976; Wicker, McGrath, & Armstrong, 1972). Historically, this concept was termed the *theory of manning*, but today it is known by the gender-neutral phrase *theory of staffing*.

In order to understand the theory, let us first define some terms proposed by Wicker and his colleagues that are related to the concept of staffing. The minimum number of inhabitants needed to maintain a behavior setting is defined as the **maintenance minimum.** The maximum number of inhabitants the setting can hold is the **capacity.** The people who meet the membership requirements of the setting and who are trying to become part of it are called **applicants. Performers** in a setting carry out the primary tasks, such as the teacher in a classroom, the workers in a factory, or the cast and supporting staff in a play. **Nonperformers,** such as the pupils in a classroom or the audience in a theater, are involved in secondary roles. Maintenance minimum, capacity, and the applicants are different entities for performers and nonperformers. For example, maintenance minimum for performers in a classroom would be the smallest staff (teachers, custodians, secretaries, deans) required to carry out the program. For nonperformers, maintenance minimum would be the smallest number of pupils required to keep the class going. Capacity for performers in a classroom might be determined by social factors (e.g., how many teachers are most effective in one setting) and by physical factors, such as the size of the room, number of lecterns, and so on. For nonperformers, room size would be the primary determinant of capacity. Whether your class contains 10 or 1,000 students depends in most cases as much on classroom size as on educational policy. For performers, applicants are the individuals who meet the requirements of the performer role and who seek to perform, as in the number of teachers available to teach a given class. Applicants for nonperformers are those who seek secondary roles, as in the number of students trying to get into the class. If students are available but do not seek to get into the class, or if teachers do not want to teach a given class, then they are not considered applicants.

AN ECOLOGICAL PSYCHOLOGY APPROACH TO MANAGING SMOKING IN PUBLIC PLACES

Smoking is banned today in many public spaces, but sometimes smokers light up in nonsmoking areas. Gibson and Werner (1994) suggested that part of the reason may not be disrespectful smokers, but rather environmental layout and cues.

One aspect of ecological psychology involves the circuitry of the setting—those elements by which it is regulated. For example, the setting program defines what is supposed to happen in the setting, and the deviation-countering circuit restores order to violations of the program. Gibson and Werner viewed smoking in a nonsmoking area as a violation of the program. The problem, they suggested, is that the program is not always obvious. Recall from Chapter 3 that legibility is an important aspect of defining the ease of cognitive mapping in a city. Lack of legibility may also be a factor in smokers lighting up in nonsmoking areas. In one study, Gibson and Werner found that smoking was much more likely in ambiguous areas than in clearly marked nonsmoking areas. In another study, they created a distinct boundary between smoking and nonsmoking areas, or they kept the boundary ambiguous by having a row of chairs cross between the two areas. In addition, ambiguity was created by sometimes having ashtrays in nonsmoking areas—all of this despite the clear presence of "No Smoking" signs. Again, they found that the more ambiguity, the more smoking in nonsmoking areas (i.e., the more violation of the setting program). In fact, when boundaries were distinct and ashtrays not present, no one ever smoked in a nonsmoking area. In a third study, these researchers found that nonsmokers' responses to an intruding smoker could be predicted by location and legibility of the setting. Nonsmokers were more likely to reprimand a smoker in a nonsmoking area (i.e., counter the deviation or defend the territory) if the violation occurred in the center of the area versus the edge of the designated nonsmoking area; deviation countering was also more likely to occur if the program of the nonsmoking area was highly legible (i.e., if the boundary was distinct and no ashtrays were present).

Thus, it seems that legibility is important in getting people to conform to a setting's program. Furthermore, increasing legibility may help manage conflicts.

If the number of applicants to a setting (either performers or nonperformers) falls below maintenance minimum, then some or all of the inhabitants must take on more than their share of roles if the behavior setting is to be maintained. This condition is termed **understaffed.** If the number of applicants exceeds the capacity, the setting is **overstaffed,** and if the number of applicants is between maintenance minimum and capacity, the setting is *adequately staffed.* Wicker (1973) has labeled an adequately staffed setting with a low number of participants as *poorly staffed,* and an adequately staffed setting with a high number of participants as *richly staffed.* Thus, we can consider a continuum of participation levels from understaffed to poorly staffed, adequately staffed, richly staffed, and overstaffed.

When conditions of understaffing exist, the consequences for the inhabitants of the setting are many. As stated earlier, inhabitants must take on more specific tasks and roles than would otherwise be the case. As a result, inhabitants have to work harder and at more difficult tasks than they would otherwise, and peak performance on any task is not as great as in an adequately staffed setting. Furthermore, admissions standards to understaffed settings may have to be lowered, and superficial differences among inhabitants may be largely ignored, whereas in adequately staffed settings these differences are highlighted to fit each person into his or her appropriate role. Each inhabitant in an understaffed setting is more valued, has more responsibility, and interacts more meaningfully with the setting. Since understaffed settings have more opportunities for the experience of failure as well as success (owing to the increase in number of experiences per inhabitant), these settings are likely to result in more feelings of insecurity than are adequately staffed settings. The consequences of understaffing are summarized in Table 4-1.

Overstaffing, on the other hand, results in adaptive mechanisms being brought into play to deal with the huge number of applicants. One obvious solution would be to increase the capacity, probably through enlarging the pres-

ent physical milieu or moving to a larger one. Another adaptive mechanism would be to control the entrance of clients into the setting, either through stricter entrance requirements or through some sort of funneling process (Figure 4-7). For example, Wicker (1979) describes how ecological psychologists implemented and evaluated a queuing (waiting line) arrangement at Yosemite National Park to alleviate overcrowding and associated disruptive behavior at bus stops. Still another regulatory mechanism would be to limit the amount of time inhabitants can spend in the setting. These three mechanisms are elaborated in Table 4-2.

In general, predictions for staffing theory have been supported by research. For example, in a laboratory study involving too many, too few, or an intermediate number of participants to run a complex racing game, those in understaffed conditions reported feelings of involvement in the group and having an important role within the group (e.g., Wicker et al., 1976). Studies of large versus small high schools (Baird, 1969; Barker & Gump, 1964) suggest that students in small schools (which are less likely to be overstaffed) are indeed involved in a wider range of activities than students from large schools and are more likely to report feelings of satisfaction and of being challenged. Similar results have been reported for colleges as well (Baird, 1969; Berk & Goebel, 1987). Even student groups within college conform to the principles of staffing: As group size declines, groups become more open to prospective and new members (Cini, Moreland, & Levine, 1993). Studies of large versus small churches (e.g., Wicker, 1969; Wicker & Kauma, 1974; Wicker, McGrath, & Armstrong, 1972; Wicker & Mehler, 1971) also indicate that members of small churches are likely to be involved in more behavior settings within the church (e.g., choir, committees) and to be involved in more leadership positions; such predictions are based on the assumption that smaller churches are more likely to be understaffed and larger churches overstaffed. Norris-Baker (1999) describes how staffing theory is useful in evaluating the effects

| **Table 4-1** | CONSEQUENCES OF UNDERSTAFFING ° |

Time and Effort

Increased effort to support the setting
Increased time to support the setting

Participation, Tasks, and Roles

Participation in a greater variety of tasks and roles
Participation in more difficult and important tasks

Responsibility, Perception of Self and Others

Assumption of more responsibility for
  behavior-setting tasks
Perception of self in terms of task-related
  characteristics
Perception of others in terms of task-related
  characteristics
Perceived by others as more important to the
  functioning of the setting

Views About Differences

Pay less attention to personality differences among
  individuals
Pay less attention to other nontask related
  differences among individuals
Lower admission standards for applicants

Performance, Success, and Failure

Accept lower levels of performance for self
Accept lower levels of performance for others
Feel insecure regarding success of the setting
Experience success and failure frequently

° *Adapted from Allan W. Wicker, personal communication, 1989*

Figure 4-7   Funneling is one way to regulate entrance into a potentially overstaffed behavior setting.

Table 4-2	MECHANISMS FOR REGULATING THE POPULATION OF A BEHAVIOR SETTING[°]

Regulating access of applicants into the setting by:
- scheduling appointments for entrance
- increasing or decreasing recruiting
- raising or lowering admission standards
- asking participants to wait in holding areas
- preventing unauthorized entrances

Regulating the setting's capacity by:
- changing the arrangements or contents of the physical milieu
- changing the duration (hours open) of the setting
- increasing or decreasing staff (performers) to handle applicants
- assigning staff (performers) to different tasks as demands of applicants increase or decrease

Regulating the time applicants or inhabitants occupy the setting by:
- admitting applicants at different rates
- changing the limits on how long people can stay
- using a fee structure based on length of stay
- establishing priorities for dealing with different classes of applicants
- changing the standing patterns of behavior to facilitate the flow of applicants

[°]*Adapted from Allan W. Wicker, personal communication, 1989*

of population decline in small rural communities. Altogether, then, these and other studies suggest that staffing theory is very useful in assessing involvement and satisfaction within a number of environments, from businesses (e.g., Greenberg, 1979; Oxley & Barrera, 1984), to psychiatric institutions (e.g., Srivastava, 1974) to schools and churches, and to "home, sweet home" (Jones, Nesselroade, & Birkel, 1991).

Assessing the Ecological Psychology Perspective

Ecological psychology has its advantages and disadvantages. It necessitates using a field observation methodology (described in Chapter 1) that gives the theory the advantage of using real-world behavior. It certainly insists on preserving the integrity of the person–environment interrelationship. However, it includes the disadvantage of not being able to study many detailed cause-and-effect relationships in the laboratory, though certainly some laboratory research on ecological psychology principles has been and will continue to be conducted (e.g, Wicker, 1987; Wicker & Kirmeyer, 1976). Studies of real-world behavior in context lead to difficulties of interpretation without scientific control of variables. For example, the observed effects of large versus small schools or churches could be due to differential group influence such as staffing demands, or to individual differences in the types of people who choose to affiliate with large versus small institutions. Here we have a theory that is so broad in its scope that specific predictions about one person's behavior become difficult to make and troublesome to confirm. Since this approach is designed to study group behavior, it does a respectable job of handling group data in the context of a given setting, but it does not handle individual behavior as well as other theories. To its credit, ecological theory does generate many valuable research questions, such as what common properties of certain behavior settings result in the same group behavior, what happens when the structure of a behavior setting changes, and what effects

one behavior setting has on behavior in another setting. Finally, the ecological approach is applicable to a large variety of settings and circumstances (Sommer & Wicker, 1991).

OTHER THEORETICAL PERSPECTIVES

We have described six major, broad theoretical perspectives used in environmental psychology. There are additional approaches that are used in a more restricted domain, particularly with regard to designing environments. We will briefly mention a few of them here and describe their application beginning in Chapter 11.

We noted in Chapter 2 that humans often develop attachment to places. This includes not just natural areas such as parks but also built environments such as our homes or communities. This attachment is often accompanied by a sense of ownership and a tendency to defend the place against intruders. The theory of defensible space (see page 356) implies that we can design places to promote attachment and make defense easier. As we will see in Chapter 7 on disasters, disturbed places have more impact on us if we are attached to them.

We also noted on page 110 that we seek an ideal level of privacy. This not only is similar to the principles of adaptation level theory, but is also one mediator of our use of space and the operation of personal space, territoriality, and crowding as means of maintaining the desired privacy level (see also Altman, 1975; Kupritz, 1998; Newell, 1998; Pedersen, 1997). In designing environments, privacy regulation is one example of the principle of congruence, which refers to the fit of an environment to the needs it is designed to meet and the behaviors it is designed to promote. To the extent that an environment is not congruent with intended behaviors (in ecological psychology terminology, is not synomorphic with its purpose), some sort of stress is likely. To the extent that congruence is present, satisfaction and attachment are likely.

We have also described a model of environmental press and competence in the box on page 112. To the extent that press and competence are matched (i.e., congruent), psychological well-being as well as attachment and satisfaction are promoted. Wayfinding as described in Chapter 3 can be facilitated by user-friendly designs, which is another example of congruence, as is designing environments for accessibility by those who have disabilities. Environments with inadequate designs in this regard are likely to foster constraints and stress, but those with appropriate designs should yield satisfaction, attachment, and competence (see also Pedersen, 1999).

Integration and Summary of Theoretical Perspectives

We will encounter still more conceptual perspectives in future chapters, and we will see repeatedly that the major theories we have described thus far are not mutually exclusive. Each theory selects one or two mediators inferred from empirical data and attempts to explain a large portion of the data using the mediator. Just because one mediator explains a particular set of data, however, does not mean that other mediators do not operate in the same set of data. It is entirely conceivable, for example, that loud noise produces information overload, stress, arousal, and psychological reactance all at the same time in the same individual, and that architectural designs that fail to attenuate noise lead to all of these outcomes. Furthermore, regardless of which of these mediators is involved (either alone or in combination), any number of coping responses are likely to result, such as flight, erecting barriers or other protective devices, ignoring other humans in need, and directly attempting to stop or reduce the stimulus input at the source. Although one particular mediator may best predict or explain which coping responses will occur in a given situation, other mediators are not necessarily excluded from that or similar situations. It is our position that all of the mediating processes discussed thus far probably occur at some time, given all the possible situations in which environmental stimulation influences behavior. Therefore, we now present an eclectic scheme of environment–behavior relationships as a summary and integration of the theoretical concepts we have discussed in this chapter.

This scheme of theoretical concepts is presented in the flowchart in Figure 4-8. Objective environmental conditions, such as population density, temperature, noise levels, pollution levels, and building designs, exist independent of the individual, although individuals can act to change these objective conditions. The scheme includes such individual difference factors as adaptation level, length of exposure, perceived control, personality, privacy preference, attachment, and competence to deal with the elements of the environment, as well as such social factors as social support and liking or hostility for others in the situation. Perception of the objective physical conditions depends on the objective conditions themselves, as well as on the individual difference factors and the attitudinal, perceptual, and cognitive processes discussed in Chapters 2 and 3. If this subjective perception determines that the environment is within an optimal range of stimulation or is congruent with intended behavior, the result is **homeostatic,** the adjective form of homeostasis, or an equalization of desired and actual input. On the other hand, if the environment is experienced as outside the optimal range of stimulation (e.g., understimulation, overstimulation, or stimulating in a behavior-constraining manner—including being overstaffed or understaffed or incongruent), then one or more of the following psychological states results: arousal, stress, information overload, or reactance. The presence of one or more of these states leads to coping strategies. If the attempted coping strategies are successful, adaptation or adjustment occurs, possibly

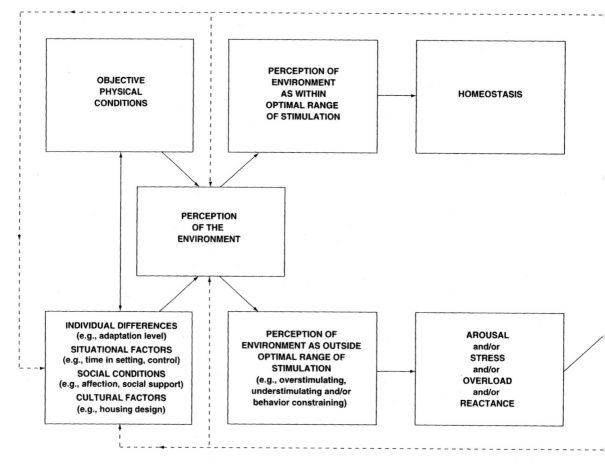

Figure 4-8 An eclectic model of theoretical perspectives.

followed by such aftereffects as lowered frustration tolerance, fatigue, and reduced ability to cope with the next immediate stressor. Cumulative aftereffects might include any of these, but would also include increased self-confidence and a degree of learning about coping with future occurrences of undesirable environmental stimulation. Should the coping strategies not be successful, however, arousal and stress will continue, possibly heightened by the individual's awareness that the strategies are failing. Potential aftereffects of such inability to cope include exhaustion, learned helplessness, severe performance decrements, and mental disorders. Finally, as indicated by the feedback loops, experiences with the environment influence

perception of the environment for future encounters and also contribute to individual differences for future experiences.

We present this model not as a completely developed environmental theory but merely as an attempt to integrate the various mediating concepts that have been applied to environment–behavior relationships. Undoubtedly, some data exist that do not support one aspect or another of this integration. However, we think this eclectic approach will help explain many of the environment–behavior relationships to be covered in the remainder of the textbook. We will continue to see this model in following chapters, where we will discuss how the physical environment (noise, weather, air

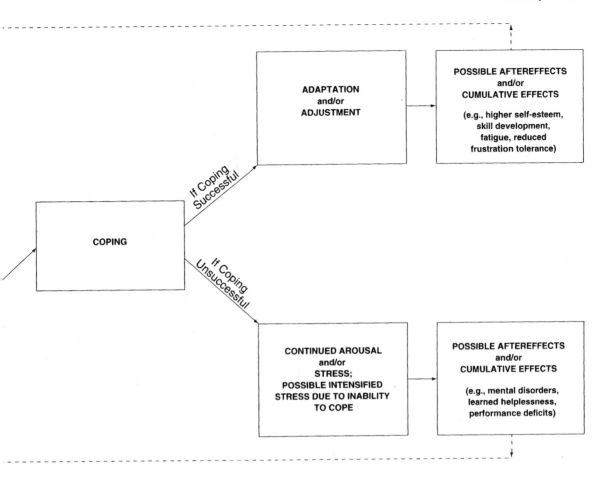

pollution), personal space and crowding, cities, and built and natural environments influence specific behaviors. When appropriate, we will point out how the various theoretical notions in this chapter help explain those specific influences.

SUMMARY

Environmental psychology, as a science, seeks to understand cause-and-effect relationships through prediction and uses publicly observable data to verify these predictions. Once enough predictions are verified, theories are constructed, which consist of a set of concepts and a set of statements relating the concepts to one another. Usually, theories infer that a more or less abstract variable mediates the relationship between one observable variable and another. Good environmental theories should predict and summarize empirical data, should offer generalizability to many situations, and should suggest ideas for research.

The arousal approach to environment–behavior relationships suggests that environmental stimulation leads to increased arousal. According to the Yerkes–Dodson law, this increased arousal will improve or impair performance, depending on whether the individual's arousal is below or above an optimal level. Other behaviors, such as aggression, also tend to follow this curvilinear relationship with arousal.

The information overload model proposes that our capacity to process information is limited and that when excessive stimulation occurs, peripheral inputs are ignored in order to give adequate attention to primary tasks. As a result, responses to these peripheral nonsocial or social stimuli are minimal or nonexistent. The understimulation approach notes that monotonous environmental stimulation leads to boredom and thus to behavioral deficiencies. Wohlwill's approach posits an individual difference variable, or adaptation level (AL), such that stimulation levels above or below this AL will bring discomfort and efforts to reduce or increase the stimulation. The behavior constraint model proposes that perceived loss of control over the environment leads to reactance or efforts to regain freedom of action. If these efforts at reassertion are unsuccessful, learned helplessness may be the result.

The stress model of environment–behavior relationships posits that once stimuli have been evaluated as threatening, coping strategies are brought into play. These strategies can be beneficial, as when their use results in learning more efficient ways of coping with stress. However, prolonged exposure to stress can lead to serious aftereffects, including mental disorders, performance decrements, and lowered resistance to stress.

Barker's ecological psychology model examines environment–behavior interdependencies and focuses on the behavior setting as the unit of study. If the number of applicants to a setting falls below maintenance minimum, performers and nonperformers in the understaffed setting must take on additional roles in order to maintain the setting.

Finally, there is no reason to assume that only one mediator operates in any given environment–behavior situation. An eclectic model is offered that attempts to integrate a number of different theoretical concepts.

KEY TERMS

adaptation
adaptation level (AL)
adjustment
aftereffects
alarm reaction
ambient stressors
applicants
appraisal
arousal
background stressors
behavior constraint
behavior setting
capacity
cataclysmic events
catecholamines
challenge appraisal
control model
coping
corticosteroids

curvilinear
  relationship
daily hassles
denial
determinism
ecological psychology
empirical
empirical laws
en masse behavior
  pattern
environmental
  competence
environmental load
environmental press
environmental stress
  model
equilibrium
extra-individual
  behavior pattern

general adaptation
  syndrome (GAS)
generalizability
harm or loss appraisal
heuristics
homeostatic
hypothesis
intervening construct
learned helplessness
maintenance
  minimum
mediating variable
model
moderator variable
nonperformers
overload
overstaffed
overstimulation
palliative

perceived control
performers
personal stressors
physical milieu
primary appraisal
psychological
  reactance
psychological stress
reactance
refractory period
repression-
  sensitization
Restricted Environ-
  mental Stimulation
  Technique (or)
  Therapy (REST)
reticular formation
screening
secondary appraisal

sensory deprivation
social comparison
social support
staffing theory

stage of exhaustion
stage of resistance
standing patterns of
  behavior

stress
synomorphic
systemic stress
theory

threat appraisal
understaffed
understimulation
Yerkes–Dodson law

SUGGESTED PROJECTS

1. Observe a behavior setting for a week. What behavior patterns are always present? Is the setting understaffed, overstaffed, or adequately staffed?

2. Keep a diary for a week or more of all the events that constrain your behavior. Do you respond with reactance, learned helplessness, or some other behavior?

3. Keep a log of your performance levels in classroom, study, and leisure situations, noting your arousal level and amount of environmental stimulation. Does your performance vary as a function of arousal level, overload, or underload?

4. Construct your own model of environment–behavior relationships. How well can you integrate the various theoretical perspectives discussed in this chapter and the previous one?

NOISE

Introduction

It was competition for the loudest sound system in the high-school parking lot. Reeves had just installed eight new 14-inch woofers in his Wrangler and wanted to blast away Harrison. It seems Harrison had what most conceded was the clear winner for the last three months.

Mr. Martinez, the science teacher, had loaned the guys a sound-level meter from time to time, and no one's amplifier had yet exceeded the 142 decibels achieved by Harrison. To top it off, the morning cruise around the school with sound systems blaring consistently yielded more neighborhood complaint calls to the principal when Harrison entered the parade.

Reeves's older brother had suggested that before he try to beat out Harrison, he might want to get his hearing checked. In a college psychology class the brother had observed class members as they tested each other on an audiometer; those with loud sound systems in their cars showed a distinct hearing deficit that they had not been aware of before seeing the audiometer's printed output. Reeves wondered about the advice from his brother, but for the moment, beating Harrison seemed more important. The audiometer could wait a few years.

For Reeves, Harrison, and their friends, the loud music was a good thing—something to work for and to be proud of—despite the possibility that it might affect their hearing. For passersby, family, and neighbors, however, it was not so desirable; a loud bass from a sophisticated car sound system might sound good to the driver but can be irritating or unwanted by others nearby. When sounds are unwanted and aversive, they are called "noise." They become stressors, capable of making people unhappy and even ill.

Distinctions among types of stressors are useful in several ways. With some stressors, like disasters (see Chapter 7), we study cataclysmic events, the most powerful kinds of stressors that humans experience. These events severely tax our ability to cope, and it is difficult to believe that anyone exposed to an earthquake, tornado, or nuclear accident would not be very aware of what was happening. Another kind of stressor, which may actually be more harmful for mood, behavior, and health, is recurrent or continuous lower intensity stressors that are "normal" and routine. They are part of the "background" in that they are always (or frequently) there. Noise is one such stressor. In many cases, noise is not readily noticeable by long-time residents of noisy areas. Noise is common and widespread and often occurs regularly or all of the time. Because it is less intense than are more dramatic stressors, noise may cause few or no acute effects. However, ambient stressors that persist may be more likely to cause long-term reactions, allowing for the possibility of effects on health and well-being.

Of the many environmental stressors, noise is one of the most thoroughly studied. And no wonder; noise is one of the most frequently mentioned stressors in surveys of what people like and do not like in their neighborhood or community. In part, this may be due to its pervasiveness in our society. Consider some of its many sources: traffic, aircraft, construction, sirens, trains, equipment at work, machinery, and, of course, other people. Everywhere we go there is noise, particularly if we live in cities (Figure 5-1). Sometimes we can adapt to noise, and may not even be aware of it as we get used to it at a certain level. However, research suggests that noise can harm us in many different ways, and studies have sought to define these consequences, identify factors that make the effects of noise more or less severe, and reduce noise levels or noise-related health problems. Regulations governing noise exposure have been put into effect, reflecting recognition of this important problem.

But how does noise affect us? Why is it that under some circumstances we can adapt to noise and under others we cannot? Why would trains that noisily pass our home every three hours be easier to get used to than an occasional aircraft passing overhead? How can it be that people talking softly but audibly during a movie can be more annoying than the loud sounds of rock

Figure 5-1   Noise is one of the most frequently mentioned stressors in surveys of what people like and do not like in their neighborhood or community.

music? In this chapter, we will discuss these issues and the kinds of problems that have been associated with noise exposure. Keep in mind as we go through this research that noise is considerably less powerful or overwhelming than are events such as disasters, but that it may be possible that noise can cause more severe and/or long-lasting consequences. How can this be? Could the repetitive or constant nature of noise be responsible for these effects, suggesting that cumulative effects over time can exceed those of very severe, acute events?

In this chapter we will briefly discuss the nature of sound and noise — how we perceive them, how they are measured, and where they come from. The effects of noise on a range of physiological, psychological, and behavioral variables are considered as well, and research on occupational exposure to noise is summarized. As we will see, noise effects on task performance and social behavior have received a great deal of attention. We also discuss some attempts to apply what we know and to evaluate the effects of noise-abatement programs.

In Figure 5-2 we show how these noise effects can be conceptualized within the eclectic environment–behavior model we presented in the previous chapter; that is, noise outside the desired level requires adaptive efforts that may or may not be successful. In the process of adapting to noise, we may experience arousal, overload, loss of perceived control, and a range of physiological and psychological concurrent effects and aftereffects.

What Is Noise?

The simplest and most common definition of **noise** is "unwanted sound." You may enjoy listening to your favorite rock group on your stereo, but if the music disturbs your roommate's studying or sleep, then as far as your roommate is concerned the rousing sound of the

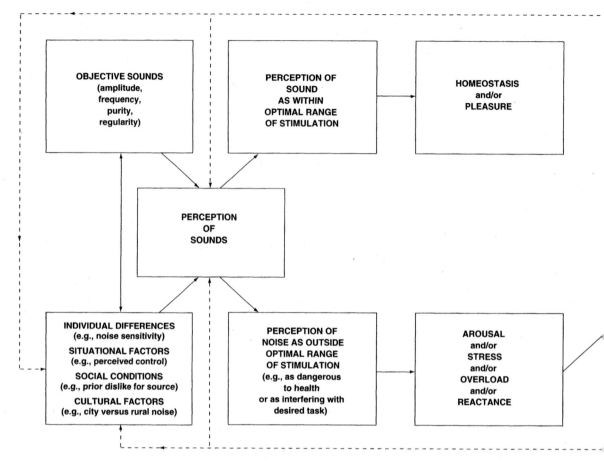

Figure 5-2   Our eclectic model of theoretical perspectives as applied to noise.

talented musicians is noise. The sound of a garbage truck making pickups early in the morning may be necessary in order to maintain healthy sanitation, and for the early riser this sound may provide a wake-up cue signaling a bright new day. But if you do not wish to be awakened so early in the morning, then the motorized contraption is making noise. Loud industrial machinery, jet aircraft, telephones, and pneumatic hammers also generate noise, but only if someone finds the sound undesirable. Of course, some sounds are more likely to be unwanted, either because they interfere with activities or because of their tone, loudness, or quality. However, few if any sounds are always or never unwanted, and nearly any sound can be experienced as noise under the right conditions.

The physical condition, sound, is necessary but not sufficient to produce noise. Thus, the concept of noise implies both a significant psychological component ("unwanted") as well as a physical component (it must be perceived by the ear and higher brain).

PERCEIVING SOUND AND NOISE

The measurement of sound is based primarily on its physical components, although the brain's interpretation of the sound is also crucial to the structure of the measuring scale. Physically, sound is created by rapidly changing air pressure at the eardrum. As air molecules are forced together, positive pressure is created, and when they are pulled apart, negative pressure results.

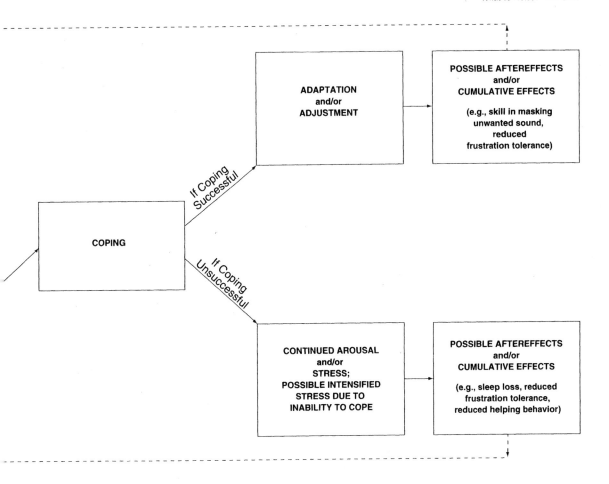

This alternating pressure can be represented graphically by waves, the peaks of the waves representing positive pressure and the valleys reflecting negative pressure (Figure 5-3). These alternating pressures cause the eardrum to vibrate. The eardrum then transmits these vibrations through the structures of the middle and inner ear to the basilar membrane in the cochlea (Figure 5-4A & 4B). Tiny hair cells in this membrane, which are activated by the noise vibrations, pass along the noise stimulation through the auditory nerve to the temporal lobe of the brain. Auditory *sensation* is initiated through activation of the nervous system by the sound stimulus. *Perception* begins somewhere between the basilar membrane and the temporal lobe of the brain, where a code we have yet to unravel completely allows the organism to interpret the sound stimulus. (This code probably involves the pattern of neuron firing—nerve cells transmitting messages—and the rate at which neurons fire.)

Examine the waves depicted in Figure 5-3 once again. Physically, the more times per second the wave motion completes a cycle (from peak to valley), the greater the **frequency** of the sound. Psychologically, frequency is perceived as **pitch** (i.e., highness or lowness). The normal human ear can hear frequencies between 20 and 20,000 cycles per second, or **hertz (Hz).** However, most sounds we hear are not a single frequency but a mixture of frequencies. Psychologically, purity of frequency is known as **timbre** or **tonal quality.** Sound

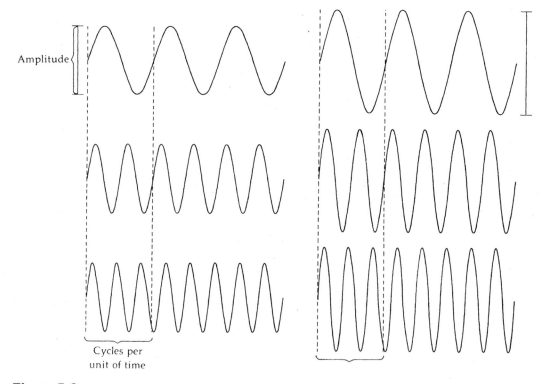

Amplitude

Cycles per
unit of time

Figure 5-3 Examples of sound waves. Frequency increases from top to bottom; amplitude increases from left to right.

stimuli that consist of few frequencies are often called **narrow band** sound, whereas stimuli with a wide range of frequencies are called **wide band.** Extremely wide-range unpatterned frequencies are called **white noise.**

Sound waves also vary in height or **amplitude,** experienced psychologically as **loudness.** The greater the amplitude of a wave, the greater the energy or pressure in the sound wave, and the louder the sound. The smallest pressure or threshold that a young adult can detect is about 0.0002 microbars, or dynes per square centimeter, where a dyne is a measure of pressure. At 1,000 microbars, the pressure is experienced more as pain than as sound. Measures of loudness must cover an enormous range of pressures, and to help us understand and study noise, a more workable scale of sound pressure has been developed that uses **decibels (dB)** as the

Figure 5-4A Schematic diagram of human ear, showing important structures associated with perception of sound.

From Gardner, E., 1975. Fundamentals of neurology, *6th ed. Philadelphia: Saunders.*

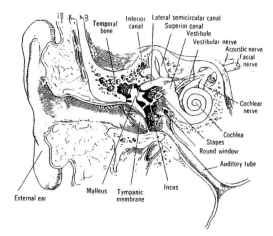

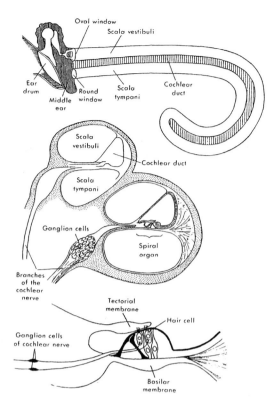

Figure 5-4B Internal structure of human auditory system.

From Gardner, 1975.

Table 5-1	DECIBEL EQUIVALENTS OF MICROBARS

Sound Pressure in Microbars	Equivalent Decibels
0.0002	0
0.002	20
0.02	40
0.2	60
2.0	80
20.0	100
200.0	120
2000.0	140

frequencies, and more intense pressure has different effects at varying frequencies. For this reason, sound intensity is sometimes measured in quantities called *phons* or *sones;* we will use the more common decibel scale when discussing noise in this text, but you may want to consult a sensation and perception text or psychophysics text for a discussion of phons and sones.

basic units of sound (decibels are a logarithmic function of microbars). Table 5-1 gives the corresponding decibel equivalents of audible ranges of pressure. Note that an increase of 20 decibels represents a tenfold increase in pressure: A sound of 80 dB is not twice as intense as one of 40 dB; it is 10 times 10, or 100 times, as intense. Figure 5-5 lists some common sounds associated with various points on the decibel scale.

The decibel scale measures the physical component of sound or noise amplitude. However, this scale does not accurately reflect the perception of loudness. That is, a difference of 20 dB means that one sound has 10 times more pressure than another, but it does not necessarily mean that the more intense sound will be perceived as 10 times louder. The human ear is differentially sensitive to sounds at different

Figure 5-5 Some common sounds associated with the decibel scale.

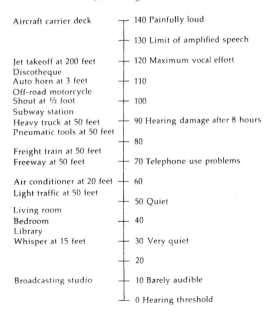

dB (A) Hearing Effect

ANNOYANCE

One of the major consequences of noise is a psychological one—annoyance. Like many stressors, noise is annoying. It makes people irritable and impatient and can make us unhappy as well. **Annoyance** is a general term that refers to these negative effects of noise and has been used to measure the adverse effects of environmental features that produce noise, such as airports, railroads, and superhighways. Airplanes flying overhead, trains thundering by on their tracks, and the constant din of trucks, cars, and horns on our highways can all contribute to general negative feelings and irritability. When this happens, the noise is associated with annoyance.

Some kinds of noise are more annoying than others. As you might guess, loud noises are often more annoying than quieter noises. However, there is usually more to it than that. Noise is a disturbing environmental phenomenon because noise is, by definition, unwanted. It is this irritating and distracting psychological component that causes noise to be a problem. Kryter (1994) and Glass and Singer (1972) in their groundbreaking work have pointed out that some types of noise are more annoying than others. Three major dimensions influencing how annoying a noise is include: (1) volume; (2) predictability; and (3) perceived control. All other things considered, loud noise, unpredictable noise, and low perceived control should be associated with greater annoyance or negative reactions to noise than is softer, predictable, or controllable noise.

How are we affected by noise at different volumes? Above 90 dB, which is the level of noise produced by a heavy truck 50 feet away, noise becomes psychologically disturbing, and after repeated periods of exposure for eight hours or more, it can be physiologically damaging to hearing. Moreover, the louder the noise, the more likely it will interfere with verbal communication, and the greater the arousal and stress associated with it, the more attention one allocates to it.

Unpredictable, irregular noise is generally more annoying than is predictable or constant noise. A constant, unbroken noise (especially if it is not loud) is not disturbing. However, once it is broken up into periodic "bursts," the noise becomes more disturbing; if we then make the bursts of noise aperiodic (i.e., coming at unpredictable or irregular intervals), the disturbing quality becomes even more pronounced (e.g., Glass & Singer, 1972). The more unpredictable the noise, the more arousing it is, and the more likely it is to lead to stress (unpredictable noises may be evaluated as more novel or threatening than predictable ones). In addition, more unpredictable noises require greater attention in order to understand and evaluate them, leaving less total attention available for other activities. Finally, it is easier to adapt to a predictable noise, since the same stimulus is presented over and over again; with unpredictable noise, adaptation is more difficult.

Noise over which we have no perceived control is also more disturbing than noise we can easily control. If you have the means to stop or muffle a noise, or if you can get away from it when you need to, you are less likely to be annoyed by it than if you cannot control it. For example, if you are using a noisy power saw, you can control the noise by stopping the saw. If your neighbor uses the saw next door, you have less immediate control over the noise, and so it is more disturbing. If you can close your window and reduce the intrusiveness of the sound, it may not bother you as much, even if you do not actually close the window. Of course, you could close it and find out the noise is still loud and disturbing. It may not always be good to test our perceived control!

From the theoretical perspectives discussed previously, we know or can propose that uncontrollable noise is more arousing and stressful, requires more attention allocation, and is more difficult to adapt to than controllable noise. Lack of control over noise can lead to psychological reactance and attempts to regain freedom of action by trying to assert control. If such efforts are unsuccessful, learned helplessness can result (see Chapter 4), in which a person simply accepts the noise and never tries to control it, even though control may become possible at a later time.

These three noise variables can, of course, occur in any combination. That is, we can have loud, predictable, uncontrollable noise, or quiet, unpredictable, uncontrollable noise, and so on. As we will see later in this chapter, loud, unpredictable, uncontrollable noise has the most deleterious effects on behavior. Although these three factors are probably the most important in determining the effects of noise on behavior, research suggests that other psychosocial factors and beliefs also influence how annoying we find noise (e.g., Green & Fidell, 1991; Miedema & Vos, 1999). Annoyance increases if: (1) one perceives the noise as unnecessary or not contributing to something we want or value; (2) those who generate the noise seem to be unconcerned about the welfare of those who are exposed to it; (3) the person hearing the noise believes it is hazardous to health; (4) the person hearing the noise associates it with fear; or (5) the person hearing the noise is dissatisfied with other aspects of his or her environment. In addition, noise sensitivity has been proposed as an individual difference variable that affects noise-related annoyance (e.g., Job, 1988; Lercher, 1996; Staples, 1996; Stansfield, 1992; Weinstein, 1978).

This last possibility, that noise sensitivity is a stable personality trait, suggests that it is related to differences in attitudes toward a variety of kinds of noise. Some people find noise more aversive than do others, in part because their attitudes about it are more negative. There is some evidence that people who are more sensitive to noise, that is, those people who react more intensely to noise or to lower levels of noise, are more annoyed by noise as well (Taylor, 1984) and are more prone to psychological disorders (Stansfield et al., 1993). In most research, correlations between sensitivity and annoyance are moderate, ranging from .25 to .45 (on a scale from .00 to 1.00) across many different kinds of noise (Job, 1988). Increases in these observed relationships may be forthcoming as we construct better measures of noise sensitivity (e.g., Zimmer & Ellermeier, 1998) and move away from single-item measures that simply ask how sensitive or negative people are about noise (Zimmer & Ellermeier, 1999). Examples of items that might be found on a noise sensitivity scale are shown in Table 5-2.

An interesting study by Staples and her colleagues (Staples, Cornelius, & Gibbs, 1999) provided mixed support for this idea and suggested that cost–benefit analyses were more useful in predicting and understanding annoyance. They studied 901 residents of areas next door to airports and administered the Environmental Noise Risk Scale (ENRS). This new scale measured how people interpret or perceive tradeoffs between the economic benefits and convenience of having an airport nearby and the adverse environmental effects of local airport development. General noise sensitivity, measured with

Table 5-2	Examples of Items That Might Be Found on a Noise-Sensitivity Scale [*]

1. I find it difficult to study in a noisy setting.
2. I cannot concentrate when people around me make noise.
3. It bothers me when others do not respect my need for quiet.
4. I can't stand loud traffic noise.
5. I find aircraft noise to be very annoying.
6. I cannot sleep when even minor noise is present.
7. It bothers me when neighbors play loud music.
8. There should be laws against loud noise after 10 P.M.
9. I would avoid living in a noisy neighborhood.
10. I believe noise is hazardous to my health.

[*] Answer on a scale from 1 = disagree strongly; 2 = disagree; 3 = neutral; 4 = agree; 5 = agree strongly.

traditional items, was not strongly related to how disturbed people were by the airport noise, but scores on the ENRS were more strongly related to disturbance. Belief that the airport benefited the community tended to decrease the strength of the relationship to disturbance.

SOURCES OF NOISE

As you might guess, noise can come from almost anywhere. Because it has a subjective component (it must be judged as unwanted), noise can come from anything that makes a sound. And, as you would also expect, the same sound may be unwanted at some times but not unwanted at other times. A dripping faucet passes as a faint whisper against the background sounds of a busy afternoon, but at night, when we are trying to sleep, it can be noisy and very disruptive! In contexts where sound is either so loud that it is considered to be a noise, or is softer but more irregular or disruptive, most people will complain about noise. We will briefly discuss two of the common settings where noise can be a problem.

Transportation Noise

Noise caused by cars, trucks, trains, planes, and other modes of transportation is of great interest for a number of reasons. First, it is very widespread. Surveys have indicated that automobile noise is the most often mentioned source of urban noise, and that opening of new highways is associated with increases in annoyance among nearby residents (Lawson & Walters, 1974). Reports estimate that 11 million or more Americans are exposed to vehicular noise at or above levels that risk hearing loss (Bolt, Beranek, & Newman, 1982). Increases in air traffic have boosted noise levels around airports, and studies have shown that half to two-thirds of those people living near airports where aircraft noise is a problem report annoyance and unhappiness about the noise (e.g., McLean & Tarnopolsky, 1977). Some evidence suggests that airport noise or noise from airplanes is associated with higher blood pressure and ear symptoms as well as annoyance (Ising et al., 1990). We will deal

with this research in more detail later in this chapter. Other sources of noise, including rail traffic and other transportation systems, are pervasive and contribute to the din of modern cities. Several studies indicate that the more exposure to transportation noise one receives, the greater the annoyance among community residents (Fidell, Barber, & Schultz, 1991).

A second characteristic of transportation noise is that it is usually loud. This is clear from the sound levels noted above, as well as from estimates of sound levels near airports (ranging from 75 to 95 dB). A quick glance at Figure 5-5 also provides evidence of this, as do EPA measurements of noise levels in third-floor apartments next to freeways in Los Angeles (90 dB; see Raloff, 1982). However, annoyance does not appear to be related to loudness itself; acoustic (volume) and nonacoustic (e.g., predictability) factors in ratings of annoyance jointly determine the mood effects of noise (Green & Fidell, 1991). For example, a study of annoyance due to airport noise near Hartsfield International Airport in Atlanta indicated that annoyance was not reduced by insulation that attenuated sound in some people's homes (Fidell & Silvati, 1991).

Occupational Noise

Noise exposure in the workplace is a second major problem and has also received a great deal of research interest as well. One characteristic of occupational noise, particularly office noise, is that it is very wide band noise, being made up of many sounds of differing frequencies. If this is extreme, it may result in a **masking** (covering up) of the noise and yield a tolerable situation (see the box on page 164). However, if not that broadband, the resulting noise may be resistant to adaptation and more likely to cause annoyance and distress (Loewen & Suedfeld, 1992). Examination of effects of office noise, as well as possible ways to reduce these effects, should consider the frequency as well as loudness of the noise in influencing psychological reactions to it.

Occupational noise is also very pervasive, and sound levels in many occupational settings are loud. More than half of United States production workers are exposed to regular noise levels above the point at which hearing loss is likely, and more than 5 million are exposed to levels above the legally permissible ceiling of 90 dB (OSHA, 1981). Construction workers may be exposed to equipment noises of 100 dB, air-craft mechanics to levels ranging from 88 to 120 dB, and coal miners to continuous levels between 95 and 105 dB (Raloff, 1982). For reasons noted above, these exposures are sufficient to cause concern and underscore the need for continued research on noise and its effects.

Effects of Noise

Regardless of the source of noise, it can have unpleasant and/or unhealthy effects on us. Sources of noise and annoyance are not limited to the major ones described in the preceding section (transportation and workplace noise). An interesting study of residential noise makes this point well. Observing that noise produced by air conditioners is often substantial, Bradley (1992) surveyed 550 people, roughly equal numbers of whom lived in noisier or quieter areas and owned or did not own an air conditioner. Noise measurements were made as well. The physical level of noise (loudness) was related to annoyance and to reports of hearing neighbors' air conditioners. The amount by which air conditioners' noise exceeded ambient noise levels was also related to annoyance. Negative reactions to the air conditioner noise were common and tended to occur most frequently in quieter neighborhoods where the air conditioner noise exceeded background levels (Bradley, 1992). As we might have expected, owning an air conditioner resulted in less annoyance from air conditioner noise, presumably because some of the noise was being generated for the person's comfort. We now examine effects of noise beyond annoyance, regardless of its source.

HEARING LOSS

The most logical or expected effect of exposure to noise is hearing loss, and this consequence of noise is an important concern for employers and regulators as well. Although very loud sounds (e.g., 150 dB) can rupture the eardrum or destroy other parts of the ear, damage to hearing from excessive noise usually occurs at lower noise levels (90 to 120 dB) because of temporary or permanent damage to the tiny hair cells in the cochlea of the inner ear (Figure 5-4). Such **hearing loss** is measured in terms of a baseline of "normal" amplitude thresholds at given frequencies. When a hearing loss occurs at a given frequency, it requires more than the normal amplitude (in dB) for a person to hear that frequency; that is, the amplitude threshold is greater. The usual index of hearing loss for a given frequency, then, is the number of decibels above the normal threshold required to reach the new threshold. Such hearing losses are generally identified as one of two types: (1) **temporary threshold shifts (TTS),** in which the normal threshold returns within 16 hours after exposure to the damaging noise, and (2) **noise-induced permanent threshold shifts (NIPTS),** which are typically measured a month or more after the cessation of exposure to the damaging noise (Kryter, 1994).

Hearing loss, which affects millions of people, is a serious problem in industrialized countries. A classic report by Rosen et al. (1962) compared the extent of the problem in the United States with a much quieter Sudanese culture, and found that 70-year-old Sudanese tribesmen had hearing abilities comparable with those of 20-year-old Americans! To avoid serious hearing loss among industrial workers, the Occupational Safety and Health Administration (OSHA) has established guidelines that allow only limited exposure to noise (e.g., eight hours for 90 dB, four hours for 95 dB, two hours for 100 dB, and so on). However, some trucks can, at some distances, emit noise of 95 dB. Thus, individuals living near heavy traffic routes are undoubtedly exposed to noise levels (for at least brief periods of time) exceeding government industrial standards.

As was the case with annoyance, absolute levels of noise alone do not determine hearing loss. For example, certain drugs may increase the damaging effects of noise (Miller, 1982). Studies with animals have indicated that administration of an antibiotic in conjunction with exposure to noise can increase the effects of the noise and cause greater hearing loss than would the drug or noise levels alone (Raloff, 1982). Other drugs, including aspirin, may also interact with noise and increase effects on hearing, but evidence remains mixed. At this point it appears that a few drugs can, in combination with noise, cause increased hearing loss, but the magnitude of effect of most is small.

College students and adolescents are frequently exposed to another damaging source of noise—loud rock music. Several studies (e.g., Lebo & Oliphant, 1968) found that rock groups playing in nightclubs are exposed to music from 110 to 120 dB for nonstop periods of up to 1½ hours. Serious hearing loss can result (the federal industrial limit for 110 dB sound is 30 minutes a day). More current rock music, as different as it is from '60s acid rock, is still loud enough to cause hearing loss.

OTHER HEALTH EFFECTS OF NOISE

High levels of noise can lead to increases in arousal and stress (e.g., Cohen et al., 1986; Glass & Singer, 1972). We might expect, then, that the incidence of diseases related to stress—hypertension (high blood pressure) and ulcers, for example—would increase as one is exposed to higher levels of more unpredictable and/or uncontrollable noise. Research evidence on this relationship is not conclusive but suggests that noise can harm health in several ways (e.g., Fay, 1991; Kryter, 1994; Passchier-Vermeer, 1993).

First, noise appears to affect immune system function in humans and animals in ways that could render people more susceptible to infection (e.g., McCarthy, Ouimet, & Dunn, 1992; Sieber et al., 1992; Weisse et al., 1990). Noise also appears to affect the gastrointestinal (GI) system, increasing GI complaints and producing organic digestive problems (Passcheir-Vermeer, 1993). Ulcers in particular appear more likely among workers exposed to a lot of occupational noise. Doring, Hauf, and Seiberling (1980) have suggested that sound can affect intestinal tissue directly, so it does not even have to be noticed to predispose a worker to digestive problems. As we shall see shortly, noise also appears to contribute to cardiovascular disease and hypertension. Sustained noise exposure is associated with constriction of peripheral blood vessels in animals (Millar & Steels, 1990). At least one study (Ando & Hattori, 1973) found an association between exposure of expectant mothers to aircraft noise and infant mortality. Other studies have linked airport noise to birth defects or lower birth weight (Ando, 1987; Jones & Tauscher, 1978; Nurminen & Kurppa, 1989). Finally, survey or correlational studies have found that frequent exposure to noise is associated with reports of acute and chronic illness and sleep problems (e.g., Bronzaft, Ahern, McGinn, O'Connor, & Savino, 1998). Exposure to loud or frequent noise (e.g., living near an airport, working in a noisy setting) is also associated with constriction of peripheral blood vessels, higher diastolic and systolic blood pressure, and increased catecholamine secretion (e.g., Cohen et al., 1986; Evans, Hygge, & Bullinger, 1995; Evans & Lepore, 1993).

One investigation of noisier and quieter industrial settings indicated that although resting blood pressures were comparable among men in noisier and quieter plants, hearing loss and heightened blood pressure were positively correlated (Talbott et al., 1990). Among men who had worked in the settings for at least 15 years, blood pressure was higher among workers in the noisier plants (Talbott et al., 1990). Data from school children attending schools near the Los Angeles International Airport has also indicated that exposure to noisy conditions at school is associated with elevated blood pressure relative to that observed among children attending quieter schools (Cohen et al., 1986). Similar data from children living near the Munich airport in Germany suggest that over a two-year period, exposure to aircraft noise was associated with greater stress (resting blood pressure, overnight urinary catecholamine levels) and poorer quality of life (Evans, Bullinger, & Hygge, 1998). Comparisons were made between noisier and quieter neighborhoods and from before to after opening of the new airport in Munich. Both analyses indicated that chronic noise exposure was associated with greater stress symptoms (Evans et al., 1998). Simulated noise situations have also been linked to elevations in epinephrine (Maschke, Ising, & Arndt, 1995), but workers exhibit lower blood pressure and lower levels of epinephrine in their urine when they wear hearing protectors that reduce the intensity of noise (Ising & Melchert, 1980). The stresslike changes associated with noise, including elevated catecholamines, cortisol, and cholesterol are found some of the time, but studies failing to find such relationships have also been reported (e.g., Cavatorta et al., 1987; Fruhstorfer et al., 1988).

One study has provided a good deal of information about how noise may facilitate the development of hypertension. Those who were already diagnosed as having moderately high blood pressure were exposed to 105-dB noise for 30 minutes, and blood pressure measurements were made during quiet and noisy periods (Eggertsen et al., 1987). During the noise there was a significant increase in systolic and diastolic blood pressure, marked primarily by an increase in peripheral vascular resistance (the force of the heart contractions actually decreased). Thus, stress due to noise exposure was associated with constriction of blood vessels (consistent with animal studies showing vasoconstriction during chronic noise exposure). This suggests a mechanism by which noise may contribute to hypertension. Research indicates that the constriction of blood vessels that is associated with noise exposure does not habituate very well when noise is loud or unexpected (Jansen, 1973). This means that, over time, very loud or unexpected noise may continue to affect blood vessels long after people have "gotten used to" the noise and other responses have diminished. Heart rate or skin conductance changes due to noise tend to be modest and to decrease with repeated exposure (Borg, 1981; Glass & Singer, 1972).

Most of these studies suggest that noise can cause a variety of *physiological changes* that may contribute to disease. The links to *actual disease,* however, are less well established—all of which is to say that despite reported increases of several stress indicators or measures of cardiovascular function when people are exposed to noise, these studies have not definitively shown a relationship between noise and cardiovascular disease or infectious illness.

One argument against the hypothesis that noise can affect health is that studies using hearing loss as an indicator of noise exposure have shown few relationships between noise and cardiovascular function or disease among Air Force aircrews (e.g., Kent et al., 1986). However, these correlational studies rely on a strong relationship between hearing loss and noise exposure and cannot address other causes of hearing loss or aspects of noise other than those linked to hearing loss. It is also possible that the self-selection biases in aircrew members might have affected these results. Other studies have examined health problems among industrial workers as a function of exposure to noise, and these studies (e.g., Cohen, 1973; Jansen, 1973) found modest relationships between exposure to high noise levels and cardiovascular disorders, allergies, sore throats, and digestive disorders.

Interestingly, younger and less experienced workers appear to suffer more from noise exposure, suggesting that more experienced workers have adapted to the noise. Unfortunately, industrial studies rarely control for other factors that may account for adverse health effects, such as factory conditions, exposure to pollutants, and stressful work activity. In one study, annoyance at work was related to noise exposure at work and to blood pressure, but the link between noise and blood pressure was stronger when job dissatisfaction and social support were considered (Lercher, Hortnagel, & Kofler, 1993).

One way of studying the health-impairing effects of noise, then, is to examine how it interacts with other stressors or behaviors. For example, exposure to some chemicals or toxic hazards in the workplace may combine with noise to produce effects or may intensify the effects of noise (e.g., Morata, 1998). We know that noise increases people's blood pressure and other signs of arousal, as does cigarette smoking. How do the two affect us if we smoke while exposed to noise? Given that smoking is described by many who smoke as an effective coping strategy that calms them down, will it cancel the arousing effects of noise or add to them? A study by Woodson et al. (1986) looked at this question in a sample of women. Forty-eight women who smoked and 12 who did not smoke participated in the study and were exposed to noise. The smokers were assigned to a smoking group or a sham-smoking group (puffing on unlit cigarettes) in which they did not actually smoke any cigarettes. The nonsmokers were assigned to a sham-smoking control group. Reported distress increased during the session, but did not increase for those allowed to smoke. The arousing effects of smoking did attenuate some noise-related arousal. Smoking appears to reduce some of the physiological responses to noise, particularly noise-induced increases in heart rate and vasoconstriction. This may have been partially due to the periodic nature of the noise exposure (noise was not continuous), since studies of continuous stressor exposure and smoking find the opposite effect or no effect at all (e.g., MacDougall et al., 1983; Suter et al., 1983).

It is also possible that noise can affect health by changing behaviors that are related to health. If people drink more coffee or alcohol, smoke more cigarettes, or fail to exercise because of noise exposure, then relationships between noise and health might be mediated by these behaviors. A study by Cherek (1985) provides some evidence of this by showing that increasing loudness of noise was associated with increased cigarette smoking. Higher dB levels

Figure 5-6 Evidence that noise has direct effects on physical health other than hearing is often equivocal. Many health effects of noise are likely associated with other factors such as industrial toxins or smoking.

of noise were associated with higher levels of smoking during an experimental laboratory session as well as with how people smoked. The louder the noise, the more puffs they took when they smoked, and the longer the average duration of each puff.

In general, adverse effects of noise exposure on health occur primarily in conjunction with other stressors (such as industrial pollutants, on-the-job tensions, economic pressures, and so on; see Figure 5-6) or are limited to those who are particularly susceptible to certain physiological disorders (e.g., Kryter, 1994). For example, in one study, noise effects on blood pressure were seen only in people with family histories of hypertension (Theorell, 1990).

NOISE AND MENTAL HEALTH

We have noted that exposure to high levels of noise leads to the heightened physiological activity typical of stress and suggested that physical health may be affected as well. Some have assumed that stress is the link between noise and health problems. Since stress is a causal factor in mental illness as well, we might expect noise exposure to be associated with mental health problems (e.g., Kryter, 1994). Industrial surveys typically report that exposure to high-intensity noise is associated with such complaints as headaches, nausea, instability, irritability, anxiety, sexual impotence, and changes in affect or mood (e.g., Bing-shuang, Yue-lin, Ren-yi, & Zhu-bao, 1997). However, a study of 2,398 men in the United Kingdom showed that traffic noise was related to annoyance but not to psychological disorders (Stansfield et al., 1993). As with surveys of physical health and noise exposure, the results of these studies must be interpreted with caution, since other stressors related to home and work are usually not fully considered or controlled.

Aircraft noise is a very common cause of complaints among those who live near airports. An interesting but controversial series of studies (reviewed by Kryter, 1990) examined the relationship between airport noise and mental health near London's Heathrow Airport (Figure 5-7). Evidence was found for a positive relationship between noise and psychiatric admissions, which are determined by other factors as well as by noise. However, assessment of

Figure 5-7 Aircraft noise is one of the most common sources of noise annoyance in the community. Some research has tried to associate airport noise with mental health problems of residents in the area. Although the findings are controversial, there is limited evidence that psychiatric hospital admissions are unusually heavy for areas surrounding airports.

psychological or emotional disturbances in noisy and quiet areas has provided no reliable evidence of noise contributing to the development of psychopathology (Stansfield, 1992). As with physical health, we should explore the possibility that noise contributes to mental illness primarily in combination with other factors that precipitate or permit mental disorders to develop. For example, in addition to influencing stress, noise exposure may lead to loss of perceived control and learned helplessness (see Chapter 4), which, in turn, increases susceptibility to psychological disorders.

EFFECTS OF NOISE ON PERFORMANCE

Effects During Exposure

Most people will quickly tell you that they make more errors in noisy than in quiet settings, but their beliefs about noise do not always match their performance (Smith & Jones, 1992; Smith & Stansfield, 1986). Laboratory research on the influence of noise on performance has shown mixed results. For detailed reviews, see Cohen et al. (1986) and Stansfield (1992). Briefly, whether noise affects performance adversely, favorably, or not at all depends on the same properties of noise discussed earlier (i.e., intensity, predictability, controllability), the type of task being performed, and stress tolerance and other personality characteristics of the individual (e.g., Baker & Holding, 1993; Cohen & Weinstein, 1982; Koelega & Brinkman, 1986). Very loud noise (more than 100 dB) appears to affect performance in some ways. In general, data from laboratory research suggest that regular noise in the range of 90 to 100 dB does not adversely affect performance of simple motor or mental tasks. However, noise in this amplitude range that is unpredictable (intermittent at irregular intervals) will interfere with performance on vigilance tasks, memory tasks, and complex tasks in which an individual must perform two activities simultaneously. Sudden, loud, unpredictable noise may momentarily distract an individual from a task and thereby cause errors if the task requires much vigilance or

concentration, yet Glass and Singer (1972) found that even these performance problems were minimal and/or overcome by individuals who perceived that they had control over the noise (i.e., could stop it if they wished).

To some extent, the kind of effects that noise has on task performance may be a matter of personality or differential sensitivity to noise. Research has indicated that noise effects on performance may depend in part on personality. Extraverts are generally underaroused relative to introverts, suggesting that extraverts should show better performance when working in noisy conditions. Indeed, this has been observed (e.g., Campbell, 1992; Dornic & Ekehammar, 1990), and some evidence suggests that extraverts prefer working in noisier settings (e.g., Geen, 1984). Age, sex, and other characteristics could also be influential. Children do not experience as much disturbed sleep, and younger individuals have been shown to have smaller physiological changes when exposed to noise than do older persons (Vallet, 1987). Some studies indicate that women are more adversely affected by noise than are men (e.g., Gulian & Thomas, 1986), while others have found no differences (Edmonds & Smith, 1985).

Exposure to loud, uncontrollable noise appears to bias retrieval of information from memory, causing more attention or greater recall of negative mood-laden items or memories (Willner & Neiva, 1986). This is not unlike observations of depressed people, who suffer from negative memory biases and seem better able to retrieve unpleasant memories (e.g., Fogarty & Hemsley, 1983). On the other hand, Bell et al. (1984) found that noise interfered with recall whether presented during a learning or recall phase. The effects of stressors on this sort of memory distortion and bias toward negative recollections represent an important area for future research.

One recent study examined the effects of background sound fluctuations from ventilation fans on how tired students became during long lectures (Persinger, Tiller, & Koren, 1999). During four consecutive lectures, data were collected after each hour reflecting student fatigue and ability to concentrate. During half of these lectures, overhead ventilation fans in the lecture room were turned on, generating sound that was more or less continuous and averaged about 60–65dB. During the remaining lectures, the fans were turned off. The effects of the additional sound were dramatic: Students reported greater fatigue when the fans were running, and this may have interfered with their ability to concentrate. Other studies confirm the effects of background noise in work settings, which we will consider shortly.

Aftereffects

Noise has more than just immediate effects on performance. In Chapter 1 we described a task designed to measure frustration tolerance in which individuals try to solve paper-and-pencil puzzles that are actually unsolvable; the number of attempts to solve such puzzles serves as an index of tolerance for frustration, or persistence (Figure 5-8). In one famous study, a group exposed to 108 dB of unpredictable and uncontrollable noise *before* attempting the task showed one-half to one-third as much tolerance of frustration. These participants also made considerably more proofreading errors compared with a no-noise control group and with groups exposed before the task to either predictable or controllable noise (Glass, Singer, & Friedman, 1969). Apparently, the aftereffects of noise can be as severe as the effects during perception of the noise. In a similar experiment, it was found that aftereffects depend on the amount of perceived control (Sherrod et al., 1977). These researchers gave some participants control over starting the noise, and still others control over both starting and stopping it. Another group had no control over the noise. Results showed that the greater the perceived control, the more persistent individuals were in working unsolvable puzzles once the noise had stopped.

Such aftereffects can also be explained by the theoretical approaches discussed previously. For example, arousal remains elevated for a time after an arousing stimulus (such as noise) has ceased. Thus, this "carried-over" arousal can account for some aftereffects. The environmental load approach also suggests that once an

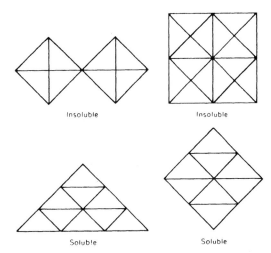

Figure 5-8   Examples of puzzles used by Glass and Singer (1972). Figures must be drawn without crossing a line or lifting the pencil.

Adapted by Glass and Singer from Feather, 1961.

attention-getting noise has stopped, a fatigue effect ensues, and it takes time to reallocate enough attention to perform a mental task. If the noise is presented with perceived control, less attention is allocated to it to begin with, so recovery time is less, and the potential for learned helplessness is decreased.

NOISE EFFECTS ON CHILDREN: WHY JOHNNY CAN'T READ?

Many children spend considerable time in homes or schools near noisy transportation corridors. Sadly, research suggests that noise from freeways, rail systems, and aircraft has substantial effects on children's classroom performance and stress levels. For example, Hambrick-Dixon (1986) studied children who attended day-care centers close to noisy elevated subways and others far from the subways. Psychomotor task performance was impaired by experimentally imposed noise in a laboratory, but children from the noisier centers performed better when exposed to noise than when not, while children from quieter centers showed worse performance under noise conditions. In another study, Damon (1977) found that children living in housing where traffic noise was high were more likely to miss school.

Cohen, Glass, and Singer (1973) theorized that urban noise may impair the educational development of children if it is severe enough. Studying a large high-rise apartment complex situated over a noisy highway in New York City (see Figure 5-9), the investigators found that noise exposure on the lower floors of the complex was more severe than on the upper floors. While carefully controlling for such factors as social class and air pollution, which might also vary with the floors of the building, the researchers found that children on the noisier lower floors had poorer hearing discrimination than children on the upper floors. Moreover, the hearing problems of children on the lower floors may have influenced their reading ability, for it was found that they had poorer reading performance than children on the upper floors. In another study, Bronzaft and McCarthy (1975) compared the reading skills of children from two sides of a school building. One side of the

Figure 5-9   This is the high-rise apartment building used in the study by Cohen et al. (1993). Note the traffic passing underneath.

building was adjacent to elevated railroad tracks, but the other side was much quieter. It was found that 11% of teaching time was lost in classrooms facing the noisy tracks. Not surprisingly, the reading skills of children on the quieter side of the building were superior to those of children on the noisy side (see also Crook & Langdon, 1974). Indeed, Evans and Maxwell (1997) concluded that much of the academic performance shortfall shown by children exposed to chronic noise, including aircraft noise, can be attributed to deficits in language acquisition. Noise can interfere with the ability to discriminate between similar sounds (e.g., the letters "b" and "d"), resulting in slower development of reading and verbal skills.

Other research also suggests that aircraft noise has marked effects on children's performance. Cohen et al. (1986) studied children attending school near the Los Angeles International Airport. Some were in schools in which aircraft noise was very loud (up to 95 dB), while others were in schools where there was considerably less noise. After controlling for the effects of socioeconomic variables and accounting for differences in hearing loss, results of a multimeasure assessment indicated that children attending noisier schools had more difficulty solving complex problems. Children attending noisy schools were also less likely to solve a solvable task than were students from quiet schools and were more likely to give up. These data suggest that students from noisy schools were simply less able to solve cognitive tasks, in part because they were more likely to quit before completing the task. These studies also examined school achievement and distractibility among the young students. Over time, the effects of chronic exposure to aircraft noise did not dissipate, that is, the effects did not decrease or go away with time. Instead, students seemed to be more distractible the longer they attended school under noisy conditions. Actual school achievement was affected more by noise levels at home than in schools.

A more recent study of chronic exposure to airport noise was conducted near the Munich International Airport in Germany (Evans,

AIRCRAFT NOISE:
Why So Loud and Disturbing?

One of the marvels of space-age technology is supersonic flight and the arrival of the supersonic transport (SST) for commercial passengers. The American version of the SST was scuttled for economic and environmental reasons. The British–French Concorde, however, went into production and has been controversial ever since, but it, too, has been scheduled to be phased out. Among its problems is noise. Engines on an SST must be slim and trim for better flight. As engine diameter decreases and speed increases, the exhaust noise becomes greater, which is why older, slimmer engines are being phased out and replaced with newer, wider ones in subsonic aircraft. Typical Concorde noise on a runway is 100 to 120 dB, depending on one's distance from the jet. This is 10 to 20 dB greater than subsonic jets. Research suggests that a single flight of an aircraft 10 dB louder than another produces the same annoyance level as 10 flights of the less noisy aircraft.

Another noise problem with the SST is sonic booms. These thunderclap sounds are produced by any supersonic aircraft. Sound travels at a speed of 334 meters per second (747 miles per hour). The SST moves faster than the noise it produces (since passengers are ahead of the sound, they do not hear it as someone on the ground does). Consequently, the sound waves crowd together, increasing their pressure and causing a sonic boom. The tail of the aircraft leaves a partial vacuum, lowering the pressure as it passes. The result is an increase in pressure followed by a decrease. These pressure changes move away from the jet in the pattern of cones (Figure 5-10), so that anyone on the ground between the two cones hears the boom. If the aircraft is long enough, the positive and negative pressure changes may be heard as two distinct sounds. The boom itself continues from the time the aircraft breaks the sound barrier until it resumes subsonic speeds, but a person on the ground hears it only 0.1 to 0.5 second. Thus, the entire area over which the SST flies at supersonic speeds will experience the sonic boom. For this reason, the SST has been forbidden to fly over most land areas at supersonic speeds, including over the United States. Since it is allowed to fly at supersonic speeds only over oceans (e.g., New York to London), this makes it very costly to operate.

Figure 5-10 Cones representing increased and decreased pressure in a sonic boom. The area where the cones intersect the ground (shaded gray) experiences the sonic boom.

From Turk, A., Turk, J., Wittes, J. T., and Wittes, R., 1974. Environmental science. Philadelphia: Saunders.

Aircraft noise is very disturbing, in part because it is loud and intermittent on the ground (Kryter, 1994). In one study, Bronzaft et al. (1998) found that about 70% of those living within the path of an airport flight corridor were bothered by aircraft noise. In Chapter 13 we will see how tourist aircraft noise is one of the most disturbing sources of complaints by visitors to popular national parks (e.g., Mace et al., 1999).

Hygge, & Bullinger, 1995). A group of 135 third- and fourth-grade students living near the airport or in quieter, urban neighborhoods was studied in air-conditioned, sound-proof trailers parked near the students' schools. Measures collected included blood pressure, levels of stress hormones, and several indices of various cognitive abilities and task performance. The children living near the airport showed clear evidence of stress when compared with the quieter neighborhood students, showing higher levels of epinephrine and norepinephrine in their urine, higher diastolic blood pressure, and greater blood pressure reactivity during a cognitive task. Chronic noise exposure did not affect performance on some tasks, as no differences were found for reaction time and a perceptual task, but memory and reading task performance was better among children from quieter neighborhood settings. Students living in noisier areas (near the airport) were less motivated, showed less tolerance for frustration, and were more annoyed.

These field studies provide strong support for those who argue that chronic exposure to noise is associated with chronic stress and impairment of cognitive performance. Other ex-

planations for these effects, such as actual auditory damage or loss of hearing, were ruled out in the Munich study, and results were consistent with other studies of airport and traffic noise exposure and classroom noise (Hygge, 1993). Issues such as how long these effects of noise exposure last once children are no longer exposed to the noise (e.g., they move or the airport is closed) are currently being examined.

Can such problems be ameliorated? Bronzaft (1985–86) found that adding sound-absorbing ceiling tiles in classrooms and noise-reducing pads on rail tracks reduced class-

Figure 5-11 Changes in environmental and job satisfaction concurrent with increased, decreased, or unchanged disturbance by noise from specific and combined sources.

From Sundstrom et al., 1994.

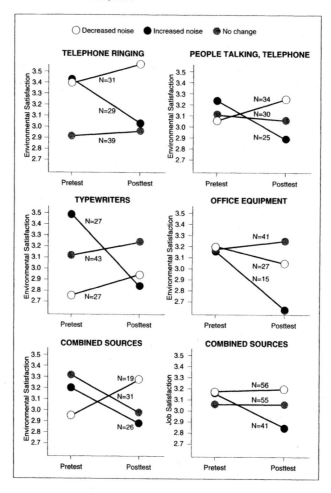

room noise and led to recovery of reading scores. Can such problems be prevented in the first place? A report by Ward and Suedfeld (1973) suggests that they can be avoided in the design phase. In response to a plan for routing a major highway next to a classroom building, the researchers played tape recordings of traffic at noise levels that simulated those of a real highway. Interference with learning was discovered before construction began, suggesting that we can plan ahead to avoid problems of this kind.

EFFECTS OF NOISE IN OFFICE AND INDUSTRIAL SETTINGS

How serious is occupational noise? One study, a survey of 2,391 employees in offices before and after office renovation, found that more than half reported that noise bothered them at work (Sundstrom et al., 1994). Irritation due to noise was also associated with dissatisfaction, and when renovated offices were noisier than they had been before renovation, job satisfaction declined (Sundstrom et al., 1994). As can be seen in Figure 5-11, changes in office design that resulted in less noise increased satisfaction with the new setting, while changes that led to more noise decreased satisfaction.

One of the most serious problems of background noise in commercial and industrial settings is its interference with communication. When a number of distinct auditory signals are presented simultaneously, it is often difficult for the human ear to distinguish or discriminate among them. This phenomenon is known as *masking,* and it accounts for our difficulty in hearing others talk in the presence of loud background noise. The background noise in the Glass and Singer (1972) research was created by combining simultaneously the sounds of a mimeograph machine (a type of copier), a calculator, a typewriter, two people speaking Spanish, and another person speaking Armenian, with the final effect being little discriminability among the various sounds due to masking. Background noise that consists of speech is especially problematic (e.g., Banbury & Berry, 1998). Apparently, we try to hear background conversation as communication, so we pay a lot of attention to

it. Nonconversational noise, however, requires less attention but does interfere with efforts to communicate.

Difficulty in hearing a communication varies not only according to amplitude and frequency of background noise (the more similar the frequency of the noise and of the communication, the worse the interference), but also according to the distance between communicator and listener. Figure 5-12 demonstrates the combined effects of ambient noise amplitude and interpersonal distance on communication. These "acceptable" levels of background noise are sometimes referred to as *speech interference levels* or SILs. Limited research (e.g., Acton, 1970) indicates that some communicative adaptation to background noise does occur, so we can learn to communicate effectively in the presence of many types of background noises. Thus, industrial workers accustomed to a noisy environment were found to be more effective in communicating against a loud background noise than were university employees accustomed to a quieter environment. Many designers and builders today use a standard of 55 to 70 dB as

Figure 5-12 Relationship between communication effort, noise level, and interpersonal distance.

Adapted from Miller, 1974.

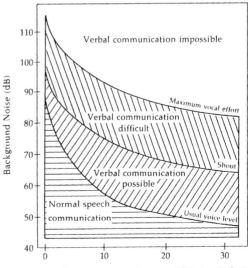

Distance between communicator and listener (in feet)

acceptable for executive offices (e.g., Fraser, 1989; Kryter, 1994).

Research on the effects of noise on productivity in industrial settings generally finds no direct effect of noise on nonauditory performance. Because such studies are conducted in field settings not subject to strict laboratory control of extraneous variables, the results are inconclusive at best. Reduction of noise, for instance, may boost employee morale (Sundstrom et al., 1994), which, in turn, boosts productivity. In other words, mediating variables, such as morale, fatigue, or communication difficulty, may be more significant than direct effects of noise on performance (e.g., Kryter, 1994). Nevertheless, if noise influences productivity even indirectly, industry would certainly want to take such influences into account by designing equipment and working space with noise levels in mind.

Concern over productivity, morale, and detrimental health effects of noise has led many industry and government officials to emphasize noise abatement factors in office and industrial settings. Among the more common abatement procedures are use of thick carpeting, suspended and acoustical tile ceilings, sound-absorbing wall materials, heavy draperies, and even plants. Other approaches involve making machines quieter in the first place, such as putting a layer of felt between typewriters and desks, enclosing computer printout equipment with felt or foam-lined covers, and producing equipment with less noisy components. Still another approach is to mask noise with constantly humming ventilating equipment or piped-in music.

How Do These Effects Occur?

Given all of the noise effects we have described, we believe it would be helpful to review some common themes in the ways these effects occur. One theory of how noise affects task performance is that it "masks" internal speech, or makes it more difficult to "hear ourselves think" (Poulton, 1977). When inner speech is blocked or cannot be used, noise has more negative effects on task performance (Wilding & Mohin-

dra, 1980). Research has found some evidence that noise masks internal speech and that this leads to poorer performance, but evidence does not indicate that this is the only way noise affects our work (Jones et al., 1979; Smith & Jones, 1992). Other theories posit memory or comprehension as the primary "victims" of noise, and studies also provide some support for this idea. Prolonged exposure to noise in the workplace is associated with deteriorating memory and recall, but attention seems to be less affected (Gomes, Martinho-Primenta, & Castelo, 1999). Noise also appears to reduce comprehension of reading material (e.g., Smith & Stansfield, 1986). Broadbent (1971) argued that noise can affect task performance by increasing the likelihood that dominant or well-learned, readily available information will be used in making decisions. In other words, noise causes people to use less of the information available and to sample or recall primarily dramatic, well-learned, or easily produced responses. Several studies have provided evidence of this narrowing of attention phenomenon, including poorer recall of irrelevant or peripheral information available during a task, or poorer ability to access uncommon or rarely used information (see Smith & Jones, 1992).

All of these models of noise effects on performance have weaknesses or parts for which the data do not provide support. Given that these and other models of noise impact do not satisfactorily account for what we have learned about noise and task performance, how can we explain why noise affects performance only in certain circumstances?

One way to answer this question is to turn to the theoretical approaches discussed in Chapter 4. It might also be useful to examine once again Figure 5-2 showing our eclectic model applied to noise. For example, adaptation level theory predicts variations in performance for different levels of skill, experience, and stimulation for each individual. Furthermore, the Yerkes–Dodson law and arousal theory suggest that noise that is arousing will facilitate performance on simple tasks, up to a point. However, high levels of arousal interfere with performance on complex tasks, and extremely high levels of arousal interfere with performance on simple tasks. Data from research on noise and performance are consistent with this explanation and suggest that arousal characteristic of stress can interfere with memory, reading, and problem-solving abilities.

Environmental load can also be used to explain many findings bearing on the relationship between noise and performance. Researchers have argued that unpredictable noise requires greater allocation of attention than does predictable noise, and it should therefore interfere more with performance. For complex tasks, even more attention is required for optimal performance, and any stimulus that distracts us or calls attention away from the task will hurt performance. Behavior constraint models can also explain some aspects of noise exposure, such as why lack of perceived control over noise hurts performance: When control is apparently lost, more effort may be given to restoring control than to attending to the task at hand, resulting in poorer task performance.

One of the ways in which noise may contribute to physical and mental health problems as well as mood and performance deficits is by disturbing sleep. Noise wakes us up and makes it harder to fall asleep. We all know from personal experience, and studies have confirmed, that noise, even in hospital-like settings, can lead to stress and poor sleep (Topf, 1992a, 1992b). In another study, objective measures of noise exposure (e.g., loudness) were not related to sleep or to reported health, but subjective measures of noise exposure (e.g., annoyance) were related to sleep (Nivison & Endresen, 1993). Most likely, people learn to ignore noise that occurs continuously or regularly and adapt to such noise without losing much sleep; however, the effort required and one's sensitivity to the noise may affect sleep and health.

NOISE AND SOCIAL BEHAVIOR

If noise has stressful, arousing, attention-narrowing, or behavior-constraining properties, exposure to it will be likely to influence inter-personal relationships that are affected by these mediators. We will now look at three specific social relationships—attraction, altruism, and aggression—to determine just what noise can do to social interaction. Much of this work consists of laboratory experiments conducted three decades or so ago.

Noise and Attraction

One might expect loud, disturbing noise to have a deleterious effect on feelings of liking toward others. That is, noxious stimuli associated with others may lead to less pleasant evaluation of those others. One way to measure attraction, as suggested by research on personal space, is to examine physical distances between ourselves and others; we stand or sit closer to those people we like than to those we dislike (see Chapter 8). Thus, if interpersonal distance is an indicator of attraction and if noise decreases attraction, we would expect noise to increase interpersonal distancing. In support of this hypothesis, Mathews, Canon, and Alexander (1974) found that even a noise of 80 dB increased the distance at which individuals felt comfortable with each other. Also, in a correlational study, Appleyard and Lintell (1972) found less informal interaction among neighbors when traffic noise was greater.

Other researchers have found equivocal results on the relationship between noise and attraction. Kenrick and Johnson (1979), for example, found that among women, exposure to aversive noise may increase attraction toward one who shares the aversive experience with the individual but may decrease attraction toward someone not actually experiencing the noise.

One explanation for some effects of noise on attraction involves environmental load: Noise affects the amount of information that people gather about another person. Perhaps in causing people to narrow their attention and focus on a smaller part of their environment, noise causes people to pay attention to fewer characteristics of other people. Thus, noise could cause a distortion in perceptions of other people. Research by Siegel and Steele (1980) suggests that this may be the case; these researchers found that noise led to more extreme and premature judgments about other people but did not cause these judgments to be more negative.

Noise and Human Aggression

Research on the effects of noise on aggression has been much more conclusive than research on noise and attraction. Theories of aggression in which arousal is posited as at least one mediator (e.g., Anderson, Anderson, & Deuser, 1996; Berkowitz, 1993) predict that under circumstances in which aggression is likely to occur, increasing an individual's arousal level will also increase the intensity of aggressive behavior. Thus, to the extent that noise increases arousal, it should also increase aggression in individuals already predisposed to aggress.

Geen and O'Neal (1969) tested this hypothesis by first showing participants either a nonviolent sports film or a more violent prize-fight film, with the expectation that the violent film would predispose individuals to aggress. Next, research participants were provided with an opportunity to aggress against a confederate "victim" by ostensibly delivering electric shocks to that person. In traditional laboratory studies of aggression, participants were given the chance to shock a confederate or stooge victim, and the shock level (intensity, duration, or number) they chose was the index of aggression. No shocks were actually administered, although the participant, until the end of the experiment, was led to believe that he or she was actually delivering shocks (Figure 5-13). During the shock phase of the experiment, Geen and O'Neal exposed half the participants to the normal noise level of the laboratory and the other half to a 2-minute burst of continuous 60-dB white noise (i.e., a broadband of frequencies). It was predicted that the 60-dB noise would increase the level of aggression of those exposed to the violent film. Results suggested that both the violent film and the added noise increased the number of shocks delivered to the victim. Furthermore, the greatest aggression occurred under the con-

Figure 5-13 An example of the apparatus used in laboratory experiments on aggression from the 1970s. Participants were told that pushing higher-numbered buttons delivered stronger electrical shocks to another individual. The level of the shock and its duration served as measures of aggression.

dition that combined the violent film with the arousing noise, as originally predicted.

Additional laboratory research on noise and aggression was conducted by Donnerstein and Wilson (1976). Recall that Glass and Singer (1972) found unpredictable noise to be more aversive than predictable noise. One would thus expect unpredictable noise to be highly arousing and consequently to lead to heightened aggression, in accordance with the dominant response hypothesis noted above. Using a similar ostensible shock procedure as described above, Donnerstein and Wilson exposed individuals to either 55 dB or 95 dB of unpredictable, 1-second noise bursts during the study and made half the participants angry through an insult from the confederate. As expected, those in the angered group delivered more intense shocks than did the nonangered participants. Furthermore, the 95-dB unpredictable noise increased aggression relative to the 55-dB unpredictable noise only for those in the angered conditioned. Noise level made no difference in the intensity of shocks delivered by those in the nonangered group.

Following Glass and Singer's findings that controllable noise is less aversive and arousing than uncontrollable noise, we would expect that perceived control over noise would make it less aversive and less likely to facilitate aggression.

Donnerstein and Wilson tested this hypothesis by conducting a second experiment in which participants worked on a set of math problems while exposed either to no artificial noise, to 95 dB of unpredictable and uncontrollable noise, or to 95 dB of unpredictable noise that they believed they could terminate at any time (i.e., over which they perceived they had control). All noise was terminated as the shock phase of the experiment began so that only the after-effects of noise could influence aggression. As in the previous experiment, participants were either angered or not angered by the victim, in this case immediately after the math task. More intense shocks were delivered by the angry than nonangry group, and for those in the angry group, unpredictable and uncontrollable noise increased aggression. The 95-dB noise had no effect on aggression, however, when participants perceived they had control over it.

The finding that noise increased aggressiveness only when people were angry suggests again that the noise served to facilitate aggression caused by anger rather than creating or causing the aggression directly. Konecni (1975) also found this to be the case—noise increased aggressiveness only for a group that had been angered.

These experiments suggest, then, that under circumstances in which noise would be expected to increase arousal or when there is a predisposition to aggress (i.e., when people are angry), aggression increased. However, when the noise does not appreciably increase arousal (as when an individual has control over it) or when the individual is not already predisposed to aggress, noise appears to have little, if any, effect on aggression. Cohen and Spacapan (1984) have argued that noise strengthens or increases aggression but does not provoke it. In order for noise to affect aggressive behavior, the behavior must be present for other reasons.

Noise and Helping

Research suggests that noise influences at least one more social phenomenon—whether or not

people help one another. It seems reasonable to assume that aversive noise that makes us irritable or uncomfortable will make us less likely to offer assistance to someone who needs help. Research in social psychology has indicated that being in a bad mood can reduce our inclination to help others (e.g., Cialdini & Kenrick, 1976; Weyant, 1978). Another reason for this expected decrease in help is offered by the environmental load approach discussed in Chapter 4. Since noise reduces the attention paid to less important stimuli, and if we are focused on an important task, then noise should make us less aware of incidental signs of distress in others. Cohen and Lezak (1977) demonstrated that the content of slides depicting social situations was less well remembered under noisy than under quiet conditions when participants were asked to concentrate initially on material other than the slides. Under such conditions, social cues in the slides were relatively unimportant, so noise interfered with attending to those cues.

Two experiments, one conducted in the laboratory and the other in the field, have suggested that noise does indeed decrease the frequency of helping (Mathews & Canon, 1975). In the laboratory experiment, participants were exposed to 48 dB of normal noise, to 65 dB of white noise piped into the laboratory through a hidden speaker, or to 85 dB of white noise from the same speaker. As participants arrived for the experiment, they were asked in turn to wait in the laboratory for a few minutes with another individual (actually a confederate of the experimenter), who was seated and reading a journal. On the confederate's lap were additional journals, books, and papers. After a few minutes, the experimenter called for the confederate who, upon getting up, "accidentally" dropped the materials right in front of the unsuspecting participant. The dependent measure of helping was whether or not the participant helped the confederate pick up the spilled materials. Results suggested a definite decrement in helping in the loud-noise conditions: 72% helped in the normal noise condition, 67% in the 65-dB condition, and only 37% in the 85-dB condition.

The Mathews and Canon field experiment revealed even more interesting results. In this study, a confederate dropped a box of books while getting out of a car. To emphasize his apparent need for aid he wore a cast on his arm in half the experimental situations. Noise was manipulated by having another confederate operate a lawnmower nearby. In the low-noise condition, the lawnmower was not running, and background noise from normal sources was measured at 50 dB. In the high-noise condition, the lawnmower was running without a muffler, putting out an 87-dB din. Once again, the dependent measure of helping was how many passing individuals stopped to help the confederate pick up the dropped books. Results showed that noise had little effect on helping when the confederate was not wearing a cast (approximately 15% helped). But when the confederate wore a cast (high-need condition), the loud noise reduced the frequency of helping from 80% to 15%! Apparently, noise led people to attend less to cues (i.e., the cast) that indicated another person needed help.

A series of studies by Page (1977) also provides evidence that noise can reduce the likelihood that people will help one another. In one study, participants encountered a confederate who, with an armful of books, had dropped a pack of index cards. One of three levels of noise was present: 100 dB, 80 dB, or 50 dB. Results suggested that high levels of noise reduced helping, although the effect was weak (see also Bell & Doyle, 1983).

A second study reported by Page (1977) found stronger results. In this one, participants saw a confederate drop a package while walking past a construction site. When the jackhammers were being used on the site, noise levels were 92 dB; when they were not being used, levels were 72 dB. People were less likely to help the confederate when noise levels reached 92 dB than at a less noisy 72 dB.

These results suggest that people who experience noise simply may not notice that someone needs help. Page (1977) conducted one more study in which people were approached and directly asked whether or not they could provide

change for a quarter. In this context, people had to pay at least some attention to the request, although the noise could certainly be distracting. Once again, noise decreased the likelihood that people would respond to the request.

The reasons noise suppressed helping behavior in these studies are not conclusively known, but the most likely explanations still appear to be the "narrowing of attention" notion and the "mood" explanation. Nevertheless, each of these has been disconfirmed by at least one study. A study that argues against the idea that noise reduces helping by putting people in bad or irritable moods was reported by Yinon and Bizman (1980). Participants were exposed to one of two noise levels (high or low) while working on a task, and then they received positive or negative feedback on their performance. Participants then encountered someone who asked them for help. One might expect the combination of the negative feedback and loud noise to dampen participants' moods and cause them to refuse to help. This is not, however, what was found. Under the high-noise condition, no differences in helping between the positive and negative feedback groups were found. Only under low noise did the feedback make a difference. Apparently, the loud noise distracted people from focusing on the feedback or provided a reason for the negative feedback. Although it is still possible that mood states were involved, their role in this study does not appear to be crucial.

We have seen thus far that perceived control over noxious noise reduces its impairment of performance and its facilitation of aggression. A study by Sherrod and Downs (1974) similarly demonstrated that perceived control can reduce the suppressing effect of noise on helping behavior. In that study, participants engaged in a proofreading task while simultaneously monitoring a series of random numbers presented on audio tape. Three conditions were established: (1) a control condition in which the numbers were superimposed on the pleasing sounds of a seashore (e.g., waves striking the beach); (2) a complex noise condition in which the numbers were superimposed over a round of Dixieland jazz and another voice reading prose; and (3) a perceived-control condition using the same tape as the complex noise condition but with individuals told they could terminate the distracting noise if they so desired. After 20 minutes in one of these situations, participants left the laboratory and were approached by an individual asking their assistance in filling out forms for another study. The most help was volunteered

by those in the seashore sound condition, for whom noise was least noxious. Those in the perceived-control condition offered more help than those in the uncontrollable complex noise condition. Thus, the effects of noise on helping behavior depend on several factors, among which are perceived control of the noise, volume of the noise, and stimulus characteristics of the person needing assistance.

Reducing Noise: Is It Really Effective?

We saw in the section on noise in office environments that reducing noise increased job satisfaction (Sundstrom et al., 1994). Reducing noise in other settings would seem to be worthwhile, such as adding noise-attenuating fences along freeways that run through residential ar-

eas or constructing bypasses to route traffic away from sensitive areas (Figure 5-14). Griffiths and Raw (1987) have argued that changes in noise levels due to such interventions can result in changes in dissatisfaction that are greater than what might be expected on the basis of the

Figure 5-14   Tall fences along freeways are designed as sound barriers to reduce the traffic noise around the nearby homes.

∞ CAN NOISE BE THERAPEUTIC?

We described the phenomenon of masking in the section on noise in office and industrial settings. Masking occurs when a wider band of white noise "covers up" the sound of a narrower band source (e.g., the voice of someone). While this may be aggravating if we want to hear the source that is being masked, an interesting application is to use masking to "cover up" an unwanted source. One effective way of reducing negative effects of workplace noise is to actually add sound to the situation. Masking white noise or other steady, uniform sound that distracts or "covers up" more disturbing noise may reduce disruptions linked to noise (e.g., Ellermeier & Hellbrueck, 1998). In Chapter 12 we will describe Alzheimer's disease, a condition characterized by memory loss, cognitive decline, and confusion that often results in placement in a nursing home. Burgio et al. (1996) describe an intervention in which agitated nursing home residents with cognitive impairments were exposed to audio tapes of the sounds of a stream running over rocks or ocean waves crashing against a shore. Those with Alzheimer-like conditions are easily distracted and confused by sounds such as footsteps, voices, or vacuum cleaners (e.g., Elm, Warren, & Madill, 1998), so the tapes were designed to mask these noises. Results showed a 23% reduction in verbal agitation due to the masking intervention!

resulting noise levels alone. In other words, if an area is characterized by ambient noise levels of 80 dB and this is reduced to 70 dB by directing traffic away, resulting annoyance or dissatisfaction may be even lower than in a neighborhood characterized by 70-dB noise to begin with. This does not appear to be due merely to contrasting conditions, as the increase in satisfaction can persist beyond the change in traffic levels for up to two years (e.g., Griffiths & Raw, 1987; Kryter, 1994).

Annoyance does not always decrease when we intervene to reduce noise, but it frequently does change. Research on the effects of acoustic barriers for reducing traffic noise has not shown strong evidence of corresponding reduction of annoyance due to the noise, but studies that consider other methods of noise reduction, such as traffic control and redistribution of noise, may show more substantial changes. In these cases, too, the reduction of noise can result in larger changes than would be expected from the absolute level of the lower noise (Kastka, 1980; Vallet, 1987). Why does this happen? When we initially described noise annoyance, we noted that annoyance is higher if we believe those responsible for the noise do not care about our

welfare or if we believe the noise is harmful to our health and well-being. We may interpret the noise reduction efforts of others to mean they really are concerned with our welfare or that our health is being protected, in which case reduced noise is likely to have a very large impact. On the other hand, if the noise is reduced and we still believe those producing it are unconcerned or that our health is in danger, we may still report high levels of annoyance (e.g., Kryter, 1994). Bronzaft et al. (1998), for example, found that about 70% of those living within the path of an airport flight corridor were bothered by aircraft noise. These respondents also reported that the aircraft noise interfered with daily activities, and they complained about sleep loss and poor health. Under these circumstances, reducing the noise may not be adequate to reduce annoyance.

We also noted earlier that Bronzaft and colleagues found that students in classrooms facing elevated subway tracks in New York had substantially lower reading ability than did students from the same school in classrooms on a quieter side of the building (Bronzaft & McCarthy, 1975; Bronzaft, 1981) and were able to convince authorities that steps had to be taken to reduce transit noise in these classrooms (Bronzaft, 1985–86). Installation of acoustic ceiling tiles in classrooms and sound-absorbing pads on the tracks reduced noise significantly and was associated with apparent recovery of reading scores; after the intervention, there were no longer any significant differences in reading ability between students from the side of the school facing the tracks and those on the other side (Bronzaft, 1985–86).

In general, the basis for noise-related public policy has been limited, and models for estimating annoyance or other untoward effects of noise "oversimplify" the extent and causes of noise effects (Staples, 1997). In particular, there is little appreciation of the social and psychological variables that affect responses to noise or of the stress-related psychophysiological consequences of noise exposure (Staples, 1997). Since psychosocial variables (e.g., sleep deprivation, belief that noise impacts health) are more closely tied to annoyance than is noise level by itself, policy makers and those responsible for decreasing the negative impact of noise need to consider such influence systematically.

This need extends to several areas, including issues related to the introduction of noise into previously quiet settings (as was studied by Evans and his colleagues as the new airport in Munich was opened) or into nonurban, sparsely populated areas (Staples, 1997). Most models are based on areas already affected by noise or were drawn from surveys of densely populated developed areas (e.g., Kryter, 1994). Further, these data are often collected from people who elect to remain in noisy areas since we would expect people who find the noise intolerable to move away. However, people cannot always move, and noise may make selling their old house more difficult; Kryter (1994) cites studies showing that noise can reduce the value of a home by 1% per dB of excess noise!

Community complaints about noise are most often due to ground traffic and aircraft. As our communities expand to outlying areas where highways and other sources of noise already exist, and as we add new sources of noise (e.g., new runways at old or new airports, construction noise from building renovation), the need to be cognizant of the impact of noise and attempts to reduce it will become more paramount.

SUMMARY

In this chapter, we discussed the nature of noise, where it comes from, and how it affects us. Often, noise is a less intense stressor than some that we will examine in other chapters; it is not as overwhelming as disasters and not as momentarily debilitating as noxious chemicals. However, it appears to have many effects. Noise can lead to increased arousal, stress, narrowing of attention, and constraints on behavior. The aversiveness of noise depends on volume, predictability, and perceived control. In combination with other stressors, noise may have adverse effects on physical and mental health. Whether noise hurts or helps performance depends on the type of noise, the complexity of the task, and individual factors such as personality and adaptation level. In the classroom, noise interferes with language skills and can elevate blood pressure. Noise interferes with verbal communication and may affect productivity. Depending on the situation and the type of noise, noise may increase or decrease attraction, facilitate aggression, or interfere with helping behavior. Whether noise has these effects, or whether we are even aware of noise, is determined by a number of factors. However, its pervasiveness demands careful study, and many interesting and complex issues remain to be explored. Reducing noise in the workplace improves job satisfaction. Whether reducing noise in the community also reduces annoyance depends in part on psychosocial factors such as perception of its effects on health.

KEY TERMS

amplitude	loudness	noise-induced perma-	timbre
annoyance	masking	nent threshold shift	tonal quality
decibels (dB)	narrow band	(NIPTS)	white noise
frequency	noise	pitch	wide band
hearing loss		temporary threshold	
hertz (Hz)		shifts (TTS)	

SUGGESTED PROJECTS

1. Interview people who live in noisy and quiet places. If you know of residence halls on campus that are noisier than others due to traffic or some other factor, that might be a good place to start. Ask residents whether they are aware of the noise, how much it annoys them, when it is bothersome, and so on. Do peoples' activities, goals, and attitudes about the source of the noise affect their reactions to it?

2. Make a tape recording of an ambiguous sound—one that could be almost anything, or at least could be from a few different sources. Tell some people each of the possible causes. For example, a high, whining sound could be interpreted as radio static or a dentist's drill. Do the labels you attach to the sound make it more unpleasant or annoying?

3. Observe the way furniture is laid out in various places around campus and see whether there are effects of noise on the ways in which people use these places and furniture arrangements. What happens in places where furniture is set up to encourage people to talk if there is loud noise coming from nearby? Explore other settings as well. What patterns do you see?

WEATHER, CLIMATE, AND BEHAVIOR

COLD TEMPERATURES AND BEHAVIOR

WIND AND BEHAVIOR

BAROMETRIC PRESSURE AND ALTITUDE

SUMMARY
KEY TERMS
SUGGESTED PROJECTS

 Introduction

It is the year 32,825 and the great city of Seltisar has changed considerably over the past 30,000 years. A geologic uplift combined with shifts in upper-atmospheric air currents has cooled the city 20° below its average 25,000 years ago. Snow occurs 300 days per year and the surrounding plants are very different from those of the Great Millennium of Heat that was recorded between 5125 and 6231. The museum shows that clothes during those times were much thinner than the layers and layers now typical. Citizens cannot imagine that track meets and soccer games were actually held outdoors back then. A popular exhibit describes the Dordellian War of 5172–5178. Humans had finally achieved a degree of world peace for 2,000 years before this war erupted. Historians disagree about its causes; some say it was aggravated by the high temperatures, others that it was fought over declining crop yields and water rights associated with the reduced flow of upland rivers. With all of the snow today, it is hard to imagine that people could have gone to war over water. What they do agree about is that the dramatic decline of the population helped end the war.

Strange as it may seem, this imaginary tale parallels the rise and fall of several earlier civilizations that archaeologists have studied, and it taps some of the relationships between weather, climate, and behavior that environmental psychologists study today. Just how do temperature, humidity, wind, and air pressure affect behavior? We are all familiar with certain consequences of exposure to these factors. When it gets cold outdoors, we behave in ways that minimize discomfort, such as putting on heavy coats. When the wind blows down the street at 50 miles per hour (80 km/h), we behave in ways that will minimize our discomfort from wind exposure, such as not riding a bicycle and walking at an angle to the ground to maintain our balance. When we travel from a cool community to a very hot one, we may restrict our outdoor activity.

Why Are Environmental Psychologists Concerned About Weather and Climate?

RECIPROCAL RELATIONSHIPS BETWEEN WEATHER, CLIMATE, AND PEOPLE

Research in the past decade has told us much more about what behavior to expect when people are exposed to abnormal levels of heat, cold, and wind. Such research lets us answer rather detailed questions about specific environment–behavior problems. For example, how do high outdoor temperatures affect the level of aggressive and violent behavior in society, as suggested by the popular notion of the **long, hot summer effect?** Or how do weather changes affect mental health and interpersonal relationships? Such questions about the influence of the physical environment on personal and interpersonal behavior are becoming more and more important for at least two reasons. First, humans are constantly exposed to natural changes in the physical environment. Parts of the world typically undergo temperature changes from $-20°F$ to $100°F$ ($-29°C$ to $38°C$) in different seasons. Some cultures exist in hot tropical climates, whereas others thrive in arctic conditions. Do such temperature differences influence behavior? What if climatological changes, which according to many climatologists are becoming more and more extreme, should result in exaggerated cold or hot temperatures? If a long, hot summer effect really does exist, and if climatological changes result in average daily summer temperatures of $110°F$ in urban centers, are we likely to see disastrous rioting and violence? Whatever the case, it becomes important for us to know the behavioral influence of extreme or even very mild natural changes in the physical environment.

The second major reason why we need to know more about the effects of the physical environment on behavior is that we ourselves are making drastic changes in the natural environment, changes that we may be motivated to correct if they can be shown to have deleterious effects on behavior. We will shortly describe research on global warming and the greenhouse effect, in which it is thought that human use of fossil fuels causes the accumulation of atmospheric gases that in turn cause surface temperatures to increase in ways that impact almost all ecological systems. In addition, waste heat from the compressors of air conditioners and heat-absorbing concrete are actually heating up cities to levels of $10°F$ ($6°C$) or more above the temperature of the surrounding countryside.

Environmental psychologists are concerned that such warming could affect people directly and indirectly. In 1991, retreating Iraqi military forces set fire to more than 700 Kuwaiti oil wells, sparking concern that regional or global weather change could result. These fires burned 4.6 million barrels of oil per day, producing sulfur dioxide emissions equivalent to 57% of that of U.S. electric utilities and carbon dioxide equivalent to 2% of global emissions. In some areas the smoke absorbed 75% of the sun's radiation. Fortunately, substantial associated weather change did not seem to occur (Hobbs & Radke, 1992). But could we possibly be adding to a long, hot summer effect by the way our daily living habits

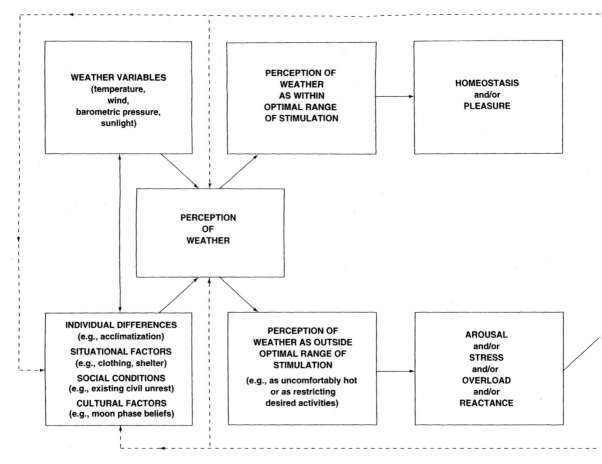

Figure 6-1   The effects of weather variables on people can be interpreted through a variation of our eclectic model of environment–behavior relationships.

alter the physical environment? As another example, if high winds have negative effects on mental health, physical well-being, and behavior, we might want to reevaluate building designs that actually increase wind speeds in pedestrian areas.

Whether the source of environmental stimulation is derived from natural or human causes, the concern of environmental psychologists is the same: What differences in behavior can be expected under different conditions in the physical environment? In previous chapters we have seen how we perceive and process information about the environment, how we can view environment–behavior relationships from several theoretical perspectives, and how environ-

mental stressors such as noise influence us. In the present chapter we will examine in detail weather and climate as important types of physical environmental factors, how they affect us, and how these effects can be explained from various theoretical perspectives. Specifically, we will look at the behavioral effects of weather variables—heat, cold, wind, and barometric pressure—as well as at the effects of climate. As we do so, perhaps it would be helpful to keep Figure 6-1 in mind as an overall framework; you will recognize this as a variation of our eclectic model from Chapter 4. The objective physical environment (e.g., heat, altitude), individual factors such as how accustomed we are to the climate, and our perception of the weather as

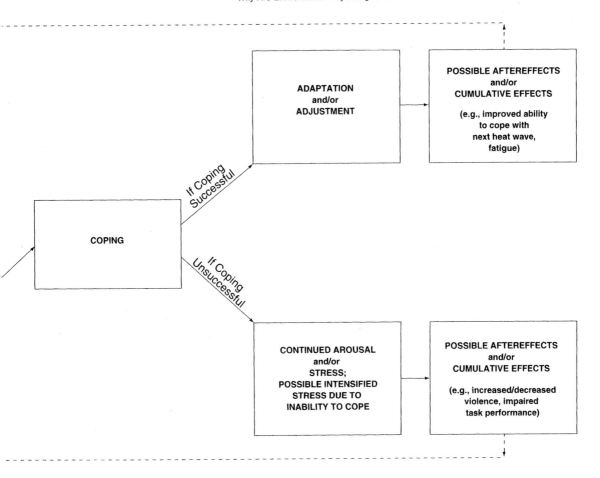

outside an optimal range all lead to mediational states (e.g., arousal) and to coping strategies such as escape or sweating; in the process, health, performance, and social behavior may be affected.

We should mention a few caveats before proceeding. Although we will primarily discuss each of the meteorological factors separately (e.g., temperature, wind, and low air pressure), in actuality they often occur together with one another as well as with air pollution. Singular effects of weather variables on mental health, outlook, and behavior rarely occur in our lives. Hot days are often associated with high barometric pressure and calm winds. Stormy days usually involve changes in temperature, wind, humidity, and air pressure. High altitudes not only re-

sult in reduced oxygen supply, but in increased exposure to solar radiation. As Suedfeld (1991, 1998) notes, individuals stationed in Antarctica face not only cold temperatures, but high altitudes, high winds, unusual light–dark cycles of the sun, and social isolation as well (details are provided in the November 1991 issue of *Environment and Behavior* on polar psychology). Moreover, low winds and temperature inversions increase the concentration of air pollutants, and high humidity can intensify the effects of photochemical smog (see Chapter 7). Greenhouse gases appear to increase global warming, which impacts local weather patterns. Thus, weather and air pollution variables are interrelated, and one can rarely conclude with certainty that a given behavior or health effect is attributable to any one of these factors (cf. Rotton & Frey, 1985).

We will have more to say on combinations of factors when we discuss climate. Closely related to this point, we should briefly note a distinction between weather and climate. Essentially, **weather** refers to relatively rapidly changing or momentary conditions, such as a cold front or a heat wave. **Climate,** on the other hand, refers to average weather conditions or prevailing weather over a long period of time. The distinction is important for environmental psychologists because the measured effects of climate on behavior are often not the same as the measured effects of weather on behavior. For example, climatological precipitation has definite long-term behavioral correlates (e.g., farming practices), but weather measures of precipitation primarily predict short-term behavior (e.g., seeking shelter in a rainstorm). Also, we can rarely study these variables alone but rather must control for effects of cultural and social factors that also affect behavior, and it is easier to control for these factors when studying climate than weather.

GLOBAL WARMING, THE GREENHOUSE EFFECT, THE OZONE HOLE, AND EL NIÑO

Global Warming and Related Concerns

A growing concern among scientists today is the phenomenon of **global warming** (IPCC, 1991; Oskamp, 2000; Taylor, 1999). The surface temperature of the Earth fluctuates somewhat regularly over millions of years. Some of the regular fluctuation corresponds with ice ages, the last one ending about 11,000 to 12,000 years ago. Independent of these natural fluctuations, the temperatures we experience change mostly with daylight and nightfall and with the annual seasons. Normally, much of the sun's heat that strikes the Earth is reflected back into space. Some of the sun's energy is also absorbed at the Earth's surface through photosynthesis—the process by which green plants grow. In a

balanced ecological system, the carbon in living organisms is recycled repeatedly as one living organism becomes the food and fertilizer of others. However, a problem occurs when this carbon recycling process becomes disrupted through at least two human-caused interventions. The first problem is the elimination of biomass as humans cut down forests (which store much carbon) to make farmland and pasture (which store less carbon). Later we convert farmland into parking lots and office buildings (which store very little carbon), and along the way turn savanna into desert and otherwise inhibit the natural conversion of carbon into plant material (e.g., Kaiser, 1999a). The second problem occurs when humans burn fossil fuels in massive quantities, as we have been doing since the Industrial Revolution with the burning of coal, oil, and natural gas. Such burning releases tremendous amounts of carbon dioxide and other "greenhouse gases" into the atmosphere. If these gases are not recycled into biomass through photosynthesis by the vegetation that would normally cover land surface (owing to humans converting forests into farmland and parking lots), then the gases build up in the atmosphere and create a **greenhouse effect.** In a horticultural greenhouse, glass panes help trap heat from the sun: What enters the greenhouse does not easily escape back into the atmosphere. Increased carbon dioxide in the atmosphere acts much like the glass in a greenhouse, trapping more of the sun's heat and warming the Earth's surface abnormally. Some 12,000 years ago at the end of the last ice age, the Earth's surface warmed about 7°F (3°C) and stayed on average at that level until recently. Over the past 100 years or so, we have warmed another 2°F (0.5°C), and estimates from the accelerating greenhouse effect suggest we may warm another 2°F to 7°F by the year 2100—a dramatic warming in a relatively very short time.

Many complicating factors are associated with global warming. There is strong evidence, for example, that aerosols and other pollutants from industrial processes moderate the warming process by increasing cloud brightness and reflecting more light back into space (Kerr, 1995;

Kiehl, 1999; Schwartz & Andreae, 1996). One study predicts that aerosols from sulfur in fossil fuel waste gases will combine with greenhouse effects to make India and China much drier—potentially impairing crop production (Mudur, 1995). On the other hand, aerosols in the form of chlorofluorocarbons (CFCs) contribute to the **ozone hole** (e.g., Kerr, 1992a). Atmospheric ozone absorbs harmful ultraviolet sunlight, among other things (see the box on page 200). It apparently is becoming rapidly depleted over Antarctica and the Arctic in a pattern called the ozone hole. The ozone-destroying process occurs heavily in stratospheric ice clouds. The greenhouse effect warms planet Earth but cools the stratosphere, causing more ice clouds to form and destroying more ozone. Although natural events such as the eruption of Mt. Pinatubo in 1991 seem to decrease atmospheric ozone (Kerr, 1993), most scientists believe the ozone hole increases skin cancer rates and is a major cause for concern (Taubes, 1993).

Another complicating factor in the global warming picture involves ocean currents (e.g., Kerr, 1998, 1999b). Deep ocean currents transfer heat across the sea as well as flush the waters of harmful levels of salinity. Changes in these currents appear correlated with global warming, and one fear is that these changes could both accelerate global warming and make large portions of the ocean uninhabitable for algae and other aquatic life. Ocean life not only provides food for humans and other organisms, but also absorbs much of the carbon in greenhouse gases.

Further complicating the global warming picture is that not everyone is convinced it is real nor that it is caused by human intervention (e.g., Kerr, 1995, 1997; Weiss, 1996). Some argue that the increase in temperature over the past century or so could be part of a natural fluctuation unassociated with human activity. Others argue that recent temperature increases could be attributable to long-term land use practices (e.g., turning prairie into farmland) or to an artifact of the measurement process (e.g., Stohlgren et al., 1998). For example, we tend to keep 100-year records of temperatures in largely inhabited areas, and over the decades these areas have

become urbanized. Urban areas in turn become "heat islands" that are warmer than the surrounding countryside because they trap heat in concrete buildings and parking lots and freeways, so if our 100-year temperature measurements are made in places that become more and more urbanized, it makes sense that average temperatures will be recorded as rising.

The **Gaia hypothesis** says the Earth self-regulates the whole thermal process (e.g., Lovelock, 1988, 1998). That is, the heating and cooling fluctuate naturally as vegetation, animal life, oceans, and the atmosphere absorb or release heat and emissions and compensate for normal fluctuations, although human intervention could cause things to get out of hand.

The great fear of global warming is that we might upset the natural balance so much that living systems will pay a heavy price. If we warm up things too much, agriculture could collapse in some parts of the world but become possible in previously cooler areas, greatly changing economic relationships as well as the ability to feed local populations. Enough heat to melt polar ice caps would flood large coastal regions, including New York City, New Orleans, immense regions of southeast Asia bordering the South China Sea, and the delta regions of Bangladesh and Egypt. Storms and drought could become more frequent and intense. Tropical infectious disease could move toward more temperate climates where people and other life forms have no built-in immunity to them (Brown, 1999; Kerr, 1999c; Oskamp, 2000; Taubes, 1997).

Can anything be done to prevent global warming? Reducing harmful emissions is one solution (Rubin et al., 1992). There is much recent work on developing practices that will sustain both civilization and the natural world on which we are dependent (Oskamp, 2000; Stern, 2000). One solution is to "green up" the Earth. Better forestry and agricultural management techniques may be part of the answer. Several studies suggest that existing forests absorb 25% to 40% of the carbon released by burning fossil fuels (Moffatt, 1997), with the implication that replanting forests and using more carbon-absorbing agricultural crops could help restore the natural balance of carbon as well as boost crop production (Culotta, 1995; Schmidt, 1998). Some 7 million square kilometers of trees (about the size of Australia) would absorb all the carbon humans burn in fossil fuels and turn it into wood (e.g., Marland, cited in Booth, 1988). There are other promising solutions involving changing human behavior that we will explore in Chapter 14. For the moment, we wish to emphasize that weather and climate variables, whether influenced by humans or not, do have an impact on our everyday lives.

El Niño and Other Oscillations

Another important phenomenon in weather and climate research is the periodic appearance of several fluctuations, most famously the El Niño Southern Oscillation, or **El Niño** for short. This is a pattern in which tropical Pacific waters warm and spread to the equatorial coast of South America (the name "El Niño" is Spanish for "The Child" and is given to the event because the arrival of this warmth often corresponds with the Christmas season). Some other oscillations include a cooling of the tropical Pacific called "La Niña" and the Pacific Decadal Oscillation

(PDO) characterized by a shift in northern Pacific waters from cold in the west and warm in the east to the opposite pattern, as well as the well-known 11-year sunspot cycle that is associated with warming and drought (Kerr, 1996, 1999a). In a classic El Niño, storm tracks that typically cross North America at mid-latitudes become replaced with warmth, which makes the northern United States warmer and drier and the southern United States colder and wetter. La Niña, on the other hand, makes the southwest United States drier and the northern states wetter. A PDO event can alter these patterns, making them more or less intense. El Niño and Arctic and Atlantic oscillations can also predict the severity of Atlantic hurricane seasons and European and African storm patterns (Kerr, 1998).

El Niño events interact with human interventions to exacerbate problems. For example, logging practices on the Indonesian island of Borneo have been shown to exacerbate forest fires associated with an El Niño drought, and logging and El Niño together can decrease the supply of seeds on which many species are dependent (Hartshorn & Bynum, 1999).

Geographical and Climatological Determinism

Much of the literature on the influence of climate on people is quite speculative, and not based on sound empirical data. Very early writings on the topic, in fact, are based on nonsystematic observation, and many of the later writings are based on flawed observations. More recent investigations, however, include careful biological measurements that show at least some physiological adaptations associated with climate, as well as archaeological revelations that show a strong relationship between climate and the rise and decline of individual civilizations. Altogether, it is instructive to examine the evidence from both the early and recent literature. In doing so, it is helpful to keep in mind three

perspectives on the degree of climate influence. Briefly, these perspectives are determinism, possibilism, and probabilism (cf. Rotton, 1986). In terms of climate, **determinism** suggests that climate forcefully causes a range of behaviors, such as heat waves causing crimes. As noted in Chapter 4, environmental psychologists take a broad view of determinism, which means that many interrelated variables contribute to predicting our behavior. Some of the climatological determinists we discuss in this section, however, proposed a more specific determinism of single-factor explanations. **Possibilism** proposes that climate sets physical limits within which behavior may vary, such as modest winds permitting sailing but high winds restricting bicycle riding. **Probabilism** falls somewhat in between these two positions, implying that climate does not absolutely cause specific behaviors but does influence the chances that some behaviors will occur and others will not. For example, snow decreases the probability that people will drive and increases the probability that people will engage in winter sports. Determinism, possibilism, and probabilism are not necessarily mutually exclusive; they may be differentially applicable to different domains of behavior. For example, climate may influence what type of farming one practices by determining which crops cannot be grown in an area, but making it possible (but not inevitable) that other crops can be grown in the same area (cf. Gärling & Golledge, 1993). Similarly, high population density seems to be associated with high average snowfall, perhaps because the higher density increases the collective ability to cope with the disruptive effects of snow (Guterbock, 1990).

It is also appropriate to note that geographical and climatological determinism are closely linked. Indeed, most of the time the two are simply called "geographical determinism." It is often difficult to separate geographic influence from weather influence, since geography plays a major role in weather. Mountains, for example, are usually associated with high altitudes, cooler conditions than surrounding lowlands, wet weather on the side of the mountains facing prevailing atmospheric movements, and dry weather on the side away from oncoming storms. Since the geography and the climate are so closely linked, it is not easy to say whether the weather or the geography is primarily responsible for associated behavior (Figure 6-2).

EARLY BELIEFS ABOUT CLIMATE AND BEHAVIOR

Suspicion that climate determines behavior has been around practically since the beginning of civilization. Sommers and Moos (1976) provide a fascinating review of early thinking about climate effects. The ancient Greeks, including Hippocrates and Aristotle, believed that

weather and climate influenced bodily fluids, which in turn influenced individual disposition. The Roman Vitruvius and the Arab Ibn Khaldun, along with Aristotle, believed that geography and climate made some people more industrious than others, some more spirited, and so on. Not surprisingly, each writer indicated that the prevailing climate in his own region led to superior civilizations! For example, Khaldun believed that moderate climates fostered superior cultures. How, then, could his own civilization on the hot, dry, Arabian peninsula be at an advantage? His answer was that cooling waters of the sea moderated the Arabian climate sufficiently to produce overall favorability (Sommers & Moos, 1976).

LATER CLIMATOLOGICAL DETERMINISM

Sommers and Moos (1976) review the writings of numerous more recent authors, most of whom make equally presumptuous and self-serving observations about climatological influences (see also Glacken, 1967). Some later geographical determinists include such theorists as Carl Ritter, Frederic LePlay, Edmond Demolins, Henry Buckle, and Ellsworth Huntington. Three of these are particularly interesting for the details of their beliefs. Buckle, for one, was the son of a wealthy London merchant and was widely traveled. In *The History of Civilization in England* (1857–1861), he posited that labor conditions and climate were closely intertwined: Cold climates inhibited work, and hot climates led to lethargy; temperate climates, however, were thought to be invigorating. With fertile soil available, then, temperate climates would lead to heavy production. Moreover, Buckle believed that the advancement of a culture was tied to the creation of a leisure class, which was possible only if some other class produced more than was needed. Thus, temperate climates in regions with fertile soil permitted the necessary overproduction, which permitted the rise of the leisure class, which theoretically, at least, enabled the entire civilization's advancement. The "proof" of Buckle's theory came from his

Figure 6-2A & 2B Some examples of the connection between geography and climate. (A) The high altitude of the mountain range pulls enough moisture out of the atmosphere to provide extensive forests on the slopes. This moisture does not fall on the relatively dry plains, which require irrigation for any farming. A similar pattern accounts for desert areas of many parts of the globe, including the Rocky Mountain region. (B) Cities, with their concrete canyons and industrial and transportation pollution, create "heat islands" such that they are several degrees warmer than the surrounding countryside.

"observations" of conditions in such diverse regions as Central America, Ireland, Egypt, and India. So inviting was the theory that it was intellectually popular for some time after its writing (Timasheff, 1967).

It is also worth summarizing the beliefs of theorist Ellsworth Huntington (1915, 1945), who held that climatic factors other than temperate conditions were necessary for the growth of major civilizations. The major ingredient was hypothesized to be seasonal change and moving storms. The change could not be too severe, but regular changes should require adaptation, and as "necessity is the mother of invention," the adaptations encouraged creative solutions, which invigorated the civilization. Huntington did indeed collect sociological data on such things as productivity, suicides, and library circulation to support his point of view that geographic bands producing these changing climate conditions were associated with advancing civilizations, although he also believed that many factors besides favorable climate influence the growth of civilizations. Similar to Huntington's ideas about climate adaptation, Markham (1947) suggested that the most important climatic factor for the development of a civilization was living in a cool enough region that technology became necessary in order to keep warm. He noted, for example, that the Romans developed a central heating system, using pipes to distribute warmth through buildings. How important is this factor? Markham noted that this heating system deteriorated shortly before the decline of Roman civilization.

While appealing in some respects, these geographical and climatological deterministic beliefs have only tenuous empirical evidence, at best, to support them. Climatological experiments on a culture are not practical, so we are left with correlational data—and there are many factors, such as war, natural resources, and technological innovation, which are difficult to measure and assess as possible explanatory variables in climatological studies. In this light, few scientists today would endorse very strong statements of climatological determinism. It is overstepping the cause-and-effect boundaries of methodology to assume that just because a civilization occurs in a given climate, the climate is a prerequisite for that civilization.

On the other hand, long-term climatological change can have substantial influences—on a civilization and even on evolution. About 55 million years ago, a dramatic global warming brought crocodiles to northern Canada; it also led to the appearance of modern mammals such as rodents, primates, and split-hoofed animals. Then, about 34 million years ago, a dramatic global cooling led to the extinction of about 60% of European mammals (Kerr, 1992b). In terms of human civilization, Henry (1994), for example, describes how climatological change was associated with shifting elevations of winter and summer camps for prehistoric humans over tens of thousands of years. Similarly, Issar (1995) describes how growing evidence indicates that ancient Mesopotamian culture collapsed around 2500 B.C. due to a warmer and drier climate that impaired agriculture by increasing salinity of soils. Drier climate change and associated resource depletion may also have led to incessant war and population decline among the Anasazi—the ancestors of Navajo, Hopi, and other tribes—in the southwestern United States about 800 years ago (e.g., LeBlanc, 1999). For 300 years before that, however, climate was favorable, resources plentiful, and warfare relatively scarce. Fischman (1996) describes research indicating that the Chumash tribe in California lived in a very egalitarian society until about A.D. 1100, when a drought led to violence and a culture that strongly differentiated among individual social status. Pringle (1997) reviews research showing that a Norse colony of 5,000 to 6,500 people had thrived for 500 years in Greenland, yet was completely wiped out in the A.D. 1300s by an increasingly cold climate (the "Little Ice Age") that destroyed agriculture. Finally, in a fascinating contemporary view of the role of geographic determinism, Diamond (1998) suggests that geography and associated rainfall patterns have repeatedly made a difference in the rise and fall of human civilizations.

CURRENT VIEWS AND DISTINCTIONS

Along with our caveat on overstating causal relationships, several additional methodological cautions are warranted. Recall that weather refers to short-term variations and climate to average weather over a longer period. Thus, on a given day, a city in the northern United States might have a higher temperature than a city in the southern part of the country, but on the average, temperatures in the south are higher. Now consider that more violent crimes occur in the south than elsewhere. Can we conclude that heat has a causal role in these crimes? The problem is that many variables, such as food preferences and ethnic mix, differ between the two regions, as well as a cultural tradition of defending one's honor with violence (e.g., Anderson & Anderson, 1996; Cohen, 1996; Cohen et al., 1996; Rotton, Barry, & Kimble, 1985). Contemplating correlations among so many climatological and sociodemographic variables, we must be very cautious about conclusions. In addition, there are seasonal differences in behavior (e.g., automobile buying, television programs, gift purchases) that probably have nothing to do with weather—although the weather certainly varies with these activities. It is the case that more assaults and homicides occur in the United States in summer than in winter, but the peak time for homicides is December. Thus, we should not conclude that meteorological variables are responsible for seasonal differences in behavior.

It is also worth noting that biometeorologists in Europe (e.g., Muecher & Ungeheuer, 1961; Tromp, 1980) have focused on the possible effects of "weather phases," or correlated patterns of changes in meteorological conditions (e.g., a storm front). North American researchers, on the other hand, look more at specific variables, such as temperature, precipitation, or barometric pressure. We have organized the remainder of this chapter primarily around the latter approach for ease of presentation, but we must keep in mind that weather variables such as wind, humidity, and barometric pressure are themselves correlated.

BIOLOGICAL ADAPTATIONS TO CLIMATE

An area of study with more convincing evidence of climatological influence is that of measuring physiological or other biological factors within a culture that is exposed to extremes of climate. For example, Frisancho (1993) reviews many lines of evidence that people living at high altitudes, such as in Tibet or Peru, in the hot climates of Africa, or in the cold environment of Lapland, may have developed special physiological capabilities for coping with these extremes. Such adaptations may even be genetic, such that ancestors with these adaptive characteristics were more likely to survive the extremes and pass their hereditary characteristics along to the next generation. For example, hearts may be larger and their walls thicker among high-altitude cultures, since hearts need to circulate more oxygen-rich blood at altitudes where oxygen is in relatively lower atmospheric concentration. Some such adaptations may well be acquired (i.e., accruing during one's lifetime) rather than genetic.

Heat and Behavior

Ambient temperature is a term used to describe the surrounding or atmospheric temperature conditions. In the natural environment, humans experience a range from arctic cold to debilitating tropical heat. As stated previously, temperature is one factor in the physical environment that humans are changing through urbanization and industrialization. Waste heat from air conditioners, industrial processes, and burning of fuel tends to accumulate in urban areas; concrete buildings and parking lots absorb more solar heat than vegetation does; and tall buildings block wind patterns that would otherwise disperse this heat, so that urban centers are typically 10°F to 20°F (6°C to 12°C) hotter than surrounding agricultural areas—the so-called heat island effect. As will be seen in this section and the next, extremes of heat and cold, regardless of the source, can have dramatic effects on people.

PERCEPTION OF AND PHYSIOLOGICAL REACTIONS TO HEAT

Perception of temperature involves physical as well as psychological components. The primary physical component is simply the amount of heat in the surrounding environment, typically measured on the Fahrenheit or Celsius scale. One psychological component of temperature perception is centered on the internal temperature of the body, known as **core temperature** or **deep body temperature.** Another psychological component involves receptors in the skin called **thermoreceptors.** Although some receptors seem to be sensitive to lower temperatures and others to higher temperatures, both types respond to change in temperature more than to absolute temperature (e.g., Stevens, 1991). This is why you may perceive even mildly warm water as very hot when your hands are extremely cold from being exposed to winter air.

Since perception of ambient temperature is largely dependent on differences between body and ambient temperatures, the mechanisms controlling body temperature have much to do with the perception of ambient temperature. Body temperature is regulated by the need to keep core temperature close to 98.6°F (37°C). Since death occurs when core temperature rises above 113°F (45°C) or drops below 77°F (25°C), maintaining it at a normal level is mandatory for survival. Without a defensive or adaptive mechanism, the body would overheat when exposed to high ambient temperatures and would "freeze" when exposed to cold ambient temperatures. Fortunately, a number of such adaptive mechanisms, under the general control of a brain center known as the **hypothalamus,** are available for use whenever core temperature is threatened by adversely hot or cold ambient conditions. When core temperature becomes too hot, the body responds by activating mechanisms designed to lose heat, such as sweating, panting, and **peripheral vasodilation.** The latter process refers to dilation of blood vessels in the extremities, especially those near the surface of the skin, which allows more blood to flow from core areas to surface regions. This blood carries with it the excess core heat, which is removed through air convection or sweating (note that peripheral vasodilation allows more sweat to reach the surface of the skin). In "heat wave" emergencies, the body may increase the supply of water available for evaporation by suppressing urine formation and extracting water from body tissues. Such dehydration causes us to become thirsty and to replenish our body's supply of water, which is another process mediated by the hypothalamus. When these adaptive mechanisms fail, a number of physiological disorders can result, including heat exhaustion, heatstroke, and heart attack (see the box on page 180). Interestingly, blood pressure may increase upon initial sensation of ambient heat, owing to a "startle" response or alarm reaction (see page 120). Once vasodilation begins, blood pressure drops. With heat stroke, blood pressure may rise again, then fall off as coma and death approach. Clearly, measuring blood pressure only once during heat exposure is not a good indication of the overall picture.

Prolonged exposure to moderately high ambient temperatures need not have disastrous consequences. Individuals who move from cool climates into very warm climates can adapt to the hot environment without too much difficulty. This adaptation process is known as **acclimatization,** and it primarily involves changes in physiological adaptive mechanisms. For instance, the body may "learn" to start sweating much sooner after the onset of high ambient temperatures (Lee, 1964). How long does it take to acclimatize when moving from a warm to a cold environment, and vice versa? Tromp (1980)

PHYSIOLOGICAL DISORDERS ASSOCIATED WITH PROLONGED HEAT STRESS

When the body's adaptive mechanisms to heat stress fail to keep core body temperature close to 98.6°F (37°C), a number of physiological disorders can occur. Among the more common are:

1. **Heat exhaustion,** characterized by faintness and nausea, vomiting, headache, and restlessness. This disorder results from excessive demands on the circulatory system for blood. Water needed for sweating, blood needed near the skin surface for heat loss through convection, and blood needed for normal or increased metabolic functioning place too much strain on the body's capacity to supply blood. Continued loss of salt and water through sweating compounds the problem. Replacement of lost water and salt, together with rest, will both prevent and cure heat exhaustion.

2. **Heatstroke,** characterized by confusion, staggering, headache, delirium, coma, and death. This disorder results from the complete breakdown of the sweating mechanism. Because body heat cannot be lost, the brain overheats. Survival or prevention of brain damage depends on quick action—the most effective being immersion in ice water. When a victim collapses from heat, the continuation of sweating implies heat exhaustion; the absence of sweating implies heatstroke.

3. Heart attack, resulting from excessive demands on the cardiovascular system due to increased need for blood by the body's cooling mechanisms. During urban heat wave conditions, most deaths beyond what would normally be expected are caused by heart attacks.

For more information on heat and cold disorders, see Kalkstein and Davis (1989).

suggests that the answer is no more than 3 to 14 days, depending on an individual's cardiovascular fitness.

Sometimes the distinction is made between acclimatization, meaning adaptation to multiple stresses in an environment (e.g., temperature, wind, humidity) and **acclimation,** meaning adaptation to one specific stressor in an experimental context (see Frisancho, 1993). In our discussion, we will use the term *acclimatization,* since in most environments we must adapt to more than one element. Frisancho (1993) indicates that acclimatization may occur through developmental changes, through genetic adaptation, or through physiological and behavioral changes following prolonged exposure to heat. The Saharan Touareg, for example, have tall, slender bodies that maximize surface cooling area in proportion to the amount of body tissue that produces heat. Behaviorally, the Touareg avoid heavy exercise during the highest temperatures of the day and wear loose, porous clothing (Beighton, 1971; Frisancho, 1993; Sloan,

1979). Keep in mind, though, that nonnative visitors to hot regions can usually acclimatize in a few days. Leithead and Lind (1964) suggest that maximum efficiency in acclimatization occurs with exposure of 100 minutes per day (Figure 6-3).

COMPLICATING FACTORS

Since perception of ambient temperature depends to some extent on the functioning of the body's thermoregulatory adaptive mechanisms, any environmental factor that interferes with these mechanisms will influence perception of ambient temperature. The primary environmental factors in this regard are humidity and wind. The higher the humidity in a hot environment (i.e., the more saturated the air with water vapor), the lower the capacity of the air to absorb water vapor from sweat. This is the reason, for example, that conditions of 100°F (38°C) and 60% humidity are perceived as more uncomfortable than those of 100°F and 15% humidity.

Figure 6-3 Acclimatization may occur through genetic changes, developmental changes, physiological changes, or behavioral changes. This desert is normally very hot and dry, but can be cool and wet at times. What would you do to acclimatize to this environment?

Thus, perception of ambient temperature is not a function of temperature alone. Psychologically, the problem of perceptual measurement can be partially solved by taking into account a comfort level that is influenced by both temperature and humidity, thus creating a new ambient environment index. One such index is known as **effective temperature.** A chart showing some effective temperatures is presented in Table 6-1. Other similar indexes exist, such as the **Temperature–Humidity Index,** or **THI** (see Tromp, 1980 for a summary).

Table 6-1	EFFECTIVE TEMPERATURE (°F) AT 0% HUMIDITY AS A FUNCTION OF ACTUAL TEMPERATURE AND HUMIDITY					
Relative Humidity (%)	**Thermometer Reading (°F)**					
	41°	**50°**	**59°**	**68°**	**77°**	**86°**
	Effective Temperature					
00	41	50	59	68	77	86
20	41	50	60	70	81	91
40	40	51	61	72	83	96
60	40	51	62	73	86	102
80	39	52	63	75	90	111
100	39	52	64	79	96	120

Since the amount of air flowing over the skin determines how much sweat is evaporated as well as how much body heat is carried off by convection, wind speed must also be taken into account in perceiving ambient temperature. On a hot day, a breeze helps carry heat off the body and thus has a cooling effect. As we will see shortly, wind on a cold day further chills the skin and thus amplifies that temperature effect, as well.

HEAT AND PERFORMANCE

Laboratory Settings

Laboratory studies of the influence of high ambient temperatures on performance have examined such varied behaviors as reaction time, tracking, and vigilance, as well as memory and mathematical calculations (see Bell & Greene, 1982; Hygge, 1992; Poulton, 1970; Sundstrom, 1986 for reviews). In general, temperatures above 90°F (32°C) will impair mental performance after two hours of exposure for unacclimatized individuals. Above this same temperature, moderate physical work will suffer after one hour of exposure. As temperatures increase, shorter exposure times are necessary to show performance decrements. Interestingly enough, some researchers find that heat has no influence on performance, others find that heat is detrimental to performance, and still others find that heat improves performance. Moreover, some studies suggest that as temperatures rise, performance first improves and then deteriorates, whereas other studies show this pattern for one task but the reverse pattern (i.e., initial decrements followed by improvements) for other tasks. Hancock (1986) notes that performance on vigilance tasks is impaired when thermal homeostasis is disturbed but improved when a new equilibrium state is reached. In general, heat impairs complex mental tasks after prolonged exposure, impairs motor tasks after fairly brief exposure, and may impair or enhance vigilance. Before examining possible explanations for these complex findings, let us first examine heat research from applied settings.

Industrial Settings

Industrialists, such as steel manufacturers, are naturally concerned about the effects of blast furnaces and other hot industrial environments on workers who are in these surroundings for eight or more hours a day. Generally, exposure to such industrial heat can cause dehydration, loss of salt, and muscle fatigue, which taken together can reduce endurance and hence impair performance. In order to overcome or avoid such problems, care is generally taken to ensure that workers have an adequate intake of water and salt, are not exposed to intolerably hot conditions for long periods of time, wear protective clothing, and, when new on the job, have adequate time to adapt to working conditions (see Fraser, 1989; Poulton, 1979; Sundstrom, 1986). When feasible, adequate air conditioning and heating systems seem to minimize physiological effects of temperature in work environments (Kristal-Boneh et al., 1997).

Classroom Settings

Temperature appears to have some effects on classroom performance. Pepler (1972) studied climate-controlled (air-conditioned) and nonclimate-controlled schools near Portland, Oregon. In nonclimate-controlled schools, academic performance showed more variance (i.e., wider distribution of test scores) as temperatures rose. However, at climate-controlled schools, such variability did not occur on the warmest days. Apparently, some students suffer more than others when heat waves hit the classroom! Support for this finding has been reported by Benson and Zieman (1981), who found that heat hurt the classroom performance of some children but actually helped the performance of others (see also Griffiths, 1975; Figure 6-4).

Military Settings

If ambient heat has any deleterious effect on performance, the consequences of moving unacclimatized troops into a tropical area (e.g., from North America or Europe to the Persian

Figure 6-4   Research suggests that weather variables, including heat and barometric pressure, may influence disruptive behavior and academic performance of children in the classroom. Interestingly, for some children heat has beneficial effects, whereas for others it has detrimental effects. In this classroom, there is no air conditioning and the only window cannot be opened. What adaptive responses would you expect when the weather gets hot?

Gulf) could be disastrous. Adam (1967) reviewed a number of British military studies that generally found that 20% to 25% of troops flown into tropical regions from more moderate climates suffered serious deterioration in combat effectiveness within three days and became in effect "heat casualties." Solutions to this problem include allowing several days for acclimatization or expanding the number of troops available to allow for heat casualties.

Interpreting the Data

How can we account for the complexity of the above research findings? Why does heat sometimes hurt performance and sometimes help it? Sundstrom (1986) notes that body temperature, metabolic cost of physical activity, acclimatization, skill level, motivation, and stress (including threat appraisal) are all factors that make a difference in the impact of heat on performance. Bell and Greene (1982) offer several other suggestions, which require an integration of several theoretical perspectives presented in Chapter 4

(see also Figure 6-1). First, arousal explains some heat effects. Initially, exposure to heat may cause a brief "startle" response that heightens arousal and hence improves performance. Heat may eventually lead to overarousal, causing performance decrements as would be predicted by the Yerkes–Dodson law (see Chapter 4). Eventually, high temperatures would result in physical exhaustion (see the box on page 180) as the body can no longer keep core temperature at a safely functioning level, so performance would completely deteriorate. A second mediator of performance, then, is core temperature. A third mediator of performance is attention, as examined in the overload interpretation of environmental stress. As heat stress increases, attention is narrowed toward stimuli central to the task at hand, so that performance on noncentral activities deteriorates. Bell (1978), for example, found that as heat increased, performance on a secondary task suffered, but performance on a primary task did not. A fourth mediator of heat effects is probably perceived control, as advocated by the behavior constraint model.

According to this interpretation, as heat stress increases, individuals feel less and less in control of the environment, and thus performance deteriorates. Greene and Bell (1980), for example, found that participants in a 95°F (35°C) environment felt more dominated by it than did those in more comfortable temperatures. Finally, each individual almost certainly has an adaptation level or maximum level of tolerance for heat (see also Greene & Bell, 1986). In sum, arousal, core temperature, attention, perceived control, and adaptation level all probably operate as explanatory mechanisms in understanding the effects of heat stress on task performance.

HEAT AND SOCIAL BEHAVIOR

Heat and Attraction

Most individuals exposed to high ambient temperatures will report subjectively that they feel uncomfortable and perhaps irritable. Ruback and Pandey (1992), for example, found that rickshaw passengers in India reported more negative feelings as the temperature became uncomfortably hot; interestingly, they found that telling people about the effects of heat gave them a greater sense of perceived control. We might expect that unpleasant feelings associated with heat or other factors would also give us an unpleasant disposition toward others. According to one model of attraction (Byrne, 1971), we should expect a decrease in interpersonal attraction when we are experiencing the unpleasant effects of either debilitating heat or cold. Griffitt (1970) demonstrated precisely this effect by asking participants to evaluate anonymous strangers who seemed to agree with them on either 25% or 75% of a set of attitudes. Participants performed this evaluation task under an effective temperature of either 67.5°F (20°C) or 90.6°F (32°C). The results indicated that high ambient temperatures decreased attraction, regardless of the degree of attitude similarity. Griffitt and Veitch (1971) reported comparable results.

However, research by Bell and Baron (1974, 1976) suggests that heat may have a relatively minor influence on attraction under other circumstances. In two experiments, these researchers found that heat did not influence attraction toward another person in the room if that person had recently complimented or insulted the individual. In this situation, the compliment or insult appears to be so overwhelming as to "wipe out" any possible influence of heat (see also Bell, Garnand, & Heath, 1984). Rotton (1983) notes that in the Griffitt studies, participants were rating hypothetical strangers who were not actually present, whereas in the Bell and Baron studies, participants rated a real stranger who was actually present in the same room. Perhaps when the stress is not shared with someone actually present, attraction decreases; but when someone is there to share the distress, the decrease may not occur (Figure 6-5).

Heat and Aggression

During the urban and campus riots of the 1960s, a popular belief arose that riotous acts of violence were in some way precipitated by the unrelenting heat of the summer months. This supposed influence of heat on aggression was popularly known as "the long, hot summer effect." It became common for television commentators and newspaper editorial writers to mention fears that "It's going to be another long, hot summer!" High ambient temperatures became even more suspect when the United States Riot Commission (1968) noted that, of the riots in 1967 on which records were available, all but one began on days when the temperature was at least in the 80s (above 27°C). A more formal study by Goranson and King (1970) strongly suggested, as evidenced in the graph in Figure 6-6, that heat-wave or near-heat-wave conditions were associated with the outbreak of the riots. More recent research shows that calls to the police do indeed increase as temperatures rise (e.g., Cohn, 1996), and violent crimes, including sex crimes, seem to increase with temperature (Cohn, 1993; Perry & Simpson, 1987; Rotton, 1993a). Even major league baseball seems susceptible to a milder form of the heat-aggression phenomenon: More batters are hit by "wild"

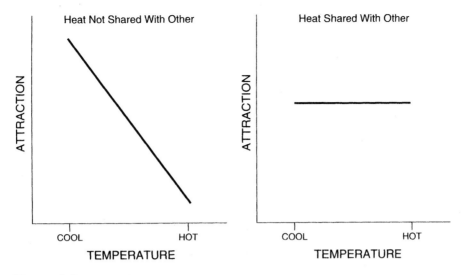

Figure 6-5 As in the hypothetical graph on the left, heat may decrease attraction when people share the discomfort of the heat. However, as in the diagram on the right, heat may not affect attraction when the two individuals do not share the discomfort.

pitches as the temperature increases (Reifman, Larrick, & Fein, 1991)!

Actually, systematic study of temperature and violence goes back a century or so. Anderson (1989) notes three types of studies that have examined this relationship over the years. *Geographic region studies* compare crime rates across different regions of a country or continent. For example, you might divide your native

Figure 6-6 Average daily mean temperatures before, during, and after riot outbreak.

From Goranson & King, 1970. Reproduced with permission.

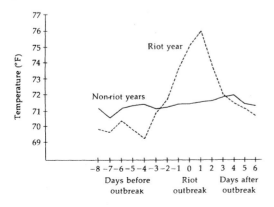

country into warm, temperate (intermediate), and cool thirds and then see whether violence is most prevalent in the hottest region and least common in the coolest region. Indeed, studies often find just such results, although sometimes the outcome is equivocal, especially when you take into account other variables such as socioeconomic status that may vary by region (e.g., DeFronzo, 1984; Rotton, 1986). *Time-period studies* make it somewhat easier to control for these extraneous variables. In a time-period study you measure violence across intervals of time such as days or months. You then see whether quantities of crimes in each day or month rise or fall with the average (or high or low) temperature for each time interval. Again, these studies often show that violence increases with temperature (e.g., Anderson, 1987; Anderson & Anderson, 1984; Cotton, 1986; Harries & Stadler, 1988; Rotton & Frey, 1985). However, in time-period studies we do not know the temperature at the time a given act of violence occurred; two violent crimes on the same day may have occurred at different times when the temperature was very cool or very warm, and we would not know that temperature had anything to do with either crime. A more accurate strategy would be to employ a *concomitant temperature study,* in which the temperature is known at the actual time of the violence. Once again, these concomitant studies often find that aggression goes up with heat. For example, Baron (1976) found that automobile drivers honked their horns (which can serve as a measure of irritation or hostility) more when temperatures were above 85°F (29°C) than when they were below. For drivers in air-conditioned cars, however, heat did not increase horn honking.

One series of laboratory concomitant studies has found a rather unexpected pattern to the temperature–aggression relationship. For example, Baron and Bell (1975) arranged for laboratory participants to be either provoked or complimented by a confederate before being given an opportunity to aggress against this individual by means of ostensible electric shock. It was found, as might be anticipated, that participants in comfortable ambient temperature conditions (73°F to 74°F; 23°C) were more aggressive toward an anger-provoking confederate than toward a complimentary confederate. However, those in uncomfortably hot conditions (92°F to 95°F; 35°C) showed just the opposite behavior: These individuals showed reduced ag-

gression toward the insulting confederate but increased their level of attack against the friendlier one. Several other studies by the same research team found the same pattern: Provoked individuals in a hot lab showed a relatively low level of violence (e.g., Bell & Baron, 1976). How can we explain this unexpected pattern?

There are many potential explanations for this and other patterns observed in the data (see Anderson, 1989). The explanation that has provoked the most attention and the most debate is what is now termed the **negative affect-escape model.** According to this explanation, negative affective feelings are the mediator in the relationship between heat and aggression. This relationship takes the curvilinear form of an inverted U, as shown in Figure 6-7. As shown on the x-axis, heat and other unpleasant events (e.g., insults) generate more and more negative affect. Up to a critical point, negative affect increases aggressive behavior, but beyond this point, stronger negative feelings actually reduce aggression, since flight behavior or other attempts to escape discomfort become more important to the individual than does aggression. Laboratory tests of the proposition of a **curvilinear relationship** between negative affect

Figure 6-7   The negative affect-escape model predicts that up to a point, the discomfort of heat increases aggression; however, extreme discomfort tends to decrease aggression because people would rather escape the situation than fight.

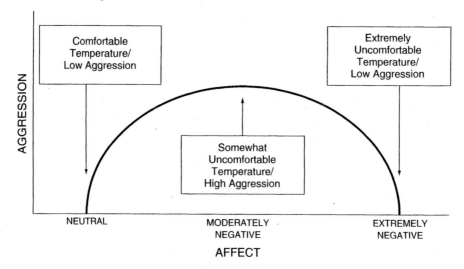

and aggression, using heat as one factor influencing affect, have been generally affirmative. What these findings suggest with regard to the relationship between high ambient temperatures and aggression is that there is a critical range of uncomfortably high ambient temperatures at which aggression may well be facilitated. On the other hand, extremely high ambient temperatures, especially when combined with other sources of irritation or discomfort, may become so debilitating that aggression is no longer facilitated and may well be reduced when individuals prefer to concentrate on escaping the heat.

Other research supports various aspects of the model. For example, the escape tendency is illustrated by Rotton, Shats, and Standers's (1990) finding that on hot days Miami pedestrians walked faster to get to their air-conditioned cars than they did to shop on the streets (see also Palamarek & Rule, 1979). Similarly, Kenrick and MacFarlane (1986) studied horn honking in both moderate and very hot temperatures (well above 100°F, or 38°C) in Phoenix, Arizona. Honking increased with temperature, and this effect was strongest for those with their automobile windows rolled down. Such results should not be surprising, since the honking was always at a car that did not move when a traffic light turned from red to green. Thus, honking was perceived as instrumental in obtaining relief from the discomfort of the heat, especially for drivers with their windows down (and presumably with no air conditioning): If the stalled car would only move, drivers could get some relief from the heat, and honking might be a way of prompting the driver of the stalled car to get on with it. In addition, some research on unpleasant odors (Asmus & Bell, 1999; Rotton et al., 1979) and on cold temperatures (Bell & Baron, 1977) also supports the model. It is also clear that high ambient temperatures generate negative affect (Anderson, Deuser, & DeNeve, 1995).

However, the failure of most geographic region and time-period studies to show any inverted-U (curvilinear) relationship between temperature and aggression has sparked a heated debate about whether extremely uncomfortable temperatures actually do decrease aggressive tendencies (Anderson, 1989; Anderson & Anderson, 1998; Anderson, Bushman, & Groom, 1997; Anderson & DeNeve, 1992; Bell, 1992; Rotton & Cohn, 1999, 2000). It may be that the lab studies showing the inverted-U do so because of a methodological artifact; for example, participants may attribute extreme discomfort to factors other than anger toward the potential victim, which in turn could lead them to decrease their aggression (see also Anderson, 1989; box on page 115). However, since carefully controlled laboratory experiments usually (but not necessarily) have more internal validity than do correlational studies, perhaps the explanation lies elsewhere. Bell (1992) and Anderson (1989), for example, note that geographic region studies, in averaging the temperature across long periods of time, probably mask the effects of very high temperatures. That is, the really high temperatures are averaged in with more moderate temperatures, and high rates of crime are averaged in with lower rates of crime, such that the averages do not show the extremes necessary to detect a potential decline in violence at very high temperatures.

A similar problem occurs with time-period studies (e.g., Rotton & Frey, 1985). The time interval studied may simply not have enough very hot days to detect a decline in violence associated with escape motives. Other variables may also be operating in time-period studies that are controlled in laboratory studies. However, controlling for humidity, wind, air pollution, and other atmospheric variables in a time-period study still results in a **linear relationship** between temperature and violence (Rotton & Frey, 1985).

Perhaps other explanations for the inconsistencies are needed. Bell and Fusco (1986, 1989), for example, suggest that on very hot days, the spread of violence widens: Some days show high violence and some low violence, whereas at cooler temperatures the spread from day to day is not as great. In essence, both linear and curvilinear trends may be in the data, but the curvilinear trend is masked by the increasing variation in violence as temperatures rise

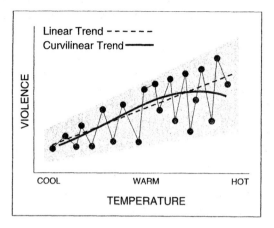

Figure 6-8 Laboratory studies tend to confirm the curvilinear relationship between heat and aggression posited by the negative affect-escape model. However, archival studies of violent crimes often show a linear increase in crime as temperatures increase. As the shaded cone shows, the variance or difference in violent crime from day to day tends to widen as the temperature increases. Bell and Fusco (1986, 1989) suggest there is room for both the curvilinear and linear outcomes within this cone, depending on such factors as alcohol consumption and the availability of escape from the heat. Rotton and Cohn (1999, 2000) report analyses that confirm this possibility.

(see Figure 6-8). Thus, escape and aggressive tendencies may both be operating under high temperatures, but other variables such as alcohol consumption, availability of escape, or attributions about the source of discomfort may determine whether escape or aggression become manifest. Depending on which of the other variables are operating, extremely high temperatures could be associated with either increased or decreased aggression. Moreover, the many types of studies conducted so far use very different measures of violence. The influence of temperature on willingness to deliver electric shock in the laboratory may be somewhat different from the influence of temperature on willingness to commit murder or rape; field studies sometimes find a temperature relationship with one type of crime but not another (e.g., Anderson, 1987; Perry & Simpson, 1987).

Recently, at least two time-period studies have found a curvilinear relationship between heat and violence, but only at certain times of day (Cohn & Rotton, 1997; Rotton & Cohn, 2000). These researchers interpret their results in terms of *routine activity (RA) theory*, which posits that people are more likely to become crime victims when they leave the safety of their homes and other familiar places, which is most likely to occur during warm periods. In addition, van de Vliert et al. (1999) found the inverted-U relationship between temperature and political violence, replicating an effect observed earlier by Schwartz (1968).

It is clear from all this research that there is indeed a relationship between heat and aggression. Much of the time, heat increases aggression and violence, yet under some circumstances heat appears to be associated with escape motives and a decline in violence. Specifying exactly what these circumstances are will require additional research.

Heat and Helping Behavior

A third type of social behavior that may be affected by high ambient temperatures is that of offering assistance to someone in need of help. Believing that weather conditions are pleasant and sunny has been shown to increase the amount of tipping for hotel room service (Rind, 1996). When people feel unpleasant, however, they are not inclined to help others unless they have reason to believe that helping will improve their own condition (e.g., Cialdini & Kenrick, 1976; Weyant, 1978). Since heat obviously produces discomfort, what might it do to helping behavior?

One study found that after leaving an uncomfortably hot experimental room, participants were less likely to volunteer their assistance in another experiment than those who had been in a more comfortable environment (Page, 1978). Another study (Cunningham, 1979) similarly found that when participants were asked to help in an interview, willingness to help declined as temperatures rose in summer months, but willingness to help increased when temperatures rose during winter months. Other research, however, has failed to find a relationship between heat and helping. For example, outdoor temperature was found to have no effect on the amount of tips left at an indoor restaurant (Cunningham, 1979). Moreover, neither high nor low temperatures reduced helping when a person: (1) using crutches dropped a book; (2) lost a contact lens; (3) dropped a sack of groceries; or (4) asked for help in a survey (Schneider, Lesko, & Garrett, 1980). Data from Bell and Doyle (1983) also failed to show that heat had any effect on helping, either during or after exposure to high temperatures. With so little data available, it is difficult (and unwise) to draw firm conclusions about the relationship of heat to helping behavior. It is possible, though, that discomfort increases helping in some cases and decreases it in others, so these two tendencies may "cancel out" each other in many instances (e.g., Cunningham, Steinberg, & Grev, 1980). In addition, many factors, including the expressions on our face, may be related to signals that the brain interprets in terms of temperature fluctuations. Thus, feeling good or bad from exposure to temperature changes—and resulting behavior patterns—may be mediated by a much more complex set of events than a simple model would imply (Zajonc, Murphy, & Inglehart, 1989).

Cold Temperatures and Behavior

PERCEPTION OF AND PHYSIOLOGICAL REACTIONS TO COLD

The physiological reaction to cold ambient temperatures (i.e., below 68°F, or 20°C) is in many ways the opposite of the reaction to heat. In contrast to overheating, when core temperature becomes too cold (as detected in part by the hypothalamus), the body reacts by activating mechanisms that generate and retain heat, resulting in increased metabolism, shivering, peripheral vasoconstriction, and piloerection. **Peripheral vasoconstriction** serves just the opposite function of peripheral vasodilation: It keeps core heat inside the body and away from the surface where it is easily lost through convection. This constriction process also makes more blood available to internal organs, which are generating more heat through increased metabolism. **Piloerection** refers to the stiffening of hairs on the skin, usually accompanied by "goose bumps." This skin reaction increases the thickness of a thin layer of insulating air close to the skin, which again helps to minimize heat loss by convection. As with perception of heat, humidity and wind influence perception of cold.

High humidity speeds heat loss, as does wind, so both amplify the perception of cold. The **chill factor** or **windchill index** shows how much colder wind makes an already cold temperature feel. For example, an ambient temperature of 23°F with a wind speed of 15 mph has the same psychological effect as an ambient temperature of −1°F with no wind. Table 6-2 depicts the broad range of the chill factor index. In very cold temperatures, it becomes extremely important for thermoregulatory survival to take

Table 6-2	WINDCHILL INDEX°		
Actual Temperature (°F) at 0 mph	**Wind Speed**		
	5 mph	**15 mph**	**25 mph**	**35 mph**
	Equivalent Temperature (°F)			
32	29	13	3	−1
23	20	−1	−10	−15
14	10	−13	−24	−29
5	1	−25	−38	−43
−4	−9	−37	−50	−52

° *Equivalent temperatures (°F) at 0 mph as a function of actual temperature and wind speed.*

PHYSIOLOGICAL DISORDERS ASSOCIATED WITH PROLONGED COLD STRESS

If cold exposure persists for long periods of time, two serious consequences can result. One danger is **frostbite**, characterized by the formation of ice crystals in the skin cells. Since the initial reaction of the body to cold stress is constriction of surface blood vessels, freezing of the skin is not uncommon. When the ice crystals tear the cell structures, cell death results and infection can occur, leading to the necessity of amputation.

Another danger of cold exposure arises when the adaptive mechanisms fail to maintain core body temperature. A decline in core temperature is known as **hypothermia.** In the initial stages of hypothermia, cardiovascular activity, including heart rate and blood pressure, is dramatically increased. As core temperature falls between 86°F and 77°F (30°C to 25°C), cardiovascular activity falls off and becomes irregular. Below a core temperature of 77°F, death due to heart attack is likely to result. At an intermediate stage of hypothermia, clouding of consciousness and coma may well occur. If the victim has not found shelter by this time, the loss of mental functioning may preclude an effort to seek warmth or assistance. Since inadequate clothing in extremes of cold is most likely to precipitate hypothermia, it is those individuals caught unprepared for cold stress, such as mountain climbers faced with sudden cold winds or shipwreck victims in arctic waters, who are most likely to suffer the disorder. Removal of wet clothing and provision of warmth are necessary to save the lives of hypothermia victims.

windchill into account. Just how critical this factor can be is illustrated by the fact that exposed human skin will freeze in less than one minute at −40°F with a 6 mph (10 km/h) wind, at −20°F with a 20 mph (32 km/h) wind, and at 0°F with a 30 mph (48 km/h) wind!

Acclimatization to cold environments may take several forms (see also Bell & Greene, 1982). For example, the Alacaluf Indians of Tierra del Fuego have an elevated metabolism that seems to keep body temperature elevated in the cold environment (Hammel et al., 1960, as cited in LeBlanc, 1975). Bushmen of the Kalahari Desert and Australian aborigines have another adaptive mechanism for tolerating very low nighttime temperatures. In these populations, shivering does not occur as it would in unacclimatized individuals, but rather core temperature actually drops at night (LeBlanc, 1975). Moreover, LeBlanc (1956) reported reduced shivering in a group of Canadian soldiers who had been moved to a cold climate, and Budd (1973) found a similar pattern for Australians of European heritage on an Antarctic expedition. Exposure to cold increases circulation in the hands for Eskimos (LeBlanc, 1975)

and for fishermen on the Gaspé Peninsula of Quebec (LeBlanc, 1962). Thus, several mechanisms are available for acclimatization to cold environments.

At least two other factors besides acclimatization modify the effects of cold temperatures. First, humans rarely have to work or interact unprotected in cold climates. We usually wear protective clothing in uncomfortably cold situations. If we do not, disease or death is not unlikely. Because of the clothing factor, performance outcomes in studies of cold environments are somewhat difficult to interpret. If workers on the Alaskan pipeline wore heavy clothing, their efficiency was not dramatically affected. This does not, however, mean that cold temperatures do not affect performance. A second factor complicating the relationship between cold and behavior is that some parts of the body may be cold, while others are not. Whether only the hands are cold, or whether the core temperature is lowered, may make a difference. In addition, in the section on heat we noted that increases of only 10°F to 15°F (6°C–9°C) above comfort levels (i.e., above about 70°F, or 21°C) often affect performance and other behavior. Comparable temperatures below comfort levels (i.e., 55°F–65°F, or 13°C–18°C) are rarely studied. Instead, research on cold tends to concentrate on behavioral effects of temperatures below 55°F (13°C). With these facts in mind, let us now examine some of the research on cold temperatures and behavior.

COLD TEMPERATURES AND HEALTH

We note in the box on page 190 that prolonged exposure to cold can lead to hypothermia and frostbite. Do people living for long periods of time in cold climates experience health effects due to the cold temperatures? The answer appears to be "probably not directly." Eskimos in North America and Lapps in northern Scandinavia do not seem to suffer prolonged disorders associated with cold temperatures. Any noticeable differences from other societies are probably the result of culture. If adequate clothing and shelter are available, cold temperatures are not all that hazardous to health. (For a review of design considerations for adequate shelter in cold climates, see Matus, 1988.) Mental health also appears not to be directly related to cold temperatures. A study on health at Antarctic stations (Gunderson, 1968) found that although residents experience insomnia, anxiety, depression, and irritability, these effects appear more attributable to isolation and work requirements than to climate. To the extent that climate is a factor, concern about it and perceived threat from it are probably more important than temperature itself (see also Suedfeld, 1998).

COLD EXTREMES AND PERFORMANCE

Humans on arctic expeditions, on military maneuvers, and in underwater diving occupations often experience extremes of cold. Research on cold stress and performance (see Poulton, 1970; Sloan, 1979 for reviews) suggests that even temperatures of 55°F (13°C) can reduce efficiency in reaction time, tracking proficiency, muscular dexterity, and tactile discrimination. As temperatures fall below this level, performance usu-

ally deteriorates further. This deterioration is at least partly due to overload and heightened arousal. That is, the body's mechanisms are heavily allocated to maintaining adequate core temperature, so there is not enough energy or attention left for optimum performance on manual and mental tasks. If the hands are exposed, loss of tactile discrimination and stiffening reduce manual dexterity. If the hands do not become cold, lowered core temperature may still hurt performance, though probably not as much. Interestingly, if the hands are kept warm, considerable cooling of the rest of the body can be tolerated without severe performance decrements (e.g., Gaydos, 1958; Gaydos & Dusek, 1958).

Whether less chilling temperatures (55°F–65°F, or 13°C–18°C) actually enhance performance cannot be stated with confidence. However, we might speculate that if the physiological reactions to slightly cooler temperatures increase arousal without overburdening the body's adaptive mechanisms, performance might be slightly enhanced.

Walking speed tends to increase in cold temperatures, presumably to try to increase body temperature and/or to hasten the escape from the cold (Rotton, Shats, & Standers, 1990). Some people appear to be more bothered by cold than others, and performance is less severely affected by cold for some individuals. Furthermore, practice on tasks in cold temperatures can improve performance, so some adaptive mechanisms seem to be at work. Adaptation level to cold almost certainly plays a role in these temperature–performance relationships.

COLD EXTREMES AND SOCIAL BEHAVIOR

Surprisingly little research has been conducted on the effects of cold ambient temperatures and social behavior. One interesting bit of laboratory evidence does suggest that "low" temperatures around 62°F (16°C) make individuals feel more affectively negative (Bell & Baron, 1977). Consequently, one might expect low ambient temperatures to influence aggression in the same

AMBIENT TEMPERATURE AND DRIVING: Can Temperature Cause Accidents?

Icy and foggy road conditions obviously contribute to winter pileups on freeways. But can temperature alone contribute to accidents? As early as 1958, Provins noted that human performance in an automobile may well be affected by ambient temperature alone. Here are several ways:

1. Temperatures below 50°F (10°C) or above 90°F (32°C) reduce grip strength and impair muscle dexterity, which could diminish control over steering, braking, and shifting gears.
2. Temperatures below 50°F (10°C) or above 90°F (32°C) also reduce tactile discrimination (sensitivity of touch), which could reduce a driver's "feel" for the road.
3. Temperatures below 55°F (13°C) or above 90°F (32°C) impair vigilance and tracking performance, possibly making a driver less cognizant of potential hazards and traffic directional or signaling devices.
4. If high or low temperatures produce irritation, drivers may become more aggressive and take more dangerous risks.

When high wind speeds (such as would be experienced by drivers of convertibles or motorcycles) are added, these temperature effects probably become more severe. Perhaps still more frightening is that increased levels of carbon monoxide in the blood of drivers—which especially occur on crowded freeways—further reduces mental responsiveness.

curvilinear fashion as high ambient temperatures in the lab. Bell and Baron (1977) found evidence for this effect: Moderately negative feelings associated with cold temperatures tended to increase aggression, but more extreme negative feelings associated with cold seemed to decrease aggression. Moreover, Rotton (1993b) found that fewer sex crimes were reported on cold days. Although such results are far from conclusive, they are supported by other research (Bennett et al., 1983) and do suggest that more studies in this area would be valuable. All of these results do bring to mind "cabin fever," the idea that people forced indoors for prolonged periods during cold weather become agitated and hostile. However, Christensen (1982, 1984) found the idea to be no more than folklore, and Rotton and Frey (1985), studying temperatures as low as 5°F (−16°C), found no increase in family and household disorder associated with cold conditions.

As cited above for heat effects, Cunningham (1979) reported a slight decrease in helping with an interview as temperatures declined in the winter, although Schneider et al. (1980) found no effects of cold temperatures on helping. Informal observation has shown that cold, harsh winters tend to increase helping behavior and to reduce crime rates. Bennett et al. (1983) affirmed these observations and suggested a cold weather helping norm as one explanation. Attributing such behavioral influence to temperature is speculative, and, as is the case for evidence on heat and helping, there is too little research available to draw any firm conclusions on the relationship of cold temperatures to helping behavior.

More generally, it is obvious that cold temperatures can restrict some types of outdoor activities but encourage others (e.g., carving snow sculptures). Li (1994) carefully documents how people adapt to colder temperatures in New York City parks and plazas by changing activities.

SUMMARY OF TEMPERATURE EFFECTS ON BEHAVIOR

The body reacts to high and low ambient temperatures by respectively losing or preserving body heat. Associated physiological activity tends to increase arousal, leading to improved performance at low levels of arousal and deteriorated performance at higher levels. Attention, perceived control, and adaptation level also play a role in the relationship of temperature to behavior. Heat has been shown to affect attraction, aggression, and helping behavior in complex ways, depending on other factors. Cold ambient temperatures appear to influence aggression in much the same way as hot temperatures and may increase helping behavior under some circumstances.

Wind and Behavior

Anyone familiar with the "Windy City" of Chicago knows how discomforting wind can be when all you want to do is walk along a sidewalk. Few areas of the world can escape this natural phenomenon, although winds tend to be more severe in certain regions. Winds formed in tornadoes and hurricanes can easily reach speeds in excess of 80 mph (129 km/h). Parts of the Rocky Mountain states, especially those regions where the mountains meet the plains, experience wind speeds of over 100 mph several times per year. Fortunately, because of the altitude and climate, such Chinook winds are so thin and dry they do little physical damage, but they do cause discomfort and inconvenience (try riding a bicycle in one!). On the other hand, in 1988, storms in Chicago blew out 200 windows in the 110-story Sears Tower, moved a refrigerator across a room, and blew furniture, briefcases, and papers onto the streets below (Johnson & Richards, 1988). Such natural winds are not all that humans are exposed to. As indicated in more detail in Chapter 10, tall buildings create uncomfortable and even dangerous winds in the hearts of our major cities (see the box on page 365). Because of the influence of buildings on natural wind patterns, these human-made (or human-altered) winds can far exceed natural winds in both *speed* and *wind turbulence* (gustiness, shifting directions). Some people attribute the Sears Tower winds to wind tunnel effects of nearby skyscrapers. As urban structures are built taller, we can expect even more exposure to these unnatural winds. Thus, we suspect that potential effects of wind on behavior will become a more important topic for future research within environmental psychology. Of course, there is a beneficial side to wind; it blows away pollution from large cities (to the misfortune of some of those downwind), and it is useful as a power source. As we describe in Chapter 14, wind could provide as much as 20% of the electricity demand of the United States (Abelson, 1993). On the other hand, wind can have immediate unpleasant effects on us, as we discuss next.

PERCEPTION OF WIND

Although the body has specialized receptors for detecting light, sound, odors, and so forth, there are no receptors designed specifically for wind detection. Thus, to detect wind we have to rely on several perceptual systems. If you are actually in a wind, pressure receptors in the skin probably tell you the most about its presence: The stronger the wind, the more pressure on exposed skin. If the wind is particularly cold or hot, moist or dry, then temperature receptors in the skin also signal its presence. Muscular effort in resisting the wind is still another clue you can use to detect the force of a wind. The sight of others being blown over or of flags whipping tells you about the force of the wind even if you happen to be in a shelter. Finally, wind makes noise as it brushes past the ears or moves around obstacles, and the intensity and frequency of these sounds gives you a clue to the wind's presence and force. One of the earliest and most widely known indexes for evaluating wind is a scale developed by Admiral Sir Francis Beaufort in 1806. The **Beaufort Scale,** depicted in

Table 6-3	BEAUFORT WIND SCALE AND RELATED EFFECTS °	

Beaufort Number	Wind Speed (mph)	Atmospheric and Behavioral Effects
0,1	0–3	Calm, no noticeable wind.
2	4–7	Wind felt on face.
3	8–12	Wind extends light flag; hair is disturbed; clothing flaps.
4	13–18	Dust, dry soil, loose paper raised; hair disarranged.
5	19–24	Force of wind felt on body; drifting snow becomes airborne; limit of agreeable wind on land.
6	25–31	Umbrellas used with difficulty, hair blown straight; walking becomes unsteady; wind noise on ears unpleasant; windborne snow above head height (blizzard).
7	32–38	Inconvenience felt when walking.
8	39–46	Generally impedes progress; great difficulty with balance in gusts.
9	47–54	People blown over by gusts.

° *The Beaufort Wind Scale contains three additional levels that involve damage to property.*
Courtesy of A. D. Penwarden.

Table 6-3, was originally devised for activities at sea, but it has been adapted to land use over the years. As can be seen from this scale, wind effects range from problems of keeping hair combed to having difficulty walking and even to being knocked off one's feet by gusts of 47 mph (72 km/h) or more. Cases have actually been reported of individuals (especially elderly persons whose agility is less than ideal) being killed by winds that blew them over.

More scientific and precise scales of wind effects on humans have been proposed by Penwarden (1973). Some of these proposed indexes include force of wind on the body (which takes body surface area into account), angle at which one can lean into a wind, and body heat loss due to various types of winds. This body heat loss index would of course be influenced by moisture content and temperature of the wind, as indicated in the previous discussion of windchill.

BEHAVIORAL EFFECTS OF WIND

Very little systematic research has been conducted to date on the specific behavioral effects of wind. A very intriguing series of wind studies, however, was reported by Poulton et al. (1975). These researchers exposed female participants to winds of either 9 mph or 20 mph (14.5 or 32.2 km/h), with varying degrees of turbulence, in a wind tunnel. These wind conditions were intended either to be just strong enough to be noticeable and cause slight discomfort or to be extremely uncomfortable and detrimental to performance. Air temperature varied between 65°F and 70°F (18°C and 20°C), with humidity at 70 to 85%. Among the findings were that high wind and gustiness (1) significantly deflected participants from walking a straight path; (2) increased the time required to put on a raincoat from 20 to 26 seconds; (3) increased the time required to tie a headscarf by 30%; (4) increased blinking to 12 to 18 blinks per minute; (5) increased the time required to pick selected words from a list and to find a circled word in a newspaper; (6) caused more water to be spilled when poured into a wine glass; and (7) increased feelings of discomfort and perceived windiness (see also Cohen, Moss, & Zube, 1979). Taken together, these results generally suggest that winds influence affective feelings and at least some types of performance. Since some of these effects can be quite disturbing subjectively, city planners and architects need to be sensitive to these wind effects when designing public spaces (Pressman, 1995).

Correlational research has examined interesting behavior patterns associated with winds

AIR IONIZATION AND ELECTROMAGNETIC FIELDS:
Mediators of Weather–Behavior Relationships?

Lightning and other factors may ionize the air. In **air ionization** the molecules in the air partially "split" into positively and negatively charged particles. Moreover, extremely low-frequency electromagnetic fields, or **ELF-EMF**, are associated with some low altitude weather disturbances. Could it be that these factors play a role in the influence of the weather on behavior? Some researchers think so, at least to some extent. For example, there is evidence that **negative ions** improve reaction time, enhance positive mood, and improve interpersonal feelings (e.g., Baron, 1987a, 1987b; Baron, Russell, & Arms, 1985; Brown & Kirk, 1987; DeSanctis, Halcomb, & Fedoravicius, 1981). Under some circumstances, however, these effects can be opposite to what others have found (cf. Baron, Russell, & Arms, 1985). **Positive ions** are associated with worsening performance and mental outlook, although some people are more sensitive to ion effects than others (Charry & Hawkinshire, 1981). Interestingly, one interpretation of the disruptive effects of the Sharav wind in Israel (see p. 196) is that this wind generates an excess of positive ions (Sulman et al., 1970).

Early research on low-frequency electromagnetic fields suggested that they may slow reaction time, impair estimation of time (constricting time), and lead to complaints of headaches and lethargy (for reviews, see Beal, 1974; Persinger, Ludwig, & Ossenkopf, 1973). In the 1980s, research suggested that when EMFs are generated by high-power electrical lines they might be associated with increased frequencies of diseases such as leukemia and cancer (e.g., Savitz & Calle, 1987; Savitz et al., 1988). The potential dangers of EMFs are hotly debated. Findings do not always replicate, and often results of one study are in the opposite direction of the results of another study. The magnetic field of a power line is hundreds of times weaker than Earth's natural magnetic field and is weaker than the pull of the moon on our bodies. Riding a bicycle through Earth's magnetic field creates at least as much electric field within the body as does walking under a power line. Moreover, the more carefully researchers control extraneous, potentially confounding variables, the weaker the effects of EMFs emerge. Furthermore, it is difficult to find a dose–response relationship; that is, the longer one is exposed to EMFs, the greater the risk should be, but studies sometimes find that a lower dose leads to higher risk. New studies do sometimes appear that suggest cause for concern, however (e.g., Oak Ridge Associated Universities Panel, 1992, 1993), and alarming publications do arise that allege conspiracies to cover up the danger to the public (e.g., Brodeur, 1993), so we can only feel confident that the debate over EMF dangers will continue. Interestingly, Holden (1995) cites two studies suggesting that EMFs may increase growth in plants. Recently, the most carefully controlled study yet found essentially no cancer risk from EMFs associated with electrical lines in the home (Linet et al., 1997), and two very thorough reviews of all the research to date concluded that EMF effects on health are minimal to nonexistent (National Research Council, 1997; National Institute of Environmental Health Sciences, 1999). One researcher who reported hazardous effects of EMFs has even been charged with fabricating the data and has withdrawn the published findings (see news report by Vergano, 1999).

Whether or not ions and ELF-EMF account for the effects of weather on behavior is unknown. The effects of ions noted above are primarily from laboratory conditions with higher levels of negative ions than would be found in natural settings (Culver, Rotton, & Kelly, 1988; Kroling, 1985; Reiter, 1985; Rotton, 1987a). As with all individual weather variables, more than one factor is operating at a time, so it is difficult to conclude that any one mechanism "causes" the observed behavior or feeling state. It is intriguing to consider the possibility, however, and we are sure that experimentation and speculation will continue in the area.

around the world, such as the Föhn, Bora, Mistral, and Scirocco in Europe, the Sharav and Chamsin in the Near East, the Chinook in Colorado and Wyoming and Santa Ana in California, and the Pomponio in Argentina (Sommers & Moos, 1976). The Föhn and Chinook are warm, dry winds that descend from mountains. It is not uncommon for residents in these regions to attribute depression, nervousness, pain, irritation, and traffic accidents to wind (Sommers & Moos, 1976). In the Middle East, some governments have been known to forgive criminal acts that are committed during the periods of disturbing winds. In an empirical study, two researchers (Muecher & Ungeheuer, 1961) measured performance on several tasks. As expected, performance was worse on days of Föhn-like weather than on less stormy days. In addition, they and other researchers (e.g., Moos, 1964) have reported that accident rates increase just before or during the approach of the winds. Rim (1975) examined performance of individuals on psychological tests during hot, desert wind (Sharav) periods in Israel and compared their scores with those taking the tests on less turbulent days. The windy days led to higher scores on neuroticism and extraversion and to lower scores on IQ tests and other measures. Other research has shown some relationship between windy days and poor classroom behavior (e.g., Dexter, 1904), yet windy days have also been found to be associated with more social interaction among preschoolers (Essa, Hilton, & Murray,

1992). High wind speed has also been associated with increased mortality rates, felonies, and delinquency (Banzinger & Owens, 1978; for a review of these studies, see Campbell & Beets, 1981), and with increased suicides (Lester, 1996).

Whether these effects are directly attributable to wind, to air pressure changes, or even to atmospheric ion changes, is subject to debate (see the box on page 195). Also, temperature and other weather changes usually accompany winds, so more than one factor may account for wind effects. For example, Cunningham (1979) found that wind was associated with increased helping in summer months and decreased helping in winter months, which suggests that the effects of wind are mediated by what wind does to perception of temperature (i.e., chilling effect in winter, cooling effect in summer). However, Cohn (1993) found that even controlling for other weather variables, wind was correlated with domestic violence; interestingly, with all the statistical controls in place, domestic violence was slightly less likely to occur as wind speed increased. Quite probably, weather conditions increase the stress one experiences, and the heightened stress leads to many of the psychological effects discussed in Chapter 4. Moreover, attention, arousal, and loss of perceived control are likely to mediate many wind effects. Further discussion along these lines is presented in the following section on altitude and barometric pressure.

Barometric Pressure and Altitude

Many people live at rather high altitudes, such as in the Rocky Mountain region of the United States, the Tibetan Plateau of southern China, the Andes, and the high plains of Ethiopia. Others of us travel to these high places. Still others experience high altitudes in aircraft or experience below sea-level conditions during underwater dives. At high altitudes we are exposed to

a variety of stresses, most notably **hypoxia,** or reduced oxygen intake resulting from low air pressure. Other high altitude stresses include increased solar radiation, cold temperatures, humidity changes, high velocity winds, reduced nutrition, and strain from negotiating rough terrain (Frisancho, 1993). In underwater environments we also experience problems from high pressures, cold temperatures, and physical exertion. Thus, it is appropriate to examine the physiological and behavioral changes associated with altitude and air pressure differences. (For more detailed reviews, see Fraser, 1989; Frisancho, 1993; Poulton, 1979; and Sloan, 1979).

PHYSIOLOGICAL EFFECTS

Normal atmospheric or **barometric pressure** at sea level is 14.7 pounds per square inch (psi) (1.022 kg/cm^2). Lower than normal pressures occur as one rises higher and higher above sea level. Under normal air pressure conditions, oxygen is taken into the body through the alveolar walls of the lungs, with the pressure difference between the atmosphere and sides of the walls being just enough to "force" oxygen into the body.

In low pressure environments, however, it becomes more difficult for oxygen to pass through the alveolar walls, resulting in reduced oxygen available, or the hypoxia noted above. Hypoxia has a number of physiological and behavioral consequences; and, it should be mentioned, hypoxia is not limited to high-altitude environments, but is also a major problem in carbon monoxide pollution, as discussed in Chapter 7. Most habitable environments are located below 15,000 feet (4,572 m), though at much higher altitudes, two special air pressure problems occur. First, above 30,000 feet (9,144 m), the pressure on the interior (body) side of the lung walls becomes so much greater than the pressure on the atmospheric side that oxygen actually passes from the blood into the atmosphere. Second, above 63,000 feet (19,203 m), air pressure is so low that water in the body at a core temperature of 98.6°F (37°C) will actually vaporize.

As stated above, the hypoxia at high altitudes has a number of physiological ramifications. Frisancho (1993) provides some interesting details. Visitors to high-altitude areas are likely to experience deeper, and perhaps more rapid, breathing to help compensate for hypoxia. As a result, more carbon dioxide is removed from the lungs,

leading to increased alkalinity of the blood. In addition, resting heart rate increases, though maximum heart rate during exercise decreases. Consequently, total cardiac output is reduced, and enlargement of the heart may occur. Red blood cell count increases, hemoglobin concentration increases, but plasma volume decreases, so total blood volume is largely unaffected. Moreover, retinal blood vessel diameter increases, and light sensitivity of the retina decreases. Also, an increased desire for sugar will likely be experienced, although hunger is suppressed and weight loss likely. Hormone production is also affected by high altitudes: Adrenal activity increases, and thyroid activity decreases. Testosterone production and sperm production decrease, and menstrual complaints may increase. In sum, initial exposure to high altitudes leads to many physiological changes.

ACCLIMATIZATION TO HIGH ALTITUDES

Fortunately, most of the physiological changes noted above are short-term responses to high altitudes, and, as Frisancho (1993) elaborates, acclimatization to the environment at these elevations does occur. For example, hemoglobin concentration levels off after six months, and testosterone production returns to normal after a week of high altitude exposure. Acclimatization is not without long-term consequences, however. Populations native to high-altitude areas do show physiological differences from lowland natives, probably as a result of developmental adaptations. For example, high-altitude natives show larger lung capacity, higher blood pressure in the pulmonary (leading to the lungs) arteries, lower weight at birth, slower growth rates, and slower sexual maturation (Frisancho, 1993). Although some of these differences may be attributable to nutrition, genetics, and culture, many of them are almost certainly tied to the hypoxic environment of high elevations.

Adaptation to high altitudes can be used to athletic advantage. Levine and Stray-Gundersen (1997) showed that runners who lived at high

A RECYCLED CYCLE:
Moon Phases and Behavior

Folklore and commonly held beliefs maintain that many aspects of our behavior are related to phases of the moon. Sexual prowess, menstrual cycles, birthrates, death rates, suicide rates, homicide rates, and hospital admission rates are among the phenomena various people claim are affected by the moon. Often, it is maintained that a full moon increases strange behavior. Surveys of undergraduates indicate that half of them believe people behave strangely when the moon is full (Rotton & Kelly, 1985a). Other beliefs are that the tidal pulls of full and new moons influence human physiology or psychic functioning, or that the moon's perigee (closest distance to the Earth) and apogee (farthest distance from the Earth) influence us in strange ways. Indeed, the word *lunacy* is derived from a belief in a relationship between the moon and mental illness.

The last full moon of the previous millennium occurred on December 22, 1999, which also corresponded to its perigee, as well as to the winter solstice or shortest day in the Northern Hemisphere. In addition, the moon on that date was almost as close to the sun as it ever gets. As a result, it appeared 14% larger than at its apogee as well as 7% brighter. No particular anomalies in human behavior were reported. But the lore says that when a similar full moon, perigee, and solstice occurred on December 20 and 21, 1866, the Sioux warrior Crazy Horse was inspired to ambush and wipe out 80 U.S. soldiers (Golden, 1999).

It is the case that lunar tides affect some marine organisms, and there is some evidence that a full moon ever so slightly increases temperatures on Earth (by 0.02°K; Balling & Cerveny, 1995). From time to time, research also appears that actually gives credence to beliefs that the moon causes drastic changes in human behavior. For example, Blackman and Catalina (1973) found that full moons were associated with an increase in the number of patients visiting a psychiatric emergency room. In another study, Lieber and Sherin (1972) reported a relationship between moon phase and homicide. Rape, robbery, and assault; burglary, larceny, and theft; and auto theft, drunkenness, disorderly conduct, and attacks on family and children have also been linked to a full moon (Tasso & Miller, 1976). At first glance, then, it would appear that science has confirmed the folklore of the ancients (see also Garzino, 1982).

Not so fast! Closer examination of the data indicates that the mysticism of the lunar cycle may be more myth than reality. Campbell (1982), Kelly, Rotton, and Culver (1985–86), and Rotton and Kelly (1985b, 1987), among others, have reviewed the available research on the topic and concluded that no firm relationship exists between any lunar variable and human behavior (see also Byrnes & Kelly, 1992). For example, studies conducted over a period of three to five years may report a relationship between the full moon and suicide or homicide for only one of the years studied. Researchers who conclude that such a relationship exists are ignoring the fact that it does not exist for the other years, or that these behaviors are actually lower during full moons for another year. Moreover, it is consistently found that crimes increase on weekends. For some periods of the year, lunar phases may coincide with weekends. Data based on only these periods will obviously show a relationship between the moon and crime, but data based on other periods will show the opposite relationship or no relationship at all. In addition, a self-fulfilling prophecy may operate: If police believe crime increases during a full moon, they may become more vigilant at these times and thus arrest more people. Altogether, the evidence suggests that positive links between moon phases and behavior are spurious and are attributable to mere chance probabilities in the data or to variables not considered by individual investigators. Why do these mistaken beliefs persist? Reasons include misconceptions about physical processes (Culver, Rotton, & Kelly, 1988), attitudes acquired from one's peers (Rotton, Kelly, & Elortegui, 1986), and cognitive biases, such as basing conclusions on only a few occurrences. Given the tenacity of beliefs in moon phases causing disruptive behavior, we suspect the lunacy of it all will continue for some time!

altitudes and trained at low altitudes took an average of 13 seconds off their time in a 5-kilometer race.

BEHAVIORAL EFFECTS OF HIGH ALTITUDES

Obviously, extreme hypoxia will lead to loss of consciousness and death. Performance impairment, however, occurs well before this extreme stage. To the extent that the body can compensate for hypoxia, high altitudes will not lead to substantial performance decrements. During strenuous work, however, the capacity of the body to compensate for hypoxia is taxed, and performance decrements are likely to be observed. Task performance can be impaired by altitudes as low as 8,000 feet (2,438 m). Learning of a new task can be impaired by rapid decompression to altitudes as low as 5,000 feet (1,524 m). In general, learning of new things is more affected by high altitudes than is recall of previously learned material (Cahoon, 1972; McFarland, 1972). We should note that such learning impairments are generally of small magnitude, and people living at high altitudes are certainly capable of learning. Also, we should note that the challenge of high-altitude mountaineering is exhilarating for some adventurous individuals; the exhilaration may even be in part a consequence of knowing that one has successfully faced the dangers of hypoxic effects. Ewert (1994) documents how exhilaration, excitement, and accomplishment motivate those who climb in Denali National Park, Alaska.

HIGH AIR-PRESSURE EFFECTS

Extremely high pressure is experienced primarily under the sea. For each 33 feet (10 m) of depth, the pressure increases by 14.7 pounds per square inch (psi) or by **one atmosphere** (1.033 kg/cm^2). Thus, at 33 feet (10 m), the pressure is 29.4 psi (two atmospheres); at 99 feet (30 m) of depth, the pressure is 58.8 psi (four atmospheres); and so on. Hazards encountered at such pressure extremes (see also Fraser, 1989; Poulton, 1979; Sloan, 1979) include

1. Increased breathing difficulty caused by reduction of maximum breathing capacity (reduced by 50% at a depth of 100 feet [30 m]);
2. Oxygen poisoning caused by breathing excess oxygen or oxygen under pressure;
3. Nitrogen poisoning caused by the narcotic effects of breathing nitrogen under extreme pressure. Symptoms include light-headedness and mental instability.
4. Decompression sickness caused by nitrogen bubbles forming in body tissues (especially in the circulatory system) when one rapidly changes from a high-pressure to a lower-pressure environment. The "bends" is one relatively acute form of decompression sickness. Permanent damage to the bones may also result from rapid decompression.

Most of these high-pressure problems can be corrected or prevented by breathing the proper mixture of air for the diving depth and by surfacing slowly to permit the gradual release of nitrogen from tissues.

Physical performance decrements in dives with scuba gear (air tank and wet suit) are not substantial down to about 200 feet (60 m), but after depths of 300 feet (90 m), performance on mental and motor tasks deteriorates noticeably. Even depths of 120 feet (36 m) can impair some memory tasks (Philip, Fields & Roberts, 1989). Divers sometimes invoke what they call "Martini's law," which holds that each 50 feet (15 m) of depth is equivalent to coordination effects of drinking one gin martini (Fraser, 1989). Performance and anxiety are not particularly impaired in dives to 1,000 feet (300 m) in a pressure chamber (Abraini et al., 1996).

MEDICAL, EMOTIONAL, AND BEHAVIORAL EFFECTS OF AIR-PRESSURE CHANGES

Low and high barometric or atmospheric pressures are not only associated with altitude. All of us, in fact, are subjected to often dramatic swings in barometric pressure associated with weather

SUNLIGHT ON MY SHOULDERS REALLY DOES MAKE ME HAPPY!

One weather variable that has significant impact on humans is sunlight. As indicated by Frisancho (1993), thermonuclear reactions within the sun convert millions of tons of hydrogen into millions of tons of helium every second, releasing radiant energy in the process. Approximately eight minutes after it leaves the sun, some of this energy reaches the Earth in various wavelengths. Ranging from short to long wavelengths, the energy takes the form of x-rays, ultraviolet rays, visible light, infrared rays, and radio waves. The wavelengths shorter than visible light are hazardous to life. Fortunately, most of these wavelengths are either absorbed by ozone, blocked by ozone, or "consumed" in the process of making ozone high in the atmosphere. **Ozone** is a form of oxygen in which three atoms are molecularly combined (O_3). As you are probably aware, there has been concern in recent years that several human-generated substances, most notably chlorofluorocarbons in aerosol propellants, destroy the layer of ozone that protects us from harmful solar radiation. When it hits certain atmospheric pollutants, sunlight leads to photochemical smog. The solar energy that does reach the surface of the Earth can also harm us through sunburn and as a factor in skin cancer. To protect us from some of this danger, the skin produces melanin and other dark pigments to act as a partial shield, a process we know as tanning (see Frisancho, 1993). Exposure to excessive midday sun has been implicated in cataracts (Taylor et al., 1988).

Sunlight is not just potentially harmful, of course, but provides us with light, heat, and through photosynthesis, food. Moreover, sunlight induces the skin to produce Vitamin D.

Behaviorally, increased hours of sunlight have been associated with increased suicide rates and crime rates (see Sommers & Moos, 1976). These effects most probably are not due directly to sunlight, but rather to increased opportunities to encounter social stress, which may lead to depression and suicide, and increased opportunities to engage in criminal activity. There are seasonal trends in suicide rates, with a peak in spring to early summer. Noting seasonal trends, Kevan (1980) concluded from a review of more than 80 studies that suicide is not related to meteorological factors.

Interestingly, two experiments by Cunningham (1979) suggest that sunlight not only leads to good moods in people, but is also associated with increased altruistic behavior! In one of these experiments, people in Minneapolis were greeted by an experimenter as they walked outdoors and were asked to answer a few brief questions. They were more willing to answer the questions the more sunshine was present, regardless of any other weather conditions in both summer and winter. In the second experiment, waitresses in a restaurant were found to receive more tips with increased sunlight. This relationship was found even though customers were indoors and were not experiencing direct sunlight at the time of leaving the tip. Moreover, the more sunlight, the more positive the mood of the waitresses. Cunningham interpreted these results in terms of mood: The more pleasant we feel, the more willing we are to be kind to and to help others. Similarly, Rind (1996) found that telling hotel guests that the weather would be sunny and warm for the day increased the quantity of tips.

In a study of office workers in southern Europe, Leather et al. (1998) found that sunlight penetrating into the workspace was associated with higher job satisfaction, decreased intention to quit, and increased feelings of well-being. Apparently, sunlight really does make us happy!

changes. Hurricanes, cyclones, and other "tropical storms," for example, are special types of low-pressure weather systems. Clear, sunny skies on the other hand, are generally associated with high pressure. Do these changes in barometric pressure affect our feelings and behavior? According to a number of researchers, the answer is "yes," although the picture is a bit cloudy (pun intended) in that: (1) the data are not always consistent from study to study; and (2) humidity, temperature, and wind variations accompanying pressure changes may account for the observed psychological changes (see also Campbell & Beets, 1977; Moos, 1976).

FEELING DOWN IN THE WINTER:
Seasonal Affective Disorder

We have noted that many human activities, including crime rates and suicide rates, vary with the seasons. For millennia, physicians have observed that depression and mania (a hyperactive state opposite of depression) often come and go with the seasons in some individuals—so much so that at one time depression was thought to be caused by cold and mania by heat (Jackson, 1986). Psychiatrists have given the name **Seasonal Affective Disorder**, or **SAD**, to a depressive cycle that varies with the seasons (Rosenthal et al., 1984). SAD usually occurs in women and begins in early adulthood, and the depressive episode typically shows excessive sleep (hypersomnia), fatigue, craving for carbohydrates, and weight gain. Since the most studied pattern is for depression to occur in the winter and a brighter mood (hypomania) to occur in summer, and since the hypersomnia is reminiscent of hibernating animals, the shortening and lengthening of daylight that goes with the seasons has been thought to be a potential causal factor. Indeed, the depression episodes may respond well to intense artificial light (e.g., Rosenthal et al., 1984; Wehr et al., 1986), and at least a dozen companies manufacture light-therapy devices (Johannes, 1996). However, carefully controlled studies suggest that light therapy may be useful for only some individuals (Terman et al., 1996) and may actually be acting as a placebo. For instance, it may work primarily because those who use it expect it to work (Teicher et al., 1995).

The SAD pattern has been tied to a substance called melatonin, which is involved in hibernation and which declines in concentration in animals exposed to bright light. However, melatonin does not seem to decrease with light therapy in humans (Wehr et al., 1986), and some people have a "reverse" cycle with the depressive episodes occurring in summer months (Wehr, Sack, & Rosenthal, 1987), so the connection of SAD to melatonin and number of daylight hours seems questionable. Another possibility is that a brain chemical called serotonin—which is related to some forms of depression—may play a role in SAD.

Although the reasons behind SAD are unclear, it has generated much clinical and research interest, and will certainly continue to do so for years to come (see also Nowak, 1994; Rosenthal, 1993). A newsletter can be obtained by writing the National Organization for Seasonal Affective Disorder (NOSAD), P.O. Box 40133, Washington, DC 20016.

In general, researchers have observed three types of effects that air pressure changes have on people: increased medical complaints, increased suicide rates, and increased disruptive behavior. With respect to medical complaints, many arthritis victims claim that their condition worsens with changes in weather (Jamison et al., 1996). A number of studies have been conducted over the years to examine the relationship of mental hospital admissions and suicide rates to weather changes. Both of these clinical occurrences show fluctuations with seasons, with the highest rates coinciding with the increased temperatures and daylight hours of spring and summer months (see Campbell & Beets, 1977; Jessen et al., 1998; Salib & Gray, 1997; Sommers & Moos, 1976; Wang & Wang, 1997). It is entirely likely that weather associations with mental hospital admissions and suicide rates reflect seasonal variations, and that the social contact that goes along with seasonal variations accounts for the behavioral pathologies (e.g., Kevan, 1980).

Finally, several studies have shown that disruptive school behavior and police dispatch calls fluctuate with weather, especially air pressure changes. As early as the turn of the 20th century, one researcher found that low barometric pressure and wind and humidity fluctuations were associated with poor behavior in the classroom (Dexter, 1904). Similar findings were reported in later work (e.g., Russell & Bernal, 1977). Also, it has been found that complaints to police and investigative activity increase with low pressure and high temperature, and that accident reports and related investigations increase with stormy weather (e.g., Cohn, 1996).

What do the above findings mean? Are our mental health and behavior helpless victims of barometric and other weather changes? Fortunately, the answer seems to be "probably not." First, the effects of weather changes on most psychological and behavioral indices are small relative to the influence of other factors, such as social conflict. Second, to the extent that weather does affect behavior, it probably does so indirectly, as an added stressor, similar to "the straw that broke the camel's back." For example, increased suicide rates associated with pleasant weather probably reflect increased time available to interact in stressful social situations and increased opportunities to worry about these social stresses (see Sommers & Moos, 1976). Similarly, weather changes may simply provide something else to worry about and cope with, adding to the strain on adaptation capacity that has been built up by other stressors. Nevertheless, the additional stresses brought about by the weather must be dealt with, and may have important consequences, especially under times of other duress. (For another viewpoint, see the box on page 201.)

SUMMARY OF AIR PRESSURE EFFECTS

Low air pressure is associated with high altitudes and stormy weather conditions. High pressure is found in underwater environments and in fair weather circumstances. At high altitudes, the major stress is hypoxia, or low oxygen intake. Adaptation to hypoxia may have short-term and long-term consequences, including respiratory, cardiovascular, and hormonal changes, as well as performance impairment. High pressure in underwater environments may lead to breathing difficulty, oxygen poisoning, nitrogen poisoning, and decompression sickness, although steps may be taken to avoid these problems. Low pressure associated with weather changes may coincide with increased medical complaints, high suicide rates, and increased disruptive behavior. These observations associated with weather may be due to weather variables other than air pressure, and are likely attributable to additional stress to go along with social stresses and other sources of duress.

SUMMARY

Weather (short-term changes in temperature, storms, and the like) as well as climate (longer-term changes) have measurable impacts on human behavior. Global warming may be human caused and could have a disastrous impact on civilization, as have past climatological changes. Climatological determinism suggests that climate and weather effects are profound, but current research suggests that the effects are quite complex. Exposure to heat or cold results in a number of physiological adaptive steps aimed at maintaining a stable core body temperature. Heat impairs performance only when it is extremely uncomfortable, and heat may actually facilitate performance if it is only moderately uncomfortable.

Uncomfortable heat tends to reduce interpersonal attraction. Moderately uncomfortable heat increases aggression. Controversy exists over whether even more uncomfortable heat increases or decreases aggression. Extremes of cold impair performance and may also influence aggression in the same manner as extremes of heat. Winds can be quite disturbing and interfere with many types of performance. Health complaints and accident rates also appear to be associated with high wind conditions. Low barometric pressure at high altitudes leads to hypoxia and other forms of physiological distress, but we can adapt to low- and high-pressure environments. Changes in air pressure associated with weather patterns can have deleterious effects on us. In general, weather effects occur in combination with one another and with pollution effects.

KEY TERMS

acclimation
acclimatization
air ionization
ambient temperature
barometric pressure
Beaufort Scale
chill factor
climate
core temperature
curvilinear
  relationship
deep body
  temperature

determinism
effective temperature
ELF-EMF
El Niño
frostbite
Gaia hypothesis
global warming
greenhouse effect
heat exhaustion
heatstroke
hypothalamus
hypothermia
hypoxia

linear relationship
long, hot summer
  effect
negative affect-escape
  model
negative ions
one atmosphere
ozone
ozone hole
peripheral vasocon-
  striction
peripheral
  vasodilation

piloerection
positive ions
possibilism
probabilism
Seasonal Affective
  Disorder (SAD)
Temperature–
  Humidity Index
  (THI)
thermoreceptors
weather
windchill index

SUGGESTED PROJECTS

1. Keep a daily record of temperature and humidity readings. Obtain crime reports from the local newspaper or from the police department. Is there a relationship between weather and crime?

2. Keep daily records of temperature, humidity, and wind conditions. Using a stopwatch, check walking and bicycle riding speeds of students as they make their way across campus. Do these speeds vary with weather conditions?

3. Using a sound-level meter, check the noise levels of winds on a breezy day. Is the noise level higher around buildings?

4. Time the length of lectures in your various classes. Are weather conditions related to length of lectures?

Disasters, Toxic Hazards, and Pollution

INTRODUCTION
NATURAL DISASTERS

What Constitutes a Natural Disaster?
Characteristics of Natural Disasters

Effects of Warnings
Summary of Disaster Characteristics

Perception of Natural Hazards
Psychological Effects of Natural Disasters

Acute Stress and Mental Health
Longer-Term Effects
Why Do These Effects Occur?
Social Support and Disaster
Children and Disasters
Age and Disaster Response

Environmental Theories and Disasters
Summary of Effects

TECHNOLOGICAL DISASTERS

Characteristics of Technological Catastrophe
Effects of Technological Disasters

The Buffalo Creek Flood
The Three Mile Island and Chernobyl Accidents

Summary and Conceptual Considerations

PRIMARY AND SECONDARY VICTIMS
EFFECTS OF TOXIC EXPOSURE

Asbestos Exposure
Living Near Toxic Waste
Radon Exposure
Sick Building Syndrome
Teratology and Behavioral Toxicology

AIR POLLUTION AND BEHAVIOR

WHERE DO WE PUT ENVIRONMENTAL HAZARDS? NIMBY AND ENVIRONMENTAL RACISM

SUMMARY

KEY TERMS

SUGGESTED PROJECTS

 # Introduction

Think of some of the famous, gripping Hollywood movies that have been released over the years about people facing disasters. Are there similarities among the different types of disasters? Is there something about disaster movies that instills us with fear yet draws us to the drama? Something inherent in the disasters themselves rather than the script and the stars? Winds of 100 mph or more, flooding, falling trees, damaged buildings, and, in some cases, death, may occur in tornadoes and hurricanes. Earthquakes and other natural disasters are similar; their presence is hard to deny, the threats they pose are intense, and they can kill or maim. Although they pose similar threats, toxic hazards and air pollution are not always visible and obvious. Sometimes they make us cough or cause our eyes to water or our drinking water to taste bad, but for much of the time we are not very aware of them. Yet, toxic chemicals can cause or contribute to cancer and can damage our heart and lungs and affect our behavior; in disaster movies, though, such slow processes are difficult to portray. Silent, constant exposure to toxic hazards may be more harmful in the long run than are more "memorable" natural disasters, yet because they are slow and silent in their effects, it seems to take a fast-paced Hollywood drama to make us really fear them. Situations such as at Three Mile Island

(TMI), Bhopal, Chernobyl, and Love Canal have sharpened our awareness of the vast possibility for toxic exposure in our world, so perhaps we don't accept movies of similar events as purely fictional. At TMI and Chernobyl, radioactivity was involved in accidents at nuclear power stations. Some radiation was released, affecting many nearby residents. At Bhopal a massive accidental release of poisonous gas killed or incapacitated many townspeople. At Love Canal, children played and families lived on top of a landfill packed with hazardous waste chemicals. The hazards were also frightening and unprecedented. These accidents or tragic hazards were dramatic and overwhelming for those people who were affected. However, exposure to potentially harmful agents is also a by-product of technology, and one could ask, as we did about autos, where would we be without plastics or power plants? Perhaps there are common elements about disasters—doom, gloom, denial, destruction, escape, recovery—that make us fear them in real life and yet draw us to movies where they become part of adventures.

In the previous chapter we discussed weather and climate and the effects that changes in them can have on us. In this chapter we consider the severely disturbed and stressful environment—weather turned into violent storms and floods, other natural disasters such as earthquakes, cataclysmic technological disasters, and slower toxic pollution releases. In discussing stress, we noted that Lazarus and Cohen (1977) distinguished among three types of events that cause stress. One type, *daily hassles,* referred to small-magnitude events that occur repeatedly — commuting to work, going to class, and so on. Another kind of stressor, termed *personal stressors,* referred to more powerful threats or losses that occur on an individual level, including loss of a loved one, loss of a job, and other personal problems. Still another type of stressor was called *cataclysmic events* to capture the intensity of these events and their potential for widespread devastation and destruction. These sudden, powerful events typically require a great deal of adaptation in order for people to cope, and large numbers of people are affected. They include war, imprisonment, relocation, and natural disaster (Lazarus & Cohen, 1977). As major events with the potential to kill, maim, disrupt, and wipe out communities, disasters are important environmental stressors. However, there are many different kinds of disasters, and although some of the characteristics of cataclysmic events are shared by most disasters, other cases are different. Sometimes cataclysmic events such as major earthquakes can precipitate other stressors such as death of a loved one (personal stressor) or disrupted transportation systems (daily hassle). In this chapter, we will consider the distinction between natural disasters — caused by natural forces — and human-made catastrophes which are due in some way to our actions or modification of the environment.

Natural Disasters

Natural disasters are relatively infrequent events, but their dramatic qualities make them memorable and seem more frequent. Few of us will directly experience more than one or a few such events in our lives, and then only if we live in areas where they are likely. Earthquakes, hurricanes, tornadoes, tsunamis, and volcanic eruptions are all instances of powerful natural events that tend to occur in certain areas of the world. Defining these events is an important place to begin our discussion of natural disasters. The massive flooding in North Carolina in 1999 or in the U.S. Midwest in 1993, the onslaught of Hurricane Andrew in 1992, or the massive earthquakes that took so many lives in Turkey and in Taiwan in 1999 are vivid examples of events that we label "disasters." Examining what they have in common may help us to better understand what makes a disaster a "disaster" (Figure 7-1). In North Carolina, Hurricane Floyd had almost benign winds but caused unprecedented flooding that completely inundated communities along rivers in the eastern part of the state and contaminated entire towns with sewage and waste from hog farms, oil tanks, and dead farm animals. In the Midwest a few years earlier, entire regions disappeared under floodwaters,

Figure 7-1   Natural disasters often leave immense destruction in their wake, as in this Owensboro, Kentucky, tornado.

Courtesy of Federal Emergency Management Agency.

and in Turkey, the worst of several earthquakes centered on the city of Izmit and killed more than 15,000 people and left more than 200,000 homeless. Buildings were totally destroyed and emergency workers and rescuers were overwhelmed by the total devastation. In the United States, earthquakes have not been as destructive or deadly for many reasons. Some quakes in both northern and southern California seemed to cause more inconvenience than suffering—although the 1989 Loma Prieta quake severely damaged parts of San Francisco, Oakland, and Santa Cruz, and the Northridge quake caused substantial death, destruction, and disruption in the Los Angeles area. The potential impact of these disasters shows how important studying them can be.

WHAT CONSTITUTES A NATURAL DISASTER?

Natural disasters have been difficult to define, not because we do not know what they are, but because specific criteria are hard to establish. The *natural* part is easy: Natural disasters are caused by natural forces and are not under human control. They are uncontrollable, because they are the product of the physical forces that govern the Earth and atmosphere, and people must learn to deal with them when they strike. Defining what constitutes a *disaster*, on the other hand, is a little trickier. Since it is obvious that not all storms are disasters, what distinguishes disasters from less serious events or series of events? We could simply list all of the events associated with disaster, such as typhoons or earthquakes, but this may not be satisfactory since these storms or events do not always cause damage—and damage or death must usually result before it is considered a disaster. How should we define or quantify damage? Should it be viewed on the individual level, as death, injury, or loss? Is there a cutoff, a certain amount of damage above which an event is a disaster and below which it is not? Or, should responses by victims (for example, if they panic) be used to index disasters? Definitional issues have posed problems for researchers interested in disaster

and extreme stress for many years (Quarantelli, 1998).

The emphasis on what makes a disaster "disastrous" seems to be on effects. The Federal Emergency Management Agency (FEMA), the U.S. agency responsible for helping disaster victims, has offered the following definition:

> A major disaster is defined . . . as any hurricane, tornado, storm, flood, high water, wind-driven water, tidal wave, tsunami, earthquake, volcanic eruption, landslide, mudslide, snowstorm, drought, fire, explosion, or other catastrophe . . . which, in the determination of the President, causes damage of sufficient severity and magnitude to warrant major disaster assistance (1984, p. 1).

This definition includes a precipitating disaster event (a storm or other natural event of great power) and a judgment of sufficient disruption or damage. The nature of the event (is it one of these disaster events?) and the extent of damage (is it bad enough?) are used to officially designate disasters. We must keep in mind, however, that this definition is used to determine whether emergency aid and relief are to be given. As a result, it focuses on issues related to taking such action. Still, it features the same assumptions and biases as most of us: Disasters are destructive and cause suffering, harm, and loss. A tornado in a desolate desert area where no one lives and no one may even be around to see it is not a disaster, but the same tornado loose in downtown Birmingham is.

Quarantelli (1998) has argued that physical indices of damage and destruction are not sufficient to define disasters. The magnitude of impact of a disaster event may be better viewed in terms of social **disruption,** the degree to which individual, group, and organizational functioning is disturbed and can no longer operate as it used to. It is possible for natural events like earthquakes to be destructive but not disruptive, though the two are frequently related. An earthquake could collapse freeway bridges, which is sure to cause considerable disruption,

Figure 7-2A   Classification of an event as a disaster depends in part on its disruptive impact. A washed out roadway will usually cause considerable disruption in daily transportation.

Courtesy of Federal Emergency Management Agency.

Figure 7-2B   It is possible to have great disruption with little visible sign of destruction. A flash flood covered parts of this university campus. A day or two later, the major visible evidence from the outside was piles of furniture and files such as seen here. In actuality, more than $100 million in damage was caused. Professors lost a lifetime of lecture and research materials, the library lost all bound journals, and utility tunnels were flooded, knocking out the heating and telecommunications systems for months.

or it could strike a more remote area in which similar damage inconveniences only a few people (Figure 7-2A). More importantly, it is possible to have little visible destruction with great disruption (Figure 7-2B). Thus, storms that cause great disruption by the threats they pose might be considered to be disasters even if they end up doing little damage. If an approaching hurricane turns out to sea at the last minute, it may do little damage to seaside cities, but if the emergency evacuations that were ordered caused a great deal of disruption or even injury or death, the storm could still be considered to have caused a disaster. An advantage of using this kind of definition is that disruption can be measured and provides an important outcome estimate by telling us how much disorganization or interruption of normal life occurs. As with recent earthquakes in desolate areas of California, little disruption may be experienced by the people living there. Conversely, disasters such as major snowstorms may cause little physical disfigurement but a great deal of disruption as cars are abandoned, accidents increase, and people experience great difficulty getting from place to place. This approach suggests that **disaster events**—those occurrences that can cause disasters—must *disrupt* victimized communities in order to be considered disasters.

This makes disasters a little easier to think about and allows a number of events, such as blizzards and droughts, to be included as disasters. However, focusing on disruption alone is not a complete solution to problems in defining disasters. The same problem of how much must occur arises as soon as we try to apply this definition, and we must decide where the cutoff may be for distinguishing between disasters and nondisasters. Consequently, most definitions remain vague about how much disruption is needed to distinguish between a disaster and just a "bad scene." We can define **natural disasters** as events caused by natural forces that disrupt the communities that they strike. We can probably also assume that this disruption must be substantial. A number of characteristics of these events distinguish some natural disasters from others, but for now, we will settle on this as a definition.

Our definition of natural disasters includes extreme weather of any kind (heat, cold, hurricanes, tornadoes, blizzards, ice storms, wind storms, monsoons, etc.). Changes in weather patterns may also be considered disasters if they cause sufficient damage and disruption, although the specific events they cause are more likely to be identified as disasters. As we saw in Chapter 6, El Niño, a weather phenomenon originating in temperature changes in the water of the southern Pacific, has caused dramatic changes in weather across North America whenever it occurs. Its most recent manifestations have been discussed extensively in the media and were associated with better weather in some areas of the continent and worse weather in others. A related ocean-based phenomenon, called La Niña, also affects the weather patterns in North America, and still other oscillations are associated with intensified disaster events in other parts of the globe. These events last only a year or so, but larger changes such as global warming have the potential to cause more extensive and long-lasting climate change and disaster. What are some likely effects of melting of the polar ice caps and a 10-foot rise in sea level, as some predict? Surely, environs from Houston to New Orleans to Melbourne to Bangladesh would experience problems.

Global warming raises a complicating issue when defining natural disasters. Although the weather phenomena and other disaster events it causes are of natural origin, the syndrome of warming may be due, at least in part, to human alteration of the environment. Industrialization and greenhouse effects from air pollution, ozone layer depletion, and other phenomena may contribute to warming. Earthquakes and volcanic eruptions, mudslides, and avalanches are natural disasters that can also be affected by human alteration of the planet. Turning dry prairie into irrigated farmland, for example, could affect underground aquifers (natural reservoirs) and add weather-altering moisture to the atmosphere. We also include floods in our definition, even though these are often caused by a combination of natural events (e.g., rain) and actions taken by people (e.g., improper use of riverbanks). Some floods are almost entirely caused by humans, as in the case of dam failures. These would be more appropriately considered as *technological catastrophes* or mishaps. Other cataclysmic events that are human made, including mine disasters, air crashes, nuclear accidents, and toxic waste contamination, among others, are generally considered technological mishaps.

Research on natural disasters is difficult to conduct for a number of reasons. To start with, these events are almost always studied after they have already occurred. It is very difficult to get information about people *before* they are exposed to disasters, nor can we usually demonstrate changes in mood and behavior from before to afterward. Even if we compare a group of disaster victims to a control group of people who were not exposed to the disaster, we cannot be sure that differences we observe were not present before the disaster. A second problem is the choice of an appropriate control group. With whom shall we compare our findings about victims of a storm or earthquake? Also, choosing measures is difficult because research must often be conducted in recently devastated, often chaotic conditions far from the researcher's laboratory. Obtaining information or recruiting participants is also problematic because recruitment and data collection for a study often must be done quickly. Many times, one cannot sample randomly and must resort to quasi-random or nonrandom sampling (e.g., selecting every third person on a given street). In studies of the impact of Hurricane Andrew, for example, researchers found that entire neighborhoods had been abandoned because of the devastation, making it more difficult to locate those who had been victimized the most. This may cause the available sample to be nonrepresentative of the entire area affected by the mishap and can limit the degree to which we can make general statements about our findings. Despite these methodological problems, research has identified several important characteristics of disasters and has begun to document effects of these events on victims.

CHARACTERISTICS OF NATURAL DISASTERS

In addition to being sudden, cataclysmic events and natural disasters are usually unpredictable. Although we may have some warning, we usually have little time to prepare or leave, and we may not know exactly where the event will hit.

Natural disasters are also typically viewed as uncontrollable. Since there is nothing we can do to guide storms to a particular area or away from another, they go where natural conditions dictate. These characteristics make disasters good candidates for being major stressors.

The destructive power of natural disasters is sometimes enormous and usually substantial. In other words, they usually do damage and sometimes wreak havoc. The sheer magnitude of some disasters makes them unique among stressors. Natural disasters are usually acute, as they often last only seconds or minutes and rarely persist for more than a few days. Heat waves, droughts, and cold spells may persist longer, but most storms, quakes, and other mishaps are over quickly. Once the crisis has passed, coping can proceed, and rebuilding and recovery can be achieved. Usually, this coping requires a great deal of effort.

Of these and other characteristics of disasters, which are the most important in determining how people will react? There is no easy way to answer this question, particularly because not all disasters share all these features. Some may be sudden, some more drawn out. Some may cause destruction, others may not. Is it possible to identify characteristics that will allow us to predict whether a disaster will have major effects? **Event duration**—how long the disaster event affects people or how long it is physically present—is one important variable (e.g., Bolin, 1985; Davidson & Baum, 1986). The longer a disaster event lasts, the more likely victims will be exposed to threat or harm, and so longer events may have bigger effects. Consider the possible effects of varying speeds with which disasters strike and subside: Some, like tornadoes, strike quickly and disappear almost as quickly as they appeared. Others are slower to develop and take longer to recede, as in the case of many floods. Which is worse? The answer to this clearly depends on several other factors such as how strong the storm was or how much destruction was done. However, the length of time a disaster lasts should affect how people caught in it react.

Related to event duration is the presence of a **low point** in a disaster, the point when things are as bad as they are going to get and will now improve over time (Baum, Fleming, & Davidson, 1983). Once past, recovery may unfold as the threat posed by the disaster recedes and attention begins to focus on secondary stressors and rebuilding. Most disasters have such a low point. After a tornado strikes and leaves, the danger has passed and threats to life quickly recede. Though victims must now confront and deal with the damage left behind, the worst is over and improvement can be anticipated. Similarly, storms and earthquakes are associated with threats and damages after which the trend is toward improvement of conditions as recovery efforts restore community services, rescue victims, rebuild homes, and so on. However, some disasters have less clearly discernible low points that are not easily seen or predicted and that increase the duration of the disaster. Earthquakes may be followed by tremors and aftershocks that may obscure the fact that the major damage has already occurred. Each aftershock is accompanied by fear, distress, and apprehensiveness about harm yet to come. Similarly, long-lasting disasters, such as crippling droughts or season-lasting floods, may never seem to "hit bottom," as things just keep getting worse. The impact of multiple events associated with disaster (e.g., a cluster of tornadoes, several earthquakes, or strong aftershocks) could in part be explained as the effect of a new threat when the previous one was over and victims thought the threat was gone. Clearly, the duration and intensity of disasters are important, and the low point may provide a way of looking at them that taps into several aspects of these stressors.

Effects of Warnings

Another characteristic of a disaster situation that appears to affect whether it has severe consequences is whether people have adequate warning of the storm, quake, or other disturbance. Half a century ago, Fritz and Marks (1954) suggested that a lack of warning can increase the consequences of a disaster. However, being warned of a disaster does not ensure minimization of consequences, as the effectiveness of the warning system, the preparedness of a community, and other factors affect the usefulness of alerting news (e.g., Drabek, 1986; Mileti & Sorensen, 1990). Obviously, the rational response to an oncoming disaster is often to run and hide—to take shelter and other precautions to adequately protect life and property. However, we do not always have enough warning to do that, and even when warnings are issued, protective behaviors are not always the first thing we implement. Some people do not take warnings seriously, particularly if the area they live in has had several "false alarms" or predicted storms or earthquakes that did not materialize (Mileti & Fitzpatrick, 1993). Some want to be spectators. In the United States when hurricanes are approaching, some people go out to "watch" them arrive. Hurricane parties are not

uncommon in areas where these storms strike. However, not everyone wants to watch the raw power of nature at its most destructive, and response to natural disasters ranges from well-planned emergency behavior to random, nonproductive activity (Figure 7-3).

This was shown in a study of response to warnings of a flash flood (Drabek & Stephenson, 1971). The effectiveness of repeated warnings in getting people to evacuate was undermined by several factors. First, when families were separated at the time of the warning, they showed more concern about finding each other than with evacuation. Unless a direct order to evacuate was given, people first sought confirmation of the danger and of the need to leave. Further, though the news media actually notified the most people, it was the least effective in producing appropriate responses.

The question of why people do or do not evacuate when faced with potential disasters has been a tough one to answer for years (e.g., Drabek, 1986; Mileti, 1999; Mileti & Sorensen, 1990). Even in instances where we can explain people's responses to impending disasters, most of the differences in actual evacuation cannot be explained. Among more than 750 adults interviewed after the destruction phase of Hurricanes Andrew and Hugo had passed through south Florida and South Carolina, risk perception, social influence (what others did or said they would do), and access to resources predicted reactions to the approach of the storms (Riad, Norris, & Ruback, 1999). Most nonevacuaters provided rationales for their decisions not to leave, some having to do with protecting their homes or their belief that the storm impact would not be "as bad as they say it will be." Determining who will evacuate and why some people resist this precaution remains an important goal as we continue to work to minimize disaster-related death and injury.

One can easily think of instances in which warnings and evacuations cause problems all by themselves. Say you are living in a coastal area and a hurricane warning is issued. All people living in a particular area are advised to leave their homes and move to higher ground for the

Figure 7-3 Response to disaster warnings varies. Some may look for family members, others may ignore the warning, and still others may take protective steps such as sandbagging levees or beaches.

Courtesy of Federal Emergency Management Agency.

storm. Clearly, some people will ignore these warnings, but many will heed them. Without adequate planning, roads may become jammed and people may not know where to go. Some people may panic as they are stuck in traffic, others may become belligerent because evacuation is taking so long, and still others may worry about their unprotected homes. Even if the hurricane turns and heads back out to sea, sparing the community, it has caused a great deal of disruption because people responded to warnings but the plans for evacuation were not clearly drawn (e.g., Drabek, 1986; Miletti, 1999).

Summary of Disaster Characteristics

Other disaster characteristics are listed in Table 7-1. Clearly, disasters can involve a number of factors—they can cause injuries or death, massive property losses, and considerable disruption. They occur in a context as well, and the extent to which communities are prepared for the disaster, conduct orderly pre- and post-disaster procedures (e.g., evacuation), and are able to pull together and rebuild quickly also affect the impact of a disaster on individual victims. When people are exposed to extreme life threat, as is the case in a number of disasters, one would expect reactions to be more extreme. Similarly, when one is injured and/or witnesses others' injury or death, when one is bereaved either before or because of the disaster, and/or when losses are substantial, we would expect

Table 7-1	CHARACTERISTICS OF DISASTERS THAT MAY AFFECT RESPONSE BY VICTIMS

- Life threat
- Injury
- Witnessing injury or death
- Death or injury of relative or friend
- Preparedness of community
- Social cohesion of community
- Financial loss
- Property/possession loss
- Separation from family

stronger, longer-lasting reactions and, perhaps, some negative consequences.

Natural disasters, then, have a number of important characteristics. They are sudden, powerful, and uncontrollable, cause destruction and/or disruption, are usually relatively brief in duration, have low points, and sometimes may be predicted. How perception of these characteristic features contributes to the ways in which disasters affect people is the topic of the next section.

PERCEPTION OF NATURAL HAZARDS

At the beginning of this chapter we discussed the fact that people might not be able to discern accurately how much risk is posed by various hazards. To some extent this is due to the dramatic nature of some hazards. Natural disasters are more dramatic than other hazards such as air pollution, and this may lead some people to assume they are riskier and more hazardous. Among the factors that influence whether individuals are aware of the potential consequences of becoming hazard victims are the crisis effect, the levee effect, and adaptation (e.g., Burton, Kates, & White, 1993).

The **crisis effect** refers to the fact that awareness of or attention to a disaster is greatest during and immediately following its occurrence, but greatly dissipates between disasters. Flood warnings, for example, may be largely ignored until there is a flood. Once the flood occurs, there may be a rush to study the problem, together with the implementation of some public works program. Efforts to prevent the next disaster, however, frequently disappear after this initial rush of activity. The same principle holds for droughts: We tend to take strong water conservation measures only when the drought arrives. We do not practice stringent measures between droughts, and we do not limit population in areas that are drought prone so that there is more water available in time of drought.

The **levee effect** pertains to the fact that once measures are taken to prevent a disaster, people tend to settle in and around the protective mechanisms. Levees are built to keep floodwaters out of populated areas. After a levee is built, however, houses and factories are constructed on what was once considered to be a dangerous floodplain. Unfortunately, levees are built with projected figures for floodwaters in mind, and projections often go wrong. Many communities along the Mississippi River are testimony to this fact. The levee effect also applies to such preventive measures as breakwaters along coasts and reservoirs in drought-prone areas: Once they are built, people flock to settle nearby.

A third factor involved in hazard perception is **adaptation.** Just as we adapt or habituate to a noise or odor, so, too, do we adapt to threats of disaster. Apparently, we can hear so much about a hazard that it no longer frightens us. Large populations in earthquake-prone regions of the world such as California, Iran, Japan, and parts of China attest to this adaptation phenomenon. Floods, mine disasters, and hurricanes follow the same principle: People in the area "learn to live with it." In learning to live with it, they generally discount the possibility that they themselves might become victims.

Several variables appear to influence adaptation to potential hazards. For one thing, when the hazard is closely related to the well-being or resource use of a community, the inhabitants are more aware of the danger. Individuals whose businesses depend on coastal tourist industries, for example, may take more precautions against hurricane damage than residents whose well-being does not depend on the tourist industry.

Personality variables may also affect how we perceive hazards, or at least what we do once they are perceived. In a famous study, Sims and Baumann (1972) noted that although the heaviest concentration of tornadoes is in the U.S. Midwest, most tornado-related deaths occur in the South. The researchers found that differences in subjective perception of danger related to the personality dimension of internal–external *locus of control.* "Internals" believe they are in control of their own fate, whereas "externals" believe outside sources, such as powerful persons, government, God, or fate control their destinies (Rotter, 1966). In interviewing residents

of Illinois and Alabama, Sims and Baumann found that the Illinois residents felt luck had far less to do with their fate than Alabama residents did. Furthermore, Illinois residents appeared to take more precautions when storms approached, such as listening for weather bulletins and warning neighbors, whereas Alabama residents paid less attention to the need to listen to radio or television bulletins. More recent research in both student and nonstudent samples confirms that those with internal locus of control, who believe that damage or injury is preventable, are more likely to take precautions and less likely to take risks (McLure, Walkey, & Allen, 1999). Among students, the belief that disasters (earthquakes) were likely also predicted low risk taking. Apparently then, personality plays some major role in determining humans' perceptions of their control over hazards. Experience with disasters and emergent attitudes about them may be also important factors in whether people prepare for disasters or take steps to minimize potential danger (Jackson, 1981; Mileti & Fitzpatrick, 1993; Norris, Smith & Kaniasty, 1999; Shippee, Burroughs, & Wakefield, 1980).

PSYCHOLOGICAL EFFECTS OF NATURAL DISASTERS

Research has produced varying findings about how natural disasters affect behavior and mental health. Some studies suggest that disasters result in profound disturbance and stress that may lead to continuing emotional problems, whereas other studies suggest that psychological effects are acute and dissipate rapidly after the danger has passed (Rubonis & Bickman, 1991).

Initial impact during the actual precipitating event (i.e., the storm or earthquake that causes the disaster) may be dramatic, and people may be frightened. Yet, as we noted earlier, the immediate response to disasters may be to search for family members or even to have a storm-watching party! The negative effects that many of us expect to appear once the danger has passed are common, but they dissipate more rapidly than many people think. In fact, some studies have found that overall effects of disaster may be positive, because of increased social cohesiveness as victims band together in local groups and help others cope.

One thing people do not do very often is panic in the face of a natural disaster (e.g., Drabek, 1986; Miletti, 1999; Quarantelli, 1998). The immediate response by some is withdrawal, and many people at first appear to be stunned after a disaster strikes. Other immediate reactions may be apathy, disbelief, grief, and a desire to talk about the experience with others. Perception of time may be affected; some people may report everything speeding up while others report that everything seemed to be happening in slow motion. Episodes in which the victim feels like a spectator, passively watching him- or herself and the surrounding carnage may also be experienced. Some antisocial behavior has been noted. For example, after the earthquake that destroyed large parts of Managua, Nicaragua, in 1972, looting was widespread, even while some survivors made efforts to rescue others, put out fires, and so on (Kates et al., 1973). Many authorities argue, however, that looting is rare in natural disasters and prosocial, altruistic responses predominate (e.g., Drabek, 1986; Mileti, 1999). Studies have also found evidence of more positive response to disaster. For example, Bowman (1964) observed the behavior of residents at a psychiatric facility after a massive earthquake near Anchorage, Alaska, on Good Friday, 1964. The residents' initial response was positive—they wanted to help with problems that arose. Bowman observed "a stimulation of all personnel, a feeling of unity, a desire to be helpful, and a degree of cooperation which I only wish it were possible to have at all times" (p. 314). Mileti (1999) also reports that community bonding and social cohesiveness is more common than antisocial actions during and immediately after the disaster event.

Disasters do disrupt organizations and communities as well as families and individuals (e.g., Wright et al., 1990). The functions once performed by larger groups may be assumed by small groups of victims. One of the most serious problems in disasters is coordination of various relief efforts. Despite the need for coordination in successful disaster management, large organizations are often hesitant to assume responsibility. This reluctance can further promote the emergence of cohesive local groups who must assume responsibility for things not being accomplished by formal organizations. A lapse of higher authority also contributes to the development of these local groups. Positive social response during or immediately after a disaster event also appears to be influenced by the needs of the community. When destruction is so vast that rescue teams and official relief efforts cannot cover all needs, locally based groups may fill the void.

Acute Stress and Mental Health

As noted earlier, natural disasters do appear to cause stress, anxiety, depression, and a range of other mood or perceptual disturbances. The extreme life threat associated with most major disasters, combined with other factors that could intensify fear and terror during or after the storm, should be expected to induce mood changes and cause at least transient mental health problems. Indeed, research suggests that disaster victims are more likely to exhibit symptoms of stress and emotional problems shortly after a disaster (e.g., Canino et al., 1990; Lima et al., 1991; Shore, Tatum, & Vollmer, 1986; Tobin & Ollenburger, 1996). For example, a study of survivors of severe mudslides and flash floods in Puerto Rico in 1985 found increased reporting of medically unexplained physical symptoms (e.g., pain, nausea) as a result of exposure to the disaster (Canino et al., 1990). Other studies, this time of the 1989 Loma Prieta earthquake in California, found that survivors of the earthquake reported more depression if they experienced difficulties during the earthquake (Nolen-Hoeksema & Morrow, 1991). They also reported about twice as many nightmares as did a control group at the University of Arizona who did not live near the earthquake (Wood et al., 1992). Most of those surveyed four months after the severe flooding in the U.S. Midwest in 1993 reported symptoms of stress, particularly those who were anxious about the flood or who viewed it most negatively (Tobin & Ollenburger, 1996).

Longer-Term Effects

As the events associated with a disaster recede, the mental health and stress-related effects we have seen should decrease as well. For the most part, studies find that intrusive thoughts, anxiety, depression, and other stress-related emotional disturbances are manifest among victims of floods, tornadoes, hurricanes, and other natural disasters. These effects have been found to last as long as a year, but often do not last that long. Further, they are not nearly as widespread among victims as one might expect. Studies rarely show more than 25% to 30% of victims suffering psychological effects months after a disaster, and it appears that people who lost the most or were otherwise affected more by the disaster are those who continue to suffer (e.g., Moore & Moore, 1996; Parker, 1977).

A review of studies of the mental health problems that follow disasters confirms that there is a small but significant association between them (Rubonis & Bickman, 1991). Depression was observed in about one-quarter of individuals in 10 studies, while nearly 40% of individuals in 15 disaster studies exhibited symptoms of anxiety and 32% exhibited phobias. However, not all studies provided measures of these and other stress symptoms, and the timing of measures was not considered in the analysis. As a result, the intensity or frequency of general effects can be estimated, but the duration of effects cannot be separately assessed. One can conclude, cautiously, that disasters have "small but consistent" effects on mood and well-being and that they can cause stress and a range of psychological problems (Rubonis & Bickman, 1991; Selten, vander Graf, van Duursen, & Gispen-de Wied, 1999).

A profound form of enduring effect of disasters is **post-traumatic stress disorder (PTSD),** an anxiety disorder characterized by having experienced a traumatic event (like a disaster) as well as frequent, unwanted, and uncon-

trollable thoughts about the event, heightened motivation to avoid reminders of the event, sleep disturbances, social withdrawal, and heightened arousal. This disorder can be thought of as an extreme outcome, as it can be debilitating and difficult to treat. Research on veterans of combat, disaster victims, rape victims, and victims of other crimes and severe stressors has contributed to a growing literature on PTSD. Among disaster victims, PTSD appears to be associated with intrusive thoughts and episodes of re-experiencing the disaster (Solomon & Canino, 1990). Other major symptoms include avoidance, numbing, and arousal (Anthony, Lonigan, & Hecht, 1999). Research also indicates that diagnosable PTSD and acute stress have been found among many victims up to four months after tornado and flood disasters (Steinglass & Gerrity, 1990). While some victims continued to experience distress, this study found substantial reductions in stress and PTSD over the year following these four-month assessments (Steinglass & Gerrity, 1990).

Why Do These Effects Occur?

Despite these findings, one can list a number of reasons why acute disaster events could give rise to long-term distress. This is important not only for explaining those long-term effects of natural disasters that have been found, but also in understanding the effects of toxic or human-made disasters discussed in the next sections. One theory suggests that intrusive thoughts about the disaster keep the event "alive"; that is, victims "relive" the disaster each time they think about it. Thoughts about frightening or threatening experiences, even if they occurred months or years before, can elicit distress and responses like those associated with the actual event (e.g., Hall & Baum, 1995). Disaster victims reporting fewer intrusive thoughts tend to exhibit fewer symptoms of chronic stress, and more ruminative styles have been associated with distress following the Loma Prieta earthquake (Baum et al., 1993; Nolen-Hoeksema & Morrow, 1991). Recent studies also suggest that worry and intrusive experiences following Hurricane Andrew and a major earthquake in California predicted stress effects such as those seen in the immune system (Ironson et al., 1997; Segerstrom, Solomon, Kemeny, & Fahey, 1998).

Alternatively, prolonged stress following a disaster could be due to the occurrence of secondary stressors or stressful events that are secondary to the disaster and that can cause problems for days, weeks, or months. Data support the idea that these secondary stressors occur as people run into problems with stores and shops being closed, phone and/or power outages, financial aspects of loss (or loss of one's job if the person's place of business is damaged), or other stressful results of the disaster. Life change is clearly greater after a disaster than would be expected if there had been no disaster (Janney, Minoru, & Holmes, 1977; Melick, 1978; Robbins et al., 1986; see Table 7-2). Further, if damage is severe as it was after Hurricanes Andrew or Hugo, considerable relocation of victims is necessary (Figure 7-4). Relocation is associated with distress, often lasting six months or more, due in part to disruption but also to crowding, lack of privacy, and deprivation (Riad & Norris, 1996).

A study of the effects of Hurricane Hugo examined the relative impact of secondary stressors or chronic stress on postdisaster distress more than a year after the hurricane (Norris & Kaniasty, 1992). Secondary stressors, measured as chronic problems related to financial, marital, parental, filial, and occupational aspects of victims' lives, were strongly associated with distress. Ecological stress (stress derived from as-

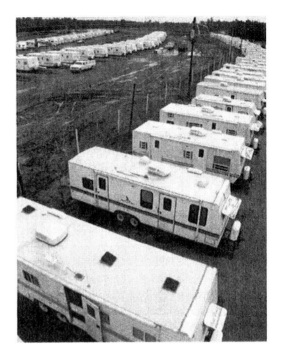

Figure 7-4 Relocation is associated with distress due to disruption, crowding, lack of privacy, and deprivation. These manufactured homes were used for temporary housing after Hurricane Floyd.

Courtesy of Federal Emergency Management Agency.

Table 7-2	CONSEQUENCES OF DISASTER THAT INCREASE DISTRESS

- Injury (being injured oneself)
- Witnessing injury or death of others (horror)
- Life threat (terror)
- Financial loss/damage
- Separation/relocation
- Disruption of the environment
- Bereavement

pects of the neighborhood, such as fear of crime) and stress due to continuing consequences of injury were also related to overall mood and distress. Effects of acute stressors associated with the disaster itself, such as loss or life threat as the hurricane came ashore, were largely explained by secondary financial, marital, filial, and physical stress. Long-term distress associated with loss, injury, terror, or other aspects of the disaster may be caused by secondary stressors that persist long beyond the disaster itself (Norris & Uhl, 1993).

This conclusion is consistent with research by Erikson (1976), who suggested that many of the symptoms of disaster survivors arise from the destruction of the community and loss of a sense of community and belonging. Disorientation and "lack of connection" are common symptoms among disaster survivors when the community has been torn.

Social Support and Disaster

We noted in Chapter 4 that social support can be of major importance in helping people cope with stress. To some extent, positive social effects of disasters may be related to effects of the disaster on social support and one's sense of where he or she fits into a social network. Social support is usually defined as a person being valued and esteemed by other people and being able to get help, emotional support, or other aid if it is needed (e.g., Cohen & Wills, 1985). People with more social support usually fare better in dealing with stress and appear to have fewer adjustment problems after disasters (e.g., Fleming et al., 1982; Norris & Murrell, 1984; Thoits, 1982). However, some stressful situations obviously change how much social support we have. Loss of a spouse, close friend, or other confidant by divorce, death, graduation, or relocation alters one's base of support, in turn contributing to problems associated with stress (Eckenrode & Gore, 1981; Rook & Dooley, 1985).

Kaniasty and Norris (1993) addressed the possibility that disasters can "deplete" available social support, rendering it less effective in buffering stress. We know that people experiencing stress usually receive support if it is available, and that stress seems to increase need for support (Cohen & Hoberman, 1983; Kaniasty & Norris, 1991). One can reasonably expect that disasters will increase people's needs for support, particularly if they are injured, left homeless, or similarly affected by the disaster. Victims of disasters should have greater need for support after a disaster than before it.

At the same time, disasters appear to decrease the amount of social support that is available. This is true for several reasons. First, the number of people in a social group who need support at the same time is likely to be greater than normal. If a group of 12 friends usually serves as a social support source for its members, the usual scenario is probably one in which a few members need support and the other 9 or 10 are there to provide it. However, after disasters, most of these people probably will need help because they have all been affected by the same major stressor. Can people who need support also provide it effectively? If they cannot, available support will decrease. At the same time, those who are available to provide support after a disaster are facing dramatically increased demand for support, and it is possible that they burn out or otherwise cannot keep up with the need around them. Together, these social processes may lead to greatly reduced available support at a time when need for support has increased dramatically (Kaniasty & Norris, 1993).

Evidence of this deterioration of social support following natural disasters has been reported. In a study of severe flooding in Kentucky in 1981 (Kaniasty, Norris, & Murrell, 1990), perceptions of social support decreased as a result of the flooding. In a longitudinal analysis of these disaster victims, evidence of decreases in social grounding and social support was found (Kaniasty & Norris, 1993). Perceived availability of support also declined, and this was related to losses from the flood. However, there appeared to be broad decreases in the extent of and need for social support as well as increased estrangement from one's social networks. The situation was one in which need for support exceeded the amount available, which contributed directly to stress and adjustment difficulties after the disaster (Kaniasty & Norris, 1993).

Children and Disasters

In general, research suggests that children respond to disasters in much the same way as adults (Garmezy & Rutter, 1985), although in some cases they appear to recover more rapidly (e.g., Green et al., 1994). A study of victims of a wildfire in California indicated that children who lived in homes that were burned (and either destroyed or seriously damaged) reported more evidence of PTSD, including more thoughts and dreams about the fire and more attempts to avoid reminders of the fire (Jones, Ribbe, & Cunningham, 1994). This is consistent with other studies (Aptekar & Boore, 1990; Lonigan,

LONGER-LASTING TRAUMA IN CHILDREN:
The Kidnapping

We have observed that stress effects of disasters usually do not last long in children. This should not be taken to mean that other stressors have only short-term effects in children.

For example, Terr (1979, 1983) studied child victims of a mass kidnapping in Chowchilla, California, in 1976. Twenty-six children, ages 5 to 14, were kidnapped in their school bus and held for 27 hours; some of this time was spent in a buried truck trailer with stale, foul air and cramped space. During the year after the kidnapping, nearly all of the children were interviewed, and all of them were found to be experiencing some symptoms of stress. Most also experienced intrusive thoughts and dreams, and some evidence of perceptual distortions, fantasies, and anxiety was found (Terr, 1979). A follow-up two to five years after the kidnapping showed that many symptoms were still experienced and that most continued to show signs of distress (Terr, 1983). This was not a natural disaster and it did involve human-caused victimization. It serves to illustrate that some stressful events can have lasting effects.

Anthony, & Shannon, 1998; Maida et al., 1989; Vogel & Vernberg, 1993). Specific fears related to the nature of the disaster (e.g., high winds if the child was exposed to a tornado), depression, anxiety, and PTSD have been found in child victims of several disasters (Vogel & Vernberg, 1993). Among more than 5,000 child and adolescent survivors of Hurricane Hugo, many cases of PTSD emerged and were generally characterized as consisting of active avoidance of intrusive thoughts, passive avoidance and numbing, and arousal (Anthony et al., 1999). When children are exposed directly to environmental disruption, they react strongly to scenes of death and mutilation if these are present (Newman, 1976). They may also regress to earlier stages of behavior. Personality measures of anxiety and emotional reactivity during a severe hurricane were found to be related to distress three months after the storm (Lonigan et al., 1994). In some cases more severe or longer lasting consequences have been seen, but these usually occur after particularly savage disasters (e.g., Yule & Williams, 1990). More commonly, evidence across several studies suggests that effects on children tend to be mild and may not last very long (Belter et al., 1991; Bromet, Hough, & Connell, 1984; Handford et al., 1986; McFarlane et al., 1987; Sullivan et al., 1991). Denial, projec-

tion, and other coping were associated with *less* distress among children within two months of a fatal lightning strike that they witnessed (Dollinger & Cramer, 1990).

Age and Disaster Response

The fact that children suffer from disasters is not really surprising, but it is of interest nonetheless. More surprisingly, age of children does not seem to have much effect on responses to disasters, although older children are able to recall more details of a disaster than are younger children (Bahrick, Parker, Fivush, & Levitt, 1998). While some symptoms may be more likely among younger and older victims, the effect of age on severity of distress depends on the symptom or situation being sampled (Vogel & Vernberg, 1993). This seems true over the life span as well. Research has been similarly difficult to interpret when focused on the elderly or on comparisons at older ages. A number of studies indicate that older people are likely to suffer distress after disasters and in some cases have suggested that worries and concerns may differ from those of younger victims (e.g., Ollendick & Hoffman, 1982). Older people may emphasize the loss of items that symbolize their lifetimes—a family photo album or a tree or a garden, for example.

However, comparisons of older and younger victims generally show comparable severity of distress or that older victims are less upset and stressed (e.g., Bell, Kara, & Batterson, 1978; Bolin & Klenow, 1982; Huerta & Horton, 1978). Failure to find reliable differences in distress, coupled with findings suggesting that older and younger victims use similar types of coping, suggests that distress following disasters is similar among most or all age groups (e.g., Craig et al., 1992). However, some studies divide victims into three rather than two groups (young/middle/older age, rather than just younger and older) and have found that middle-aged flood victims are more likely to exhibit distress than the younger or the older groups (e.g., Gleser et al., 1981; Shore et al., 1986).

An analysis and study by Thompson, Norris, and Hanacek (1993) may help explain these effects. There are several ways to look at the differences in how people of different ages might respond to disasters. For example, Thompson and her colleagues suggest that older victims are likely to receive the greatest exposure to disasters or their effects; they are more likely to be injured, less likely to evacuate, and so on (e.g., Bolin & Klenow, 1982). Alternatively, one could argue that older victims have fewer resources with which to cope. Either would lead one to predict that older people would show more distress. A third way to think of this, however, would favor the hypothesis that younger people are more adversely affected: If coping efficacy and scope of one's coping increases with age, older victims should be able to cope more efficiently, negating and reversing any advantage associated with resources. Finally, if one considers burden, or the extent to which people are in caregiving or provider roles (and assuming that burden makes them more vulnerable to disasters and disruption), our predictions would be more in line with findings suggesting that middle-aged people (who are taking care of children and, possibly, their parents) would be most affected.

These four kinds of predictions were systematically evaluated in a study of 831 adults from four areas affected by Hurricane Hugo but varying in severity of storm impact (Thompson

et al., 1993). The effect of age on distress was studied 12, 18, and 24 months after the storm. A number of exposure variables (variables that increased or decreased how much people were affected) were related to postdisaster distress. Injury, life threat, financial loss, personal loss, and scope of exposure all had effects on distress, with the effects of exposure greatest among middle-aged victims. These findings led the investigators to conclude that, regardless of age, the disaster had a substantial impact on mood and mental health. However, the burden hypothesis, positing a more profound impact of disaster on middle-aged caregivers, was supported by the finding that the stressor effects of each exposure variable were greatest among this middle-aged group (Thompson et al., 1993).

ENVIRONMENTAL THEORIES AND DISASTERS

As we have already observed, stress formulations as well as other theories or approaches to the study of environmental events can be used to understand the phenomena described in this chapter. Destruction of a community can involve behavioral constraint, as options for activity are reduced and behavioral freedom is limited. People may be forced to leave their homes and move to large shelters where behavior must conform to emergency rules. Water and power service may be disrupted, further limiting what people can do, and plans are unavoidably changed by the sudden impact of the event. When our behavior is constrained, we may react negatively to the loss of freedom, feel bad, and act in ways to reestablish our freedom. Continuous constraints on our behavior that cannot be removed may eventually cause us to experience helplessness. Fortunately, once the emergency is past, constraints are reduced and gradually disappear. People return to their homes and normal services are restored. However, for those who are made homeless or who have lost a family member or close friend, constraints and their negative effects may continue. Rebuilding or relocating is necessitated, and choices of activities are further limited by losses and the need to cope with them.

In communities that have lost many people in a disaster, ecological perspectives may help explain the effects of the event. You will recall that *staffing* refers to the number of people in a setting relative to the roles that need to be played. Over- and understaffing, in which too many or too few people are present in the setting, are viewed as negative states that cause problems. Communities are settings, and there are roles that must be played in them. When a large number of the members of the community suddenly die, the community may become understaffed, and those remaining are forced to assume multiple roles. This can cause strain and, coupled with problems caused directly by specific losses, can help to explain the effects of disaster. However, understaffing can also increase cohesiveness among those left, and thus produce positive effects.

A relatively new approach to disaster impact is based on the theory of **conservation of resources (COR)** that was proposed by Hobfoll (1989). The COR proposes that the extent to which people lose important resources (social and psychological resources as well as material resources) or are able to minimize this loss will determine how much stress is experienced. Resources are anything that can help people achieve important goals, and they include tangible or material assets (such as money), social resources (family roles, work roles), as well as personal strengths (optimism, coping skills;

Freedy et al., 1992). Loss or threatened loss of resources should exacerbate stress, while maintaining stable resource bases should minimize stress. Similarly, reestablishing resources after they have been lost should reduce stress, and significant loss in any of these may cause problems. A study of the effects of Hurricane Hugo (Freedy et al., 1992) suggested that resource loss in these domains was associated with distress following the disaster. Two to three months after the hurricane, resource loss was strongly related to distress and was the strongest predictor of postdisaster outcomes. The experience of intense resource loss was linked to clinically significant episodes of distress. This COR theory is also applicable to a number of other natural disasters and may prove to be an excellent model for predicting and understanding distress following natural disasters. Figure 7-5 shows how all of these theories can fit our eclectic environment–behavior model.

SUMMARY OF EFFECTS

Natural disasters are clearly stressful, limit freedom and behavioral options, deplete resources, and may cause a shortage of people, leading to the breakdown or disruption of a community. People seem capable of coping with disasters, and serious long-term effects of disasters—especially with social support—are not extensive. In many ways, these events are similar to disasters caused by human-made aspects of the environment, which we examine next.

Technological Disasters

To a large extent our perceived dominance of the natural environment and our adaptation to its hazards have been achieved through advances in technology. When we come upon a problem or a threat to continued survival and well-being, we build machines or otherwise fabricate the tools to solve the problem. Improvements in the quality of life, prolongation of life,

mastery over disease, and the like are based on a broad technological network we have created. These machines, structures, and other human-created additions to our environment share unparalleled responsibility for supporting our way of life. For the most part, they accomplish this goal and work well under human control. However, this network occasionally fails,

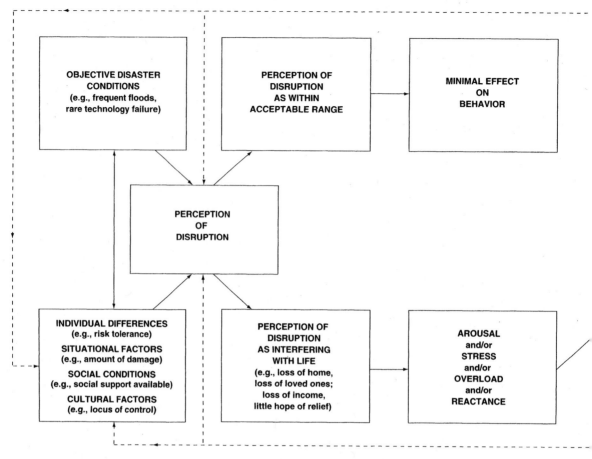

Figure 7-5 Our eclectic model of theoretical perspectives applied to disasters.

and something goes wrong. As a result, we have blackouts of major cities, transportation accidents, leakage of toxic chemicals from waste dumps, dam failures, and bridge collapses. Many of the characteristics of these catastrophes appear to differ from those of natural disasters, and the effects of being in one may differ as well. We will first consider the characteristics of technological catastrophes, comparing them with natural events, and then discuss their effects.

CHARACTERISTICS OF TECHNOLOGICAL CATASTROPHE

What are the differences between these catastrophes and the natural disasters we have already discussed? For one thing, they are human

made. They are not the product of natural forces, but rather are caused either by human error or miscalculation, as some part of our extensive technological net fails. The duration of technological accidents is variable. They may be acute and very sudden, as in the case of a dam failure or blackout. When a dam fails, the resulting wave of water assumes most of the characteristics of a natural disaster, striking swiftly and continuing on its way. Major power failures can plunge entire cities into darkness in a matter of minutes, but are usually quickly fixed. These technological mishaps are usually brief, and the worst is soon over. However, other technological catastrophes are more chronic and may not have clear low points. The discovery of contamination at Love Canal began a chronic period of distress for area residents, as did the nuclear accident at Three Mile Island. In both of these cases, sources of threat (toxic contamination and radioactivity) remained in the area for many years and required long-term coping with the threats that were posed. For people affected by these events, the worst was not over quickly, nor was it easy to identify the point at which things began to improve. A great deal of uncertainty can accompany such events.

Interestingly, technological catastrophes are often more likely to threaten our feelings of control than are natural disasters. This is somewhat paradoxical, since natural disasters are inherently uncontrollable and we never really expect to be able to control their occurrence. Technological catastrophes, on the other hand, represent the consequences of occasional loss of control over something we normally control

quite well. If this loss of control is intermittent, temporary, and not indicative of an entire collapse, why is it so disquieting?

It is possible that because technological catastrophes are losses of control we are supposed to have, they shake our confidence in our ability to control events in the future. These events are never supposed to happen—technological devices are designed never to fail unpredictably, and to warn us when they are worn out. Thus, nuclear power plants are not supposed to have accidents, ever, and toxic waste dumps are not supposed to leak. But these things do happen, and they often appear to occur at random. Instead of saying, "No accidents will happen," we may often find ourselves wondering, "Where will the next explosion occur?" "Which plane will crash?" "Which waste dump will leak?" and so on. While the above analysis is somewhat speculative, it provides some feel for the complexities of people's responses to technological catastrophes. By reducing our sense of control over technology usually taken for granted, these disasters may reduce expectations of control in other areas of our lives. These kinds of events can reduce general perceptions of control and lead to stress (Davidson, Baum, & Collins, 1982).

While natural disasters often cause a great deal of destruction, human-made catastrophes are often characterized by a lack of visible destruction. Natural disasters are more familiar to us, occurring at fairly regular rates around the world in almost predictable ways. The chances are that during some seasons in some regions, tornadoes are likely and that at other times, in other areas, hurricanes are likely. Natural disasters also begin very quickly and, to some extent, can be forecast. They are powerful, among the most threatening stressors we know about, and pass very quickly. Once they pass, rebuilding and recovery can begin, and once these are complete, a sense of closure may be gained.

Technological disasters are less familiar to us, seem to occur less often, but are potentially more widespread. Natural disasters are selective in where they occur; hurricanes tend to be coastal and tornadoes inland. Technological accidents are not predictable at all: One cannot forecast the breakdown of something that is never supposed to break down in the same way as one can forecast a storm. The onset of technological catastrophes is usually sudden, with little warning. The speed with which these events unfolds often makes them difficult to avoid. Those in the path of a flood following a dam break have little time to get to safety.

Some technological mishaps, such as factory explosions, train accidents, and mine accidents, have a well-defined low point. In these cases, coping with the disaster may be similar to that of natural disaster recovery. However, it appears that some of the most powerful technological disasters may also be those without a clear low point. For example, situations in which individuals believe that they have been exposed to toxic chemicals or radiation (e.g., Love Canal

and TMI) involve long-term consequences connected with the development of disease many years after exposure. There may be considerable uncertainty about this, and for some technological disasters there is no clear low point from which things will gradually get better. The worst may be over, or it may yet surface. Thus it could be difficult for some persons to return to normal lives after the accident has ended.

Another possible difference between natural and human-made disasters is the nature of the postdisaster community response. As we have noted, several studies have found that positive as well as negative effects of disasters are apparent, and that social cohesiveness or feelings of social bonds may be stronger afterward (e.g., Cuthbertson & Nigg, 1987). Following natural disasters, these social changes may provide a crucial resource in aiding recovery from loss and disruption. While these developments do not occur in all areas affected by natural disasters, they are noted often enough to be considered a possible outcome of a natural disaster. But what about human-made disasters? Are the same changes likely after a human-made accident or hazard has occurred? Anecdotal evidence suggests that controversy and conflict among neighbors may be more likely, leading one to question the postimpact similarities between natural and technological catastrophes.

Cuthbertson and Nigg (1987) studied two human-made disasters to determine whether socially supportive postdisaster groups developed in their wake. One involved asbestos contamination at a trailer park, the other spraying of pesticides near residential areas. In both cases, victims included those who were worried about having been exposed to toxic substances (the asbestos and pesticide) and those who thought the exposures were nothing to worry about. Their differences of opinion, common following events involving questionable toxic exposures with long-term, uncertain effects, was the basis for conflict and disagreement among neighbors. Just as at Three Mile Island, where some are strongly opposed to nuclear energy including the TMI plant while others are strongly in favor, those who were worried and those who were not were in conflict. Cuthbertson and Nigg (1987) found evidence of anger, frustration, resentment, helplessness, defensiveness, and a polarization of attitudes about the hazards and did not find any evidence of the development of supportive, cohesive groups.

While similar in many ways, technological and natural disasters do appear to be different. These differences may be partially responsible for the greater preponderance of chronic distress among victims of technological accidents that are discussed in the next section.

EFFECTS OF TECHNOLOGICAL DISASTERS

The *immediate* effects of human-made disasters are often similar to those of natural disasters. This is particularly true when technological catastrophes are like natural ones in duration, suddenness, and so on. Thus, when Fritz and Marks (1954) studied several human-made accidents, including an air disaster in which a plane plunged into a crowd of air show spectators, they found the same types of responses that have been found in studies of natural disasters. Panic did not often occur, and when it did, it was usually seen as an attempt to escape immediate threat. Less than 10% of victims interviewed reported that they felt they were "out of control" during the disaster, and while many were confused and disoriented, others behaved in constructive, "rational" ways.

Fewer similarities in response to natural and technological mishaps have been observed when the impact of the technological stressor is longer lasting. When people are told that they have been exposed to toxic chemicals or believe that they have been irradiated, the perceived threat to life and limb may be no less than when a plane is about to crash into a crowd. Clearly, it is less intense, partly because it is longer lasting and usually slower to unfold. Unlike an air crash, where one might have a minute or two at most to decide what to do, people exposed to toxic hazards may have months or years to think about what is happening. Weil and Dunsworth (1958) observed the reactions of townspeople to a coal mine cave-in at Springhill, Nova Scotia. While rescue efforts were in progress, the panic,

grief, and anxiety was punctuated by mood swings to euphoria when some miners were rescued, but after rescue efforts had ceased, response appeared to be more suppressed.

Some argue that the consequences of technological catastrophes are more complex and/or longer lasting than those caused by natural disasters (Baum, 1987; Baum, Fleming, & Davidson, 1983; Gleser, Green, & Winget, 1981). Adler (1943) reported on the effects of a human-caused fire at the Coconut Grove nightclub in Boston. The fire killed 491 patrons of the club and was characterized by great terror. More than half the survivors developed psychiatric symptoms such as anxiety, guilt, nightmares, and fear a year after the fires. Interestingly, of those who did not develop psychiatric problems, 75% had lost consciousness during the fire, most remaining unconscious for more than an hour. Of those who did exhibit psychological difficulties, only half lost consciousness, mostly for less than an hour. Remaining conscious through the fire appeared to contribute to emotional distress. Unconsciousness, and therefore less exposure to the terror and horror during the fire, was associated with more positive mental health outcomes.

Long-term psychiatric distress, including mood disorders, is common in victims of technological disasters. One clear consequence of disasters is an increase in the experience of intrusive thoughts and memories. Several studies have linked intrusive thoughts and PTSD with human-made disasters (e.g., Smith et al., 1990), including a study of survivors of a ship disaster in which nearly 4 in 10 passengers died (Thompson, Chung, & Rosser, 1994). Intrusive thoughts and neuroticism were elevated among survivors, relative to norms. Many of the emotional effects of disasters are associated with PTSD, and recent research suggests that acute PTSD constitutes or co-occurs with the majority of psychiatric disorders following disaster or trauma (North, Smith, & Spitznagel, 1994). In other words, PTSD was the most frequently seen disorder and when other disorders, such as depression, were present, PTSD was likely to be diagnosed as well. This may suggest that the symptoms and sources of distress in PTSD are fundamental in initial response and coping after disasters, and that continued or prolonged PTSD could facilitate the development of depression or other disorders. Intrusive thoughts or memories may prove to be one of the more "lethal" aspects of disasters for stress and stress-related consequences. Not only do they appear to contribute to chronic stress following disasters (Baum et al., 1993), but they also may affect the likelihood that someone who has been exposed to very stressful events will seek treatment and/or counseling.

Though the number of studies of human-made disasters has increased and we have learned more and more about them, the Buffalo Creek Flood and the Three Mile Island and Chernobyl nuclear accidents remain vivid and well-studied examples of these types of catastrophes. Considering them in some detail will help us understand how human-made disasters differ from natural ones and how they affect mood and behavior.

The Buffalo Creek Flood

Perhaps the most intensively studied disaster with a human cause is the dam break and flood at Buffalo Creek in West Virginia. On February 26, 1972, a dam constructed by a mining company, which had been dumping coal slag in the creek, gave way and unleashed a wave containing millions of gallons of water. The flood washed away houses, automobiles, and everything else in its path, careening off the walls of the valley and killing 125 people. When the wave finally spent its rage and drained into a river at the foot of the valley, it left behind a scene of death and devastation. Five thousand were left homeless, and the valley was disfigured and permanently altered.

The Buffalo Creek flood was clearly due to failure of a human-made device—the dam holding back the creek. As with other technological mishaps, the flood was never supposed to happen. As a result, it was even less predictable than the storms that had swollen the creek behind the dam. And, as we have suggested, this disaster, partly because of its human origins, appears

to have had more chronic effects on the victims than ordinary floods or natural disasters. While many of the specific effects are similar to those observed in studies of natural disasters, they seem to have had more lasting consequences.

Research at Buffalo Creek has identified a number of problems occurring as late as two years after the flood (e.g., Gleser, Green, & Winget, 1981; Lifton & Olson, 1976; Tichener & Kapp, 1976). These problems included

1. Anxiety. Fears about the disaster and about the changes in lifestyle that came in its aftermath were common.
2. Withdrawal or numbness. Almost all researchers at Buffalo Creek noted apathy and blunted emotion after the flood.
3. Depression. Many survivors lost everything they had worked a lifetime for and became sad and subdued.
4. Stress-related physical symptoms. Almost all somatic or bodily symptoms, including gastrointestinal distress, aches and pains, and so on, were increased.
5. Unfocused anger. Survivors found themselves angry and upset. When disasters are human made, the rage tends to be worse. This is due, in part, to the fact that although there was a culpable agent, identification of a specific person to blame was difficult.
6. Regression. Children often regressed to earlier stages of behavior.
7. Nightmares. Dreams about dying in the disaster and about dead relatives occurred frequently. Sleep disturbances were common as well.

Tichener and Kapp (1976) noted that traumatic neurosis was evident in more than 80% of the sample they studied. Anxiety, depression, character and lifestyle changes, and maladjustments and developmental problems in children occurred in more than 90% of the cases. Anxiety, grief, despair, sleep disturbances, disorganization, problems with temper control, obsessions and phobias about survival guilt, a sense of loss, and rage were some of the symptoms. Lifton and Olson (1976) listed several characteristics of the flood at Buffalo Creek that intensified the reactions to it: the suddenness, the human-cause factor, the isolation of the area, and the destruction of the community. Survivors were aware of the symptoms. They were surprised at how long they had survived and afraid that recovery was impossible (Lifton & Olson, 1976).

There are many reasons why the Buffalo Creek flood appears to have caused more extensive, longer-lasting psychological distress than do most floods. In fact, follow-up research showed that symptoms of anxiety and depression were still elevated among flood survivors 14 years later (Green et al., 1990a, 1990b). The human cause is but one of these potential reasons, focusing anger on the mining company and affecting the ways in which the disaster was experienced. In addition, the flood was unusually severe, washing away an entire community and causing immense destruction. Recovery was inhibited by delays in removing debris, and it was weeks before homeless survivors were provided with temporary homes. Trailers, brought in to house victims, were not assigned so as to allow friends and family to live together, further disrupting the sense of community that had characterized the valley. All of these factors are likely contributors to the enduring effects of the disaster.

The Three Mile Island and Chernobyl Accidents

Research at Three Mile Island (TMI) illustrates the kinds of effects that can occur over a long period of time following a technological catastrophe. In March 1979, an accident occurred in Unit 2 at the TMI nuclear power station. Through a number of equipment failures and human errors (see Chapter 13), the core of the reactor was exposed, generating tremendous temperatures. The fuel and equipment inside the reactor was damaged, and by the time the reactor was brought back under control, some 400,000 gallons of radioactive water had collected on the floor of the reactor building. In addition, radioactive gases were released and

remained trapped in the concrete containment surrounding the reactor. During the crisis, which lasted several days, there were a number of scares. Some people feared a nuclear explosion, others a meltdown, and still others massive radiation releases. Information intended to reduce fears often increased them because it was contradictory or inconsistent with other information that had been released. An evacuation was advised, and this probably contributed to the chaos and fear of the moment (Figure 7-6).

Without a doubt, the accident at TMI caused stress. During the crisis period there was a good probability that threat appraisal would occur; research suggests that most people living near the plant were threatened and concerned about it (Flynn, 1979; Houts et al., 1980). Immediately after the accident, studies found greater psychological and emotional distress among nearby residents than among people living elsewhere (Bromet, 1980; Dohrenwend et al., 1979; Flynn, 1979; Houts et al., 1980).

Despite the fact that the severe threats associated with the accident disappeared relatively quickly, it does not appear that the kind of recovery that characterizes the aftermath of many disasters was evident at TMI. The potential danger of radiation release remained long after the reactor was brought under control. The radioactive gas remained trapped in the containment building for more than a year after the accident. For some area residents the potential for exposure from this source remained a threat, accentuated by occasional leaks of small amounts of the gas. Approximately 15 months after the accident, the gas was released in controlled bursts into the atmosphere around the plant. The radioactive water remained in the reactor building, and decontamination of the reactor required many years.

Research on the chronic effects of living near TMI suggests that stress persisted among some area residents up to six years after the accident. Bromet (1980), for example, has reported evidence of emotional distress among young mothers living near TMI a year after the accident and found evidence of more persistent distress among these TMI area residents as

Figure 7-6   Like other technological mishaps, the accident of Three Mile Island has had long-lasting effects.

well (Dew et al., 1987). A series of studies of men and women also identified stress effects among some TMI area residents, starting about 15 months after the accident (Baum, Fleming, & Davidson, 1983; Baum et al., 1993; Gatchel, Schaeffer, & Baum, 1985) and lasting until 1986 and beyond, more than six years later (e.g., Davidson & Baum, 1986; McKinnon et al., 1989). Studies also suggest persistence of sleep-related difficulties and of arousal related to stress (Baum & Fleming, 1993; Davidson & Baum, 1986; Davidson, Fleming, & Baum, 1987). People living near the damaged TMI reactor reported more bothersome symptoms, were more easily awakened at night, and took longer to fall back to sleep than did a control group (see Table 7-3). Interestingly, urinary norepinephrine levels among TMI area residents were higher both while individuals were awake and asleep, and whereas controls showed normal differences between sleeping and waking levels, TMI area residents did not (Davidson, Fleming, & Baum, 1987). Some differences in immune response were also found, with TMI area residents showing some evidence of fewer numbers of some immune cells and less effective control of latent viruses (McKinnon et al., 1989). Though the intensity of this chronic stress appears to be moderate, the fact that it persisted for so long is unusual.

Table 7-3   PERSISTENCE OF STRESS AT THREE MILE ISLAND THREE YEARS AFTER THE ACCIDENT

Group	Total number of symptoms reported	Being awakened at night (1–7 scale)	Time to fall back asleep (minutes)	Urinary Norepinephrine (mg/ml)	
				Awake	*Asleep*
TMI	32	3.7	18.0	31.3	35.9
Control	17	2.5	16.3	18.5	13.8

These studies also reported effects of several variables that we have considered as influencing stress. Not all TMI residents seemed to be stressed. Fleming et al. (1982) found that TMI area residents who reported having lower amounts of social support exhibited greater evidence of stress than did those who had a great deal of support. Differences along coping style dimensions were also found, as TMI area residents who were more concerned with palliative coping (managing their emotional response) showed fewer stress symptoms than did TMI residents who were more concerned with taking direct action and manipulating the problem (Collins, Baum, & Singer, 1983). Finally, the continued uncertainty at TMI appears to have suppressed feelings of personal control among TMI area residents, and those who reported the least confidence in their ability to control their surroundings exhibited more symptoms of stress than did residents who were more confident (Davidson, Baum, & Collins, 1982).

The 1986 accident at the nuclear power plant at Chernobyl, in the Ukraine, was considerably larger and more hazardous than the TMI accident. It is generally considered the worst known nuclear accident in history and undoubtedly involved much more radiation release and exposure than did TMI. It also affected a larger number of people and appeared to last longer as well. However, research has suggested similarities in acute and more sustained distress around the Chernobyl plant 3, 6, 12, and 20 months after the accident. The MMPI (an instrument containing scales of clinical symptoms) was administered to the Chernobyl workers and a control group of workers from another nuclear power plant (Koscheyev et al., 1993). Workers at the Chernobyl plant showed more symptoms of distress than a control group, and the percentage of workers in the Chernobyl group with at least one elevated clinical scale increased over time, from 18% to 33%, while the control group had about 10% showing at least one elevated clinical scale (Koscheyev et al., 1993).

SUMMARY AND CONCEPTUAL CONSIDERATIONS

In addition to causing distress, technological catastrophes also involve processes related to the other theoretical orientations discussed in Chapter 4. With few exceptions, technological mishaps lead to behavioral constraint, loss of control, and the problems associated with these states. Evacuation, whether temporary or more permanent, disrupts and limits what we can do. People may find it more difficult to sell their homes if they live near a damaged reactor or hazardous waste site and thus may be limited in their freedom to move. At Buffalo Creek, the destruction of almost everything in the valley also severely limited what people could do and required almost complete attention to a circumscribed set of recovery options. People could, for example, rebuild their homes and places of business, move to a "safer" nearby area, or "call it quits" and leave altogether. Very few realistic options may be available following such an event.

Staffing levels may become important when a community loses many members, but for some technological catastrophes this is not the case. At TMI, the number of people living in the community did not drastically decline, while in the Buffalo Creek flood, many people died. Staffing theory provides useful predictions primarily when losses have occurred.

Primary and Secondary Victims

There are many victims of disasters, often not people directly threatened by the storm, earthquake, or accident. Those people who are directly affected, called *primary disaster victims*, include people who live or work in the path of a tornado, the epicenter of an earthquake, the landfall of a hurricane, or near a toxic waste site. Primary victims also include those living near nuclear plants that malfunction or those riding in aircraft that crash. These victims are the ones we usually think of as disaster victims, people whose lives are directly threatened, and/or those who experience the disaster event in all its fury. *Secondary disaster victims* are also affected by disasters, but not as directly. They are not in the disaster, they do not experience the fury of the event firsthand, nor do they survive any direct threats to life associated with the disaster event. They include people who own property that is destroyed by a disaster but who are not there during the event; people with friends and/or family who are primary victims and may have been killed or injured; or people who live nearby, escape the wrath of the disaster, but suffer later from shortages, disruption, and other latent effects of disasters.

Distinctions between primary and secondary victims are sometimes difficult to make, and it can be hard to draw a clear line between those who were directly affected and those who were indirectly affected. For example, when a tornado strikes a community, it approaches, strikes some part of the town or countryside, and moves on. Most people living in the community are directly threatened, and many will take shelter in anticipation. However, tornadoes strike only parts of communities, perhaps going down one side of a street or meandering aimlessly through a neighborhood. Many of us have seen photos of communities where one house has been demolished by a tornado but the house next door stands untouched. In such cases, one could consider all the people in the community as primary victims, or we could decide that only those living where the twister actually touched down should be considered to be primary victims. If we found that everyone was responding in the same way after the disaster, it would favor the community case, but if we found that those more directly affected were more upset, were more stressed, or exhibited more mental health problems, it would favor the narrower categorization.

One of the more interesting victim groups is rescue and recovery workers who follow disasters into a community and work at the disaster site. Initially, these disaster workers are most concerned with rescue efforts, searching for survivors, pulling them from rubble, and providing them with life-sustaining support until they can receive needed medical or other help. As time passes and the likelihood of finding survivors declines, attention typically turns to recovery of personal property, bodies, and debris. Some disasters, like motor vehicle accidents, require only a small team of rescue workers—police, paramedics, firefighters. Other disasters, such as air disasters or earthquakes, usually require larger numbers of medical, emergency, and other workers, such as military personnel. In larger disasters, many of these workers may be volunteers and may not be prepared for the difficult work they must perform. In some cases, such as in recent earthquakes in Turkey and Taiwan, flooding in Venezuela, or recent air disasters at sea, a veritable army (or navy) of rescue and recovery workers is often needed (Table 7-4).

There are many sources of stress for disaster workers. First, they usually have to work under difficult conditions, for long hours, with considerable pressure on them. For example, after destructive earthquakes or tornadoes, debris must be cleared away in order to rescue survivors buried in the rubble, and seriously injured people must often be rescued and provided medical care. After air crashes, the bombing of the Murrah building in Oklahoma City, or other human-caused disasters, similar scenes played out, and too soon attention turned from rescue to recovery. Once it is deemed unlikely that any

Table 7-4	DIFFERENCES BETWEEN TYPES OF DISASTER VICTIMS	
Disaster Victims	**Disaster Experience**	**Major Stressors**
Primary victims	Directly affected by event	Life threat, injury, or loss
Secondary victims	Indirectly affected; no life threat	Bereavement, loss
Workers	Clean-up, body recovery, rescue	Response to death, grotesque disfigurement, fatigue

Figure 7-7 Disaster workers may show their own signs of stress. Encountering death and body parts has particularly distressing effects.

Courtesy of Federal Emergency Management Agency.

more survivors will be located, disaster workers begin the sometimes gruesome task of locating and transporting bodies of victims. In some cases, this activity can be particularly disturbing, especially when victims are children or when parts of bodies rather than complete remains are found (Figure 7-7). This process is hard work and because of the urgency of the situation, it often demands around-the-clock effort, giving rescuers and emergency workers little time for rest. The sources of stress for disaster workers are summarized in Table 7-5.

Evidence is mixed, however, about how much stress is actually experienced by disaster workers. Even though many disaster workers are volunteers, many have experience with emergency or medical work, and most select professions or positions in which these stressors regularly occur. It is possible that a combination of personality characteristics, coping skills, and experience or training makes some people more resistant to the upsetting aspects of disaster work. Regardless, studies generally indicate that disaster workers experience some distress as

Table 7-5	SOURCES OF STRESS FOR DISASTER WORKERS PARTICIPATING IN RESCUE OF SURVIVORS, RECOVERY OF VICTIMS, AND OTHER EMERGENCY ACTIVITIES

- Exposure to death/dead bodies
- Identification with victims
- Long hours and difficult working conditions
- Urgency of situation
- Fatigue
- Injury

a function of rescue and/or recovery work, particularly when exposure to death is part of these efforts (Ursano, McCaughey, & Fullerton, 1994). Rescue workers responsible for the cleanup of a DC10 air crash in Antarctica that killed 257 passengers and crew showed substantial distress for years after the experience (Taylor & Frazier, 1982). Firefighters and other emergency workers fighting major fires also exhibit long-term symptoms of stress (Hytten & Hasle, 1989; McFarlane, 1987). Rescue and cleanup activities following natural and industrial disasters have also been linked to symptoms of PTSD, including intrusive thoughts and dreams as well as anxiety and negative mood (Durham, McCammon, & Allison, 1985; Ersland, Weisaeth, & Sund, 1989; McCammon et al., 1988). Demoralization, helplessness, de-

pression, and arousal have also been found among rescuers following a variety of disasters (Fullerton et al., 1992; Markowitz et al., 1987; Raphael et al., 1983). Stress-related immune system changes, mood changes, and intrusive thoughts have also been reported among emergency workers following a major air disaster (Delahanty et al., 1997; Dougall et al., 1999). In this study of disaster workers who responded after an airliner crashed and exploded on contact, exposure to death was not the most important predictor of distress. Instead, experiencing unexpectedly gruesome scenes and intrusive thoughts that seemed to "come out of nowhere" were associated with greater distress (Delahanty et al., 1997; Schooler, Dougall & Baum, 1999).

Effects of Toxic Exposure

We know that toxic substances such as radiation, dioxin, and chemical wastes can cause physical health problems, but we are not as well informed about how people respond psychologically to known exposure. How do people feel when they believe that they have been exposed to toxic substances, and what do they do? Can psychological reactions be understood in terms of beliefs about the toxicity of the substance? What is it about these substances that evokes strong reactions in most of us? Why does exposure to toxic substances as a result of accidents or leaks seem to arouse greater response than toxic exposure that results from air or water pollution?

Consider the case of radiation. Experts argue about what levels of exposure are dangerous. Of course, at very high levels of exposure people die, but the consequences of long-term, low-level exposure are debated. The possibility of being exposed to radiation evokes strong emotional responses in many of us, and we tend to view nuclear power plants as more risky than do experts (Brown, 1992; Slovic, 1987; Slovic, Fischhoff, & Lichtenstein, 1981). In part this is due to

the dramatic nature of the nuclear accidents that have occurred. The invisible threat posed by radiation and the possibility of being exposed to it without even knowing about it add to the threat. Finally, the effects of radiation may take many years before they can be detected. Cancers and birth defects, two possible consequences of exposure to radiation and to many toxic chemicals, take years to develop or become evident. Long after TMI, Chernobyl, or Love Canal, concerns about possible future exposure may be compounded by worry and fear about effects that have already been set in motion (Figure 7-8).

The belief that one has been exposed to toxic substances, regardless of whether one has actually been exposed, seems to be sufficient to cause a stress reaction. In many cases, the extent of real exposure is unclear but is generally thought to be low. However, some area residents believe that they were exposed to dangerous levels of radiation, and this may have contributed to chronic stress. The lack of early warning signs of toxic effects on health, lack of clear information about

Figure 7-8 The effects of toxic exposure may not appear for years. Exposed individuals may therefore perceive that the risk is low, or they may worry about the future effects, which increases stress.

Courtesy of Federal Emergency Management Agency.

whether one was actually exposed to dangerous levels of the toxic substance, and severity of the long-term consequences of exposure may contribute to uncertainty and distress. The very belief that one has been exposed to toxic substances may cause long-term uncertainty and stress as well as pose a threat to one's health (Baum, 1987). Like many environmental stressors, however, the extent to which this occurs is determined by a number of situational and psychological factors. One important factor is trust. If people who may have been exposed to dangerous substances maintain trust in responsible officials and agencies, distress may be minimal. Often, however, trust problems emerge from toxic incidents and transport (e.g., Binney et al., 1996; Williams, Brown, & Greenberg, 1999).

Research has considered several different types of toxic exposure. Occupational exposure occurs at work. In many cases workers identify toxic hazards like radiation as a cause of health problems and experience stress as a result (e.g., Madsen, Dawson, & Spykerman, 1996). Other types of toxic exposure occur at home or in one's neighborhood. Still others occur before birth or during childhood. Let us examine some of the more publicized types of exposure.

ASBESTOS EXPOSURE

One source of toxic exposure that receives much attention is asbestos. Used because of its durability and resistance to heat, asbestos was common in many industries, and it has been estimated that over a period of 50 years, more than 13 million workers were exposed to asbestos (Lebovits, Byrne, & Strain, 1986). Some asbestos contamination may be found in schools and old housing, but for the most part, exposure occurs in occupational settings. When inhaled, asbestos fibers lodge in the lungs, where they may be coated by bodily defenses and left there. These particles can then cause damage to the lungs and result in pulmonary diseases, including lung cancer. Asbestos is a particularly risky toxicant to people who also smoke cigarettes. Regardless, the diseases and damage caused by asbestos require long periods of time to develop, so much harm can be done before an individual recognizes that there is anything wrong. Therein lies a major problem: Because the damage takes years to develop and become evident, workers are resistant to taking recommended safety precautions.

A study of asbestos workers who were first exposed to asbestos at least 20 years earlier provides some insight into reactions to being exposed to this hazard (Lebovits et al., 1986). Most had not been aware of the dangers of asbestos when they started working with it, though they had learned of the risks many years earlier than a control group of people who did not work with asbestos. Asbestos workers were also very aware of the consequences of asbestos: Nearly 80% of them had known four or more co-workers who had developed asbestos-related illnesses (less than 10% of the control group knew anyone with such disease). Consequently, asbestos workers reported greater perceived risk of developing cancer and heart disease than did control workers (Lebovits et al., 1986).

Interestingly, the sample of asbestos workers, compared with a control group, did not exhibit any more depression, anxiety, or other mental health problems, used mental health services infrequently, and reported similar perceptions of perceived control as did the control group. The asbestos workers also reported that they had not taken preventive precautions, such as wearing masks, visiting their doctor, and so on. A third of them continued to smoke even though they were aware of the special risks of doing so. Apparently denying the risks associated with their occupation and behavior, these workers showed little evidence of distress (Lebovits et al., 1986).

LIVING NEAR TOXIC WASTE

Known exposure to toxic waste in the home or neighborhood can result in serious consequences. Living near Love Canal and being exposed (or thinking one was exposed) to the toxic waste there appears to have generated fears about developing illnesses while reducing trust in officials responsible for the situation (Gibbs, 1982; Levine, 1982). In two studies of two toxic accidents, one involving pesticide exposure and the other toxic smoke from an explosion in a toxic waste facility (Markowitz & Gutterman, 1986), perceived threat to health was associated with psychological distress.

Several studies have zeroed in on the reactions to living near hazardous toxic waste sites. Love Canal is clearly the most infamous, and studies of people affected by the situation there suggest some evidence of long-term distress (Levine, 1982). Problems began when a chemical company dumped thousands of tons of toxic waste in the canal. The same land was later sold, and an elementary school was built on the site. A neighborhood of several thousand people grew up around the canal area. About 20 years later, it was discovered that hazardous waste was leaking from the canal, and in 1978 area residents were alerted by state officials (Levine & Stone, 1986). Toxic vapors were detected in some homes, increased miscarriage rates were discovered, and offers to move some of the affected residents were announced. The resolution of problems dragged on for years, and the full extent of the consequences of Love Canal remains to be assessed.

We do know that the Love Canal hazards were stressful for area residents. Of those interviewed in a study reported by Levine and Stone (1986), nearly 90% viewed the situation as a problem, and the nature of the problems posed ranged from uncertainty and threat to health, to financial and practical concerns. Residents felt that their health had worsened as a result of living near the canal and reported feelings of lost control and helplessness. Though some positive changes were reported, most evidence suggested a long-term state of distress and worry (Levine & Stone, 1986).

These findings are consistent with studies of other, less well-known toxic waste hazards. Stress, worries, uncertainty about exposure and its apparent or invisible effects, depression, anger, mistrust, and anxiety are common symptoms of those whose neighborhoods are exposed to toxic waste (e.g., Baum & Fleming, 1993; Davidson, Fleming, & Baum, 1986; Eyles et al., 1993; Fleming, 1985; Gatchel & Newberry, 1991; Gibbs, 1986).

Radon Exposure

Another hazard that may have severe consequences is *radon*, a colorless, odorless gas that comes from uranium deposits in the ground. It is a naturally occurring gas, and small exposures are both normal and of little apparent consequence. However, some people's homes have been found to have radon levels far in excess of safe or normal levels, and the health problems that can result from this kind of exposure are extreme. Apparently, radon problems are more widespread than was previously believed, and it has been recommended that everyone test his or her home for radon levels (Figure 7-9).

Not unexpectedly, people appear to overestimate the risk of radon problems when they do not have dangerous levels in their homes, though when radon is found, risk is underestimated (Sandman, Weinstein, & Klotz, 1987). As with many sources of danger, people's estimates of risk do not correspond to those made by experts, nor are they based on readily apparent criteria (Bostrom, Fischhoff, & Morgan, 1992). The invisible nature of radon compares with radiation hazards and toxic chemical contamination. Radon does no damage to buildings and can exist for many years without being detected. However, unlike these other hazards, radon is not human made but rather is a naturally occurring phenomenon. Thus, it is a natural hazard rather than a technological one, though human factors such as siting and insulation of homes can exacerbate problems (in well-insulated homes, radon may get trapped, and levels may increase due to lack of ventilation).

Figure 7-9 Radon, particularly in homes, has emerged as a widespread and serious problem. Occurring naturally, radon is radioactive and thus shares characteristics of many types of hazards.

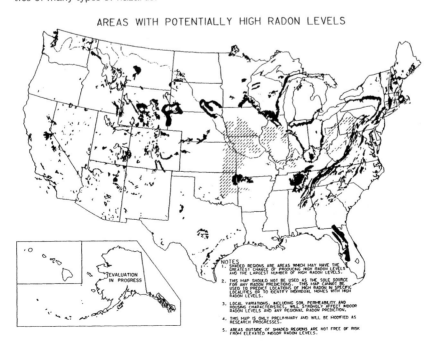

AREAS WITH POTENTIALLY HIGH RADON LEVELS

SICK BUILDING SYNDROME

An increasingly common workplace problem is the development of chronic complaints about impaired health and well-being attributed to foul air or toxins in construction materials (e.g., adhesives in floor or ceiling coverings). A number of developments in several fields are relevant here. Studies suggest complex interactions of toxic and behavioral variables that can cause a variety of problems. Actually, there are two distinct syndromes to consider: building-related illness, which involves diagnosable illnesses linked to toxic exposure or some other aspect of an indoor environment, and **sick building syndrome,** which involves symptoms and discomfort but no clear disease (Oliver & Shackelton, 1998; Thorn, 1998; Woods, 1988). Patterns of symptoms often suggest a cause in a particular building or job. For example, symptom distress often diminishes over weekends or after work, and symptoms could be the result of environmental conditions (e.g., headache, eye and nose irritation), stress (lethargy and fatigue), or both (Burge et al., 1987). These reactions can be widespread (up to 80% of building occupants have been shown to experience sick building symptoms; Burge et al., 1987) and can be debilitating for those affected.

Causes of these syndromes vary, but are often related to ventilation systems, heating systems, and building design. The prevalence of sick building syndrome is higher in air-conditioned office buildings than in naturally ventilated buildings (Mendell & Smith, 1990; Teculescu et al., 1998) and symptoms are often attributed to poor air quality (Hedge, 1984). However, these explanations frequently fail to account for symptoms, and efforts to clean up buildings may not result in decreases in symptoms (Kildeso, Wyon, Skov, & Schneider, 1999). Investigations of many "sick" buildings have not found any indoor air pollutants or air quality problems that might cause these symptoms (Hedge, Mitchell, & McCarthy, 1993).

In other cases, dampness and ammonia build up under building floors and cause nasal symptoms and eye irritation (e.g., Wan & Li,

1999; Wieslander, Norback, Nordstrom, Walinder, & Venge, 1999). An experimental study of perceived air quality, sick building syndrome symptoms, and productivity found some evidence of joint effects of psychological factors and environmental conditions (Wargocki, Wyon, Baik, Clausen, & Fanger, 1999). In this study, a removable source of pollution was used, a 20-year-old carpet that was contaminated during its use. The carpet was systematically introduced and removed from a work environment in which small groups of workers were exposed. Exposures lasted about four hours. Dissatisfaction with air quality was greater when the carpet was present than when it was absent, and more complaints of headache, less effort directed toward tasks, and poorer work performance were also associated with introduction of the pollution source (Wargocki et al., 1999). These data suggest that environmental conditions are a primary source of sick building problems and that improving them can have positive effects.

Others have proposed psychological causes and discussed *mass hysteria* or *mass psychogenic illness* explanations for these symptoms (e.g., Colligan & Murphy, 1979). Studies have suggested that these syndromes reflect preexisting or underlying psychological disturbances, stress, behavioral contagion, or problems at work (Bjornsson, Janson, Norback, & Boman, 1998; Brodsky, 1983; Stahl & Lebedun, 1974). Former uses of buildings, as in prior manufacture of pharmaceutical products, may contribute to symptoms when no traces of pollution or prior contamination remain (Engelhart, Burghardt, Neumann, Ewers, Exner, & Kramer, 1999). Some have suggested that women are more likely than men to report sick building syndrome symptoms, but these reports of greater sensory irritation and negative indoor climate appear due to the fact that women often work in more undesirable conditions (Bullinger et al., 1999). However, some studies have not found predicted relationships between personality and susceptibility to behavioral contagion or between distress and complaints associated with sick building syndrome (e.g., Bauer et al., 1992; Eysenck, 1975). Evidence points to the conclusion that many, if not most, cases of building-related illness or sick building syndrome are caused by real exposure to environmental contaminants (Cooley, Wong, Jumper, & Straus, 1998; Hodgson & Morey, 1989; Lyles et al., 1991). Smoking, stress, and other behavioral factors appear to play a role in the intensity of symptoms and reporting of discomfort (e.g., Muzi, Abbritti, Accattoli, & dell'Omo, 1998), but the likelihood that these are purely psychological in nature seems small (e.g., Oliver & Shackelton, 1998).

Researchers have compiled an impressive list of symptoms of general malaise that have been linked to "sick" buildings. These include eye, nose, and throat irritation (dryness, pain, hoarseness); skin irritation (itching, dry skin, pain); somatic symptoms (headache, nausea, sleepiness, fatigue); nonspecific allergic reactions (runny eyes, nasal congestion, asthma-like symptoms); and complaints about sensory changes (bad odors, bad taste; see Hedge,

Erickson, & Rubin, 1994). Correlational studies of office workers in many office buildings suggest that job stress and use of video display terminals (VDTs) are associated with sick building syndrome complaints (Burge et al., 1987). Other studies have found relationships between symptoms and factors such as air temperature (hotter was associated with more symptoms), allergy history, hours worked each day or week, photocopying, and job satisfaction (less satisfaction was associated with more symptoms; see Hedge, Mitchell, & McCarthy, 1993). While some studies do not find these relationships, others have suggested that job stress or use of VDTs alone is enough to cause some of the symptoms reported as sick building syndrome (e.g., Frese, 1985; Knave et al., 1985; Smith, Cohen, & Stammerjohn, 1981).

Consistencies in findings from studies of sick building syndrome have led to an eclectic model based on several interacting factors (see Figure 7-10). Studies of 18 office buildings

Figure 7-10   A model of factors that cause sick building symptoms; these factors interact to determine workers' experiences and distress.

After Hedge et al., 1989.

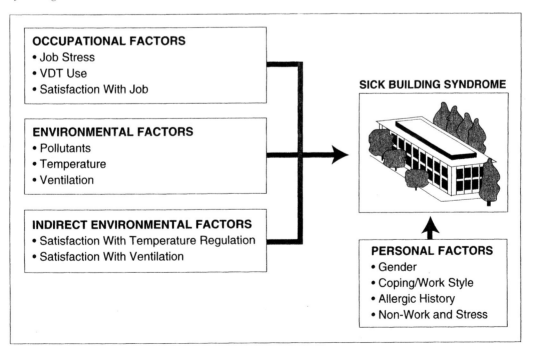

(Hedge, Mitchell, & McCarthy, 1993) and of 46 office buildings in the United Kingdom (Hedge et al., 1989) as well as studies of schools (Thorn, 1998) provide support for multifactorial causes of symptoms and suggest that environmental conditions, together with job stress, perceived comfort of environmental conditions, VDT use, and job satisfaction jointly predict sick building syndrome symptoms.

TERATOLOGY AND BEHAVIORAL TOXICOLOGY

The development of research in behavioral toxicology is also important in understanding the impact of toxic exposure. Studies of the effects of chemicals on development of organisms (**teratology**) and examination of behavioral effects of chemical exposure (**behavioral toxicology**) have identified a number of symptoms and syndromes associated with chemical exposure. For example, Spyker (1975) suggested 11 assessment categories for studying the effects of chemical exposure before or after birth (see

Table 7-6). In other words, there are at least 11 different kinds of problems that can occur and a variety of symptoms and ages at which they appear. To evaluate the presence and/or extent of the effects of prenatal chemical exposure (e.g., while pregnant, the mother drank alcohol or took drugs that can cause birth defects), one must consider a complex array of physical and behavioral problems.

In addition to these major effects of chemical or toxic exposure, a number of more subtle behavioral effects are associated with intelligence, developmental processes, and maturation. For instance, several studies have indicated that lead "poisoning" or substantial exposure to lead during childhood is associated with poorer school performance and lower IQ scores (e.g., Chang, 1999; Needleman, Shell, Bellinger, Leviton, & Alldred, 1990; Taylor, 1990). Exposure to chemicals at doses that do not produce symptoms of "poisoning" can still cause problems, and asymptomatic mothers exposed to lead, mercury, or other toxic substances are more likely to deliver infants who

Table 7-6	ASSESSMENT CATEGORIES IN THE EVALUATION OF TERATOGENIC EFFECTS[°]	
Assessment Category	**Illustration**	**Age of Testing**
1. Morphological characteristics	Limb or facial anomalies	Birth to maturity
2. Physical characteristics	Discoloration of the skin; facial swelling	Birth to maturity
3. Maturational landmarks	Preterm birth (neuromuscular and physical immaturity)	Birth
4. Growth	Small birth size; depressed postnatal growth	Birth to maturity
5. Reflexes	Poorly organized sucking; depressed reactions	Birth to maturity
6. Activity levels	Hypo- and hyperactivity; clinical assessment of apathy	Birth to maturity
7. Neuromuscular and sensory motor capacities	Poor hand–eye coordination (swimming in mice)	Postbirth to maturity
8. Sensory and attentional functions	Deficits in visual, auditory, or olfactory functions; numbness	Birth to maturity
9. Learning ability	Alternation and reversal learning deficits in monkeys; low IQ	Postbirth to maturity
10. Autonomic regulation	Depressed response to stress; emotional lability; tremulousness in infants	Birth to maturity
11. Sexual development	Reproductive failure; menstrual irregularities	Maturity

[°] *Adapted from Spyker (1975), Fein et al. (1983).*

experience symptoms of the mother's exposure (Harada, 1977). Further, consequences of maternal exposure may not show up immediately in infants and may appear later in development (Fein et al., 1983). These concerns have resulted in legislation that requires removing lead from paint and gasoline.

Research in these newer areas of behavioral investigation in environmental science will produce new and important information about human susceptibility to toxic exposures. New methods of ways of measuring sick building syndrome are appearing, suggesting that some sources of inconsistency will be overcome (e.g., Citterio et al., 1998; Kildeso et al., 1999; von

Mackensen, Bullinger, & Morfeld, 1999). The possibility of discovering effects of low-level exposure to toxic substances, even at levels thought to be safe, underscores the importance of this kind of work, as do the many costs of the syndrome (Lynch & Kipen, 1998). Issues regarding multiple chemical sensitivity and sensitization by exposure to toxic substances are also important (e.g., Bell, Baldwin, Russek, Schwartz, & Hardin, 1998). As the number and frequency of toxic compounds and sources of contamination increase and change, we must systematically evaluate the effects of these exposures on both behavioral and biological processes.

Air Pollution and Behavior

Toxic exposure occurs in many contexts and is not new (e.g., Eiser et al, 1998). Study of Greenland ice cores has shown that atmospheric lead levels were four times higher than natural concentrations from 2,500 to 1,700 years ago. These levels resulted from Greek and Roman mining and smelting of silver and lead (Hong, Candelone, Patterson, & Boutron, 1994). Over the past few decades, air pollution has become one of our primary environmental problems. Acid rain— the acidification of rain and the lakes it collects in and land it falls on due to the tendency of rain to "cleanse" the air of pollutants—is a major issue, as is depletion of the ozone layer of the atmosphere. Recent evidence indicates that air pollution actually reduces the amount of rainfall over a region (Rosenfeld, 2000). We often walk around in air that is filled with toxic particles generated by exhaust gases from automobiles, aerosol spray emissions, and factory discharges, as well as gaseous and solid airborne particles from industrial wastes. Even the smoke from cigarettes, forest fires, and cozy fireplaces in the home can have seriously adverse effects on health. Among the most common pollutants are carbon monoxide, sulfur dioxide, nitrogen dioxide, particulate matter, hydrocarbons, and

photochemical pollutants formed from the reaction of other pollutants with light and heat. Fortunately, with increased environmental awareness and responsibility and passage of legislation such as the Clean Air Act, we are well on our way to reducing many types of air pollution. A report released by the U.S. Environmental Protection Agency in 2000 stated that since 1970, air pollution had been cut by a third and acid rain by 25% (EPA, 2000). Nevertheless, the air is still being contaminated and will continue to be for many years to come. In this section we will examine some of the available research on air pollution and then examine the health effects, performance effects, and social effects of air pollution.

PERCEPTION OF AIR POLLUTION

Perception of air pollution depends on a number of physical and psychological factors. What do you think of when you hear the term *air pollution?* Probably, you think of two bad things —bad odors and smog-like conditions. Unfortunately, we depend primarily on our sense of smell and on atmospheric visibility to perceive air pollution. We say "unfortunately" because many of the most harmful types of air pollution are not detectable in these ways. Carbon monoxide, for example, is both odorless and colorless. Moreover, the atmosphere in airtight homes designed to restrict heat loss may be two or three times more polluted than outside air (e.g., Guenther, 1982).

Perception of air pollution is also likely to be affected by factors such as stress or annoyance (e.g., Crawford & Bolas, 1996). Attitudes toward the source of pollution or the attractiveness of this source may, for example, affect our awareness of pollution or how we report it when asked. Winneke and Kastka (1987) found that pollution from a chocolate factory produced less annoyance among nearby residents than did emissions from a brewery, a tar oil refinery, or an insulation plant. It is possible that exposure to odor was perceived as less intense near the chocolate factory. Other studies suggest that perception of pollution or annoyance caused by it does not always correspond to physical levels of pollution. This means that physical levels of pollutants provide only rough estimates of actual exposure or consequences that are "indirect, behavioral impacts of air pollution" (Evans & Jacobs, 1982, p. 117).

When pollution is detectable through smell, how do we perceive the smell? The answer appears to be chemically, through the *olfactory membrane*. The olfactory membrane lies at the top of the nasal passage, just behind the nose. This membrane, which is similar to the basilar membrane in the cochlea, is lined with hair cells. Gaseous chemicals stimulate these cells as they pass by, sending signals to the brain, which interprets the signals as various odors. Several factors determine whether the olfactory membrane detects a specific odor. For one thing, the chemical stimulating the membrane usually has to be heavier than air. Also, sufficient quantities of the chemical have to be present. This is one reason "sniffing" the air helps you detect odors. To the extent that pollutants are capable of stimulating the olfactory membrane, humans can detect air pollution through smell. Procedures have been developed to measure human detection of odorants. For example, with the *dynamic olfactometer*, a group of people smell different dilutions of the odorant, and the threshold (minimum level for detection) is the dilution at which the majority of the group detects the odor. With the *scentometer*, foul air is passed through a carbon filter and various dilutions are presented to an individual; the ratio of filtered air to unfiltered air at detection defines the threshold (for a review of odor detection, see Berglund, Berglund, & Lindvall, 1987). Some individuals are especially sensitive to chemical odors and score distinctively on a measure of sensitivity known as the *Chemical Odor Intolerance Index*, or *CII* (Szarek, Bell, & Schwartz, 1997). This may be related to a controversial condition known as *multiple chemical sensitivity*, or *MCS*, a condition that can be associated with psychopathological symptoms (e.g., depression, panic disorder) and may contribute to complaints of sick building syndrome.

In addition to noticing smells, most of us infer air pollution from smog-like conditions. That is, we use visual perception to determine the presence or absence of pollution. If a scene looks hazy, especially if the haze is brown, we perceive that pollution is considerable. Two early studies suggested that visibility is the primary cue that average citizens use to detect air pollution (Crowe, 1968; Hummel, Levitt, & Loomis, 1973). These researchers asked the open-ended question, "What do you think of when you hear the term 'air pollution'?" There was a strong tendency for respondents to specify effects of pollution, such as smoke or smog, rather than to specify causes, such as factories or automobiles (see Table 7-7).

We might also detect air pollution indirectly, through its consequences or sources. To detect particulate pollution, we may, of course, observe dust on our belongings; for some pollutants, eye and respiratory irritation are cues. Pollution experts often base their judgments of pollution on the concentration of automobiles, the absence of rain (rain cleanses the air), the presence of tall buildings (which block winds), and the presence of stop-and-go traffic as opposed to freeway traffic (idling and accelerating automobiles produce more pollution than automobiles moving at a constant speed; Hummel, Loomis, & Hebert, 1975; see Figure 7-11).

Table 7-7	CLASSIFICATION OF TYPICAL DEFINITIONS OF AIR POLLUTION°

Component of Definition	Percentage of Respondents Using Each Component	
	Urban Sample	*Student Sample*
Specific manifestation (smoke, haze)	43	14
Causative source (cars, industry)	22	43
Effects (health or property damage)	18	19
Combination (two or more of the above)	17	34

°*Note that causes are specified less than half the time.*
From Hummel, Levitt, & Loomis (1973).

Perceptual awareness of pollution may change with our exposure to it and may also depend on other factors (Cameron, Brown, & Chapman, 1998; Eiser et al., 1998; Evans & Jacobs, 1981). Interestingly, we tend to think "the other guy" has more pollution than we do. That is, we think our own immediate geographic area is less polluted than adjacent areas (De-Groot, 1967; Rankin, 1969).

Stressful life events, or having experienced a good deal of recent stress, appears to be related to perceptions of pollution as well (Jacobs et al., 1984). In addition, anxiety appears to be related to how we perceive air pollution (Navarro, Simpson-Housley, & DeMan, 1987). Because anxiety is characterized by perception of the environment as more threatening or harmful, it should not be surprising that anxiety is linked to perception of pollution; but this anxiety can also lead people to take more positive action to reduce pollution (Navarro et al., 1987). Whether or not one changes his or her behavior for protection from the effects of pollution appears to depend on beliefs about the nature and dangers of air pollution and other health beliefs (Skov et al., 1991).

Does prior exposure to air pollution decrease or increase our awareness of it? Unfortunately, the evidence is mixed on this question. For example, Wohlwill (1974) compared two

Figure 7-11 Nonexperts typically judge the presence of air pollution by its consequences (e.g., smog, eye irritation), whereas experts more often judge its presence by causes (e.g., automobile traffic, smoke-stacks).

HOGWASH!

A growing source of foul-smelling air is large-scale livestock operations, particularly feedlots and hog farms. While those in rural areas are accustomed to livestock odor, two trends are causing growing concern. First, the size of the individual livestock operation is increasing. Large, factory-like hog farms can produce half a million to several million hogs per year. Such facilities produce far more odorant in far greater concentration than traditional farming. Second, formerly small rural communities may expand such that the town's residents move closer to the large-scale operations. The manure from these operations is stored in open pits, or "lagoons." Hogs produce two to four times the amount of waste as humans. One planned facility for 2.5 million hogs is expected to produce as much waste as the city of Los Angeles. Regulations for this waste, unlike human waste, are often minimal or nonexistent. In addition to the odor, there is great concern that groundwater can become contaminated and that large volumes of nitrogen, ammonia, and methane are released into the atmosphere, damaging nearby plant life. Also, microbes associated with human disease can be found in the waste. One of the devastating effects of Hurricane Floyd in 1999 was that in North Carolina, tens of thousands of hog carcasses from these large-scale operations were carried away in the flood, and waste lagoons were washed out into local streams. Several states and local communities are now trying to balance the economic and environmental issues associated with this growing phenomenon. For more information, see Asmus (1998).

groups of people in one location: those who had moved there from a highly polluted region, and those who had moved there from a relatively unpolluted area. The current location was considered more polluted by those from the unpolluted area than by those from the highly polluted area; this suggests that the two groups were using different adaptation levels in making their assessments. In essence, these findings suggest that the more people were familiar with pollution, the less they were bothered by it (see also Evans, Jacobs, & Frager, 1982). Data from Lipsey (1977) and Medalia (1964) support the opposite position: The more people encounter pollution, the more concerned about it they become. For example, Medalia (1964) found that the longer people had lived near a malodorous paper mill, the more aware they were of its pollution. Asmus (1998) reported that long-term residents of a community with a chronic environmental odor problem were neither more nor less sensitive to the odor than newer residents, suggesting that adaptation to the odor did not occur. Thus, there is no consistent evidence indicating that humans adapt to persistent air pollution.

AIR POLLUTION AND HEALTH

The hazardous effects of air pollution on health are becoming well known (see Table 7-8). From time to time, very high concentrations of pollutants have been known to increase the death rate for urban areas, as in the December 1952 disaster in London, in which 3,500 deaths were attributed to excessive levels of sulfur dioxide (Goldsmith, 1968). Such disasters, however, are rare. More worrisome are adverse health effects of high concentrations of pollutants that occur more frequently (Bullinger, 1989; Fraser, 1989; Holgate, Samet, Maynard, & Koren, 1999). As early as the 1970s in the United States, for example, 140,000 deaths were attributable to pollution each year (Mendelsohn & Orcutt, 1979). Carbon monoxide (CO), the most common pollutant, prevents body tissues (including those of the brain and heart) from receiving adequate oxygen, a condition known as *hypoxia* that we described in Chapter 6 as a potential consequence of high altitudes. The primary sources of CO include motor vehicles, coal and oil furnaces, and steel mills. Prolonged exposure to heavy concentrations of CO can lead to very

Table 7-8	WHAT YOU CAN'T SEE IN THE AIR CAN HURT YOU: THE MAJOR COMPONENTS OF AIR POLLUTION HAVE A VARIETY OF HEALTH EFFECTS [a]
Respiratory symptoms:	Ozone, formed in sunlight when nitrogen oxides and hydrocarbons combine, aggravates respiratory problems by damaging epithelial cells in the trachea.
Skin problems:	Arsenic, produced by furnaces, can cause skin cancer. Depletion of the ozone layer of the atmosphere may also contribute to skin cancer.
Nervous system diseases:	Arsenic and lead can disrupt development in children or cause central nervous system problems.
Liver:	Lead can cause liver disease.
Reproductive difficulties:	Cadmium can retard development of the fetus.
Eyes:	Hydrogen chloride causes irritation; carbon monoxide and ozone affect eye–hand coordination.
Heart:	Carbon monoxide can reduce blood's ability to carry oxygen and cause symptoms of heart disease.
Lungs:	Almost all particulates and metals accumulate in lungs and can contribute to cancer.

[a] Based on Holgate et al. (1999).

serious health problems, including visual and hearing impairment, epilepsy, headache, symptoms of heart disease, fatigue, memory disturbance, and even retardation and psychotic symptoms. Particulates can cause respiratory problems, cancer, anemia, and neural problems, among other things. Photochemical smog can cause eye irritation, respiratory problems, cardiovascular distress, and possibly cancer. Oxides of nitrogen and sulfur, also produced by cars, trucks, and furnaces, impair respiratory function and may lower resistance to disease. Finally, furnaces, smelters, wood stoves, dry-cleaning plants, and petroleum refineries produce arsenic, benzene, cadmium, and hydrocarbons, all of which can cause irritation or illness. For most pollutants, the elderly and the ill are the most likely victims. The list of ailments aggravated, if not caused, by air pollution seems endless.

Some research has considered behavioral and mental health consequences of air pollution (e.g., Shusterman, 1992). Data suggest, for example, that people are less likely to engage in outdoor recreational activities when air quality is poor, but these effects are not as clear as one would expect, partly because subgroups of people were not considered. Evans, Jacobs, and Frager (1982) reported that people who exhibited more internal locus of control and who

were newly arrived from low-pollution areas more clearly reduced outdoor activities during periods when air quality was poor. Feeling that one is in control of the situation has also been shown to reduce the effects of malodorous pollution on frustration (Rotton, 1983). Data suggest that air pollution can increase hostility and aggression and reduce the likelihood that people will help each other (Cunningham, 1979; Jones & Bogat, 1978). Finally, there is some evidence that psychological disturbance can follow exposure to air pollution. Symptoms of depression, irritation, and anxiety have been observed after indoor air pollution exposure (e.g., Weiss, 1983), and epidemiological studies have revealed correlations between air pollution levels and psychiatric hospital admissions (Briere, Downes, & Spensley, 1983; Strahilevitz, Strahilevitz, & Miller, 1979). Perceptions of smog were related to symptoms of depression in a survey of randomly sampled residents of Los Angeles (Jacobs et al., 1984). Rotton and Frey (1984) found evidence of increased emergency calls for psychiatric problems when air pollution levels were high, and Evans et al. (1987) found that poor air quality increased the likelihood of distress following major life stressors.

Recall that having experienced stressful life events was related to perceptions of pollution.

The more "stressed" people are, the more likely they are to be irritated by pollution. Of concern, then, is whether stress influences how people respond to air pollution. Are people who have experienced other stressful events (like moving, starting school, or separation from home and family) more likely to show negative effects of air pollution? A study of 500 Los Angeles residents over a three-year period found that having experienced stressful life events was related to more symptoms of emotional distress and mental health problems, and interacted with perceived pollution levels to predict distress (Evans et al., 1987). The highest levels of distress were observed among people experiencing stressful life events at medium levels of pollution. These findings suggest that air pollution has similar effects as other stressors when experienced in combination with other problems, and that people who are experiencing stress are more vulnerable to effects of air pollution.

AIR POLLUTION AND PERFORMANCE

Most available research on air pollution and performance involves studies of carbon monoxide (CO). Because CO produces hypoxia, the higher the level of CO, the greater the effects on performance. Concentrations of CO at 25 to 125 parts per million (ppm) are typical on freeways at rush hour. In one early study, Beard and Wertheim (1967) exposed volunteers to concentrations of CO ranging from 50 ppm to 250 ppm for various periods of time and asked them to make discrimination judgments about time intervals. It was found that 90 minutes of exposure to CO at 50 ppm significantly impaired performance on the time judgment task. As CO concentration increased, shorter periods of exposure were required for similar levels of impairment. Such research suggests that air pollution on major traffic arteries may impair driving ability enough to increase the frequency of automobile accidents. This possibility is supported by results of a study by Lewis et al. (1970), in which participants were exposed to "clean" air or to air drawn 15 inches (38 cm) above ground at a traffic site handling 830 vehicles per hour. Performance decrements occurred in three out

of four information-processing tasks for those breathing the polluted air.

AIR POLLUTION AND SOCIAL BEHAVIOR

Research has shown that malodorous air pollution influences at least several types of social behavior. First, recreation behavior in particular, and outdoor activity in general, are restricted by pollution (Chapko & Solomon, 1976; Peterson, 1975). Second, Rotton et al. (1978) examined the effects of ammonium sulfide and butyric acid on interpersonal attraction. In one experiment, it was found that ammonium sulfide increased attraction for similar others with whom participants thought they were interacting. That is, when individuals were exposed to an unpleasant odor, attraction toward others who were also exposed to the odor increased. In a second experiment, however, the same researchers found that those who did not expect to interact with others evaluated the others less favorably if exposed to ammonium sulfide or butyric acid than if not exposed. It seems that the unpleasant affective states associated with pollution led to decreased attraction if not shared with others, but to increased attraction when odor exposure was the same for all parties. Malodorous air pollution affects more than just attraction to people. Interestingly, foul odors also reduce liking for paintings and photography (Rotton, 1983). Asmus and Bell (1999) found that foul odors made participants feel more unpleasant, reduced willingness to help, increased anger, and increased flight behavior.

In another experiment, Rotton et al. (1979) investigated the effects of exposure to ethyl mercaptan and ammonium sulfide on aggression. Using the shock methodology common in aggression research (see page 160), the researchers ostensibly allowed participants to shock a confederate. In accordance with the research on heat and aggression, it was anticipated that exposure to a moderately unpleasant odor (ethyl mercaptan) would increase aggression, but that exposure to an extremely unpleasant odor (ammonium sulfide) would decrease aggression. Consistent with these predictions, it was found

SECOND-HAND CIGARETTE SMOKE:
More Concern Than Ever

For a number of years we have known that tars and nicotine can have major effects on the health of smokers. There is now considerable evidence that nonsmokers breathing the air in a room where others are smoking may also suffer ill effects. In a study of 78 urban nonsmoking high school students, a third showed evidence of nicotine metabolites in their urine, and this evidence of passive smoke exposure was associated with altered lipids and heightened risk for premature heart disease (Hardoff et al., 1997). Cigarette smoke has been shown to contain significant quantities of carbon monoxide and probably some degree of DDT and formaldehyde as well. A nonsmoker inhaling the air around a person who is smoking may experience an increase in heart rate, blood pressure, and breathing rate (Luquette, Landiss, & Merki, 1970; Russell, Cole, & Brown, 1973). These effects may be particularly pronounced or important in children exposed to cigarette smoke by smoking parents (e.g., Martinelli, 1999). Further, the 1986 Surgeon General's report on involuntary smoking concludes that passive smoking can cause disease, that children of smokers have more respiratory infections than do children of nonsmokers, and that we must act as a society to minimize exposure of children and nonsmoking adults to others' tobacco smoke (Koop, 1986). Passive smoking has also been described as a major cause of premature death (Goldman & Glantz, 1998).

Research has certainly confirmed that nonsmokers are disturbed by cigarette smoke. In one study (Bleda & Sandman, 1977), smokers were evaluated negatively by nonsmokers if they smoked in the presence of the nonsmokers. Other research indicates that nonsmokers have increased feelings of irritation, fatigue, and anxiety when exposed to cigarette smoke (Jones, 1978). In another study (Bleda & Bleda, 1978), it was found that persons sitting on a bench in a shopping mall fled faster if a stranger next to them smoked than if he or she refrained from smoking. Finally, cigarette smoke may not just lead to feelings of irritation and dislike, but to overt hostility as well. Both feelings of aggression (Jones & Bogat, 1978) and hostile behavior increase in the presence of others' cigarette smoke.

Opposition to smoking in public places has been growing for many years (e.g., Brenner et al., 1997), and a number of states have adopted laws prohibiting or restricting smoking in elevators, stores, and some restaurants (e.g., Aakko, Remington, Dixon, & Ford, 1997; Brauer & t'Mannetje, 1998). There is some evidence that creating smoke-free environments encourages people to cut back on their smoking or give it up altogether. For example, several studies find evidence of decreases in daily smoking (the number of cigarettes consumed in a day) as smoke-free environments increase (Chapman, Borland, Scollo, Brownson, Dominello, & Woodward, 1999). Smoke-free environments are responsible for an annual decrease of more than 600 million cigarettes (1.8% of all that might otherwise be consumed; Chapman et al., 1999). Efforts to minimize tobacco smoke exposure also extend to the home, where passive exposure is a particularly important issue for children's health (Ashley & Ferrence, 1998; Ashley, Cohen, Ferrence, Bull, Bondy, Poland, & Pederson, 1998). Some programs to reduce tobacco smoke exposure among asthmatic children have had unusually good success over a two-year period (e.g., Wahlgren, Hovell, Meltzer, Hofstetter, & Zakarian, 1997).

that relative to a no-odor control group, the moderate odor increased aggression. In addition, there was suggestive evidence (though not statistically reliable) that the stronger odor decreased aggression.

Two other studies show that air pollution affects our social behavior and the way we feel about other people. Rotton and Frey (1985) found that complaints about household disturbances, including child abuse, were elevated when ozone levels were high compared with when they were low. Pollution may contribute to such behavior by making us more hostile or depressed, a possibility suggested by a study showing that a combination of high life stress and high air pollution predicts how hostile or depressed people feel (Evans et al., 1987). Regardless of how these effects occur, the research on pollution, malodor, and the sick building syndrome, which was described earlier, suggests complex relationships among foul odor, annoyance, distress, and social behavior (Hempel, 1997; Hoppe & Martinac, 1998; Rotton & White, 1996).

Where Do We Put Environmental Hazards? NIMBY and Environmental Racism

One of the thorniest issues facing society today is the siting of industry, or where we build manufacturing plants and energy producing plants. If we accept the fact that we need energy and we need to have materials (e.g., plastics, steel) that produce pollution or other hazards during their production, then we must decide where to locate these hazards or potential sources of harm. Do we put them near where people live so that it is convenient for workers and transportation to reach them, or do we locate them far from where people live? Of course, the latter is not always possible, and if we decide to place a power plant or chemical plant, oil refinery, or other manufacturing plant near people, how do we decide where to put it? Places like eastern Texas, southwestern Louisiana, northern New Jersey, northern Indiana, and others seem to have concentrations of refineries or chemical plants, and examples of potentially dangerous power plants being located near where people live are many (for example, Three Mile Island). In West Virginia there is an area sometimes called "chemical valley" because of its concentration of chemical plants, again near a moderately sized population. Some of these areas grew up around the industries that were placed there first, but in other cases industries were located near existing communities.

Most people do not want such hazards or sources of pollution to be located in their communities. There is a substantial literature on the effects of where such facilities are located and how environmental changes affect residential satisfaction (e.g., Craik & Feimer, 1987). The acronym **NIMBY** ("not in my back yard") has come to symbolize resistance to locating undesirable plants or other facilities in people's neighborhoods or communities, and opposition to locating energy or manufacturing plants, high-energy electric transmission lines, oil and gas pipelines, highways, and even schools has taken this position. Often fought in local courts and agencies, these issues can be larger and come to represent the age-old struggle between the welfare of the larger group and that of a small group, in this case communities (Figure 7-12).

There are many criteria that are used in decisions about where to locate new plants or highways (and their associated pollution). Environmental impact analyses and complex hearings and other legal proceedings are usually involved. Some people believe that an "unwritten" criterion is one based on the ethnic or national origin of a group of people or on their income or educational level. In other words, the disadvantaged in society are thought to bear an unfair burden. Hazardous plants are rarely built near affluent communities but are instead more likely to be found in poorer areas in which minority groups are more likely to live.

At times, the circumstances surrounding the siting and remediation of environmental hazards arouse suspicion. For example, in 1979 in Houston, six out of eight incinerators were sited in African-American neighborhoods. A white blue-collar community in Globe, Arizona, and a middle-class African-American community in Texarkana, Texas, were both built on contaminated soil, at about the same time. Yet it took only two years of complaints for the Globe residents to receive a settlement of $80,000 each, compared with 20 years for the Texarkana residents to receive half that (Taylor, 1999). In numerous cases, it appears that legal protections against environmental hazards are implemented

Figure 7-12   The issue of where to site sources of potentially toxic wastes can become part of a community battle. The acronym NIMBY ("not in my back yard") symbolizes the pressure to keep such potentially hazardous sources out of individual neighborhoods. As a result, environmental hazards are often disproportionately located in disadvantaged neighborhoods.

inequitably across various groups (e.g., Lavelle & Coyle, 1992).

The **environmental justice** movement that we introduced in Chapter 2 aims to correct prior and prevent future abuses of the environment by framing the issues in terms of "justice" due (e.g., Taylor, 2000). Advocates from minority groups sometimes frame the environmental justice cause in the context of past injustices suffered by disadvantaged groups, and may use the term **environmental racism** to describe the disproportionate exposure of minorities to environmental hazards (for a review, see Taylor, 1999). The tendency for hazards to be located in minority communities may be as much a matter

of discrimination against lower socioeconomic status (SES) as well as membership in a minority group. Some people believe these are mere coincidences, due in part to the availability of low-cost land on which to build these plants and to zoning laws that limit where they can be located. Others argue that those exposed are insufficiently informed about the dangers to be as concerned as they should about hazards or where they are sited (e.g., Bullard, 1993; Taylor, 1989). Still others suggest that minority group members may have different environmental concerns and that local siting issues may not be as important as other concerns (e.g., Arp & Kenny, 1996). Some people rebut this by arguing that the siting of human-made hazards reflects an inherent bias in our society. It could also be that neighborhoods where hazards are sited simply do not have the political power of wealthier citizens to fight the site plan. Alternatively, wealthier people may have the means to buy homes elsewhere, and less well-off residents may be able to afford housing only on sites whose value is depressed because they are located near toxic sites.

These issues are not new and may reflect an important source of stress effects on mental and physical health. The correlations between minority status and SES are high, and it is often the case that lower SES and minority status are associated with more dangerous, crowded, and/or polluted environments and with chronic stress (e.g., Baum, Garofalo, & Yali, 1999). Economically disadvantaged people typically live in areas characterized by more crime, more traffic, poorer health care, more crowding, more noise, and poorer quality of services and goods than do more affluent people. These conditions, together with discrimination and the greater likelihood of environmental hazards posed by industrial complexes, may affect a range of mental and physical health conditions. Careful decision making and reduction of the hazardous aspects of having factories or other sources nearby will help, but the issue may require more comprehensive social and cultural changes as well. In the process, all citizens of · the world might benefit from the *Principles of*

Environmental Justice as adopted by the People of Color Environmental Leadership Summit in October 1991. As cited by Taylor (1999, p. 56), the first principle reads:

> Environmental justice affirms the sacredness of Mother Earth, ecological unity and the interdependence of all species, and the right to be free from ecological destruction.

In Chapter 14 we will have more to say about achieving environmental change for the benefit of all.

SUMMARY

Stressors may be categorized as cataclysmic events, personal stressors, or daily hassles and background stressors. This categorization is based on both the power of the event and the number of people affected. At one extreme, cataclysmic events are potent events that demand a great deal of effort from people trying to cope with them. However, they often pass quickly. Disasters are a primary example of such events. At the other extreme are daily hassles: minor, transient stressors that by themselves demand little effort to cope with them. However, they are repetitive or constant, and over time they may exact a substantial cost. Air pollution is one of these background stressors.

Natural disasters are powerful, destructive events that require a great deal of adaptation. They involve threat to life or loss of property but are usually brief in duration. Most disaster events last for less than a day, some for only a few minutes. Once the disaster has passed, various means of coping may be directed at recovery and a number of effects may occur. Beneficial effects due to the formation of cohesive groups in the face of adversity have been found when these groups stay together long after the disaster. Though there is some concern for long-range effects of these events, research evidence of this is equivocal, and most research suggests only short-lived stress reactions.

Technological catastrophes are similar to natural disasters but are also different in several ways. They are caused by human actions rather than by natural forces, are not necessarily as destructive, are less predictable, and lack a clear low point, after which recovery may begin.

These accidents and disasters seem to have few, if any, positive effects and have longer lasting negative consequences, including stress, negative mood, and uncertainty. One reason for this is that the events are human caused, providing a better focus for blame than in natural disasters. Another is that technological disasters occur when control is lost over something previously under control (natural disasters are rarely controllable). A final possibility is that technological catastrophes often also involve toxic substances, and we have seen that the possibility of exposure to toxic substances can cause many long-term problems. Victims of disasters include those directly affected, their friends and family, and rescue workers, who face their own forms of stress. Toxic exposure, such as to asbestos and radon, often goes unnoticed because the consequences take years to appear. Living near toxic waste dumps can increase worry and long-term stress. Sick building syndrome may result from a combination of toxic exposure and psychological distress.

We have also discussed a very prevalent but often ignored source of toxic exposure, air pollution. Air pollutants cause or aggravate a variety of ailments, most notably respiratory and cardiovascular problems. Carbon monoxide in sufficient quantities impairs performance. Malodorous pollutants can increase or decrease aggression and attraction, depending on other factors. Disproportionate exposure to environmental hazards often occurs among the disadvantaged. Framing attempts at amelioration in terms of justice due is part of the environmental justice trend.

KEY TERMS

adaptation	crisis effect	event duration	post-traumatic stress
behavioral toxicology	disaster event	levee effect	disorder (PTSD)
conservation of	disruption	low point	sick building syn-
resources (COR)	environmental justice	natural disaster	drome
theory	environmental racism	NIMBY	teratology

SUGGESTED PROJECTS

1. Interview some people who have lived through severe storms, tornadoes, floods, or earthquakes. Ask them how they felt during and just after the event, how they coped, and what they worried about. Then interview some people who experienced some kind of technological accident. Are there differences in their reports?

2. Talk with homeowners in your area. Have they ever heard of radon? How much do they know about it? How concerned are they about it, and how likely are they to test for it? Is knowledge about radon related to fear of it and likelihood of remedial action?

3. Interview the environmental safety officer on your campus. What are the sources of con-cern about toxic wastes (e.g., the chemistry labs, cleaning agents)? What special steps must be taken to dispose of these substances?

4. Keep a log for two weeks in which you enter your local air pollution and pollen indexes each day. Also, write down your observation about your and others' apparent mood, behavior, and level of activity. Is pollution related to any of these measures of mood and behavior? If so, are people aware of it? Be sure not to look at the pollen and pollution measures until just before you go to bed each night, so that the knowledge of these levels does not affect your mood or how you observe others throughout the day. Could you guess about the pollution or pollen levels from your observations of people's behavior?

CHAPTER 8

PERSONAL SPACE AND TERRITORIALITY

INTRODUCTION

PERSONAL SPACE

FUNCTIONS OF PERSONAL SPACE
    Size of Personal Space

METHODS OF STUDYING PERSONAL SPACE

SITUATIONAL DETERMINANTS OF PERSONAL SPACE
    Attraction and Interpersonal Distance
    Effect of Other Types of Similarity on Interpersonal
        Distance
    Type of Interaction and Interpersonal Distance

INDIVIDUAL DIFFERENCE DETERMINANTS OF PERSONAL
    SPACE
    Cultural and Ethnic Determinants of Personal Space
    Gender Differences in Personal Space
    Age Differences in Personal Space
    Personality Determinants of Spatial Behavior

PHYSICAL DETERMINANTS OF PERSONAL SPACE

INTERPERSONAL POSITIONING EFFECTS

SPATIAL ZONES THAT FACILITATE GOAL FULFILLMENT
    Optimal Spacing in Learning Environments
    Optimal Spacing in Professional Interactions
    Optimal Spacing to Facilitate Group Processes

CONSEQUENCES OF TOO MUCH OR TOO LITTLE PERSONAL
    SPACE
    Predicting the Effects of Inappropriate Distances
    The Consequences of Inappropriate Spacing

CONSEQUENCES OF PERSONAL-SPACE INVASIONS
    The Effects of Being Invaded on Flight Behavior
    The Effects of Being Invaded on Arousal

Introduction

To help you get a "feel" for the content of this chapter, imagine you are suddenly whisked off to a faraway land, a land full of seemingly normal people with a strange disregard for the need to maintain a personal space "buffer" around their bodies. In this land, it is not uncommon for a complete stranger to sit down right next to you on an empty bus; to stand right in front of you, knees touching yours, and ask you for directions; or to stand an inch away from you in a swimming pool even though there is nobody else around. In all types of settings, it is not uncommon for people to "brush" against each other, even in intimate ways, and nobody minds at all. If you go to the checkout counter in a department store, rather than spreading out and using the space, people in line pack together tightly, so that you can barely fumble for your checkbook. When you go to the beach, you find that people choose
to put their towels right next to yours, restricting your movement and comfort, while the rest of the beach remains vacant.

Before long, you discover another peculiarity about these people. You realize that they have no concept of territoriality or territorial behavior. People move randomly from one dwelling to another and have no place they call "home." They wander into your home without knocking and proceed to eat your food, use your toothbrush, and sleep in your bed, before walking out and taking some of your things with them. You go out to a fancy restaurant for a romantic evening, and a complete stranger pulls up a chair and begins sampling your food and conversing with your date. The inhabitants of this place seem not to mind that they have nowhere to call their own, or that at any moment a complete stranger might walk in on whatever they are doing.

What would life be like in such a place? How would you feel, and what would you be able to accomplish in life, if you had to live there? While this "fantasy" may seem extreme, it should make you aware of your need to maintain a portable personal space "bubble" between yourself and others, and of the importance of territorial functioning. In this chapter we will discuss personal space and territoriality, two ways in which people create different types of boundaries to regulate their interactions with others in their environment. The importance of such boundaries is probably quite evident to you from our description of what life would be like without personal space and territoriality (Figure 8-1).

Thus far in this book we have examined how we perceive the environment and how various aspects of it affect us. As we move toward an examination of the "built" environment in the later chapters of this book, this chapter and the next mark a transition in which we describe spatial relationships between humans and their environment. As we will see, these spatial relationships have behavioral implications, design implications, as well as commonalities with the environment–behavior relationships we have examined thus far.

Personal space is defined as a portable, invisible boundary surrounding us, into which others may not trespass. It regulates how closely we interact with others, moves with us, and expands and contracts according to the situation in which we find ourselves. In contrast, **territories** are relatively stationary areas, often with visible boundaries, that regulate who will interact. The person is always at the center of his or her personal space — it is always with him or her. On the

Figure 8-1 In addition to constant spacing between various groups, members of individual groups maintain relatively constant personal space from one another.

other hand, territories, which often center on the home, can be left behind. While territoriality is more of a group-based process, personal space is more of an individual-level process. Let us now take a more comprehensive look at these two spatial phenomena.

Personal Space

FUNCTIONS OF PERSONAL SPACE

The term *personal space* was coined by Katz (1937). The concept is not unique to psychology, and also has roots in biology (Hediger, 1950), anthropology (Hall, 1968), and architecture (Sommer, 1959). Both popular and scientific interest in personal space have been intense during the past several decades. In fact, about 1,000 published experiments have been conducted in this area since 1960.

What is the function of the personal-space bubble we maintain around ourselves? A number of conceptual explanations have been suggested, some of which correspond to the theoretical formulations we reviewed in Chapter 4. Briefly, an *overload* interpretation of why we maintain personal space between ourselves and others stipulates that it is necessary to avoid overstimulation. According to this notion (Scott, 1993), too close a proximity to others causes us to be bombarded with too many social or physical stimuli (e.g., facial details, olfactory cues). An alternative formulation, the *stress* interpretation, assumes that we maintain personal space to avoid various stressors associated with too close a proximity. Further, the *arousal* conceptualization suggests that when personal space is inadequate, people experience arousal. When this occurs, we attempt to understand why we are aroused (e.g., is it because someone we love is close to us, or because someone we fear is close?), and the type of explanation we come up with determines how we respond to inadequate personal space. A fourth conceptual perspective, the *behavior constraint* approach, suggests that personal space is maintained to prevent our behavioral freedom from being taken away because others are too close to us.

In addition to these theoretical approaches, which we have discussed in detail earlier in this book, other explanations (e.g., Hayduk, 1994) have been suggested for why we maintain personal space. One, proposed by anthropologist E. T. Hall (1963, 1966), conceptualizes personal space as a form of *nonverbal communication.* According to Hall, the distance between individuals determines the quality and quantity of stimulation that is exchanged (e.g., tactile communication occurs only at close proximity). Distance also communicates information about the type of relationship between individuals (e.g., whether it is intimate or nonintimate), and about the type of activities that can be engaged in (e.g., lovemaking cannot occur between individuals who are far apart).

Another theory, proposed by Altman (1975), views personal space (and territoriality as well) as a boundary regulation mechanism to achieve desired levels of personal and group **privacy.** Privacy is an interpersonal boundary process by which people regulate interactions with others. Through variations in the extent of their personal space, individuals ensure that their desired and achieved levels of privacy are congruent. When it is impossible to regulate these boundaries so that privacy is within desired levels, negative consequences and coping will occur.

Related in some ways are the intimacy–equilibrium model proposed by Argyle and Dean (1965) and the comfort model formulated by Aiello (1987). According to these approaches, in any interaction (or relationship) people have an optimal level of intimacy they want to maintain (Gibson, Harris, & Werner, 1993). Lovers aspire to more intimate relations than friends, and so on. Intimacy is a function of personal space and other factors such as eye contact, facial gestures, and the intimacy of the topic under discussion. If the level of intimacy in an interaction becomes too great (e.g., the people are interacting about too intimate a topic and too much eye contact is being maintained), **equilibrium** will be restored through **compensatory behaviors** in some other modality (e.g., moving physically farther away). If the level of intimacy is too small, equilibrium will be restored as well (e.g., by moving closer or maintaining more eye contact). While nonoptimal levels of intimacy will typically prompt compensatory behaviors, Aiello (1987) suggests that these may not result from small deviations from an optimal level of intimacy, and that really large deviations may cause people to completely lose interest in the interaction.

A final perspective on personal space is provided by **ethological models** (cf. Evans & Howard, 1973). These assume that personal space functions at a cognitive level and has been selected by an evolutionary process to control intraspecies aggression, to protect against threats to autonomy, and thereby to reduce stress. We should note, however, that in contrast

IS PERSONAL SPACE REALLY A BUBBLE?

Researchers, including the authors of this text, have a tendency to liken personal space to a bubble of sorts that surrounds us and fulfills a number of functions. While the bubble analogy gives us a concrete image to visualize and may thereby aid our understanding, if taken too literally it has some drawbacks that may lead to misunderstanding. First, it has been suggested that the notion of a personal space "bubble" emphasizes the protective function of personal space more than the communicative function (Aiello, 1987). Second, one might begin to think that if personal space is analogous to a bubble, it is the same size for all individuals and in all situations, and this is *not* the case. As we will see in this chapter, people have varying spatial zones, and the amount of space we desire between ourselves and others expands and contracts depending on the situation. Personal space is really an *interpersonal distance continuum*.

In addition to the bubble analogy, there are also problems with the term "personal space." One could easily get the idea that since it is *personal*, personal space is somehow attached to an individual in all situations. This is also not true; personal space has meaning only with respect to another individual, and does not apply to distance between people and desks, for instance. Also, the label "personal space" may emphasize the idea of *space* and thus suggest that researchers are concerned only with distance. As we will see in this chapter, those studying personal space must also focus on other behaviors, such as body orientation and eye contact, in order to get a complete understanding of spatial behavior (Aiello, 1987; Knowles, 1978). The latter behaviors are interpersonal rather than personal. Overall, the personal, as well as the space, components of the term "personal space" may be misleading.

Due to these possibly misleading aspects of the term "personal space," should we consider replacing it with something else? Some researchers, such as Aiello (1987), have suggested that a term such as **interpersonal distance** might be better to use than personal space. Such a term implies that distance is a continuous dimension in which a variety of behaviors can occur. However, even Aiello admits that the concept is probably too well engraved in our minds—and in the literature—to be readily abandoned!

to the assumption of ethological models that personal space evolved naturally, most researchers would probably argue that it is more a product of learning. However, after it is learned, our spatial behavior seems to be governed unconsciously (i.e., we do not have to "think" about how to position ourselves in different situations).

If all the conceptual perspectives we have described are integrated, personal space may be seen as an interpersonal boundary regulation mechanism that has two primary sets of purposes. First, it has a *protective* function and serves as a buffer against potential emotional and physical threats (e.g., too much stimulation, overarousal leading to stress, insufficient privacy, too much or too little intimacy, physical attacks by others). The second function involves *communication*. The distance we maintain from others determines which sensory communica-

tion channels (e.g., smell, touch, visual input, verbal input) will be most salient in our interaction. To the extent that we choose distances that transmit intimate or nonintimate sensory cues and that suggest a high or low concern with self-protection, we are communicating information about the quality of our relationship with other persons (i.e., the level of intimacy we desire to have with them).

Size of Personal Space

What determines the size of the personal space we want to maintain between ourselves and others? The distance we maintain must be appropriate to fulfill the two functions of personal space—protection and communication. One determinant of the amount of space necessary to accomplish these functions is the situation

Table 8-1	TYPES OF INTERPERSONAL RELATIONSHIPS, ACTIVITIES, AND SENSORY QUALITIES CHARACTERISTIC OF HALL'S SPATIAL ZONES

	Appropriate Relationships and Activities	Sensory Qualities
Intimate distance (0 to 1½ feet)	Intimate contacts (e.g., making love, comforting) and physical sports (e.g., wrestling)	Intense awareness of sensory inputs (e.g., smell, radiant heat) from other person; touch overtakes vocalization as primary mode of communication.
Personal distance (1½ to 4 feet)	Contacts between close friends, as well as every-day interactions with acquaintances	Less awareness of sensory inputs than intimate distance; vision is normal and provides detailed feedback; verbal channels account for more communication than touch.
Social distance (4 to 12 feet)	Impersonal and businesslike contacts	Sensory inputs minimal; information provided by visual channels less detailed than in personal distance; normal voice level (audible at 20 feet) maintained; touch not possible.
Public distance (more than 12 feet)	Formal contacts between an individual (e.g., actor, politician) and the public	No sensory inputs; no detailed visual input; exaggerated nonverbal behaviors employed to supplement verbal communication, since subtle shades of meaning are lost at this distance.

(i.e., whom we are with and what we are doing). Certain relationships and activities demand more distance than others for appropriate communication and adequate protection. Situational conditions are not the only determinants of the size of our personal space, however. Some individuals always preserve minimal personal space zones, while others maintain relatively large personal space zones. Individual differences in spatial behavior probably reflect different learning experiences concerning the amount of space necessary to fulfill the protective and communicative functions. Individual differences that affect spatial behavior include gender, race, culture, and personality.

One of the first observational studies of the effect of **situational conditions** and **individual difference variables** on spatial behavior was conducted by E. T. Hall. Hall (1963, 1966) suggested that depending on situational conditions, people use one of four personal space zones in their interactions with others. The particular zone that we use depends on situational conditions such as our relationship with the others and the activity we are engaged in. The four zones (which are labeled *intimate distance, personal distance, social distance,* and *public distance*) vary in terms of the quality and quantity of stimulation that is exchanged (see Table 8-1).

With respect to the effect of individual differences on spatial behavior, Hall observed in cross-cultural investigations that cultures vary widely in terms of spatial behavior, an observation that has been corroborated (Aiello & Thompson, 1980b). Cultural differences were attributed by Hall to different norms regarding the sensory modalities seen as appropriate for communication between people who are interacting. Hall (1968) observed that these differences can lead to awkward social interactions with unintended inferences if people from cultures with disparate personal-space norms interact. Someone from a "close" culture may approach too closely into the space of someone from a "far" culture, who may then back off to a more comfortable distance but as a consequence appear aloof and withdrawn.

METHODS OF STUDYING PERSONAL SPACE

While Hall's studies were primarily observational and qualitative in nature, many *experimentally based* investigations have considered the effect of situational and individual difference variables on personal space. In this research, several different methodologies have been employed. Many of the experimental studies exploring factors that affect personal space have used *laboratory methods*. Some of this work involves real interaction between people. Here, the personal space between participants is measured with a tape or similar standard as a function of experimental conditions (e.g., degree of mutual attraction). Other laboratory methods include *simulation methods*—techniques in which participants manipulate the personal space between dolls or symbolic figures, or approach an inanimate object under various experimental conditions. Another set of studies has used *field methods*. These involve observing and experimenting with interpersonal positioning in naturally occurring situations, as a function of individual difference or situational variables. Distance might be estimated from photographs, or the dependent measure might be flight behavior from too close an interaction.

Fortunately, many of the important relationships between situational and individual difference variables and the size of personal space have been corroborated using different experimental approaches. While this has occurred with some regularity (Knowles, 1980), in other cases (e.g., Eaton, Fuchs, & Snook-Hill, 1998; Wann & Weaver, 1993) there have been dissimilar findings for studies that used different methods to assess the same spatial relationship. Knowles and Johnson (1974) suggested that although there is sometimes a general association between the various methods used to measure personal space (i.e., they can be considered to be indicators of the same dimension), the level of convergence is only moderate. Others (e.g., Aiello, 1987; Hayduk, 1983, 1985) have suggested that the different measures of personal space are not measuring the same thing. In general, it appears that laboratory and field methods that involve actual interaction between participants are better measures of our spatial behavior than simulation techniques (Aiello, 1987; Hayduk, 1983; Love & Aiello, 1980).

SITUATIONAL DETERMINANTS OF PERSONAL SPACE

Attraction and Interpersonal Distance

Love songs often suggest that the greater the attraction between individuals, the more physically close they want to be. There is some truth to this, but the relationship between affection and personal space is somewhat more complex.

Classic studies (Allgeier & Byrne, 1973; Byrne, Ervin, & Lamberth, 1970) indicated that when males and females interact, increased attraction is associated with closer physical distance. In one such study, Byrne and his colleagues manipulated attraction by sending male–female pairs, who were similar or dissimilar on a variety of personality traits, on a brief date. From research in social psychology, we know that similar individuals tend to be more attracted to each other than dissimilar individuals (Berscheid & Reis, 1998). When the "matched" or "mismatched" couples returned from the date, the experimenter measured their degree of mutual liking, as well as the distance between them as they stood in front of his or her desk. The "matched" couples liked each other more and stood closer together than the "mismatched" ones. Another classic study examined whether the attraction–proximity relationship for opposite-sex dyads occurs because the male moves closer to the female, because the female moves closer to the male, or because both the male and the female move closer to each other. This study suggested that the smaller distances between close friends of the opposite sex were primarily attributable to females moving closer to males to whom they were attracted (i.e., females respond more to attraction by their spatial positioning than do males; Edwards, 1972).

If the spatial behavior of females is primarily responsible for the attraction–proximity relationship, then the distance between female–female pairs should be determined by their degree of mutual attraction, while the distance between male–male pairs should not. In line with this assumption, early work showed that while female–female pairs position themselves closer together with increased liking, positioning does not vary with liking for male–male pairs. In one experiment (Heshka & Nelson, 1972), pairs of adults were unobtrusively photographed by researchers as they walked down the street. The use of a standard in each of the pictures permitted a fairly accurate estimate of the distance between the people in the dyad. After taking the photograph, the experimenter approached the unknowing "participants" and asked them what type of relationship they had. It was found that female–female pairs interacted at closer distances as their relationship became closer, while distance between male–male pairs did not change as a function of friendship.

Why is it that the attraction–proximity relationship holds for females but not males? One explanation derives from socialization differences between the sexes, which could be reflected in spatial behavior with liked others. For males, who may be socialized to be fearful of homosexual involvement and to be more independent and self-reliant overall (Crawford & Unger, 2000), and who have less experience with intimate forms of nonverbal communication (Crawford & Unger, 2000; Deaux & LaFrance, 1998; DePaulo & Friedman, 1998), spatially immediate situations with liked males or females are ambivalent. Close distances with liked males may trigger concerns about homosexuality, close distances with liked females may evoke concerns about dependency, and for males, physical closeness and its attendant high degree of sensory stimulation is generally somewhat foreign. On the other hand, females may be socialized to be more dependent, to be less afraid of intimacy with others of the same sex, and generally to be more comfortable in affiliative situations. They also have more experience as senders and receivers of intimate nonverbal messages (Crawford & Unger, 2000; Deaux & LaFrance, 1998; DePaulo & Friedman, 1998). Thus, it is not surprising that they have less difficulty responding spatially to liked others (see also Bell, Kline, & Barnard, 1988).

The research demonstrating that in some cases people in dyads interact at closer distances with increasing friendship (cf. Bell et al., 1988;

Holmes, 1992) suggests that closer personal space is an outcome of increased attraction. Indeed, observers viewing people interacting at close range infer higher degrees of attraction, and people interacting at closer distances are judged to have a more positive interpersonal relationship than individuals interacting at farther distances (Haase & Pepper, 1972; Wellens & Goldberg, 1978).

Effect of Other Types of Similarity on Interpersonal Distance

We have noted that personality similarity leads to attraction, which elicits closer interpersonal positioning. Since other types of similarity have been shown to affect attraction, similarity on these other dimensions should also lead to closer interpersonal positioning. This has been found to be true in a number of studies. Closer distances are maintained between individuals of similar rather than dissimilar age (Latta, 1978; Willis, 1966), sex (Kaya & Erkip, 1999), race or subculture (Aiello, 1987), religion (Balogun, 1991), sexual preference (e.g., heterosexual versus bisexual; Barrios et al., 1976), and status (Lott & Sommer, 1967). One setting where status is highly salient is in the military. When initiating an interaction with a superior, the greater the similarity between the initiator and the other in terms of rank, the smaller the interpersonal distance maintained (Dean, Willis, & Hewitt, 1975). Finally, it is both interesting and unfortunate that those without disabilities or stigmatizing diseases prefer to interact at closer interpersonal distances with similar others than with people who have disabilities or stigmatizing diseases like AIDS (Mooney, Cohn, & Swift, 1992; Rumsey, Bull, & Gahagan, 1982).

Why should similarity and attraction lead to closer interpersonal distances than dissimilarity and dislike? People generally anticipate more favorable interactions with similar (liked) than with dissimilar (disliked) others. Since one of the functions of personal space is protection against perceived threats, we *should* be willing to interact at closer distances with similar than dissimilar others because we anticipate fewer threats from them (cf. Skorjanc, 1991). Maintaining close interpersonal distances with liked others is also a means of fulfilling the communicative function of personal space. By choosing closer distances, we convey information to liked others that we are attracted to them and that we expect to communicate more intimate sensory cues to them.

Type of Interaction and Interpersonal Distance

If expectations of pleasant interactions lead to closer interpersonal positioning, then situational qualities (e.g., type of interaction, discussion topics) that can be placed on a pleasant–unpleasant dimension should also affect the size of our personal space. Indeed, studies have shown that negatively toned situations precipi-

tate larger spatial zones (Albas & Albas, 1989; Sinha & Mukherjee, 1996; Strayer & Roberts, 1997). In one study, participants in a stressful interaction maintained more distance than those in a low stress condition (Ugwuegbu & Anusiem, 1982). An analysis of the literature further suggests that women, as opposed to men, are especially apt to react to threatening situations by expanding their personal space (Aiello, 1987; Figure 8-2A–2D).

While it seems that affectively negative situations generally lead to more distant interactions, there is a special case in which contrasting results are sometimes found. When participants are angered as a result of personal insults, they may show closer interaction distances than non-angered participants. This may be interpreted as a retaliatory stance that facilitates communication of anger. However, some studies (O'Neal

Figure 8-2A–2D Interpersonal distance and such nonverbal behaviors as eye contact, body angle, and facial expression vary with the affective tone of the interaction context. Can you suggest the affective tone of the interaction situation and relate it to personal space and nonverbal behavior in each of these pictures?

et al., 1980) suggest that anger, like the other negative affects, may also produce farther distances. Additional situational conditions probably determine when anger leads to closer distances (for retaliation) or to farther distances (for protection).

INDIVIDUAL DIFFERENCE DETERMINANTS OF PERSONAL SPACE

In addition to situational conditions, differences between individuals or groups that reflect diverse learning experiences or cultural or subcultural norms also determine the size of personal space (Aiello, 1987; Kaya & Erkip, 1999). Although we will find that there *are* consistent relations between individual difference variables and personal space preferences, some findings are inconsistent, perhaps because of methodological differences across studies.

Cultural and Ethnic Determinants of Personal Space

Cross-cultural variations in spatial behavior have been observed, although the patterns are not always consistent (Aiello, 1987; Remland, Jones, & Brinkman, 1991). Hall (1966) proposed that in highly sensory "contact" cultures (e.g., the Mediterranean, Arabic, and Hispanic cultures), where individuals use smell and touch as well as other sensory modalities more, people should interact at closer distances. In contrast, more reserved "noncontact" cultures (e.g., northern European and Caucasian American cultures) should exhibit larger interaction distances. This hypothesis has received support (Aiello, 1987; Remland, Jones, & Brinkman, 1995). Thus, although the research is not entirely consistent and many cultures have yet to be studied (Aiello & Thompson, 1980b), various cultural groups may need different distances to fulfill the protective and communicative functions of personal space.

The research on *subcultural differences* in spatial behavior within North American culture is more confusing (Hayduk, 1983). As we said earlier, subcultural groups tend to interact at closer distances with members of their own subculture than with nonmembers (Aiello, 1987). Also, it seems as if Hispanic-Americans interact more closely than Anglo-Americans (e.g., Aiello, 1987). Unfortunately, findings for other subcultural differences (e.g., differences between African-Americans and Caucasians) have often been inconsistent. It has been suggested (Hayduk, 1978) that socioeconomic status may be a better predictor than subculture of learning experiences related to spatial behavior. Although members of a particular subculture may vary greatly in their living conditions, those in a particular socioeconomic group tend to live under relatively similar conditions. However, support for the view that socioeconomic status should have consistent effects on spatial behavior has also been mixed (Aiello, 1987).

Gender Differences in Personal Space

In terms of same-sex others, female–female pairs maintain closer distances than male–male pairs (Aiello, 1987; Barnard & Bell, 1982). These findings may reflect a stronger female socialization to be affiliative, more experience by females with intimate nonverbal modalities (Crawford & Unger, 2000; Deaux & LaFrance, 1998; DePaulo & Friedman, 1998), and a greater male concern about not being intimate with others of the same sex (Berscheid & Reis, 1998; Maccoby, 1990). The tendency for women to interact more closely than men does not hold for all situations. While it occurs in affiliative situations, in contexts that imply threat, women interact at greater distances than men (Aiello, 1987).

When dyads are of mixed sex, distancing depends on the relationship of the people who are interacting. Acquaintances maintain an intermediate distance (between that used by female–female and male–male pairs), while mixed-sex dyads who are in a close relationship maintain closer distances than either male–male or female–female pairs (Aiello, 1987). Interestingly, some research suggests that a woman's point in the menstrual cycle affects the personal space she maintains with opposite-sex others. Females' personal space zones tend to be larger during the menstrual flow than during the middle of the cycle (Gallant et al., 1991). This has been interpreted as reflecting the midcycle peak in sexual desire. In effect, hormonally determined sexual receptivity may affect personal space in opposite-sex interactions.

Age Differences in Personal Space

Personal-space norms appear to develop between ages 45 and 63 months, or perhaps a little later (Duke & Wilson, 1973; Eberts & Lepper, 1975; Meisels & Guardo, 1969), and then change somewhat as children mature. Children less than 5 years old show inconsistent spatial patterns, and after age 6 (grade one in Figure 8-3), the older the child (until adulthood), the greater the preferred interpersonal distance (Aiello, 1987; Hayduk, 1983). This pattern holds across cultures (e.g., Lerner, Iwawaki, & Chihara, 1976; Lomranz et al., 1975). Adult-like spatial norms are first exhibited around the time of puberty (Aiello, 1987). In addition to personal space becoming larger with age, the distance adults maintain from children becomes greater as the child's age increases (Larson & Lowe, 1990; Sigelman & Adams, 1990).

Figure 8-3 Mean interaction distances of male and female dyads at six grade levels.

From Aiello, J. R., & Aiello, T. (1974). The development of personal space: Proxemic behavior of children 6 through 16. Human Ecology, 2, 177–189. *Reprinted by permission.*

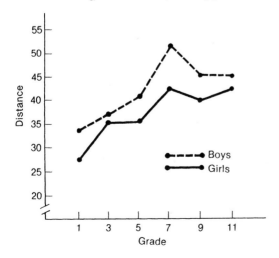

Personality Determinants of Spatial Behavior

Since personality represents one's way of looking at the world and reflects learning and experience, it seems reasonable that personality orientations should be reflected in spatial behavior. One of the *individual personality* traits that has been explored in this regard is **internality–externality.** The theory of internality–externality views an individual's orientation (internal or external) as a reflection of past learning about internal or external causation of events. *Internals* view reinforcements as under the control of the self; *externals* view reinforcements as controlled by external sources. Duke and Nowicki (1972) found that externals desired more distance from strangers than internals. It seems that if past learning leads to the belief that one is in control of a situation, he or she feels more secure at close distances with strangers than if past learning leads to the belief that events are controlled externally. It has also been found that anxious individuals maintain more personal space than nonanxious people (Karabenick & Meisels, 1972; Patterson, 1977), that introverts maintain more space than extraverts (e.g., Cook, 1970), and that those who work in relative isolation (e.g., at computer terminals) require more personal space, even outside of the work setting, than those who do not work in isolation (Gifford & Sacilotto, 1993). Also, those with high self-esteem maintain smaller personal space than those with low self-esteem (Frankel & Barrett, 1971), people high in need for affiliation prefer closer distances than those low in need for affiliation (Mehrabian & Diamond, 1971a), and field dependent persons (reliant on external cues) maintain closer distances than field independent ones (reliant on internal cues; Kline, Bell, & Babcock, 1984).

Other studies have found that spatial behavior differs as a function of psychopathology. Srivastava and Mandal (1990) compared the spatial needs of schizophrenics and "normals" and found that schizophrenics require more space, and Akhtar (1990) found that those with borderline personality disorder have difficulty maintaining appropriate personal distances. This research notwithstanding, many of the studies that have attempted to relate individual personalities to spatial behavior have not been terribly enlightening and have resulted in conflicting findings (cf. Aiello, 1987; Hayduk, 1983). Patterson (e.g., 1978) has proposed a procedure that may be more fruitful than the individual personality trait approach. Rather than studying single personality traits and their relationship to spatial behavior, Patterson conceptualized personality dimensions in more general terms. He looked at *clusters* of personality variables related to a general approach tendency for social situations (e.g., need for affiliation, extraversion) and related these to personal space preferences. His research demonstrated that such a strategy may yield better, more stable predictors of interpersonal distancing than focusing on individual traits. Another way to maximize the possibility of observing relationships between personality variables and personal space was suggested by Karabenick and Meisels (1972). Specifically, it may be necessary to study such relationships in situations in which the personality trait in question is salient (e.g., studying the relationship between aggressiveness and personal space in an anger-provoking situation), ·as opposed to the neutral contexts generally employed.

PHYSICAL DETERMINANTS OF PERSONAL SPACE

Although we have focused primarily on situational and individual difference determinants of personal space (as has past research), studies also suggest some interesting physical determinants of interpersonal spacing. First, a number of architectural features affect personal space. For example, Savinar (1975) found that males had more need for space when ceiling height was low than when it was high. White (1975) reported that personal space increased with reductions in room size and decreased with increases in room size, and Daves and Swaffer (1971) found that individuals desire more space in a narrow than a square room. Also, Baum, Reiss, and O'Hara (1974) suggested that in-

stalling partitions in a room can reduce feelings of spatial invasion. Do we maintain closer distances with others when "in the dark" than when there is light? Gergen, Gergen, and Barton (1973) reported that we are more likely to touch others (the ultimate in closeness), which may make many people uncomfortable, when it is dark than under more typical lighting conditions. Perhaps because touching is more apt to occur in the dark, maintaining a close personal space in a place that is dark causes more discomfort than being close when there is full illumination (Adams & Zuckerman, 1991).

People exhibit greater personal space when in the corner of a room than when in the center (Tennis & Dabbs, 1975). Also, it seems that we maintain closer distances when standing than while seated (Altman & Vinsel, 1977). With respect to spatial differences as a function of being indoors or outdoors, Cochran and Hale (1984) found that participants kept more distance between themselves and others when indoors than when outdoors. Similarly, we prefer greater distances in crowded than in uncrowded conditions (Jain, 1993). The "corner–center," "sitting–standing," "crowded–uncrowded," and "indoor–outdoor" effects may reflect differences in availability of escape; when we know we can get away, we are content with less space.

INTERPERSONAL POSITIONING EFFECTS

Individual difference and situational variables also affect the body orientation that we maintain between ourselves and others. Whereas males prefer to interact with liked others in an across (i.e., face-to-face) orientation, females prefer to have liked others adjacent to them. In two related studies, Byrne, Baskett, and Hodges (1971) manipulated the attraction between a participant and two confederates so that the participant liked one of the confederates but disliked the other. The participant was then asked to join the confederates in another room where his or her choice of seats with respect to the liked and disliked confederates was recorded. In the first experiment, which involved side-by-side seating, females sat closer to the liked confederate than to the disliked one, while males showed no preference. In the second experiment, which involved face-to-face seating, males sat closer to the liked confederate, while females showed no preference.

The cooperativeness or competitiveness of the interaction situation also affects spatial positioning. In an initial study, Sommer (1965) observed the spatial arrangement of individuals who were cooperating or competing and found that cooperating pairs sat side-by-side, while competing pairs sat across from each other. A second study found corroborative results. Participants anticipated either a cooperative or a competitive interaction and sat opposite a decoy in competitive conditions and adjacent to him or her in cooperative conditions. In addition to affecting interpersonal orientation, Sinha and Mukherjee (1996) reported that people maintain closer distances in more cooperative than less cooperative contexts.

SPATIAL ZONES THAT FACILITATE GOAL FULFILLMENT

What distances lead to the best results between teacher and student, therapist and client, or physician and patient? This important question shows the applied significance as well as the design implications of research on personal space. For example, research by Latané, Liu, Nowak, Bonevento, and Zheng (1995) found that generally, the impact of a communication will decline as the physical distance between the communicator and the target increases. In three different studies spanning three diverse cultures, the researchers observed a consistent finding such that the social influence from a communicator declined greatly with distance (in mathematical terms, impact was an inverse-square function of distance from the target of the communication). Does this mean there is an optimal distance for teaching and healing?

Optimal Spacing in Learning Environments

In general, teacher–student interactions at Hall's closer zones may lead to better perfor-

mance by a student. For example, in a study by Skeen (1976), a learner performed a serial learning task either 6 inches (intimate distance) or $3\frac{1}{2}$ feet (personal distance) from the experimenter. For tasks of varying levels of difficulty, the learner's performance was better at the personal distance than at the intimate one.

What about typical classroom situations, where there are many students present? In reviewing research on this question, Montello (1988) concluded that when seats are assigned by the instructor, seating position has little if any effect on academic outcomes (e.g., grades), but especially when you are free to choose your seat, where you sit makes a difference in the quality of the educational experience. Where is the best place to sit? It seems as if the middle-front section of the classroom is a relatively high communication zone. Sitting there promotes verbalization (except for those who are very low verbalizers) and facilitates attention (Koneya, 1976; see Figure 8-4). It has been found that people who choose middle-front seats have the highest self-esteem (Hillmann, Brooks, & O'Brien, 1991), show higher participation and more positive attitudes toward the educational experience (Montello 1988), and get the best grades in the class (Becker et al., 1973; Sommer, 1972). While the relationship between seating position and grades is only correlational and could be due to students choosing to sit in more

Figure 8-4 Where you sit in a large lecture hall can make a big difference.

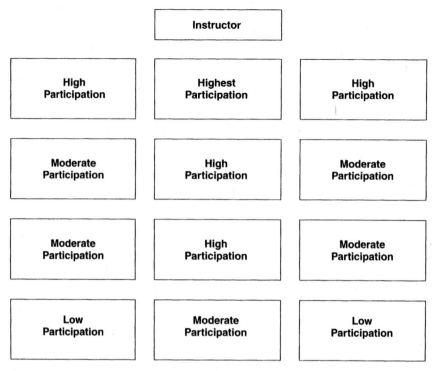

Figure 8-5 Typical student participation in class as a function of seating position. These participation rates correlate positively with grades in the course; however, this relationship could be due to effects of seating on participation, or to higher-performing students selecting the higher-participation seats, or both.

or less favorable locations, there is also some experimental evidence that partially supports it (Stires, 1980; Figure 8-5).

Optimal Spacing in Professional Interactions

An interesting question concerns the distance at which people feel most comfortable disclosing personal information about themselves to clinical psychologists. Generally, an intermediate distance is preferred for a counseling situation (Brokemann & Moller, 1973), and psychiatric patients talk most about their fears and anxieties at that distance (Lassen, 1973). This pattern of effects also holds for college students. When Stone and Morden (1976) had students discuss personal topics with a therapist at a distance of 2 feet, 5 feet, and 9 feet, they found that students volunteered the most personal information at the 5-foot distance. Since this distance is cul-

turally appropriate for such communications and is expected for them, these data support Hall's (1968) prediction that deviation from the appropriate distance elicits unintended negative consequences. However, these data may not generalize to self-disclosures between two strangers in nonclinical interactions (cf. Skotko & Langmeyer, 1977).

How far should a physician position him- or herself from a patient so that the patient's compliance with medical regimen will be highest? According to available evidence, the answer depends on whether the doctor is delivering basically "accepting" or "neutral" evaluative feedback for the patient's self-disclosures. In a study by Greene (1977), close physical proximity strengthened adherence to dieting recommendations when "accepting" verbal feedback was offered, but lowered compliance when "neutral" verbal feedback was given. If a closer distance implies a more accepting attitude and a farther distance a more neutral one, then when the intimacy of both the verbal and environmental "channels" was consistent, more positive effects occurred than when there were inconsistencies.

Optimal Spacing to Facilitate Group Processes

Suppose that an environmental psychologist wants to promote interaction within a group. This calls for **sociopetal** spacing (spacing that brings people together, such as the conversational groupings found in most homes), rather than **sociofugal** spacing (spacing that separates people, like the straight rows of chairs found in airports or bus terminals). In an early study, Sommer and Ross (1958) examined conditions at a Saskatchewan hospital, where a newly opened ward with a lovely, cheerful decor seemed to be having a depressing and isolating effect on patients. They observed that chairs were lined up against the walls, side by side. All the chairs were facing the same way, and rather than seeing one another, people just gazed off into the distance. When Sommer and Ross rearranged the chairs into small, circular groups, the frequency of interactions among patients almost doubled. Other studies have similarly found that arranging space so that people face each other more directly results in greater interaction between group members (Mehrabian & Diamond, 1971b). A nonfacing orientation may elicit longer pauses, more self-manipulative behaviors and postural adjustments, and perhaps even more negative ratings of group interaction (Patterson et al., 1979).

How could one manipulate his or her spatial positioning in a group in order to become its leader? It seems that in small group settings, people direct most of their conversation to the person sitting across from them (i.e., the one who is the most highly visible; Michelini, Passalacqua, & Cusimano, 1976). Also, people who occupy a central position in a group initiate the most communications (Michelini et al., 1976). This suggests that you could become highly influential merely by choosing a central spatial orientation where others eye you directly. This assumption has received some support. Those

who choose the end of a rectangular table are more likely to be elected foreman in simulated jury studies (Strodtbeck & Hook, 1961) or to otherwise dominate group interaction. Of course, this could be due to the fact that dominant individuals *choose* to sit at the "head" of the table, or it could be a reciprocal relationship.

CONSEQUENCES OF TOO MUCH OR TOO LITTLE PERSONAL SPACE

What happens when we are forced to interact with another person under conditions of "inappropriate" (i.e., too much or too little) personal space? For example, imagine an interaction with a door-to-door salesperson who insists on extolling the virtues of his or her product at an inappropriately close distance (e.g., 3 inches) or an inappropriately far distance (e.g., 10 feet). Would you be likely to buy anything from this person? Since personal space serves some important functions, we can assume that inappropriate distancing often has negative consequences for the interactants.

Predicting the Effects of Inappropriate Distances

The effects of inappropriate positioning can be described in the context of the eclectic environment–behavior model introduced in Chapter 4 and presented in Figure 8-6. Before we discuss the model, however, recall that we have observed throughout our coverage of research on personal space that situational conditions and individual differences determine optimal interpersonal distances. It is evident in Phase One of the model that whether we perceive our personal space as optimal or nonoptimal at a particular objective distance from another person depends on situational conditions (e.g., attraction) and individual differences (e.g., personality). If we perceive our personal space as within an optimal range, homeostasis is maintained. If we perceive it as outside this range, a variety of responses may occur.

What is the nature of our response to nonoptimal personal space? The same conceptual

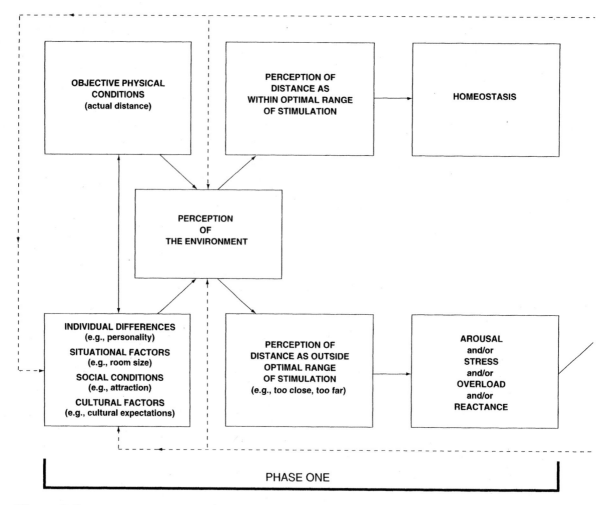

Figure 8-6   Eclectic environment–behavior model adapted to conceptualizing reactions to inappropriate personal space.

formulations used earlier to explain why we maintain personal space (e.g., overload, arousal, and behavior constraint) predict the effects of inappropriate interpersonal distancing. For example, *overload* notions predict that stimulus overload occasioned by inappropriate interpersonal distancing should cause performance decrements and elicit coping responses in order to restore stimulation to a more reasonable level. In terms of the *stress* approach, inappropriate positioning leads to a stress reaction, which may have emotional, behavioral, and physiological components. Coping responses are directed at reducing stress to a more acceptable level.

The *arousal* conceptualization assumes that being too close leads to overarousal and negative attributions, and suggests coping mechanisms designed to lower arousal. According to *equilibrium* and *comfort* models, distances that are too close or too far will lead to compensatory reactions in other modalities (e.g., changes in body orientation or eye gaze) and, when these are impossible, to a loss of interest in continuing the interaction. Altman's (1975) **privacy regulation model** implies that inadequate personal space will elicit attempts to "shore up" boundary control mechanisms and thus ensure privacy. Finally, the *behavior constraint* approach sug-

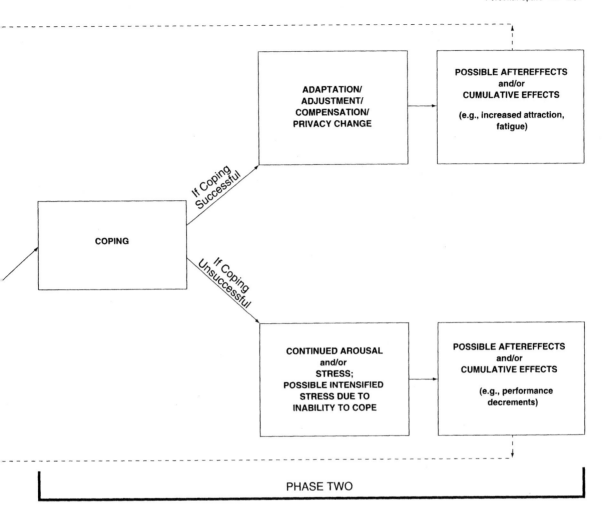

gests that inadequate personal space frequently leads to an aversive feeling state and to coping responses that attempt to reassert freedom.

We noted earlier that Hall (1966) proposed a model based on *communication properties* to explain personal space and that ethological models have also been applied (cf. Evans & Howard, 1973). In terms of Hall's formulation, it might be predicted that inappropriate distance constitutes a negative communication and leads to negative attributions and inferences. The *ethological approach* makes still another set of predictions. It assumes that when personal space is inadequate, fear and discomfort are experi- enced due to feelings of aggression or threat (cf. Evans, 1978).

How can we integrate all these formulations? Although each of the approaches proposes a somewhat different reaction to inappropriate positioning, we should not view them as competing with one another. Rather, it is probable that inappropriate personal space may at times lead to each of the responses we have described. Further, the predictions of all of the conceptual schemes may be integrated into the sequence of events shown in Phase Two of Figure 8-6. When personal space is perceived to be inadequate, which may be due to a combination of objective physical distance and situational and social conditions (Zakay, Hayduk, & Tsal, 1992), various types of coping responses are employed, which may or may not be successful. When coping is successful, it leads to adaptation or habituation, and aftereffects are less likely. If coping is unsuccessful, inappropriate positioning can lead to aftereffects such as dislike for the other, poor performance, and so on.

The Consequences of Inappropriate Spacing

Several studies have shown that when an environmental setting forces two people to interact in an inappropriate spatial zone, unfavorable feelings and inferences are elicited. In an experiment on the effects of distance on persuasion, Albert and Dabbs (1970) hypothesized that negative feelings and attributions would be elicited if a communicator and a target person were positioned more or less than 5 feet apart, an appropriate distance for such interpersonal contacts. Accordingly, the communicator and the target person were constrained to interact at the "appropriate" distance of 5 feet (1.5 m), or at one of two inappropriate distances (i.e., 2 feet or 15 feet; 0.6 m or 4.6 m). Several effects were measured, and the findings basically supported the prediction. Participants paid more attention to the communicator and rated him or her as more of an "expert" at the 5-foot distance than at either of the other distances.

Boucher (1972) found parallel results using schizophrenics as participants. First, interviewers sat down with patients at a distance that was inappropriately close, appropriate, or inappropriately distant. The distance manipulation was accomplished by fastening both chairs to the floor at one of the three ranges so that people in "inappropriate" positions could not adjust their proximity to a more comfortable zone. Following a 10-minute interview under these conditions, the patient's attraction to the interviewer was assessed. It was found that more attraction was expressed for the interviewer at the appropriate distance than at distances that were inappropriately close or far.

Several studies suggest that maintaining inappropriate interpersonal distance is associated with considerable stress. For example, Dabbs (1971) found that a persuasive communicator who was positioned too close caused participants to feel more pressured, unfriendly, and irritated than they did when a more appropriate distance was maintained. And when Aiello and Thompson (1980a) had participants converse at either a comfortable distance or an uncomfortably far one, participants who sat too far apart not only felt ill at ease, but blamed the other for their discomfort, even though the other was clearly not responsible! Patterson and Sechrest (1970) reported that participants evidenced more positive feelings when interacting with a confederate at a moderate distance (4 feet; 1.2 m) than at either a closer distance (2 feet; 0.6 m) or a farther distance (8 feet; 2.4 m).

In another study, Hayduk (1981) found a linear relationship between the degree to which a spatial arrangement was inappropriate and the amount of people's discomfort. Also, he observed that participants with smaller personal space zones responded more positively to an inappropriately close distance than participants with larger zones. A recent study with nursing-home residents found that insufficient personal space was associated with lower levels of resident satisfaction (Sikorska, 1999). Interestingly, a study by Fisher (1974) suggests that inappropriate distances with a similar (liked) other led to less negative reactions than the same distances with a dissimilar other.

According to Argyle and Dean (1965), nonverbal compensatory coping reactions should occur to restore a comfortable "equilibrium" when the physical distance between two individuals is too close or too far. In one study (Albas, 1991), when an interviewer moved uncomfortably far from a target person during an interview, the target reestablished equilibrium by coming closer. In other studies changes in eye contact and body orientation were in line with the equilibrium hypothesis (e.g., Kline & Bell, 1983; Rosenfeld et al., 1984); with too much proximity, body orientation became less direct and percentage of eye contact decreased.

COMPENSATION VERSUS RECIPROCATION:
Too Close Is Not Always Too Bad

Up until now, we have suggested that inappropriate interpersonal distancing results in negative consequences (e.g., dislike) and compensatory reactions (e.g., indirect body orientation). However, we have also noted that there is some conflicting evidence. One conceptual formulation (Patterson, 1976, 1978) suggests a way of integrating both sets of data. Patterson hypothesizes that when two individuals are interacting, a sufficient change in the intimacy of one of them (e.g., moving too close) produces a changed state of arousal in the other. Depending on cognitions or attributions about the situation, this arousal may be labeled as either a positive or a negative emotional state by the other person. If the arousal is negatively labeled, a compensatory response (such as moving farther away) will occur. On the other hand, if the arousal is positively labeled, a **reciprocal response** (moving still closer to the other) will occur. This model makes an important point: the situation should determine whether the effects of interacting at a very close range will be negative (i.e., eliciting compensatory reactions and dislike) or positive (i.e., eliciting reciprocal reactions and liking). For example, reciprocity may occur when two people like each other, while compensation may occur when they are unsure about their relationship or dislike each other (Firestone, 1977; Ickes et al., 1982).

A study by Storms and Thomas (1977) supports the notion that the situation determines whether interacting at close range is positive or negative. In this study participants interacted with another who was either friendly or similar, or unfriendly or dissimilar, at a very close or normal distance. The other was liked more when he or she sat close than at a "normal" distance in the friendly or similar conditions. In effect, when the situation is positive, closeness may facilitate a desire for reciprocal intimacy. On the other hand, the participant was liked less when he or she sat closer than farther away in the unfriendly or dissimilar conditions. Closeness in this situation promoted disliking and a desire for a compensatory response.

While Patterson's model received some support from research, it has been criticized on theoretical grounds (e.g., Hayduk, 1983). Hayduk has suggested that it is often difficult to make predictions with the model since it does not specify what causes a positive or negative evaluation of a change in intimacy. Attempts to modify the model have met with some success (Andersen & Andersen, 1984).

Other studies consistent with the equilibrium hypothesis have shown that decreased directness of body orientation leads to greater proximity among individuals in the situation (e.g., Felipe & Sommer, 1966). In addition, the longer participants interact under inappropriate conditions, the greater the degree of compensation observed (Sundstrom & Sundstrom, 1977). However, research has not always supported the predictions of the equilibrium notion (cf. Altman, 1973), and some studies find opposite results (i.e., closeness begets closeness). One way of resolving this apparent conflict is indicated in the box on this page. Others have suggested modified equilibrium theories (i.e., the "comfort" model reviewed earlier), which better correspond to certain experimental findings (e.g., Aiello, 1977; Aiello & Thompson, 1980b).

CONSEQUENCES OF PERSONAL-SPACE INVASIONS

What happens when a person is sitting alone minding his or her own business, with no intention of interacting with anyone, and a stranger sits down at an uncomfortably close proximity?

The Effects of Being Invaded on Flight Behavior

A famous early study of the effects of personal-space invasions was conducted by Felipe and Sommer (1966). At a 1,500-bed psychiatric

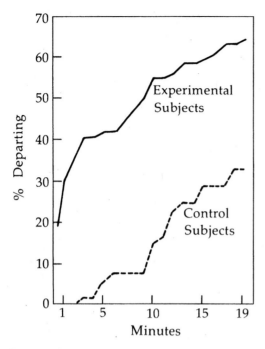

Figure 8-7 Cumulative percentage of patients departing at various intervals.

Based on data from Felipe & Sommer (1966).

institution where patients spent a great deal of time outdoors, a stranger (actually an experimental confederate) approached lone patients at a distance of 6 inches (15 cm). If the patient attempted to move away, the confederate moved so as to maintain a close positioning. The flight behaviors of the "invaded" group were compared with those of patients who were not invaded but who were watched from a distance. As can be seen in Figure 8-7, after 1 minute, 20% of the experimental targets and none of the control targets had fled. After 20 minutes, 65% of the experimental targets had left their places, and only 35% of the control targets displayed such a reaction.

Similar results for flight behavior after personal-space invasion were reported by Konecni et al. (1975). In this study (see Table 8-2), it was observed that both male and female pedestrians crossed the street more quickly as personal-space invasions became more severe. Smith and Knowles (1979) reported the same thing, and also observed that invaded pedestri-

ans formed more negative impressions of the invader and experienced more negative moods than those in control conditions. In another setting, Patterson, Mullens, and Romano (1971) reported that "invaded" targets turned away, avoided eye contact, erected barriers, fidgeted, mumbled, and displayed other compensatory and coping reactions more than "noninvaded" control targets. Such reactions are especially common in individuals who choose not to escape altogether, or who do not have the option of escape. In a study by Terry and Lower (1979), it was found that in the latter group the more severe the invasion, the more intense the attempts at perceptual withdrawal. In research with children, it was found that personal-space invasions caused behavior to become more primitive and to be characterized by increasing movements (e.g., fidgeting; Bonio, Fonzi, & Saglione, 1978). In nursing-home residents with dementia, personal-space invasions were found to be a significant cause of agitated behaviors among the invasion "victims" (Ragneskog, Gerdner, Josefsson, & Kihlgren, 1998).

The Effects of Being Invaded on Arousal

Does invasion of your space increase your arousal? A very ingenious—and as we noted in Chapter 1, controversial—study by Middlemist, Knowles, and Matter (1976) bears directly on this question. The setting for the study was, of all places, a three-urinal men's lavatory! The unknowing participants were lavatory users who were "invaded" by a confederate at either a close

| Table 8-2 | TIME IN SECONDS TAKEN TO CROSS THE STREET BY EXPERIMENTAL CONDITION[°] |

Sex of Participants	Experimenters' Lateral Distance From Participants (in feet)			
	1	*2*	*5*	*10*
Male	7.65	8.45	9.09	9.08
Female	8.94	8.95	9.41	9.79

[°] *Based on data from Konecni et al. (1975).*

Figure 8-8A & 8B   Observation apparatus that was used to study the effects of arousal from personal-space invasions in a men's room. As you can see, the periscope is quite unobtrusive when hidden by the stall.

or a moderate distance. In the control condition, the confederate was not present. Since research indicates that stress delays the onset of urination and shortens its duration, it was reasoned that if closer invasions cause stress, greater delay of onset and shorter duration of urination should result. Accordingly, an experimenter stationed in a nearby toilet stall with a periscope and two stopwatches recorded the delay of onset and persistence of urination. As can be seen in Figures 8-8A and B and in Figure 8-9, results confirmed the assumption that personal-space invasions are stressful. Close interpersonal distances increased the delay and decreased the persistence of urination. No wonder installing partitions should add to people's comfort in lavatories!

One implication of the arousal elicited by personal-space invasions is its effect on task performance. In line with the Yerkes–Dodson law (page 104), available evidence suggests that the consequences of invasion-induced arousal for performance depend on the complexity of the task. With simple tasks, performance does not seem to be negatively affected by having another person too close. With more complex tasks, invasions take a noticeable toll (Evans & Howard, 1972).

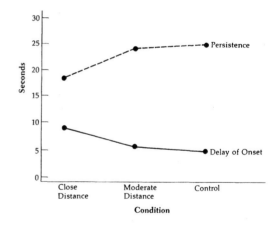

Figure 8-9   Mean persistence and delay of onset for urination at three levels of personal-space invasion.

From Middlemist, R. D., Knowles, E. S., & Matter, C. F. (1976). Copyright © 1976 by The American Psychological Association. Reprinted by permission of the author and publisher.

Other Effects of Being Invaded

If personal-space invasions are aversive for the target, they should elicit a host of additional behavioral reactions. In some situations, it would not be unreasonable to expect that personal-space invasions could lead to aggression (Ryden,

TOO CLOSE FOR COMFORT: Gender Differences in Response to Invasions of Personal Space

As you recall from earlier in the chapter, it has been found that males prefer to position themselves across from liked others, while females prefer to position themselves adjacent to liked others (Byrne, Baskett, & Hodges, 1971). On the basis of these findings, Fisher and Byrne (1975) reasoned that for each sex the spatial position most favored for "liked" others should be the one least favored for an invading stranger. Specifically, it was hypothesized that females should respond more negatively than males to side-by-side invasions of personal space, while males should respond more negatively than females to face-to-face invasions.

Those invaded were males and females who were sitting alone at tables in a university library. As they attended to their business, they were "invaded" by a male or female "invader" from either a face-to-face or an adjacent position (the invader simply sat in a chair at the same table either across from or next to the "victim.") After 5 minutes, the invader appeared to have concluded his or her work and left the area. Three minutes later, an experimenter arrived, claiming to be a student who was conducting a study of people's impressions of various stimuli for an introductory psychology class. The experimenter also claimed to have noticed that someone had been sitting at the target's table and wondered whether he or she would rate impressions of that person as well as impressions of the library environment on questionnaires. The questionnaires specifically tapped affective state, attraction toward the invader, perception of the aesthetic quality and crowdedness of the environment, and the positiveness of motivation attributed to the invader.

How did invasion victims respond to the questionnaires? Regardless of the invader's gender, males responded negatively on all measures when the invader sat across from them but were not affected by an adjacent invasion. Females responded negatively when the invader sat adjacent to them but were not affected by one who sat across from them. It is as if special significance is associated with face-to-face positioning for males and adjacent positioning for females, and invading these "special" zones leads to particularly negative reactions.

The results of the study led Fisher and Byrne to make a simple prediction that, if confirmed, would lend additional support to their findings. It was assumed that if males dislike face-to-face invasions, they should place their books and personal effects between themselves and facing seats in a library, and if females dislike adjacent invasions, they should place their possessions between themselves and adjacent seats. To test these hypotheses, an observer was sent into the library to record where males and females placed their possessions. The hypotheses were confirmed: Males erected barriers primarily between themselves and facing positions, while females erected barriers between themselves and adjacent positions.

Why is it that the sexes seem to attribute special significance to different spatial positioning arrangements? One possible explanation lies in the socialization process, with males taught to be relatively competitive and hence more sensitive to competitive cues, and females taught to be relatively affiliative and more sensitive to affiliative cues (Crawford & Unger, 2000; Deaux & LaFrance, 1998). Adjacent seating (which occurs in affiliative situations) may signal affiliative demands to females. Females like to have someone they "feel safe with" in this relatively intimate affiliative position and react negatively when it is occupied by a stranger. On the other hand, facing seats (which occur in competitive situations) may signal competitive demands to males. Males like to have a trusted (and nonthreatening) friend in this competitive position.

It is rather humorous, but the gender differences observed in the Fisher and Byrne research may be the source of considerable miscommunication between the sexes. A female who wants to befriend an unknown male may be surprised to find that a nonthreatening (to her) eyeball-to-eyeball approach causes consternation and alarm. In the same way, a male who attempts to ingratiate himself with an unknown female by sitting adjacent to her in a nonthreatening (to him) position may be surprised to find he elicits a "Miss Muffet" reaction.

Bossenmaier, & McLachlan, 1991). In fact, sometimes (e.g., during demonstrations in which both pro- and antiabortion protestors are present) it is necessary to keep people from the two factions apart to prevent personal-space invasions from occurring, which helps to avoid aggression (Hern, 1991).

If your personal space were invaded by someone, would you be less likely to help him or her, if given the opportunity? Two sets of studies have looked at the effect of personal-space invasions on helping and have reached conflicting conclusions. In one group of studies (Konecni et al., 1975; Smith & Knowles, 1979), a confederate first violated the target's personal space and then dropped one of several objects. It was found that when the personal space violation was severe, victims failed to retrieve even objects that seemed to be important. In addition to failing to help the *invader*, Smith and Knowles (1979) reported that the reluctance to help on the part of those who had experienced "severe" invasions also generalized to an unwillingness to assist others in need of aid. A second set of studies (e.g., Baron & Bell, 1976) found just the opposite: Personal space invasions facilitated helping. However, in these studies, confederates asked the target for help when at either an "invaded" or an appropriate distance. These conflicting findings can be readily resolved. In the first set of studies, the victim may have attributed the invasion to negative intent, dismissed the invader as a "nasty" person, and refused to help him or her. In the second, the invader's violation may have been attributed to the importance of the request rather than to negative personal qualities and therefore resulted in greater helping.

Smith and Knowles (1979) suggest an interesting possibility for integrating the mechanisms behind some of the findings presented thus far, in line with Patterson's theory of compensation versus reciprocation (see the box on page 269). Smith and Knowles imply that our initial response to a personal-space invasion involves *arousal*. As we know, arousal can have many behavioral consequences in and of itself. However, Smith and Knowles argue that arousal is also followed by a secondary, cognitive response (e.g.,

attributions). In effect, our arousal response draws our attention to the invader and causes us to try to understand why we are aroused, and why the invader behaved as he or she did. Characteristics of the invader and the situation affect the explanations at which we arrive. These, in turn, determine our attributions about the invader, our liking for him or her, whether we will aggress against the invader, help the invader if he or she is in need, and so forth.

An interesting question remains: Do all personal-space invaders elicit the same reactions in their victims, or is it more aversive to be "victimized" by some people than others? The model proposed by Smith and Knowles (1979) would suggest the latter. The more negative our attribution for why someone invaded our space, the more uncomfortable the invasion will make us. It appears that people flee more slowly after having their personal space violated by an attractive rather than an unattractive confederate (Kmiecik, Mausar, & Banziger, 1979). Also, it may be more upsetting to be invaded by a male than a female. Specifically, Murphy-Berman and Berman (1978) and Bleda and Bleda (1978) observed that male intruders were evaluated more negatively and elicited more movement in their victims than female intruders. Perhaps this is because we attribute more negative motives to male than female invaders. Other research suggests that more negative reactions to male invaders may be limited to ambiguous settings (Aiello, 1987). In less ambiguous settings, such as on the beach or in a bar, a male's invasion of a female's personal space may elicit a favorable response (Skolnick, Frasier, & Hadar, 1977). Additional studies suggest that the degree of choice invaders have in their action affects how negatively they are evaluated (Murphy-Berman & Berman, 1978), and it seems that invaders who smoke elicit more flight reactions in their victims than those who do not (Bleda & Bleda, 1978).

What about the effect of invaders of different ages? Fry and Willis (1971) had children who were 5, 8, and 10 years old stand 6 inches (15 cm) behind adults in theater lines. Five-year-olds were given a positive response, 8-year-olds were ignored, and 10-year-olds were given a cold reaction. Thus, as children get older, they are treated more like adult invaders. Finally, in a study that examined whether the status of the invader affects reactions to him or her, Barash (1973) varied the clothing confederate invaders wore and found that those who wore "faculty-like" attire evoked faster flight than those who wore casual clothing.

How could one ameliorate some of the negative effects on the victim, if forced to invade someone's personal space? Sundstrom and Sundstrom (1977) suggest that asking permission can make a difference. But perhaps the best way to avoid torturing those you invade comes from research by Smith and Knowles (1979, p. 449), who suggest that "negative reactions occur only when there is no immediately apparent and appropriate reason for the invader to be standing close." So, as long as you behave so that your victim believes you have a good reason for your invasion, you may be able to avoid inflicting pain on others.

Are there gender differences in reaction to personal space invasions? It is interesting to note that males generally react more negatively to invaders than do females (Patterson, Mullens, & Romano, 1971), although there are exceptions (Bell, Kline, & Barnard, 1988). On airplanes men use the common armrest three times as much as women and are more apt to get annoyed if their neighbor uses it. In part, the larger spaces taken up by men and their more negative reactions to invasions may be responsible for the fact that women are typically approached more closely than are men (Kaya & Erkip, 1999; Long, Selby, & Calhoun, 1980). Overall, women have more tolerance for distances that are inappropriately close than do men (Aiello, 1987).

Invading Another's Personal Space: Effects on the Invader

We have spoken about how it feels to be the *victim* of a personal-space invasion, but what happens when people are placed in dilemmas that require them to become personal-space "invaders"? Several studies have found that people do not even like to *approach* the personal space of others. In one study (Barefoot, Hoople, & Mc-

Clay, 1972), a lone confederate was stationed 1, 5, or 10 feet (0.3 m, 1.5 m, or 3.0 m) from a water fountain. Fewer passersby approached the fountain when doing so would violate the confederate's personal space (i.e., at the 1-foot distance) than when it would not (i.e., at the 5- or 10-foot distances). While people will avoid a water fountain when the setting is uncrowded, it becomes easier to invade someone's personal space (and take a drink) under crowded conditions (Thalhofer, 1980). Surrounded by a crowd we become "overloaded" and are less attentive to social cues (e.g., that we may cause another person discomfort). Another study demonstrated that it may be aversive to approach the personal space of a group, as well as a lone individual. Knowles and Bassett (1976) positioned groups of varying sizes on a hallway bench and observed "deflection" in the walking patterns of passersby as they walked past the seated confederates. As the number of confederates on the bench increased, passersby were "deflected" farther away. Thus, it appears that approaching the personal space of either lone individuals or groups is a threatening experience, to be avoided if possible. Design features that allow people to avoid invading others' personal space would probably be appreciated by many of us.

Several other experiments have looked at the invader's reactions to physically penetrating, rather than merely approaching, the personal space of interacting dyads. One study suggests that for females it is easier to invade the personal space of someone who is smiling than of someone displaying a neutral face, while for males the reverse is true (e.g., Hughes & Goldman, 1978; Lockhard, McVittie, & Isaac, 1977). And, at least for males, it may be still easier to violate the space of one who has his or her back toward the invader (Hughes & Goldman, 1978). Efran and Cheyne (1974) found that passersby are less likely to "invade" if the individuals in the dyad are conversing, if they are occupying Hall's personal-space zone rather than social distance, and if they are of the opposite sex. Similar results were reported in other studies (e.g., Bouska & Beatty, 1978), which also found that

HOW DO GROUPS RESPOND TO PERSONAL-SPACE INVASIONS?

In general, we have restricted our attention to the effects of personal-space invasions on lone individuals. In a very interesting study, Knowles (1972) extended this line of research to an exploration of how *groups* of people respond to personal-space invasions. His findings suggest that groups, like individuals, engage in compensatory responses when their space is invaded, constituting evidence for a group analog to personal space.

On a city street Knowles had a confederate approach a pair of pedestrians walking in the opposite direction. The invader walked so it appeared that he or she intended to walk right between the two pedestrians. Over half the pairs moved together to avoid an intrusion, and some reprimanded the invader; this suggests that groups try to maintain their personal space—even in the face of invasion. Further, group-level avoidance of intrusion was more frequent when the pedestrians consisted of a male and a female rather than individuals of the same sex. In a later study, Knowles and Brickner (1981) found that the more cohesive the dyad, the more it resisted the intrusion, i.e., protected its "group space."

an "invasion" was less likely if the interactants appeared to be of high status (e.g., a businessman, a priest). When it is necessary to invade, the status of the victims also determines how the invader treats them. High-status individuals receive "positive deferential" behaviors (e.g., signals of appreciation), while those with low status receive signs of negative deference (e.g., derogation) from the invaders (Fortenberry et al., 1978). Finally, dyads comprising individuals who are African American may be more likely to be invaded than Caucasian or mixed-ethnic dyads (Brown, 1981).

Even though it is easier to invade some people's personal space than others, it is generally aversive to violate the personal space of others. In a finding that shows just how taxing it is to invade interacting dyads, it was observed that participants forced to invade tended to look at the floor rather than ahead and to close their eyes (Cheyne & Efran, 1972; Efran & Cheyne, 1974). Further, Efran and Cheyne (1974) reported that the act of invading interacting dyads has affective consequences: Participants in "invasion" conditions displayed more negative moods and more hostile facial responses than noninvading control participants. How does the invader react to penetrating a group larger than a dyad? Knowles (1973) created target groups of two or four persons who were interacting in a

hallway so that passersby had two choices: to violate the group space or to go around the interactants. Fewer people penetrated the four- than the two-person group, and low-status individuals were invaded more often than high-status individuals. Thus, not only individuals and dyads but larger groups appear to be aversive to invade. For the invaders, the permeability of interacting targets depends on such factors as status, group size, and gender composition. Further, it is apparent that groups, like individuals, are recognized as having a sort of personal space. (Note that the study reviewed in the box on this page, in which groups responded as a unit to a confederate invader, also supports the idea of personal space at the group level.)

SUMMARY OF PERSONAL SPACE

Personal space is an invisible, portable boundary that regulates how closely we interact with others. Many conceptual perspectives may be applied to suggest different functions of personal space. A combination of these approaches implies that personal space serves two major functions: protection and communication.

Personal space expands and contracts depending on situational conditions, and as a function of individual differences. People interact more closely with similar than with dissimilar others, and in pleasant than in unpleasant interaction situations. Individual differences that affect personal space preferences include gender, certain cross- and subcultural differences, age, and personality factors (e.g., internality–externality, anxiety, introversion–extraversion). Physical factors (e.g., ceiling height, position in room) also affect personal space preferences.

Some of the same factors, such as gender and interpersonal attraction, that have an impact on the amount of personal space we prefer to maintain also affect our interpersonal positioning. In addition, the appropriateness of spacing (e.g., too close versus too far) affects goal fulfillment. Inappropriate spacing also leads to negative affect and to compensatory responses.

Similarly, personal-space invasions elicit negative affect, arousal, negative inferences, and compensatory reactions. The intensity of negative reactions varies as a function of situational conditions and individual differences. Finally, being placed in the role of personal-space invader is aversive, and is avoided if possible. With this capsule summary of personal space in mind, we move on to a discussion of territorial behavior.

Territorial Behavior

In contrast to personal space, *territory* is visible, relatively stationary, visibly bounded, and tends to be home centered, regulating who will interact (Maher & Lott, 1995; Sommer, 1969). Also, territories are generally much larger than personal space; and whether or not we are on our own territory, we still maintain a personal-space zone.

One way of viewing *territories* is as places that are owned or controlled by one or more individuals. Anyone who has ever been on the sending or receiving end of a statement like "Don't you ever set foot on my property again!" has confronted the concept of *territoriality* head on. In addition to the notion of demarcation and defense of space, territories also play a role in organizing interactions between individuals and groups, can serve as vehicles for displaying one's identity, and can be associated with feelings, valuation, or attachment regarding space (Figure 8-10).

For us, *human* **territoriality** *can be viewed as a set of behaviors and cognitions a person or group exhibits, based on perceived ownership of physical space.* Our definition of territoriality in humans is representative of "mainstream" views in the field (for commentaries on the definitions used by others, see Altman & Chemers, 1980;

Brown, 1987; Maher & Lott, 1995; Taylor, 1988; Taylor & Brooks, 1980). Perceived ownership as used here may refer either to actual ownership (e.g., as with your home) or to control over space (e.g., you may control but not own your office if it is part of a building owned by another). Territorial behaviors serve important motives and needs for the organism and include occupying an area, establishing control over it, personalizing it, having thoughts, beliefs, or feelings about it, and in some cases defending it (Brown, 1987; Harris & Brown, 1996; Omata, 1995; Taylor, 1988). Note that the concepts of "territory" and "territoriality" illustrate the interdependent nature of human–environment transactions. Without a territory there would be no territoriality, and vice versa (e.g., Taylor, 1988).

According to Altman and his colleagues (Altman & Chemers, 1980), three types of territories are used by humans, and this distinction has been supported in research by others (Taylor & Stough, 1978). These differ in their importance to the individual's or group's life— *primary* territories are most important, followed by *secondary* and *public* territories. They also differ in the duration of occupancy, the cognitions they foster in the occupant and others (e.g., the extent of perceived ownership), the

Figure 8-10   Fences and signs are among the many ways people demarcate and defend their territories.

amount of personalization, and the likelihood of defense if violated (these differences are highlighted in Table 8-3). As we will discuss later, different types of territories provide different benefits for individuals. For example, primary territories such as a bedroom promote privacy and control and allow for the expression of one's identity—functions not promoted in a public territory. Therefore, based on the type of activity we want to engage in and the needs it poses, we choose a particular type of territory (Taylor, 1988).

Table 8-3	TERRITORIAL BEHAVIORS ASSOCIATED WITH PRIMARY, SECONDARY, AND PUBLIC TERRITORY °	
	Extent to Which Territory Is Occupied/Extent of Perceived Ownership by Self and Others	**Amount of Personalization/ Likelihood of Defense if Violated**
Primary Territory (e.g., home, office)	*High.* Perceived to be owned in a relatively permanent manner by occupant and others.	*Extensively personalized;* owner has complete control and intrusion is a serious matter.
Secondary Territory (e.g., classroom)	*Moderate.* Not owned; occupant perceived by others as one of a number of qualified users.	*May be personalized to some extent during period of legitimate occupancy;* some regulatory power when individual is legitimate occupant.
Public Territory (e.g., area of beach)	*Low.* Not owned; control is very difficult to assert, and occupant is perceived by others as one of a large number of possible users.	*Sometimes personalized in a temporary way;* little likelihood of defense.

° Based on Altman (1975).

THE ORIGINS OF TERRITORIAL FUNCTIONING

Territorial behavior is practiced by humans and other animals. Some researchers consider human territoriality to be *instinctive,* some consider it to be *learned,* and some consider it an *interaction* of the two (cf. Brown, 1987; Taylor, 1988). According to the instinct view, there is an innate drive to claim and defend territory (e.g., humans and nonhumans mark off their turf to keep others out and respond with vocal warnings and bodily threats to invaders; Ardrey, 1966; Lorenz, 1966). Since Earth has a limited amount of space, and we are all driven to make and defend territorial claims, conflict is inevitable. Needless to say, this set of beliefs makes some fairly pessimistic predictions regarding the future of humankind. However, few investigators hold that territorial behavior is entirely instinctive.

The position that territoriality is *learned* suggests that in humans it results from past experience and from culture. For example, people learn through socialization that certain places are associated with particular roles. And the patterns of learning that occur depend on culture (e.g., some cultures are nomadic and relatively aterritorial, while others are highly territorial). Some proponents of learned territorial behavior assume that such learning is limited to humans, and that in nonhumans territoriality is instinctively driven. For them, even when humans and other animals exhibit similar behaviors, such as aggression against an intruder, the same mechanisms may not be responsible.

Finally, from another perspective, human and perhaps even nonhuman territorial behavior may result from an *interaction of instinct and learning.* This view holds that both processes contribute to territorial actions. The exact way in which this may occur is yet to be specified (Altman & Chemers, 1980). However, it is quite possible that we are predisposed toward territorial behaviors through instinct, but that learning determines the intensity and form of our territorial actions. Alternatively, it has been proposed that instinct guides some types of elementary territorial behaviors, while learn-

ing is responsible for more complex ones (e.g., Edney, 1974).

Some research suggests replacing the instinct perspective on nonhuman territoriality with a more complex conceptualization. Based on this work, the notion of a territorial instinct in nonhuman animals that is "unresponsive to learning and driven to expression" is becoming less accepted (Brown, 1987, p. 508). Instead, nonhuman territoriality is viewed more as an adaptive mechanism that is responsive to ecological considerations (e.g., resource availability) and that is flexible across time and different types of settings (Brown, 1987). Given that the conceptualization of nonhuman territoriality has become less biological, the argument that human territoriality is strictly biologically based has come to hold even less weight than before.

FUNCTIONS OF TERRITORIALITY

Territorial functions vary among species (e.g., Maher & Lott, 1995). Animals maintain territory for such important functions as mating, dispersing the population more evenly, food gathering and protecting food supplies, finding shelter, rearing young, and minimizing intraspecies aggression. Thus, territories are often quite essential to their survival. Also, animals tend to defend territories vigorously when violations occur (Edney, 1976; Maher & Lott, 1995), though this depends on resource distribution and competition.

The human species is flexible with respect to the use of territories for functions such as those mentioned above. For us, many of the purposes territories serve are not as closely related to survival, and they may be seen primarily as "organizers" on a variety of dimensions (e.g., they promote predictability, order, and stability in life; e.g., Edney, 1975; Brown, 1987). For example, territories allow us to "map" the types of behavior we can anticipate in particular places, whom we will encounter there, what someone's status is, and so forth. In this way they help us plan and order our daily lives. Territories also contribute to order due to their relationship to social roles (e.g., the boss controls his or her office, the company lounge, the lunchroom, etc.).

Table 8-4	THE ORGANIZING FUNCTIONS OF HUMAN TERRITORIES IN SOME EVERYDAY SETTINGS °

For People in . . .	Organizing Function of Territory
Public places (e.g., a library, the beach)	Organizes space; provides an interpersonal distancing mechanism.
Primary territories (e.g., a bedroom)	Organizes space by providing a place that promotes solitude; allows intimacy; expresses personal identity.
Small face-to-face groups (e.g., the family)	Clarifies the social ecology of the group and facilitates group functioning; may provide home-court advantage.
Neighborhoods and communities	Promotes an "in-group" who "belongs" and can be trusted; differentiates it from an "out-group" who doesn't belong and can't be trusted. In some urban areas, territorial control makes a space safe to use.

° *After Taylor (1988).*

Precisely *how* territories function to "organize things" depends on the particular space in question (for some examples, see Table 8-4). In addition to their organizing function, territories may lead to feelings of distinctiveness, privacy, and a sense of personal identity (Harris & Brown, 1996). People may experience a higher self-concept due to the territories they possess and the ways they have personalized them. They may even proudly refer to themselves as "the person who lives in the big red house on Oak Street." In sum, social, cultural, and cognitive elements are characteristic of human territoriality. While nonhuman territoriality is rooted in survival needs, human territoriality is also associated with "higher-order" needs (e.g., self-image, recognition; Gold, 1982).

How do species differ with regard to territorial defense? In general, humans very rarely resort to aggressively defending their turf. When they must deal with territorial invasions, their defense is typically based on laws that defend territorial rights, rather than brute force (Brown, 1987). This is not to suggest that relatively dramatic forms of territorial defense never occur in humans. Indeed, many international problems (Coakley, 1993) as well as frequent interpersonal difficulties (e.g., fights with a roommate over use of the refrigerator) involve territorial issues and associated aggression. However, one reason for the typically lower degree of territorial defense in humans than other species is that people generally recognize and avoid one another's territory. This varies, of course, with the type of territory (as depicted in Table 8-3 on page 277; e.g., Schiavo et al., 1995). In addition, humans routinely entertain others on their turf without aggression. When human territorial aggression does occur, it often takes a different form from that of other species. While human territory-related fighting tends to occur more often at the group level (e.g., one nation versus another), fighting in other species is more frequently at the individual level, though it occurs at the group level as well. Unfortunately, humans now have the capacity to destroy one another's territory without physically invading, through the use of long-range weapons.

A broad conceptualization of the functions of territory for humans may be achieved through an analysis in terms of the environment–behavior theoretical formulations (e.g., arousal, overload) discussed in detail in Chapter 4, and earlier in this chapter in the context of personal space. For example, in terms of the *overload* approach, clearly defined territories reduce environmental load by lending a sense of order that lowers the amount and complexity of incoming stimulation and makes life easier to cope with. In effect, territories afford role organization (e.g., the host has one role and the visitor another); allow us to assume continuity in the future (e.g., we will always be able to sleep in our house); and afford us control over inputs from the outside world (e.g., "No Trespassing" signs keep out extraneous inputs). *Stress formulations* view territories as functioning to reduce stress by controlling the amount of stressful stimuli with which we must contend. According to the *privacy regulation model*, territories are used to maintain a consistency between desired and achieved levels of privacy. From the *arousal* perspective, territories hold down arousal (e.g., by moderating the amount of stimulation we are exposed to). In the context of the *ethological* conceptualization, territories may be seen as minimizing aggression and affording identity. Finally, in line with the predictions of *control* models, the fact that territories facilitate unhindered performance of chosen behaviors should be quite beneficial. Territories should also have favorable effects because the "owner" of a territory controls access to it and what goes on there.

METHODS OF STUDYING TERRITORIALITY IN HUMANS

Research on human territoriality has looked at territorial behavior in both groups and individuals, employing methodologies that range from controlled laboratory and field experimentation to naturalistic observation. In some cases these methodological approaches have inherent problems when applied to human territorial behavior.

Laboratory experiments are difficult to perform with humans because territoriality implies a strong attachment between an individual and a place, which is not easy to create under artificial laboratory conditions. Introducing experimental manipulations (e.g., territorial invasions) into real-world settings (e.g., dormitories) where people do perceive a degree of territorial "ownership" avoids this problem, and has provided some rich data. In addition, many researchers have relied on nonmanipulative (and nonexperimental) field observation of behavior in naturally occurring territories.

Unfortunately, nonmanipulative observation is often fraught with interpretive problems. For example, Vinsel et al. (1980) reported an interesting relationship between the way in which students personalized their dorm room (or primary territory) with decorations, and whether or not they dropped out of college during the following year. Students whose decorations showed diversity and commitment to the university setting were more likely to survive the rigors of college than those whose decorations did not. However, what these data mean is very unclear. It could be that the way nondropouts personalized their territory led to feelings of security, which promoted success in school. On the other hand, personalizing one's dorm room may reflect commitment to it, and lack of personalization may reflect a sense of alienation that one might expect in someone planning to drop out of school.

RESEARCH EVIDENCE OF TERRITORIAL BEHAVIOR

Territorial Behavior Between Groups

In a classic study, Suttles (1968) observed the territorial actions among various ethnic groups on Chicago's South Side (public territory). Each group claimed and defended a separate territory, and there were some "shared territories" in which certain community resources were used separately by each ethnic group in a prescribed fashion. Different groups would use them, but never at the same time. Another interesting example of group territoriality stems from an analysis of street gang behavior in Philadelphia (Ley & Cybriwsky, 1974a). It was found that street gangs are highly territorial, often taking their names from a street intersection at the center of their territory. Each gang demarcates its territory, and territorial domains are recognized by gang and nongang youth. Outsiders usually avoided the in-group's turf, and were greeted with hostility when they entered.

What functions does territoriality between groups serve? Such actions tend to facilitate trust *within* the group. Sharing a territory can lead to feelings of group identity and security, perhaps because people in the same territory share common experiences (Taylor, 1988). And the security afforded by a territory is important: In some areas of a city, having territorial control of a space makes it safe to use (Taylor, 1988). However, the in-group cohesion resulting from territories can have negative effects (e.g., the formation of gangs). It may also cause "outsiders" to be viewed with suspicion. Both of these consequences could elicit aggression.

Territorial Behavior Within Groups

Group members often adopt certain areas as "theirs." In primary territories, families have territorial rules that facilitate the functioning of the household (Ahrentzen, Levine, & Michelson, 1989; McKinney, 1998; Omata, 1995). These support the social organization of the family by allowing certain behaviors by some members, in particular areas (e.g., the parents can engage in intimacy in the bedroom undisturbed). In one study of territoriality in family life, it was found that people who share bedrooms display territorial behavior, as do individuals at the dining table (e.g., through seating patterns). Territorial divisions in the home depend on the particular activities of family members, as well as whether or not the mother works outside the home (Ahrentzen et al., 1989). Family members generally respect one another's territorial markers, such as closed doors (Altman, 1975), and a violation of territorial rules often leads to punishment of the one at fault (Scheflen, 1976).

Territorial behavior within groups is not limited to primary territories. Lipman (1967) found that residents of a retirement home made almost exclusive claims to certain chairs in the day rooms. They defended their "territory" despite considerable psychological costs and physical inconvenience. Even students stumbling into their 8:00 a.m. class display territorial behavior. Haber (1980) found that in formal style (e.g., traditional lecture) classes, about 75% of the students claimed a particular seat and occupied it more than half the time. In informally run classes, this occurred for only 30% of the students. Also, of those students who claimed a seat to be their territory, 83% chose the one that they occupied during the first, second, or third class period. In addition to choosing a seat as their territory, many students used markers to delineate their turf. "Marking" (e.g., placing books and possessions to defend one's turf) is also frequent in libraries and cafeterias, among other places (e.g., Fisher & Byrne, 1975; Taylor & Brooks, 1980).

A number of researchers have investigated whether some members of intact groups are more territorial than others. It is reliably found that males are more territorial—have larger territories—than females (Mercer & Benjamin, 1980). In addition, some studies with nonhumans show a strong relationship between dominance within a group and territoriality, generally finding that more dominant animals are more territorial. This finding, however, depends on resource scarcity and competition. Although research with humans has sometimes demonstrated mild support for the dominance–territoriality relationship found in other species, in some studies opposite results have been observed. It seems that the relationship between dominance and territoriality is quite complex (Brown, 1987; Taylor, 1988).

Perhaps whether more dominant individuals will display higher or lower territorial behavior depends on the situational context. In environments offering only a few desirable places (e.g., private rooms in a home for delinquent boys), dominant individuals should end up with them and hence appear to be highly territorial. In contrast, when a setting has no areas that are more desirable than others, dominant individuals should roam over large amounts of space and appear to be very low in territoriality. The dominance–territoriality relationship probably also depends on group composition and social organization. Adding and removing group members or changing the social organization of the group can significantly affect the nature of dominance–territoriality relationships (e.g., Sundstrom & Altman, 1976). Although studies of the dominance–territoriality relationship have shown mixed results, researchers (e.g., Taylor, 1988) suggest that where a dominance–territoriality relationship does exist, it should facilitate group functioning. Investigators propose that if those with high dominance are recognized as having access to the best space, this helps clarify the ecology of the group and thus reduces conflicts within it.

Territorial Behavior When Alone

Territoriality also exists for individuals who are alone. In fact, research suggests that people may feel a stronger ownership of a setting when alone than when part of a group (Edney & Uhlig, 1977). Thus, members of a family or roommates may feel lower responsibility to maintain their turf, and may individually exert less surveillance over it, than single occupants. This also implies that there may be more vandalism, theft, and other similar acts in group than in individual residences.

Signals of Territoriality: Communicating Territorial Claims

What do a backyard fence, a chair with a coat on its back, a nameplate on an office door, and a blanket at the beach have in common? All are ways of communicating territorial ownership to others as well as, perhaps, reassuring oneself regarding ownership or propriety over something (Barber, 1990). We engage in these types of behaviors differently in primary, secondary, and public territories. In certain public territories (e.g., an airport, bus station), rather than providing protection from invasion, valuable territorial markers (e.g., a suitcase, a fur coat) are apt to be stolen (Brown, 1987; see Figures 8-10 and 8-11).

Sommer (1969) conducted classic studies that looked at the relative effectiveness of various strategies for warding off territorial invaders in libraries. At low levels of overall density, people were less likely to sit at tables with any kind of marker (e.g., a sandwich, a sweater, books) than at tables without such personal effects. However, under conditions of high density, it appears that potential invaders take an attributional approach to interpreting whether or not particular markers really represent someone who intends to return. To the extent that markers are personal and valuable (a coat, a notebook with a name on it), territorial "ownership" tends to be respected. However, when attributions of intent are not clear, as when the marker is a library book or a newspaper, the resulting uncertainty coupled with the fact that only a few seats are available tends to lead to

invasions. Research also indicates that "male" markers are much more effective at territorial defense than "female" markers, and that territories belonging to males (e.g., a man's versus a woman's desk) are less apt to be invaded (Haber, 1980; Shaffer & Sadowski, 1975). It should be noted that although we and others consider such markers as books and coats to be territorial indicators, there is debate as to whether they function mainly as territorial markers or as interpersonal distance maintainers.

Ley and Cybriwsky (1974a) suggested another interesting means of indicating turf ownership. They found that in Philadelphia, wall graffiti offer an accurate indication of gang territorial ownership. As a general rule, gang graffiti (i.e., graffiti that include a gang's name) become denser with increasing proximity to the core of the gang's territory. These graffiti are readily accepted by neighborhood youth as an accurate portrayal of each gang's area of control. It was also found that often, when street gangs invaded each other's territory, they spray-painted their name in the rival gang's turf. The "invaded" gang generally responded by adding an obscene word after the rival gang's name! Gangs that were not respected (or feared) generally had turf covered with a large amount of graffiti put there by neighboring gangs.

Finally, in addition to traditional (e.g., books, coats) and nontraditional markers, such as graffiti, it has been suggested that nonverbal markers may be used to communicate territorial claims. Studies have found that restaurant patrons touch their plates when they have reason to assert a territorial claim (Truscott, Parmelee, & Werner, 1977), and in a video arcade, Werner, Brown, and Damron (1981) found that standing close to or touching a video game protected it from violation by others. In addition, people were more apt to touch their video game to assert ownership when others were approaching.

Personalizing Territories

In addition to staking territorial claims, people tend to *personalize* their territory. Some means of personalizing territory (e.g., working on one's lawn or garden, making improvements to one's

Figure 8-11   Note the various forms of territorial defenses that people employ.

property) may provide opportunities for neighbors to get to know each other better, to become more cohesive (Brown & Werner, 1985), and to learn to distinguish between residents and strangers. This may lead to more surveillance and fewer problems with outsiders (Taylor, Gottfredson, & Brower, 1981). Personalization may also elicit greater feelings of attachment to a place and instill the feeling that it is "comfortable" and "homelike" (Becker & Coniglio, 1975). In addition, personalizations often reflect the self-identity of the owner. For example, decorative complexity of housing interiors correlates with materialistic values (Weisner & Weibel, 1981). Further, observers may form impres-

sions of others' idealized self-images (Sadalla, Burroughs, & Quaid, 1980), of their commitment to their home territories (Harris & Brown, 1996), of their degree of sociability (Werner, Peterson-Lewis, & Brown, 1989), or of their ethnic identities (Arreola, 1981) from their personalizations.

The literature suggests that women engage more in personalization and have greater feelings of attachment to their homes than men (Sebba & Churchman, 1983; Tognoli, 1980). Nevertheless, these feelings of person–place attachment are important for both genders and are one of the many losses experienced by the growing members of homeless people (Rivlin,

1990). For more on homelessness, see Chapter 10 on cities.

TERRITORY AND AGGRESSION

One of the most interesting aspects of territoriality is the relationship between territory and aggression. Although it is not always realized, territory may serve either as an instigator to aggression or as a stabilizer to prevent aggression. The function it serves depends on a number of situational conditions. One factor that affects the relationship between territoriality and aggression is the status of a particular territory (i.e., whether it is unestablished, disputed, or well established). When territory is unestablished or disputed, aggression is more common. Ley and Cybriwsky (1974a), for example, found that street gangs engaged in more intergang violence when territorial boundaries were ambiguous or unsettled than when they were well established. Parallel evidence is available for nonhumans: It has been found that other species fight more when territories are being established or are under dispute than after territorial boundaries have been well drawn (e.g., Lorenz, 1966).

While unestablished or disputed territory promotes aggression, established territorial boundaries often lend stability and lead to reduced hostility in humans as well as in other species. For example, Altman, Nelson, and Lett (1972) observed that confined groups that established territories early in their confinement evidenced smoother interpersonal relationships and were more stable socially than groups that failed to establish territories early. O'Neill and Paluck (1973) reported a drop in the level of aggression in groups of developmentally delayed boys after the introduction of identifiable territories. What are the dynamics of the process by which territorial boundaries decrease aggression? We mentioned earlier that territorial behavior serves an organizing function, indicating what is "ours" and what is "theirs." Thus, well-established territories should be less subject to intrusion and should therefore decrease aggression.

When territorial invasions do occur, what are the consequences? Predicting reactions to territorial invasions is complex because our responses appear to depend on situational conditions. For example, some have suggested that the likelihood of territorial defense is the result of cost–benefit analysis, involving the perceived costs of resisting the intruder versus the benefits of maintaining the territory undisturbed (Brown, 1987). In addition, Altman (1975) proposed that the attributions we make for a violation will determine our response, and that we will consider aggression only when we feel the other's behavior was malicious. And generally, we try other verbal adjustive responses (e.g., warning the individual to leave, threatening him or her), as well as physical ones (e.g., putting up a fence or a "No Trespassing" sign) first, resorting to aggression only when these are unavailable or unheeded. In addition, Edney (1974) suggested that for humans many forms of "appropriate" territorial invasion exist (e.g., when guests are present) that do not elicit aggression.

One additional factor that may determine whether invasion leads to aggression in humans is the location of the territory along a primary territory–public territory dimension (Brown, 1987). Invaders of primary territories are likely to elicit the most intense aggression (see Table 8-3). By definition, primary territories are more central to the owner's life, symbolize his or her identity, and are associated with more legitimate feelings of control than public territories. Invasions of primary territories (e.g., homes) are also more apt to be intentional and to involve a deliberate crossing of boundaries or markers than invasions of secondary or public territories. Thus, invaders of primary territories are seen as more threatening and hence are dealt with more harshly. The intensity of the territorial invasion–aggression relationship for primary territories is reflected in the ambiguity of many local laws dealing with the prosecution of a homeowner accused of killing an intruder. One means some homeowners use to prevent invasion of primary territories is to erect markers of territorial defense (e.g., "No Trespassing" signs). Edney (1972) compared homeowners who displayed such markers with those who did not. He found that individuals who erected forms of territorial defense had lived in their houses longer and intended to stay longer than people without territorial markers. Further, residents who displayed markers answered their doorbells faster, which may be interpreted as a sign of defensive vigilance.

In contrast to the defensive posture assumed by holders of primary territory, because people have minimal "rights" to public territories, they may respond to invasions of public territories by retreating (Brown, 1987) or do nothing at all. However, even in public territories, people may exhibit defensive behaviors toward intruders. Haggard and Werner (1990) suggested that people will ask intruders to leave a public territory if information cues (e.g., signs telling others to keep out) and environmental cues support privacy regulation. In such situations, individuals are likely to cite aspects of the situation (e.g., the signs), or to give other "excuses" (e.g., the difficulty of performing their task with the territorial invader present) as their reason for asking the invader to leave. Attempts to regain public territory are generally preceded by expressions of surprise or intimidation (Taylor & Brooks, 1980). Also, as the value of the invaded public territory increases (e.g., a library carrel versus a seat at a table), the likelihood of its defense rises (Taylor & Brooks, 1980). Even under some of the conditions that maximize the likelihood of defense of public territories, several studies suggest a strong reluctance on the part of participants to defend them (e.g., Becker, 1973; Becker & Mayo, 1971). It seems that flight is the most frequent reaction to invasions of public turf (Brown, 1987).

One means of defending a public territory about to be invaded would be simply *not* to yield to the apparent demands of the territorial invader. Ruback and Snow (1993) studied the behavior of individuals who were drinking at a public water fountain and whose territory was about to be invaded. The results showed evidence of nonconscious racism: White participants left the drinking fountain faster when they were intruded by a white territorial invader than in a control condition. In contrast, African Americans stayed at the fountain *longer* when they were invaded by a white confederate than in a control condition. In a second study, it was found that racially dissimilar, "would-be" territorial invaders waited longer to invade the water fountain than did same-race invaders, and that water-fountain drinkers stayed longer (i.e., defended their territory longer) following cross-race territorial invasions than following same-race invasions. Other studies have also shown that under certain conditions, people will refuse to yield public territory. Ruback, Pape, and Doriot (1989) reported that individuals using public telephones spent more time on the phone (i.e., defended their territory more) when threatened by a territorial invasion (another who was encroaching on them and who wanted to use the phone) than in a control condition. Similar observations have been made by Werner, Brown, and Damron (1981). More recently, Ruback and Juieng (1997) found that drivers leaving public parking spaces at a mall who were "intruded upon" by another car wanting to take their space took longer to vacate the space than drivers whose parking spaces who were not intruded upon.

TERRITORY AS A SECURITY BLANKET: HOME SWEET HOME

We have suggested that many properties of territories are associated with positive effects. The truth of the phrase "Home Sweet Home" has been assessed in a number of experiments. In a study that also supported the assumptions of Altman's (1975) conceptual distinction among primary, secondary, and public territory, Taylor and Stough (1978) found that participants reported the greatest feelings of control in primary territories (e.g., residence hall rooms), followed by secondary territories (e.g., a fraternity house) and public territories (e.g., a bar). In a great deal of research, feelings of control are related to a sense of well-being as well as other positive effects (e.g., beneficial implications for health). A study by Edney (1975) using Yale undergraduates highlights additional benefits of being on one's turf. The experiment took place in the dormitory room (primary territory) of one member of the pair, where the other member was a "visitor." Participants who were in their own territory were rated by visitors as more relaxed compared with how residents rated visitors, and residents rated the rooms as more pleasant and private than visitors did. Residents also expressed greater feelings of passive control. In a related study, Edney and Uhlig (1977) reported that participants induced to think of a room as their territory felt less aroused and found the setting to be more pleasant than others in the control group.

While being on one's own turf is typically associated with enhanced perceived and actual control, in offices at least, this may depend in part on status. Katovich (1986) had participants role-play a conversation between an employer and an employee that took place either in the boss's or the employee's office. It was found that the office holder always initiated the handshake at the start of the interaction but that the power to invite the "visitor" to enter depended on status. While the boss invited the employee to enter when the meeting occurred in the boss's office, the boss sat down without waiting for an invitation when in the employee's office! Thus, in certain places, one's territorial "rights" depend on one's status.

An additional advantage of being "at home" is that under conditions that do not promote liking (e.g., competition, disagreement, or unequal roles), the resident has a "home-court" advantage that allows him or her to dominate the visitor. Martindale (1971) reported that dormitory residents were more successful at a competitive negotiation task on "their own turf" than were visitors. Similarly, Conroy and Sundstrom (1977) found that when resident–visitor dyads held dissimilar opinions (conditions that cause disliking), residents talked more and exerted more dominance over the conversation than visitors. When the two had similar opinions (conditions that promote liking), visitors talked more and dominated the conversation. The authors interpreted residents' allowing this as a sort of "hospitality effect." In addition, Taylor and Lanni (1981) have shown that residents have an ad-

vantage under conditions that do not facilitate liking in triads as well as dyads, and for both low- and high-dominance individuals. The effect is even true of larger groups and in settings other than primary territory (see the box on page 287).

SOME DESIGN IMPLICATIONS

Our review thus far suggests that if the design and administrative policies of psychiatric facilities, nursing homes, prisons, and other institutions allow occupants to personalize their quarters and otherwise stake out territory, the social atmosphere of the unit should improve and more positive feelings toward the environment should be observed (e.g., Holahan, 1976; Holahan & Saegert, 1973). Designing space so that it appears to look like someone's turf should have other advantages as well. When spaces have clear boundaries that signal they "belong" to somebody, there is evidence that less crime and vandalism occur. In a study of low-cost urban housing developments, Newman (1972) found that public areas having no clear symbols of ownership were more likely to be vandalized than those with well-marked boundaries. (Expanded coverage of this relationship and the factors that may account for it is provided in Chapter 10.) Although Newman's findings have been subjected to methodological criticism (cf. Adams, 1973), supportive evidence is provided in a famous study that observed the locations where cars were vandalized in inner-city Philadelphia (Ley & Cybriwsky, 1974b). It was suggested that more vandalism took place near "public" places such as factories, schools, and vacant lots than in areas that signaled territorial ownership, such as private dwellings and small businesses. While the nature of the research precludes a definitive statement, it seems that people tend to respect properties that can be identified as someone's territory more than properties that cannot be easily identified.

An interesting study by Brown (1979) identified a number of specific characteristics of residential areas in general, and homes in particular, that are associated with burglary. Before we tell you what she found, take a look at the

AN ANALYSIS OF THE "HOME-COURT ADVANTAGE"

Just how pervasive is the "home-court advantage" in professional and college sports? A study by Schwartz and Barsky (1977) looked at the outcomes of 1,880 major league baseball games, 182 professional football games, 542 professional hockey games, and 1,485 college basketball games that took place in 1 year. They assumed that in the absence of a home-court advantage, about half of a team's total wins for the season should occur at home and half "on the road." As seen in Table 8-5, for all sports they found a decisive home-court advantage. This varied somewhat according to the sport in question, ranging from professional baseball, where 53% of the total wins occurred at home, to professional hockey, where 64% of the wins during the season occurred at home. The analysis of basketball records, which employed slightly different techniques (and is therefore not incorporated into the table), suggested that still a higher proportion of college basketball contests are won on the home court. This implies that the advantage of the home team becomes more pronounced for indoor than for outdoor sports.

Other studies suggest that the "home-held advantage" may be greater for better teams. While even mediocre teams benefit, the better the team, the greater the benefit (James, 1984). However, when pressure to succeed is very high, being at home may be a disadvantage. While being on home turf is beneficial to teams in the first few games of the World Series, when the series goes to a "sudden death" seventh game, home teams win less than 40% of the time (Baumeister, 1985)! The fact that the home-turf advantage is lowered when the pressure is on is supported by other studies. Often, teams play better at home during the regular season than during the championships or playoffs (Baumeister & Steinhilber, 1984; Heaton & Sigall, 1989).

Table 8-5	PERCENTAGE OF GAMES WON BY HOME TEAM IN BASEBALL, FOOTBALL, AND HOCKEY IN A GIVEN YEAR[*]			
Home Team Outcome	**Sport**			
	Professional Baseball	*Professional Football*	*College Football*	*Professional Hockey*
Win	53	58	60	64
Lose	47	42	40	36
Total	100	100	100	100

[*]*Ties are excluded. After Schwartz & Barsky (1977).*

Schwartz and Barsky found that the underlying factor in the home-court advantage was that while superior offensive play occurs at home compared with "on the road," there were no differences for defensive play. How strongly should the home-court advantage be "weighed," compared with factors like team quality? Strikingly, analyses of the data suggested that the advantage from just being on one's own turf can actually be as significant in determining the outcome of a game as the quality of the team!

Schwartz and Barsky (1977) felt that home audience support—the applause for the home team and jeers for the visitors—is an important determinant in the home-court advantage in sports. This may also be why the home-court advantage is greater for indoor sports, where sounds do not get lost in the air. Also, distance between noisy fans and players is greater outdoors than indoors. And as baseball fans have been taught, batting averages are higher and home runs more frequent in domed stadiums (Goodman & McAndrew, 1993), regardless of the home team!

houses in Figure 8-12. Which do you think you would rob if you were a burglar? Brown found that signs of defensibility, occupancy, and territorial concern were different in a sample of homes that were not burglarized than in a corresponding sample that were. Specifically, burglarized homes differed in that the **symbolic barriers** they possessed were public, as opposed to private. For example, burglarized homes had fewer assertions of the owner's

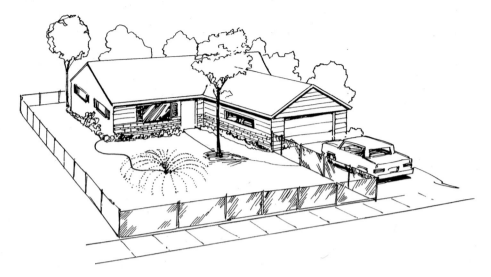

Figure 8-12A A nonburglarized house on a nonburglarized block.

From Brown (1979); reprinted by permission.

Figure 8-12B A burglarized house.

From Brown (1979); reprinted by permission.

private identity (e.g., name and address signs), more signs of public use (e.g., public street signs in front of them), and fewer attempts at property demarcation from the street (e.g., hedges, rock borders). They also had fewer *actual barriers* (e.g., fewer locks or fences to communicate a desire for privacy, and deter public access). Also, on streets where burglaries occurred, there were fewer *traces* (e.g., signs of occu-

pancy) that showed the presence of local residents. Burglarized houses had fewer parked cars and fewer sprinklers operating, and residents were less apt to be seen in their yards by the researchers. In this regard a garage was significant, since it often made it ambiguous whether or not people were home (e.g., a garage without windows can disguise the absence of the car). More burglaries occurred in homes without

TERRITORIAL BEHAVIOR AND FEAR OF CRIME IN THE ELDERLY

We have seen that when symbols of ownership are present, less crime and vandalism may occur. Are people who display more territorial markers (e.g., "No Trespassing" signs, fences, external surveillance devices) less fearful of being victims of crime than people who do not display such markers? In an interesting study, Patterson (1978) explored this problem with an elderly population in central Pennsylvania. Since fear of crime is a major source of anxiety for older citizens (some studies have shown it to be greater than fear of illness), determining the effectiveness of territorial markers in ameliorating such fears is important from both an applied and a conceptual perspective. Patterson had interviewers approach the homes of elderly citizens to record unobtrusively any territorial markers.

After gathering these data, the interviewer approached the homeowner and conducted an interview. The interview consisted of several sets of questions, including fear of property loss (e.g., "When I am away, I worry about my property") and fear of personal assault (e.g., "There are times during the night when I am afraid to go outside"). It was found that displaying territorial markers was associated with less fear of both property loss and assault, especially for males (see Figure 8-13.)

What do these data mean? It is clear that there is an important relationship between reduced fear of crime in the elderly and territorial behavior. However, since this study is correlational, the mechanism by which territoriality is associated with reduced fear is not clear. One possibility is that erecting territorial markers gives one perceived and perhaps actual control and thus leads to feelings of safety. A study by Pollack and Patterson (1980) as well as research by Normoyle and Lavrakas (1984) tentatively supports this interpretation. If this is the case, there is a clear design implication: Encourage people to display territorial markers to enhance their feelings of security. But there is another explanation that cannot entirely be ruled out: Those elderly homeowners who feel sufficient mastery of the environment to erect territorial boundaries are also those who would feel secure from victimization in any event.

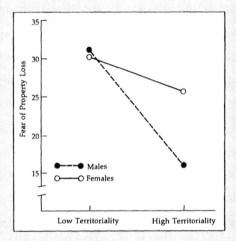

Figure 8-13   Fear of property loss by males and females high and low in territoriality.

From Patterson, A. H. (1978). Territorial behavior and fear of crime in the elderly. Environmental Psychology and Nonverbal Behavior, *3, 131–144.*

garages, perhaps because for these homes the absence of cars made it likely that the house was empty. In houses where burglaries occurred, *detectability* (the potential for exercising surveillance) was also lower, and neighboring houses were less visually accessible (see also Brown & Bentley, 1993; MacDonald & Gifford, 1989). This finding of low "surveillability" increasing the likelihood of burglary has also been found with student apartments (Robinson &

Robinson, 1997). While these findings are *consistent* with defensible-space theory (Newman, 1972), other findings are not as supportive. In *contrast* to defensible-space theory, MacDonald and Gifford (1989) found that a sample of individuals who had actually burglarized homes would *not* avoid houses that looked especially "cared for" (e.g., that had signs of the owner's identity, hedges, or sprinklers). Instead, they seemed to assume that such houses might have valuables inside that would make them an especially good target. Additional findings from this study challenged some of the other findings by Brown (1979; e.g., that traces of occupancy and actual barriers necessarily deter burglary).

Designers should consider the above findings. Too often spaces do not communicate the types of territorial messages they should, or are ambiguous with regard to their territorial status,

due to designed-in characteristics. In addition, some territories do not promote the sorts of activities for which people use them (e.g., the value of a primary territory may be hampered due to a lack of adequate soundproofing, or too much visual access). This often leads to lack of use, to use by the wrong parties, or to various types of misuse. Care during the design process could prevent this.

SUMMARY

Personal space is invisible, mobile, and body centered, regulating how closely individuals interact. It has two purposes: protection and communication. The size of the spatial zone necessary to fulfill the protective and communicative functions changes according to situational variables (e.g., attraction, the activity occurring) and individual difference variables (i.e., ethnicity, personality). Individuals find it aversive (1) when they are constrained to interact with another person under conditions of inappropriate (too much or too little) personal space; and (2) when their personal space is "invaded" by others. Interacting at inappropriate distances leads to negative affect and negative inferences; personal-space invasions precipitate withdrawal and compensatory reactions.

Territory is visible, stationary, and home centered, regulating who will interact. It serves somewhat different functions in humans and other species; in humans it serves a variety of organizational functions. Human individuals and groups exhibit territorial behavior and have adopted a variety of territorial defense strategies that vary in effectiveness. Territorial invasion by others may or may not lead to aggressive responses by the target, depending on the situation. Further, being on one's own turf has been shown to have a number of advantages and elicits feelings of security and improved performance. Finally, areas that appear to be someone's territory are less likely to be vandalized.

KEY TERMS

compensatory behaviors	internality– externality	privacy regulation model	sociopetal
equilibrium	interpersonal distance	reciprocal response	symbolic barriers
ethological models	personal space	situational conditions	territoriality
individual difference variables	privacy	sociofugal	territories

SUGGESTED PROJECTS

1. A scale called the C.I.D.S., or Comfortable Interpersonal Distance Scale (Duke & Nowicki, 1972) permits us to diagram the shape of our personal space without even getting out of our

seat! It works like this. Imagine that Figure 8-14 represents an imaginary round room, for which each radius is associated with an entrance. You are positioned at dead center, facing position

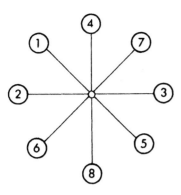

Figure 8-14   Diagramming the shape of personal space.

From Duke, M. P., & Nowicki, S. (1972). Diagramming the shape of personal space: A new measure and social learning model for interpersonal distance. Journal of Experimental Research in Personality, 6, 119–132.

number 8. For each of the 8 radii, respond to an imaginary person approaching you by putting a mark on the radius indicating where you would prefer the stimulus person to halt (i.e., the point at which you think you would begin to feel uncomfortable by the individual's closeness). After you have marked all 8 radii, connect the points you have marked, and you will know the shape of your personal space.

The C.I.D.S. may also permit you to verify some of the relationships that we have described between situational and individual difference conditions and personal space without leaving your chair. For example, imagine that the approaching individual is a friend, and mark the radius; then do the same imagining that he or she is a stranger. Does your experiment confirm the results of the experiments we reviewed which found that people maintain smaller personal-space zones for friends than for strangers? Do the same for an approaching individual who is ethnically similar or dissimilar to yourself or for any of the other relationships we have discussed. You will see how the C.I.D.S. is a useful means of assessing the effects of many factors on spatial behavior. But before you begin, we have one note of caution. Be sure to remember that because the various measures of personal space are not perfectly related (see the discussion on page 257), failure to replicate studies with the

C.I.D.S., which originally used different methods, does not necessarily mean that the original measures are invalid. Thus, while a replication with the C.I.D.S. of earlier findings that employed other methods is valuable supportive evidence, failure to replicate should not be seen as terribly damaging.

2. Your nearby library offers an opportunity for you to study the fine art of territorial defense. Before you go, make some hypotheses about the relative effectiveness of various types of territorial markers for repelling potential invaders. Consider the possible effects of a wide range of markers, including some you expect to be highly effective and some you expect to be less effective. Collect the necessary materials and report to the library for your experiment.

When you arrive at the library, make mental note of the overall level of population density, and place each of the artifacts you brought at a separate empty table, trying not to use too many of the study tables in any one room for your experiment. After you have distributed all your artifacts, "make the rounds" of all your experimental tables at 15-minute intervals, noting which markers are more effective and which are less so in preventing territorial invasion. Repeat the procedure using different levels of population density.

3. Position yourself and a same-sex other on either side of a busy doorway through which many people must pass. Face each other and engage in a lively conversation for 15 minutes. Watch the reactions of the passersby, whom you have placed in the role of personal space "invaders." What do they do? Do they force their way through, wait for you to invite them to pass, or look for an alternative exit? Next, remain in your positions, but stop conversing for 15 minutes and note whether the reactions of passersby change. Is your personal space more or less difficult to invade when you are talking? Finally, follow the same procedures with an opposite-sex other positioned across the arch from you. Do you find that passersby find it more difficult to violate the personal space of same- or opposite-sex dyads?

HIGH DENSITY AND CROWDING

INTRODUCTION

EFFECTS OF POPULATION DENSITY ON NONHUMAN ANIMALS

PHYSIOLOGICAL CONSEQUENCES OF HIGH DENSITY FOR NONHUMANS

BEHAVIORAL CONSEQUENCES OF HIGH DENSITY FOR NONHUMANS

CONCEPTUAL PERSPECTIVES: ATTEMPTS TO UNDERSTAND HIGH-DENSITY EFFECTS IN NONHUMANS

SUMMARY

EFFECTS OF HIGH DENSITY ON HUMANS

METHODS USED TO STUDY HIGH DENSITY IN HUMANS

FEELING THE EFFECTS OF DENSITY: ITS CONSEQUENCES FOR AFFECT, AROUSAL, AND ILLNESS

Affect

Physiological Arousal

Illness

Summary

EFFECTS OF DENSITY ON SOCIAL BEHAVIOR

Attraction

Withdrawal

Prosocial Behavior

Aggression

Summary

EFFECTS OF HIGH DENSITY ON TASK PERFORMANCE

PUTTING THE PIECES TOGETHER: CONCEPTUALIZATIONS OF DENSITY EFFECTS ON HUMANS

BASIC MODELS

THE CONTROL PERSPECTIVE

A SUMMARY PERSPECTIVE ON HIGH-DENSITY EFFECTS

ELIMINATING THE CAUSES AND EFFECTS OF CROWDING

 Introduction

Suppose you are put in charge of a large experimental device called a "mouse universe." You begin with eight mice, four males and four females, in the apparatus and are told female mice can bear a litter of four to eight pups once a month. You are instructed that your job as keeper of the mice is to create a veritable mouse paradise for them, in which they can live, bear young, and do whatever mice like to do, protected from their natural enemies. As an incentive for you to do your job well, your employer offers you a large bonus if the population of the "mouse universe" increases dramatically while you are in charge. To ensure that you get the bonus, you decide to do whatever can be done to provide a utopian setting for the mice. You supply them with unlimited food, water, and nest-building materials. Furthermore, although you do not enjoy it, you clean the mouse droppings that accumulate on the floor of the "universe" frequently to minimize disease.

After creating your "ideal" environment, you sit back and watch, thinking of how you will spend your new riches. At first, things run smoothly; the males are establishing territories and mating with the females in their areas. The females are constructing nests, bearing young (very quickly, to your satisfaction), and successfully rearing them to weaning. However, when the population begins to get larger, you observe that the mice start behaving quite differently
from before. Their odd behavior increases as the population increases. Although some animals still maintain their normal lifestyle, most males no longer function well as territorial defenders and procreators (two of their major roles in life), and most females no longer function well as bearers and rearers of young. The birthrate declines rapidly. In addition, the mortality rate of the young becomes extremely high, and some animals become hyperactive and cannibalistic. You envision your bonus disappearing and wonder why these ungrateful creatures are doing this to you. You gave them everything they could possibly need—or did you?

Although the strange behavior of the mice as the population increased may seem bizarre, it is not. Experiments have shown that if animal populations are allowed to multiply unchecked, high-density conditions lead to disease and behavior disorders, even when other aspects of the environment (food, water) are ideal (Calhoun, 1962; Dyson & Passmore, 1992; Judge & deWaal, 1993; Pearce & Patterson, 1993). In fact, our chapter-opening description of high-density behavior in mice is based on the results of actual experiments. This research, as well as other, similar work with nonhumans provided a strong reason for studying the effects of high density on humans, which is the focus of this chapter.

Another initial impetus to research on how high density affects humans was the environmental movement (e.g., the writings of Barry Commoner, 1963; Paul Ehrlich, 1968) and the awareness of upcoming global overpopulation that these writers generated. Concern about global overpopulation continues today. We are going to live in a world characterized by higher and higher population densities, which makes the importance of studying the effects of high density on humans paramount. For example, the present population of the world is about 6 billion, and it is increasing by approximately 78 million annually (United Nations Population Division, 1998). In about 11 years, the world's population will increase by 1 billion—which equals the present population of Europe and North America combined (United Nations Population Division, 1998). For environmental psychologists, this population growth is significant for the demands it will place on the world's resources, as well as for the direct effects of high density on people and accompanying implications for design (Harvey & Bell, 1995; Figure 9-1).

Will the expected **high density** in humans lead to negative behavior as it did with the mice in our imaginary mouse universe? Early correlational work by sociologists, which examined the relationships between human population density and behavioral and health abnormalities, suggested that it might. As you may remember from Chapter 1, correlational research does *not* allow us to draw strong conclusions concerning causes and effects. However, some of the initial studies of this type strongly suggested that increases in human population density may be associated with pathology.

In this chapter, we will discuss research and theory on the effects of high density. Our discussion will start with a brief overview of experimental and theoretical work on the reactions of other species to high density. We will then move to the primary focus of the chapter—the effects of high density on humans. After a review of this area, some theories that try to explain human responses to high density will be discussed. The last section of the chapter will focus on several ways to alleviate the causes and effects of high density. ▨

Figure 9-1   Crowding refers to the way we feel when there are too many people and/or not enough space.

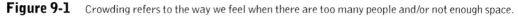

Effects of Population Density on Nonhuman Animals

Two major types of research methods have been used to study the effects of density. These include naturalistic observation, which tends to be descriptive, and laboratory methods, which tend to be experimental. In laboratory research methods with any species, density is manipulated in one of two ways. **Social density** manipulations vary group size while keeping area constant. A manipulation of social density might entail putting 15 rats (or people) in a given area in the low-density condition and putting 75 rats (or people) in the same area in the high-density condition. Or, it might entail starting out with 15 rats in a constant-sized space and observing behavior change as they reproduce, and as population density increases. In contrast, **spatial density** manipulations vary area while keeping group size constant. To manipulate spatial density you might place 15 rats (or people) in a relatively large area in the low-density condition, and 15 rats (or people) in a relatively small area in the high-density condition.

As we will see later in this chapter, there is quite a bit more than what we have said so far to the distinction between social and spatial density. They are not *simply* ways of manipulating density, and they are not interchangeable. Rather, they reflect different conditions with different problems and consequences. Under high social density the primary problem is too many other individuals with whom one must interact; under high spatial density the primary problem is too little space. Is it better to manipulate social density than spatial density, or vice versa? The answer is "it depends on what you are studying." Both manipulations are somewhat imperfect. Social-density variations include the confounding of group size and space supply (i.e., group size and space per individual are changed at the same time). In contrast, spatial-density manipulations confound room (or experimental apparatus) size and space per individual (i.e., room size and space per occupant are changed at the same

time). One thing we will find in our literature review is that the way in which density is manipulated sometimes affects the results obtained (i.e., social- and spatial-density manipulations do not always yield the same results).

In contrast to laboratory methods in which social or spatial density is typically manipulated by the experimenter, studies employing naturalistic observation assess how naturally occurring density variations affect behavior in "real-world" settings. A classic example of naturalistic observation is provided by the work of Dubos (1965), who found that when Norwegian lemmings become overpopulated, they migrate to the sea, where many drown. He attributed this to density-induced malfunctions of the brain. In this study, as in others using naturalistic observation, changes in density occur "on their own"— the researcher does not manipulate them.

Given these brief examples of how research with nonhuman species is conducted, let us turn to what has been found in this research. Our discussion will first highlight some of the more consistent physiological and behavioral effects that occur when animals are "densely packed" and will then describe some conceptual perspectives we can use to understand these effects.

PHYSIOLOGICAL CONSEQUENCES OF HIGH DENSITY FOR NONHUMANS

Past research demonstrates that when animals interact under high population density, they experience negative physiological consequences. Many of these effects parallel the reactions in Selye's General Adaptation Syndrome, discussed in Chapter 4. For example, studies show hormone abnormalities in animals under high density (Chapman, Christian, Pawlikowski, & Michael, 1998), as well as depressed immune systems (Kingston & Hoffman-Goetz, 1996). Other studies show that high social and spatial density lead to abnormalities in endocrine functioning, which is an indicator of stress (e.g., Chaouloff & Zamfir, 1993).

One important effect of high density on endocrine functioning is that it leads to decreased fertility in both males and females (e.g., Ostfeld, Canham, & Pugh, 1993). For example, it has been found that male rats living under high density produce fewer sperm than those under low density. With females, estrus cycles of "high-density" animals begin at a later age, occur less frequently, and are shorter than those of "low-density" animals (e.g., Ostfeld, Canham, & Pugh, 1993). Given such differences, it is not surprising to find both smaller litter sizes and less frequent births in crowded populations.

BEHAVIORAL CONSEQUENCES OF HIGH DENSITY FOR NONHUMANS

Some interesting studies have found that high-density manipulations can significantly disturb normal social organization in animals (Calhoun, 1962; Chapman et al., 1998; Dyson & Passmore, 1992; Judge & deWaal, 1993). The pioneering work of John B. Calhoun in this area serves as an excellent example of how high density affects nonhuman animals. Calhoun studied both rats and mice, but his most startling study employed rats. He placed a small number of male and female rats in the apparatus pictured in Figure 9-2 and allowed them to bear young and eventually overpopulate (just like in our chapter-opening

Figure 9-2   The "universe" used by Calhoun (1962) to study the effects of high density on rodent behavior. There are no ramps between pens 1 and 4, which means that they are essentially "end" pens. This eventually precipitates a behavioral sink in pens 2 and 3.

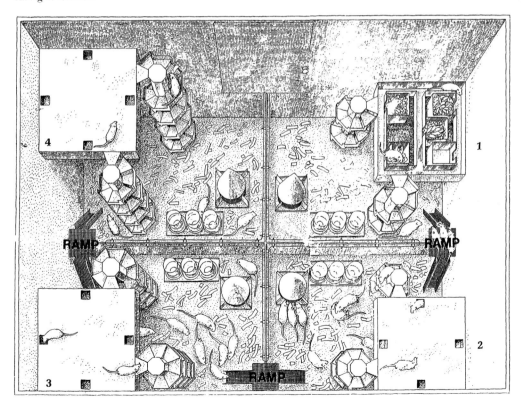

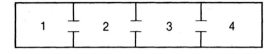

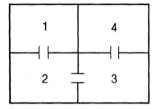

Figure 9-3 If the arrangement of pens 1, 2, 3, and 4 is changed from the one in the top diagram to the one depicted below it, pens 1 and 4 remain "end" pens with only one entrance/exit.

introduction). The "apparatus," which can comfortably handle 48 animals, consists of a 10 foot by 14 foot (300 cm by 420 cm) platform divided into four cells, each with a capacity of 12 animals. One of its important features is that ramps connect all the pens except the two "end" pens, which eventually causes many animals to crowd into the two central pens. Pens labeled "1" and "4" take on the role of end pens, while the other two are more central (see Figure 9-3).

Before they become extremely crowded, "average" male rats busy themselves accumulating a harem, mating with members of the harem, and defending their territory. They do not fight much and do not mate with females in other harems. Females occupy themselves with building nests and rearing their young. They do not fight, and they resist advances from males outside their harem. How do rats behave under high density? Calhoun observed that under high density the normal social order disintegrated, and a new one emerged.

Allowing the animals to overpopulate had negative effects on the social behavior of all the occupants of the apparatus, and these effects were particularly negative in pens 2 and 3, where high density was acute. (Calhoun calls this extremely crowded area, which is described in the box on page 299, a **behavioral sink**). In pens 1

and 4, males and females still attempted to enact their normal social roles. Females engaged in nurturant behaviors, and dominant males guarded the sole entrance and maintained a semblance of territorial behavior. But in the behavioral sink (pens 2 and 3), neither males nor females carried out their roles effectively. Although females in the less crowded pens tried to nurse their young, to build nests for them, and to transfer them in the event of harm, none of these behaviors was effectively engaged in by mothers in the behavioral sink. This accounts, at least in part, for the fact that the infant mortality rate in the behavioral sink was extremely high, with 80% to 96% of all pups dying before being weaned. In contrast (but nothing to be pleased about), only about 50% of the pups in the less crowded "end" pens suffered this fate. While dominant males protected estrous females in the less crowded pens, packs of socially deviant males in the crowded inner pens relentlessly pursued estrous females, who were unable to resist their advances. This led to a high rate of mortality from diseases in pregnancy (almost half of the females in these pens died by the 16th month of the study), which was not experienced by female residents in the less crowded pens.

Within the bizarre setting of Calhoun's "universe," several social classes emerged, varying in the extent and type of their pathological behavior. There were four groups of males. First, there was a group of dominant males that generally lived in the less crowded pens. These were the most "normal" animals in Calhoun's "universe." They were also the most secure, since the majority of the other animals were victims of almost continuous aggressive attacks. The second group consisted of pansexual males. These animals made advances to females who were not in estrus as well as to males. The third class of males was completely passive and ignored other rats of both sexes. The fourth and most unusual group of males Calhoun termed "probers." These animals lived in the behavioral sink and were hyperactive, hypersexual, pansexual, and cannibalistic. Classifying the female rats was relatively simple.

⌒ WHAT IS A BEHAVIORAL SINK?

At this point, you probably have the general (and correct) impression that a "behavioral sink," such as existed in pens 2 and 3, is an area in which the negative effects of high density are intensified. However, up to now we have not discussed the dynamics by which behavioral sinks are formed. According to Calhoun (1967), a behavioral sink develops when a population that is uniformly distributed becomes nonuniformly distributed in groups far exceeding optimal size. Two processes are involved in "behavioral sink" formation. First, some aspect of the environment or the behavior of the animals makes population density greater in some places than in others (the absence of ramps connecting pens 1 and 4 did this in Calhoun's study). Second, animals come to associate the presence of others with some originally unrelated activity. For example, in Calhoun's studies, animals came to associate food (a reinforcer) with the presence of others, which caused them to be attracted to areas where there were many animals. As we have seen, the intense crowding and the need to make accommodations to so many others was highly detrimental to the social order. Can you think of any areas in the human environment that would qualify as "behavioral sinks"?

Given an understanding of the dynamics of the behavioral sink, how can this condition and its associated pathology be remedied? One approach taken by Calhoun was to substitute granular food (which can be eaten very quickly) for the hard food pellets (which were very time consuming to eat) used in his earlier studies. Consequently, it took much less time for animals to eat, decreasing the probability that two or more animals would be eating simultaneously and that the conditioning process described previously would occur. In these studies, behavioral sinks failed to develop, and the pathological behavior associated with high density was less intense, although by no means low. Can you think of a similar means of eliminating the human behavioral sinks you thought of?

One group (which lived in the behavioral sink) was completely abnormal, could fulfill no sexual and maternal functions, and "huddled" with the male rats. The second group (which lived in the less crowded pens) behaved much more like "normal" rats.

While Calhoun's work is clearly important, it is not without criticism. Because of the design of his apparatus, some have claimed that in addition to studying high density, Calhoun was also manipulating territoriality. Due to design features, some rats became territorial, while others were kept from having territories. The fact that rats in the high-density pens were also less territorial suggests that both high density and lack of territoriality may be responsible for the negative effects. Others have criticized Calhoun's work for having low ecological validity. In the wild, rats are not penned in as in Calhoun's apparatus and tend to emigrate when density

becomes too high (Archer, 1970). The latter critics suggest that the way the apparatus was designed, "behavioral sinks" were inevitable. They imply that while the research provides a look at how things could be in the worst of all possible worlds, it may not portray a completely accurate picture of rats' behavior under high density.

Nevertheless, Calhoun's work has been central to the study of the behavioral effects of high density in nonhumans, and some of his results find parallels with several species. For example, Dyson and Passmore (1992) found similiar results for aggression in frogs. Other studies have found density-related increases in withdrawal rather than aggression among monkeys and pigs (Anderson et al., 1977; Pearce & Patterson, 1993). Still other researchers have found, like Calhoun, that crowding affects the sexual and aggressive behavior of rats and mice in high-density settings (e.g., Chapman et al., 1998; Williams, McGinnis, & Lumia, 1992).

Having discussed the effects of high density on animals' social behavior and physiological responses, one might ask what other consequences high density has for animals. While a complete discussion is beyond our scope, a final effect worthy of mention is that high density is associated with decrements in learning and task performance (Pearce & Patterson, 1993).

CONCEPTUAL PERSPECTIVES: ATTEMPTS TO EXPLAIN HIGH DENSITY EFFECTS IN NONHUMANS

Several attempts have been made to explain the negative effects of high density on nonhuman animals. In a sense, these perspectives view the negative responses to high density as adaptive mechanisms that act to prevent extinction due to overpopulation. Although none of these conceptual schemes have received unqualified scientific support, they are useful in adding to our understanding of the effects of high density. At this point, we should consider the various viewpoints as "possibilities" and expect the eventual explanation to be an integration of these approaches.

One conceptualization of the effects of high density was proposed by Calhoun (1971), based on his classic work. This formulation can be used to explain both the extremely negative consequences that occurred in the behavioral sink and the relatively less severe negative effects that occurred elsewhere. Calhoun assumes that species of mammals are predisposed by evolution to interact with a particular number of others. This is termed their "optimal group size." It leads to a tolerable number of contacts with others each day, some of which are gratifying and some of which are frustrating. Calhoun suggests that as the group increases beyond the optimal size, the ratio of frustrating to gratifying interactions becomes more unfavorable. Further, interruptions in necessary periods of solitude increase, and these are experienced as aversive. This state of affairs becomes extremely

Figure 9-4 Calhoun suggests that animals are evolutionarily predisposed to interact with a particular number of others. When more than the "optimal group size" are present, interactions become aversive, and at twice the optimal size, conditions may become debilitating.

Courtesy of Bob Swerer.

debilitating when the number in the group approaches twice the optimal number, and a sustained period under such conditions produces the sort of effects observed in Calhoun's "rat universe" (Figure 9-4).

Another conceptualization of the negative effects of high density is social stress theory (Christian, 1955). From this perspective, it is assumed that the social consequences of high density (e.g., increased social competition, effects on social hierarchies) are stressful, and that stress produces an increase in the activity of the adrenal glands as part of a stresslike syndrome. (Recall the evidence described earlier that high density is associated with changes in endocrine functioning.) It is believed that increased adrenal activity is responsible for many of the negative physiological and behavioral effects associated with high density. Interestingly, social stress theory predicts that glandular activity may also moderate a population *increase* when populations are very small. Since its social consequences are not stressful, low density does not elicit negative physiological and behavioral effects, and thus facilitates higher birthrate, longer lifespan, and so on. Social stress theory involves an endocrine feedback system that keeps density at an acceptable level.

An explanatory framework based on territorial behavior has been proposed in the work of Ardrey (1966) and Lorenz (1966). It assumes that the negative effects of density are caused primarily by aggression induced by territorial invasions. These writers suggest that as population density increases beyond an optimal level, violations of territorial "rights" increase, precipitating high levels of aggression. (The relationship between territorial invasions and aggression was discussed in Chapter 8.) Such aggression results in the negative physiological and behavioral effects described earlier as associated with high density. Under conditions of low density, territories are not violated, aggression is low, and the population can increase toward the optimal level. This formulation, however, cannot explain the effects of density on species that are relatively nonterritorial.

A way to integrate these conceptualizations, as well as others, was proposed by Wilson (1975). He assumes that there is a tendency for populations to return to an optimal level of density. How does this occur? It is accomplished by "density-dependent controls" (e.g., by varying levels of aggression, stress, fertility, emigration,

predation, and disease). According to Wilson, such controls operate through natural selection. For example, at high density, selection may favor an aggressive organism, which will bring the population into decline. At low density, aggressive organisms would be at a disadvantage, and the gene frequencies would change to favor more gentle behavior, permitting the population to expand. Such a process would protect the species from extinction caused by under- or overpopulation.

SUMMARY

Nonhuman species experience severe negative physiological and behavioral reactions to high density. These include changes in body organs, glandular malfunctions, and extreme disruption of social and maternal behavior. Calhoun's research with rodents powerfully demonstrates many of these effects and shows that they are intensified when behavioral sinks develop. Although the findings we discussed in this section are quite consistent, it is important to note here that there are variations among species in the reactions that occur. Finally, several different conceptual schemes have been proposed to explain reactions to high density.

Effects of High Density on Humans

After reviewing studies on the physiological and behavioral effects of high density on other species, as well as some conceptual frameworks in which to view them, it is tempting to speculate about whether this pattern of effects does generalize to humans (Figures 9-5 and 9-6). Scientists and philosophers have puzzled over the differences between humans and other animals for centuries, and endless arguments have emerged. We should view nonhuman animal studies as important in their own right for what they say about the impact of density on animals and as a rich source of hypotheses concerning how humans may respond to high density. For

example, the notion of unwanted interaction and social regulation that forms the basis of Baum and Valins's (1977) studies of college dormitories (page 308) was directly derived from Calhoun's notion of balancing frustrating and gratifying interactions. As with nonhuman populations, Baum and Valins and others have found that exposure to large numbers of others has negative effects, as does a lack of social structure. The value of nonhuman animal work as a source of hypotheses—and sometimes generalizable data—about human reactions to high density is enhanced by at least two methodological strengths of research with nonhumans: (1) Since

Figure 9-5 While it is generally the case that animals respond negatively to high density, for some it constitutes "standard operating conditions" and does not lead to negative consequences. Optimal population densities for some species may appear quite crowded to us.

other animals bear young more quickly than humans, it is possible to observe the cycle in which they reproduce and overpopulate in a much shorter period of time; and (2) it is easier to study physiological and behavioral responses of other species without disturbing the process being monitored.

Differences between humans and other animals notwithstanding, most early research on human crowding assumed that for us, high density would lead to uniformly negative effects. To the surprise of everyone, this was not the case. While for nonhumans high density is generally aversive, for humans it depends more on the situation. For the most part, the effects of high densities on people are neither severe nor uniform (Baum & Paulus, 1987). Our discussion of human response to high density will first highlight representative research findings and then integrate them in terms of the general environment–behavior model presented in Chapter 4. Before proceeding, however, we will pause to consider the methodologies used to study human reactions to high density.

Figure 9-6 In general, humans evidence more variable reactions to high density than animals do. Sometimes we like it; sometimes we do not.

METHODS USED TO STUDY HIGH DENSITY IN HUMANS

The method most often used to study high density in humans is laboratory experimentation. As discussed in Chapter 1, laboratory experiments have a number of advantages over other techniques. For exploring high density in humans, however, they have two disadvantages worthy of mention. First, creating high-density conditions in the laboratory is somewhat artificial, which may affect generalizability to the "real world." Second, laboratory experiments can explore only very short-term high-density effects, which is a serious problem. In attempts to remedy these deficiencies, researchers have increasingly turned to field research techniques. These offer greater realism than laboratory experiments and permit us to study longer-term high-density effects. However, while field experiments permit us to make causal inferences, correlational field studies do not. Quasi-experimentation, a field study technique that permits one to more closely approximate a causal inference through the use of certain types of research designs, allows both the realism of field settings (e.g., prisons, dormitories) and some ability to infer causality.

Correlational research involves relating different measures of population density to the frequency of various abnormal behaviors. These studies have generally looked at correlations between pathology and two types of density: **inside density** (e.g., number of persons per residence or per room) or **outside density** (e.g., number of persons, dwellings, or structures per acre). Unfortunately, early correlational research failed to control for a number of variables that may vary along with density (e.g., income, education), and the results are of questionable value. A "second generation" of studies has statistically controlled for these confounding variables.

Although these later studies represent an improvement over earlier correlational research, several important weaknesses remain. First, so many different indices of inside and outside density are used that meaningful comparison among studies is difficult. Second, it appears that the most fruitful of the studies have focused on relating pathology to smaller-scale indices (e.g., persons per room) rather than larger-scale indices (e.g., persons per acre; Gove & Hughes, 1983). Third, while they can tell us whether various disorders are associated with density, correlational studies can give us very little information about the specific cause of the pathology. We will review some of the research employing a correlational approach in this chapter, but since much of it focuses on the relationship between levels of urban density and urban pathology, this work will be considered more fully in Chapter 10.

Having reviewed the methodologies used in human research on high density, we now turn to the research itself. Our discussion will be organized into conceptually related areas: how density makes us feel, how density affects our social behavior, and how density affects task performance. While we will discuss the effects of density on each of these areas separately, it is important to keep in mind that density may have simultaneous effects in several of these areas, and that these effects may be interrelated.

FEELING THE EFFECTS OF DENSITY: ITS CONSEQUENCES FOR AFFECT, AROUSAL, AND ILLNESS

Affect

One of the most common assumptions that people make about crowding is that it makes people "feel bad." Not surprisingly, studies have reported that high *social* density may cause negative affective states. One field study had people perform a series of tasks in either crowded or uncrowded settings. It was found that they reported more anxiety in the dense than in the nondense conditions (Saegert, MacIntosh, & West, 1975), although this probably does not surprise anyone who has ever had to perform a task with hordes of others "breathing down their neck." A study by Baum and Greenberg (1975) found that even the mere anticipation of being in high social density conditions causes a negative mood.

Before concluding that crowding invariably leads to negative moods, however, we should consider some evidence suggesting that negative feelings caused by high *spatial* density may be stronger in males than in females. Several studies (e.g., Freedman et al., 1972) found that while males experience more negative moods in high than in low spatial-density conditions, the reverse is true for females. One way to explain these effects is the finding in the personal space literature (see Chapter 8) that males have greater personal space needs than do females. Alternatively, these findings may reflect a female socialization to be more affiliative (and therefore to have more of an affinity for others at close range), and a male socialization to be more competitive (and thus to view others at close proximity as sources of threat; Deaux & LaFrance, 1998). Research also indicates that women may sometimes approach high-density settings in more cooperative ways than do men (Karlin, Epstein, & Aiello, 1978; Taylor, 1988).

It is important to note that the studies that have found uniformly negative moods in response to high density (i.e., no gender differences) are primarily studies of high *social density.* In contrast, those that reported gender differences in affective response are studies of high *spatial density.* Recall that we said earlier that social and spatial density refer to more than just methodological differences—they reflect very different kinds of problems. It is possible that high social density is equally aversive to men and women, but that high spatial density is bothersome only for males (Figure 9-7).

Physiological Arousal

If high density affects our feelings, can it also lead to physiological effects, such as increased heart rate? The answer is "yes" (Baum & Paulus, 1987). In one experiment, Evans (1979a) had mixed-sex groups of five males and five females participate in a 3½-hour study in either a large

Figure 9-7   Several factors determine whether high density leads to negative consequences. For example, males are generally more affected than females by high density, confined laboratory settings often yield more negative density effects than unconfined real-world settings, and friendships and social support reduce negative effects of high density.

or a small room. Participants' heart rate and blood pressure were recorded both before the experiment began and after three hours. Results indicated that in high-density conditions, participants showed higher pulse rate and blood pressure readings than in more spacious conditions. Evans, Lepore, Shejwal, and Palsane (1998) reported that Indian males from homes characterized by chronic high density experienced higher levels of blood pressure than males from homes characterized by lower levels of density. Similarly, research by D'Atri et al. (1981) found that increasing levels of population density in prisons were associated with higher levels of blood pressure. When prisoners were transferred back to lower density accommodations, these effects were reversed. Schaeffer, Baum, Paulus, and Gaes (1988) also reported that crowded inmates had higher levels of urinary catecholamines than less crowded inmates.

Several other physiological measures of arousal have been found to be affected by high density (cf. Baum & Paulus, 1987). Skin conductance (a measure of arousal) increases significantly over time for those in high but not low spatial-density conditions (Aiello, Epstein, & Karlin, 1975a), and exposure to a large number of others leads to arousal as measured by palmar sweat (Saegert, 1975). Finally, Heshka and Pylypuk (1975) compared cortisol levels (indicative of stress) of students who spent the day in a crowded shopping mall and those who stayed on a relatively uncrowded college campus. When compared with the control group, males who had been in the high-density shopping conditions had elevated cortisol levels indicative of higher stress, but females did not. (Does that say anything about possible gender differences in feelings about shopping?)

Field studies in Sweden have also investigated stress-related arousal in high-density settings (e.g., Lundberg, 1976; Singer, Lundberg, & Frankenhaeuser, 1978). Lundberg (1976) studied male passengers on a commuter train, comparing their response to trips made under high- and low-density conditions. Despite the fact that even under the most crowded conditions there were seats available for every-

one, negative physiological reactions increased as more people rode the train. Lundberg collected urine samples from individuals and found higher levels of epinephrine after high-density trips than after low-density ones (epinephrine is an endocrinological marker of stress-related arousal).

Other results, however, qualified the nature of these findings. Regardless of how densely packed the train was, riders who boarded at the first stop experienced less negative reactions and had lower levels of epinephrine in their urine than passengers boarding halfway to the city. Despite the fact that their ride was considerably longer (72 minutes versus 38 minutes), those boarding at the first stop entered an empty train and were able to choose where to sit and with whom they traveled. For example, groups of commuters who were friends could be assured of finding seats together. In this way, they could buffer themselves from the high density that would occur by structuring the setting before it became crowded. Apparently, the control afforded initial passengers reduced the effects of high density, while the lack of control associated with boarding an already crowded train resulted in increased arousal (Lundberg, 1976; Singer et al., 1978).

Illness

It would seem reasonable that if high density leads to negative feeling states and to physiological overarousal, living under such conditions would have negative health consequences. High density can contribute to illness due to stress, but it can also be associated with poor health because disease can "spread" more quickly in high- than low-density settings (Baum & Paulus, 1987; Paulus, 1988). There is evidence in prison settings to support the assertion that high density is associated with decrements in health (e.g., Baum & Paulus, 1987; Cox, Paulus, McCain, & Karlovac, 1982). McCain, Cox, and Paulus (1976) reported that in a prison setting inmates who lived in conditions of low spatial and social density were sick less than those who lived in high densities. Requests for medical attention

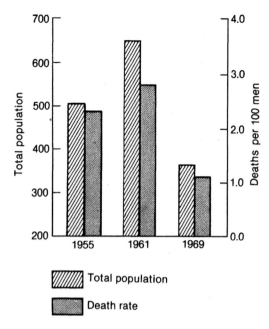

Total population

Death rate

Figure 9-8 As population size increased in prison settings, the death rate in the prison also increased. Decreased population size was associated with lower mortality. These findings controlled for a number of factors, including violent deaths. The correlation between death rates and population size was .81.

Based on data in Paulus, McCain, & Cox (1978).

by inmates were also related to absolute levels of density (Wener & Keys, 1988). Studies also indicated that high density was related to blood pressure increases in inmates and even to increased death rates (Baum & Paulus, 1987; Cox, Paulus, & McCain, 1984; Wooldredge & Winfree, 1992; see Figure 9-8).

In other studies, Baron et al. (1976) found evidence of more visits to the student infirmary by residents in high than low social-density dormitories, and Dean, Pugh, and Gunderson (1975, 1978) reported associations between high density and illness complaints aboard naval vessels. In the same vein, a study by Konarski, Riddle, and Walker (1994) studied the relation between the census of a facility for mentally retarded individuals and the rate of injuries to occupants. They found that as population density decreased, the number of health-related in-

juries decreased markedly. Other correlational studies on the relationship between high density and illness show mixed results and collectively suggest that high density may not be an important factor in medical pathology (Fuller et al., 1993; Ruback & Pandey, 1991), although these mixed results may be due to methodological inadequacies.

Summary

Having considered the effects of high density on affect, arousal, and illness, we can draw several tentative conclusions. First, it appears that high density leads to more negative affective states (especially in males) and to higher levels of physiological arousal, as measured on a wide variety of indices. Further, there is evidence (although somewhat inconsistent) that high density is associated with illness and injury. With this capsule summary in mind, we now turn our focus to the effects of high density on social behaviors such as interpersonal attraction, withdrawal, prosocial behavior, and aggression.

EFFECTS OF DENSITY ON SOCIAL BEHAVIOR

Attraction

Will we tend to like a stranger more if we meet him or her in a crowded subway car or in a more spacious setting? Generally, it seems as though high density leads to decrements in attraction whether we are merely anticipating confinement, are confined for a relatively short period, or are confined for a long time. For example, Baum and Greenberg (1975) found that merely expecting to experience high social density elicited dislike; students who were told 10 people would eventually occupy a room liked those with whom they waited for the experiment to begin less than did those who were told only 4 others would be present. In a study of short-term high density confinement, groups of eight males who were together for an hour attributed more friendliness to other group members under low than high spatial density (Worchel & Teddlie, 1976). Looking at long-term density ef-

Table 9-1	SATISFACTION WITH ROOMMATE UNDER CROWDED AND UNCROWDED CONDITIONS °

	Uncrowded	Crowded
Satisfaction with roommate	4.9	3.7
Perceived cooperative-ness of roommate	4.7	3.9

° *Higher numbers indicate more positive responses.*
Based on data from Baron et al. (1976).

fects, Baron and his colleagues (1976) reported that dormitory residents living in "triples" (three students in a room built for two) were less satisfied with their roommates and perceived them to be less cooperative than students living in "doubles" (Table 9-1). We discuss other studies on "tripling" and elaborate on the above findings later in this chapter.

Although it appears that high density leads to lower attraction, there is evidence (as noted earlier for affective state) that for high spatial density, this response is more characteristic of males than females (cf. Baum & Paulus, 1987). For example, in an experiment by Epstein and Karlin (1975), males and females participated in same-sex groups of six. Consistent findings on a variety of measures indicated that while males responded more negatively to group members in high than low spatial-density conditions, females liked group members more under high-density conditions (Table 9-2). We speculated

Table 9-2	RATINGS OF PERCEIVED SIMILARITY UNDER CROWDED AND UNCROWDED CONDITIONS °

Sex	Crowded	Uncrowded
Male	5.7	4.4
Female	4.2	5.7

° *Lower numbers indicate greater perceived similarity.*
Based on data from Epstein & Karlin (1975).

earlier that gender differences in response to high spatial density may be due to different sizes of personal space zones or to a more cooperative socialization of females and a more competitive socialization of males.

Epstein and Karlin (1975) suggest another possibility. They state that while both males and females experience arousal from high spatial density, social norms may permit females to share their distress at being "packed like sardines." It is also possible that females may experience more social support from other females under high-density conditions (Evans & Lepore, 1993; Ruback & Riad, 1994). Either of these responses would lead to greater liking and cohesion. In contrast, the same norms prohibit males from sharing distress, which may cause a more negative reaction. In a follow-up experiment (Karlin et al., 1976), it was found that when females were not permitted to interact with one another, their positive reactions to high spatial density were attenuated. However, support for this interpretation has been limited (cf. Keating & Snowball, 1977).

Withdrawal

In support of Baum and Valins's observation that withdrawal may be associated with high levels of social contact (see the box on page 308), studies have reported that withdrawal may function as an anticipatory response to high density, as a means of coping with ongoing high density, and as an aftereffect (Baum & Paulus, 1987). For example, the mere expectation of high social density elicits withdrawal responses, including lower levels of eye contact, head movements away from others (Baum & Greenberg, 1975; Baum & Koman, 1976), and maintenance of greater interpersonal distances (Baum & Greenberg, 1975). Withdrawal also occurs during ongoing high-density interactions: People are less willing to discuss intimate topics under high-density conditions (Sundstrom, 1975), and both children (e.g., Loo, 1972) and psychiatric patients (Ittelson, Proshansky, & Rivlin, 1972) interact less frequently as room density increases.

Withdrawal due to high density may have a very important consequence: It may disrupt the

HIGH DENSITY IN THE DORM:
Where Would You Like to Live Next Year?

One of the most often studied residential environments is the college residence hall. Some important effects of high residential density in this setting were reported by Baum and Valins (1977). These investigators performed studies comparing the responses to high density of students assigned to *suite-style* dormitories and students assigned to *corridor-style* dormitories. Corridor residents shared a bathroom and a lounge with 34 residents on the floor; suite residents shared a bathroom and a lounge with only four to six others (see Figure 9-9A and B). All students shared a bedroom with one other student. While the suite and corridor designs were identical in terms of space per person and number of residents per floor, as you might guess, they led to dramatic differences in the number of others that residents encountered constantly.

What were the behavioral effects of the greater number of interpersonal contacts in corridor-style dormitories? Corridor residents responded differently from suite residents in a number of ways. They perceived their floors to be more crowded, felt they were more often forced into inconvenient and unwanted interactions with others, and indicated a greater desire to avoid others. Corridor residents were also far less sociable, perceived less attitude similarity between themselves and their neighbors, and were less sure of what

Figure 9-9A & B Floor plan of corridor-style residence hall (below) and suite-style residence hall (right).

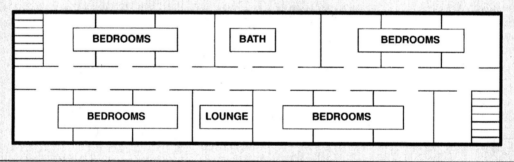

very social support networks that we rely on to cope with negative life events (Evans et al., 1989; Evans, Lepore, & Schroeder, 1996; Lepore, Evans, & Schneider, 1991). This may leave people who live under high-density conditions with even fewer resources to deal with critical stressors. Further, Evans and Lepore (1993) have found that individuals from crowded homes were actually less likely to seek social support from a confederate when they needed it, and rated the confederate to be less supportive, than individuals from less crowded homes. Individuals from crowded homes were also less apt to offer social support to another in need. Evans, Lepore, and associates (Evans et al., 1989; Evans & Lepore, 1993; Evans, Lepore, & Schroeder, 1996; Lepore, Evans, & Schneider, 1991) found that negative effects of residential crowding were due, in part, to this breakdown in individuals' social support systems. Further evidence for withdrawal in high-density environments is reviewed in Chapter 10, which considers the effects of city life.

Prosocial Behavior

If high density leads to lower attraction and to withdrawal responses, how might it affect helping? For example, suppose you lost something of value. Where would you be most confident their neighbors thought of them. Not surprisingly, a significantly lower number of corridor residents reported that the majority of their friends lived on the same floor.

It was also found that living in a suite- or a corridor-style dormitory led to different behaviors in other places and with other people. For example, Baum and Valins reported that corridor residents looked less at confederates and sat farther away from them while waiting for an experiment. Corridor residents also performed significantly worse than suite residents on tasks under cooperative conditions, although they performed better under conditions that inhibited personal involvement with an opponent. In another study, Reichner (1979) found that when ignored in a discussion, residents of corridor-style dorms felt less bad than those living in suite-style dorms.

What do these data mean? It may be that corridor residents find themselves "overloaded" by their high level of interaction with others, or that they experience frequent *unwanted interactions,* and their withdrawal responses may be interpreted as coping strategies that prevent such involvement. Baum and Valins suggest that high-density living in suites and corridors may be considered as a type of social conditioning process. Obviously, this process results in a more positive orientation to others in suite- than in corridor-style dormitories. Subsequent studies have linked this social conditioning process to later differences in prosocial behavior, to differences in interpersonal bargaining strategies, and to differences in response to violations of social norms (e.g., Reichner, 1979; Sell, 1976).

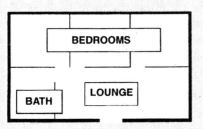

Is there any way to make life in corridor-style dormitories more tolerable? Baum and Valins found that membership in small local groups, when it occurred, tended to reduce many of the negative effects of corridor-style dormitory living. And Baum and Davis (1980) found that an architectural intervention—dividing the long corridor into two shorter ones by adding a door in the middle—reduced overload and eliminated many negative outcomes. How does your own experience as a dormitory resident correspond to these observations?

of finding someone who would help you look for it: in a higher- or a lower-density building? In a cafeteria that is full or one that is empty? Interestingly, most research on how density affects prosocial behavior has been conducted in field settings like these (Baum & Paulus, 1987).

In studies that explored how helping is affected by building density, it was found that greater density leads to less helping. For example, Bickman et al. (1973) compared prosocial acts in high-, medium-, and low-density dormitories. Envelopes, which were stamped and addressed, were dropped in the dormitories, and helpfulness was measured by the number that were picked up and placed in the mail. The results showed that 58% were mailed in the high-density condition, 79% in the medium-density condition, and 88% in the low-density condition. In an interesting study, Jorgenson and Dukes (1976) observed the effect of social density on compliance with a prosocial request (printed on signs) for cafeteria users to return their trays to designated areas. It was found that fewer users complied during high-density periods. A final set of studies that address the effects of high density on helping has compared prosocial behavior in urban and rural areas (Baum & Paulus, 1987). These studies are reviewed in Chapter 10.

Aggression

If high density can make us less likely to help others, does it also make us more apt to hurt them? One approach to this question has explored the effects of density on aggressiveness of children's play. This strategy has led to inconsistent results. In an attempt to resolve this controversy, Loo suggested that density may affect children's aggression in a curvilinear fashion. This was supported by a study that observed that moderately high density led to increased aggression in males, while very low and very high density led to decreased aggressiveness (Loo, 1978). Subsequent research, however, has shown increases in aggression under conditions of very high social density among boys (Loo & Kennelly, 1979). In another attempt to resolve this controversy, Baum and Paulus (1987) suggested that competition over scarce resources may be a major determinant of children's aggression in high-density situations. If there are more children than toys and each child wants a toy, aggression is more likely than if there are enough toys to go around. This relationship was supported by Smith and Connolly (1977), who found that increased aggression occurred during play if playground equipment was made more scarce.

Other research has suggested that children's responses to high density change with continuing development (Aiello et al., 1979; Loo & Smetana, 1978). Since children are presumably less restrained and more outwardly aggressive than adults, it may be that high density has more subtle effects on adult aggressiveness. Several studies have addressed the aggression-enhancing effects of high density among adults. Often, it appears that increased density leads to aggression in adult males but not in females, a familiar pattern in high-density research. For example, Stokols and his associates (1973) studied same-sex groups under high and low spatial density and found that males rated themselves as more aggressive in the small room, while the reverse was true for females. Freedman et al. (1972) also found that increasing spatial density was associated with increasingly aggressive be-

havior among men but not women. When Schettino and Borden (1976) used the ratio of people in a classroom to the total number of seats as an index of density, they found that density was significantly correlated with self-reported aggressiveness for males but not females.

Baum and Koman (1976) found gender differences in aggressive response to *anticipated* crowding as well, but *only* when spatial density increased. Men in small rooms who expected to be crowded behaved more aggressively than did women in the same situation. Further, men were more aggressive in a smaller room than when a larger room was used. However, increases in *social* density did not produce increased aggression. In fact, under conditions where participants expected large numbers of people rather than limited space, they tended to withdraw rather than act aggressively.

From the studies reviewed thus far, it appears that the aggression-enhancing effects of density may be related more to spatial and resource-related problems than to issues created by the presence of too many people. It also seems that the magnitude of the effect of density on human aggressiveness is less than overwhelming (Baum & Paulus, 1987). The studies that have found increases in aggressiveness during high density have reported them primarily among men, and the effects have been mild at worst. However, it is important to keep in mind that the measures of aggression employed in this research (e.g., people sentencing a hypothetical criminal to a longer prison term) have been artificial. They are *not* the sort of aggression one finds in "real-world" crowded environments. One reason for this artificiality is the settings in which the research has been conducted. Most investigations of adult aggression during high density have been conducted in the laboratory, where participants are confined only briefly, and where the measures one can use to assess aggression are limited. Overall, the weaknesses of these research methods pose serious limitations to our understanding of the density–aggression relationship.

These problems are only partly resolved by the correlational research that has been attempted. This research involves finding the association between long-term high-density confinement and "real-world" measures of aggression (e.g., crime). Unfortunately, this body of research contains methodological flaws and shows a tenuous relationship between high density and various indicators of crime (e.g., Bagley, 1989; Galle, Gove, & McPherson, 1972). However, the more "fine grained" the measure of density used in the study (room density as opposed to people per acre), the higher the correlation with aggression (e.g., Palmstierna, Huitfeldt, & Wistedt, 1991). In addition, there is evidence that high density is even more strongly associated with fear of crime than with actual victimization (Gifford & Peacock, 1979).

An exception to some of the above criticisms of research on the density–aggression link is research that has been conducted in prisons and in other "real-world" institutional settings. Here, people are confined for relatively long periods of time under high density, "real-life" aggression does take place, and there are accurate records of both population density and aggressive behaviors (Pontell & Welsh, 1994). In prisons, Paulus, McCain, and Cox (1981) observed that increases in disciplinary infractions were associated with increased population density, and studies by Cox, Paulus, and McCain (1984) found extremely high correlations between prison density and inmate aggression. In one prison, a 30% decrease in the census resulted in a 60% decrease in assaults. When a 20% increase in the census occurred, it was followed by a 36% increase in assaults! Similar results are reported by Ruback and Carr (1984). It should be noted, however, that some prison studies have been less conclusive (Bonta, 1986). In addition, the extent to which this body of research could be expected to generalize to other populations living under high density is unclear. Obviously, prison inmates are different from the "person on the street," and prisons are not a typical high-density setting. Nevertheless, research by Morgan and Stewart (1998a) found that increased social and spatial density were associated with increased instances of disruptive behavior in elderly residents with dementia in a long-term care facility, and Teare et al. (1995) reported that as the number of youths in a crisis shelter increased, the probability of problem behavior occurring on a given day increased as well. While this research points to the possibility of stronger relations between density and aggression in the general population than occurred in the more artificial laboratory studies, the extent of its generalizability is still somewhat uncertain.

Summary

Our discussion of the effects of density on social behavior (i.e., attraction, withdrawal, helping, and aggression) allows us to draw several tentative conclusions. First, it appears that high density leads to less liking of both people and places, and that this relationship is stronger for males than for females. High density also causes withdrawal and less helping behavior in a variety of situations. Concerning aggressive behavior, the findings are somewhat inconsistent, but for certain populations there seems to be a relationship between high density and aggression. The differences between social and spatial density also appear to be important. With these ideas in mind, we turn our focus to the effects of high density on a final and extremely important dimension—task performance.

EFFECTS OF HIGH DENSITY ON TASK PERFORMANCE

One of the most critical questions that can be asked about high density is whether it affects task performance. The answer has important implications for the design of all types of living and working spaces (e.g., schools, workplaces). Most early studies used tasks that were relatively simple to perform and were consistent in finding no performance decrements under high social or spatial density (Baum & Paulus, 1987). For example, Freedman and his associates (1971) reported that density variations did not affect performance of any of a series of tasks.

Later work, generally using more complex tasks, supports a somewhat different conclusion

Table 9-3	ERRORS IN MAZE PERFORMANCE AS A FUNCTION OF SPATIAL DENSITY AND SOCIAL DENSITY °	
Low spatial density		34.20
High spatial density		37.44
Low social density		32.13
High social density		39.50

° *Based on data from Paulus et al. (1976).*

(Baum & Paulus, 1987). As shown in Table 9-3, Paulus et al. (1976) found that both high social and spatial density led to decrements in complex maze task performance, but these decrements were more pronounced under conditions of high social density. In a field setting, Aiello, Epstein, and Karlin (1975b) observed decrements in complex task performance over time in residents of overcrowded dormitory rooms (three persons in a room built for two), compared with less crowded rooms (two persons in a room built for two). Evans (1979b) also found poorer complex task performance under high-density conditions but no impairment in simple task performance, and Klein and Harris (1979) reported poorer complex task performance in individuals who were anticipating crowding. Finally, Knowles (1983) reported decrements in maze learning under conditions of high social density when all the individuals in the room were watching the target person perform, but increased retention of the task, once learned.

How can we reconcile our findings of high-density decrements on some tasks but not on others? One explanation centers on the fact that high density leads to arousal (cf. Evans, 1979a; Worchel & Brown, 1984). The Yerkes–Dodson law (see Chapter 4), a formulation that relates arousal to task performance, states that arousal *should* interfere only with complex task performance. In terms of this law, our observation that high density causes decrements only in complex task performance would be expected, rather than discrepant. Since the Yerkes–Dodson law has been supported in numerous research con-

texts in the psychological literature, it seems quite tenable as an explanation here. In addition, Baum and Paulus (1987) offer other suggestions for why density has not consistently affected task performance. They conclude that such factors as the psychological salience of the others present, the feelings of being evaluated, and the number of tasks participants must perform may be important as well.

An alternative explanation for the past inconsistent findings has been offered by Heller, Groff, and Solomon (1977). They propose that many studies of high density have focused only on the physical aspects of a setting at the expense of the kinds of interactions that are typical of high-density situations. For example, some studies occupy people with tasks so that interaction is minimized. Heller et al. suggest that this kind of procedure reduces the likelihood of finding effects of high density on task performance. In support of this, they showed that decrements in task performance occurred only under conditions characterized by high density and interaction among participants. High-density settings in which people did not interact very much did not produce task performance decrements (see Figure 9-10).

Figure 9-10 Interaction is necessary for density-related task performance deficits to occur.

Adapted from Heller, Groff, & Solomon (1977).

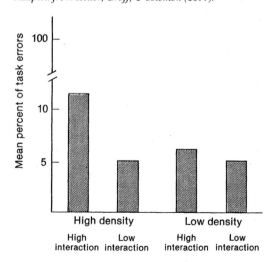

Another potential explanation is provided by a study reported by Schkade (1978). She manipulated spatial density and expectancy (how well people thought they would do on the task). Results showed that the poorest task performance occurred when density was high and expectations were low—that is, when people did not expect to do well on the task. Problems with task performance under high-density conditions may be evident primarily when negative outcomes are anticipated.

A final and very important question is whether high density can cause *aftereffects,* as well as immediate effects, on performance (Baum & Paulus, 1987). As you will recall, noise has been linked to consequences for performance, occurring *after* exposure, and some studies suggest that crowding has lingering effects as well. For example, it was found that people exposed to high density later showed less persistence at working on unsolvable puzzles than those exposed to low density (Evans, 1979a; Sherrod, 1974). In an attempt to determine whether perceived control would lessen aftereffects, Sherrod (1974) gave some individuals the option of leaving a crowded room to complete the study in a spacious setting (i.e., perceived control). Although no one took advantage of the option, this group showed fewer aftereffects than a group that was not offered an opportunity to leave. The similarity of these findings to those observed by Glass and Singer with noise (reviewed in Chapter 5) further suggests that noise and high density may affect people in similar ways.

Putting the Pieces Together: Conceptualizations of Density Effects on Humans

Up to this point, we have explored a number of density–behavior relationships, finding that high density may lead to various negative effects. As suggested in the general environment–behavior model from Chapter 4, we have seen that high density, like other potential stressors, may lead to (1) immediate effects such as physiological arousal and negative affect; (2) coping responses (e.g., withdrawal); and (3) aftereffects and cumulative effects (e.g., illness). However, high density does not always have negative consequences. For example, it affects task performance in some situations but not in others. Overall, the most appropriate conclusion might be that density negatively affects some of the people some of the time in some ways.

BASIC MODELS

What is it about high density that causes those negative effects that do occur? Several different theoretical approaches, which we have reviewed earlier and have discussed in detail in Chapter 4, have been applied to answer this question.

Briefly, the *overload notion* posits that high density can be aversive because it may cause us to become overwhelmed by sensory inputs. When the amount and rate of stimulation occasioned by high density exceeds our ability to deal with it, negative consequences occur. In contrast to the overload approach, the *behavior constraint* approach views high density as aversive because it may lead to reduced behavioral freedom (e.g., fewer behavioral choices, more interference). Thus, whether or not we will experience negative effects depends on what we want to do and whether high density constrains us. Finally, the *ecological* model assumes that high density can have negative consequences since it may result in insufficient resources for people in the setting. Resources are broadly defined and include anything from materials to roles. When density causes resources to become insufficient, negative effects occur.

Not surprisingly, additional explanations have been offered to account for how density affects us. *Arousal theory* (Evans, 1978; Paulus & Matthews, 1980) suggests that high density may increase arousal. As we noted in Chapter 4, arousal has effects on performance in and of itself. Also, arousal may be attributed by the person in an arousing situation to various factors, depending on situational and cognitive cues. For example, Worchel and Teddlie (1976) argue that personal-space violations associated with high-density settings cause arousal, which results in a negative experiential state attributed to others being too close. If arousal is misattributed (e.g., is attributed to something *other than* others being too close), the likelihood of a negative emotional state being linked to high density is lessened (Aiello et al., 1983b).

Baum and Valins (1977) build upon an overload framework and propose that the negative consequences of high density are caused by *unwanted interaction.* While too many contacts (overload) may be distressing, this is not always the case—sometimes a large number of social interactions may be bearable or even fun. However, when these interactions are unwanted, problems are more likely. Thus, difficulties in regulating when, where, and with whom one may interact can lead to too many unwanted interactions and, eventually, to stress. In support of this notion, Baum and Valins (1977) consistently found that "tripled" dormitory residents complained about unwanted contacts with neighbors.

Another explanation, the **behavioral interference** formulation (e.g., Schopler & Stockdale, 1977), asserts that when inadequate space or large numbers of people interfere with goal-directed behavior, negative effects are experienced. This explanation is loosely derived from the behavior constraint model discussed earlier and is supported by studies showing that interference increases the negative effects of high density (Heller, Groff, & Solomon, 1977; Sundstrom, 1975). Unwanted interaction can be subsumed by this model, since it, too, can disrupt or prevent goal achievement. Also consistent with the model are findings that the presence of social structure or rules governing conduct (which lessen interference) reduce the negative consequences of high density (Baum & Koman, 1976; Schopler & Walton, 1974). Finally, studies have

directly linked the type of interference that occurs and the importance of blocked goals to the intensity of stress (Evans & Lepore, 1993; Lepore, Evans, & Schneider, 1991; McCallum et al., 1979; Morasch, Groner, & Keating, 1979).

Related in some ways is Altman's (1975) **privacy regulation model.** According to Altman, high density has negative effects when breakdowns occur in the achievement of desired levels of privacy. As we stated in Chapter 8, privacy is an interpersonal boundary process by which a person or group regulates interactions with others (Altman, 1975, p. 67). When achieved privacy is less than desired, control of social interaction is inadequate, and the person cannot regulate his or her level of interaction with others. Under these conditions, there may be negative consequences of high density. According to Altman, people cope with the inadequate privacy characteristic of high density by using stronger or additional privacy control mechanisms.

So, we now have a number of different views concerning what the critical determinants are of when high density will lead to negative effects (see Table 9-4). Without doubt, all of them are relevant, and there are probably other ways of conceptualizing the negative effects of high density as well. How can we resolve these competing explanations? Or, do we actually need to resolve them as much as to combine them into a more unified perspective? Despite the fact that the different formulations are presented as competing with one another, it is probably true that too much stimulation, overarousal, too many constraints on behavior, inadequate privacy, excessive unwanted social contact, interference, and resource inadequacy each accounts for some negative effects of high density (Baum & Paulus, 1987).

Some conceptual efforts have focused on more parsimonious explanations for why high density has the effects that it does. One of these, the *control perspective,* has been used in this regard because it crosses the lines of the models in Table 9-4 and unifies diverse theoretical currents. We will discuss the control formulation in some detail here.

Table 9-4	SUMMARY OF THEORETICAL PERSPECTIVES ON CROWDING [*]		
Conceptual Approach	Critical Cause(s) of Crowding	Primary Coping Mechanisms	Reference
Social overload	Excessive social contact; too much social stimulation	Escape stimulation; prioritize input and disregard low priorities; withdrawal	Milgram, 1970; Saegert, 1978
Behavior constraint	Reduced behavioral freedom	Aggressive behavior; leave situation; coordinate actions with others	Stokols, 1972; Sundstrom, 1978
Ecological	Scarcity of resources	Defense of group boundaries; exclusion of outsiders	Barker, 1968; Wicker, 1979
Arousal	Personal space violations plus appropriate attributions	Lower arousal to more optimal level	Evans, 1978; Paulus & Matthews, 1980
Unwanted interaction	Excessive unregulable or unwanted contact with others	Withdrawal; organization of small primary groups	Baum & Valins, 1977; Calhoun, 1970
Interference	Disruption or blocking of goal-directed behavior	Create structure; aggression; escape	Schopler & Stockdale, 1977; Sundstrom, 1978
Privacy regulation	Inability to maintain desired privacy	Privacy control mechanisms	Altman, 1975

[*] *Adapted from Stokols (1976).*

THE CONTROL PERSPECTIVE

As you recall from Chapter 4, *perceived control* is a potent mediator of stress. When we believe that we can control a stressor or other aspects of a situation, its aversiveness appears to be reduced. On the other hand, even if no other problems are apparent, losing or not having control can be stressful. Several researchers have proposed that high density can cause a loss of control (or prevent someone from ever having control), and that this loss of control is a primary mechanism by which density causes stress (Baum & Paulus, 1987; Evans & Lepore, 1992; Lepore, Evans, & Schneider, 1992).

Can we really explain many of the negative effects of high density as a loss of control? All of the theories outlined above—with the exception of the arousal theory—can be readily subsumed by the concept of control. *Overload models* assume that under conditions of high

density we are bombarded by more stimuli than we can process—a situation where a loss of control is likely. And Baum and Valins's notion of *unwanted interaction* is a control-based perspective. Negative effects are the result of contact that is too frequent, which makes control over when, where, and with whom people interact difficult to maintain. As a result, interactions become unpredictable and frequently unwanted. The *behavior constraint* notion is also control based, viewing high density as eliminating behavioral options and reducing freedom to behave as one might like. For example, having inadequate space can constrain our behavior by making it impossible to control the nature of interaction with others (Figure 9-11).

In addition, *privacy regulation models* are related to the concept of control. We can typically control the degree of intimacy in one-on-one interactions by adjusting the distance we stand from people, but in a very high-density

Figure 9-11   As a ski lift line suggests, whether crowding leads us to experience negative affective reactions or positive responses depends in part on whether we perceive control over the circumstances. While some do not like lift lines, others find them to be good opportunities to socialize.

room, we may find ourselves with no choice— we must stand close to people whether we know them well or not. In this manner, we can lose control over intimacy regulation. *Interference* can also be viewed as a threat to control, because our attempts to achieve one goal or another are repeatedly blocked or disrupted. And resource problems, the focus of *ecological* models, can limit our choices and restrict our ability to exercise control. Thus, to some extent, many of the "consequences" of high density that have been related to negative effects do cause a reduction in control. But is there any direct research evidence for the relation between density and a loss of control (see Figure 9-12)?

Research examining the links between high density and loss of control has taken two different tacks. The first has been to manipulate personal control (i.e., provide some individuals, but not others, with perceived control) and to see whether this has any effect on experience in

high-density settings. Rodin, Solomon, and Metcalf (1978) attempted to manipulate whether or not people riding in a crowded elevator had control. First, they observed people's response to riding in crowded elevators and found a tendency for them to gravitate toward the floor selection panel (a control panel, if you will). In effect, people attempted to "take control" in a high-density elevator by standing near the panel that regulates entry, exit, and floor selection. Next, Rodin and colleagues manipulated whether people were able to stand near the control panel (high-control condition) or not (low-control condition). The results indicated that those allowed to stand near the panel (i.e., those who were given control) felt better, and thought the elevator was larger, than those not near the panel. Rodin and co-workers (1978) also examined the effects of control on the experience of high density in a laboratory context. Again, people were provided with varying degrees of control over the setting, and those with control felt better than those without.

A somewhat different approach to demonstrating the importance of control was taken by Sherrod (1974). Recall that in research on noise (see Chapter 5), Glass and Singer (1972) found that the negative aftereffects associated with exposure to noise were reduced if people had control over it. Sherrod conducted a similar study with density instead of noise as the environmental stressor. As in research on noise, negative aftereffects were associated with exposure to high density only when control was not available. Individuals who had perceived control did not exhibit negative aftereffects following exposure to high density.

Taken together, the above studies all demonstrate that high density associated with loss of control is more aversive than high density with control, and that introducing control can reduce the potential negative effects of high density. They do not, however, demonstrate that high density itself has consequences similar to those associated with loss of or lack of control.

Fortunately, that bit of evidence has been reported by other researchers. Rodin (1976) conducted two studies in field settings characterized by chronically high residential density. She attempted to ascertain whether living under high density was associated with helplessness-like behavior. *Learned helplessness* is a syndrome in which people who are exposed to uncontrollable settings learn that they cannot control the setting, and hence stop trying to do so (Seligman, 1975). This manifests itself in reduced motivation and cognitive activity. Does chronic exposure to high density result in learned helplessness? Rodin's participants, who were children and adolescents, showed symptoms of learned helplessness that were associated with the high density in their homes (see the box on page 318). Similar results have been obtained in children living in India by Evans, Lepore, Shejwal, and Palsane (1998).

Figure 9-12 As this festival photo suggests, whether or not crowding leads to negative consequences depends in part on perceived control and whether we believe we can escape the density if we want to. It may be that the density intensifies the fun. Imagine, though, what it would be like to live in this kind of density for months!

⌒ CROWDING IN THE HOME AND IN THE SCHOOLS

Imagine yourself growing up in a small apartment with five other people who are continuously interacting with one another and with you. With so many people in so little space, you may grow up to feel the world is a complex place in which you have little power to influence events (cf. Evans et al., 1998). What consequences does this have? Seligman (1975) has demonstrated that when we come to believe we cannot control our outcomes by responding appropriately (as may result from living in high-density conditions), we no longer perform effectively in a number of situations. This syndrome is called "learned helplessness."

Two interesting and provocative studies by Rodin (1976) and a study by Evans, Lepore, Shejwal, and Palsane (1998) demonstrated the relationship between residential density and susceptibility to "learned helplessness." In the former experiments, Rodin hypothesized and found that children who lived in high-density conditions were less likely than those in low-density conditions to try to control the administration of rewards they were to receive. In a second experiment, she exposed individuals from both groups to an initial frustrating task on which responses and outcomes were noncontingent. Rodin found that only children from high-density homes did significantly worse on a subsequent task for which outcomes were contingent. Thus, it appears that density in one's home is an important determinant of both the use of control and of performance after frustrating noncontingent reward situations.

A fourth study, by Saegert (1982), also focused on the consequences of residential density for children from low-income families, but this time on its effects for school performance. Children from high-density homes were more apt to be rated as behavior problems by teachers, and they exhibited more evidence of distractibility and hyperactivity than children from low-density homes. In addition, reading scores were lower and vocabularies less developed for children from high-density homes. Very similar effects of high-density living conditions on behavior at school and academic performance were reported by Evans et al. (1998). In this study, children growing up under high density also reported less supportive relationships with parents than children from lower-density homes.

Evans et al. (1998) point to aspects of parent–child relationships characteristic of high-density homes as responsible, in part, for the effects of high density on students' learned helplessness, school performance, and behavior. They suggest that in high-density homes parents are more distressed, have poorer interpersonal relationships with others (including spouses), receive less social support, and ultimately may have impaired parent–child relationships. It is this factor, they believe, that most directly causes children to suffer negative outcomes. In effect, they argue that the link between residential crowding and children's general well-being and school performance is crowding induced, strained parent–child relationships.

Besides the home environment, what else could be responsible for the negative effect described above? With respect to "learned helplessness," one important culprit may be the schools themselves. Baron and Rodin (1978) suggest that as class size increases, learned helplessness training begins to occur. They argue that larger classes lead to lower student expectations for control of reinforcement, because teacher feedback concerning student work becomes less discriminative. For example, as class size goes up, individualized student–teacher interactions decrease, and generalized (rather than individualized) praise and criticism increase. Clearly, such conditions could lead to a state of learned helplessness and its negative consequences for performance. What have your experiences been in small and large classes?

Baum and Valins (1977) also found symptoms of helplessness among people exposed to high density in a residential environment over a prolonged period of time. While neither Rodin (1976) nor Baum and Valins (1977) linked help-lessness *directly* to loss of control in residential settings, this has been accomplished by Baum, Aiello, and Calesnick (1978) and Baum and Gatchel (1981). It was established in these studies that as people in high-density situations

MORE THAN 57 VARIETIES OF CROWDING

The research we have reviewed thus far suggests that crowding has situational antecedents, an emotional component, and, of course, behavioral consequences. The theoretical notions introduced have pinpointed some of the situational factors associated with crowding, and the research we have reviewed has highlighted some of its affective and behavioral consequences. While we have "pieced the crowding story together" from a variety of sources, a study by Montano and Adamopoulous (1984) had individuals rate how they would feel and act in a variety of crowded situations, applied sophisticated statistical techniques, and yielded a picture of crowding quite consistent with the themes of this chapter—all in a single study.

The researchers specified four major situations in which people felt crowded, three major affective consequences, and five typical behavioral responses. The situations in which people became crowded were: (1) feeling that one's behavior was constrained; (2) being physically interfered with; (3) being uncomfortable due to the mere presence of many others; or (4) having high density cause expectations to be disappointed. What types of affective responses did crowding elicit? It was associated with negative reactions to others and the situation, and under certain conditions, positive feelings. When is positive affect associated with crowding? According to Montano and Adamopoulous, this occurs only when people feel that they have coped successfully with it. Regarding the types of behaviors caused by crowding, the researchers identified five: (1) assertiveness; (2) rushing to complete activities so that one can flee to less dense environs; (3) physical withdrawal; (4) psychological withdrawal; and (5) adaptation—making the best of a bad situation. While this research did not involve people engaged in "real-world" experiences with high density, it is useful as corroborative evidence for earlier findings, and it suggests some new approaches. By crossing the four situations in which crowding was found to occur with the three affective and five behavioral responses, the result implies that crowding comes in at least 60 different varieties!

relinquished their beliefs that they could control their environment, their behavior became increasingly like that associated with helplessness.

Overall, there is compelling evidence that control is involved in the negative effects of high density (Baum & Paulus, 1987). When control is available, high density has less impact on people than when control is not available. Further, chronic exposure to high density appears to be associated with learned helplessness. Additional evidence of the links posited by the control model is needed, but it is fairly clear that the effect of high density is at least partly determined by people's perceptions of control.

A SUMMARY PERSPECTIVE ON HIGH-DENSITY EFFECTS

Given that high density involves an array of potentially disturbing elements (e.g., loss of control, overarousal, overstimulation), how can we explain the fact that it only *sometimes* influences

our behavior? Many researchers have addressed themselves to this issue (Baum & Paulus, 1987). A conceptualization of the effects of density on humans is presented in Figure 9-13. As you will notice, the conceptual scheme is a special case of the general environment–behavior model presented in Chapter 4.

In Phase One of our conceptualization, an important distinction is made that explains why high density is sometimes stressful (leading to negative effects) and other times is not. In terms of this distinction (first proposed by Stokols, 1972), *high density* is viewed as a physical state involving potential inconveniences (e.g., loss of control, stimulus overload, lack of behavioral freedom, resources, or privacy), which may or may not be salient to a person in the situation. Whether or not these conditions are salient depends on (1) individual differences between people (gender, personality, age); (2) situational conditions (what the person is doing; time in the setting; presence of other stressors); and

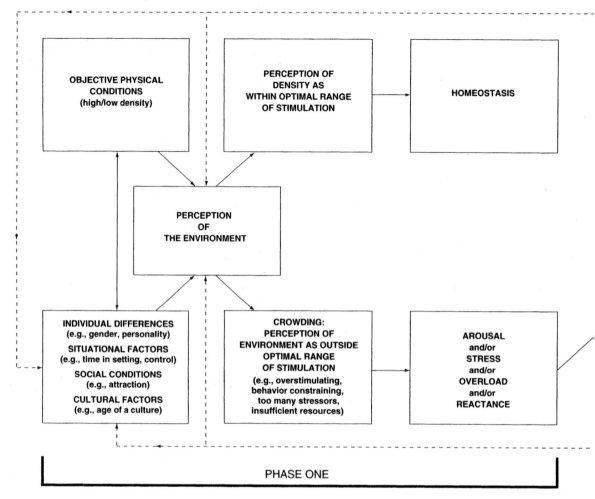

Figure 9-13 A conceptualization of the effects of high density on behavior based on our eclectic environment–behavior model

(3) social conditions (relationships between people, intensity of interaction). If the negative aspects of high density are *not* salient, the environment is perceived as being within an optimal range, homeostasis is maintained, and no negative effects occur. If the potential negative aspects (e.g., overstimulation, behavior constraint) of high density are salient, crowding occurs. **Crowding** is conceptualized as a psychological state characterized by stress and having motivational properties (e.g., it elicits attempts to reduce discomfort).

Having incorporated the density–crowding distinction into our model, we turn now to Phase Two, which specifies the consequences of the psychological state of crowding. As in other stressful situations (see Chapter 4), it is assumed that the stress associated with crowding involves coping responses that are directed toward reducing stress (e.g., withdrawal). Interestingly, the overload, behavior constraint, and ecological approaches, as well as the others, each predict qualitatively different types of coping responses (see Table 9-4 for a description of these varying responses). Regardless of these minor differences in the *types* of coping, the sequential links specified in Figure 9-13 among stress, coping, adaptation, and aftereffects conform to the general environment–behavior model found in Chapter 4. It is assumed that when coping is

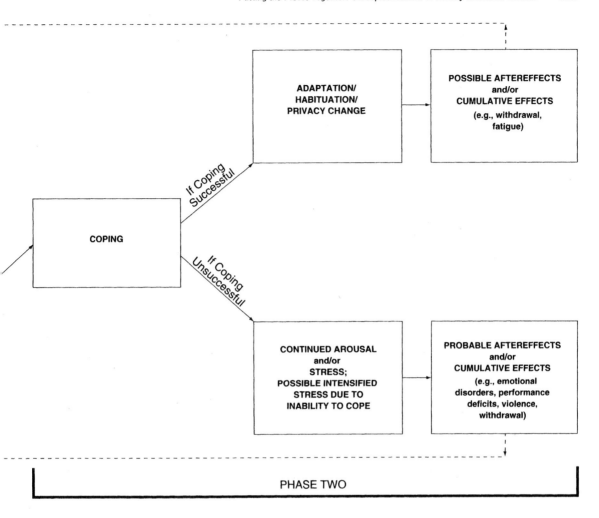

successful in handling stress, adaptation or adjustment occurs, and the individual is less likely to experience aftereffects or cumulative effects. If coping is unsuccessful, the stress continues, and the individual is extremely likely to experience aftereffects and cumulative effects (e.g., illness).

Coping is an important part of any model of crowding for two reasons. First, it is usually directed at reducing the causes or effects of crowding, and second, it is a continuous process. From the moment that crowding is first experienced or anticipated, people attempt to deal with it. These attempts are dynamic, continuously unfolding until adaptation is achieved, the crowding dissipates, or fatigue makes further coping impossible. The notion of dynamic coping underlying crowding suggests that responses to crowding change with the situation. Such a coping process has been addressed by research examining adjustments to high spatial and social density. For example, Greenberg and Firestone (1977) observed adjustments in verbal and visual behavior in contexts where other forms of coping were blocked. Greenberg and Baum (1979) reported continuing adjustment and readjustment of social behaviors among participants who anticipated changing degrees of crowding in their experimental session.

Eliminating the Causes and Effects of Crowding

At this point in our discussion, you might be wondering how we can eliminate the causes and the effects of crowding. We will explore the predictions derived from our general environment–behavior model and apply them to address this issue.

PREDICTIONS FROM OUR GENERAL ENVIRONMENT–BEHAVIOR MODEL APPLIED TO HIGH DENSITY

One extremely valuable feature of our model is that it provides a framework for speculation and research about how to moderate the causes and effects of crowding. This can be of great conceptual and applied significance. As you recall, the model specifies that individual differences among people, situational conditions, and social conditions determine whether or not high density is perceived as "crowding." Research has supported the assertion that these three sets of factors can produce the experience of crowding, which is associated with negative consequences. We will briefly discuss representative individual differences, situational conditions, and social conditions found to affect our reactions to high density.

Identifying *individual difference* variables that determine whether high density is experienced as crowding is of practical value, since it allows us to select those individuals who will be most and least sensitive to the constraints of limited space. For example, we have found that in a variety of situations, males are more apt to experience crowding than females, and have suggested several explanations for this. There is some evidence, however, that this pattern of gender differences may be limited to laboratory settings, where there is no possibility of escape (Aiello, Thompson, & Brodzinsky, 1983a). Under such conditions, women seem to handle stress better, perhaps because they are more apt than men to share their distress with others. In long-term high-density contexts, however,

women may cope more *poorly* than men. For example, in dormitory crowding studies (e.g., Aiello, Baum, & Gormley, 1981a), women sometimes report more crowding and negative effects. This may be because men cope with high density by leaving their rooms, whereas women are more involved with their roommates and spend more time in their room, which results in increased stress (Aiello, Thompson, & Baum, 1981b). Additional long-term crowding studies conducted in real-world settings similarly suggest that women may react more negatively than men in certain ways to long-term high density (e.g., Booth & Edwards, 1976; Ruback & Pandey, 1996), though some studies of long-term high density have found no gender differences (e.g., Lepore, Evans, & Schneider, 1991) or mixed results (e.g., Ruback & Pandey, 1996).

In addition to gender, the amount of personal space people desire to maintain between themselves and others constitutes an individual difference variable that may affect the degree to which crowding is experienced. For example, Aiello et al. (1977) found that those with preferences for large interpersonal distances were more adversely affected in a high-density setting than those with smaller preferred distances. Individuals who liked to sit far away from others showed greater physiological arousal, discomfort, and poorer task performance than those who preferred to sit closer.

Another important determinant of our reactions to high density is our level of social support. Studies by Evans and Lepore (e.g., Lepore, Evans, & Schneider, 1991) found that individuals who were experiencing high density and who had low social support had more negative psychological reactions than those with high social support. After very long exposure to high density, this buffering effect of social support on the negative consequences of high density disappeared, because long-term exposure to high density disrupted the very social networks that had protected people from its negative effects!

⌒⌒ AN ALTERNATIVE APPROACH:
The Density-Intensity Model

We have summarized a model that distinguishes between density and crowding and that specifies a number of factors that may cause crowding and its attendant effects. Freedman (1975) takes another position, which has been a source of great debate among environmental psychologists for a quarter of a century. In general, his **density-intensity** model does not support the density-crowding distinction accepted by most researchers.

He also argues that density intensifies reactions that would occur in any case in a particular situation. High density heightens the importance of other people and *magnifies* our reactions to them. Thus, for Freedman, high density will intensify the pleasantness of positive situations and intensify the negativeness of aversive ones. From this viewpoint, any number of factors can cause a negative reaction in a high-density situation.

In research to support his notion, Freedman provided a link between density and the intensity of contagion, which occurs when the behaviors or emotions expressed by one person spread rapidly throughout a group of people. In a study using different room sizes and group sizes to vary density, Freedman, Birsky, and Cavoukian (1980) observed people's reactions to humorous films. After viewing the films, a confederate began to applaud, and the spread of this reaction throughout the group was noted. As one would expect from the density-intensity notion, contagion was more extensive in high-density groups. Freedman and Perlick (1979) similarly found intensification of contagion with high density; and Freedman (1975) and Aiello, Thompson, and Brodzinsky (1983a) reported additional evidence consistent with the general model. On the other hand, several studies have found *reduced* appreciation of humor under high-density conditions (Prerost, 1982; Prerost & Brewer, 1980). In addition to the possible effects of high density on intensifying our appreciation (or lack of appreciation) of humor, other studies have found that high density may intensify our reactions to more negative stimuli (e.g., stressors). Lepore, Evans, and Palsane (1991) reported that social hassles in the home were associated with psychological symptoms *only* among people living under crowded conditions. In effect, crowding may intensify the effects of other stressors one may be experiencing. Lepore, Evans, and Schneider (1992) suggest that this may be due, in part, to lower perceived control, because crowding makes it impossible to avoid or escape social hassles.

There is no doubt that one of the effects of density *is* the magnification of responses to various situational variables. At baseball or football games, excitement is often intensified by larger crowds, while negativity may be amplified if the home team loses and the drive home is in bumper-to-bumper traffic. Yet, this is only one of the many effects of density. Freedman and co-workers (1980) are careful to point this out, and it remains clear that high density can exert independent effects as well as intensifying the feelings that would otherwise be present. Identification of those cases in which intensification is the primary mechanism underlying individuals' response to high density and when it is not is an area for future research.

Research has examined whether *personality characteristics* moderate our reactions to high density (Baum & Paulus, 1987). Some of this work has focused on locus of control (i.e., whether people believe they, or outside forces, control their outcomes). It has typically but not always been found (cf. Walden, Nelson, & Smith, 1981) that internals (individuals who feel they control their fate) display a *higher* threshold of crowding than externals (individuals who feel events are controlled by outside forces; Schopler & Walton, 1974). In addition, people who are highly affiliative are more tolerant of high density than those who are less affiliative (Miller & Nardini, 1977). In fact, high-affiliative individuals experienced more stress in a low- than a high-density dormitory living situation (Miller, Rossbach, & Munson, 1981).

When considering individual differences, it is important to note that the characteristics of a particular high-density setting may affect how an individual difference variable will affect coping (Baum & Paulus, 1987). For example, Baum et al. (1982) found that people who screen themselves from interaction and organize their surroundings were better able to cope with high *social* density than individuals who did not screen themselves. One would expect that this "screening" variable would be less important under conditions of high *spatial* density.

We should also keep in mind that the individual differences we have discussed were found for North American samples and may not hold cross-culturally. Indeed, the personal-space literature (see Chapter 8) and work on crowding suggest that we may expect cultural differences in reactions to high density. Research corroborates this assertion. For example, studies find that high density is related to social pathology in some places but not in others (e.g., Fuller et al., 1993; Lester, 1995). Similarly, Nasar and Min (1984) predicted and found that Mediterraneans would respond more negatively than Asians when placed in a small, single dormitory room. Other studies, however, show few cross-cultural differences (e.g., Vaske, Donnelly, & Petruzzi, 1996).

What differentiates the cultures where high density may be more, and less, associated with pathology? One factor could be the *age* of the culture. Young, as opposed to older, cultures may have had less time to develop means of coping with high density. According to this logic, as cultures evolve, ways are developed to cope with density, and negative effects may decrease (Gifford, 1997). Some suggestive data support the above line of reasoning, although more research is clearly needed. One society that is very old, in which people may cope especially well with high density, is the Chinese culture. It has been suggested that the Chinese may have become so familiar and comfortable with high density that, when given the choice, people often opt for high- as opposed to low-density conditions (Aiello & Thompson, 1980). Further, it has been suggested that the Chinese have developed an elaborate set of norms, rules, and coping strategies to support them in a "densely packed" existence (Ekblad, 1996). There are rules about access to space, a low level of emotional involvement is expected with others, and interaction between different groups (e.g., men and women; high- and low-status individuals) is regulated. Further, sounds that others might view as noise are regarded as acceptable (Aiello & Thompson, 1980). Similar practices are found in other cultures that have adapted successfully to high density (Iwata, 1992; Munroe, Munroe, & Vutpakdi, 1999).

However, other research casts doubt on the assumption that the Chinese have an affinity for high density. In a study of people living in Chinatown in San Francisco, Loo and Ong (1984) found that residents view crowding as undesirable and even harmful. The experience of crowding was also a major reason why they thought they might want to move. Overall, this research suggests that the Chinese like crowding no more than anyone else and presents a forceful challenge to work implying that they bear up especially well under high density.

A final related individual difference variable that has been linked to reactions to high density is one's adaptation level from past experience under high-density conditions. Some investigators hypothesize that people with a history of high-density living are less likely to experience crowding in a novel situation than those with a history of isolation. In support of this hypothesis, it has been found that the Japanese, residents of Hong Kong, and the Logoli (all of whom live under extremely high density) have developed social mechanisms that may be viewed as adaptive for high-density living. Further, Zhou, Oldham, and Cummings (1998) reported that employees from high-density settings (childhood residences or urban communities) responded more favorably (e.g., in terms of performance) to high-contact work environments than those from other types of residences and communities. While several additional studies (e.g., Gove & Hughes, 1983; Sundstrom, 1978) support this "high-density experience-adaptation" hypothesis, Paulus and his colleagues (1975) found that the longer an inmate was imprisoned, the *lower* his or her tolerance for crowding became. Inconsistent evidence is also reported by Lepore, Evans, and Palsane (1991), by Rohe (1982), and by Loo and Ong (1984). Further, work by Maxwell (1996) suggests that when children who grew up in crowded homes were placed in high-density classrooms, they were doubly negatively affected. Other research by Webb and Worchel (1993) suggests that in addition to level of past experience under high density, expectations regarding present levels of density, and whether they have been confirmed or disconfirmed, may play a role in determining individuals' reactions to high-density confinement. This research may help to explain some of the inconsistencies in past research in this area.

In addition to being moderated by individual difference variables, reactions to high density are affected by *situational conditions*. An important situational condition is the degree of control we have. Complementing the studies we discussed earlier, additional research has found that allowing people increased control over a situation leads to less perceived crowding (Langer & Saegert, 1977) and to fewer negative effects (Baum & Fisher, 1977; Langer & Saegert, 1977). The applied potential for introducing control into high-density situations is great. For example, providing individuals who live under high density with training that enhances control (e.g., giving pointers about how to share space and ensure privacy) may help alleviate crowding and its negative effects. Schmidt and Keating (1979) have identified three types of control that might be introduced: cognitive control (accurate information), behavioral control (ability to work toward a goal), and decisional control (having choices available). They suggest that providing one or more of these to people in high density could ameliorate crowding stress.

Other situational conditions may affect reactions to high density as well. For example, at a constant level of density, men experience more negative effects when others are touching them than when this is not the case (Nicosia et al., 1979). This supports Knowles's (1983) assertion

that in addition to the more traditional measures of social and spatial density, the physical proximity of others is important to consider. All current measures of density assume that people are evenly distributed across space. However, it may be not only how many square feet per person there are in a room, but how close the others are to you that counts.

Does the amount of time we are confined to high-density conditions moderate our reaction? The relationship between time under high density and crowding is somewhat unclear, but it is probably fair to suggest that the longer the period of confinement, the more aversive the response (Loo & Ong, 1984). Of course, as we noted earlier, very prolonged confinement under high density may affect one's adaptation level, and for some such individuals, density may become less problematic. Whether we are in a primary environment (like a home) or a secondary environment (like a restaurant), and the extent to which other stressors (e.g., noise) are involved may also affect crowding. When we are in a primary environment (Stokols, 1976, 1978) and when other stressors are operating, we may be more likely to experience crowding. This will be especially true when we view others as responsible for our distress (Sundstrom, 1978). Finally, there is suggestive evidence that we are more likely to experience crowding when engaged in work than when engaged in recreation (Cohen, Sladen, & Bennett, 1975).

In addition to individual differences and situational conditions, *social conditions* can be manipulated to affect whether or not we are crowded. These variables make up the social "climate" of a high-density situation (e.g., the degree of friendship and the level of social interaction and interference). For example, our relationship with the people we are with may determine how crowded we feel: Less crowding is experienced with liked rather than disliked others (Fisher, 1974; Schaeffer & Patterson, 1980), with others who engage in activities we approve rather than disapprove of (Gramann & Burdge, 1984; Womble & Studebaker, 1981), and with acquaintances rather than strangers (Cohen et al., 1975; Rotton, 1987b). To the extent that we experience social interference (interruptions) by others (especially if these are perceived as intentional; Stokols, 1978) or experience excessive proximity or immediacy (too-direct eye contact or body orientation; Sundstrom, 1975), crowding is more likely. Finally, crowding is more often experienced in unstructured than in structured task situations (Baum & Koman, 1976).

High density within primary social groups (e.g., families) has repeatedly shown fewer negative effects than in other groups. For example, when density *within* apartment units is examined, it appears to be negligible as a factor associated with illness or behavior difficulties (e.g., Giel & Ormel, 1977). Degree of acquaintance with others and one's relative position in a group's dominance hierarchy also affect crowding: The presence of friends or the possession of high status tends to reduce the aversiveness of large numbers of people or cramped spaces (Arkkelin, 1978).

The dynamic way in which social conditions can affect crowding may be illustrated by considering, once again, studies on the effects of overassignment of student residents to dormitory rooms. You recall that this research assessed the consequences of having three students live in a room designed for only two. Initial study of this phenomenon (e.g., Baron et al., 1976; Karlin, Epstein, & Aiello, 1978; Walden et al., 1981) revealed that the "tripling" of dormitory rooms was associated with negative mood, increased health complaints, and suppressed task performance. These findings made sense, given the increased difficulties of sharing resources, coordinating activities, and achieving privacy created by the addition of a third roommate. Yet the question remained—was this a problem of too many people or too little space? Baum et al. (1979) reasoned that it was neither. Going back to the social psychological literature on groups, they found confirmation of a notion that many of you already know: Three-person groups are very unstable and susceptible to coalition formation such that two people get together and exclude the third (e.g., Kelley & Arrowwood, 1960). Given this, it seemed possible that the primary problem in "tripling" was not that there were too many roommates or insufficient space. Instead, it was that there were three roommates, one of whom was likely to feel left out and, as a result, to have less control over the shared bedroom. This "isolate," when compared with the other two roommates, would have less input into how the room was arranged and used, feel generally more "left out," have greater difficulty achieving privacy, and feel more crowded.

Research examining the formation of coalitions in "tripled" dorm rooms provided support for this interpretation (e.g., Aiello et al., 1981a; Gormley & Aiello, 1982; Reddy et al., 1981). These studies indicated that students living in tripled rooms were especially likely to feel "left out" by roommates. Those who felt like isolates reported more problems related to using the room and more perceived crowding. Tripled residents who did not feel "left out" reported experiences and moods more like students living in doubled rooms (Baum et al., 1979).

An extension of this research examined the effects of tripled and quadrupled rooms (Reddy et al., 1981). If the instability of three-person groups was responsible for the effects of "tripling," one would expect residents of four-person rooms to feel less crowded than residents of three-person rooms. If, on the other hand, the primary problem was the absolute number of roommates, then quadrupled rooms would be associated with greater crowding. Results indicated that isolates were more likely in the tripled than in quadrupled rooms and that nonisolate residents of tripled rooms reported experiences similar to quadrupled residents. Isolates, on the other hand, reported more problems with crowding than either of the latter two groups. From this research, we can see the importance of considering social processes in attempting to understand crowding.

ARCHITECTURAL MEDIATORS OF CROWDING

Now that we have tried to provide you with a feeling for some of the conditions that moderate the experience of crowding, it should be interesting to consider how we can modify existing environments or plan new ones so that crowding is less of a problem (for a complete discussion of the design process, see Chapter 11). What would you do if you were a planner charged with evaluating (and possibly modifying) some of the plans for a building that might affect the level of crowding residents are likely to experience? First, you would probably assess objective physical conditions (i.e., space allotted to each resident) in terms of its adequacy for the type of functions to be performed in that space. Next, you would estimate how spatial needs would be affected by anticipated situational conditions (e.g., how well the individuals occupying the space could be expected to get along) and individual differences (e.g., adaptation level). From your evaluation of objective physical space plus situational, social, and individual difference conditions, you would have an idea of how much of a crowding problem there would be. If you had anticipated that crowding would be a problem, you could institute some of the architectural modifications we will describe below.

How can environments be designed or modified to alleviate crowding and its consequences? A number of studies suggest alternatives that can be incorporated into existing structures or planned into new ones. For example, for males, greater ceiling height is associated with less crowding (Savinar, 1975), and it has been found that rooms with well-defined corners elicit less crowding than rooms with curved walls (Rotton, 1987b). In addition, rectangular rooms seem to elicit less crowding than square rooms of the same area (Desor, 1972), and rooms that contain visual escapes (e.g., windows and doors) are rated as less crowded than similar areas without such escapes. The latter findings suggest that in some cases, the *design* of a building affects how crowded people feel in a constant amount of objective space. Rapoport (1975) makes the important point that the level of density that people perceive, rather than the actual level of density, is apt to determine their behavior. Therefore, designs that lessen perceived density could be expected to be associated with less crowding and negative effects.

One type of design that seems to lessen perceived crowding is low- as opposed to high-rise buildings. High-rise buildings are associated with greater feelings of crowdedness and less perceived control, safety, privacy, and satisfaction with relations with other residents, than low-rise buildings (McCarthy & Saegert, 1979). Some research suggests that residents of higher floors in high-rise buildings are less crowded than those on lower floors (Nasar & Min, 1984; Schiffenbauer, 1979), but other studies are equivocal on this point (e.g., Mandel, Baron, & Fisher, 1980). A study by Evans, Lepore, and Schroeder (1996) suggests another determinant of the effects of high density. Specifically, the farther into a setting one resides (i.e., the more spaces in the setting one must pass through to get there), the less likely people are to withdraw or to become psychologically distressed in high-density settings.

Clearly, the above types of features would be fairly difficult to change in a structure that has already been built. However, there are simple things that can be done to lower crowding in such settings. For example, positioning furniture in the center of the room, as opposed to having it arranged at the side of the room, was perceived as more crowded (Sinha, Nayyar, & Mukherjee, 1995). A number of studies (Baum, Reiss, & O'Hara, 1974; Desor, 1972; Evans, 1979b) provide evidence that adding flexible partitions to rooms lessens feelings of crowding. In one study, "privacy cubicles" surrounded by high partitions and containing desk and storage space were placed in dormitory-style prison rooms. Inmates having the cubicles in their dormitories had more positive reactions to their environment and lower rates of noncontagious illnesses (McGuire & Gaes, 1982). Parallel results were reported by Schaeffer, Baum, Paulus, and Gaes (1988). Similarly, segmenting large dormitory rooms in prisons into smaller rooms by building a lounge area in the middle lowered illness complaints (Baum & Paulus, 1987), and adding enclosed sleeping areas had a similar effect (Cox, Paulus, McCain, & Karlovac, 1982). These studies suggest, incidentally, that at least in prisons, the number of people one must have

SOCIAL VERSUS SPATIAL DENSITY:
Which Is More Aversive?

Should a designer faced with the unenviable choice worry more about creating a design with high social density or high spatial density? Each type of density offers different problems to individuals experiencing it (Baum & Paulus, 1987). Increasing numbers of people bring with them more interaction, more need for social structure, more social interference, and greater threats to control. Too little space, on the other hand, may be associated with physical disruption, loss of intimacy regulation, spatial invasions, and physical constraints (Baum & Paulus, 1987).

Not surprisingly, then, research on high density suggests differences in the effects of high social and spatial density. And, based on a careful analysis, researchers have tentatively concluded that manipulations of social density are more aversive than manipulations of spatial density (cf. Baum & Valins, 1979; Baum & Paulus, 1987). Specifically, they have found that high social density will produce negative effects more consistently than high spatial density, and that while social density manipulations are generally aversive, spatial density manipulations are often problematic only to males in same-sex groups (Baum & Paulus, 1987). In natural settings, social density is clearly a more serious concern for people (Paulus, 1980). For example, Cox et al. (1984) report that while high social density led to very negative effects in prisoners, high spatial density had few negative effects. They conclude that the best way to house prisoners would be in small, single rooms. Others report consistent findings. For example, Ruback and Carr (1984) reported that prisoners who lived in single rooms liked their accommodations more, had higher perceived control, and experienced less stress than those in accommodations characterized by higher social density.

Why might an overabundance of others be more distressing than too little space? The answer is unclear, but there are some hypotheses. One explanation put forth by Baum and Valins suggests that people are more immediately aware of problems created by large numbers of others than by spatial limitations. Also, the loss of control that results when too many people are in a room is frequently more serious than that caused by being in too small a room. In addition, people may be threatened by the presence of many others. What are the theoretical consequences of the assertion that social and spatial density may affect us differently? While we should not draw the conclusion that the effects of spatial limitations are inconsequential, we should develop predictive frameworks that account for the differences in the two manipulations.

contact with is a more important determinant of outcome than the amount of space one has. For a discussion of whether it is generally worse to experience high social or spatial density, see the box on this page.

It has also been found that brightness (provided by wall and accent colors or appropriate light sources) leads to less perceived crowding (Mandel et al., 1980; Nasar & Min, 1984; Schiffenbauer, 1979), and that the presence of visual distractions (e.g., pictures on walls, advertisements on transportation vehicles) leads to more perceived space (Baum & Davis, 1976; Worchel & Teddlie, 1976). In addition, sociofugal seating arrangements (when people face away from

each other) are associated with less crowding than sociopetal ones (when people face each other; Wener, 1977). However, this may not be the case when relations between interactants are good.

INTERVENTIONS IN HIGH-DENSITY SETTINGS

While many intervention strategies can be derived from the various models of crowding, only a few have actually been implemented. Some have attempted to prevent crowding from occurring in the first place (e.g., by providing people with information on how to exert "control" over the situation, or by modifying high-density environments in ways that would help people to cope, thereby reducing or preventing crowding). Others have focused on treating the consequences of crowding (e.g., dealing directly with the negative mood created by high density).

Preventing Crowding From Occurring

One form of intervention has involved providing what is often referred to as heightened *cognitive control* to people in high-density situations. Cognitive control is the increased sense of predictability or controllability that people gain when given prior warning or information about a situation. As we noted in Chapters 4 and 5, perceived control can reduce the aversiveness of stress. Some important work has found that increasing cognitive control is beneficial in high-density situations. Langer and Saegert (1977) reported an experiment in which information about crowding was given to some participants, but not others, before they entered grocery stores varying in actual levels of density. The information people were given focused on how they would feel if the store became crowded. All participants were given a task to perform that required them to move around the store to find a number of items. Not surprisingly, the results suggested that when density was higher, task performance was poorer. In addition, those who had been given prior information about crowding performed better and reported a more positive emotional experience than those who did not receive information. Having information about how they might feel allowed individuals to better select appropriate coping strategies and to behave more confidently (Langer & Saegert, 1977). This pattern of effects has been replicated in both laboratory and field settings (Baum, Fisher, & Solomon, 1981; Fisher & Baum, 1980; Paulus & Matthews, 1980). In one related study, Wener and Kaminoff (1983) introduced informational signs into the crowded lobby of a federal correctional center. Visitors reported less perceived crowding, discomfort, anger, and confusion.

Additional strategies for preventing crowding have involved architectural, as opposed to cognitive, interventions. Baum and Davis (1980), for example, reported a successful architectural intervention in high-density dormitories (see the box on page 308). By altering the arrangement of interior dormitory space, they were able to prevent residents from experiencing crowding stress. Other strategies are sure to arise. It is important to understand that architectural interventions in high-density settings can be effective only if they consider the specific dynamics of the situation they are addressing.

Another thing to keep in mind in planning any intervention is that high density is not invariably negative, and one must consider the complexities of the situation before deciding whether it is even desirable to intervene. For example, a study by Szilagyi and Holland (1980) revealed that when an organization moved into a new building characterized by higher social density, employees reported less job autonomy but greater feedback about their job performance from others, increased friendship opportunities, and work satisfaction. While it is not clear that higher social density is uniquely responsible for the reported effects, there may be situations in which, rather than decreasing social density, one might actually want to increase it!

Treating the Consequences of Crowding

In addition to interventions that determine whether or not crowding occurs, a second set of interventions focuses on moderating the effects of crowding when it does occur. Karlin, Rosen, and Epstein (1979) reported a study that sought to lower the anxiety and arousal associated with crowded transportation settings. Three therapeutic interventions were used to treat participants in a laboratory analogue of a transportation context. Individuals were given training in *muscle relaxation, cognitive reappraisal* (in which they were told they could improve their mood by focusing on the positive aspects of the situation), or *imagery* (in which they were instructed to concentrate on a pleasant, distracting, pastoral image). A fourth group received initial instructions to relax but was given no other training. Responses to crowding among individuals in these four groups provided mixed support for the value of therapeutic intervention. Those given cognitive reappraisal instructions showed more positive responses to the setting than those in the other groups. The effectiveness of the muscle relaxation and imagery treatments in reducing the impact of high density was less marked.

SUMMARY

Considerations of global overpopulation make the study of the effects of high density particularly important. This area of investigation has become increasingly popular and has included research with both human and nonhuman populations. Two types of density manipulations are commonly used: varying *spatial density* (in which space is manipulated and group size held constant), and varying *social density* (in which group size is manipulated and space is held constant). With nonhuman animals, it appears that the physiological and behavioral effects of high density are almost uniformly negative. It has been found that other species experience changes in body organs and glandular malfunctions that affect birthrate and also experience severe disruptions of social and maternal behaviors. A number of conceptual schemes have been developed to account for reactions to density.

Human reactions to high density often depend on the particular situation. While density does not have a totally consistent negative effect on humans, it leads to aversive consequences on a variety of dimensions. Concerning its effect on feeling states, high density leads to negative affect (especially in males) and to higher physiological arousal. There is also some evidence that it is associated with illness.

In terms of effects on social behavior, high density has been found to result in less liking for others (especially in males), and it is associated with withdrawal from interaction. Also, there is suggestive evidence that high density leads to aggression and to lower incidence of prosocial behavior. Finally, for task performance, it leads to decrements for complex but not for simple tasks, and it may also be associated with aftereffects.

Overall, high density causes (1) immediate effects on behavior; (2) coping responses; and (3) aftereffects. A number of explanatory schemes (e.g., overload, behavior constraint, and ecological models) attempt to explain why high density is aversive, and each stresses a different element of density as critical. Why does high density not always lead to negative consequences? A model that differentiates between high density and crowding accounts for this finding. It is suggested that while high density contains negative aspects, individual differences and situational and social conditions determine whether these are salient and whether "crowding" occurs. The model specifies a progression of effects that follow when crowding is experienced and also offers ways to eliminate the causes and effects of crowding.

KEY TERMS

behavioral interference	density-intensity	privacy regulation model
behavioral sink	high density	social density
crowding	inside density	spatial density
	outside density	

SUGGESTED PROJECTS

1. In our discussion about eliminating the causes and effects of high density, we stated that individual differences, situational conditions, and social conditions moderate perceived crowding. A simple procedure allows us to verify this relationship and many more that have been highlighted throughout the chapter. The procedure is the "model room technique." First, get a shoe box or a somewhat larger size carton. Modify it a bit so that it looks something like a room. (You may be creative and include elaborate windows, draperies, and so on if you like, but be sure to leave the top off.) Next, take a large number of clothespins, small blocks, pieces of styrofoam, or the like, which can be modified to stand up and to look something like people. Now you are ready to start doing "model room" experiments.

How do you begin? First, decide what relationship you want to test. Let's assume you want to test the assertion that people will feel more crowded in primary environments than secondary ones. Have a willing participant imagine that the box is his or her living room (a primary environment). Tell the individual to place figures in the box up to the point at which he or she feels the room is crowded. Count the number of figures in the box, and then remove the figures. Next, tell the individual the box is a restaurant (a secondary environment) and ask him or her to place figures in the box until the space seems crowded. Determine how the number of figures placed in the box varies, depending on whether it is described as a primary or a secondary setting. You now have data concerning the "threshold" of crowding in primary and secondary environments. If you find a lower threshold of crowding in primary than in secondary environments, you may have supportive evidence for Stokols's (1976) assertion that we experience crowding more readily in a primary than in a secondary setting. Some other hypotheses to test using this procedure are listed below:

a. Test Baum and Valins's (1977) assertion that people who live in corridor-style dormitories avoid social contact situations more than people who live in suite-style dormitories. (To do this, you will need two groups of participants, one living in corridor dormitories, the other living in suite dormitories.)

b. Test whether different personality types have different thresholds of crowding by first administering personality tests and then relating the test results to the number of figures the individuals place in the box.

2. By comparing other people's reactions in high- and low-density natural settings, you can get a feeling for how high density affects you. Select a "real-world" setting that varies over time in the number of people who are present. For example, a bus or train that becomes more crowded as it approaches the end of the line would be ideal. Your school cafeteria or library, which varies in terms of density over time, would also work. Then, pick a number of dimensions (e.g., friendliness of people toward one another, eye contact, defensive postures, object play) to observe for behavioral changes as population density increases. Compare how people respond on these dimensions in high- and low-density situations. By employing this procedure in a variety of settings, you can gain firsthand experience about how people react to high density.

3. We cited some evidence showing that people who like each other respond more favorably to high density than people who dislike each other. One reaction to high-density confinement with a disliked other should be a variety of coping strategies. Check whether high density with a disliked other leads to coping by making an informal study of residents on your residence hall floor. First, list five residents who like their roommates and five who do not. Look into the rooms of both groups, and note the arrangement of furniture. If confinement with a disliked other leads to coping, furniture should be arranged so as to block interaction and ensure privacy.

4. Make several comparisons concerning environmentally destructive behavior between high-rise and low-rise (i.e., high- and low-density) residence halls on your campus. Compare graffiti, damage to furniture and public telephones, and so on to assess whether aggressive behavior accompanies higher levels of density.

THE CITY

Introduction

Maria anxiously clutched her daypack as the big jet descended in its final approach to Boston's Logan Airport. It was not that the flight was responsible for her anxiety; the succession of miniature farms, villages, and waterways out the window had been far more entertaining than the slick magazine in the plane's seat pocket. What she feared was not the trip, but the destination— Boston. Of course Maria knew (and kept telling herself) that Boston was not a really big city compared with the packed towers of New York or the concrete sprawl of Los Angeles. And she would not be alone. Her cousin Jon would meet her near his office, just a short subway ride from the airport. But Boston was certainly too big for Maria. It was far too big for a person who had grown up in the pine ridges of northwestern Nebraska. "Cities should never grow larger than Scottsbluff, Nebraska, or Casper, Wyoming," she thought. "Why would anyone want to live with filth, crime, gangs, and noise? And how will I ever be able to handle traveling alone on the subway?"

A week later Maria was again at the airport, preparing to leave her favorite city. Boston had become a rich blend of new memories. Some were centered on the suburbs—the home shared by her cousin and his family, trips to the town beach, and sightseeing excursions to the historic towns of Lexington and Concord. Memories of the city itself were just as important, and just as

fond. Her introduction had begun with a long walk down Commonwealth Avenue, a shady parkway between old brownstone apartments. She and Jon sat next to a statue and watched the city squirrels race among the trees. They strolled on through the Public Garden (and rode on one of the swan boats) and across the expanse of Boston Common. History was everywhere they walked. It colored the facades of the homes on stately Beacon Hill, added texture to the old headstones in a little graveyard tucked into the financial district, and creaked in the rigging of Old Ironsides down by the waterfront. She could almost feel the presence of John Hancock and Paul Revere. One morning was spent along the Charles River. They visited the science museum that perches over the water, then walked the banks of the Charles to a little playground. They continued their walk through the bustle of city traffic to the Faneuil Hall Marketplace. There they found food and music, and shops galore. Everything was so alive!

As the jet banked and began its climb over the Charles, Maria strained to make out the waterfront aquarium, the Federal Center, and old Trinity Church sitting beneath a glassy city tower. She was already planning her return.

No environment more clearly shows the hand of humanity than cities. When some groups of humans gave up their nomadic life to settle permanently in groups, they began the trend toward concentration of people and services that eventually resulted in large cities in Mesopotamia, China, Egypt, and Europe. In the New World, Aztec, Inca, and Mayan cities were eventually replaced by huge urban centers like Mexico City and New York. In contrast with nature-dominated landscapes, cities are our creations — perhaps our delights, and perhaps our nightmares. Nowhere is there such a diversity, novelty, intensity, and choice as in cities. They provide an immense variety of cultural and recreational facilities, such as concert halls, museums, sports stadiums, educational facilities, and all types of restaurants. Further, there is a much wider variety of services available to the average city dweller than to the resident of a small town. Contrary to what most of us probably imagine, cities even provide habitat for hundreds of different species of small animals (Kloor, 1999). On the other hand, cities are, quite clearly, dangerous. They seem to attract crime, avarice, and noise. The city as a place is characterized by multiple and contrasting realities. Within the city both ends of almost any continuum (e.g., excitement and boredom; safety and danger) can and do exist simultaneously. Cities can pull people apart or bring them together, yield opportunities for us and in other ways constrain our behavior. There are good and bad, rich and poor, isolation and integration within the city's limits. And, of course, cities are not just one place, but a series of interconnecting, sometimes hierarchical districts (Bonnes et al., 1990).

However you choose to view them, cities are an environmental feature that you will probably have to contend with throughout your life. Cities and their suburban outgrowths are where most North Americans live. The U.S. Census Bureau provides a fascinating Web site (www.census.gov) with regular updates to maps and tables that chronicle changes in the population and makeup of both urban and rural areas. As British colonies, both the United States and Canada were heavily rural. Today, about 80% of the population of Canada and the United States lives in urban areas (U.S. Bureau of the Census, 1990), compared with 6% in 1800 (Gottman, 1966).

In 1940 the United States had only 11 Major Metropolitan Areas (those cities and suburbs with more than 1 million residents each), and these accounted for about one-fourth of the nation's population. Now the number of MMAs and their populations have approximately quadrupled. Population change is uneven, however, with far greater percentage increases in the western states and actual declines in some northeastern urban areas (U.S. Census Bureau, 1998, www.census.gov/). A key feature of the new century will be worldwide "megacities" such as Mexico City, with populations in the dozens of millions. On the other hand, simply saying that metropolitan areas are growing obscures important differences between the older central cities and the surrounding suburbs. As we shall see, much of the dramatic growth experienced by metropolitan areas in Canada and the United States has occurred in the suburbs, not in the city cores (e.g., Garreau, 1991; Southworth & Owens, 1993). Growth and other changes present different opportunities and different challenges to the suburbs and their parent cities (Figure 10-1).

In Chapter 9 we observed that population density can have significant effects on people. Interestingly, some urban areas have very dense populations, whereas in others people are more spread out. For example, 1990 census data showed population densities of more than 23,000 individuals per square mile in New York City, and fairly high densities in Chicago, Philadelphia, and Boston. Jacksonville, Florida, on the other hand, had a 1990 population density of only about 800 people per square mile, and even larger cities such as Phoenix and Houston had densities of less than 15% of those reported for New York.

Our chapter will begin with a reconsideration of some theoretical perspectives we first encountered in Chapter 4. These theories offer broad explanations that have been proposed to explain the effects of urban life. We will then move on to a discussion of some of the negative consequences of city life. Finally, we will

Figure 10-1 City living will probably be an inevitable fact of life for most of the world's population in this century. Simply saying that metropolitan areas are growing obscures important differences between the older central cities and the surrounding suburbs.

highlight various solutions that have been proposed to capitalize on the city's opportunities, while moderating some of its negative consequences. As we will see, some of these "solutions" create problems of their own. Nevertheless, as the world becomes more and more urbanized, the need to "humanize" the city is a challenge we can hardly avoid — a challenge very amenable to applications of environmental psychology.

Effects of Urban Life on the City Dweller: Conceptual Efforts

Before we attempt to broadly describe several of the major theoretical views of the impacts of city life, we should reiterate a point that we emphasized in Chapter 6 on climatological and geographic determinism. As attractive as it is to believe that we can exert control over our environments, it is usually an oversimplification to speak of any environment as *determining* behavior. For instance, although cities do have disproportionate rates of crime, many of their residents live fulfilled lives without being driven to vandalism, isolation, or despair. Instead of operating on people directly, as implied by deterministic theories, cities may actually be experienced as **urban villages** (Gans, 1962)—relatively homogeneous, small social structures of neighborhoods or businesses. In other words, the day-to-day life of the average city dweller may not require him or her to deal with the monolithic city itself, but rather, a small part of it centered on his or her home or workplace. Perhaps one primary effect of the city is to provide the "critical mass" to allow various ethnic and social groups to establish enclaves. With this caution in mind, examine again our eclectic model applied to cities, as depicted in Figure 10-2, then join us in seeing how each perspective applies individually to city environments.

OVERLOAD NOTIONS

One of the most deterministic formulations is a form of *overload* theory (see Chapter 4). Stanley Milgram (1970) hypothesized that urban existence involves being exposed to a profusion of stimulation. These stimuli are frequently more than we can deal with, so we employ coping strategies in order to lower stimulation to a more reasonable level. These could include setting priorities on inputs so that we attend to only important stimuli (which may result in ignoring those in need of certain types of help), erecting interpersonal barriers (e.g., behaving in an unfriendly fashion), establishing specialized institutions (e.g., welfare agencies) to absorb inputs, and shifting burdens to others (e.g., requiring exact change on buses). Even successful coping may be costly, leading to such aftereffects as exhaustion, fatigue, or disease. When successful coping does not occur, the individual will be subject to continued overload and is extremely likely to suffer serious physical or emotional damage.

ENVIRONMENTAL STRESS

A number of researchers (e.g., Glass & Singer, 1972) have applied the *environmental stress* approach to understanding and predicting reactions to urban life. In general, this approach views the presence of particular negative stimuli (e.g., noise, crowding) as critical for the negative effects of city life, as opposed to the overload assumption that too much stimulation per se is the critical element. The negative elements of city life may be experienced as threatening and may elicit stress reactions, which have emotional, behavioral, and physiological components. Stress reactions lead to a variety of coping strategies, which may be either constructive (e.g., using reasonable means to control the stressor) or destructive (e.g., aggression). If coping is successful in eliminating threat, adaptation occurs, and long-term consequences of the

stressor are often prevented. If coping is unsuccessful, long-term costs are likely to result.

BEHAVIOR CONSTRAINT

This formulation assumes that city dwellers experience constraints on their behavior (such as those caused by fear of crime, or getting "stuck" in traffic jams) that are not generally shared by people who live in rural areas. Such constraints often determine whether or not people can achieve their goals in a setting (Stokols, 1978).

Initially, behavior constraint notions predict that individuals experiencing this situation will evidence a negative feeling state and will make strong attempts to reassert their freedom. However, predictions of the consequences of long-term adaptation may be more pessimistic. If our efforts at reasserting control are repeatedly unsuccessful, or if we are overwhelmed by too many uncontrollable events, we may be less likely to attempt control of urban settings even when it is actually possible to control them. In effect, we may experience *learned helplessness.* While city life does impose many constraints on behavior, it should be noted that in some ways it is less constraining than small town life. For example, urbanites probably have more control over the information others obtain about their activities than those living in small towns, and are less frequently constrained by unavailable resources.

ADAPTATION LEVEL

Although overload, environmental stress, and behavior constraint theorists emphasize that the high level of stimulation characteristic of the city will have negative effects, the adaptation level approach (see Chapter 4) implies that this is not necessarily the case. The effects will vary across persons depending on their past experiences and personalities. For certain people the city may offer an optimal level of stimulation. The city offers so much diversity (e.g., quiet parks, busy streets) that somewhere within its environs it could harbor an optimal level of stimulation for everyone. Those not accustomed to the city often find it to be too noisy, too crowded, or

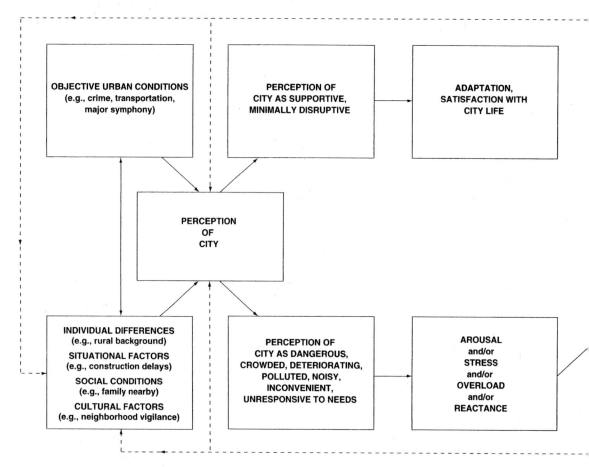

Figure 10-2 Our eclectic model applied to urban environments.

too uncomfortable in general, but after adaptation, they tolerate or even prefer more complex stimuli.

THE CITY AS A BEHAVIOR SETTING

You may recall from Chapter 4 that Roger Barker emphasized the importance of behavior settings, a point of interaction between individuals, their physical setting, and standing patterns of behavior. Two different dimensions of Barker's theory have made it increasingly important in modern urban studies. First, the behavior setting approach takes a very molar or broad view of the interaction of humans and environments. Increasingly, researchers treat the city as a place or series of places. Places represent the nexus

of setting and experience—a complex amalgam of memories, feelings, and more direct effects. This multivariable, broad approach to understanding transactions between humans and their surroundings was anticipated and at least partly inspired by Barker's behavior setting analysis.

More directly, Barker's approach gave birth to staffing theory (Wicker, McGrath, & Armstrong, 1972). As you recall from Chapter 4, overstaffing occurs when the number of participants exceeds the capacity of the system. A brief look at any city is sufficient to convince us that we are looking at an overstaffed environment. In terms of staffing theory, city dwellers should respond to such conditions by experiencing feelings of competition and marginality, by establishing priorities for interaction, and by at-

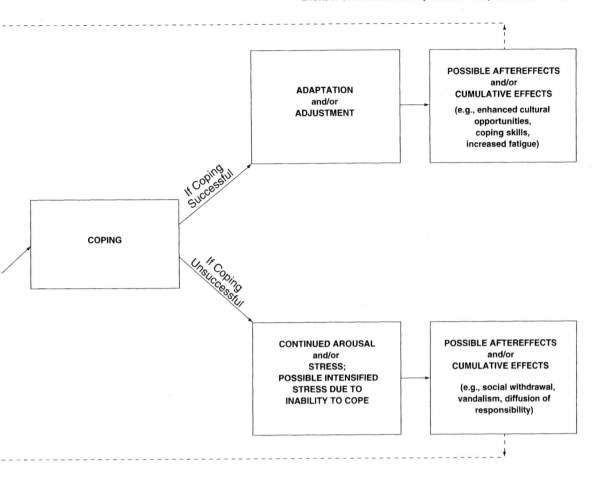

tempting to exclude others from their lives. If overstaffing is chronic, these behaviors may come to characterize everyday existence. However, it should be kept in mind that cities offer high-diversity behavior settings and numerous behavior settings overall, which could have positive effects.

INTEGRATING THE VARIOUS FORMULATIONS

It is probably the case that too much stimulation, too much stress, too many behavioral constraints, and overstaffing each accounts for some of the negative effects that may result from an urban existence. Furthermore, many of these same theoretical approaches account for the desirable stimulation and rich opportunities afforded by city life. The ways in which people are assumed to cope differ for the various models we have discussed. Nevertheless, when coping is successful in handling stress, adaptation or adjustment occurs, and the individual is less likely to experience aftereffects or cumulative effects. If coping is unsuccessful, the distress continues, and the individual is likely to experience aftereffects and cumulative effects (e.g., illness).

The balance of this chapter will reflect the discrepancy between the good and bad aspects of the city. First we will examine the negative effects of cities on residents and visitors, then we will turn to attempts to take advantage of opportunities of city life while minimizing its negative outcomes.

Negative Effects of Urban Life on the City Dweller: Research Evidence

Research on the effects of urban life has relied on two methodological approaches. The *single-variable approach* attempts to synthesize a picture of urban life from studies of how various individual stressors present in the city (e.g., noise, pollution) affect urbanites' behavior. These studies have often been conducted in "real-world" settings, where the stressor under study varies naturally, rather than at the will of an experimenter. For instance, to explore the effect of urban noise on psychiatric disorders, one might compare the mental health of the residents on two streets that differ only in their closeness to a noisy factory. This strategy may allow us to approximate a cause-and-effect relationship between a potential urban stressor and behavior, although the nonrandom assignment of individuals to conditions may lead to problems in causal inference.

While the single-variable approach may allow us to synthesize a picture of urban life as the sum of the separate effects of various stressors, it does not allow us to understand the city as a place—the result of a multitude of individual elements. A realistic view of the consequences of urban stressors may come from considering how they affect us collectively. Research using this approach generally compares cities (which obviously contain a full range of urban stressors) with nonurban areas on various dependent variable dimensions. You will find that many of the foundational studies for both of these research approaches are several decades old, yet their implications still hold true today.

STRESS

Clearly, urban areas differ from one another, and some are much better places to live than others. Nevertheless, comparisons of urban and rural areas generally suggest that cities contain more stressful environmental features. Furthermore, as city size increases, so do the risks of both physical stressors such as noise and pollution,

and social stressors like crowding and divorce (Dillman & Tremblay, 1977; Frug, 1999).

The pace of life seems faster in the city. "Pace" in this case may include actual physical movement. For example, residents' walking speed increases with the size of the community (Bornstein, 1979). The effect of population size is not always strong, however. In fact, there is greater variation in speed at different locations within the same city than between two cities (Sadalla et al., 1990), and speed may be influenced by gender, time of day, or even weather (Walmsley & Lewis, 1989). Sadalla et al. (1990) argue that more subjective, psychological measures of tempo are more representative of the perceived differences between cities than actual physical speed. That is, certain cities seem "fast paced" because of a complex mix of psychological factors, only some of which actually involve physical movement.

It is sad to note that the urbanites most intensely exposed to urban stressors are those with other problems as well. The poor are probably exposed to more urban crowding, noise, and crime. Given the fact that these individuals are already vulnerable to stress, adding the environmental stressors characteristic of urban life can be especially problematic.

With all of the evidence indicating that living in the city is probably experienced as more stressful, do urban and rural dwellers really perceive different levels of stress? Although only a few studies directly compare urbanites and rural people, a famous study of individuals who had recently migrated to the city or to a rural area suggested that this may be the case (Franck, Unseld, & Wentworth, 1974). Investigators interviewed a sample of students who were newcomers to either a small town or a large city. It was observed that the urban newcomers reported experiencing significantly more tension when living in the city than in their previous residence; the reverse was observed for rural newcomers. When sources of stress were broken down into those associated with the physical environment and those associated with the social environment, some additional differences emerged. For physical stressors (e.g., pollution, noise, crowding), urbanites reported being affected far more adversely than rural dwellers. For social stressors, results depended on the particular stressor. Public social stressors (e.g., slums, aversive individuals one must deal with) were experienced more strongly by urban newcomers. However, some rural newcomers complained about the lack of cultural diversity in their environment. Personal social stressors (stressful personal relationships) did not differ significantly for the two groups. These findings are suggestive and should be interpreted with caution, as should all studies involving urban and rural comparisons.

Perhaps the most complete study of commuting as a source of stress (Figure 10-3) was conducted by faculty and students in the Social Ecology Program at the University of California–Irvine (Novaco et al., 1979; Stokols & Novaco, 1981). The research was a longitudinal field experiment, using urban commuters traveling varying distances to work. Individuals were tested twice in their work settings— 18 months apart—to determine the effects of commuting stress. The data mostly corroborate earlier studies but give a more complete picture. Conditions that interfere with a commuter's movement (e.g., congestion) elicit stress reactions such as physiological arousal, negative mood, and performance deficits, and the intensity of these responses depends on personality characteristics of the commuter. Also, when people view commuting in a negative light they attempt to change the situation (e.g., move closer to work, try other routes). Importantly, such coping often makes them feel better at the psychological level. (See the box on page 342 for an examination of another aspect of commuter stress.)

AFFILIATIVE BEHAVIOR

On a number of dimensions, city life seems to be associated with a decreased desire for affiliative behavior. This ties in well with several of the conceptual notions we have discussed as well as with conceptual formulations discussed elsewhere (e.g., Wirth, 1938) but can also be explained in terms of reinforcements derived from past experience with city life (e.g., more experiences with crime). In one study suggesting a lower degree of affiliation, Newman and McCauley (1977) found that individuals' eye contact with strangers who looked them in the eye was relatively rare in center city Philadelphia, more common in a Philadelphia suburb, and very common in a rural Pennsylvania town (e.g., 15% versus 45% versus 80% of passersby made eye contact with the stranger in the three communities). In a study extending these findings, McCauley, Coleman, and DeFusco (1977) showed that commuters were less willing to meet a stranger's eye when they arrived at a downtown terminal than when they were in a suburban train station. Finally, Milgram (1977) reported that when undergraduate students approached strangers on the street and extended their hands in a friendly manner (as if to initiate a handshake), only 38.5% of city dwellers reciprocated, compared with 66% of small-town dwellers.

While this pattern of effects suggests that city dwellers are apt to avoid contact with strangers, it is important to assess whether this behavior extends to friends and acquaintances.

Figure 10-3 Commuting can be a source of urban stress. Road rage seems to be at least partly triggered by cues in the environment.

ROAD RAGE:
When Commuter Stress Gets Deadly

We have noted that commuter stress can negatively affect physiological arousal, mood, and performance. A related phenomenon that is gaining increasing concern is road rage—acts of violence committed by drivers who apparently "can't take it anymore." *Road rage* is a fairly rare occurrence (but receives much media attention) that involves an actual attempt to cause physical harm to another person, and as such is a criminal offense. This is in contrast to *aggressive driving*, which is a more common occurrence involving speeding, the cutting off of other drivers, profanity, horn honking, and general disregard for the rights of other drivers to share the road. Goehring (1999) reviews several surveys of the problem. In one, 62% of respondents said the behavior of another driver had been a threat to them in the past year. In another, 10,037 incidents of aggressive driving were examined between 1990 and 1996. A total of 12,610 people were injured and 218 killed in the traffic incidents reported. In yet another survey, 90% of respondents reported seeing aggressive driving within the past year, 80% within the past month, and 50% within the past week—and two out of three admitted aggressive driving themselves within the past year.

What causes aggressive driving? Are we dealing with a generation of disturbed personalities on our freeways? A study by Ellison-Potter et al. (2000) suggests the culprit is more our environment than our innate dispositions. College students in a driving simulator drove more aggressively and had more accidents (and killed more simulated pedestrians!) when told they were in a convertible automobile with the top up such that other drivers could not identify them. Participants drove much more safely when told the convertible top was down and others could identify them. Also, aggressive bumper stickers and road signs (e.g., "My Kid Beat Up Your Honor Student") increased aggressive driving, but a personality measure of driving anger was largely unrelated to aggressive driving.

What can be done to stop the carnage of road rage and aggressive driving? The results of the Ellison-Potter et al. study suggest that making drivers identifiable (perhaps through readily recognizable license plates) and reminders to be courteous may help. In another simulation study, Parsons et al. (1998) report that adding attractive, nature-dominated landscaping to roadways helps recovery from prior stress and reduces stress responses to subsequent unpleasant events.

In an experiment designed to test this hypothesis, McCauley and Taylor (1976) asked small-town and large-city residents about yesterday's telephone conversations with friends and acquaintances. Phone conversations in the city were just as likely to be with close friends and as intimate in subject matter as conversations in small towns. This pattern is corroborated by additional research (see Korte, 1980).

PROSOCIAL BEHAVIOR

Overload notions suggest that the overstimulation of urban life leads us to filter out less important inputs—a needy stranger, for instance. Does the urbanite's lack of desire to affiliate with strangers extend to a disregard for strangers who are in need? When a child claiming to be lost asked for aid in New York City and in several small towns, he or she was more likely to be the recipient of **prosocial** (helping) **behavior** in the smaller towns (Milgram, 1977). In a similar vein, Milgram found that willingness to allow a needy individual into one's home to use the telephone was higher in a small town than in a large city. Also, in Milgram's study 75% of all city respondents answered the person in need by shouting through a closed door, while 75% of all rural respondents opened the door, reinforcing our earlier conclusion about urbanite avoidance of affiliation with strangers. A statistical summary of all past research on the subject by Steblay (1987) strongly supports the nonurban–urban difference in helping, though decreases in helping were found to begin at a higher threshold (i.e., communities with a population of 300,000 or more) than previously thought (cf. Amato, 1983). According to Levine et al. (1994), population density (the number of people in a restricted area) is more likely to be related to unhelpful behavior than population size (the total population of a city). In an ambitious study of 36 cities in different regions of the United States, Levine and his colleagues compared the correlation between population size and density and a group of six helping behaviors. Overall, population density was the strongest and most consistent predictor of helping, particularly in situations that required a fast, more spontaneous response.

Among a number of possible explanations for the lower helpfulness of urbanites, we first should examine overload theory. As that model would predict, the high levels of stimulation characteristic of the city make passersby less attentive to novel stimuli, such as someone needing help (Korte, 1980). Second, Fischer (1976) suggested that the diversity of appearance and behavior characteristic of others in urban areas may make people feel insecure and thus less likely to help, and third, Wirth (1938) proposed that being brought up in an urban as opposed to a rural area elicits an "urban personality," which is simply not characterized by prosocial behavior. Overstaffing theory could offer a fourth explanation, and diffusion of responsibility notions (cf. Latané & Darley, 1970), a fifth. Work on **diffusion of responsibility** suggests that when there are many people around who could help (as would occur more in cities than in small towns), perceived responsibility to help lessens, which affects the likelihood of giving aid. Interestingly, meta-analytic work by Steblay (1987) suggests that it is the urban context rather than personality factors that is responsible for lower levels of urban helping. This discounts Wirth's "urban personality" theory, while lending a measure of support to several of the others.

Although research suggests that there may often be less prosocial behavior in cities, other studies imply some important moderators of this pattern (Forbes & Gromoll, 1971; Korte, Ypma, & Toppen, 1975; Weiner, 1976). Korte and his colleagues suggested that urban and rural settings may lead to differences in helping

only insofar as environmental input level (i.e., amount of incoming stimuli) is higher in cities. Their findings led them to conclude that input level may be the critical determinant of helping, rather than the urban–rural distinction per se. Similarly, others (e.g., Kammann, Thompson, & Irwin, 1979) suggest that pedestrian density in the area where help is to be given, rather than city size, is the major factor in whether or not aid will occur.

Some studies have even found more helping in urban than in nonurban contexts. Weiner (1976) and Forbes and Gromoll (1971) found greater helping by individuals reared in cities than by those reared in small towns. Interpreting her findings, Weiner posited that different patterns of social-perceptual learning in the city and the country may cause urbanites to be more socially effective in certain circumstances. In effect, she suggests that the experience of growing up in the city allows one to learn skills that may be particularly adaptive in certain dependency situations.

THE FAMILIAR STRANGER

One thing our discussion has suggested is that urbanites are less likely to acknowledge strangers (e.g., by shaking hands and making eye contact) than rural dwellers. Some extremely interesting research by Milgram (1977) indicated that while city dwellers fail to display such amenities in everyday situations, they may show their feelings in other ways. Milgram and his students found that many city residents have a number of people in their lives who may be called **familiar strangers.** What is a familiar stranger? It is someone we observe repeatedly for a long period of time but never interact with, probably because of overload. Milgram found that commuters to New York City had an average of four individuals whom they recognized but never spoke to at their train station, and that 89.5% of the commuters had at least one "familiar stranger." How did the researchers find this out? They took pictures of groups waiting for a train at the station and had individuals tell them how many of those present met the definition of a "familiar stranger."

WHAT MOTIVATES VANDALISM?

Vandalism is the willful destruction or defacement of any public or private property. The results of a vandal's work may seem completely senseless, but the intent may reflect deep personal frustrations or even a calculated political expression (consider the Boston Tea Party). Overall, the resulting cost to our schools, parks, recreation areas, public housing, and transit systems is estimated at billions of dollars per year, and these costs are increasing rapidly.

Preventative strategies could focus on increasing territorial control (e.g., designing areas to promote defensible space). Aesthetic factors associated with an object's appearance (e.g., physical beauty) and the extent to which a site is hardened (made difficult to vandalize; see Figure 10-4) also affect the level of vandalism (Pablant & Baxter, 1975). Just as aesthetic variables affect how much we enjoy socially acceptable interactions with an object, they affect the pleasure we experience from vandalizing it. Objects that break in aesthetically interesting or pleasing ways may be more apt to be vandalized than those that break in dull, uninteresting ways (Allen & Greenberger, 1980; Greenberger & Allen, 1980). Designing objects (e.g., street lights) that will not break in a satisfying manner may be another way of decreasing vandalism.

The presence of graffiti may encourage new "artists" (Samdahl & Christensen, 1985; Sharpe, 1976). Thus, one priority of both parks and cities is to clean up the signs of vandals as quickly as possible. Other factors have been implicated as causes

Figure 10-4   "Hardening" features of the environment, such as this trash can, may help prevent breakage from vandalism.

What is the difference in urbanites' behavior toward "familiar strangers" and other strangers? First, many passengers told the researchers they often think about their familiar strangers and try to figure out what kinds of lives they lead. According to evidence, urbanites are more likely to help a familiar stranger in need than an ordinary stranger. Finally, Milgram found that under some circumstances, familiar strangers do interact with each other, although it is rarely in the place where they usually meet. He suggested that the farther they are away from the scene of their routine encounter (a foreign country, for instance), the more likely they are to interact.

CRIME

Reasons for High-Crime Prevalence

Studies of victimization suggest that even what could be thought of as trivial crimes may have long-lasting consequences for victims' well-being (Greenberg & Ruback, 1984), and there is ample evidence that crime is more prevalent in urban than rural areas. For instance, the rate of violent crimes per person is almost six times greater in the large metropolitan cities than in rural areas (U.S. Census Bureau, 1998).

An amusing anecdote related by Zimbardo (1969) suggests the intensity of crime in and around many cities. While repairing a flat tire of vandalism. Allen and Greenberger (1980) suggest that low perceived control (see Chapter 4) will under certain conditions elicit vandalism. When we come to believe we cannot control our outcomes, we sometimes resort to vandalism as a way of showing ourselves and others that we can control at least certain things (Warzecha, Fisher, & Baron, 1988). This would suggest that the greater people's control over a setting, the less vandalism there will be. Also, it has been suggested that vandalism results when there is a "lack of fit" between the person and the environment; for example, school vandalism may be due to poor congruence between personal characteristics of students and the social or physical environment of the school. Increasing the goodness of the fit through social and/or environmental means could help lower vandalism, according to this conceptualization. Work by Richards (1979) has identified peer relationships (i.e., associations with antisocial peers) and adult–child conflict as major causes of vandalism by middle-class adolescents. Vandalism may also occur due to financial need, in the pursuit of social causes, due to nonmalicious play, or due to poor achievement (cf. Cohen, 1973; Sabatino et al., 1978).

One model encompasses many of the reasons suggested above for why vandalism occurs, under the concept of *perceived inequity* (Baron & Fisher, 1984; Fisher & Baron, 1982). What is perceived inequity? Equity theories in social psychology imply that we are socialized to believe we should treat others fairly (or equitably) and should be treated equitably by others. When this does not occur (i.e., when we perceive we are being inequitably or unfairly treated), we become upset and try to restore equity. Fisher and Baron's model views vandalism as a way to restore equity by responding to one type of perceived rule-breaking unfairness with another type (i.e., disregard for another's property rights). In effect, some vandals seem to say, "If I don't get any respect, I won't give you any either." Fisher and Baron suggest that inequity will result in vandalism only when the person who feels inequitably treated has low perceived control—that is, little likelihood of influencing whether equity will be restored. When we have high perceived control, we restore equity within the system (e.g., complain to the authorities), and when we have very low perceived control, we become helpless and simply accept our fate. But when we have moderate to low control, we are likely to opt for a way of restoring equity. For some, this method is vandalism—an immediate, low-effort, and certain means of paying society back. Fisher and Baron suggest that increasing perceptions of control, decreasing perceived inequity, or both, can be effective means of lowering vandalism.

alongside a highway in Queens, New York, a motorist was startled when he observed that his car hood was being raised, and a stranger was removing his battery. "Take it easy, buddy," said the thief to his assumed car-stripping colleague, "you can have the tires—all I want is the battery!"

Why is there more crime in cities than in small towns? Although these findings can be interpreted in terms of overload, stress, behavior constraint, or overstaffing notions, several other explanations have been offered. One is the theory of **deindividuation.** It was used by Zimbardo (1969) to explain why an "abandoned" car he left in New York City was stripped of all moving parts within 24 hours, while a similar car left in Palo Alto, California, was untouched. According to this theory, when we feel we are an anonymous member of a crowd (i.e., deindividuated), our inhibitions against antisocial behavior are released. This is partly because we feel it is very unlikely that we will be identified and punished. Under such conditions, criminal behavior is clearly less costly and more likely to occur. Other explanations for the higher levels of crime in urban areas include a lack of employment opportunities, the greater number of antisocial role models available, and the fact that there may be fewer prosocial models available than in nonurban areas. Another explanation is that there are simply more possible victims, more goods to steal, and more outlets for stolen goods in cities than elsewhere. Individuals who want to pursue crime may even migrate to the city.

Fear of Crime

Imagine being afraid to go outside of your apartment to buy food or to cash a check, or of opening the door in terror when someone knocks, hoping he or she is not a criminal. Fear of crime and associated stress are major problems in urban areas. Interestingly, it has been found that fear of crime is increasing faster than actual crime rates (Taylor & Hale, 1986). In fact, in some cases, fear of crime in a subpopulation is not related to the true likelihood of being victimized (Maxfield, 1984).

Various aspects of the urban environment may have an impact on fear of crime, which varies from neighborhood to neighborhood (Maxfield, 1984). Teenage loitering, which can be facilitated or inhibited by environmental features, can elicit crime stress (Lavrakas, 1982; Lewis & Maxfield, 1980). It has also been suggested that physical decay of the environment and signs of urban "incivilities" (e.g., reports of crime; vandalism, graffiti, litter) can imply to people that the social order has broken down, and elicit fear of victimization (Lewis & Maxfield, 1980; White et al., 1987). This is especially likely when residents attribute the cause of the incivilities to factors residing "within the neighborhood" (Taylor & Hale, 1986). Importantly, studies have shown that perceptions of incivilities are more strongly related to fear than the objective number of incivilities (Taylor & Hale, 1986). Areas where attackers could be concealed are particularly fear provoking (Nasar & Jones, 1997). Finally, perceived loss of territorial control appears to be associated with fear of crime (Taylor & Hale, 1986).

Some people seem to fear crime more than others. Those who are most concerned are those with lower incomes, females, minorities, the aged, and residents of the inner city (Clemente

& Kleiman, 1977; Gordon et al., 1980; Keane, 1998). Interestingly, in areas with the highest crime, age is not related to fear of crime. Where crime is a regular feature of daily life, the physical vulnerability associated with age may be a less important determinant of fear than other factors (Maxfield, 1984).

People with high fear of crime feel they must restrict their activities greatly to avoid being victimized (Keane, 1998; Lavrakas, 1982). Environmental designs that help promote social cohesion among residents (e.g., defensible space; Lavrakas, 1982, p. 343) may moderate fear and make people feel more comfortable "moving about." Also, having supportive neighbors who are accessible may act to quell fear of being victimized (Gubrium, 1974; Sundeen & Mathieu, 1976). For the aged, this seems to occur more often in socially homogeneous living situations (e.g., retirement communities) than in other settings. Being able to move away from high-crime areas increases feelings of safety (Varady & Walker, 1999).

HOMELESSNESS

Although homelessness occurs in rural areas, it is disproportionately an urban malady (see Desjarlais, 1997; Da Costa Nunez, 1994; Jencks, 1994). Homelessness is also more than simply being "houseless." That is, **homelessness** is not just the lack of shelter, but also the loss of security and social or economic support that is associated with unreliable shelter or the prospect of losing shelter (Bunston & Breton, 1992). There have always been homeless people, but their number has increased dramatically in the past few years, and their plight has become the focus of more and more public attention (Figure 10-5).

Formal definitions for experiences like homelessness sometimes obscure insight. Rivlin (1990) provides a particularly dramatic description:

> Picture a day when you cannot be certain
> where you will sleep, how you will clean

Figure 10-5 Homelessness is one of the saddest urban problems.

yourself, how you will find food, how you will hold on to your belongings (in settings where they can be stolen with ease), how you will be safe, how you will dress properly to go to work (for many homeless people do, indeed, work), and for some, how you will fill up the long hours of the day and do so in places that will tolerate your existence (p. 50).

There is an ongoing debate about the relative importance of structural versus individual causes of homelessness (Main, 1998). *Structural explanations* lay the blame outside the individual and suggest that the homeless are largely competent victims of unemployment or inadequate social policies. Shinn (1997), for instance, emphasizes that homelessness is not a trait;

homelessness is usually a temporary state that results from a confluence of environmental or personal challenges, rather than a personality trait or motivational deficit. In contrast, others place more emphasis on the role of *individual causation* such as mental illness or substance abuse. Certain personality traits may help one's chances of overcoming homelessness. For example, a sense of self-efficacy (being able to control one's own destiny) and an orientation that focuses on the future may reduce the time a person remains homeless and increase the likelihood that he or she will seek out additional education and other coping strategies (Epel, Bandura, & Zimbardo, 1999). Thus, as Main suggests, homelessness can probably be attributed to both structural and individual causes.

So why are so many people homeless? Many individuals are homeless because their incomes have failed to keep up with sharply increased housing costs. Especially for those with few close family ties who have marginal incomes, the loss of even a few days' pay due to an injury, losing a job, or similar misfortune can quickly result in homelessness. Urban renewal and gentrification can also cause homelessness. **Gentrification** occurs when middle- and upper-income people move back to the city and occupy and improve areas formerly lived in by poor people. Unfortunately, the return of the "gentry" is often accompanied by displacement of the poor.

Another factor in the homelessness problem is society's failure to provide adequate community-based housing and care for people. Certain health problems tend to cause homelessness (e.g., major mental illnesses such as schizophrenia). In fact, some studies suggest that a very large percentage of the homeless have psychological or addictive disorders (Bassuk et al., 1986; Baum & Burnes, 1993) or diseases like AIDS, which may render one unable to pay rent and/or to be undesirable to landlords and even to family due to stigma. In many instances there is not just one cause. Many homeless families are "multiproblem families" (Bassuk, Rubin, & Lauriat, 1986) with fragmented social networks and difficulty utilizing available public welfare services.

Although researchers often find considerable sympathy for the homeless (e.g., Barnett, Quackenbush, & Pierce, 1997), homelessness bears a stigma even greater than that attached to poverty in general (Phelan, Link, Moore, & Stueve, 1997). One adaptation to homelessness may be so-called shelterization, characterized by low self-esteem and a dependency on the system (Grunberg & Eagle, 1990). From their study of homeless women in Toronto, Bunston and Breton (1992) concluded that the simple provision of shelter alone overlooks important needs for autonomy, programs to foster independent living, and education and job skills to break the cycle of homelessness. Although efforts are not yet well coordinated, some regions are beginning to provide follow-up services to support the transition from shelters to more permanent housing (Da Costa Nunez, 1994).

The characteristics of the homeless differ dramatically from place to place, and estimates of the proportions of men, women, and children who are homeless vary. Overall, the composition of the homeless is changing: Middle-aged men now make up a shrinking percentage of the homeless, and families with small children are the fastest growing segment of this population (Da Costa Nunez, 1994). In fact, families now account for about one-third of the homeless (Berck, 1992) and an even higher percentage in places like New York City (Da Costa Nunez, 1994). As you can imagine, homelessness disrupts almost every facet of a family's life. For children, homelessness usually means changing schools, leaving teachers and friends behind. Not surprisingly, many children become restless, aggressive, or listless when they must move to a shelter (Neiman, 1988). Interestingly, most homeless are long-term residents of particular cities, which negates some public officials' arguments that if they do more to help them, increased numbers of homeless will come to their city (Committee for Health Care for Homeless People, 1988).

In addition to being a cause, health and psychological problems can also result from homelessness. Homelessness may increase the risk of developing many diseases, the likelihood of incurring trauma, and the risk of being victimized (rape for women; violent assault for both sexes; Kelly, 1985). For those with medical problems, homelessness makes treatment extremely difficult. How can people be on "bed rest" when they do not have a bed? The mortality rate for the homeless is three times that of the total U.S. population, and the homeless die 20 years earlier than expected (Baum & Burnes, 1993). Homelessness is also associated with other risks (e.g., the fires used by street people to keep warm often cause burns). The trauma of being homeless has negative psychological consequences and is associated with anxiety and depression. Homelessness can also contribute to alcoholism or drug addiction as an attempt to "medicate" the psychological pain of not having a place to live.

Environmental Solutions to Urban Problems

Not surprisingly, many people and businesses have attempted to find true happiness by escaping from urban areas, a fact that has resulted in population declines for the urban cores of many North American cities even as the suburbs and metropolitan areas grow (e.g., Garreau, 1991; Southworth & Owens, 1993). Those who remain behind in the decaying city core are often individuals whose social or economic position makes them incapable of departing. This has left cities in deteriorating physical condition, with a dwindling tax base and high levels of unemployment and attendant social problems, such as crime. How can this situation be ameliorated? Many significant social, economic, and physical changes are needed, but cities are an unwieldy aggregate of corporate and private interests only loosely controlled by overburdened governments. Perhaps the earliest and simplest approach was to add new amenities such as museums, parks, or playgrounds without attempting to change the basic structure of the city.

A LITTLE PIECE OF NATURE: PARKS AND URBAN GARDENS

One cure for what ails cities might be a little piece of the country. We saw in Chapter 2 that natural scenes have many psychological benefits, so perhaps going back to nature would help cities. Indeed, the presence of trees and maintained grassy areas increases feelings of safety in inner-city neighborhoods (Kuo, Bacaicoa, & Sullivan, 1998). In a study of inner-city children and outdoor spaces, it was found that level of play, supervision by adults, and creativity in play were approximately doubled in areas with green grass and trees compared with barren areas (Taylor, Wiley, Kuo, & Sullivan, 1998).

Somewhat before medicine provided an explanation, citizens of Europe and North America began to suspect that outbreaks of diseases like cholera were tied to the unsanitary living conditions and close quarters of big cities. People needed to breathe, they said. In London, for example, the large open squares were believed to have positive health effects, and former royal parks across European cities were opened to the public in the new spirit of populism (Schuyler, 1986). As their new nation became more urban, citizens of the United States began to seek relief from urban noise, disease, and confusion. Initially many people went on outings to cemeteries, the first public or semipublic gardens in cities (Kostof, 1987). In 1857, 25-year-old Frederick Law Olmstead became the Superintendent of Construction for what was to become New York's Central Park (see Cranz, 1982; Hayward, 1989; Hiss, 1990; Kostof, 1987; and Schuyler, 1986, for reviews of urban park design). With his associate Calvert Vaux, Olmstead created Central Park, North America's first great public park. Olmstead was inspired by the *English Romantic style*, a literary and artistic return to nature that was popularized by the European elite (see Chapter 2). Olmstead and Vaux sought to create an illusion of nature through careful manipulation of topography, water, and plant materials (see Figure 10-6). Parks were a form of landscape art meant to represent nature not as it really was, but as it ideally might be. In spite of modern intrusions and neglect, Central Park retains many of the romantic landscapes that first attracted visitors when the park opened in 1859. The ideas Olmstead and Vaux developed in Central Park were refined in Brooklyn's Prospect Park and repeated by the Olmstead firm across North America (hundreds of Olmstead parks remain: San Francisco's Golden Gate Park and Montreal's Mt. Royal are examples). In Boston, Olmstead proposed and built an integrated necklace of interconnected parks along waterways, and his hand is also apparent in the design of a number of college campuses (see Chapter 11).

The 19th-century park was a natural "pleasure ground," intended for the most part to

Figure 10-6   Illusion of nature created by Olmstead and Vaux in New York's Central Park.

provide quiet vistas and opportunities for re-flection in the midst of expanding cities (Cranz, 1982). Open lawns in some cities supported herds of sheep, deer, or even reindeer, with the dual purpose of adding to the pastoral or pictur-esque landscape and keeping the lawns clipped. Parks were separated from the neighboring city with berms and barriers of vegetation, and al-though drives were provided for carriages and horses, pedestrian pathways were generally seg-regated. The park was also seen as a place where different ethnic groups could experience leisure and reflection together. Less charitably, Kostof (1987) suggests that one primary goal was to wean the working classes from their ethnic neighborhoods in order to make "good" Ameri-cans of them.

Even in the 1800s, parks supported some organized group activities (ice skating became popular in Central Park, for instance), but at the turn of the century, North Americans found themselves with more leisure time and a height-ened desire to participate in organized activities. In many parks, open pastures were replaced by ballfields and tennis courts.

Neglect, fences, crime, and conflict left many parks at the end of the 20th century with many of the problems of the surrounding city and anything but romantic reputations. Yet his-toric preservation efforts and the environmen-tal movement have led to a park renaissance. Some efforts concentrate on the construction of new, specialized parks and cultural centers such as botanical gardens, zoos, and aquariums (see Chapter 13 for a discussion of some of these dual learning/leisure environments). Others seem aimed at refurbishing more traditional parks. In the latter case, it is unclear how well historic parks address the needs of modern city dwell-ers. Olmstead's vision may have been on the mark, but if not, to what degree should an old park evolve to meet the needs of a changing so-ciety? Part of this tension is between the need for environments to support both organized ac-tivities such as sporting events and somewhat more passive pursuits.

At a more personal level, private garden-ing may provide nature-based benefits for urban residents (Kaplan, 1984, 1985; Kaplan & Kaplan, 1987; Lewis, 1973). For example, Lewis (1973) researched the effects of providing recreational gardening environments for residents of run-down urban areas in New York City. The New York City Housing Authority sponsored recreational gardens. Groups of tenants wanting to garden could apply to the Authority, which gave them a garden site close to their project, turned over the ground for them, and also provided money for seeds or plants and a gardening manual.

Providing garden plots for residents has had many beneficial effects. Lewis (1973) reported that recreational gardening by inner-city residents led to pride in accomplishment, to increased self-esteem, and to reduced vandalism outside as well as inside the buildings. Social factors are important as well. In Kaplan's (1985) study cited above, residents who had adequate access to gardens found their neighbors to be more friendly and felt a stronger sense of community. Gardens add social cohesion in the community by providing a meeting place and a chance for people to work together toward a common end (Lewis, 1973). Gardening may also increase the proprietary sense of territoriality, and thus make nearby space more apt to be defended and defensible (see Chapter 8).

Why is recreational gardening beneficial? In addition to their natural beauty and potential as a food source, gardens may provide a restorative experience that allows people to recover from the stresses of day-to-day life (Kaplan & Kaplan, 1987). The chance to be outside, to labor, to see things grow, and to experience a diversion from the routine involves many of the same benefits observed in wilderness recreation (R. Kaplan, 1984; Talbot & Kaplan, 1986).

DESIGNING URBAN PLAYGROUNDS

One fact of nearly every childhood is play. Most researchers believe that play activity has great significance for children and serves as an important vehicle for learning about the world. Children play in a variety of environmental contexts, including recreation rooms, museums, vacant lots, and streets. In fact, street play may be a universal phenomenon, and it is especially important to children's lives in countries with emerging economies (Abu-Ghazzeh, 1998). According to Abu-Ghazzeh, in some respects streets are superior to designed playgrounds. They are smooth enough for riding bikes and playing games with balls, but their most important attribute is their immediacy for close-to-home activities.

Because there is less space available for children to appropriate, playgrounds may be better used and more important in cities than in small towns (Moore, 1989). The beginning of the playground movement in America can be traced to 1885, when a pile of sand was provided for a "sand garden," a play area for the children living near a mission in Boston (Dickason, 1983). This structured play experience was well supervised and was apparently intended to "Americanize" the children of immigrants by enticing them to a site where they would be subject to instruction or propaganda. Although this goal may seem rather heavy handed, it represents an early recognition of the usefulness of formal play facilities in creating effective educational environments.

Researchers (e.g., Brown & Burger, 1984; Hayward et al., 1974) generally distinguish between three broad playground styles: traditional, adventure, and contemporary playgrounds (see Figures 10-7A, B, and C). Perhaps you are most familiar with the traditional and contemporary playground types. *Traditional playgrounds* contain the standard apparatus (e.g., swings, monkey bars, jungle gyms). This is still a widely spread playground type in North America and seems primarily geared toward exercise. Unfortunately, these playgrounds are often dangerous places with a variety of metal parts, chains, and (with a curious insensitivity to life and limb) concrete or asphalt paving. Aesthetics are not necessarily ignored, but seldom are central.

Contemporary playgrounds include many of the same elements found in traditional designs, but with a flair for aesthetics and abstract shapes (e.g., the slide may extend from a multileveled wooden structure with a variety of ladders, balancing beams, ramps, bridges, and other delights). Note that in a contemporary playground a single apparatus often serves multiple rather than single play functions.

Figure 10-7   Playground styles (A) Traditional (B) Adventure (C) Contemporary.

Hart (1987) remarks that highly manicured outdoor settings are usually controlled by adults, with the loose parts of scrap wood, dirt, and other materials systematically removed. These environments afford children few opportunities for fantasy and spontaneous design. On the other hand, snow in the wintertime and the "odds and ends" that accumulate in untended lots or rural areas provide rich opportunities for children. *Adventure playgrounds* (at least the "official" versions) began in Denmark during World War II and encouraged youngsters to use scrap wood and other castoff materials to build their own world of fantasy and dirt. Instead of traditional play equipment, scraps of materials such as wood (i.e., "loose parts") and tools like hammers, nails, and saws are supplied. Children are encouraged to build structures and to modify old ones as time goes on and interests evolve. Clearly these playgrounds require careful supervision, and the presence of adult supervisors makes possible activities such as cooking and gardening

in addition to less structured digging and hammering. Moore (1989) concludes that adventure playgrounds support more fantasy and richer cognitive experiences. Of course, some members of the community complain about their unplanned nature and unattractive appearance.

Different playgrounds attract different clientele. In the famous study by Hayward et al. (1974), practically no preschool children attended the adventure playgrounds, whereas older children were more likely to patronize them. According to Moore's (1989) summary, girls are more often seen on traditional apparatus such as swings, whereas boys seem more attracted to climbing apparatus and to ball games. On the other hand, Moore believes that settings dominated by nature attract a more equal mix of boys and girls and encourage more cooperative play.

The activities engaged in at the three playgrounds differed as well (Hayward et al., 1974). In the traditional playground, swinging was the most common activity; but at the contemporary playground, children engaged in a continuous mode of activity that included playing on varied equipment. This result was probably due in part to the equipment at the contemporary playground (e.g., there was much more "multi-purpose" equipment). At the adventure playground the most popular activity was playing in the "clubhouse," an option that did not exist in the other two settings.

Overall, for the three playgrounds, it may be seen that the opportunities and constraints provided by the environment predict the predominant activities engaged in by children. Such opportunities and constraints also affect how the children play (e.g., alone or in groups) and the focus of their interaction (e.g., on the "here and now" versus fantasy). For example, fantasy play was least common in the traditional playground.

Such pretend or imaginative play is of particular theoretical interest because it is thought to foster divergent, creative thinking. One recent study (Susa & Benedict, 1994) specifically examined the link between playground design, pretend play, and creativity. As expected, more pretend play occurred on a complex contemporary playground than on a somewhat simpler traditional playground. In addition, the researchers

asked the children they had been observing on the playgrounds to try to think of different uses for a large wooden cable spool. At least for this measure, creativity was positively correlated with pretend play, leading to at least a tentative conclusion that playground design may foster creativity.

A number of researchers believe that environmental designers do not sufficiently weigh the preferences or concerns of children and parents (e.g., Iltus & Hart, 1994; Moore, 1989). The reasoning that goes into play space design often involves untested assumptions about the nature of children and play (Brown & Burger, 1984; Hayward, Rothenberg, & Beasely, 1974). Consequently, we should ask how well the resultant play spaces actually meet the needs of the user population. Moore (1989) and Iltus and Hart (1994) emphasize the importance of community involvement. Especially in small and medium cities, new playground construction is often a community project involving teachers, parents, public officials, and children themselves. The playground in Figure 10-8, for instance, was constructed as a community project under the supervision of a commercial design firm. During the course of the construction, hundreds of different individuals volunteered

Figure 10-8   A playground constructed as a community project.

time and materials, and subsequent postoccupancy evaluation (see Chapter 11) resulted in the construction of two additions to ease congestion on the popular playground and to make it more accessible for physically challenged children. Children were involved in the construction and planning too, but in this and other playground projects children sometimes are constrained to a role that allows them only token participation or an opportunity to decorate what is really an adult project. Hart (reported by Iltus & Hart, 1994) advocated a more difficult but more genuine approach whereby children consult with adults, participate in construction, or even initiate playground projects.

Fears for children's safety and the reduction in park and playground staff have resulted in increases in private playgrounds for middle-class children. Unfortunately, relatively little attention has extended to low-income families (Gaster, 1992; Iltus & Hart, 1994). Iltus and

Hart report a successful program in New York City that involves children in most of the stages of design participation we will outline in Chapter 11. The project capitalized on the children's creativity by involving them in drawing and model building. Capable children even participated in some of the negotiation and compromise decisions that eventually accompany almost every design. Additionally, some of the playgrounds were combined with community gardens, thus increasing adult supervision and involvement.

Many playgrounds become abandoned when warm fall days give way to chilly winds. Careful site planning, however, can create microclimates that extend the play season (Pressman, 1994). Winter can even be an asset. Each year a snow and ice playground is built as part of the Winterlude celebration in Canada's Capital Region (which includes Ottawa, Ontario, and Hull, Quebec; see Figure 10-9).

Figure 10-9   Winter can even be an asset in playground development. Each year a snow and ice playground is built as part of the Winterlude celebration in Canada's Capital Region.

REVITALIZING ENTIRE URBAN DISTRICTS: URBAN RENEWAL

Urban renewal can be defined as an integrated series of steps taken to maintain and upgrade the environmental, economic, and social health of an urban area (Porteus, 1977). It is not really a new idea. More than a century ago at the Colombian Exposition in Chicago (a world's fair), Daniel Burnham set out to create an entire city based on a coherent master plan that was to be free from the ills of unplanned urban disorder. The fair was a success. Although no new cities resulted, Burnham's plan led to what was called the *City Beautiful Movement*. These grand master plans featured classical architecture, arranged in an orderly fashion around public open spaces having plazas and pools. City Beautiful plans were adopted and at least partially executed in Chicago, Duluth, Cleveland, and San Francisco. For United States citizens, the most familiar of Burnham's projects may be his contribution to Washington, DC's Mall, with its flanking monuments and museums (Kostof, 1987).

A more modern initiative for urban renewal focused more clearly on the decaying residential and commercial sections of post–World War II cities. The assumption was that renewal would provide better housing, safer neighborhoods, and revitalized business districts. In effect, urban renewal was to be a panacea for many urban problems. For instance, there was assumed to be a causal relationship between poor housing and a "grab bag" of social ills. In fact, aside from studies relating poor housing to problems with physical and mental health (e.g., Duvall & Booth, 1978), there is little evidence to support this assumption.

The latter half of the 20th century saw massive physical changes in the name of renewal: Houses and neighborhoods were razed and replaced by tall apartment complexes, and thousands of residents were relocated. Planners hoped to attract wealthy individuals and businesses back into the city, to destroy eyesores, and to keep the city sufficiently attractive so that people would make use of its cultural resources (Porteus, 1977). Especially in the early years of urban development, they tended to replace slums with luxury apartments and office buildings, forcing residents to move elsewhere. Unfortunately, the people who were relocated often did not perceive their area as a slum at all but as a pleasant neighborhood (Fried & Gleicher, 1961).

What are the psychological consequences of demolishing neighborhoods and forcing people to relocate? Clearly, these depend on a large number of situational conditions, including attraction to the former neighborhood and family conditions, but relocation often has negative consequences. Destroying a neighborhood not only eliminates buildings, it can destroy a functioning social system and sense of identity for neighborhood residents. According to Gans (1962), slum areas not only provide cheap housing, but offer the types of social support people need to keep going in a crisis-ridden existence.

Boston's West End served as the location for an intensive study of the effect of renewal on a well-liked Italian working-class residential area. What were the consequences of relocation? Loss of home, neighborhood, and daily interactions with well-known neighbors caused an upheaval in people's lives and disrupted their routines, personal relationships, and expectations. This led to a grief reaction in many of those who were displaced, especially in those people who had been most satisfied with the status quo. Among women who reported liking their neighborhood very much, 73% displayed short-term reactions to extreme grief, including vomiting, intestinal disorders, crying spells, nausea, and depression. About 20% of the residents were depressed for as long as two years after moving.

REVITALIZING RESIDENTIAL AREAS

If "wiping out" entire urban districts in an attempt to "save" them generates new problems, are there environmental steps we can take to target residential areas for revitalization? In this regard, let us examine defensible space, social networks, public housing, and gentrification.

Defensible Space

Newman and his colleagues (Newman, 1972, 1975, 1995; Newman & Franck, 1982) have focused on how physical aspects of a setting may affect resident-based control of the environment and ultimately lead to lower crime rates. Their ideas are captured in the concept of defensible space. **Defensible spaces** are clearly bounded, or semiprivate, spaces that appear to belong to someone; that is, a visitor is likely to recognize them as someone's territory. Defensible spaces should also allow surveillance by providing visual accessibility. Newman argued that if we create such spaces through design, they would lead residents to feel ownership over them, foster informal surveillance, and promote social cohesion between neighbors. These behaviors should reduce certain types of crime and antisocial acts and elicit improved social relations among urbanites.

Defensible spaces could lead to lower crime rates for several reasons (Taylor, Gottfredson, & Brower, 1984). First, they could have a direct effect. It may be that spaces that look "defensible" lead potential offenders to assume that residents will actively respond to intruders, a notion that has been supported in work by Brower, Dockett, and Taylor (1983). Second, as suggested by Newman, defensible space may cause the formation of local ties among residents. This may occur because it makes people feel safer, which causes them to use the space more, to come into increased contact with neighbors, and ultimately, to develop more common ties. Individuals with more ties are more apt to intervene to "defend" their neighborhood, are better able to discriminate neighbors from strangers, and, because shared norms develop, are more likely to know what types of activities should go on and what types should not (Taylor & Brower, 1985). The latter analysis was supported in research by Taylor et al. (1984). Finally, defensible space could lessen crime, since it may strengthen people's territorial functioning (i.e., because areas characterized by defensible space are well bounded and more defensible, they may elicit more proprietary attitudes).

Does the concept of defensible space have the predicted effects? There is definite support for at least part of the model. Newman (1972) compared two public housing projects in New York, one of which was high in defensible space, the other of which was low. The latter project had more crime and higher maintenance costs, and this could not be explained by tenant characteristics. However, while increased defensible space was associated with less crime, whether this was due to greater cohesion among neighbors and stronger territorial attitudes and behaviors, as suggested by defensible space theory, is unclear since this mediating link was not measured.

In addition to physical changes to enhance defensible space, other interventions that will increase neighborhood cohesion or feelings of "ownership" should also lower crime. These could include increasing the extent of home ownership in an area, assisting neighborhoods in the development of local social ties, and similar initiatives. For example, assistance with implementing block organizations and neighborhood clean-up and beautification contests could help (Taylor et al., 1984). All of these may affect some of the same types of social processes that defensible space theorists hope to achieve through physical design changes.

Taylor and his associates (e.g., Taylor et al., 1980) believe defensible space research and application should draw more heavily on the concept of territoriality. They suggest that some critical environmental features for controlling crime are signs of defense, signs of appropriation, and signs of incivility (Hunter, 1978). *Signs of defense* are symbolic and real barriers directed toward strangers that keep unwanted outsiders away. *Signs of appropriation* are territorial markers suggesting that a space is used and cared for. *Signs of incivility* are physical and social cues (e.g., environmental deterioration) that indicate a decay in the social order. These territorial signs give information to other residents and to strangers, which in turn affects whether or not crime occurs. Taylor et al. believe that territorial signs that deter crime are more common in homogeneous neighborhoods and

PUBLIC HOUSING:
The Bad Example

Pruitt-Igoe was a public housing project built in inner-city St. Louis in 1954. In this project, 12,000 persons were relocated into 43 buildings 11 stories high, containing 2,762 apartments, and covering 57 acres. The buildings contained narrow hallways with no semiprivate areas where people could congregate—a design that was praised in *Architectural Forum* (April, 1951) for having no "wasted space." The project was expensive to build but very institutional in nature, containing such "features" as institutional wall tile (from which graffiti was easily removed), unattractive (but indestructible) light fixtures, and vandal-resistant radiators and elevators.

In spite of the construction expense, within a few years Pruitt-Igoe was a shambles. Take a walk with us through the project several years after it opened. First, there is a display of broken glass, tin cans, and abandoned cars covering the playgrounds and parking lots. Some of the building windows are broken; others have been boarded up with plywood. Inside, you smell the stench of urine and garbage. The elevator is in disrepair, and the presence of feces indicates it has been used as a toilet. Next, you notice that plumbing and electrical fixtures have been pulled out of apartment and hallway walls. When you come upon a resident and ask her about Pruitt-Igoe, she says she has no friends there; there is "nobody to help you." She also tells you that gangs have formed and that rape, vandalism, and robbery are common. Since crime frequently took place in elevators and stairwells, the upper floors have been abandoned.

These conditions destroyed Pruitt-Igoe. By 1970, 27 of the 43 buildings were vacant; and the project was totally demolished decades ago. Why did Pruitt-Igoe fail so miserably? One explanation was proposed by Yancey (1972), who centered his argument around the lack of semiprivate, sociopetal spaces or other facilities that could promote social interaction and the formation of a social order. Typically "slums" are made up of low-rise tenements, narrow streets, and lots of doorways to businesses in which to stop and talk. The design of many urban renewal projects is far less successful in providing for such social interaction. This architectural failure may impair cohesion among residents and promote conflict and crime.

Yancey also contended that the high-rise architectural design of the project was greatly to blame. It put children beyond their parents' sight and control whenever they were outside their apartment and gave them many hidden areas, such as stairwells and elevators, in which to cause mischief. Such areas also provided sanctuaries for teenagers and adults to engage in illicit activities almost anonymously. As one resident said, "All you have to do is knock out the lights on the landings above and below you. Then when someone comes . . . they stumble around and you can hear them in time to get out" (Yancey, 1972, p. 133).

---

in areas with strong local social ties. As opposed to Newman's model, then, these authors suggest that sociocultural variables and social conditions, in addition to design, determine territorial cognitions and behaviors and ultimately the level of crime in a neighborhood. This model has been tested and has received support (e.g., Newman & Franck, 1982; Taylor et al., 1980, 1984).

A fairly recent means of promoting territorial defense in cities is to barricade streets or erect fences and gates to keep nonresidents out (Carvalho, George, & Anthony, 1997; Crowe, 1991). A cul-de-sac design in a residential area serves a similar function: Traffic access is reduced so that residents may be more vigilant and criminals more conspicuous. By barricading strategic points in the neighborhood, residents hope to make cruising through the neighborhood by criminals more difficult. Barriers can be barrels filled with sand and linked by long boards, concrete construction barriers, or attractively designed brick walls with landscaping. Atlas and LeBlanc (1994) report on one study of Miami Shores, a community in Florida that erected barriers over a period of years. Compared with other communities, some crime rates were lower after installation of the barricades.

Barricades and gated communities are not without controversy; often, lower-income residents or those in mostly minority neighborhoods contend that barricades are designed to keep them out of well-to-do majority residential areas (Blakely & Snyder, 1995; Carvalho et al., 1997). On the other hand, not all gated communities are so exclusive. The Five Oaks neighborhood of Dayton, Ohio, is a modification of Newman's appro▓▓▓▓▓▓▓▓ gates were left ▓▓▓▓▓▓▓▓ order to create s▓▓▓▓▓▓▓▓ borhoods, streets and alleys were closed or blocked, and speed bumps were installed to slow traffic. Instead of one large residential neighborhood, Five Oaks was divided into smaller "mini-neighborhoods" by the modifications. At least according to preliminary results, traffic decreased by 67%, and crime was reduced by 26% since implementation of the modifications. According to the Department of Housing and Urban Development (HUD), the community is not closed off and only criminals are excluded. Apparently, the plan is being reviewed as a possible model for urban neighborhoods and public housing (*Newsweek*, July 11, 1994).

Part of the reduction in crime observed in gated communities may have been due not to the barriers themselves, but to the need for neighbors to work together to implement a plan (Atlas & LeBlanc, 1994). Thus, social factors may be as important as architectural effects, a topic we will explore next.

Social Ties and Community

"Sense of community," "place attachment," "neighborhood social networks." All three concepts address a sense of emotional connection between people and their communities (see Altman & Low, 1992; Mesch & Manor, 1998; Wilson & Baldassare, 1996). Whether the residential environment is a suburb or the urban core, neighborhood social networks could play a significant role in fostering the ability to cope with urban problems. The presence of social networks helps regulate access to an area by strangers, leads to less reliance on police for

dealing with disturbances, and reduces the need for a "get tough" policy by police when their help is required (Frug, 1999; Taylor, 1988). When social cohesion is absent, urban decay can get a strong foothold. Under these conditions, diffusion of responsibility effects (Latané & Darley, 1970), in which people assume it is "someone else's" responsibility to deal with social problems, may occur. In addition, deindividuation (Zimbardo, 1969), in which people feel "lost in the crowd," unrecognizable, and therefore not responsible for their antisocial behavior, may occur.

Viable social networks are most likely to occur in certain situations. "Neighboring" in urban environments is greater when there is ethnic similarity, shared socioeconomic status, psychological "investment" in a neighborhood, satisfaction with conditions there, and a positive sense of well-being (Unger & Wandersman, 1983). Social networks can also be fostered or inhibited through environmental means. Research by Newman (1972, 1975) and Newman and Franck (1982) has also shown that defensible space works to facilitate local social network formation. Designing small spaces where people can meet and socialize informally, particularly spaces with nicely landscaped greenery, seem very helpful in building social networks and community cohesion (Abu-Ghazzeh, 1998; Coley, Kuo, & Sullivan, 1997; Kweon, Sullivan, & Wiley, 1998; Skjaeveland & Gärling, 1997; Figure 10-10).

One way in which social networks can function effectively for the good of an urban area is to form local organizations (often called **block organizations**). These work for improvements such as better lighting, police protection, street repairs, or other common goals. Residents' participation in block organizations may be predicted by several factors: how important the block environment is to the individual, whether a person believes he or she could perform the behaviors necessary to participate, the perceived existence of common needs among residents, and how much a person generally participates in activities with other residents (Florin & Wandersman, 1990).

Figure 10-10 Providing spaces for informal social interaction promotes social ties and community cohesion. Spaces with natural greenery are particularly associated with these effects.

Low-Income Housing: Some Favorable Alternatives

What factors are associated with satisfaction by residents of low-income housing? A study by Rent and Rent (1978) surveyed residents from many housing projects in South Carolina. Those who lived in single-family or "duplex" dwellings liked their residences much more than others. This satisfaction probably occurred, in part, because these residences were more often owned, which is another predictor of satisfaction in low-income housing. For other reasons, too (e.g., greater privacy), such dwellings produce more satisfaction. Not surprisingly, then, 75% of those surveyed said they would prefer to live in a single-family dwelling, and 83% wanted to own one. Another important predictor of housing satisfaction was having friends in the neighborhood (often these turn out to be neighbors). Generally, the more satisfied one was with his or her neighbors, the greater the attraction to the living situation. An interesting finding was that overall life satisfaction was associated with liking one's residence. The happier one was with his or her life, the more satisfied one was with living arrangements. Overall, then, social as well as physical factors may be important determinants of housing satisfaction among low-income individuals.

Fortunately, some of the recent trends in government housing assistance have more elements associated with residential satisfaction than earlier project housing. When government assistance is provided, the U.S. government has more or less stopped building "high-rise" projects for low-income families. No "public housing" by name has been built for over a decade. Instead, people are placed more often in townhouses or small apartment buildings. There has also been increased government assistance with home ownership, direct housing subsidies for the poor (e.g., rent vouchers), and attempts to renovate or preserve current housing instead of demolishing it. This serves the admirable function of "fixing the building and leaving the people." Another recent innovation is **urban homesteading,** where abandoned urban property is given to individuals who agree to rehabilitate it to meet existing housing codes and occupy it for a prescribed period of time. It has sometimes been quite successful, but in other instances the practical problems of having low-income families with limited resources play the role of "general contractor" have been overwhelming. There has, unfortunately, been one "backlash" from earlier fiascoes with public housing such as Pruitt-Igoe (see the box on page 357): Some municipalities refuse altogether to have any form of it within their boundaries.

When people must be moved due to urban renewal types of projects, are there some means of accomplishing this in a more humane way? One possibility would be to move people to a new setting in established social groups. This would maintain the social cohesion of the former neighborhood (Young & Willmott, 1957). It could also be maintained, to some extent, by moving people to redeveloped areas near their old neighborhood. Another important factor is citizen participation in planning the move and the new setting in which they will live (e.g., Arnstein, 1969). Designers and planners should encourage participation and be especially sensitive to cultural or subcultural differences in housing preferences.

Gentrification

While urban renewal has had an effect on cities for many years, a more recent trend has been gentrification. Gentrification, as we described in the section on homelessness, can be defined as the emergence of middle- and upper-class areas in parts of the inner city that were formerly deteriorated (London, Lee, & Lipton, 1986). Frequently, this follows renovations to buildings that were once attractive and desirable but that have fallen into disrepair. After the renovation, wealthier tenants move in, and those who lived there before the renovation must find alternative housing. While gentrification is good for cities in many ways (e.g., it encourages "resettlement" by people with greater means, raises the tax base, and improves the environment), like urban renewal, it can be "bad news" for poor residents of the city. In addition, while gentrification and urban renewal continue to occur, we should keep in mind that a major threat to cities is still the disintegration and abandonment of urban housing (Henig, 1982).

Several conceptual perspectives have been proposed to explain the emerging trend of gentrification. Demographic explanations suggest that gentrification is due to population changes. For example, as increasing numbers of "baby boomers" reached adulthood in the 1970s and 1980s, they put demands on the housing supply. Other factors include the declining birth rates in Canada and the United States and the increasing number of women in the workforce. Affluent, childless, working couples are not discouraged by the poor reputation of inner-city schools and may want to live in the city, close to their jobs and recreational opportunities (London et al., 1986). In contrast, ecological approaches suggest that the ecology of the setting determines whether or not there will be gentrification (London et al., 1986). From this perspective, cities high in white-collar businesses, low in manufacturing, low in noxious land use, and that have long commuting distances should be most apt to experience gentrification (e.g., Lipton, 1977). A third approach, the sociocultural explanation, assumes that changing values, attitudes, and lifestyles are responsible for gen-

trification. Whereas the values of most North Americans may have formerly been antiurban (Allen, 1980), this may be changing. In fact, it may even be becoming "in vogue" to live in the city among some population subgroups (e.g., "yuppies").

Finally, political-economic explanations may take several forms. One implies that the decreasing availability of suburban land, rising transportation costs, inflation, the low cost of urban, inner-city dwellings, and antidiscrimination and school desegregation laws are all conspiring to encourage gentrification. A similar perspective suggests that economic interests and political factors are responsible for gentrification, and that, in some sense, it has been willfully planned. The most cynical view is that powerful interest groups allow the city to deteriorate, mindful that gentrification could later yield major profits. They pursue gentrification for their own benefit, with little regard for individuals who would be displaced by it (London et al., 1986).

Do only the wealthy benefit from gentrification? While at first it may appear so, a closer analysis suggests that this may not be entirely correct. It has been found that while the owners of gentrified housing are "urban gentry" (e.g., young, highly educated professionals), the "renters" of such housing typically have much lower incomes and pay a large proportion of them for rent. Thus, two types of people are moving into gentrified areas (DeGiovanni & Paulson, 1984).

There are various "costs" of gentrification. It has been found that gentrification often results in an increase in violent crimes (Taylor & Covington, 1988), as well as an increase in larceny and robbery (Covington & Taylor, 1989). This may occur, in part, from the close juxtaposition of the "haves" to the "have nots" in gentrifying areas. One of the major urban trends of the past two decades, in fact, is an increasing gap between the very poor households and the other households in urban areas. This is the source of many current urban problems (e.g., violence and crime) and will probably play an even greater role in the future.

Another negative aspect of gentrification is that the poor who originally lived in the "slums" are often pressured to move out. Because they have few political advocates and little power, they are in a difficult situation. Henig (1982) argues that collective mobilization or other forms of protection for the victims of gentrification (e.g., those whose rents or property taxes are raised, or who are pressured or forced to relocate) is even more important than for victims of urban renewal. Much of the responsibility for dealing with those displaced is left to state and local officials. When these officials weigh their concerns over displacement of existing residents with their desire for an increased tax base, they often find it difficult to support the former (Henig, 1982). In addition, the private sector is resistant to policies to limit displacement due to gentrification. Not surprisingly, as with urban renewal, it has been found that being forced to relocate due to gentrification is associated with threats to health and well-being (Myers, 1978).

REVITALIZING COMMERCIAL AND BUSINESS DISTRICTS

North American cities are mere infants compared with many of their Asian and European counterparts, and North American cities (with the possible exception of Mexican archeological sites) have no opportunity to show the rich history of Paris or London, let alone the ancient cities of Athens or Rome. Moreover, until recently North Americans have shown a disregard for preserving the variety of different architectural styles, materials, and building sizes that once marked the core of cities (Day, 1992; Gratz, 1989; Kostof, 1987). In a study conducted in a small city in The Netherlands, Oppewal and Timmermans (1999) found that the attractiveness of city shopping districts depended on maintenance of streets and storefronts, public pedestrian space, greenery, the presence of coffee shops, and cafes. People preferred a moderate level of occupancy, disliking either very crowded or uncrowded public spaces. As Kostof (1987) complains, urban renewal often resulted in the indiscriminate destruction of the rich texture of old neighborhoods, replacing them with corporate towers. One part of what was lost was formal history—the architecture associated

with important people and events. A different, but perhaps related loss is the sense of place created not by age per se, but by the sense of meaning created by architecture (see also Chapters 2 and 12 for other discussions of place). One current goal of many downtown revitalization projects is not to erase the past, but to preserve or even recapture it. For example, many cities have "rediscovered" their old farmer's markets (Sommer, 1989). Such *placemaking* may not be a simple task, because it relies not on a building's power to determine behavior, but on its ability to cue the memories that create the personal meaning and experience of place (Day, 1992). As we shall see, the results of these efforts are at least as varied as their goals, and projects that are successful on some counts may exacerbate other urban problems.

Festival Marketplaces

The proliferation of so-called **festival marketplaces** provides an example of both success and the potential dangers of urban revitalization projects. In the late 1960s a Boston architect, Benjamin Thompson, proposed a plan for restoring an historic area of downtown Boston by creating a new retail marketplace. The proposed site between the waterfront and the Boston financial district was three old market buildings adjacent to Faneuil Hall, a Colonial meetinghouse (Gratz, 1989). Thompson's plan for a revitalized marketplace ran counter to the conventional wisdom that downtown retail districts were destined to fail at the hands of suburban shopping malls. Thompson's plan called for a new retail emphasis to turn the old market into something not unlike a shopping mall, but with a sense of history. Funding was difficult, even when the project attracted the successful developer James Rouse. But funds were obtained, and the Faneuil Hall-Quincy Market district became the prototype for dozens of other festival marketplaces (see Figure 10-11). These marketplaces are busy retail and tourist centers that combine retail space, leisure, and a bit of theater. The so-called *Rouseification* of downtown Boston, Baltimore, and other North American cities invokes what Hall (1988) calls the "city-as-stage."

Figure 10-11 The festival marketplace in Baltimore's Inner Harbor.

That is, the new city center, like theater, presents a wholesome, sanitized, and not quite real vision of urban life. Part of the mix includes a bow to history. In Boston the elements of history were Faneuil Hall, which was treated as an historic landmark, and the three old granite buildings of Quincy Market, which were renovated to serve as homes for retail outlets and food vendors. This rehabilitation and recycling of old structures for new uses is now known as *adaptive reuse* (Hall, 1988). In Baltimore's Inner Harbor, on the other hand, the old wharf and warehouse district was removed altogether and replaced by new retail pavilions, plazas, and a museum. Some connection to the historic waterfront was maintained (or recovered) by acquiring the three-masted frigate the SS *Constellation* and mooring it permanently near the pavilions (Kostof, 1987).

How can designers marry the architecture of the past with modern construction techniques and requirements? Much of the effect can be created by visually extending old facades through new edifices and by maintaining compatible size and mass (e.g., Hedman & Jaszewski, 1984).

Such projects require massive funding. Baltimore's Inner Harbor area, for instance, attracted $180 million in federal funds, $58 million from the city, and only $28 million in private funds (Hall, 1988). At their best, such downtown attractions revitalize decaying older industrial cities. The Inner Harbor area of Baltimore, for instance, attracts 22 million visitors a year (Hall, 1988). In Boston, the market area and a similar revitalization of the waterfront are part of a long pedestrian corridor through the heart of old Boston that stretches from the harbor, through the market, to the Boston Common and Public Garden area, and beyond to the gentrified brownstones of Commonwealth Avenue and Back Bay (this cohesive corridor is partly a result of Kevin Lynch's cognitive mapping studies reviewed in Chapter 3).

Despite these apparent successes, festival marketplaces and other so-called postmodern projects are not without critics. As we said, they preserve not history, but an idealized presentation of urban life as it never was. At their worst, such developments (or their poor imitations) make little attempt to preserve real history, but instead treat historic structures as commodities and cater more to collective nostalgia than to a genuine understanding of the historic landscape (Kostof, 1987; Roberts & Schein, 1993). Their apparent spontaneous mix of sights and smells are, in fact, anything but accidental, and to some they represent a publicly subsidized trendy superficiality. At their best, however, they may be delightful celebrations of the city.

Design Review

Both historic preservation and large-scale downtown revitalization projects operate on a massive scale. Most buildings are not part of such massive projects, of course. Unfortunately, when individual developers, landlords, and architects attempt to construct a new building, there may be a conflict between their individual goals or taste and those of the surrounding district (Devlin & Nasar, 1989; Nasar, 1994). Other projects are smaller still (a plan to refurbish an aging business facade, for instance) and may not even require the owner to hire an architect or other design professional. To ensure that building appearance is not discordant with the good of the community, cities may adopt **design review,** a case-by-case examination of proposed new projects that attempts to ensure that they will remain harmonious with both the architecture and the ongoing social fabric of a district. New York City's Times Square may provide examples of both the good of design review and the bad that can result from lack of oversight. When Times Square became the target of revitalization plans in the middle 1980s, some observers reacted with mixed emotions. On the one hand, the plans signaled a return to health for the decaying theater district. On the other hand, some feared that huge new buildings would visually overwhelm the more modest historic district. According to Hiss (1990), an environmental simulation demonstrated the need for buildings to

be set back at about the 6th floor to allow continued use of the traditional gaudy signs, and the New York City Planning Commission adopted both this suggestion and zoning regulations requiring developers to set aside space for businesses such as costume shops that serve the theater industry. Unfortunately (according to Hiss), the planning commission did not also require a second setback at approximately the 12th floor to maintain historic levels of sunlight and sky.

Pedestrian Environments: Shopping Malls and Pedestrian Plazas

Pedestrian movement through a city has characteristic patterns. People select paths and avoid obstacles in regular fashion. By understanding the development of these patterns, we can obtain important information about the design of several kinds of settings (e.g., Whyte, 1980). For example, designers sometimes construct pedestrian malls to eliminate some of the traffic congestion, noise, and clutter of an urban area. To the extent that we can make pedestrian movement easy and convenient, we also reduce the pollution associated with motorized travel.

Some shopping malls attempt to become "indoor cities" that combine recreational and community functions with the more obvious retail-oriented activities. Often the design parallels the urban tradition with broad avenues, ersatz streetscapes, and plaza fountains (Uzzell, 1995). The Mall of America near Minneapolis envelopes an entire indoor amusement park, and the West Edmonton Mall in Alberta includes a water park, miniature golf course, ice rink, and hotel. Large suburban malls act as magnets, drawing people from large regions. Unfortunately, the effect of large suburban malls may be to promote the economic decline and decay of traditional downtowns. Eventually downtowns may be stripped of retail businesses and restaurants, leaving behind only office towers and streets that empty at the close of business. Because malls are privately owned, and because their goals are ultimately commercial, Uzzel suggests that they will never reproduce the genuine places of public business districts. Perhaps the sterility of the suburban mall has created a thirst for place and urban history.

For several decades, researchers have systematically studied pedestrian behavior in plazas, shopping malls, and similar areas, and in doing so have described a number of patterns that seem to apply (e.g., Low, 1997; Oppewal & Timmermans, 1999; Preiser, 1972, 1973). One basic rule of thumb is that people choose simple, direct routes, whether formalized as landscaped paths or freely chosen, as in walking across lawns. The well-worn paths across lawns on most college campuses, despite the presence of nearby sidewalks, illustrate this principle. Another observation about movement patterns regards the speed with which people walk. Generally, people conform their speed to that of people around them. Larger crowds appear to move more slowly, and people walk slower on carpeting than on bare floors. Moreover, pedestrians match their speed somewhat to the pace of background music (cf. North & Hargreaves, 1999).

Traffic and large crowds can inhibit our movement and change our patterns. The presence of people in one's way leads to frequent changes in speed and deviations from the most direct route one can take. For example, people will walk around a small group of persons who are standing and talking rather than following a direct route between or through a group (Cheyne & Efran, 1972; Knowles et al., 1976). Preiser (1973) has incorporated all of these influences on pedestrian movement into a *friction-conformity model*. That is, "frictions" such as those mentioned above impede pedestrian flow, and conformity pressures (e.g., the speed of others) exert additional influence on movement.

Knowles and Bassett (1976) considered social cues that people use in deciding whether to stop or move in crowded settings. Their perception of whether a group is an interacting entity or a casual gathering of strangers appears to be important in determining behavior. When pedestrians encountered a group of people talking to one another, they moved on. When they encountered a casual group of people who were

simply standing and looking up in the air, they were more likely to stop and join in the gazing.

Sometimes we like to watch others pass by us. "People watching" occurs when people seek out benches or seats where they can watch others. As we noted previously, designing common areas for people to sit and socialize helps build community attachment and cohesion, and one function of these common areas is to allow people watching (e.g., Coley et al., 1997; Kweon et al., 1998). Similarly, adolescents often use an area shopping mall as a "hangout," sitting for hours watching people pass by, looking for friends, and visiting with those they find.

One innovative development in cities is the use of *pedestrian malls* to enhance city life. A street in a commercial downtown area is blocked off and turned into a plaza for pedestrians—no automobiles are allowed. The idea is to reduce traffic congestion, beautify the area, and encourage commerce in previously deteriorating areas. Whyte (e.g., 1980) has observed that pedestrian plaza areas can indeed liven up the environment. Food vendors, sunny areas, places to sit, and fountains promote the habitability of such spaces. Amato (1981) even observed that people were more likely to help another in need along an area converted to a pedestrian mall compared with when it was a "normal" street (Figure 10-12).

Figure 10-12 Pedestrian malls such as this one block off motorized traffic to increase commerce and social opportunities. However, these areas can also involve elements that are opposite to the desired effects. Another variation of a pedestrian mall can be seen in Figure 3-1 on page 58.

DESIGN RESEARCH FOR PEDESTRIAN WIND DISCOMFORT

No matter how efficient public transportation becomes, some pedestrian movement will be needed to get people to their final destination. Modern cities usually have heavy concentrations of pedestrian movement around business areas that consist of numerous high-rise buildings or skyscrapers. These buildings make it possible to locate many activities, such as work, shopping, entertainment, and living quarters, in a relatively small geographical area. However, for decades there has been evidence that concentrated areas of high-rise buildings can alter ground level climate and pollution conditions because of the effects of building design on wind patterns (Cermak, Davenport, Plate, & Viegas, 1995).

Engineers are now able to test the wind effects of proposed building designs in elaborate simulations that make use of wind tunnels (e.g., Cermak et al., 1995). Such tests involve fitting models of proposed and existing buildings with pressure-sensitive recording devices. When the models are subjected to simulated wind levels typical of that city, researchers can examine structural stress effects as well as possible wind problems for pedestrians. Two common problems occur if a smokestack on a tall building is too short, or if a tall building is located too close upwind from a short one. Resulting wind patterns can force pollutants toward the ground and trap them there, causing a variety of discomforts for pedestrians. Another problem arises when high-speed winds 30 or more feet (9 m) off the ground strike a tall building. Typically, these winds are forced straight down. If the building has an open passageway at ground level, the winds rush through it, causing a wind tunnel effect. Pedestrians, especially those carrying opened umbrellas, may be literally sucked through the passageway. Design alternatives that include wind deflectors are one means of solving this problem (Figure 10-13A & B).

Wind tunnel simulations enable researchers to test potential wind effects on entire city blocks. Design solutions to anticipated problems can also be tested. In one case, tests revealed that high winds in a plaza could be avoided by erecting partial walls at the entrance. The beauty of this type of research is that design alternatives can be tested before construction commitments are made.

Figure 10-13A & B Even though this is an attractive building complete with waterfall on the face of the left tower, the design amplifies the wind speed for pedestrians who need to walk through the passageway. Wind deflectors such as the V-shaped projection on the next building can reduce the wind tunnel effect and increase comfort for pedestrians.

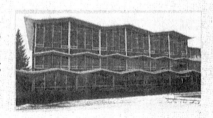

On the other hand, pedestrian malls can be a disappointment. Grossman (1987) summarized some of the problems that can occur. A mall in Eugene, Oregon

. . . became a wasteland. Pedestrians stayed away, partly out of fear that the mall's many trees and fountains were hiding muggers. Motorists skirted the area, confused by the reconfigured street routes. With sales much slower than expected, merchants departed and storefronts were vacant. Downtown Eugene began to rebound two years ago when one block of the mall was reopened to traffic. Now the city is considering reopening two more of the mall's original eight blocks to vehicles. If it does,

Eugene will have spent more than twice the mall's original $1 million development cost on revamping the mall (p. 27).

Grossman observed that Galveston, Burbank, Minneapolis, Grand Rapids, Chicago, and Little Rock faced similar decisions with pedestrian malls. What went wrong? For one thing, shoppers accustomed to suburban shopping malls did not have the convenience of free park-ing, a variety of stores, and short drives to shop in the evening. Derelicts, delinquents, drug dealers, and "boom boxes" could find their way to the malls and scare away customers. What can be done to change the situation? Solutions include allowing some motorized transportation along the malls, building hotels there to provide a ready shopping population, removing obstructions where muggers could hide, and sponsoring festivals along the mall to attract crowds.

Escaping to the Suburbs

The great American dream appears to be to leave the city. In addition to crime, urbanites cite overcrowding, pollution, housing, traffic congestion, and noise as major reasons for leaving. Where would people rather live? A 1985 survey asked a representative American sample, "If you could live wherever you wanted, would you prefer a large city, a suburban area, a small town, or a rural area?" The results were: city, 9%; suburbs, 29%; small town, 37%; rural area, 25% (ABC News/Washington Post Survey, February 22, 1985). However, it does not appear that the city is being totally rejected: Many of those expressing a preference for suburban or rural areas still wanted to be near a medium-size or large city (Figure 10-14).

This explains the massive move to suburbia, but what exactly are suburbs? They are areas within a metropolis that are relatively distant from the historic city center. Suburban living has increased dramatically, especially since World War II, and at present, more Americans live in the suburbs than in the center city or non-metropolitan areas (Garreau, 1991). Why is this happening? Quite simply, because suburban living offers an answer to a number of urban problems. As one moves farther from the city, he or she is subjected to fewer crowds and to less dirt, noise, and pollution. In addition, although sub-urban crime rates are increasing, they are still much lower than in the city.

According to Garreau, suburbanization occurred in three phases. In the first phase, which began shortly after World War II, affluent city dwellers moved to new residential developments in the suburbs, and governments built highways to allow them to travel to and from the central business district.

In the 1960s, the second phase of suburban growth saw malls and other commercial establishments following the outflow of people. Finally, offices and corporate headquarters moved to the newly established *edge cities* that now ring the old urban core. Edge cities account not only for most metropolitan areas' population, but for most of their office and commercial space as well. Unlike the old city core, however, edge cities are largely the product of individual developers who try to make their particular developments spacious, attractive, and, of course, profitable. The results are mixed. Internally, each development can be well organized, but the fast pace of building and the lack of a central political authority has also meant that adjacent neighborhoods, streets, and commercial zones are often not well integrated with one another (Southworth & Owens, 1993). The edge city consumes far more land for each unit of activity

Figure 10-14 Escaping to the suburbs seems to be preferred by a majority of North Americans. Many people preferring suburban or rural areas still want to be near a city.

than did the old city core, and its vastness commits us to the automobile. Still, the edge city seems destined to remain the home of most North Americans well into the 21st century.

What are the individual consequences of the move to suburbia? There is some evidence that the move has a positive effect. Suburbanites are generally happier with their housing, their communities, and their lives than city dwellers, even when socioeconomic status and other differences between urban and rural populations are statistically controlled (Fischer, 1973; Marans & Rodgers, 1975). Also, people who move to the suburbs are much less afraid of crime victimization and restrict their behavior less due to fear of crime (Lavrakas, 1982; Skogan & Maxfield, 1981). However, not all is well in suburbia. The price of typical suburban houses is rising tremendously, and it appears that fewer and fewer people will be able to afford or to maintain a suburban lifestyle in the future. Further, as more and more people escape the city for the

suburbs, crowding, pollution, and other urban problems are becoming suburban problems. As noted earlier, crime in the suburbs is increasing, and the use of drugs in suburban schools is cause for great concern. All this leaves one wondering whether the suburban areas of today will be characterized by a full complement of "urban" problems in the future. But perhaps the worst problem created by the move to suburbia has its roots back in the cities. Cities are experiencing decreased populations, populations that are poorer, and populations that are more minority dominated than ever before. There is a declining tax base and an increasing demand for city services (e.g., police protection). This has occurred at a time when federal support has decreased. Unfortunately, the trend toward abandoning urban areas and moving to suburbia has jeopardized all that the city has to offer (Flynn, 1995; Garreau, 1991). We will have more to say about urban–suburban living in Chapter 12.

STARTING FROM SCRATCH:
Urban Utopias

As we have seen, cities are amalgams of opportunity and despair, poverty and conspicuous wealth, fond attachments and fear. They seem to have evolved (or mutated) as a series of localized events embedded in a whole that is largely beyond the control of any rational process. But what if someone could build an entire city to take advantage of all of the excitement and variety that can be part of urban life while avoiding urban ills? Urban planners and architects have long been intrigued by utopian possibilities. Several quite different visions of the city of the future offer contrasts in implementing some of the design interventions we have covered so far. Our discussion is largely informed by Fishman's excellent review, *Urban Utopias in the Twentieth Century* (1994).

Perhaps the most famous U.S. architect, Frank Lloyd Wright, was determinedly individualistic himself and designed a city to maximize its inhabitants' opportunity for individuality. Wright wished to eliminate the distinction between rural and urban. His Broadacre City was huge in scale and the extreme of decentralization. He proposed a patchwork of individual homesteads connected and organized by an efficient network of roadways. Families could be somewhat self-sufficient on their homestead farms, and daily life would center on the home. Nevertheless, the transportation network would maintain coherence. Festive recreational and cultural facilities and a "roadside market" would always be a short drive away to provide the communal antidote to isolation.

Le Corbusier lies at the opposite land-use extreme. Whereas Wright thought existing cities were far too dense, Le Corbusier thought they were not dense enough. His Radiant City glorifies the centralized metropolis of glass and steel. Residences would be high-rise apartment blocks, and the business center would feature great skyscrapers. All of this would be tightly organized by a strong central planning hierarchy, because individuals could not be trusted to harness the challenges of greed and uncontrolled growth. Le Corbusier's desire to concentrate the city resembles the plans of another urban visionary, Paolo Soleri, who has begun construction of a demonstration town.

Repulsed by urban sprawl, Soleri (1970) drew huge futuristic megastructures that could contain entire cities of hundreds of thousands of residents. Instead of low-density housing spread across vast acreage,

SUMMARY

The city is a salient environmental element in almost everyone's life. How does the urban setting affect individuals who live in it? A number of conceptual formulations have been derived to understand and predict the effects of the city on individuals; these include overload, environmental stress, behavior constraint, and overstaffing notions. Although they are often presented as competing concepts, it is probably true that overload, stress, constrained behavior, and insufficient resources each explains some of the consequences of an urban existence. Further, the predictions of each of the models can

be integrated into the general environment–behavior formulation presented in Chapter 4. While this model has not been tested explicitly in research on cities, many of its assertions have been supported.

What are the results of experiments on the effects of city life? Two methodological perspectives (the "single variable" approach and the "urban versus rural" approach) have been used in past research. Each has its strengths and weaknesses. Overall, such urban stressors as noise, pollution, heat, crowding, and "extra demand" have at least moderately detrimental effects on

Soleri built upward, minimizing the city's "footprint." Using a living organism as a metaphor, Soleri emphasizes that living systems are actually tightly packed collections of organized and interdependent systems. Some recoil at the apartment-like density. Soleri argues that high density is an asset rather than a liability when it is coherent and recognized as a facilitator of connectedness. Soleri wishes to capitalize on the "urban effect" that occurs when the many amenities and opportunities of city life are organized and miniaturized (1993).

In 1970 Soleri began construction on Arcosanti, an urban laboratory at the edge of a mesa north of Phoenix (Wilson, 1999). Built primarily of poured concrete by mostly untrained workshop participants and residents, the project is funded by tuition and the sale of ceramic and bronze wind-bells. Perhaps the most distinctive design features of Arcosanti are the large domed apses (1/4 spheres) and vaults that capture the warmth of the low winter sun but provide shade in the heat of the summer. Roads and cars are unnecessary because homes, workplaces, and public facilities are gathered so tightly by Soleri's notions of miniaturization. The roof of one person's dwelling may be a public plaza, or a window may look down into a ceramic studio. Some produce is already grown and sold, but a series of roof-top water collection channels and cisterns will one day connect with more massive gardens spilling down the side of the mesa.

Finally, Ebenezer Howard's Garden Cities were based on the assumption that humans profit from contact with both the natural countryside and one another. His were small cities, limited to about 30,000 persons. Each city was to be roughly circular and surrounded by rural green space or an agricultural belt that could help supply the city with food. At the center was a large central park, surrounded by civic buildings and a shopping arcade. These were pedestrian cities where radial avenues would spread out to connect neighborhoods both to the commercial and recreational core and to industries, which were located on the city's periphery. Howard designed to facilitate cooperation and community. Howard would find American suburbs too uncontrolled and too sprawling, but his emphasis on moderate density, greenbelts, and a balancing of individuality and community has inspired a number of developments.

city dwellers; the effects of homelessness and crime are much more severe. Further, when cities and nonurban areas are compared, there are urban–rural differences in terms of affiliative behavior, prosocial behavior, crime, stress, coping behavior, and long-term aftereffects. On most of these dimensions, urbanites come out on the short end. However, on dimensions not often studied by researchers (e.g., ability to adapt to diverse situations), urbanites may come out ahead.

Finally, a number of solutions have been tried to alleviate urban problems. One major attempt has been urban renewal. Unfortunately, this has often involved a conflict of interest between slum dwellers and city planners, with the former being forced to relocate. Forced relocation into public housing sometimes has disastrous consequences, which might be ameliorated by proper design of public housing. Gentrification is a more recent trend. Individuals return to the city and renovate housing that was formerly in bad condition. While this improves the urban area, it again causes relocation of original residents. The dream of most urbanites, however, is suburbia. This is attainable only for those whose socioeconomic level permits it. Research on suburban living shows that it offers a solution to some of the negative aspects of the city for those who can make the move.

KEY TERMS

block organizations
defensible space
deindividuation
design review

diffusion of
    responsibility
familiar stranger
festival marketplaces

gentrification
homelessness
prosocial behavior
urban homesteading

urban renewal
urban villages

SUGGESTED PROJECTS

1. One assumption we have made is that cities differ from small towns on a number of dimensions. To test this hypothesis, first consult copies of a few newspapers from large cities and small towns. Compare the following sections: entertainment, sports, and reports of local crime. Next, locate some telephone directories from large cities and small towns. Compare listings for the following: medical specialists, tradespeople, specialized restaurants of diverse nationalities, museums, religious institutions, educational facilities, and theaters. What pattern of urban–rural differences emerges on these various dimensions?

2. Think of three cities you have visited, and attempt to rate them in terms of "atmosphere." Consider factors such as legibility, amenities, pace, and architecture. Which of these or other physical or social aspects most affect your assessments?

3. What are your views of urban life? Has our assessment led you to become more positive or more negative toward cities than before? Write down your views and compare them with those of classmates.

4. Discuss this project with your instructor before you begin because many states have very strict rules regarding observation of children. Thus prepared, get permission from school officials to visit and observe a school playground

(they may be eager to learn from you!). Would you classify the playground as traditional or contemporary? How do children use the playground? For instance, what playground equipment is most popular? Do children congregate in enclosed spaces? What are the functions of loose parts?

5. Try to replicate the studies Newman and McCauley conducted on reciprocation of eye contact (which signals accessibility for interactions). In a small town and then in a city, position yourself near a doorway. When people passing by are a few feet away, initiate eye contact. Record the number of reciprocal gestures you receive in both settings. Do your results replicate those of Newman and McCauley?

6. Broadly outline the planning principles you would employ for an urban utopia. Is yours a dense city like Arcosanti or spread out like one of Howard's garden cities?

7. Create a short list of questions to ask of at least three friends from urban backgrounds and three from small towns or rural areas. Ask standardized questions to determine the size of town or city in which they would most like to live. Also assess their view of the relative merits of small towns and cities. List those factors that were most often volunteered by your friends. Are there differences between those from urban or rural homes?

CHAPTER 11

PLANNING AND DESIGN FOR HUMAN BEHAVIOR

Introduction

It is your first weekend visit to a friend at another campus. You park your car in a lot marked "Visitors." Beside the lot stands a stately stone building. Green ivy twines up the old columns, and a worn brick walkway leads to a sign bearing a campus map. Oriented now, you make your way across a lawn searching for your friend's residence hall. And there it stands, a beautiful old stone building with grand wooden doors. What a classy school; no wonder people want to come here!

Two days later you zip your toothbrush into your bag and sit on the foot of the bed mentally replaying episodes of the weekend visit. Your friend returns from an errand and asks, "Do you want to see my favorite place before you go?" Of course. Together you walk past the new poured-concrete library. "It is big, impressive, and somehow out of place," you remark. You continue past the attractive building housing the most inefficient cafeteria you've ever seen. "Who designed this place?" your friend wondered at lunch. Ahead is the Arts Center, locally as famous for its stairways that end nowhere as the studios it houses. Cutting across a lawn on a well-worn path that short-cuts the sidewalk, you arrive at a battered old park bench underneath a bare maple. "Here we are" says your friend, and seeing a question in your eyes adds "I watch the squirrels." Your friend continues, "On a warm fall day when the sun sets the leaves look like they are burning. I sit here and watch the squirrels. Sometimes I study chemistry. Sometimes I daydream. And sometimes I write to you. I don't know why really, but the place is special." And so it is.

In Chapter 1 we defined environmental psychology as "the study of the molar relationships between behavior and experience and the built and natural environments." Although much of this text is devoted to establishing an understanding of how humans are affected by different environmental experiences, our definition also recognizes that humans are part of the environment and that our behaviors often change it. Indeed, as we explained in Chapter 2, the ability to manipulate or temper the environment is one characteristic of our species—a characteristic that has allowed us to inhabit most terrestrial environments and even outer space.

Critics may properly argue that it is our species' conceit to assume that we can predictably construct a more efficient and comfortable world by relying on technology. Regrettably, many of the environmental problems discussed in Chapter 14 result, in part, from this hubris. Nevertheless, modern humans live, work, and often play within human-modified landscapes and buildings. Good design of these constructed environments will foster better comfort, safety, and productivity, whereas bad design may foster feelings of powerlessness or stress (Evans & McCoy, 1998). (See Figure 11-1.)

Perhaps the case for a marriage between design and the behavioral sciences is an easy one. Designers know how to manipulate spaces and wish to use these skills to change the affect or behavior of the people who use the environments they create. Psychologists and other behavioral scientists study behavior. Voila! Unfortunately, the "marriage" between design and behavioral science has not always been productive (e.g. Philip, 1996). As Philip points out, there is more than one manifestation of psychology with which designers might partner. Philip terms one psychological suitor "hard psychology" and attributes it to the behavioristic or scientific tradition. Philip also identifies another kind of psychology. He labels it "humanistic," but to humanism we might add other influential but nonempirical theoretical positions such as Freudian or Jungian psychologies. The authors of this text are most comfortable with the scientific understanding of psychology, but a visit to the "psychology/self-help" section of any bookstore should demonstrate the proliferation of the other construction.

How can environmental psychology contribute to better planning and design? Clues can be found throughout this text. For example, Chapter 3 suggests ways in which we may create more legible environments that facilitate wayfinding. Similarly, Chapter 8 examines human territorial needs and offers a theoretical framework for understanding important variables that can maximize spatial or social comfort. Other chapters document the effects of ambient stressors such as noise, foul air, and crowding. The contribution of psychology to design remains modest, however, but the future of the collaboration is promising.

Architectural training is quite different from the schooling of psychologists (Sancar & Eyikan, 1998). Architects and other design professionals face many challenges, only some of which lie in the domain of behavioral science. An architect is expected to have both the sensi-

bilities of an artist and the utilitarian skills of an engineer. Most psychologists do not presume that their expertise extends to these domains. The professional nexus between behavioral science and design, of course, centers on human behavior. In fact, it might be more useful to acknowledge not one focus, but two behavioral foci. Perhaps the most obvious is the body of scientific evidence regarding the effect of environmental features, such as the size, shape, and amenities of a room, on its inhabitants. In this sense, behavior is the result of design (although we will critique the extent of this conclusion). Less evident, perhaps, but potentially as valuable is the utility of psychological knowledge in facilitating the decidedly social and often political processes of problem solving and decision making that characterize design. ▨

The Physical Environment as a Behavior Setting: Extent of Design Influence

An assumption we have made in this textbook is that the environment is an important contributor to behavior. Similarly, an underlying assumption of design is that we have at least moderate influence over environmental variables. Nevertheless, researchers, as well as designers, have had difficulty agreeing on the extent of architecture's influence on behavior.

ARCHITECTURAL DETERMINISM

One of the early conceptualizations of the architecture–behavior relationship was architectural *determinism*. The general concept of environmental determinism receives more complete treatment in Chapter 6, but in this context, architectural determinism holds that the built environment directly shapes the behavior of the people within it. In its most extreme form, the physical environment is seen as the only—or at least the primary—cause of behavior. It has become clear, however, that such a view is too simplistic to adequately account for the effects of

design. Franck (1984) criticizes this extreme determinism on several counts. First, this view exaggerates the importance of the physical environment by underestimating the importance of social and cultural factors. For example, refer to almost any building shell as a "cathedral" and the behavior of those inside is likely to be muted. Conversely, the behavior that occurs in the same structure's "fellowship hall" can become quite raucous on other occasions. Second, determinism overlooks the importance of indirect environmental effects and interactions among several environmental variables acting in combination. Although variables can be independently manipulated in laboratory experiments, such control and isolation is rare in real-world settings. Finally, determinism ignores the fact that people engage in transactions with the environment—that is, they are not passive, but influence and change the environment as it influences and changes them. Figure 11-2 illustrates a geology laboratory on a college campus. Although the "flume room" was certainly not

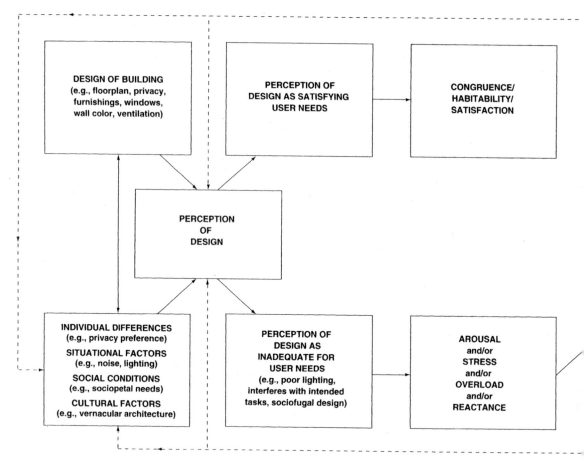

Figure 11-1   Our eclectic model applied to design.

Figure 11-2   Behavioral traces reveal that this laboratory has been adapted and adopted by students.

designed as a social space, over the years students have gradually appropriated space and modified the laboratory to make it a popular "hangout."

Roger Barker's approach to behavior settings outlined in Chapter 4 provides a useful framework for critiquing determinism. Barker was one of the first and most articulate voices calling for psychology to shift its focus to include the context of behavior rather than to eliminate it. In an important assertion, Barker and his colleague Herbert Wright concluded that the immediate environments of children were more important in determining their behavior than were individual characteristics of their personalities. That is, two children in the same place behaved more consistently with each other than did one child in two different places (Barker &

```
                    ┌─────────────────┐        ┌─────────────────────┐
                    │                 │        │ POSSIBLE AFTEREFFECTS│
                    │   ADAPTATION    │        │      and/or         │
                    │    and/or       │───────▶│  CUMULATIVE EFFECTS │
                    │  ADJUSTMENT     │        │                     │
                    │                 │        │ (e.g., user feedback to│
                    │                 │        │ improve design process)│
                    └─────────────────┘        └─────────────────────┘
                  If Coping
                  Successful
┌──────────────┐
│              │
│   COPING     │
│              │
└──────────────┘
                  If Coping
                  Unsuccessful
                    ┌─────────────────┐        ┌─────────────────────┐
                    │ CONTINUED AROUSAL│       │ POSSIBLE AFTEREFFECTS│
                    │    and/or        │       │      and/or         │
                    │    STRESS;       │──────▶│  CUMULATIVE EFFECTS │
                    │ POSSIBLE INTENSIFIED│    │                     │
                    │ STRESS DUE TO    │       │ (e.g., conflicts between users,│
                    │ INABILITY TO COPE│       │ reduced productivity,│
                    │                  │       │ increased fatigue, burnout)│
                    └─────────────────┘        └─────────────────────┘
```

Wright, 1951). Casually we can observe that all but the shyest child is spirited on a playground, and all but the most irreverent child is relatively quiet in church. Nevertheless, Chapter 4 makes clear that Barker realized that the physical environment is just one part of a behavior setting. He emphasized that a behavior setting results from the interdependency or transactions between the physical milieu and the standing patterns of behavior (sometimes called a *program*) that typically occur in that setting. Thus, a design can support or facilitate certain behaviors, but it will not reliably determine them. For example, virtually every college campus includes locations that seem to function independently of, or even in spite of, the goals and assumptions of their design. As with the "flume room" you can probably name a plaza or lounge that is seldom occupied, or an "ugly" unplanned space that has become a social focal point.

The Designer's Perspective

For our discussion of architecture practice our framework will be Lang's (1987) outline of the relationship between behavioral science and design. Our primary focus will be on the potential to improve the process by which designers gather information and make design choices. As you will notice, some of the literature concerning this collaboration consists of examples or recommendations rather than empirical conclusions. Nevertheless, we hope to outline some of the issues as they are perceived by designers, particularly those designers who are most eager to embrace environmental psychology.

According to Lang (1987, 1988) there is general agreement among architects that buildings and other designed environments must fulfill three basic purposes: commodity, firmness, and delight. **Commodity** refers to the functional goal of a design (What is the building to be used for? Does it facilitate human performance?), **firmness** refers to the structural integrity or permanence (will it last?), and **delight** encompasses aesthetic concerns. Differ-

ent architects may place different emphasis on these interrelated dimensions, but in each case the designer must draw on his or her professional expertise in facing the challenges of a specific design. An architect must practice in a broad domain that includes both engineering and art. Of course, this need not be a lonely task, and there are a number of reasons why it may be wise for designers to collaborate, both with other professionals and with the eventual occupants of a design.

Many authorities describe design as a problem-solving process. In seeking solutions, designers must draw upon an accumulation of organized data and ideas that are loosely organized as a system of theories or models (Figure 11-3). Although this is a casual use of the term *theory*, it is consistent with our definition in Chapter 4. In this instance, theory provides a

EARLY NORMATIVE ARCHITECTURAL INFLUENCE ON THE COLLEGE CAMPUS

A familiar example of positive and normative theory operating together can be found in the history of the college campus. Paul V. Turner's historical account of American campus planning (1984) offers an opportunity to ponder even broader normative influences, beginning with the establishment of the first colleges in eastern North America. According to Turner, the founders of the early English colonies placed considerable emphasis on higher education and moved quickly to establish colleges in America. Harvard College was founded by 1636, only six years after the settlement of the Massachusetts Bay Colony. By 1640, it was occupying quarters in Cambridge (named for the English university that counted many of the Harvard founders among its alumni), and by the time of the American Revolution, there were nine degree-granting institutions spread across the colonies. The importance of these institutions is reflected both in their number and their size. During most of the colonial period, the biggest buildings in America were built for higher education (Turner, 1984). The same American emphasis on education manifested itself more than a century later, when the western migration of Americans and a new desire for practical education led Congress to pass the Land Grant Act, establishing colleges in states across the new western frontier.

American colleges drew much of their architectural inspiration from the British colleges at Cambridge and Oxford. Unlike some of their counterparts on the European continent, these British institutions provided residences for their students. Throughout their early histories the individual colleges at Oxford and Cambridge enclosed inward-looking quadrangles modeled after their heritage from medieval monasteries. By the early 17th century, however, Cambridge had built several colleges following a more open style that was enclosed on only three sides. By then the British schools were also guided by an egalitarian desire to provide educational opportunities for those who could not have previously afforded them and a general increase in enthusiasm for higher education among aristocrats. Thus, enthusiasm for higher education was peaking in Britain just as the first European settlers were establishing colonies in North America. The architectural style of American colleges reflected both the enthusiasm for higher education and some of the values of Oxford and (particularly) Cambridge.

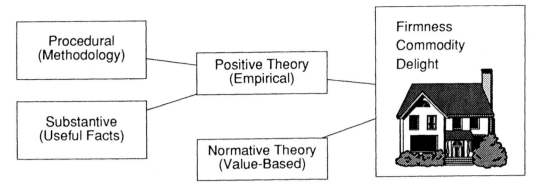

Figure 11-3 The relationship between normative and substantive theories and design.

source of information and organization to which a designer can refer. Lang (1987, 1988) distinguished between positive and normative theory. **Positive theory** attempts to discover predictable relationships between variables, in this instance, the effect of modifications of the physical environment on commodity, firmness, and delight. **Normative theory,** on the other hand, is based upon what *ought* to be done according to a set of design values or style. Normative

The British college template found a new expression in America, however. The three-sided quadrangles of Cambridge were opened up even more and replaced by unattached individual buildings at Harvard and at William and Mary. Eventually, many American colleges spread themselves across parklike settings, such as the one at Princeton (where the term *campus* was first used), establishing a unique pastoral setting that was quite different from their European counterparts. Some, built more or less symmetrically, created malls like the one at the University of Virginia (which still is true to the original design by founder Thomas Jefferson and was adapted by the designers of the University of Nevada at Reno) that are found in some altered form on many modern campuses (Figure 11-4). In the 19th century, Land Grant universities spread across the midcontinent. Frederick Law Olmstead, the founder of American landscape architecture, was responsible for many of their typically parklike designs. As the 19th century advanced, more and more women were admitted at some colleges and universities, athletic facilities became more important, and the curriculum broadened to embrace new disciplines. Some of these changes were echoed in an explosion of new buildings or resisted by revivals of classical design. But ultimately, both the changes and the challenges became integrated into the architectural record still visible both at individual schools and in college architecture generally.

Figure 11-4 The malls of college campuses reflect a Jeffersonian influence. This one is at the Manoa campus of the University of Hawaii.

theory may express itself in design manifestos or identification with a particular design movement. Greek Revival, Romanesque, or postmodern buildings subscribe to different design norms, but they may all be attractive and durable and meet user needs in a fashion that is consistent with positive theory.

As scientists, our discussion will emphasize empirical or positive theory, but we would be foolish not to recognize the importance of the normative influence of history and culture on design (for a familiar example of the influence of cultural values on architecture, see the box on pp. 376–377). Of course, even the choice to emphasize positive theory represents the normative stance of behavioral science.

Lang (1987, 1988) also differentiates between **procedural** and **substantive theory.** In introducing this chapter, we noted that behavioral research can contribute to our understanding of useful relationships between environments and people within them. The accumulation of useful facts about variables such as color, privacy, territory, or furnishings and the ability of a design to provide commodity, firmness, and delight forms what Lang would call "substantive" information. Our introduction also noted that the process of design typically entails complex decision making, involving multiple interests in an atmosphere that may be politically charged. Thus, procedural issues inform the process of gathering data or making group decisions. We believe that psychology can make both procedural and substantive contributions to design.

Before we examine **modern design** issues, however, we must acknowledge that design takes place within the normative framework of our culture. Design expectations reflect not just a rational optimization of a design–behavior relationship, but also the nature of the culture in which we participate.

HISTORY, CULTURE, AND DESIGN PROCEDURES

If someone were to ask you to name five things you associate with ancient Egypt, it is likely that among them you would name the pyramids or

Figure 11-5   An example of the grand design tradition.

the Sphinx. For ancient Greece a similar list might bring the Parthenon to mind, and symbols of modern Paris are likely to include Notre Dame, the Arc de Triomphe and the Eiffel Tower. Surely these are impressive (and relatively permanent) examples of design, but do they really represent the environments that most affect the average citizen of their societies? For that matter, Rapoport (1969) noted that monuments and other "important" buildings represent a self-conscious attempt by the designer (or his or her patron) to impress—what Rapoport referred to as the **grand design tradition.** By intent, these constructions are unusual, specialized, and not representative of the variety of environments experienced by the common person (Figure 11-5). On the other hand, the **folk design tradition** (as expressed, for example, in the home of the common person) is a more direct expression of the day-to-day world of people as they live, shop, and work; it is the design of the everyday home or neighborhood of the typical citizen. Environmental psychology should concern itself with both the monumental architecture of public buildings and the more personal design of individual dwellings. Rapoport's summary of folk design across cultures will serve to highlight some of the broad issues confronting both designers and environmental psychologists in a variety of contexts.

Within the folk tradition Rapoport distinguished between **primitive** and **preindustrial vernacular architecture.** In so-called primitive societies there is little specialization, and nearly everyone is capable of building his or her own shelter according to time-honored techniques that result in a standard style across all dwellings within the culture. The term *primitive* does not imply unsophisticated. Indeed, these shelters have evolved over time under the unforgiving challenges of survival, and they represent successful integrations of unique cultural and environmental demands. Given the resource and cultural constraints, a modern de-

signer would be hard-pressed to create a more durable and portable dwelling for a family than the Cheyenne teepee, or a more successful adaptation to Northern winters than the snow igloo. Furthermore, these and other examples of primitive architecture are sensitive adaptations to climatic conditions such as temperature, wind direction, and moisture rather than attempts at overcoming them with the huge energy expenditures needed to make the standard, modern North American frame home habitable from Florida to Alaska (see Figure 11-6).

As construction methods become more complex, a society may begin to rely on the knowledge and assistance of specialists or tradespersons. Rapoport referred to this as *preindustrial vernacular architecture,* characterized by slightly more individual variation in the design of individual buildings and by the

Figure 11-6 Shelters of so-called primitive societies have evolved into very effective solutions to the challenges of the environment. This Fiji dwelling has a high-pitched, thatch roof that sheds heavy tropical rain and provides natural cooling in a hot, humid climate.

addition of the tradesperson who has specific building knowledge. Again, however, design in these societies is based upon an evolved variation on an established and time-tested theme.

As Figure 11-7 illustrates, building design in industrial nations differs from design in traditional societies on a number of dimensions. For example, the designer is likely to be an architect or some other professional rather than a member of the family, the design is less constrained by climatic conditions, and the changes in building styles and construction techniques are likely to occur at a dramatically faster pace. Furthermore, shelter and survival are expectations rather than concerns among the middle-class or upper-class citizens who commission designs and purchase homes. Therefore, in addition to obvious structural concerns such as durability and safety, modern criteria for building design also include aesthetics, comfort, and efficiency for the people living and working within a construction. Particularly in the industrialized nations, a comfortable lifestyle has led to a high dependence on technology. We use energy to fuel our furnaces, power our air conditioners, and light our homes and factories. Unfortunately, the most common sources of energy are nonrenewable and contribute to pollution problems (see Chapter 14). It is also easy to cite instances of technological mishaps as the price of our dependence. Later in this chapter we will discuss a few ways to avoid such mishaps,

Figure 11-7   Modern buildings (as with Canada's National Gallery of Art) emphasize aesthetics, individuality, and changing technology.

but in our technologically ambitious society, some errors are inevitable. (See Chapter 7 for a discussion of human reactions to technological catastrophes such as nuclear power plant accidents.)

The Process of Design: Fostering Communication

The architect is faced with quite a challenge in attempting to design structures that address the needs of his or her clients. With hindsight it is probably easy for you to think of instances in your own home or campus environment wherein building design does not fit your needs as the user. Ironically, the premium our society places on originality and the explosion of building technology makes errors almost inevitable (e.g., Alexander et al., 1975; Rapoport, 1969).

One of psychology's most important contributions may be insight into the complex process of information gathering and decision making that occurs in design (Lang, 1987; Zeisel, 1981). Design is an evolutionary process of problem solving that involves selecting from a variety of alternatives in search of **congruence** or fit between buildings and their use. Arrangements of space inevitably restrict behavioral options (we cannot walk through a wall unless a door is there), and to the extent that these restrictions inhibit preferred ways of behaving, users become dissatisfied and negative reactions become manifest.

In our discussion of behavior settings in Chapter 4, we emphasized that a particular physical setting might support a number of different behaviors depending on the specific program or occasion. One way to achieve greater congruence or habitability is to "design in" flexibility, thereby ensuring that the space can support a variety of behaviors (Zeisel, 1975). Flexibility might occur at several levels. At the most global, flexibility might include design provisions to allow an entire building to change its function—from administrative to classroom, for instance. Within a building, flexibility might be enhanced with modular partitions, adjustable lighting, and movable furniture. At an even smaller scale, flexibility might include provisions allowing individuals to modify their own microenvironment. For example, O'Neil (1994) documented the importance of allowing workers to adjust their workplace furniture. Such personal adjustments not only allow the individual to "fine tune" his or her environment, but should increase perceived control (see Chapters 4 and 5).

The ability of a designer to adjust the physical environment to achieve congruence is determined in part by the number of potential **design alternatives** (or different ways we can think of to design or redesign a setting). In a given setting there may be a large number of design alternatives, but as different criteria are brought to bear, more and more alternatives will be ruled out. For example, some may be too expensive, others may be inappropriate due to their behavioral effects, and, of course, some may simply be out of style. The process of determining the proper design alternatives and weighing the importance of various criteria forms the heart of the design process. This is a complex undertaking since there are many interrelationships among design alternatives as well as many different social, economic, artistic, and cultural pressures (see Figure 11-8).

Figure 11-8   Flexible design allows buildings to change to meet a variety of functional needs.

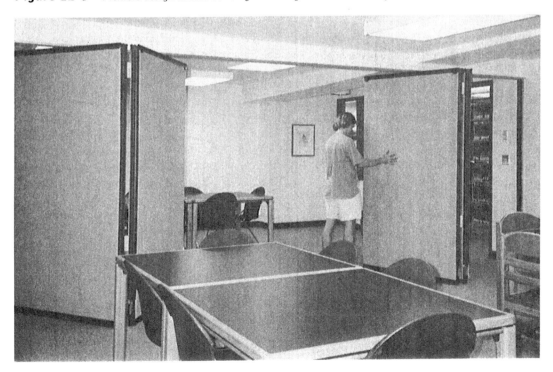

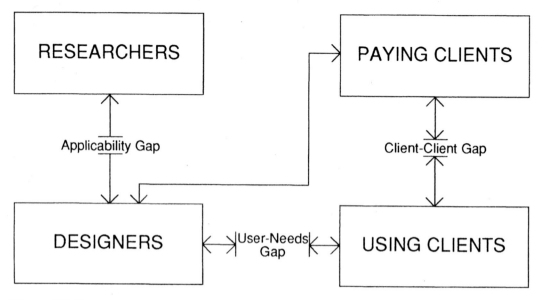

Figure 11-9 An illustration of the gaps between the paying and using client, between designers and clients, and between behavioral scientists and designers.

THE GAPS

In most introductory treatments of behavior-based design it has become common practice to speak of "the **gap**." In fact, there are several gaps (see Figure 11-9), but the term most commonly refers to the discontinuity between designers and those who will eventually live and work in their buildings (Theil, 1994; Zeisel, 1981). Unlike primitive or preindustrial vernacular cultures, the eventual users of modern architecture (using clients) are often never directly consulted in the design process. Often the architect is the only party who has direct contact with the paying client—perhaps a corporation or government agency. The problem is easy to illustrate. Were you (or some other student before you) consulted by the architect responsible for the classrooms in which you learn, the office in which you work, or the house, residence hall, or apartment in which you dwell?

Increasing communication between designers and their clients may seem to be an unremarkable suggestion, but successful communication is not accidental. The reason to hire a designer in the first place is that the owner is incapable (or unwilling) to design alone. On the other hand, it may be that design education and training shapes and changes the expert's perceptions of the environment (Devlin, 1990; Groat, 1982; Hubbard, 1996; Kaplan & Kaplan, 1982; S. Kaplan, 1987; Sancar & Eyikan, 1998; Wilson, 1996). For example, Groat (1982) reported that accountants were more likely to classify buildings according to preference or building type, whereas architects were more likely to include judgments of building style, form, design quality, or historic significance. Devlin (1990) compared the content of published architectural reviews of two buildings versus on-site interviews with nonarchitects. Architects apparently paid more attention to design ideas and concepts, whereas nonarchitects were more likely to make general affective (emotional) judgments or descriptions of the physical features of a building. Finally, Hubbard (1996) concluded that planners emphasize the physical aspects of design such as materials and the architect's normative approach to design (style), whereas the public is more likely to base its judgments on cognitive constructs such as what the building reminds a person of or a global assessment of preference.

Apparently, the divergence between the judgments of laypeople and architects results from the unique experiences of architectural training (Wilson, 1996). Although design training may result in a general tendency to react to buildings in terms of their normative style, whichever style is preferred differs based upon the exposure to specific styles. Graduates of one school may generally prefer postmodern architectures, whereas those who were trained through a different program may prefer more purely modern forms. Oddly enough, designers may feel their views are not adequately weighed in the politicized negotiations of selecting design alternatives. Fahriye Sancar and Baris Eyikan (1998) surveyed studio instructors of architecture and landscape architecture at more than 60 schools in the United States. They concluded ". . . the professional identity of designers is being transformed from that of isolated creative individuals to that of politically active professionals" (p. 379). Although the transformation is incomplete, these authors believe that designers must become better able to communicate with laypersons in order to participate in increasingly politicized negotiations. Furthermore, instead of emphasizing training based on critiques of architect-acclaimed buildings, design studio courses need to incorporate social and environmental design research.

The **applicability gap** (Russell & Ward, 1982; Seidel, 1985) is a name given to the miscommunication between psychologists trying to understand the needs of architects, and designers who try to come to grips with the data and the implications of social and behavioral sciences. Examining the applicability gap prompts questions regarding research methodology and philosophy of science, and highlights an unresolved tension in the field of environmental psychology. Altman (1973, 1975) has emphasized that the design process must reflect the different approaches of the various people involved

in environmental design. In particular, he feels that practitioners, such as architects, are inclined to attend primarily to design criteria and to particular places or settings. Researchers, on the other hand, are more likely to stress ongoing behavioral processes, such as privacy, territoriality, or personal space. Academicians often value their independence, that is, their ability to choose to investigate almost any question that interests them. They are also typically (rightfully) committed to cautious interpretation of their data. Unfortunately, at least for those who wish to see the early application of behavioral design, these goals of science sometimes result in a situation that encourages psychologists to ask simple research questions that may show statistical elegance but that hold little promise for application. Furthermore, the implications of the research literature can be difficult for design professionals to extract from the jargon and statistical descriptions of research journals. If behavioral science is to offer designers useful data, it should be responsive to practical questions of design. On the other hand, the whole point of statistics and careful research design is to ensure that conclusions are drawn carefully and objectively. Environmental psychology will suffer painfully from sloppy or misinterpreted findings. This balance between caution in interpretation and relevance in research topics is likely to pose one of the thorniest problems for environmental psychology in the years to come.

On a different issue, we might add that if a psychologist wishes to serve as a liaison between professionals and their clients, he or she would profit from a basic understanding of the graphics and technical references used by designers. In a sense, a successful liaison must be able to translate the dialects spoken by both the professional and lay participants. The training of most psychologists is at least as inadequate in preparing them to understand design as a designer's training is in preparing her or him to interpret scientific research. How eager are designers to integrate relevant behavioral information into their designs? Hopeful signs come from interdisciplinary design conferences. For example, each year the Environmental Design Research Association (EDRA) holds a conference that attracts behavioral scientists and designers who share an interest in behaviorally based design.

FOSTERING PARTICIPATION

We must acknowledge Wilson's (1996) assertion that the best design does not necessarily result from a process that restricts designers to architectural styles that the average user already appreciates. Designing for the familiar and "average" may simply result in blandness. Nevertheless, facilitating **participation** by users remains an important goal, and there are several ways to improve the chances that the needs and wants of the using client are incorporated in a new design. One way to close the gap would be to train experts to be more sensitive to people's concerns (Kaplan & Kaplan, 1982). Even more straightforward, the users could be included in the actual design process (e.g., Kaplan & Kaplan, 1982, 1989; Kaplan, Kaplan, & Ryan, 1998; Theil, 1994). Although design experts are likely to recognize the complexity and ambiguity of many design questions, they are asked to make quick, confident, and cost-effective decisions. In the short term at least, quick decisions conflict with the laborious process of gathering and assimilating data from the public (Dalholm & Rydberg-Mitchell, 1992; Kaplan & Kaplan, 1982).

Kaplan and Kaplan (1982) suggested several themes that seem to characterize instances of successful participation:

1. Involving the public at an early stage in design so that its suggestions can be fairly integrated into design alternatives. The public will rightfully feel offended if its participation is invited only when most of the decisions have already been made.
2. Availability of several concrete alternatives to react to. The designer's expertise can demonstrate the scope of design alternatives and present options that allow straightforward responses.
3. Presentation of possibilities in a format that is comprehensible. In particular, the use of visual or spatial material can make it possible for laypersons to visualize design alternatives.

To summarize, a concerted effort needs to be made to accurately communicate design alternatives to laypersons so that they may make substantive and informed decisions. Designers may have developed ways of visualizing design alternatives and communicating with other professionals that may not be comfortable for laypersons. Simulations such as models or drawings are frequently employed by designers in communicating with clients and other designers. Certainly new technology, especially new computer visualization programs and equipment, will allow designers an unprecedented ability to model design alternatives (Decker, 1994; Theil, 1994), although the complexity of design software can make it difficult for laypersons to use such programs to express their own ideas. Not all of what an architect reads in a drawing is necessarily perceived by laypersons, however. For instance, the symbols used by architects for windows, doors, and other features may be unfamiliar to laypersons, and the scale may be distorted (Dalholm & Rydberg-Mitchell, 1992). Asking laypersons to make drawings can reverse the process. Users seem able to communicate relationships between the desired size and location of large spaces; however, bathrooms, doors, closets, and hallways are often drawn too small, leading Dalholm and Rydberg-Mitchell (1992) to advocate the use of small-scale or even full-scale models in addition to drawings. On the other hand, detail and exactness in simulation may actually be counterproductive. A model that pretends to be a perfect replica of an actual design is likely to be expensive and may simply activate the human tendency to try to find all of the minor discrepancies between the model and reality (Kaplan & Kaplan, 1982).

Substantive Contributions

We have emphasized the importance of communication for gathering information about user needs and wants. Needs and wants form the basis of the program of a proposed construction. In order to support commodity, firmness, and delight, designers must match the program to materials, construction techniques, and spaces that support it. In other words, designers must consult a body of what Lang (1987, 1988) called *substantive theory*. Of course, much of this knowledge is beyond the domain of behavioral science. Firmness is a very desirable characteristic of buildings (it prevents the roof from falling on our heads), but engineering and structural materials are not domains of psychology.

PRIVACY

One of the most important aspects of the design of interior space is the amount of privacy it provides (Evans & McCoy, 1998; Kupritz, 1998; Sundstrom, 1986). Altman (1975) has defined privacy as the "selective control of access to the self or one's group." This definition has two important parts. The first is the notion of privacy as an ability to withdraw or separate ourselves from other people. In effect, this refers to the desire for seclusion. Both Altman and Ittelson et al. (1974) recognize a second important aspect of privacy—the ability to personalize spaces in order to present information about ourselves. Thus, privacy represents a dynamic process of openness/closedness to others (Altman & Chemers, 1980). Personal space and territoriality are behavioral mechanisms that regulate privacy; crowding and the loss of perceived control represent failures to achieve it. Perhaps all humans have a desire both to communicate and to keep some aspects of our personalities or thoughts to ourselves. Architectural features or policies that encourage personalization will support the presentation of personal information, whereas other characteristics of the physical environment (the lack of walls in an office or

school, for instance; see Chapters 12 and 13) may prevent us from regulating what other people find out about us. Designs that optimize privacy have to consider both elements of Altman's definition, as well as the fact that privacy means different things to different people. For example, dormitories that house students one instead of two to a bedroom promote greater privacy. Likewise, the use of barriers around one's work area may increase the sense of privacy. Often, then, privacy adjustment is centered on the structures that partition interior space (Figure 11-10).

Figure 11-10A & 10B   Privacy adjustment may be established with physical or even psychological barriers.

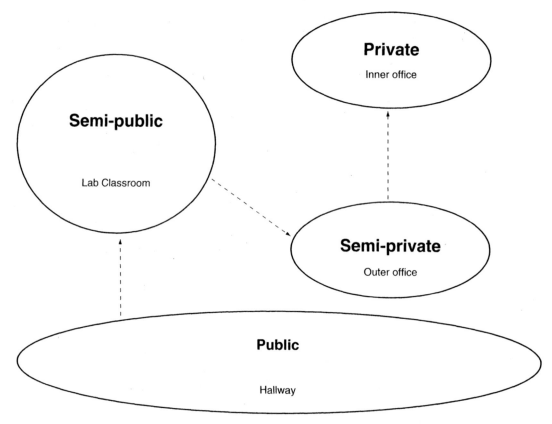

Figure 11-11   Privacy gradients in an environmental psychologist's office.

Research has indicated that barriers that block views of other people decrease the impact of these people, while barriers that do not obscure the view (e.g., clear panels) do not reduce their impact (e.g., Baum, Reiss, & O'Hara, 1974; Desor, 1972). Interestingly, some research suggests that the lack of auditory privacy may be even more troubling than losses of visual privacy (Kupritz, 1998). In particular, we are troubled by environments that make it likely that private conversations will be overheard (Sundstrom et al., 1994).

One common way to conceptualize privacy is in terms of a spatial hierarchy (Alexander, 1972; Evans & McCoy, 1998). Doorways and walls between spaces orchestrate a sequence of increasing levels of privacy (see Figure 11-11). Spaces deeper in the hierarchy allow more regulation of social interaction (Hillier & Hanson, 1984), but at an extreme, may result in isolation.

Materials and Color

Color is one of the most easily manipulated dimensions of environmental surfaces. In a business setting, for example, a coat of paint is far less costly than structural remodeling. Popular articles abound, but there is very little recent empirical research directed specifically at the effectiveness of various manipulations of environmental color. We are forced to join those who conclude that the literature addressing the application of color is surprisingly limited (e.g., Read et al., 1999; Sanders & McCormick, 1993).

For example, does the color of the walls in the room in which you are now reading affect your perception of the room's temperature? Many aspects of color are tangled in a web of symbolism. Simply asking people to report color–temperature associations may tap only an individual's ability to repeat these learned, perhaps wholly symbolic associations. As a rule, investigators who have adequately controlled the tendency for people to try to respond "correctly" with the usual color–temperature symbolism have failed to find useful effects (e.g., Berry, 1961; Greene & Bell, 1980; Fanger, Breum, & Jerking, 1977). An early study by Berry (1961) neatly makes our point. Berry chose to investigate the effects of color on the comfort of people seated in a heated room. Participants were ostensibly performing a task designed to determine the effects of colored light on performance in an automobile driving simulator. They were told that the lights generated a great deal of heat and that they should notify the experimenter when they became uncomfortably warm. Berry concluded that the color of the illumination (green, blue, yellow, and amber) did not affect the level of tolerable heat. Subsequently, however, Berry asked the same participants to rank samples of the experimental colors in terms of the amount of heat they transmitted. Thus cued, most ranked the colors in the conventional hue–heat order: warmth with amber, coolness with blue and green.

There is little question that individuals have color preferences. It is likely that being unable to escape from a disliked color threatens perceived control. Unfortunately, those investigations that have focused on the relationship between ambient color and mood or performance (e.g., Ainsworth, Simpson, & Cassell, 1993; Jacobs & Blandino, 1992; Stone & English, 1998) often conclude that there is no conclusive evidence of a useful relationship. Some promise for application has come from studies showing that perception of spaciousness, or conversely, crowding, may be influenced by color. For instance, Acking and Kuller (1972) had participants rate a series of slides depicting rooms that varied in color. Results indicated that lighter rooms were seen as more open and spacious. Similarly, Baum and Davis (1976) found that different intensities of the same color affected people's responses to model rooms. Light-green rooms appeared larger and less crowded than identical rooms painted a darker green.

Our discussion of color introduced an important question about the effect of visible build-ing materials. Are differences in the impressions of various materials based on direct differences in sensation or differences in utility (weather resistance, for instance), or are they symbolic? Sadalla and Sheets (1993) found that the materials covering the facades of houses may be perceived as indicators of the interpersonal style, creativity, and social class of the homeowner, especially if the homeowner was perceived as having a role in choosing the material. For example, wood coverings were associated with more "emotional," "weaker," and "tender" owners.

ILLUMINATION

Adequate light is clearly necessary for successful performance on some tasks requiring good visual acuity or in making color discriminations (e.g., Galer, 1987). Popular reports further suggest that fluorescent lighting that is similar to natural daylight (full-spectrum lighting) leads to both higher productivity and better health. Despite the popularity of this view, and not insignificant amounts of money spent trying to correct supposed imbalances, there is little evidence that full-spectrum lighting has any appreciable effect on mood, performance, or health (Boyce & Rea, 1994; Veitch, 1997). There is also a dearth of information regarding the effects of lighting on mood, cognition, and social relationships, and some of the research seems contradictory (Knez, 1995). Two commonly held beliefs are that low levels of light lead to both greater intimacy and to quieter or reduced conversation (e.g., Feller, 1968; Saunders, Gustanski, & Lawton, 1974). Several studies support these beliefs. For example, as we saw in Chapter 8, Gergen, Gergen, and Barton (1973) reported that when college students who were strangers to each other were placed in a dark room for several hours, considerable verbal and physical intimacy occurred between them. Darkness and anonymity had apparently removed some customary barriers to intimacy.

Butler and Biner (1987) used a questionnaire format to determine the lighting preferences of a large sample of college students. Participants in the studies reported their preferred lighting levels, the importance of these lighting levels, and the degree to which they desired control over the lighting levels. Having the proper lighting level was rated as most important in instances in which individuals reported preferences for either a rather dark (e.g., during a romantic interlude) or very bright (cutting vegetables with a knife) lighting level. Predictably, control over the level of lighting is more important in some environments than others, and particularly important for those expressing strong lighting preferences (Butler & Biner, 1987; Veitch & Gifford, 1996).

WINDOWS

Research (e.g., Butler & Biner, 1989; Heerwagen & Orians, 1986; Nagy, Yasunaga, & Kose, 1995) confirms that people like windows, but why? Butler and Biner's study of office workers assembled a list including: ventilation, a source of weather information, ability to see people, and a way to regulate temperature. We might add that windows can also serve as portals for passing objects in and out (as in a fast-food drive-through window) and can provide opportunities to psychologically escape from crowded or unpleasant rooms (see Chapter 9). Windows also serve as decorations, both for those inside a building (consider the decorative effects of stained glass) or for those on the exterior who view the architectural statement in patterns of windows and lintels. Without denying the importance of these roles, we will focus on two additional and complementary functions: first, the window as a source of sunlight, and second, the window as a visual connection with the outside environment (Leather, Pyrgas, Beale, & Lawrence, 1998).

The sun is, of course, an important source of illumination. For most of our species' existence, supplemental lighting was limited to flickering fires. Even at the time of the Industrial Revolution adequate task lighting in factories required large windows and narrow buildings to allow light penetration (see Chapter 13). Now, however, artificial lighting can easily provide illumination levels well in excess of those needed for task performance, so researchers have turned to

qualitative differences between sunlight and artificial illumination (Leather et al., 1998).

People believe that sunlight has beneficial effects over artificial illumination (Heerwagen & Heerwagen, 1986; Veitch & Gifford, 1996), a view that is at least partially confirmed by our understanding of the causes of seasonal affective disorder (SAD; see the box on page 201). In looking at reports that sunlight improves job satisfaction and general well-being, Leather et al. (1998) point out that it is not the general level of interior sunlight that is important, but rather, the size of the sunlit area.

Perhaps the most persuasive support for the generous use of windows as a source of views comes from the growing literature documenting therapeutic effects of hospital windows overlooking pleasant landscape views (e.g., Ulrich, 1984; Verderber, 1986). For example, as we have noted in other chapters, Ulrich (1984) reported that patients with views of pleasant landscapes outside their hospital rooms had shorter postoperative hospital stays, required lower doses of painkillers, and had fewer negative evaluative comments from nurses. Even posters of outdoor scenes can make a workplace more pleasant (Stone & English, 1998).

FURNISHINGS

Furniture, its arrangement, and other aspects of the interior environment are also important determinants of behavior. In classroom settings, for example, it appears that the use of nontraditional seating patterns can influence student performance; horseshoe arrangements, circular patterns, or other less formal departures from the standard "rows of desks facing the teacher" seem to generate more student interest and participation (Sommer, 1969). Some evidence even implies that within traditional classroom arrangements there are differences in performance according to where people sit. These findings are discussed in Chapter 8.

Many studies of furniture arrangements have been conducted in institutional settings. Reusch and Kees (1956) noted that the way in which patients arrange their furniture expresses their feelings regarding interaction in their space. Some arrangements (called sociopetal) are open and welcome interaction, while others (called sociofugal) are closed and discourage social contact. Sommer and Ross (1958) described the relation between furniture arrangement and behavior in a geriatric hospital. When chairs were arranged in rows along the walls, patients did not interact very much. This arrangement was simply not conducive to talking; it did not suggest that interaction was appropriate. When Sommer and Ross changed the arrangement, clustering the chairs in small groups, people began to talk to one another. The new juxtaposition facilitated conversation while the old one seemed to inhibit it. Holahan (1972) found the same kind of effect in a psychiatric hospital—patients seated around the table talked to one another more than patients seated in rows against the walls (for a thorough discussion of design in selected institutional environments, see Chapter 12).

Furniture arrangements can be used to help structure the preexisting architectural layout of a setting. In most environmental contexts, the walls, the location of the doors, and so on are fixed—they are rather difficult to move. To some extent, these elements do structure the space inside a building. However, the placement of furniture often provides additional organization. For example, if you have a large living room, you may arrange the furniture to suggest two rooms. Or, you may arrange it to unify the room.

Arrangement is not the only aspect of furnishings that can affect mood and behavior. The quality of the furnishings is also important. Later, we will discuss studies of the effects of "pretty" and "ugly" rooms and discover that being in a pretty room can sometimes have beneficial psychological effects. Unfortunately, studies varying the quality of single pieces or sets of furniture have not been conducted systematically. However, research has also been conducted on the effects of large-scale improvements in furnishings. Holahan and Saegert (1973) reported on a large-scale refurbishing of a psychiatric hospital admissions ward, comparing it with another ward that was not redone. The refurbishing included bringing in new furniture, repainting, and creating different types of space. These improvements in the quality of the environment led to increases in social activity on the ward and demonstrated that the quality of an environment can influence mood and behavior. However, because the improvements were so extensive, it is difficult to know what was primarily responsible for the observed results.

In our discussion of windows we noted the importance of a connection to the exterior world, particularly to plantings and the "green" landscape. If exterior lawns, trees, and shrubs can have a beneficial effect, one might expect a similar role for potted plants in offices and homes. Larsen et al. (1998) investigated the effect of either a moderate or high number of office plants on measures of productivity, pleasantness, and comfort. In this study the office was perceived as both more comfortable and more attractive when it was decorated with plants. Surprisingly, performance on a simple and repetitive task was not enhanced, and was actually better when the office was devoid of plants.

Ultimately, any decision about furnishings will be based on several criteria, including cost, aesthetics, and the function of the setting. This latter criterion is often the most difficult to · evaluate. Sometimes a given space is expected to facilitate communication among employees working near one another, to serve as a meeting room on occasion, and to impress clients who come for consultations. Some of these functions are at odds with each other, and the choice of arrangement of furnishings must be accomplished with these complex issues in mind.

ARCHITECTURAL AESTHETICS

One of the primary goals of a design is to evoke a pleasurable response from people viewing the finished setting. The study of **aesthetics** in architecture is an attempt to identify, understand, and, eventually, create those features of an environment that lead to pleasurable responses. The problem is that aesthetic considerations in design may operate contrary to behavioral ones. Some of the most beautiful structures are also among the most impractical. However, one cannot simply dismiss aesthetic quality as less relevant than the behavioral effects of design. Indeed, there is evidence that aesthetics may be important in determining behavior (e.g., Nasar, 1994; Steinitz, 1968).

Some authors differentiate between two kinds of aesthetic design characters (Lang, 1988; Nasar, 1994). *Formal aesthetics* includes dimensions such as shape, proportion, scale, complexity, novelty, and illumination. On the other hand, *symbolic aesthetics* is affected by different sorts of meaning. Some meanings are denotative—for instance, a building's function (bank or prison) or style (postmodern, Gothic). Other symbolic meanings may be connotative. Is your campus library's architecture friendly? Imposing?

According to Nasar's review (1994, p. 384), primary formal variables include enclosure (openness, spaciousness, density, mystery); complexity (diversity, visual richness, ornamentation, information rate); and order (unity, order, clarity). As we saw in Chapter 2, well-defined spaces that balance openness and enclosure are preferred to either wide-open or highly enclosed landscapes. Complexity reflects diversity and visual richness.

Although not specifically discussed by Nasar, the Gestalt rules of perceptual organization (see Chapter 2) may also provide clues to formal application. According to Lang (1987), formal aesthetics has traditionally been heavily dependent on the **Gestalt theory** of perception, which views the organization of elements of visual form as units that can be perceived as either simple or complex (see Chapter 3). Although Gestalt theory is now considered to apply only to a rather limited number of situations, it appeals to the designer's need for an understanding of visual forms at a broad, holistic level. For many designers, the implication is that environments ordered according to these principles of "good form" will also be good environments, whereas other designers (e.g., Venturi, 1966) deliberately violate Gestalt principles as a means of obtaining visually richer environments.

Nasar (1994, p. 389) also outlines sources of symbolic aesthetics: naturalness, upkeep, inten-

sity of use, and style. We will review the influence of naturalness in more detail in Chapter 13, but a number of authors have confirmed the importance of naturalness in aesthetic preference (e.g., Kaplan & Kaplan, 1989; Ulrich & Parsons, 1992). Style, which incorporates meaning and normative dimensions, has also been well documented as a source of aesthetic appreciation. Although style preferences may vary depending on the function of a particular building and among sociodemographic groups, vernacular architecture or architecture with an old and genuine "feel" may be appreciated more than modern styles (Herzog & Gale, 1996).

Research has indicated that the aesthetic quality of a room—the extent to which it is pleasant or attractive, for instance—may affect the sorts of evaluations we make while in that setting. In their classic study, Maslow and Mintz (1956) compared participants' ratings of a series of photographs of individuals in a "beautiful" room (well-decorated, well-lit, etc.), an average room (a professor's office), and an "ugly" room (resembling a janitor's closet). Their results showed that participants rated persons depicted in the photos most positively if they had been in the beautiful room, and most negatively if they had been in the ugly room.

Attractive environments also make people feel better. Research has shown that decorated spaces make people feel more comfortable than ones that have not been decorated (Campbell, 1979). Also, the good moods that are associated with pleasant environments seem to increase people's willingness to help each other (Sherrod et al., 1977). People feel more like talking to one another in pleasant settings (Russell & Mehrabian, 1978). Research has also suggested that decoration may be distracting (e.g., Baum & Davis, 1976), but whether this is necessarily a problem appears to depend on other factors (Worchel & Teddlie, 1976).

PLACE

In Chapter 2 we introduced the concepts of place and place attachment. Try to imagine a few of your favorite places within the everyday world of your campus or community. Some of

Figure 11-12   Settings and personal history combine to form a sense of place.

these places may be quite idiosyncratic. Your place may be the room you decorated with personal mementos, or a little hideaway in an academic building where you go to read. Other places on your list may be more public, and for these, both the space and the sense of place they create can be shared (see Figure 11-12).

Evoking an intuitive understanding of place by analogy or example is quite easy. Unfortunately, constructing a formal definition proves more difficult, and an adequate investigation of the phenomenon challenges the tools of empiricism that have been adopted by behavioral scientists. At an extreme, phenomenologically oriented philosophers, geographers, and architects (e.g., Nordberg-Schultz, 1979) propose that some places manifest a *genius loci*, a personality that is almost independent of humans. On the other hand, *placemaking* is a standard topic in the training and vocabulary of architects and designers (e.g., Dober, 1992). We will examine placemaking again in Chapter 12.

Selecting Alternatives: The Design Cycle

Whether the design criteria are behavioral or structural, normative or substantive, we see a need for a reliable model for integrating program, data, and design. Thus, design is seen as an example of problem solving. As a student of psychology, one thing you may have learned about problem solving is that it is not rational. We know that humans generally fall far short of the optimal solution to a problem, accepting instead solutions that *satisfice*—that is, solutions that are "good enough" (Kaplan & Kaplan, 1989; Simon, 1960). Because there is such a variety of

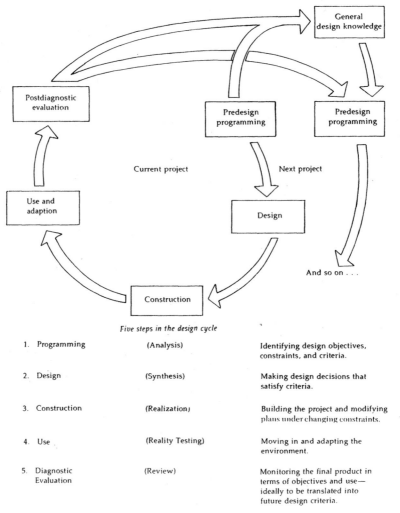

Figure 11-13   Zeisel's design spiral.

Adapted from Zeisel (1975). Sociology and architectural design: Social science frontiers, No. 6. New York: Russell Sage Foundation.

apparently acceptable solutions for most projects, and because even poor designs may appear successful until after a great deal of effort has gone into developing them, one of the design's most difficult problems may be in deciding when to stop design and begin construction. Instead of proceeding in a smooth, directed path, Zeisel (1981) suggests that a spiral metaphor is a better representation of the design process (see Figure 11-13). In separate cycles, architects propose, test, and refine possible solutions to sets of related problems. The result of each cycle will

be a possible response to a particular design problem, but this decision may limit the alternative solutions to another set of problems. By testing ideas, finding conflicts, then retesting, the spiral will gradually narrow until it lies within the domain of acceptable solutions, at which point construction will begin.

STAGES IN THE DESIGN PROCESS

Commonly, design is conceptualized as a continuous cycle of design planning and evaluation

Figure 11-14 A simplified version of the ideas of the client and designer merging into a final construction.

(Zeisel, 1975, 1981). The process begins with an awareness of general building needs. During the programming phase, all of the parties to the design process work to create an extensive specification of the requirements of the building (Bechtel, 1997). Often the process begins with a series of questions. Who will use the building and how can their behaviors be supported? Must the building support specialized machinery? Are there important aesthetic concerns? How can the building be made more accessible to users who are physically challenged? Although the process varies somewhat, one place to begin is to review other similar projects to glean examples of notable successes and failures. Usually the program has already become quite mature before the design team begins to draft its first conceptual sketches. The programming phase is also the time in which the principles of environmental psychology are most valuable. Here behavioral scientists may suggest design alternatives based on substantive environment–behavior research. Here, too, psychological expertise in managing group processes and in decision making has potential for improving the process of design. Finally, the program becomes more complete, and the designer's drawings be-

come less conceptual and more specific, resulting at last in construction specifications. In a very simplified illustration, this process is depicted in Figure 11-14, where the client for and the designer of a simple construction bring their ideas together.

After all of the work that has gone into programming, designing, and building a project, it would be a shame if other projects could not profit from all of the creative ideas and solutions generated by the design process. Indeed, one of the most important contributions of environmental psychology has been the **post-occupancy evaluation (POE).** POEs were intended to become a cumulative archive of design solutions, a kind of tool kit that would be available to designers as they began each new project (Bechtel, 1997; Zimring & Reizenstein, 1980). Unfortunately, the realities of architecture as a business often mean that a design firm has no real motivation to assess the success of past designs for which they are no longer being compensated. Ideally, however, the **design cycle** is completed and begun anew by using knowledge gained from previous design experiences to improve upon the next one.

North American College Campuses: An Example of Design Dynamics

Modern college campuses illustrate most of the issues outlined by Lang. What factors weighed most heavily in your decision to attend your

college or university? Was your choice purely rational, or was it influenced by taste or even whimsy? According to one survey, 60% of college-bound students rank the visual environment as the most important factor in choosing a college (Carnegie Foundation as reported by Gaines, 1991). The physical layouts of North American colleges and universities reflect both their European heritage (particularly British) and the unique value Americans place on higher education. In spite of what seems to be a self-conscious desire to be perceived as unique or special, many common themes can be identified on North American college campuses. As Gaines notes (1991), many have a core space or quadrangle such as:

The Oval	Michigan State University
The Heart	Earlham
The Yard	Harvard
Prexie's Pasture	University of Wyoming
The Horseshoe	University of South Carolina
The Lawn	University of Virginia
The Meadow	Mills College
The Plaza of the Americas	University of Florida

The buildings and grounds of campuses represent a rich physical record of changing design innovations and architectural norms. Architectural features are likely to become symbols for the university or college, or a state or province. Figure 11-15A shows The Old Well, a popular meeting spot at the University of North Carolina at Chapel Hill, and Figure 11-15B is the view of the capitol of Wisconsin from Bascom Hill on the University of Wisconsin campus. As an aside, we can't help but mention that the latter vista was threatened briefly in the late 1980s when a miscalculation nearly resulted in an addition to the main library that would have blocked part of Bascom Hill's view toward the state capitol building. Plans were redrawn, the building was shortened by several stories, and the new poured-concrete stairwell (already visible in the distance in Figure 11-15B) was removed. We see this as an example of both the

fallibility of design and the importance of landscapes and vistas to college campuses.

At many institutions the campus buildings and grounds reveal periods of stability and periods of growth, the establishment of new programs and the abandonment of those that have lost favor, and the effect of evolving architectural fashions from Georgian to postmodern (see the box on pp. 376–377). For example, before the late 1960s most coeducational colleges and universities housed men and women in separate dormitories, or at the very least, on separate floors (also see Chapter 9 for an additional discussion of residence halls). As students began to demand less segregation between men and women, many residence halls mixed men and women's bathrooms, but often without investing in the expense of replacing the old sex-typed plumbing. The 1970s and 1980s introduced an era in which new regulations promoting safety, energy conservation, and accessibility for the physically challenged are now leaving their marks. A little detective work will probably reveal a variety of physical traces that reflect normative changes on a campus familiar to you.

Although there may be some common campus design prototypes (the green parklike campus, for instance), there are clearly many exceptions and differences in expression of these prototypes. Unfortunately, one problem shared by campuses, urban and rural alike, is a fear of crime. Fear of sexual assault concerns students, faculty, and staff and results in both worry and a restriction in freedom of movement, particularly at night (Day, 1994; Koss et al., 1987). The degree to which sexual assault can be reduced with changes in grounds design is unclear. For instance, the actual number of cases of assault is unknown. It may be that the frequency of sexual assault has been intentionally obscured on some campuses to preserve an image of safety. Furthermore, many assaults occur indoors and many assailants are acquaintances. Nevertheless, some factors associated with fear on campus include the presence of features such as dense vegetation that provide an assailant a place to hide, low potential for escape, and poor lighting (Day, 1994; Fisher & Nasar, 1992; Kirk,

Figure 11-15A The Old Well, a popular meeting spot at the University of North Carolina, Chapel Hill.

Figure 11-15B View from Bascom Hill at the University of Wisconsin, Madison.

1988; Nasar & Fisher, 1992b). Ironically, the same features that make a landscape attractive during the daylight hours (vegetation, enclosed space) make the landscape feared at night (Day, 1994; Shoen, 1991).

PLANNING FOR THE FUTURE

Most colleges and universities have a *master plan*, a map or series of maps with supporting documents that seek to coordinate future building projects. The goal of these plans is to prevent haphazard growth and isolated constructions that lack coordination with other campus facilities or design styles (Dober, 1992, 1996). The box on pages 396–397 describes an example of the integration of psychology in the planning efforts of one small college.

The layout of campus buildings sometimes results from immediate needs or a particular college administration's artistic sense. Some (such as Thomas Jefferson's University of Virginia) continue to reflect the cohesive vision of their past (Gaines, 1991; Turner, 1984). However, our society has become even more pluralistic since Jefferson's time, and to succeed aesthetically it may now be necessary for a campus to reflect the input of planners, social scientists, officials, and naturalists (Dober, 1992, 1996; Gaines, 1991). One of the best-known proposals for user participation in college design was developed by Christopher Alexander and his colleagues (Alexander, 1979; Alexander, Ishikawa, & Silverstein, 1977; Alexander et al., 1975). Although this approach is quite normative, it does directly address the three themes outlined by the Kaplans

∾ COGNITIVE MAPPING AND CAMPUS PLANNING

Figure 11-16 indicates the mean (average) ratings of pleasantness for a small college campus and was obtained by electronically overlaying individual maps drawn by several hundred students. As you can see, the most pleasant area appears in the lower left corner of the map, which is the area depicted in the photograph in Figure 11-17. The map is actually one of a series of cognitive mapping exercises that have traced changing evaluations of the small campus (Greene & Connelly, 1988; Greene & Warden, 2000; Mace & Greene, 1997). Could these data be useful for more than just academic purposes? As it happens, in the mid-1980s the university employed an architectural and planning firm to help create a new master plan to guide development for the next two decades. Architects for the project used some of the early results to determine

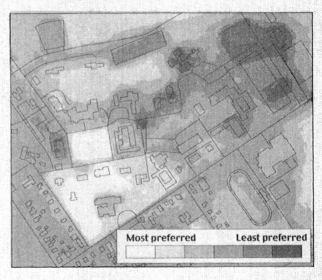

Figure 11-16 Pleasant and unpleasant areas of a college campus as rated by students.

preference zones for students and faculty. They also requested traffic-pattern data (collected in a similar manner by asking students and faculty to draw their routes across campus on a typical day), as summarized

(see page 384). Like Rapoport, Alexander believes that modern design has lost many of the advantages of participation and slow evolution that were characteristic of primitive and preindustrial cultures. In particular, Alexander attacks the modern approach in which roads are built by engineers, buildings by architects, and tract housing by developers. In Alexander's view, the average citizen has lost the ability to affect design, and designers have lost touch with the needs of those they serve. *The Oregon Experiment* (Alexander et al., 1975) illustrates the participative process as Alexander would implement it on a college campus (in this case the University of Oregon).

Alexander advocates participative planning but finds master plans themselves to be too rigid; too likely to constrain growth consistent with patterns devised decades earlier; and too likely to overlook the need for small, evolutionary renovations. Instead of a master plan, Alexander favors piecemeal, *organic growth* in which buildings are initiated and designed by their users and built by architects and contractors. Change and adaptation are constant, but piece-by-piece growth achieves a long-term order as each new space is carefully fit into the existing environment. In order to allow laypersons such unusual prominence in design, Alexander created what he calls a **pattern language** (Alexander et al., 1977) that contains prescriptions for hundreds of different design problems or situations that are designed to be understandable to both architects and laypersons.

Figure 11-17 Canopied section of the campus that is rated as
highly pleasant in Figure 11-16.

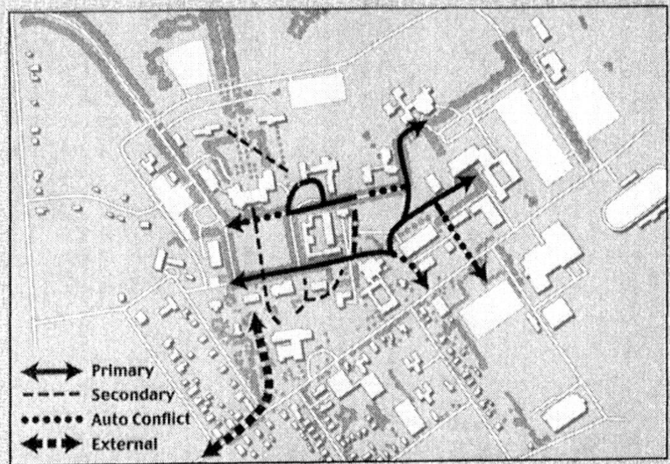

Primary
Secondary
Auto Conflict
External

Figure 11-18 Portions of the campus master plan that resulted partly from the analysis in Figures 11-16 and 11-17.

in Figure 11-18. Finally, they used their professional expertise and on-site observations to create a new campus master plan that, among other things, removed parking lots and roads from the central campus.

SUMMARY

We began this chapter with a comparison between the design traditions of the industrialized
nations and other less technological societies.
We concluded that there is frequently a gap between modern designers and those who will
eventually use or live in their constructions. We
suggested that user participation may be one
avenue to bridging this gap and that a more responsive relationship between designers and researchers provides another.

The behavioral effects of several different
individual features of built settings were another
focus. We considered some effects of features
such as privacy, materials, aesthetics, and furnishings. We also examined formal models of the
design process. For any given building there are
several important criteria: cost, durability, aesthetic quality, and the like. Among these are behavioral considerations such as the congruence
of fit between design and user needs. These criteria are used to decide among different design
alternatives in a rather complex process. As an
example of many of the issues of design, we investigated the American college campus as both
a product of culture and a fertile environment
for planning.

KEY TERMS

aesthetics	firmness	participation	primitive architecture
applicability gap	folk design tradition	pattern language	procedural theory
commodity	gaps	positive theory	substantive theory
congruence	Gestalt theory	post-occupancy	
delight	grand design tradition	evaluation (POE)	
design alternatives	modern design	preindustrial vernacu-	
design cycle	normative theory	lar architecture	

SUGGESTED PROJECTS

1. Ask several friends to list their most and least favorite buildings on your campus and to indicate the reasons for their choices. Is there any agreement among their likes and dislikes? Do you think these preferences are based upon firmness? Commodity? Aesthetics? Does what you have learned about environmental psychology offer any insights into the reasons for their opinions or ways to improve the least-liked buildings to facilitate human behavior?

2. The design process is complicated and sometimes resistant to behavioral input. Assume that your college is building a new student center. How would you approach the possibility of providing information for the process? What design alternatives would you suggest?

3. Many college campuses have inspired someone (often a graduate) to write an account of their growth and development. Visit your college library to see whether such a book has been written about a college campus familiar to you. With the book as a guide (if it is available), tour the campus. How does the architecture and use of campus buildings reflect the changes in cultural and educational values suggested by Turner's history of American college campuses?

4. Make two photocopies of a map of your campus on overhead transparencies. You will also need enough copies on plain paper to give two copies each to a sample of friends or classmates. Ask them to draw lines indicating the paths they take during typical school days. Collect these maps and trace each person's daytime paths to one transparency and their nighttime routes to another. Does this mapping technique identify commonly used travel corridors? Would this information be useful in considering the new location of a building, garden, or information kiosk? What differences do you observe between daytime and nighttime routes?

CHAPTER 12

DESIGN IN RESIDENTIAL AND INSTITUTIONAL ENVIRONMENTS

INTRODUCTION

RESIDENTIAL ENVIRONMENTS

ATTACHMENT TO PLACE

PREFERENCES

SATISFACTION WITH THE HOME ENVIRONMENT

USE OF SPACE IN THE HOME

NEIGHBORHOOD AND COMMUNITY ENVIRONMENTS

Propinquity: The Effect of Occupying Nearby Territories

Sense of Community

SUMMARY OF RESIDENTIAL ENVIRONMENTS

INSTITUTIONAL ENVIRONMENTS

HOSPITAL SETTINGS

Designing for Hospital Visitors

PRISON DESIGN AND BEHAVIOR

DESIGNING FOR THE ELDERLY

Noninstitutional Residences for the Elderly

Residential Care Facilities for the Elderly

Specialized Facilities for the Cognitively Impaired: Alzheimer Units

SUMMARY

KEY TERMS

SUGGESTED PROJECTS

Introduction

Your first breaths are taken in a sterile room with very institutional surroundings. Your parents take you "home" and place you in a strange container that looks like a cell with bars on two sides (which they call a "crib"), and when you look around, you see stuffed toys, a changing table, and a rocking chair. When you get a bit older you begin to explore your apartment and continue to be affected greatly by its environment. After 2 years, your parents buy a house and are very proud of their "very own home." Soon it is time for you to attend school. On your first day you are amazed by the large number of desks, their arrangement, and the whole educational environment. In some ways the classroom setting is stimulating, but in other ways it constricts your behavior—you must get permission to move from your personal work station to another part of the setting. More time passes, and you go to work. The work environment bears some similarity to the school setting, but in many ways it is different. You work in a large "open" office and do most of your tasks at a computer terminal. While you like the fact that you have easy access to your co-workers, you also feel you have insufficient privacy. You value your annual vacation and enjoy spending time in recreational environments, away from the work setting. As you move up the organizational ladder, you experience other work environments and note their positive and negative effects on yourself and your co-workers. After you retire, you live in a large retirement community. Although you appreciate having other retired people as well as medical facilities nearby, you miss aspects of the more heterogeneous environment you lived in before. No matter what the nature of your home has been, you have generally been satisfied with it.

In previous chapters we have examined how specific aspects of the environment interact with our behavior. In Chapter 11 we saw how we can use knowledge of these environment–behavior relationships to design environments that will facilitate the behavior we want to occur. In this and the next chapter we will see how these principles can be brought together in specific environments. That is, we will select some settings that have been studied extensively by environmental psychologists and show how these particular environments influence behavior, and how the design of these environments can be modified to achieve desired effects. This chapter will examine environmental psychological research on residential settings, hospitals, prisons, and facilities for the elderly. The next chapter will continue the same theme by examining work environments, learning environments, and leisure settings. We should caution that numerous books and articles have been written on behavior in each of these environments and that we have room to discuss only an outline of the relevant material on each of these topics. What we will emphasize is how some of the principles of environmental psychology can be applied to each setting.

Once again, we will find it useful to apply our eclectic model from Chapter 4 to our current discussion, as depicted in Figure 12-1. Our perceptions of residential and institutional settings are influenced by individual differences (e.g., housing preference) as well as the physical setting itself, and these perceptions in turn influence our favorable or unfavorable interactions with the setting. Two concepts in particular will be common threads in this chapter. First, **person–environment congruence** is paramount: The setting facilitates the behaviors and goals appropriate to the setting—a major proposition in Barker's ecological psychology. To the extent that congruence does not hold, arousal, overload, reactance, and other responses will occur, and we will attempt to change the setting and/or change our behavior as we adapt to the conditions present. Aftereffects, in turn, may well occur. A second common thread in this chapter is that perceived control (or lack thereof) is exceptionally important in the settings we will study: Our behavior in the setting is in part a function of the degree of perceived control the environment offers. Moreover, we will see that we can add design features to enhance personal control. ▣

Residential Environments

We have discussed elements of residential environments in early chapters—such as residence halls in Chapter 9—and will consider them in more detail here. It is no surprise that they have been mentioned so often—residential settings are so familiar and important to most of us that there has been a great deal of research interest in residential design and improvement (for reviews and bibliographies, see Altman & Werner, 1985; Cooper Marcus & Sarkissian, 1985; Rullo, 1987; Tognoli, 1987; van Vliet, 1998). We will begin our discussion with the concept of attachment to place, then examine satisfaction, preference, and use of space in the home, and finally move to the larger neighborhood and community, where we will discuss proximity and the sense of community.

ATTACHMENT TO PLACE

In previous chapters we observed how architectural and social factors contribute to a sense of place; such is certainly the case for our homes. Homes are important for reasons other than shelter. They also provide meaning and identity in our lives. For example, they signify status (e.g., Duncan, 1985), they structure our social relationships, they afford a location for major activities of daily living (e.g., eating, bathing), they are centers of regular and predictable events, and they trigger many of the memories central to our formative past, all of which contribute to a form of psychological bonding with this environment (Werner, Altman, & Oxley, 1985). These bonds can extend beyond the household to the neighborhood and larger regions. Environmental psychologists refer to this bonding as *attachment to place* or *place attachment* (Altman & Low, 1992; Giuliani & Feldman, 1993).

When discussing life in the city in Chapter 10, we noted that attachment to home and neighborhood can be very strong, even for those who live in slum or near-slum areas. In a study of citizens of Senegal, Ireland, and the United States, Newell (1997) found that 38% identified their home as their favorite place. In a landmark investigation, Fried (1963) studied a group of families who were forced to move from the West End as part of a Boston urban renewal project. Although the new housing was a physical improvement, the loss of social bonds between friends in the neighborhood caused considerable grief. In fact, those with a weaker attachment to the old place had an easier time adapting to the new one. Thus, social bonds can play a significant role in place attachment. In general, attachment to place includes an affective or emotional bond to the place (which bond may be mediated by social ties), memories and other cognitive interpretations that provide meaning to our experience with the place, and a sense of anxiety associated with potential removal from the place. The greater the attachment, the greater the distress can be if separation from the place is forced. Holman and Silver (1994), for example, found that residents of the Los Angeles area who lost homes to earthquakes and fires showed more distress the higher their attachment. If you attend a college or university away from home, the feeling of homesickness may in part be a result of your attachment to home (Burt, 1993). For a discussion of definitions of place attachment, see Giuliani and Feldman (1993).

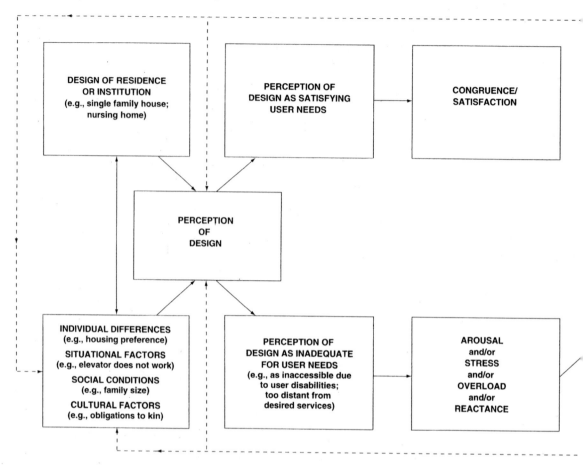

Figure 12-1   Application of our eclectic environment–behavior model to residential and institutional environments.

The extent of residential place attachment can be influenced by bonds to more than just the home and associated social ties. Furnishings, antiques, heirlooms, and other objects can be part of the attachment (Belk, 1992), as can automobiles. In a study of two Bern, Switzerland, neighborhoods, Fuhrer, Kaiser, and Hartig (1993) found that home and neighborhood attachment was often associated with vehicles; moreover, lower transportation mobility was associated with higher attachment to place. If we live in a mobile society, then, does that mean that we are unlikely to form place attachment bonds? Perhaps the attachment to a specific place is weaker, but Feldman (1990, 1996) argues that attachment to a type of settlement

(e.g., suburb versus city) remains strong even in the face of mobility (cf. Lalli, 1992). In a study of a Caracas, Venezuela, barrio, Wiesenfeld (1997a) presents strong evidence that attachment is not just to the home itself but to the neighborhood and community. What happens when we do move from a place where we have become attached? Brown and Perkins (1992) describe three stages we go through in such disruptions: predisruption, disruption, and postdisruption. Adequate preparation in the predisruption phase (e.g., having previous experience with a move) and having a means of dealing with grief in the postdisruption phase can help reduce the stress of relocation. Since attachment to possessions can play a key role in attachment

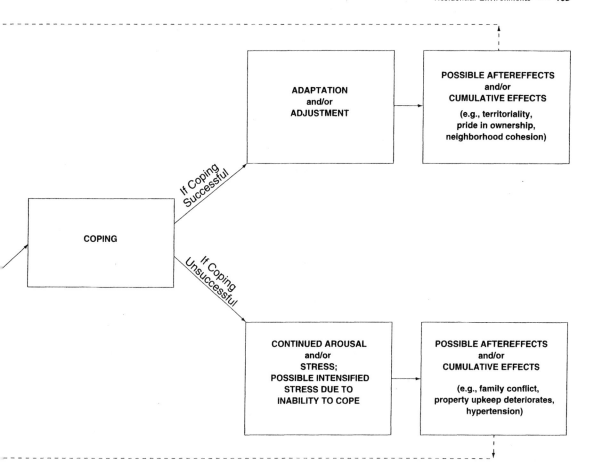

to the home, taking your possessions with you also helps you adjust to the new place. If you lived in parts of Indonesia, a move would mean taking the whole house—wood structure and thatched roof—with you, much as mobile homes are transported in the Western world (Waterson, 1991). Attachment to the new location is facilitated if it contains features that promote its preference over the old (Churchman & Mitrani, 1997).

Other contributors to place attachment include institutions such as religion. Religious rituals in the home as well as in other sacred spaces can add to the bonding to place (Mazumdar & Mazumdar, 1993, 1999). Indeed, the variety of factors bearing on place attachment is

such that even the homeless often show many of the characteristics of attachment to the place where they currently live (Bunston & Breton, 1992). Among nonhuman species at least, place attachment seems to have biological roots (Figure 12-2). Taking advantage of all these aspects of place attachment, Cooper Marcus (1996) shows how attachments and disruptions to attachments in the home can be used to guide psychotherapy and improve outlook on life experiences.

Attachment to place occurs over time (Hay, 1998). From our earliest childhood (Chawla, 1992) to our senior citizen years (Rubinstein & Parmelee, 1992), we form new attachments— and sometimes break old ones. Have you ever

Figure 12-2 There may be a biological basis to some forms of place attachment. Some varieties of salmon migrate hundreds of miles before returning to spawn at the site of their own origin. Canada geese mate for life and return to the same nesting place year after year. During the nesting season they are extremely territorial.

known someone who moved into your community and yet maintained strong attachment to an athletic team from the old community? As we form new bonds we use many mechanisms to maintain attachment to our former places. Baird and Bell (1995), for example, describe a case of a 26-year-old leukemia patient coping with treatment and, eventually, death. Attachment to a hospital room became such that the patient requested it upon repeated admissions. Attachment to home—including family bonds—was such that the patient insisted on returning home to die. The hospital room had offered a view of a cemetery, and the patient became attached enough to it to choose a final gravesite within that view. Place attachment takes many forms, indeed.

PREFERENCES

As we saw in Chapter 10, there are a number of different kinds of residential settings. The most common type, at least away from the core of the city, is the single-family detached house that many of us grew up in. In urban settings, row houses or two- or three-story apartment buildings may be the rule, while in many areas townhouses have become a predominant form. Also, of course, there are high-rise apartment buildings. In many parts of the world, "squatter" communities are commonly constructed out of cast-off building materials. With all of these different forms of housing, it has become important to examine residential preferences. Research seems to suggest that people in North America and the British Commonwealth tend to describe the detached, single-family house as the "ideal" home (e.g., Thorne, Hall, & Munro-Clark, 1982). This tendency does not appear to be a matter of socioeconomic class—people seem to reject apartment settings and prefer suburban homes regardless of ethnic or social background or the kind of housing environment they occupy.

Although you may have suspected that most people want a "nice little house in the suburbs," the reasons for this preference are not clear. Although very young and very old adults prefer living closer to the center of the city, most adults prefer to be farther away (Lindberg et al., 1992). The growth in popularity of the suburban detached house is attributable to many factors, including, among others, pride in home ownership, government incentives (such as for tax deductible mortgage interest), and the growth of transportation systems leading into and out of large cities (e.g., Jackson, 1985). To environmental psychologists, however, the most compelling explanation is the way these settings structure space. We saw in Chapter 2 that legibility and cohesiveness are important dimensions determining preference for natural landscapes. The same applies to housing: People prefer well-structured space typical of the single-family house (Herzog, 1992)—space with clear boundaries that form a distinct whole.

LIFE IN THE SUBURBS:
Is It Really What You Think?

We have noted that preferences fairly consistently show the majority of people want a single-family house and that most envision the suburbs as the best location for such an arrangement (see also Chapter 10). Schools, transportation, and recreational opportunities have some role in this vision. But just how ideal are suburbs? Jackson (1985) describes the history of the suburban movement in America in a book appropriately titled *Crabgrass Frontier.* Although development on the outskirts of cities has been around since cities began, most of us think of "contemporary" suburbs as they have developed after World War II. The "prototype" planned development by William Levitt is alleged to have started the suburbanization boom. Said Levitt, reflecting the atmosphere of the times in 1948, "No man who owns his own house and lot can be a Communist. He has too much to do." Jackson observes that in seeking our private, suburban ownership, we also evolve along a sameness dimension—everyone has the same dreams, same fence, same yard, same cars, same commuting route, same "little boxes," and it still seems preferable to the city!

But is all well in paradise? Lublin (1985), among others, has summarized some of the ills befalling suburbia. Homeowners must worry about flight paths for new airports. Office buildings, commerce, and industry are retreating to suburbs to cut costs and eliminate commuting. Growth, traffic snarls, pollution, noise, strains on public services, and lack of workers to fill low-end jobs accompany these developments. Housing becomes more expensive. Developers buy houses to put in commercial projects. Restricting the heights of these buildings to preserve views means development spreads out and traffic congestion becomes worse. The 1980 census showed that 27 million Americans actually commute from one suburb to another (Lublin, 1985). Ah, life in the suburbs!

The ways in which space is distributed in areas dominated by single-family homes allows residents to avoid intense interaction with neighbors. Urban areas are sometimes characterized by close neighborhood ties, extended families, and extensive social interaction. The single-family house appears to permit residents to avoid or control these social factors to a greater degree. Single-family housing is generally regarded as family oriented. Thus, people seem to want the family privacy afforded by these settings. Not all areas, of course, have close neighborhood ties and extensive social interaction (e.g., Altman & Wandersman, 1987). In these situations, as well, the single-family house seems to offer a sense of security along with perceived control over relationships.

Choice of type of housing may be restricted by economic factors that may make it difficult for young families to purchase single-family houses, in which case they may opt for townhouse or condominium settings or apartment living (cf.

van Vliet, 1983). A number of factors appear to determine the location that one lives in—where the house, townhouse, or the like is situated. Again, economic factors are important, since some areas of a city or its suburbs may be more expensive than others. Choices are often made between locations and types of housing. A family may be able to afford a single-family house in suburb A, but only a townhouse in suburb B. A number of things are ordinarily considered before making choices between locations and housing type (e.g., Shlay, 1985). The status communicated by housing types is very distinguishable, even for homes of 100 years ago (Cherulnik & Wilderman, 1986). Within housing types, building materials also convey status, with stone and brick having the highest status associations (Sadalla & Sheets, 1993).

Given the basic preference for the detached house, why might a family opt for a townhouse in a different suburb? Status may be an issue, if one of the two suburbs is very high or very low in prestige. Security and crime rates may be another—the detached home in suburb A may also be closer to high crime areas. Commuting time, closeness to and quality of schools, and availability of shopping and services may also be important. Clearly, preferences for and actual choice of housing is complex, with factors about the residential environment and its surrounding area being considered in each instance (see also Cook, 1988).

SATISFACTION WITH THE HOME ENVIRONMENT

Despite the general preference for single-family houses in the suburbs, are those who live in other types of arrangements satisfied with their current housing, even though the ideal might be something else? As we will see, it appears that psychological factors are very important in determining satisfaction—factors that are often present no matter what the style or location of the home. Amerigo and Aragones (1997) describe multiple studies of low-income women in Spain and how four factors consistently predict housing satisfaction: (1) the physical aspects of the housing, including quality of construction and how well the design supports basic living needs; (2) how the housing structures space, particularly with respect to crowding and privacy, and again how these affect the ability to support basic living needs; (3) safety, including safety within the unit and in the neighborhood; and (4) relationships with neighbors and the sense of community. These same factors seem consistent with the results of other studies on housing satisfaction (e.g., Bruin & Cook, 1997; Carvalho, George, & Anthony, 1997; Tognoli, 1987), as we will see in our further discussion.

As we become accustomed to a specific residential setting, we develop more and more satisfaction with our ability to perform basic tasks in it. The more easily and conveniently these functions can be performed, the more satisfied we usually become. The better we are able to adapt to the features of our residence (i.e., perform the desired functions despite less than favorable design features), the more satisfied we

are (Tognoli, 1987). Also recall that in the previous chapter we mentioned design alternatives as components of the design process. If an alternative dwelling is available for comparison that makes our own seem superior, we are more likely to be satisfied than if our own seems inferior (Tognoli, 1987). The size and floor plan of rooms, not having enough room to work, having too many rooms to clean, and being too close to noisy areas of the house can affect our ability to perform needed tasks in the intended areas. Galster and Hesser (1981) found that certain physical or environmental factors were associated with dissatisfaction. Poor plumbing, heating, or kitchen facilities were strongly related to dissatisfaction, as were neighborhood characteristics such as racial makeup, high density, or condition of the structures in the area (see also Michelson, 1977). Similarly, Kaitilla (1993) found that small size of houses, small living/dining areas, badly designed kitchen and bathroom facilities, and lack of storage space were associated with dissatisfaction in New Guinea public housing.

Physical and social factors are clearly interdependent in determining satisfaction (Anthony, Weidemann, & Chin, 1990; Weidemann & Anderson, 1982), but there are cross-cultural differences (e.g., Hourihan, 1984; Tognoli, 1987; Zube et al., 1985). Ross (1987) notes a particularly interesting problem in satisfaction among Australian aborigines who move into modern homes. Traditionally, aboriginal culture places high emphasis on taking care of relatives in need—bringing them into the home if necessary. Moreover, control over social pathology is handled by moving away—a relatively simple feat with a small, portable house and meager possessions. But a larger modern home is inviting to kin and makes moving difficult, such that residential satisfaction can be low; some aborigines even prefer smaller houses for this reason. Perceived control over social life is clearly related to residential satisfaction across cultures. Among residents of Canadian cooperatives, Cooper and Rodman (1994) found social control to be more important than control over physical aspects of the home in determining

Figure 12-3 Neighborhood ties are often stronger on cul-de-sacs than on through streets, as sometimes reflected in holiday decorations.

satisfaction. As another example, Pruchno et al. (1993) examined perceptions of space and satisfaction among American adults and children who were living in a house with a disabled elderly person. The more time the elderly person spent in space shared with the family, the more negative the perceptions.

Social ties also appear to be important in determining residential satisfaction. This is especially the case for residents of low-income housing (e.g., Amerigo & Aragones, 1997; Bruin & Cook, 1997). In urban areas, social ties appear to contribute to a sense of neighborhood and the sharing of outdoor space by residents. Greenbaum and Greenbaum (1981) found that group identity in a neighborhood is related to territorial personalization and social interaction. When people were able to establish social bonds with those around them, they took more care in decorating the exteriors of their residences, and the neighborhood took on the aura of group-owned territory. Neighborhood ties are stronger on cul-de-sacs than on through streets. This effect is often reflected in more decorations at Christ-

mas and Halloween on cul-de-sacs (Brown & Werner, 1985; Oxley et al., 1986; Figure 12-3).

Even teenagers show differences in neighborhood evaluations. Van Vliet (1981), for example, found that teenage Canadian residents of suburbs, relative to counterparts in a city, were more satisfied with neighborhood safety, "nice looks," friendliness, and quietness. City dwellers, however, rated their environment as having more things to do.

Other social factors can be significant sources of satisfaction or dissatisfaction for many people. We have already mentioned privacy regulation as an important consideration in the design of environments. How a residence is designed can affect the ease with which we achieve privacy. Individuals differ, however, in the amount of privacy they want. Privacy in the single-family home, as reported by the male members of the household, could take two forms (Altman, Nelson, & Lett, 1972). One type of family controls privacy without using physical features of the environment as a means of control. In these families, bedroom doors are rarely closed, and few areas of the home are considered the domain of one family member. A second family type is more likely to use environmental controls over privacy. These families help to ensure privacy by designating rooms as specific territories for individual use. Thus, one cannot use a single residential design and expect both types of families to be satisfied. Variety of interior design of homes helps to ensure that individual family styles can be accommodated (cf. Morris, 1987; Oseland & Donald, 1993; Shlay, 1987).

Clearly, then, psychological and social factors are at least as important as physical factors in determining residential satisfaction. This point is especially evident in a study by Paulus, Nagar, and Camacho (1991) comparing U.S. army families of comparable socioeconomic status who rented either apartments or mobile homes. Table 12-1 details just how important

Table 12-1	COMPARISON OF APARTMENT AND MOBILE HOME LIVING °	
Apartments	**Mobile Homes**
Environmental Quality	
More Positive on	*More Positive on*
Attractiveness	Noise level
Other people in complex	Crime risk
Closeness of services	
Adequacy of recreation facilities	
Reason for Choosing Housing	
Fire safety	Distance between units
Weather safety	Lower noise levels
	Can have own place
	More space
	More privacy
	Better for raising children
Satisfaction High for Both Types of Housing Because of:	
High level of perceived choice in selection process	
Future expectation of housing quality improvement	
Current housing compares favorably to past housing or friends' housing	

° *Paulus et al. (1991).*

nonphysical qualities were for both types of housing in determining satisfaction; note the psychological reasons satisfaction was high for both types of housing.

USE OF SPACE IN THE HOME

Despite differences in the ways in which people arrange their homes, consistent space-use patterns emerge. Home interior designs especially mediate social interactions (Bonnes et al., 1987; Werner, 1987). Bedrooms are, for one thing, intended to be private space, and thus are likely to be set off from less private areas. They may be located down a hallway from the living and kitchen areas, or on a different floor altogether. Bedrooms are also supposed to be for sleeping and therefore must be quiet, but they are also more likely to show signs of personalization and informality than living rooms, where guests are to be received (Figure 12-4A & 4B). Yet the master bedroom and master suite have increased substantially in size in American homes over the past decade or so, accommodating more and more functions (eating, paying bills, using computers) we are likely to perform there (Hasell & Peatross, 1990).

The bathroom is an especially interesting design problem. Consider the many functions or purposes of a bathroom. Kira (1976) identified more than 30 functions, including, among others: brushing teeth, rinsing mouth, gargling, expectorating, cleaning and soaking dentures, vomiting, treating skin blemishes, cleaning ears, applying cosmetics, shaving, defecating, urinating, bathing, washing wounds, applying bandages, taking medicine, and inserting contact lenses. Each one of these functions can serve as the basis for bathroom design criteria. To the physical hygienic functions that we are all familiar with, we can also add some social functions. Many people use the bathroom as a "sanctum" for privacy. Social conventions frown on people interrupting one another while in the bathroom and, as a result, one can often escape there for a moment of peace and quiet. Thus, even though the bathroom can take on attributes of shared space, it can also serve as a place where privacy can be achieved on a transient basis. Of course,

Figure 12-4A & 4B   Bedrooms are more likely to show signs of privatization and informality than are living rooms.

this is not the case with large bathrooms, such as those in a college residence hall. There, the small stalls within the bathroom may serve the same privacy function as the entire bathroom. The role of the bathroom in affording privacy should not be underestimated. Inman (cited in Meer, 1986) surveyed 200 households in Indiana. Regardless of how much space the rest of the house had, about half of all families with only one bathroom felt stressed because of a perceived lack of living space, compared with about 20% of those with more than one bathroom. She cautioned, however, that having more than three bathrooms increased stress because of problems related to cleaning and stocking supplies!

Kira (1976) also examined the relationship between the design of bathrooms and their functions. Many of his suggestions are intended to increase convenience, including changes in sink and vanity design and placement of shower control knobs, electrical outlets, and so on. Other suggestions were more adventurous, and one goes completely against the notion of privacy in the bathroom: the creation of a living room bath that could be used for entertaining guests. He viewed such an unusual situation as a logical response to economic pressures. When residential settings become smaller because of energy and land costs, designers often include multipurpose space in homes. Because bathrooms are not used all of the time, they are prime candidates for new functions, and Kira argued that they could be used as living rooms as well as bath areas (Figure 12-5). While we're discussing new functions for the bathroom, we should mention a high-tech toilet introduced by the Japanese firm Matsushita in 1999. Not only does it serve the usual functions, but it also weighs you, checks body fat, and analyzes sugar content of urine automatically (Holden, 1999a).

We noted that master bedrooms have increased in size in recent years. Hasell and Peatross (1990) suggest that part of the motive behind this change is to accommodate the functions of two adults who work outside the home: As women more commonly enter the workforce, the master bedroom and bath need to allow both partners to get ready at the same time. Other design adjustments with the same root cause include larger and more open kitchens to allow more sharing of functions, and a separate

Figure 12-5 Kira argues that bathrooms can be used for less privatized functions. Would you feel comfortable that way? (Photo courtesy of Harleen Alexander)

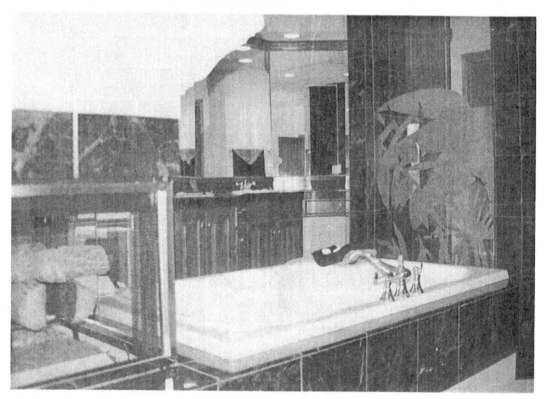

HOMES AND THE PRIVACY GRADIENT

Understanding people's social values and practices is extremely important for developing a home design that will provide desirable levels of privacy. Adequate privacy in the home also enables people to develop a sense of community with neighbors (Wilson & Baldassare, 1996). In the United States, a number of modern designs are based on open architecture that includes the kitchen as part of the area for entertaining guests. In many cultures, a **privacy gradient** exists that restricts certain areas in terms of entertaining. Zeisel (1975) describes a Peruvian arrangement under which formal friends and acquaintances are permitted only in the room intended for social activities. As guests become better known, they may be invited into other areas, but only those closest to the homeowner are ever permitted into the kitchen. Alexander (1969) suggested that a Peruvian house should be designed along a privacy gradient that places the *sala* (room for entertaining) at the front and the kitchen at the rear. Bechtel (1997) describes how the traditional Arab and Spanish home is designed with a windowless wall of the house along the street and the layout focused on an inner courtyard, encouraging privacy from outsiders. In French upper-middle-class homes, privacy is marked by distinct barriers. Carlisle (1982) observed that residents of these homes isolate intimate areas of the house with hallways, doors, grills, or curtains.

Why do we place so much emphasis on home ownership? Although there are many obvious economic and status reasons, Tognoli (1987) suggests that ownership implies less permeable boundaries (i.e., more control over privacy) than renting.

private space for the woman of the house (comparable to the den that was stereotypically for the man of the house). Another trend today is that more and more people are working at home. Ahrentzen (1990) found that maintaining a separate workspace and restricting access to it, as well as rescheduling activities, were mechanisms by which those who work at home accommodate such activities in the home.

So far, our discussion of space use has been for a typical middle-class American setting. Scheflen (1971) provides an alternative view of space use in the urban ghetto. Because many children share the same room, each can claim perhaps only one drawer and a quarter of a closet in the bedroom—if there is a separate bedroom at all. The kitchen is 9 feet by 12 feet (2.7 m by 3.7 m), but a bed, cabinets, and closets take up much of this space. The refrigerator is across the room because the original space for it was designed in 1920 and is too small for any refrigerator available today. How does one adapt to such a kitchen where two adults cannot maneuver at the same time? Scheflen noted two options: (1) All dining occurs in the living room, or (2) an end table or child's play table is the

kitchen table, so adults do not eat in the kitchen. There is one sofa and one chair in the living room, and a dominance hierarchy develops around their use, which varies from household to household. Territorial defense within the apartment is not achieved with physical partitions, but rather with behavior: Extending an elbow or leg keeps others a little farther away, and contours of posture help block visual access. Noise and lack of privacy are such problems that studying is difficult at best and frequently nonexistent. Although many factors contribute to the social ills of the inner city, these adaptations to spatial restriction should be considered by anyone exploring solutions (cf. Merry, 1981, 1987; Oxman & Carmon, 1986).

We might note also that there are considerable cultural differences in the use of space at home. Kent (1991) classifies more than 50 cultural groups worldwide based on their use of space. Some, such as the Mbuti Pygmy and the Navajo, rarely segment their living space for different tasks (e.g., eating, sleeping, entertaining). Others, such as Euramericans and Saudi Arabians, segment their space considerably. Interestingly, the congruence principle seems to hold, as design matches the culture: Those with little segmentation of functions tend to have homes with few barriers such as interior walls, and those with more segmentation of function have clearly defined, segmented spaces. What happens when a low-segmentation family is placed in a home with high physical segmentation and vice versa? Kent (1991) observes that under these circumstances, remodeling occurs: Low-segmentation Navajos occupying a three-bedroom western-style house were found to sleep and eat in the living room; high-segmentation Euramericans set up additional barriers in an otherwise low-segmentation house. Westernization in general tends to bring segmentation. Omata (1992), for example, observes how modern Japanese homes tend to have more private space than they previously did (Figure 12-6).

Figure 12-6A & 6B These housing units have the same amount of floor space. The one at the top has relatively high segmentation, whereas the one below is much less segmented.

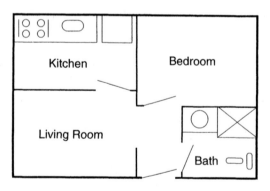

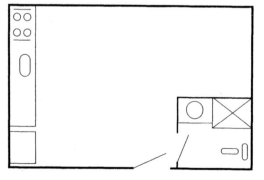

NEIGHBORHOOD AND COMMUNITY ENVIRONMENTS

Thus far, we have concentrated on environment–behavior properties of the individual household. Much of our residential life, however, centers on the neighborhood or community—our interactions with those proximal to us. But having multiple households in the same vicinity sets up a tension between desire for privacy and peace and quiet and the need to cooperate with others for mutual protection and social support. We conclude our section on residential environments by examining two important considerations: proximity and sense of community.

Propinquity: The Effect of Occupying Nearby Territories

Propinquity refers to "nearness" between places people occupy. How close you are to other residents in a housing development, an apartment building, or even a dormitory or an office building will affect your social outcomes with them. The more people know nearby friends and neighbors, the more attached they are to the neighborhood (Mesch & Manor, 1998).

Two types of propinquity have been found to lead to favorable social outcomes. First, it has been observed that the closer the objective physical distance between two individuals (what we called *euclidian* distance in Chapter 3), the more likely the individuals are to be friends. The classic study was conducted by Festinger, Schachter, and Back (1950), who investigated friendship patterns of apartment dwellers in Westgate West. When residents (who were randomly assigned to apartments) were asked, "Which three people do you see most often socially?" it was found that people were friendliest with those who lived near them. In fact, residents were more likely to be friendly with a neighbor one door away than with a neighbor two doors away, and so on. The same principle holds if you have ever been assigned to seating in a classroom based on alphabetical order of last names: You are more likely to become friendly with those whose names are close to yours in the alphabet (e.g., Segal, 1974).

Objective physical distance is not the only predictor of attraction, however. In Chapter 3 we introduced the concept of functional distance. It has been found that *functional distance,* defined as the likelihood of two individuals coming into contact, also predicts whether people will become friends or like each other (Ebbesen et al., 1976; Festinger et al., 1950). Functional distance becomes a more accurate predictor of friendship than objective physical distance when architectural features of a building constrain individuals whose apartments or offices are physically distant from frequent interaction. For example, the concept of functional distance would best predict attraction between two individuals who live five floors apart in an apartment building (distant in an objective sense) but who have adjacent mailboxes in the lobby (Figure 12-7).

Why does propinquity lead to friendship? Baron and Byrne (2000) offer some fairly convincing explanations. First, it is impossible to find grounds for friendship with someone we have never met, and those who are close to us in terms of physical or functional distance are clearly more readily accessible to us and to each other than individuals who are more distant. Second, since we have to continue to interact in the future with others who live in close proximity to us, perhaps we try a bit harder to "see the good side" of them and exert ourselves a bit harder to "make it work." Third, continued interaction with individuals obviously leads to a feeling of predictability and to a sense of security, which may make friendship more likely. Fourth, familiarity in and of itself may lead to attraction (Moreland & Zajonc, 1982).

Propinquity is more likely to lead to attraction under cooperative conditions where there is equity between individuals than under competitive conditions where there is inequity. Studies of functional and objective distance and friendship formation often examine homogeneous population groups—often newcomers needing help adjusting. Under these conditions, friendships are likely to form out of propinquity. On the other hand, propinquity can also create enemies. Ebbesen et al. (1976) found that more disliked than liked others lived close to participants of the study. They interpreted their results in terms of an **environmental spoiling hypothesis:** Positive social relationships follow from frequent contacts (which may or may not stem from propinquity), but the activities of

Figure 12-7 The distinction between functional distance and objective physical distance. The A and B units that share the same number are back-to-back and very close in objective physical distance. The functional distance for walking from Unit A-3 to B-3, however, is much greater. As a result, occupants of Unit B-3 are likely to be closer friends with occupants of any of the B units than with the occupants of any of the A units.

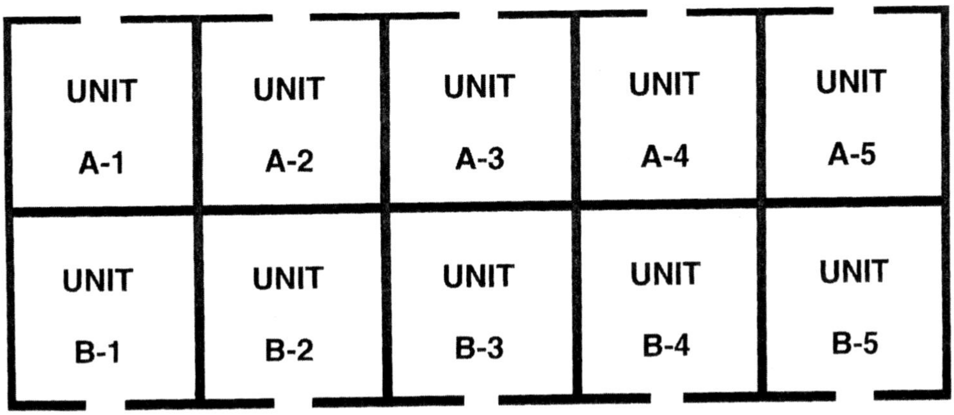

some can spoil the perceived quality of the living environment.

Satisfaction with one's neighborhood seems closely tied to this spoiling notion; spoiling, in turn, is a direct result of propinquity, since we would not object so much to what our neighbors did if their activity were far enough away that it did not disturb us. Merry (1987), for example, examined sources of neighborhood conflict. The most annoying complaints were social in nature—neighbors creating nuisances. Noise from neighbors, dogs making messes, vandalism, barking dogs, children playing in the street, children harassing adults, fights over street parking spaces, and trespassing were all sources of complaints—much more so than factors such as city services. As we indicated previously, greater ability to control these sources of aggravation is a major reason people prefer a single-family house in the suburbs; such homes typically have more distance between neighbors, fences to demarcate territory, private parking, and clearly defined areas for children to play such that they are less likely to be intrusive. This proximity/annoyance connection seems fairly universal. For example, lack of control over neighborhood noise was shown to be especially annoying in French neighborhoods (Levy-Leboyer & Naturel, 1991), and high neighborhood density was associated with low satisfaction in an Italian study (Bonnes, Bonaiuto, & Ercolani, 1991).

Sense of Community

In Chapter 10 we described the benefits of a sense of community in coping with urban life. Given that most of us must live within considerable proximity of our neighbors, what can we do to build a sense of community among our neighbors? A Dutch study suggests that neighborhood cohesion consists of two factors: neighboring (friendliness, looking out for one another's interests, providing social support) and sense of community. The relatively low cohesion found in higher-density multifamily housing was not due to lack of neighboring, but rather to lack of a sense of community (Weenig, Schmidt, & Mid-

den, 1990; but cf. Keane, 1991). Research shows that one clear benefit of social cohesion and a sense of community is reduced crime (Sampson et al., 1997), so designers should be very interested in fostering community feelings.

One interesting design feature that may build a sense of community is a front porch on the dwelling. Several locales are experimenting with ordinances requiring front porches large enough for multiple seating in order to encourage more informal interaction with neighbors, to learn who belongs and who does not, and to build more sensitivity to annoyances. The front porch also serves as a place to be away from other members of the household (Brown, Burton, & Sweaney, 1998).

Another design feature that enhances community feelings is intentionally designed informal spaces for social interaction (Abu-Ghazzeh, 1998; Skjaeveland & Gärling, 1997). Nicely landscaped, green outdoor spaces seem especially helpful in this regard (Coley, Kuo, & Sullivan, 1997; Kweon, Sullivan, & Wiley, 1998).

Another factor in a sense of community is ownership. Signs of physical decay (litter, dilapidated housing, abandoned cars) are associated with fear of crime (Perkins, Meeks, & Taylor, 1992). Removing such signs of incivility may improve a sense of community, but the owner of the problem often does not live in the neighborhood. Leavitt and Saegert (1989) describe how residents of Harlem reclaimed abandoned buildings and turned them into cooperatives; that is, the residents became owners of their buildings. Cooperation, caring, protection of common resources—all factors in a sense of community—were enhanced in this collaborative process. Similarly, Rohe and Basolo (1997) found that programs to assist low-income families in purchasing homes positively affect life satisfaction and participation in neighborhood meetings.

Neighborhood meetings and organizations can build a sense of community. Neighbors can begin with existing organizations, or a facilitator can help build a new one, such as a community improvement committee, a garden club, or a crimewatch program. Such organizations

COHOUSING:
Combining Privacy and Community

We have noted that apparent tension exists between privacy and community in housing design; the single-family home tends to maximize privacy but can minimize interactions with neighbors, whereas multi-family housing sacrifices privacy but can increase a sense of community. An alternative type of housing may be a good blend of the two. **Cohousing** communities, originally developed in Denmark, provide each family with its own detached unit but have a common building for dining and larger-scale entertaining. Variations are spreading in North America. The common building is designed to house a large kitchen and dining hall, laundry facilities, and entertainment/recreation facilities. Much of the rationale is to share larger, energy-consuming appliances. Adding solar and wind-generated power is an option, and the heating source for individual units is also shared. Privacy is afforded in separate, detached family units that contain bedrooms, baths, smaller living areas, and a small kitchen. Often, these individual units surround a common greenbelt space where children can be jointly supervised away from automobile traffic. Cohousing, then, attempts to offer privacy for private functions but to share facilities for functions where community interests are more of a concern.

promote getting to know neighbors, improve communication, and give loosely connected neighbors a common purpose (Wandersman, 1981; Wandersman & Hess, 1985; Wiesenfeld, 1997b).

SUMMARY OF RESIDENTIAL ENVIRONMENTS

Homes provide more than shelter; they organize much of our individual and social life and provide bonding. Attachment to place includes a sense of bonding, memories about the place, and anxiety in the face of potential separation from the place; attachment occurs to both the home and community. People occupy many different styles of homes, although the preference tends to be for a single-family home in a quiet area away from city problems; privacy and control over annoyances seem to be behind this preference. Satisfaction with the home, on the other hand, seems related not so much to style or location but to whether the home facilitates desired functions and meets expectations. Space within the home can serve single or multiple functions; there are large cultural differences in how domestic and other functions are accomplished in the home, and the operation of the home has much to do with privacy expectations and accommodations for privacy. Propinquity can lead to friendship formation or to perception of environmental spoiling; a sense of community promotes friendship and minimizes spoiling.

Institutional Environments

Outside the home we can observe environment–behavior relationships in numerous settings. Some of these have received much attention by environmental psychologists, and we will look at a few of them in the rest of this and the following chapter. As with residences, congruence or person–environment fit is a theme in our descriptions. Perceived control and privacy issues are also paramount. Although some of the issues are the same, institutional environments differ somewhat from residential environments in several common ways. For one, they are shared with many more people than is typical of a home. That makes privacy more difficult to achieve and territorial defense less likely—institutional settings tend to involve public territories (see Chapter 8). Second, many of the people who enter an institutional environment on a given day are there for the first time, so orientation and wayfinding aids become significant adaptive tools. Third, institutional settings usually serve multiple functions—learning, entertaining, shelter, and so on—such that designs must accommodate multiple demands. Fourth, in part because of the multiple demands and in part because of sheer size and complexity, overstimulation is often a problem in these settings —too much noise, confusing paths of exploration, and/or distracting activities. Finally, all of the above can lead to loss of perceived control in such settings, so many of the design features promoting satisfaction involve restoration of control. Let us see, then, how design influences behavior in some of these settings. The remainder of this chapter will be dedicated to institutions where people also reside 24 hours per day—hospitals, prisons, nursing homes—so we will see some direct relevance to the section on residences. The next chapter will look at work, learning, and leisure environments—settings where people typically spend less time than in a residence.

Hospital Settings

Much of what we know about design in hospital settings is derived from research on acute care and psychiatric hospital environments (Reizenstein, 1982). Many data-based investigations have studied psychiatric patients, perhaps because of greater ease and accessibility in using these individuals. For several decades researchers have methodically studied advantages and disadvantages of various hospital design fea-

tures. Three more recent volumes summarize some of these findings and the methods used to assess innovations (Carpman & Grant, 1993; Halpern, 1995; Moran, Anderson, & Paoli, 1990). We offer here a brief overview of some of the historic work (see also Malkin, 1992).

One aspect of hospital settings that has received attention is the low control or "low choice" forced upon patients and visitors. Hospitals typically have a great number of rules and allow patients only minimal control over the small spaces that they use. Olsen (1978, p. 7) pointed out that hospital designs can communicate this message—that people are "sick and dependent and should behave in an accordingly passive manner." Provision of greater spatial complexity or providing more options or variations in design can improve the situation and lead to more positive emotional responses. Originally, psychiatric hospitals resembled prisons more than hospitals (Figure 12-8A and 8B). Today, the design of psychiatric facilities tends to resemble that of a dormitory or hotel built around a central nurses' station and lounge (Figure 12-9A and 9B).

Do certain designs and locations of nurses' stations promote more efficient patient care? In a very influential study, Trites et al. (1970) investigated nurse efficiency and staff satisfaction with three different hospital ward designs (Figure 12-10). In general, a *radial ward design* was found to be the most desirable (relative to *single* and *double corridor designs*), both in terms of saving unnecessary ward travel and of increasing time with patients. Moreover, members of the nursing staff indicated a preference for assignment to the radial ward. The fact that nursing staff in the radial unit had more free time was interpreted as an indication that more patients could be housed on the ward.

Another issue that has received attention is the degree to which different designs affect social interaction among patients (see Devlin, 1992, for a review). Designs that protect privacy yet encourage appropriate social interaction would seem to be most therapeutic. We noted in previous chapters that some types of furniture

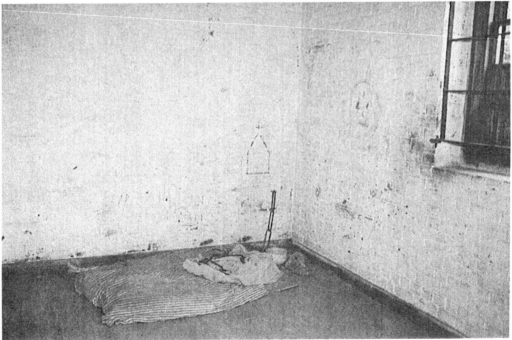

Figure 12-8A & 8B Exterior and interior of the first hospital in the United States built specifically for the mentally ill. This facility was built in Williamsburg, Virginia, in 1773.

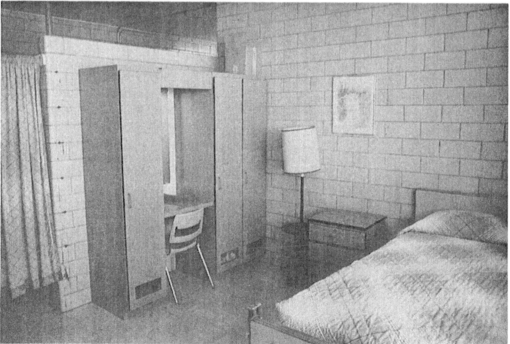

Figure 12-9A & 9B   Exterior and interior of a contemporary mental health institute.

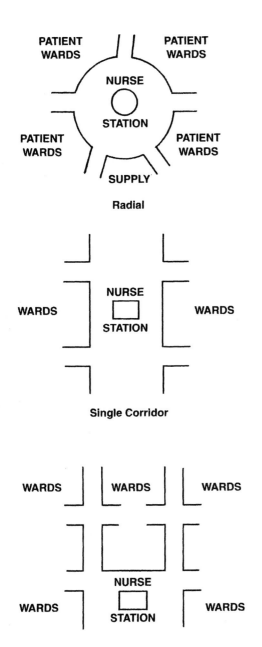

PATIENT WARDS | PATIENT WARDS

NURSE STATION

PATIENT WARDS | PATIENT WARDS

SUPPLY

Radial

NURSE STATION

WARDS | WARDS

Single Corridor

WARDS | WARDS | WARDS

WARDS | WARDS

NURSE STATION

Double Corridor

Figure 12-10 These designs represent three types of hospital wards. The radial unit appeared to be the most desirable in terms of staff satisfaction and amount of time spent with patients.

Adapted from Trites, D., Galbraith, F. D., Sturdavant, M., & Leckwart, J. F. (1970). Influence of nursing unit design on the activities and subjective feelings of nursing personnel. Environment and Behavior, 2, 303–334.

arrangements (i.e., sociopetal versus sociofugal) facilitate patient interaction more than others. In addition, keeping down the number of beds on a ward promotes social interaction and reduces withdrawal (Ittelson, Proshansky, & Rivlin, 1970).

Several case studies in which psychiatric facilities were modified are presented by Cherulnik (1993). In one, a corridor design and a suite design were compared with a control condition in which the previous institutional layout was unchanged. In the corridor design, one or two residents lived in each room with floor-to-ceiling walls and lockable doors. In the suite design, one to three residents were housed per room, with partitions used to separate sleeping areas and homelike furnishings supplied. Residents appeared most alert and purposive and interacted most with other residents (and less with staff) in the suite design; these desirable behaviors were least apparent in the traditional, unchanged control condition. In another case, dormitories were partitioned into two-room modules with added sound insulation and improved lighting; in addition, chairs were arranged in a semicircle sociopetal configuration. Social interaction among the residents doubled after the design modification (see also Halpern, 1995; Sommer & Ross, 1958).

Still other environmental features have been shown to have direct impact on well-being. Uncontrollable noise in the hospital is a problem for patients and staff (e.g., Topf, 1992a). Windows with a natural view are particularly therapeutic. Patients in rooms with a natural view have more favorable outcomes (Ulrich, 1984; see also Verderber, 1986). In describing attachment to place we mentioned a case study of a young leukemia patient coping with treatment and death (Baird & Bell, 1995). So powerful was the window effect for this patient that she preferred a hospital room with a view of a cemetery to a room with no view at all. As we described in Chapter 2, opportunities to view natural scenery seem to promote restorative experiences. Ulrich (1997) describes a number of opportunities to incorporate natural scenery directly into the hospital setting in order to promote a more favorable medical and psychological outcome.

A series of studies by Rivlin and colleagues examined a number of design variables and their effects on behavior in a children's psychiatric hospital. One important variable turned out to be bedroom size and occupancy. Design of play space in psychiatric facilities for children has also received attention (e.g., Rivlin, Wolfe, & Beyda, 1973). In general, little consideration was given to age level differences in planning play space in these facilities. Younger children, especially, seem to have difficulty adapting to highly controlled hospital ward space. Rivlin and her colleagues suggested that the behaviors younger children display in handling their disorientation to ward space may be interpreted by staff members as part of their disorder. In reality, the younger child may be reacting as any child might to restrictions on play caused by inadequate space. Consequently, special care facilities should be designed not only for treating illnesses, but also for encouraging the normal activities of a particular age group.

Designing for Hospital Visitors

Thus far we have discussed some considerations for hospital patients and staff. Zimring, Carpman, and Michelson (1987) point out that hospitals should attend to design needs of visitors, as well. They note that in this day of cost-saving efforts, visitors can often perform basic caregiving tasks instead of the staff, such as adjusting pillows, feeding the patient, or providing psychosocial support. Yet, many visitors are themselves worried about the condition of the patient, and they are in an unfamiliar environment. Numerous design flaws—many of which are readily rectified—add more stress for the visitor.

For example, how easy is it to find your way through all the corridors? Most visitors have a difficult time with wayfinding (e.g., Carpman, Grant, & Simmons, 1983–1984). Good signs in everyday language instead of medical terminology would help, as would frequent "you-are-here" maps, or paper maps visitors could carry. Other examples are cited in Zimring et al.

(1987). Visitors find hospital noise disturbing, so sound-absorbing materials (carpeting, furnishings) would help. Television sets are often positioned in patient rooms so that patients can see them from the bed, but visitors have to sit in uncomfortable positions for a good view. Privacy is also an issue: Visitors have no place to go for a private conversation. Conference rooms accessible to visitors or screened portions of waiting areas could alleviate this design problem. The list of design concerns for visitors includes many other dimensions we have discussed previously in other contexts, such as lighting, segregation of smokers and nonsmokers, odors, and furniture arrangements in waiting areas and patient rooms. Indeed, all of these components communicate whether or not the visitor has been considered when the design was developed.

PRISON DESIGN AND BEHAVIOR

In the United States and some other countries, there is a dominant attitude that those who break criminal laws deserve prison sentences. So pervasive is this view that as of June 1997 more than 1.7 million persons were incarcerated in the United States (e.g., Haney & Zimbardo, 1998). Yet imprisonment seems to have such a negative outcome that researchers suspect prison design itself could have important consequences. Recall from Chapter 9 that research by Paulus and his colleagues has led to some important conclusions regarding the behavioral effects of prison design, including consequences for behavior and health. Grouping prisoners together in large numbers is less healthful than grouping them together in small numbers, and single- or double-occupancy cells are better (i.e., judged as less crowded) than cells with more prisoners. Further, if large groups are "broken up" by partitions or segmentation of space (so that the large group becomes several smaller ones), psychological and physical health is improved (Cox et al., 1982; Cox, Paulus, & McCain, 1984; Schaeffer et al., 1988).

These findings provide strong evidence that the design of prisons influences mood and behavior of prisoners. At one level, this is not surprising. Prisons are usually designed along functional criteria and are not built for aesthetic reasons. Space is designed to facilitate order and regimentation. Bars and walls are deliberate attempts to constrain behavior, and economy is usually an important factor in determining the use of space. However, it is also clear that prison designs can have unintended demoralizing effects, and new prisons have been increasingly designed to avoid some of these consequences. This has occurred in spite of objections based on differing penal philosophies and increased costs.

The ethic that guided traditional prison design evolved from 19th-century concepts of correctional activity. Treatment was seen as being best accomplished by isolation from society, both physically and symbolically. To some extent, prisoners' clear separation from society may be traced to this ethic. Also, most people see prisoners as nonproductive elements of society, so we place a special emphasis on economy. As a result of these pressures, prisons are usually large, located in remote areas, and surrounded by high exterior walls that deny inmates visual access to the outside world (e.g., Haney & Zimbardo, 1998).

"New" prison designs usually seek to do several things. First, they attempt to provide more "humane" environments, replacing gun towers with natural barriers, adding color and lighting to otherwise drab settings, and increasing opportunities for privacy by providing more single-occupancy cells and by designing windows so narrow they do not need bars (Figure 12-11A and 11B). There has also been an attempt to build more cells into exterior walls so that inmates can have a view of the outside world. Control over the environment has also been heightened in some prisons, with inmates gaining control over heating, lighting, and even privacy in their own cells. Rather than separate observational booths or rooms for guards, the guards are placed directly in the living quarters of the inmates to increase interaction. These improvements have been made in conjunction with changing philosophies in dealing with people who break the law. Wener, Frazier, and Farbstein (1985, 1987) have indicated that such designs help remove the fear of violence at the hands of other inmates, resulting in 30% to 90%

Figure 12-11A & 11B The top picture displays a traditional prison gun tower and security wall. The bottom picture is a more contemporary facility with windows so narrow they do not require bars.

reductions in violence, similar drops in vandalism (e.g., to mattresses and light fixtures) and graffiti, and the virtual disappearance of homosexual rape (see also Williams et al., 1999; Figure 12-12).

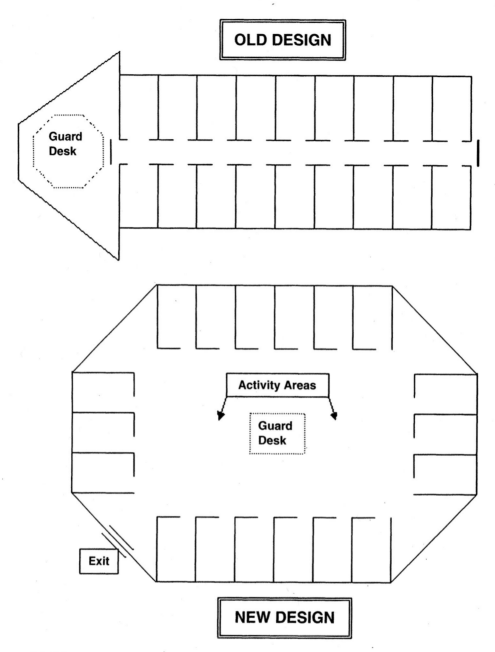

Figure 12-12 Conceptual floorplans of "old" and "new" prison designs. A critical issue in prison design is maximizing surveillance to promote security, yet doing so tends to eliminate a humane sense of privacy from any of one's daily activities. In "old" prison designs, walls of cells are simple bars so that guards and other prisoners can see everything that happens within. Guards are stationed in an adjacent room separated from the row of cells by a security door. Space for activities (exercise, visitation) is in another part of the facility and accessible only when the schedule allows. Exterior walls may not have windows, and interior lights are controlled by guards. In "new" designs, cells have solid walls for privacy, windows to the outside, and lights controlled by the individual inmate. The guard also serves as a counselor and is stationed in the open area surrounded by the cells. Inmates have free access to the guard's desk and activity areas within the unit.

DESIGNING FOR THE ELDERLY

Those over age 65 are the fastest growing segment of the population in the industrialized world, and those over age 85 are the fastest growing subgroup of the elderly. In the United States, 12% of the population was over 65 in 1987, and that figure will rise to 20% by 2030. Japan is the most rapidly aging country in the world, where the elderly population will grow from 10% in 1987 to 23% by 2020 (Dickson, 1987). As a greater proportion of our population becomes aged, providing specialized short- and long-term residential care facilities for them becomes very important. Certain characteristics of the elderly should be kept in mind when designing such environments. Especially important to consider is that the elderly are a heterogeneous lot. Too often, designers have assumed that the elderly are a homogeneous group, when actually they have only age and certain health problems in common. In fact, there is probably no other segment of the population with such a broad diversity of individual problems and needs. Some elderly citizens have trouble hearing, others have difficulty with vision, and still others have difficulty with locomotion. Many have no physical disabilities at all. Some elderly people suffer from psychological difficulties (e.g., withdrawal, distorted thought processes) while others do not. Another source of diversity in the elderly is the fact that individuals have established long-term behavior patterns, and these differ greatly among people. All of these considerations argue for "designing in" flexibility in any facility for the elderly. In general, these facilities can be discussed as either institutional on noninstitutional, and together with in-home services such as "Meals on Wheels," they make up what is called a *continuum of care*. Institutional facilities are nursing homes and similar settings that provide a relatively high level of care for residents, such as nursing care (e.g., administration of medications) and what is termed *custodial care*, or care for everyday needs such as meals and laundry. When the emphasis is on custodial care rather than medical care, the setting is often called *intermediate care, residential care,* or *assisted living,* meaning the individual can handle some aspects of daily living but needs assistance with others. When the emphasis is on a high level of medical care (e.g., rehabilitation for a broken hip or for a stroke), the setting is often called a *skilled nursing facility,* or *SNF* (pronounced "sniff" in nursing home parlance). Despite our stereotypes, only 5% of those over age 65 live in nursing homes or similar facilities. Noninstitutional settings include retirement villages or similar housing where residents provide most of their own daily care. What is paramount is to match the needs of the individual to the setting.

In Chapter 4 (see the box on page 112), we described Lawton's model of environmental press and competence in the elderly, which holds that if the press of a specific environment is within the competence of the elderly to handle it, positive adaptation will occur. Design principles can be applied to institutional and noninstitutional facilities for the elderly to help them maintain competence (see also Altman, Lawton, & Wohlwill, 1984; Carp, 1987). If we use suitable designs to compensate for disabilities, we can enable the elderly to live as independently as possible in their own homes if feasible, and otherwise in suitable supportive settings. As their needs increase, accompanying design modifications along the continuum of care can help them maintain the highest degree of competence and control that their physical and cognitive condition will allow (Gilderbloom & Markham, 1996; Slangen-de Kort et al., 1998). As we will see in our subsequent discussion, perceived control is paramount in designing successfully for the elderly (Schulz & Heckhausen, 1996).

Noninstitutional Residences for the Elderly

In many ways, it is probably better if one can stay at home or in a relatively "homelike" setting without having to endure certain almost inevitable problems associated with institutional environments—certainly the elderly prefer it that way. Accordingly, special residential housing facilities for the elderly have been planned and built. Also, many services (e.g., "Meals on Wheels," home-based health care) are now made available to elderly citizens who are living "at home."

What type of residential housing for the aged seems to be best? Various elements (e.g., high and rising rents for people on fixed incomes, long-term residences being turned into condominiums) can make it difficult for the elderly to find a decent place to live at a reasonable price. Overall, providing planned housing specifically for the elderly seems desirable (for reviews, see Baltes et al., 1987; Birren & Schaie, 1996; Carp, 1987; Moos & Lemke, 1985; Parmelee & Lawton, 1990). Is it best for the elderly to live in age-segregated or in heterogeneous environments? The confluence of evidence suggests that the elderly prefer age-segregated housing. Living with others of one's own age is associated with housing satisfaction, neighborhood mobility, and positive morale, perhaps because more similar others are in close proximity when the neighborhood is age-segregated, and both similarity and propinquity elicit attraction. In addition, age segregation probably results in more activities that are appropriate to an elderly population, specialized services (e.g., geriatric medical care) may be more easily targeted in an age-segregated setting, and annoyances such as certain forms of noise can be reduced.

What other factors should be taken into consideration when planning noninstitutional residences for the elderly? One important element in residential living for the aged population is adequate transportation. Planners should be certain there are bus routes running through areas where seniors are concentrated, and these should include stops at places where the elderly need to go (i.e., medical complexes, shopping areas). Buses should also be accessible to elderly citizens (and others) with disabilities. Designers should also explore means of making other aspects of the community accessible (e.g., Evans et al., 1984). Many elements in the environment (e.g., exterior stairways, nonautomatic doors) may be discouraging and dangerous to an elderly population. On the other hand, Kweon

et al. (1998) report evidence that green outdoor common spaces encourage social ties and a sense of community among the elderly.

Residential Care Facilities for the Elderly

How should designers approach the task of planning residential care facilities for this population? They should attempt to compensate as much as possible for the physical and psychological difficulties that some elderly individuals have, without unduly constraining the lives of people who have no particular problems (e.g., Baltes et al., 1987; Parmelee & Lawton, 1990). That is, designs should foster adaptation. A variety of environmental options should be provided so that people can continue to engage in the same activities as they did prior to institutionalization. Also, it is important for designers to attempt to view things from the perspective of aged residents, which may differ from the ideas and needs of the staff of the facility or those of the designer. Too often, facilities for the elderly are designed in accord with an architect's vision (which may not be sufficiently informed about the elderly) or are planned so that they make life easy for the nursing, cleaning, or maintenance staff of the institution. A study by Duffy et al. (1986), for example, found that designers and nursing home administrators were biased in favor of nursing home designs that emphasized social interaction, whereas residents who were not cognitively impaired actually preferred designs that fostered privacy.

A number of design features would seem to be useful to incorporate in a short- or long-term residential care facility for the elderly. First, it is important for the environment to provide for safety and convenience. To promote safety, a facility should permit sufficient staff surveillance to avoid accidents or to detect them when they do occur, while not eliciting the feeling that there is no privacy. In addition, specific design features should be included to prevent accidents (e.g., handrails in halls, "nonslip" surfaces), and aspects of the design certainly should not cause accidents. For example, entrances should be protected from the elements: Many elderly are not steady on their feet, and snow, ice, rain, or wind around entrances can be extremely hazardous. Also, elements should be included that permit clients to notify staff if they have a problem in a private area (e.g., call buttons in bathrooms). The design should promote convenience by providing orientation aids (e.g., color-coded floors, cues to differentiate halls), as well as by affording comfort (e.g., chairs should be easy to get in and out of). Convenience is also fostered to the extent that the setting is "barrier free" and allows a large proportion of the population to move about independently. Finally, important facilities (e.g., bathrooms, communal areas) should be within easy access of rooms. In many ways, facilities conforming to the above criteria will promote feelings of personal control, prevent helplessness, and elicit positive outlooks in residents (e.g., Rodin, 1986).

The design of a residential care facility for the elderly should also foster choice (and in doing so, feelings of control). The location of the facility should be sufficiently close to a community to allow residents to choose among a variety of available services (e.g., grocery stores, movie theaters; Smith, 1991). Choice is also facilitated when the design contains various types of spaces that can be used for special purposes (e.g., recreation, privacy, dyadic as opposed to large-group communication). Recreation areas should be designed to elicit communication (i.e., should be *sociopetal*), but some areas should afford privacy (i.e., should be *sociofugal*). It is very important that there be a range of social and recreational choices available to each resident. Also, each resident should have access to both a bathtub and a shower, and it is preferable for each room to have individual heating controls. Bathrooms with "grab bars" and accessibility to wheelchairs improve safety and reduce dependence. Without adequate degrees of choice being promoted by physical design, the environment can promote loss of perceived control and helplessness.

In addition to providing choices, objective physical conditions of the facility should be adequate and appropriate. Rooms should be of sufficient size, there should be enough recreational space for the resident population, and the construction should be of reasonable quality. Objective physical conditions affect patient behavior in many ways. When large sitting areas are occupied by relatively few residents, there seems to be a low level of physical interaction. Designs including long corridors appear to discourage resident mobility. When physical arrangements cause residents to be grouped in areas closely accessible to staff, some positive outcomes occur (e.g., the staff has more surveillance over accidents and danger, and interacts more with residents). However, some negative outcomes also occur under these circumstances (e.g., the staff may behave in ways that encourage patient dependency). It is also important for higher-functioning residents to have their own kitchen facility; otherwise, residents tend to depend on staff members even to get a cup of coffee. This situation encourages helplessness.

In addition to the physical environment, the social environment is extremely important to the well-being of the institutionalized elderly. A great deal of research has found that when the social environment fosters perceived choice and personal control, the well-being of the elderly is enhanced (cf. Rodin, 1986; Rowe & Kahn, 1987; Woodward & Wallston, 1987). Unfortunately, both the social conditions under which many people arrive at institutions and institutional life itself typically promote a loss of control (cf. Crozier & Burgess, 1992). The new resident is often stripped of his or her accustomed relationships and satisfactions and must give up personal property that has served as a means of self-identification. Also, people frequently come to a residential care facility after having problems with illness, financial setbacks, and family difficulties, all of which foster a sense of loss of control. Often, the family decides that the person cannot remain at home any longer and to which institution the person will go. The very character of institutional life (e.g., one must submit to rules and regulations, to authority, to "standardized" schedules and procedures) adds to one's feeling of loss of control.

One's response to being relocated and entering a long-term residential care facility is more positive if he or she is afforded a degree of control over the process. Reactions are more favorable when the person has chosen to be institutionalized and has picked the particular facility that he or she will live in, and when the difference in control between the pre- and post-relocation environments is not great (Imamoğlu & Kiliç, 1999). In Chapter 4 we noted a study by Langer and Rodin (1976) in which one group of institutionalized elderly was treated in a way designed to increase feelings of control. In the group where control was fostered, residents were happier, their conditions had improved somewhat after several months, and they showed more activity (e.g., were more apt to attend a movie and to participate in a contest) than in the condition where control was not encouraged. In addition to providing control, one way to increase predictability (and hence resident well-being) is to give people preparatory information about their forthcoming move. Another helpful procedure is to familiarize the resident with the building prior to the move by means of a three-dimensional model and slides of the various rooms and corridors (see Chapter 3). Among other things, this procedure aids wayfinding (and cognitive mapping) once the resident moves into the building (Hunt, 1984).

One problem for residents is that with subsequent declines (or improvements) in health, further relocation may be necessary. In order to minimize the negative effects of this movement, multilevel facilities—those that offer many levels of care and supervision in one place—are becoming popular. These are beneficial because they minimize the effects of relocation by enabling subsequent moves to be from one part of the facility to another, rather than from one facility to another. People can still have access to their friends and can be moved easily if their condition again improves or deteriorates. In essence, people can move to an area whose design and staff support a level of care required by the circumstances (Figure 12-13A and 13B).

Specialized Facilities for the Cognitively Impaired: Alzheimer Units

Although most elderly are not cognitively impaired, about 10% or so of those over 65 have **Alzheimer's disease,** which is characterized by loss of memory, confusion, impaired judgment, and progressive decline to more and more dependent states of existence. It is the fourth leading cause of death in the United States, afflicts at least 4 million Americans, and accounts for half of all nursing home admissions (e.g., Cook-Deegan, 1987). We have emphasized the need to permit choice among nursing home residents. What happens when residents are so cognitively impaired that free choice can be dangerous? One answer is the creation of specialized Alzheimer units or dementia units within nursing homes or other suitable settings. **Dementia** is a medical term meaning long-term loss of cognitive capabilities. Although Alzheimer's disease accounts for two-thirds of all dementias, there are many causes of dementia, and the behavioral consequences of these diseases are such that the environmental interventions for them are essentially the same.

Alzheimer units are designed with the idea that those with dementia display diminished judgment capabilities and progressive confusion. These residents are very prone to wandering off and can easily become lost. Traditionally, dementia victims have often been physically or chemically restrained to prevent wandering and other potentially dangerous behavior. An Alzheimer unit attempts to minimize the need for these restraints: The setting is designed to make life safe and less restrictive by adjusting the environment to the behavior. For example, the units have locked access so that a key (or combination) is needed in order to leave the unit. Such an arrangement may sound cruel, but if it is carefully designed, it permits minimal use of restraints: The residents can wander within the unit all they want without wandering off or otherwise endangering themselves. Most of the time the unit includes a secured outdoor area with open access for the residents, so they may

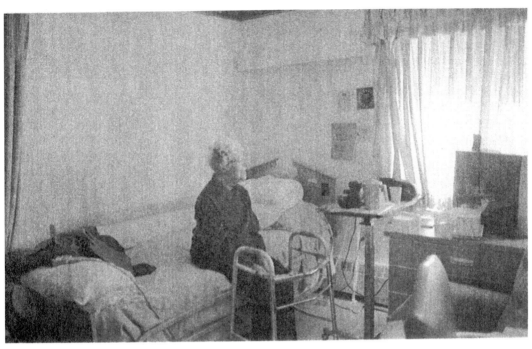

Figure 12-13A & 13B   Providing several levels of care in a facility for the aged can minimize difficulties associated with relocation but can also reduce the morale of healthier residents. An appropriate continuum of care provides an environment that supports the remaining capabilities of the residents. (Photos courtesy of Columbine Health Systems)

⌒ RETURNING TO EDEN IN THE NURSING HOME

One innovative design feature for dementia and nondementia nursing home units is called the Eden Alternative (Thomas, 1996). The philosophy behind this approach is that many of the elderly were raised in rural areas or in times when they were surrounded by animals and plants. Any bonds with these nostalgic times are thought to increase a sense of responsibility, control, and caring, and to be therapeutic. The institutional unit, then, contains dozens of animals (dogs, cats, birds, fish) and hundreds of plants for the residents to observe, care for, and handle (see Figure 12-14). The evidence suggests that this opportunity to "commune with nature" can be very therapeutic. Benefits include reduced medication use, reduced staff turnover, and improvements in resident satisfaction and self-sufficiency (e.g., Bruck, 1997; Stermer, 1998).

Figure 12-14   In the Eden Alternative facility, residents interact with numerous dogs, cats, and birds and have responsibility for caring for the animals as well as for dozens or even hundreds of plants. (Photo courtesy of Columbine Health Systems)

wander in and out of the building but never away from the unit. A very important feature of these units is extra staffing and thorough training of the staff in behavioral management techniques, such as the use of reminiscence and diversion. Also, activities to keep the residents occupied are specially planned. Environmental considerations include nonglare floors (glare can increase confusion about orientation); extra orientation aids (such as pictures indicating locations of toilets, names of residents in large letters on the room door); pictures firmly affixed to walls (to prevent their inadvertent falling when touched out of curiosity); absence of "busy" interior decorations (which can add to confusion); and a lounge area with a Dutch door (i.e., a door with a bottom half that can be closed separately from the top half). The reason for this door is that at

night, some residents will not be able to sleep, a phenomenon in Alzheimer's known as *day–night reversal* or "*sundowning*." If sleepless residents are forced to stay in their rooms, they will likely wake others. If permitted to roam in the lounge (with only the bottom half of the door closed to permit staff supervision), they typically do not disturb others. Although there are numerous variations of these dementia unit designs, a representative one is shown in Figure 12-15. How well do Alzheimer units work? Evidence suggests they have some merit (e.g., Ohta & Ohta, 1988). For example, Martichuski and Bell (1993) explain how a well-designed facility with well-trained staff and a continuous activity program can reduce *excess disability*— the disability that occurs over and above the physiological cause of the dementia due to the

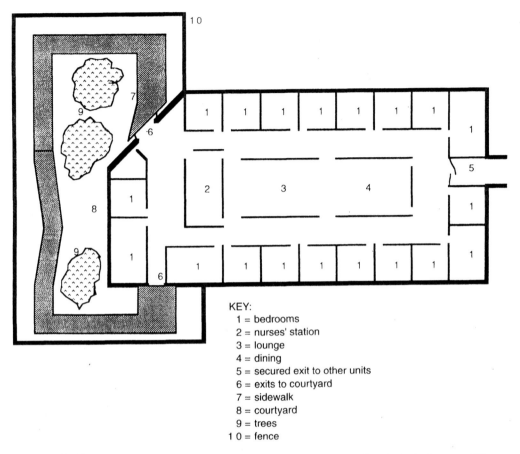

KEY:
1 = bedrooms
2 = nurses' station
3 = lounge
4 = dining
5 = secured exit to other units
6 = exits to courtyard
7 = sidewalk
8 = courtyard
9 = trees
1 0 = fence

Figure 12-15   An example of a floor plan for a dementia unit. Note that the locked access still permits wandering through the facility, indoors and out.

environmental design not supporting the person's daily activity needs. From a preliminary perspective, at least, the design of dementia units may have some desirable consequences (for additional reading on these units, see Calkins, 1987; Morgan & Stewart, 1998b; Ohta & Ohta, 1988; Sloane et al., 1995; Sloane & Mathew, 1991).

SUMMARY

In North American culture, strong preferences exist for single-family detached homes, apparently because such housing provides high control over social interaction. Economic and other factors, however, often lead to choices of other types of housing.

The more easily we can perform given tasks in a setting, the more satisfied we are with it. Other factors affecting residential satisfaction include noise, ease of cleaning, and adequate plumbing, heating, and kitchen facilities. For inner-city residents, satisfaction with a neighborhood is closely tied to social bonds.

Privacy is a significant mediator of activities in residences but is not a simple process. Whereas some families have a very open structure, others use design within the home to structure privacy for all family members. Another important issue is functional criteria in residential design. One current consideration is to combine several functions in one room, such as using a sunken bathtub as a conversation pit in a living room. Cultures differ greatly in segmentation of functions within the home.

Propinquity involves both objective physical distance and functional distance, or the likelihood that two individuals will come into contact. Both types of propinquity facilitate the formation of friendships. Propinquity can also lead to annoyance and dissatisfaction; building a sense of community tends to maximize friendship and minimize annoyance.

Hospitals are often oriented toward a high-control, low-choice atmosphere to facilitate staff functioning. This tendency, however, reduces perceived control on the part of the patient, as well as privacy. Designs that restore control and foster social interaction can help in this regard and can also facilitate patient recovery. A radial design of wards around a nursing station can improve staff efficiency and increase the amount of time staff spends with patients. Design features that could help hospital visitors include orientation maps, private areas, low-noise designs, and considerations for comfort.

Modern prison designs attempt to provide outdoor views and increased opportunities for privacy. Such designs are associated with reduced violence and vandalism. ·

Care facilities for the elderly need to consider the fact that characteristics and needs of the elderly vary widely, so designs should allow for flexibility. For those outside of institutions, planned housing in age-segregated areas seems to enhance satisfaction and morale; adequate transportation and shopping are also important for most of those in both noninstitutional and institutional settings. Whatever the setting, safety and convenience, choice and control, and physical conditions are important considerations. Whereas proximity of residents to staff facilitates surveillance, it may also encourage dependence. Providing several levels of care in one facility minimizes negative effects of relocation. Clearly, providing perceived control is one of the most effective interaction strategies for the institutionalized elderly. With dementia units, the idea is to adapt the environment to the special behavioral characteristics of the cognitively impaired.

KEY TERMS

Alzheimer's disease	environmental	person–environment	privacy gradient
cohousing	spoiling hypothesis	congruence	propinquity
dementia			

SUGGESTED PROJECTS

1. Tour some model homes in the community. Compare the privacy available in the bedrooms, kitchens, and bathrooms. What behavioral adaptations do you anticipate once someone occupies the homes?

2. Observe the activities of an entire floor of a dormitory. In what ways do the behavioral adaptations resemble those of a private home? In what ways do they resemble those of the inner city described by Scheflen?

3. Visit your local hospital and assess the following: (1) adequacy of orientation/wayfinding aids; (2) views from windows; (3) location of nurses' station relative to patient rooms. What improvements can you suggest?

4. Visit a nursing home or retirement home and note the location of lounge areas. Look for lounge areas that have a lot of interaction and those with little interaction. What factors account for the differences? Can you identify design features intended to promote competence?

CHAPTER 13

WORK, LEARNING, AND LEISURE ENVIRONMENTS

Introduction

Harry and Sue were looking forward to the weekend. Harry's job at the factory was becoming more and more unpleasant. The machines were extremely loud and the company doctor had informed him last week he was losing his hearing in the sound frequencies most necessary for conversation. What was worse, relationships among his co-workers were deteriorating. Management was pressing for increased productivity, and the only way for him to produce more was to make faster trips between the supply room and his workstation. Other workers were doing the same thing, though, and they were all getting in each other's way. If his workstation could be closer to the supply room, at least one problem would be solved.

Sue was equally hassled. Her firm had just moved into a new office without interior walls. This change was supposed to reduce maintenance costs and increase ease of communication, with everyone in one large room. She could not

stand it, though. She had to reprimand a secretary yesterday, and there was no place to do it except at her desk, where everyone else could hear the conversation. Her new workstation was attractive enough, and it bristled with the latest in computer technology. Everything was efficient. The computer word processing program eliminated much of the time spent in rewriting. Company records were carefully filed in a database, and the computer could easily "talk" to those in various branch offices. Still, Sue seemed to be more tired lately. Her eyes hurt from staring at the computer screen, and she wondered whether the pain in her back was simply from spending too much time hunched over the keyboard. To make matters worse, her doctor had called today with unfavorable news on some lab tests, and it seemed as if everyone in the office heard at least her end of the conversation. Noise was a problem for her, too. She worked hard on a marketing report due today, but it took much longer than necessary because of all the distraction from phones ringing and everybody else talking. The chatter of the old copier 10 feet from her desk did not help much, either.

To get away from these headaches, Harry and Sue decided to go camping in the state park in the next county. After all, the convenience of the park was one reason they had chosen Rockport as a home. Arriving at the park entrance, the ranger informed them they were just in time to get one of the last two campsites available. They felt fortunate, though when they pulled into the campground it was discouraging to see that one of the remaining campsites was muddy and the other was next to a group of adolescents having a loud party. This was getting away from it all?

Have you had experiences similar to those described above? Unfortunately, they occur more often than we would like. Environmental psychologists have asked whether or not environmental design and management can make a difference in our lives of work, learning, and play. Are there ways of designing the factory, the classroom, the office, and the leisure setting so that undesired effects are minimized and desired results are maximized?

Admittedly, work, learning, and leisure seem worlds apart — unlikely companions in a chapter, you might think. As we will see, however, the difference between work, learning, and leisure is often neither the activity we engage in nor the setting in which it takes place. The relative proportion of human time devoted to either work or leisure has varied from culture to culture (Csikszentmihalyi & Kleiber, 1991). Even in our time, it is hard to make an accurate estimate. Let's assume that most full-time job holders are required to work 8 hours a day, 5 days a week. If each worker gets 8 hours of sleep a night (56 hours a week) and works 40 hours, that should leave 72 hours per week for leisure or other activities. As Csikszentmihalyi and Kleiber (1991) point out, however, determining people's useful leisure time is not nearly so simple. Most people spend about 40 of the 72 "free" hours on nondiscretionary activities like driving to and from work, shopping, household chores, and getting dressed. Many of the remaining 30 hours are likely to be committed to "getting ahead" at work (who really works just 40 hours, anyway?) or family responsibilities like coaching or watching athletic events. Both the large percentage of clearly allocated time devoted to work and learning and the scarcity (but importance) of leisure time argue for the importance of these environments in understanding the human condition. 🔲

Work Environments

In previous chapters we have discussed the effects of noise, temperature, and territorial identification on behavior. In the present section we will examine how these and other components of the environment can be incorporated into the process of designing the work environment, which is sometimes called the *workspace*. For those who wish to study the design of the work environment more thoroughly, detailed reviews of previous research exist elsewhere (Becker,

1981; Sundstrom, 1986, 1987; Wineman, 1986). For our purposes, we will highlight some of the important findings of this research. As we begin, you might find it helpful to examine how our eclectic model applies to behavior in work settings (Figure 13-1). Briefly, we may perceive the work environment as fitting or not fitting our needs; the adaptive responses we employ and resulting potential aftereffects (e.g., fulfillment, exhaustion upon arriving "home") may be

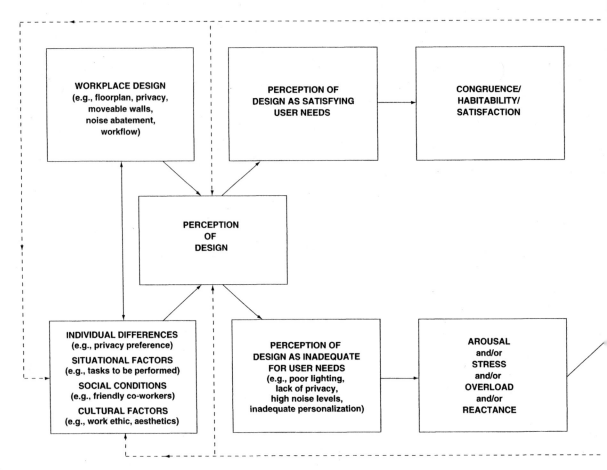

Figure 13-1   Our eclectic model applied to the workplace.

mediated by processes such as arousal, overload, or stress. The model is similarly applicable to learning and leisure settings.

A BRIEF HISTORY OF WORKPLACE DESIGN

Before the Industrial Revolution, nonfarm work was typically done in small spaces, often in the craftsperson's or businessperson's home. Of course, even in early times, some specialized products required the labor of several persons, and the specialized workplaces sometimes took a shape dictated by their product. Ropewalks, for example, were buildings in which rope was woven. Before the introduction of modern coiling machines, the maximum length of a rope was

dictated by the length of the building in which it was manufactured (Kostof, 1987). Hence, ropewalks were simple, but extremely long; in fact, one built in Charlestown, Massachusetts, in 1838 was a quarter of a mile long. Early factories were limited by their reliance on water for power and the sun for lighting. Because they required swiftly flowing streams, often they had to be constructed on remote sites far from the populated (but flat) coastal strip. Mill towns became self-contained communities, often wholly owned by one partnership (Kostof, 1987). The reliance on water to power these early mills and factories also dictated their shape. A water wheel outside the factory turned a long shaft that extended inside the building. Off this shaft ran a series of belts that powered the factory's

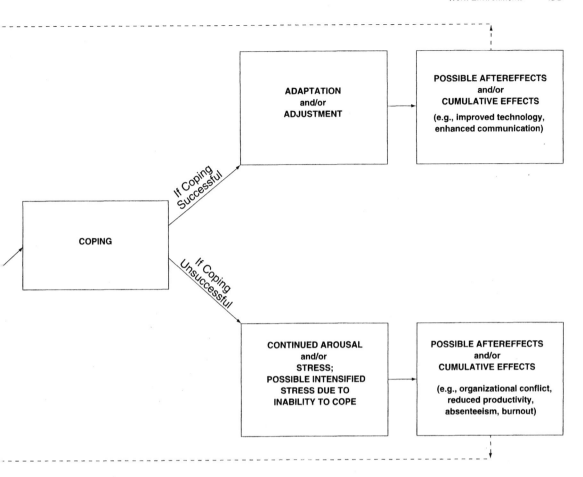

machinery. Thus, the technology of the power source dictated a long building, and the need for sunlight limited building width to about 60 feet, or 18 meters (Sundstrom, 1986).

Working conditions in early factories were miserable, a situation that persisted through the early 1900s in North America. At that time several factors combined to cause rapid improvements (Sundstrom, 1986). First, a tragic disregard for human safety led the popular press to pressure for reforms that eventually led to a variety of laws to protect worker health and safety. Second, there was a growing belief that comfortable working conditions would lead to increased productivity. Finally, managers began to discover Frederick Taylor's (1911) **scientific management.** This was a management philos-

ophy that emphasized the importance of optimal pay, the analysis of work activities to improve efficiency, and the development of work environments to support efficient movement.

Like the factory, the office building was restricted by construction technology. Stone construction and the absence of elevators meant that buildings could not be more than 6 to 10 stories high, and the need for adequate lighting through windows dictated a fairly narrow building. Two developments in technology changed both the factory and the office. First, iron and later, structural steel, combined with concrete, made it possible to span larger spaces as well as to build higher and higher. (With stone construction, walls had to be so thick at the base for support of upper floors that tall buildings were impractical.) Second, commercially available electricity allowed for more extensive indoor lighting and elevators in tall buildings, as well as the ability to separate manufacturing machines from a central power shaft. It then became technologically possible to set up a factory or office in an almost infinite number of ways (Sundstrom, 1986; Figure 13-2).

What will be the workplace of the future? The emergence of high technology, particularly the computer, has had a profound effect on the workplace. It may be that computers will take

Figure 13-2 Early large-scale work environments were long, narrow, and only one or two stories tall because of power sources and limitations of stone construction. With modern construction techniques and the flexibility of electric power, buildings can be tall, workspaces may or may not have windows, and many alternatives are available for the spatial layout.

over tasks that were once routine, increasing both productivity and the quality of worklife. On the other hand, office automation may result in loss of employment, lowered skill requirements, and an inhumane workplace (Turnage, 1990). What will be the effects of the workplace in determining our quality of life as we begin a new century?

Ambient Work Environments

As a student of environmental psychology you will not be surprised to learn that the physical environment can affect productivity and satisfaction in work environments. Prior to the 20th century, choking fumes and deafening noise, for example, were considered part of the normal manufacturing process. Concern for the safety and health of workers, along with studies showing that productivity and accident rates could be influenced by physical working conditions, has led to standards for lighting, ventilation, noise, and so on. Workplace design can provide greater safety and reduce threats to productivity and job satisfaction. Most of us expect to work in an environment that is well lit, not too warm, not too cold, and free from physically damaging noise levels. Of course, for those in outdoor occupations, the environment cannot be as carefully controlled, and some other jobs typically require work in extreme conditions (for example, boiler tenders and airline ground crews). For these occupations, the ambient environment is the direct source of stress. We refer you to Chapters 4, 5, 6, and 7 for reviews of both theories and examples of individual stressor effects. Ambient conditions such as lighting and the presence or absence of windows are also important (see Chapter 11). In this chapter we will restrict ourselves to what may be a more typical situation in which the heating, ventilating, and cooling systems work, and the lights function.

Sound

You may recall that in Chapter 5 we defined noise as unwanted sound. Unlike visual distractions, noise cannot be easily avoided by turn-

A HAWTHORNE IN THE SIDE OF
ENVIRONMENTAL PSYCHOLOGY?

In the early 1900s, Frederick Taylor proposed a management system based on the assumption that workers are primarily motivated by economic incentives. It was assumed that production would be greatest when pay was adequate and production techniques and the work environment were optimal. This approach was labeled *scientific management* and led to several investigations of environmental qualities such as heating and lighting. It also led to routine and standardized jobs and what some consider a rather unflattering picture of workers and their motivations.

One of the first real breaks with scientific management was promoted by a series of studies that took place beginning in the 1920s in the Hawthorne plant of the Western Electric Company near Chicago (Roethlisberger & Dickson, 1939). One of the early questions addressed by this project, which became known as the Hawthorne studies, was the effect of illumination on productivity. According to many reports, lighting levels were systematically varied for an experimental group, whereas a control group worked under constant illumination. Amazingly, both the experimental and control groups increased production. In follow-up observations, an experimental group was reported to have maintained the initial level of performance in spite of the fact that illumination had been reduced by 70%! Eventually, the researchers only pretended to change the illumination level, yet workers continued to increase their production and to express pleasure with what they perceived to be better illumination. These bizarre results were interpreted at the time as suggesting that workers' performance increased as a result of novelty and the fact that workers knew they were being observed, rather than because of experimental changes. This interpretation came to be known as the *Hawthorne effect*.

In hindsight, the true meaning of the Hawthorne studies is open to discussion. A number of reviewers have pointed out methodological flaws that cast doubt on many of the conclusions drawn by the researchers (e.g., Franke & Kaul, 1978; Landesberger, 1958; Parsons, 1978). Indeed, the illumination studies that are so often recounted were never formally published and seem to have served mainly as an impetus for subsequent investigations of work schedules, supervision, and work-group functions (Parsons, 1978).

Nevertheless, the illumination studies and the others that followed have had powerful effects and, in fact, probably led to a revolution in industrial/organizational psychology. In particular, the Hawthorne studies are often cited as the beginning of the human relations movement in American management (Landy, 1989; Saal & Knight, 1988). On the other hand, in focusing on the importance of the placebo-like "Hawthorne effect," these studies may have shifted research attention away from environmental variables such as illumination and delayed the development of what we now know as environmental psychology.

ing one's head. From the perspective of the environmental load approach, workers in noisy offices are forced to process not only their particular tasks, but all ambient sound information as well (Loewen & Suedfeld, 1992). Under these conditions workers must attempt to filter irrelevant inputs and focus on relevant information. Researchers have concluded that one of the most distracting noises is overheard speech (presumably speech is meaningful, automatically attended to, and difficult to ignore (e.g., Loewen & Suedfeld, 1992; Sundstrom, 1987; Sundstrom

et al., 1994). Preliminary evidence suggests that noise may act as a dissatisfier; that is, job satisfaction goes down in noisy conditions, but a corresponding increase in job satisfaction does not necessarily follow noise reduction efforts (Sundstrom et al., 1994).

Music is another source of sound in the workplace—one that is technically considered noise only if someone does not like it. Do you like to read or study with music playing in the room? Does your answer depend on the type of music? Several of the theoretical models we discussed in Chapter 4 might be used to predict the effects of music (see also Figure 13-1). If it raises your arousal to some optimal level without either overarousing you or creating a distracting source of information, we might expect your performance to improve. Sundstrom (1986) reviewed the evidence on music in the workplace and reported that although much of the research is the private, inaccessible property of firms that sell music systems to businesses, published research may or may not support the idea that music enhances the work environment. At one time it was actually thought that singing and/or listening to music with a steady, somewhat upbeat rhythm improved productivity. Later it was felt that pleasant music made employees cheerful and the environment enjoyable. Research indicates that in factories, music may or may not slightly improve productivity, but employees like it anyway. In offices, music may facilitate vigilance tasks (e.g., where an employee must monitor a screen), although it can be distracting for some. At any rate, employees often report that music helps provide a pleasant atmosphere, which may ensure that it will always be found in some work settings.

Furniture and Layout

The layout and design of the workplace may be an important determinant of people's impressions of the company or organization (for a closer look at furnishing and faculty offices, see the box on pages 446–447). The use of a desk as a "barrier" between the office occupant and a visitor can communicate a desire for physical and psychological distance, as well as status differences. Joiner (1971) observed that high-status office occupants were more likely to use a closed desk arrangement (the desk sits between the visitor and the office occupant) rather than an open placement (in which the desk is placed against the wall). Furthermore, desk arrangement can also have implications for the pleasantness of the interaction and the visitor's level of comfort (Morrow & McElroy, 1981). According to Zweigenhaft (1976), seating arranged at right

angles is perceived as facilitating cooperation and affiliation. In one investigation of photos of reception areas, organizations judged by students and executives as the most considerate and likable had upholstered couches and chairs at right angles and prominently displayed floral arrangements. Firms rated as moderately considerate had four chairs surrounding a coffee table, contemporary artwork, and either one or three plants. Finally, the firms judged least considerate lacked artwork and had chairs placed directly facing one another across a coffee table (Ornstein, 1992).

TERRITORIALITY AND STATUS IN THE WORK ENVIRONMENT

We discussed territoriality in Chapter 8 as it relates to many of our relationships with the environment. Some researchers also believe that territories are important in work environments. Often the concept relates to assignment of a specific area or machine to a worker and is termed **assigned workspace** (Sundstrom, 1986). It is often believed, for example, that if a large machine in a factory is assigned to one worker, that worker will take better care of it than if all workers roam from machine to machine. The same concept is often called the **fixed workspace.** Sundstrom (1986) suggests that the right to treat a workspace as a territory might lead to more personal attachment to it, more perceived control over it, and thus more of a sense of responsibility for it and more signs of personalization of the workspace. Whether workers in fact prefer clearly defined territories and whether territories improve job satisfaction or productivity is open to question. Most likely, territories become more important to workers the higher the rank they have in the organization. At higher ranks, territories may become symbols of status (Sundstrom, 1986).

Status symbols in the office or factory may be important in several ways. For example, they communicate status and power to others, they compensate employees as a nonmonetary benefit, and they serve as props or tools (such as larger desks, filing cabinets, computer terminals), which the worker is privileged to use on the job (Sundstrom, 1986). In addition to furnishings such as desks and size and comfort of chairs, typical status symbols include amount of floor space, the capacity to regulate privacy and accessibility (e.g., through an enclosed office), and the right to personalize the workspace. One large firm, for example, provides carpeting, a bottled water dispenser, and plants as one moves up the corporate ladder. Apparently, the more one can attach status to the office space, the more satisfied one is with the job (Konar et al., 1982).

HUMAN FACTORS: ENGINEERING FOR HUMAN EFFICIENCY

Can careful design of the tools, machines, and workspaces in an office, kitchen, or factory decrease accidents and increase productivity? Psychology's interest in facilitating human performance in different environments, especially our interactions with machines, has a long history that predates the establishment of the field of environmental psychology (see the box on page 437). Historically this is the domain of **human factors** psychology (human factors psychology is also sometimes referred to as *engineering psychology*, or, especially in Europe, as **ergonomics**). According to one popular textbook: "Human factors [psychology] discovers and applies information about human behavior, abilities, limitations, and other characteristics to the design of tools, machines, systems, tasks, jobs, and environments for productive, safe, comfortable, and effective human use" (Sanders & McCormick, 1993, p. 5).

Many of the early human factors efforts focused on the role of specific tools or procedures in increasing worker efficiency. For example, Taylor (1911) demonstrated dramatic improvements in efficiency when steel workers were issued the optimal-sized shovel for each shoveling task—smaller ones for heavy iron ore and larger ones for ashes. More recent human factors research emphasizes either the interactions between humans and machines (computers, aircraft, nuclear power plants), the workstation, or the ambient environmental conditions such as noise, light, and temperature that we have discussed throughout much of this text.

Unfortunately, it is not difficult to find equally dramatic examples of instances in which human factors-related errors have catastrophic effects. A well-meaning safety check resulted in the Chernobyl nuclear power plant disaster, for instance. In Chapter 7 we discussed another dramatic nuclear incident. On March 29, 1979, an accident at Three Mile Island Unit 2 resulted in a crisis that lasted for several days, costing the plant owner over $1 billion and subjecting people living nearby to persistent stress. During the incident, operators searched frantically to discover what was wrong with the reactor. Of the 1,600 windows and gauges in the control room (some 200 of which were flashing), several critical displays were in out-of-the-way locations, hidden by maintenance tags, or absent altogether. You may recognize this as a situation of information overload, discussed in Chapter 4. Resulting investigations revealed that many of the human errors that contributed to the accident resulted from grossly inadequate control room design (Smither, 1988). The incident sparked a flurry of interest in human factors, both in the nuclear power industry and in the workplace in general. Unfortunately, Chernobyl, airline accidents, and major power failures remind us that we are not yet truly the masters of our machines; many of us who interact with computers daily suspect the power in the relationship goes the other way.

Communicating With Machines

In spite of their obvious similarity, the historical development and focus of human factors and environmental psychology are somewhat different. Although nothing in the definition requires that human factors psychologists restrict themselves to work settings, many of them do, and their field is often thought of as a subdiscipline of industrial/organizational psychology. A small

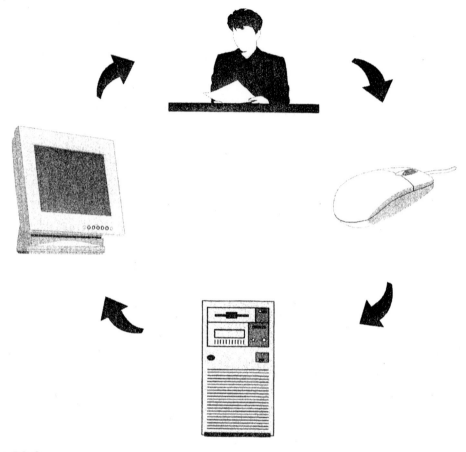

Figure 13-3   Humans and machines interact in a cycle of communication.

oversimplification may capture some of the flavor of the field. In general, human factors psychologists focus on the human–machine system, perhaps most easily understood as a communications cycle (see Figure 13-3). The human being (pilot, driver, operator) makes decisions and communicates them to a machine through controls such as knobs, levers, steering wheels, or pedals. As it functions, the machine also communicates to the human operator through displays such as dials, gauges, or warning lights. No doubt, all of this seems simple enough, and it is unlikely that any engineer would knowingly design a machine that was impossible to control. The problem is in making the human–machine system as efficient and error free as possible.

Humans vary greatly in their size, strength, perceptual abilities, and skills. Like the machines they command, humans may be injured or wear out.

Just as with human-to-human interactions and communication, an important consideration in human factors design is feedback. Have you ever borrowed a friend's car and found yourself searching for the windshield wiper controls? Perhaps your first effort turned on the lights, then you signaled both a left and then a right turn before your search finally activated the wipers. In this case, the lights and wipers, when activated, gave you quick feedback that identified your error. In other circumstances, feedback is much slower. A misadjustment in the thermostat in your residence probably does not result in uncomfortable temperatures for many minutes or even hours.

The human operator cannot adjust controls if the machine is out of reach. Sanders and McCormick (1993) list some principles that apply to the placement of the components of a workstation (the area of a factory or office assigned to an individual). Those components of a system that are most important or most used should be placed in the most convenient locations. According to the *functional principle*, controls or displays that are functionally related to each other (temperature controls and displays, for instance) should be placed together. *Link analysis* is a term applied to the systematic investigation of the number of times a movement is made to adjust a control or read a display. Proper positioning of workspace elements will minimize the distances required for common or sensitive activities and reduce errors and fatigue (Sanders & McCormick, 1993).

Mapping Control–Display Relationships

One of the most important principles in facilitating human–machine communication is what Norman (1988) calls mapping. **Mapping** refers to the relationship between the actions of an operator and those of a machine. Sometimes mapping is "natural," that is, consistent with physical or cultural analogies. For example, to move an object up, you should move the control up. Similarly, in order to unambiguously report an increase in altitude, perhaps an airplane's altimeter indicator should go up on a vertical scale (traditionally altimeters do not follow this advice, and the resulting errors in their use may account for at least some accidents; see Sanders & McCormick, 1993). In Chapter 3 we mentioned a powerful (but frequently violated) natural mapping known as *forward-up equivalence* (e.g., Levine, 1982)—what is up on a fixed information map in a mall or airport should be forward in the environment. Other mappings may at first seem just as obvious, but they may not be as clearly tied to natural events. To raise the volume of sound equipment, it is probably natural to raise a control lever, but most of us are also comfortable with a knob that requires a clockwise movement (Figure 13-4A). Does this seem unremarkable and natural? Now look at the knob in Figure 13-4B. What direction should you turn this knob to water the garden? In North America, water faucets turn left to open (releasing more water) and right to close.

Consistent mapping relationships between human actions, controls, and displays are also sometimes referred to as *population stereotypes*. These can become quite complex. Note the

Figure 13-4 A   A volume knob (turning right increases the volume).

Figure 13-4 B   A water faucet (turning left increases the volume).

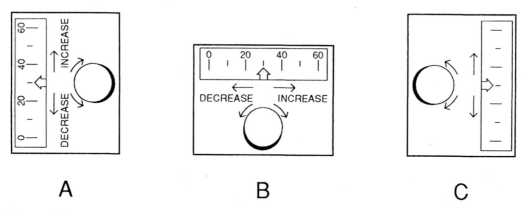

A   B   C

Figure 13-5A, 5B, & 5C   Warrick's principle. If you turn the knob in A to increase pressure, in which direction should you turn the knob in B? In C?

display–control pairs in Figures 13-5 A, B, and C. You might think of the display as a pressure gauge and the knob as the control that increases the pressure in some fictional machine. Which direction would you turn the knob in Figure 13-5A to increase the machine pressure? Which direction would you turn the knob in Figure 13-5B? Although most people would turn the knob clockwise in the first two instances, many would switch to counterclockwise for Figure 13-5C. In the final instance, one population stereotype conflicts with another. According to the *clockwise-for-increase principle,* people will turn a knob clockwise to increase the value on the display. On the other hand, *Warrick's principle* (see Sanders & McCormick, 1993) states that a pointer on a display will be expected to move in the same direction as the part of the control nearest it.

Machine or Human: Who Is in Charge?

Technological advances have allowed us to assign more and more tasks to machines, a process known as *automation.* Automation may decrease workload or replace humans in dangerous or boring jobs. However, not all of us make optimal use of automation. Sometimes we fail to use an efficient system; sometimes we rely on it too much (see Parasuraman, 1997 for a review). Like human–human interactions, one impor-

tant dimension of the human–machine communication process might be termed "trust." There are a number of instances in which an operation can either be carried out by a person or left to the automatic control of a machine. When does an operator trust the machine, and when does he or she override the automatic functions (Lee & Moray, 1992)? The autopilot in an aircraft is one familiar example. Does the flight officer trust the plane's autopilot? Even more familiar might be the spell checker in a computer word processing program. Do you trust your computer's spell checker? In the latter example, a failure to use the spell-checking function is likely to reveal a number of instances in which you are less reliable than the machine. On the other hand, blindly trusting the software program may result in errors with homonyms, proper names, and other spelling nuances.

When automated systems perform poorly, it often becomes critical for human operators to quickly diagnose the source of the problem and take corrective action. In order to quickly "come up to speed" in emergencies, operators must maintain **situational awareness,** or **SA** (Gilson, 1995). Machines often offer a variety of modes that provide the human operator with options for performing essentially the same task under different operating conditions. For example, modern computer operating systems often have two, three, or even more ways to select files or run programs. For instance, one might "click" on a program icon or use the keyboard to type a command. Automated cockpit systems provide a persuasive example of some of the difficulties that may arise. For instance, pilots can choose from at least five different levels of automation to change altitude (Sarter & Woods, 1995). Although this system provides flexibility, the price is that pilots must monitor not only the activities of the aircraft, but keep in mind which system is controlling them. In high workload or emergency situations cognitive demands build, resulting in a dramatically increased cognitive load.

Sadly, when SA is not maintained the result may be an accident (Sarter & Woods, 1995). Establishing SA for teams may be even more complex than for individuals. Another aviation example illustrates the problem (Jentsch, Barnett, Bowers, & Salas, 1999). As you may know, cockpit crews on commercial airliners alternate responsibility for actually flying the aircraft. Although the captain is usually more experienced and has higher status, about half the time the copilot is responsible for takeoffs or landings. However, the captain is always the person officially in charge of making major strategic decisions and handling emergencies. Thus, the captain has a dual role that sometimes includes actually flying the aircraft but always includes certain managerial decisions. Consistent with our understanding of cognitive overload, captains more often lose situational awareness. Thus, captains make more errors when they are at the controls than when they are not, sometimes even more errors than the less-experienced copilots.

THE ELECTRONIC OFFICE

Advances in computer technology have made dramatic changes in the workplace (e.g., Carlopio & Gardner, 1992; Grandjean, Hunting, & Pidermann, 1983; Kleeman, 1988). Clearly the Internet has revolutionized access to information (and the potential to generate an information glut). Soon trips to the business office, registrar, mailbox, or filing cabinet will be unnecessary, and we will have the "convenience"

(some might say "misfortune") of never having to leave our desks. Whether this technology is a blessing or a curse may depend in part upon our ability to exploit the computer's strengths and adjust to its demands. With the proliferation of video display terminals (VDTs), workers began complaining of eyestrain, headaches, back pain, and fatigue (Carlopio & Gardner, 1992; Kleeman, 1988; Menozzi, Von Buol, Waldmann, Kuendig, Krueger, & Spieler, 1999; Stellman et al., 1987; Turnage, 1990).

Glare from lights and windows is likely to cause discomfort, so careful location of the display, adjustable workstation furniture, and occasional breaks to rest the eyes and hands are recommended (Garcia & Wierwille, 1985; Kleeman, 1988; Rose, 1987). Furthermore, job-related stress may occur in part because of the indirect effects of VDTs on jobs (Turnage, 1990). Computers reduce the need for workers to move around the office to accomplish their jobs, and this, combined with the possibility of computer monitoring of a worker's progress, may usher in an era of greater workloads, fewer breaks, and less autonomy (Sundstrom, 1986; Turnage, 1990). Unfortunately, not every business has the time or money to systematically investigate the adequacy of each employee's workstation. However, Menozzi et al. (1999) report on a Swiss company's success in establishing a training program whereby employees were trained to assess and improve the condition of their own workplaces.

The same technology is capable of allowing exciting new freedoms to workers. For some, a trip to the "office" involves walking into a different room in the house. Communications networks now allow workers to perform many of their chores from home-based workstations, or an electronic cottage (Toffler, 1980). As Sundstrom (1986, 1987) notes, such an arrangement might eliminate much of the cost and inconvenience of commuting, but such a decentralized workforce will put new demands on supervisors and weaken some of the organization's control. Indeed, the role of the employee may change to that of a contractor, and the office building may become essentially a conference center for those meetings that require face-to-face encounters rather than telephone or video conferences.

DESIGNING THE OFFICE LANDSCAPE

We have mentioned that the use of structural steel in buildings has allowed architects to design larger and larger open spaces. With previous construction methods, the need for support walls required a building to be separated into smaller rooms. Accordingly, a relatively small number of workers shared an office. Typically, a manager or executive would have a totally private office, and several clerical workers would share an adjoining space. Such office designs are still common today, but there is also an alternative permitted by a large open space. The principle of **workflow** is that workstations should be arranged in the order in which paperwork or products are to move consecutively among them. To facilitate workflow in the large office, workstations for 100 or more workers could be placed in the same large room, with supervisors' offices along the sides of the large room. With

the growth of the human relations movement in the 1950s, more open communication between workers and managers was encouraged, employees were allowed and even encouraged to participate in decision making, and barriers of status and authority became less prominent. These developments encouraged what is now known as the **landscaped office** or **open-plan office** (Figure 13-6). This concept probably originated in Germany with work by the Schnelle brothers and their Quickborner Team consulting firm (Sundstrom, 1986). Basically, the idea involves arranging desks, filing cabinets, and other office furniture in such a manner as to make maximum use of the large open space but still provide for efficient workflow. The office landscape design typically places a supervisor very near workers and arranges work areas close together or far apart so that workflow and communication among related areas is unimpeded by myriad enclosed offices. In some schemes, portable screens are used to set areas off from others, or shelving and filing cabinets may accomplish the same purpose. As you might ex-

Figure 13-6 An example of an open office design. Compare this with the traditional offices in Figure 1-3.

pect, such an office design has a number of advantages, but also carries with it a number of disadvantages (e.g., Becker et al., 1983; Brooks & Kaplan, 1972; Oldham, 1988; Oldham & Brass, 1979; Sundstrom, 1986). Let us examine some of these separately.

Advantages of the Open-Plan Office

We have already mentioned one advantage of the open-plan office: It provides for a more efficient flow of work and communication. In addition, it often costs less because there are no internal walls to construct, and lighting and ductwork can be shared by several workspaces. Maintenance costs may also be reduced due to less painting and faster cleaning of work areas, and more people can be accommodated in the same interior space without walls. Moreover, it is easier to make changes in the design of the office when new jobs are added or eliminated or the number of people working on a project changes, because there are no fixed walls to move or add in order to change spatial arrangements. Also, the open office permits easier supervision of workers. That is, a supervisor can see all workers from his or her desk without having to walk through several offices. Furthermore, there is evidence for social facilitation in nonprivate offices (Block & Stokes, 1989): A number of researchers have demonstrated that the mere presence of others improves performance, at least for simple tasks.

Disadvantages of the Open-Plan Office

Typically, changes in any environment cannot be made without trade-offs, and open offices are no exception. For all the potential advantages of the open office, it carries disadvantages that fall into two major categories: increased noise and distraction, and lack of adequate privacy. When offices are separated by walls, the noise from keyboards, phones, and copy machines in one area seldom penetrates into the next office. With the open-office plan, however, the noises may be very distracting to those in neighboring workstations. Similarly, conversation travels, and as we saw in Chapter 5, noise that is interpret-

able as conversation is quite distracting. Indeed, although open offices may facilitate social conversations, there is little evidence that organizationally relevant communication improves (Wineman, 1982). Movement of people as they walk about doing assigned tasks is also more noticeable in the open-office plan, and adds still another source of distraction. Solutions to the problem of noise and distraction include office machinery designed to be quieter, carpeting and other treatment to muffle sounds, and the use of portable barriers (partitions, shelving, cabinets) to help screen out the distraction. One less obvious solution to the distracting quality of overheard conversations is to actually increase the overall level of noise by adding a source of so-called white noise. White noise, as we saw in Chapter 5, is made up of all audible sound frequencies and might be described as a hissing or humming sound. Played at a constant level, it may mask the distracting character of conversations and other meaningful office sounds (Loewen & Suedfeld, 1992).

Loss of privacy is also very noticeable in open offices (Kupritz, 1998; O'Neil, 1994; Sanders & McCormick, 1993; Sundstrom, 1986). Personal conversations are easily overheard and communication between supervisors and workers becomes more difficult to keep confidential. Every move a worker makes is open for public view. Phone calls with family members are overheard. Errors and embarrassing behavior are there for all to see, and personalizing the workspace with artwork or mementos may be discouraged. The use of portable barriers may help solve the privacy problem in the open office, but these barriers cannot provide the privacy of an enclosed, individual office (e.g., O'Neil, 1994; Sundstrom, 1986). Employees who move from a conventional office to an open arrangement may complain of loss of privacy and suffer from reduced job satisfaction (Oldham & Brass, 1979). Although enclosure (closed versus open design) may not be as important as other factors such as the ability to adjust one's workspace (O'Neil, 1994), Oldham (1988) demonstrated the therapeutic effects of moving employees from an open office to either a partitioned office or a

THE FACULTY OFFICE

One environment ripe for investigation, and one with which you are probably familiar, is the university faculty office. Have you strolled down a hallway of professors' offices and speculated on the character of their occupants? Is a neat office the sign of a neat mind, or is it sterile and cold? Do decorations or living things make workplaces more hospitable? Does a particular office convey a sense of distance or welcome? Research on faculty offices has centered on three issues: the placement of the desk, aesthetics, and overall neatness.

Desk Placement

Much of the interest in desk placement probably began with a survey of 10 London firms conducted by Joiner (1971). He found that higher status individuals tended to place their desk between themselves and the door rather than against a side or back wall. This finding led Zweigenhaft (1976) to hypothesize that faculty who placed their desks between themselves and visiting students would be using the desk as a physical barrier (perhaps inadvertently) and would be perceived as more behaviorally distant than those who used a more barrier-free arrangement. Consistent with Joiner's observations regarding status, Zweigenhaft's survey found that senior faculty (full professors and associate professors) were more likely to use the closed desk (desk-between) arrangement. Furthermore, those faculty members who used an open desk arrangement, in which the desk did not separate the faculty person and his or her visitors, were more likely to be rated positively by students. Other researchers have also observed open desk placement to be associated with more positive student feelings (Morrow & McElroy, 1981) and also with positive evaluations by other faculty members (McElroy, Morrow, & Wall, 1983). On the other hand, at least one attempt at replication failed to confirm the effect of desk placement (Campbell & Herren, 1978), and another found only a weak effect (Campbell, 1979). Perhaps a study by Hensley (1982) clarifies the situation. Hensley hypothesized that although there is a relationship between faculty desk placement and student evaluations, instead of the office arrangement causing more positive evaluations, both the evaluations of the teacher and the desk placement result from the influence of a third variable, the professor's formality or attitude toward education. Thus, a formal teacher is also likely to choose a more formal office, and an informal teacher is likely to have an open,

lower density open-plan office that allowed more space per employee.

In sum, open-office plans provide both advantages and disadvantages. The increased opportunity for communication may facilitate some flow of work, but also increases distraction and reduces privacy. Depending on the functions to be accomplished in a given office, the disadvantages may outweigh the advantages.

JOB SATISFACTION AND THE WORK ENVIRONMENT

In addition to productivity, managers and others have become concerned that design of the work environment can influence job satisfaction. As it turns out, both satisfaction and performance are very complex variables, and there is only a slight positive relationship between satisfaction and performance (e.g., Riggio, 2000). In general, employees do list physical conditions as important for job satisfaction, although the physical environment is not as important in this regard as such factors as job security, pay, and friendly co-workers (e.g., Crouch & Nimran, 1989). One influential theory in industrial psychology and management suggests that an adequate work environment does not substantially enhance job satisfaction, but that a substandard environment definitely leads to dissatisfaction (Herzberg, 1966; Herzberg, Mausner, & Snyderman, 1959). In Chapters 4 and 5 we emphasized the informal desk arrangement. Hensley's results support his contention in that more traditional educational philosophies were more often associated with a closed desk arrangement. In addition, the desk arrangement was also affected by the number of advisees a professor had. Perhaps because a more open desk arrangement facilitates activities such as reviewing records or completing schedules, even traditional professors with a large number of advisees tended to adopt an open desk placement.

Aesthetics

Campbell (1979) used slide photographs to investigate the effects of presence or absence of living things (four potted plants and two aquariums with fish) or art objects (four wall posters and a macrame hanging). Students associated these decorations with feelings of welcome and comfort and expected the professor to be friendly and unhurried.

Neatness

Overall neatness may be even more important than decorations (see Figure 13-7). Very messy offices make the occupant appear to be busy and rushed and to make visitors report that they would be less comfortable and welcome (Campbell, 1979; McElroy et al., 1983; Morrow & McElroy, 1981). Morrow and McElroy also introduced an intermediate level of tidiness they refer to as "organized stacks." Interestingly, the organized stacks level of tidiness was evaluated as significantly more friendly, welcoming, and comfortable than either the messy or extremely neat office conditions (McElroy et al., 1983; Morrow & McElroy, 1981).

Figure 13-7 What is the effect of a messy faculty office on students' perceptions of the occupant?

importance of perceived control in moderating the effects of environmental stressors. In that light, it may not be surprising than one postoccupancy evaluation of a successful office relocation emphasized the importance of targeting a limited number of small-scale (personally meaningful) improvements and enlisting employee participation in planning change (Spreckelmeyer, 1993).

SUMMARY OF DESIGN IN THE WORK ENVIRONMENT

In general, work environments can be designed to maximize productivity by providing acceptable ambient conditions such as lighting, and by optimizing interactions between humans and machines. The nature of the workplace is changing. Fewer employees are exposed to extremes in temperature and noise. On the other hand, changes such as the mass introduction of personal computers have the potential to become either new sources of work-related stress or sources of a more involving and pleasant workplace. Although job satisfaction is related to quality of the work environment, other factors such as the social environment are usually more important. The ability to treat a workspace as a territory, to adjust it, and to personalize it serves as a form of status and may increase job satisfaction, especially at higher ranks in the organization.

Learning Environments

Education is not just for young people; in today's changing world, education continues for a lifetime. Accordingly, the effect of the design of learning environments on the activities within them has been of great interest to researchers. These environments may range from small dormitory study areas that we examined in Chapter 9, to a large formal library or a museum setting such as we study in this chapter, to even larger settings like an entire university campus that we examined in Chapter 11. If a design feature impedes learning, we can ameliorate the problem, and if a design change can increase the effectiveness of education, we can implement it. Let us now look at several design applications in a variety of educational settings.

CLASSROOM ENVIRONMENTS

We noted in Chapter 5 that noise can be particularly problematic in classrooms, in Chapter 6 that heat can interfere with classroom goals, and in Chapter 8 that seating arrangements can influence classroom performance. Intentional changes in classroom environments have been made more or less continuously since we abandoned the one-room schoolhouse. However, as we shall see, physical factors no longer bind us to traditional designs, and research has indicated that changes in classroom design can result in more positive student attitudes and greater participation in class (Gump, 1984, 1987; Rivlin & Wolfe, 1985; Sommer & Olsen, 1980). Let us consider the successes and failures of some of these innovations.

Windowless Classrooms

One innovation, the building of **windowless classrooms,** has not proven overwhelmingly successful. Originally designed to reduce distraction in the classroom as well as to reduce heating costs, these new school buildings typically contain few, if any, windows. Research has suggested that the absence of windows in classrooms has no consistent effect on learning (some students improve, and others show worse performance), but that it does reduce the pleasantness of students' moods (Ahrentzen et al., 1982; Weinstein, 1979). One study in Sweden suggested that a windowless classroom may be associated with reduced growth and lower concentration (Küller & Lindsten, 1992); because this study examined only two windowless and two windowed classrooms, the results must be interpreted with caution, although following the study, school authorities did add windows to the rooms.

The Open Classroom Concept

The traditional design of classrooms, rectangularly shaped with straight rows of desks, dates back to medieval times, when the only source of light was natural light that came in through windows. Modern buildings, of course, do not rely solely on sunlight, so new design alternatives are possible. **Open classrooms,** like open offices, are designed to free students from traditional barriers, such as restrictive seating. In such settings, students should have more opportunity to explore the learning environment (Figure 13-8).

Research evaluating these designs is confounded by the fact that the environment is typically not the only difference between open and traditional schools. That is, an "open education" philosophy implies freedom for students to move around and less structure in class activities. However, these could occur in a traditionally designed classroom and do not necessarily occur in open classrooms. In many open classrooms, students behave much as they do in traditional classrooms, and teachers often do not use all of the space provided. As is the case in traditional settings, students spend a great deal

Figure 13-8 A variation of an open classroom.

of time engaged in solitary tasks such as reading and writing. For example, Rothenberg and Rivlin (1975) found these percentages of total activities observed in one open classroom: writing, 26%; arts and crafts, 11.8%; talking, 11.3%; reading, 6.3%; working at projects, 5.7%; and teaching, 4.3%. However, Gump (1974) observed that students in open classrooms spend less time in directed activity than students in traditional settings, and that groups in open classrooms show greater variability in size. Such heightened flexibility in open-plan rooms is often accompanied by greater activity than in the traditional classroom.

Two serious problems with open-plan designs are that they provide inadequate privacy and foster too much noise (e.g., Ahrentzen et al., 1982). These are the same problems we saw with open-plan offices. The flexibility provided by the open space can cause coordination problems, and frequently teachers do not know how to arrange furnishings so as to get the most use out of the space provided. Variable height partitions can reduce noise but still give the open feeling (Evans & Lovell, 1979). It is conceivable that by combining aspects of traditional and open-design classrooms, better environments may be created. At present, however, reviews of this kind of physical design are mixed. Overall, data-based studies indicate that open classroom designs are noisy, provide undesirable distractions, and do not foster adequate educational benefits to outweigh these problems (Bennett et al., 1980).

Environmental Complexity and Enrichment

What is the proper amount of environmental complexity in an educational setting? As we have seen elsewhere in this book, studies have indicated that the complexity of an environment can affect arousal and performance in that setting. Too many stimuli may distract students, create overload, or increase fatigue. However, extremely simple settings may be boring and equally detrimental to performance (cf. Sommer & Olsen, 1980).

Some researchers believe that classrooms should tend more toward the complex rather than the simple. Having more stimuli and opportunities for environmental exploration provides an enriched environment that facilitates learning. Others disagree, arguing that complex learning environments are distracting and make it difficult for the student to concentrate on school work. For example, Morrow and Weinstein (1982) found that having a quiet library corner increased reading activity by kindergartners. Of course, classrooms serve more purposes than just learning content relevant to a specific topic. They also involve learning how to learn, learning social responsibility, and acquisition of cultural values. Different classroom environments may facilitate one of these purposes but not the others. Working for the right fit between pupil and learning environment is probably the most desirable approach (e.g., Ahrentzen et al., 1982), a view consistent with adaptation level theory as presented in Chapter 4. Because things change with time, it is important to evaluate classroom design modifications continually. Wong, Sommer, and Cook (1992), for example, studied a University of California at Davis classroom that had been modified in the 1970s to provide a "softer" environment that would facilitate interaction among the students and instructor. Satisfaction had declined since the renovation, but remained higher than in a prerenovation evaluation. Concerns were expressed about a 1970s color scheme and a faded carpet. Still, the design was perceived to enhance interaction.

Density

Whether the classroom is open or closed, windowless or windowed, or complex or simple, educators, parents, and students are concerned about the number of students in the class, or in the term more common to environmental psy-chologists, the *density* in the classroom (see also Chapter 9). In general, high density has minimal effects on learning of simple concepts appropriate to a lecture format, but interferes with learning of complex concepts and with activities that require students to interact (e.g., Ahrentzen et al., 1982; Weinstein, 1979).

Day Care and Preschool Settings

As more and more families are characterized as having dual-career parents or as being single-parent families with that parent working, finding quality day care for children has become a paramount concern. Similarly, preschool programs are becoming fairly commonplace. A major goal of these settings is to teach scholastic skills (the alphabet, counting) as well as social skills (taking turns, sharing). Environment and behavior specialists have noted particular concerns with design features of these settings that impact the very youngest children—children whose age means they have very short attention spans and are easily distracted by visual movement and by noise. Design features that address these special concerns include carpeting to reduce noise and screens to reduce visual intrusion. Interestingly, Neill (1982) found that although carpeting increased teacher–pupil interaction, screens increased teachers' time on paperwork and decreased their interaction with pupils. Also, activity areas that are separated from one another and traffic paths that reduce intrusion into these areas promote better learning behavior and may actually improve math and language skills and creativity (Nash, 1981; Weinstein, 1981).

LIBRARIES

Multiple Functions

Library designers have a number of unique problems with which they must deal. One familiar problem at university libraries is that patterns of use for study and reading areas move through periods of over- and underuse. Because underuse wastes space that could be used for books, one design alternative for a university library is to reserve the library for the storage and dispensation of materials. Reading and study areas would be eliminated from this setting and dispersed to other areas on the campus. For many students, such a separation of library and study functions would mean a major change in work style and would sometimes prove inconvenient. After all, the campus library often is the one place where students know they can get schoolwork done. Another problem libraries encounter is that more and more materials are acquired each year, yet older materials are not discarded, so more and more space is needed. One design alternative is a space-saving strategy that eliminates the usual aisles between rows of bookshelves. The rows of shelves are mounted on tracks and move apart to create access aisles as needed (Figure 13-9).

Orientation and Wayfinding

Finding a book in a library is partly a problem of orientation to a large setting. Where do we start? Where do we go for help? Do we ask for information or try to find our way by reading signs? Pollet is one librarian who, decades ago, (e.g., Pollet & Haskell, 1979) showed interest in helping libraries improve their orientation aids. One of the most important observations she made is that library patrons must cope with information overload. Adding signs to help people find their way around contributes even more information to the environment. In particular, Pollet noted that clustering many signs together makes orientation information ineffective. People who are already receiving too much information are not apt to stop and look at a cluster of signs.

Libraries are learning to reduce the number of signs used and to experiment with critical

Figure 13-9   To save space normally taken up by aisles, these rows of library bookcases move on tracks to open up aisles as users need them. Sensors prevent the bookcases from closing when someone is in the aisles.

locations of signs throughout the building. People need information at the point of making a decision about where to go next. One helpful technique is to use a specific color for orientation information. No matter where people are, they can look for that color and become oriented. However, Pollet concluded that using too many colors for different areas simply adds more information to be processed and can cause disorientation (see also Chapter 3 on wayfinding).

As found in similar museum studies, many patrons will not ask for help in libraries. Pollet advocated a good sign system that would give patrons a sense of control over the environment instead of relying on staff to answer questions. She also commented that it is hard to find workers who can put up with answering the same questions all day. While library patrons may be experiencing stimulus overload, information staff may experience understimulation, which can leave them bored and irritable.

VISITOR BEHAVIOR IN MUSEUM ENVIRONMENTS

If you visit a museum today, you probably go for a little entertainment, a little enlightenment, and a chance to get away from your normal routine and perhaps spend time with family or friends on an outing for the day—experiences consistent with those we will shortly discuss for participating in leisure activities. In a museum, you probably also think you have complete control over the decisions as to which exhibits to explore and how much time to spend viewing each one. Most assuredly, your visit to a modern museum is not a random event, but what happens during your visit is not as much under your complete control as you might think. Rather, a team of environmental psychologists and related professionals has probably designed each exhibit hall and each exhibit in order to influence your behavior in very specific ways. Today, museums, zoos, visitor centers at national parks, and similar facilities have functions that go beyond preservation and recreation. Indeed, a major purpose of these settings is to educate the visi-

tor about history, nature, and culture. Together, environmental psychologists refer to this science as the study of *visitor behavior,* and to these settings as *informal learning environments.* These environments are a type of bridge between our previous look at traditional classrooms and the leisure settings we will explore at the end of this chapter. Indeed, museums can have the qualities of a *restorative environment* that we associate with leisure settings (Kaplan, Bardwell, & Slakter, 1993), as we mentioned in Chapters 2 and 4. Since we use museums less regularly than classrooms, the museum environment is somewhat more novel to us. Museums are also usually larger and do not provide a home base, such as a desk does in a classroom. In addition, the primary mode of activity in museums is exploration, as we make our way through halls and rooms, past endless displays and exhibits. In this section we will examine some of the design principles that apply to informal learning environments. To save space, we will concentrate on museums, but the principles apply to zoos and similar settings as well. The entire July 1988 issue of *Environment and Behavior* is devoted to this topic as it relates to zoological parks, the November 1993 issue of the same journal is devoted to the environmental psychology of museums, and Loomis (1987) provides in-depth coverage if you would like to read more details about design and behavior in such settings.

Wayfinding

The ability to find things in a museum is related to wayfinding in any setting. Museums that are confusing or hard to explore may result in less satisfaction with the visit. If you miss the exhibits you came to see because you could not find them, or if you find yourself constantly backtracking and going in circles, you probably have less fun than if everything were simpler. However, the complexity of museum environments is an almost inherent feature of their purpose—to display as many exhibits as possible.

One way of overcoming this inherent complexity is to provide aids for finding one's way through the museum. Many people prefer to consult signs and maps and are uncomfortable if they have to ask museum employees for help, just as we saw for libraries. Maps that clearly depict a setting and identify the viewer's location on the map in relation to the setting seem to be particularly helpful. Such *you-are-here maps* show the position of the viewer and how to get from "here" to other parts of the setting (Levine, 1982; Levine, Marchon, & Hanley, 1984; see also Chapter 3). Not surprisingly, orientation aids such as maps and suggestions for what to see appear to increase satisfaction with the environment, and the simpler the map, the better (Talbot et al., 1993).

Exploration

Research has also addressed the ways in which people explore museums. For example, people appear to have a right-handed bias; upon entering a gallery in a museum, they typically turn right and move around the room in that direction (Melton, 1933, 1936; Robinson, 1928). Once inside a museum, people usually stop at the first few exhibits and then become more selective, stopping at fewer the longer they explore (Melton, 1933). For example, Melton (1933) found that 49% of visitors to an art gallery looked only at the art on the left or right wall, and only about 10% made a complete circuit through the gallery to examine all the artwork. The more likely visitors are to explore a given exhibit, the higher its **attraction gradient.** Exits to other exhibit rooms are also important because people tend to use the first exit they see. Museum researchers refer to this "pull" of exits as the **exit gradient.** Due in part to attraction gradients and exit gradients, most people see only a part of each exhibit room rather than seeing everything before moving on.

Fatigue in Museum Exploration

Predictable though it is, the pattern of physical movement within a museum shows some signs of being maladaptive. Walking in a museum

should facilitate exploration of the environment. Yet, as we have seen, visitors frequently move past much of the exhibit without stopping or looking at it, thereby missing many of the rewards to be gained from a museum visit. Why is exploratory movement not more complete? The most popular museum exhibits are those that are of moderate complexity (Melton, 1972; Robinson, 1928). Museums may create overload if they are too complex or if it is difficult to get around inside (see Figure 13-10A and B). Perhaps fatigue from this overload interferes with completing more thorough patterns of exploratory behavior.

Robinson (1928) first studied fatigue in museums many years ago. In spite of his work being old, many of his observations on exploratory fatigue are still important. He concluded that fatigue was not due just to physical exertion but also to the visitor growing tired of maintaining a high level of attention. Borrowing from Gilman (1916), Robinson used the term **museum fatigue** to describe the phenomenon.

In a clever laboratory study, Robinson was able to demonstrate that museum fatigue was more than just physical exertion. He had persons seated at a table look at a series of copies of paintings from a gallery, presented in the same order as they hung in the gallery. Attention time for each painting was recorded and compared with the attention time observed in the gallery itself. It turned out that participants seated at the table and looking through the stack of pictures began to show a drop in attention at about the same point in the sequence as visitors walking through the museum. Robinson concluded that museum fatigue was due to psychological satiation or boredom as well as to fatigue from physical activity. He did not mean that visitors were bored by the exhibits. Rather, he noted that after visitors concentrated on several stimulating exhibits for a long period, they became so satiated with the museum's environment that additional exhibits were relatively unstimulating. Recall from our discussion of information overload in Chapter 4 that when we receive the massively complex stimulation typical of many

Figure 13-10A & 10B Monotonous rows of display cases (top) can create overload. Modern museums (bottom) recognize the problem of environmental complexity and fatigue by creating exhibits that pace the amount of complexity so as to reduce fatigue and orient the visitor.

Figure 13-11 The attraction gradient of an exhibit can be increased by adding features such as movement, buttons to push, flip-up labels, or sensory experiences such as sound and touch. This exhibit on air pollution includes a dial to turn, levers to push that reveal answers to questions, and a phone line for weather and pollution data. (Photo courtesy of Denver Museum of Natural History. Used with permission.)

museum environments, we tend to ignore less important cues in order to attend to more important ones. This is the sort of phenomenon that occurs with museum fatigue: We become so satiated with complex information that we spend less and less time looking at the details of various exhibits.

Museum fatigue can be alleviated somewhat by building what Robinson called discontinuity into the design of an exhibit. **Discontinuity** refers to a change of pace in the stimuli presented. For example, a series of paintings might be broken up with a piece of sculpture or an arrangement of furniture. The number of paintings or objects displayed can also be reduced, since a single gallery may contain a collection large enough to tax the attention span of the most ardent art lover. Alleviating museum fatigue helps visitors gain more satisfaction from their exploration of the museum environment. It certainly helps to have labels with large print and interpretive explanations that are brief enough to avoid overload (Bitgood & Patterson, 1993). In addition, movement in an exhibit, such as a rotating wheel or a swinging pendulum, will boost the attraction gradient; however, movement in a previous exhibit may mean that lack of movement in a subsequent exhibit contributes to fatigue. An interactive exhibit is one solution:

Encouraging the visitor to push a button that lights up a section of the display case or to lift a cover to see the answer to a question (called a "flip" by exhibit designers) helps maintain interest. Another solution is a **space surround environment,** in which the features of the exhibit entirely surround the visitor, as opposed to a series of separate display cases. In general, a space surround environment is superior in maintaining visitor interest (e.g., Thompson, 1993). As you can infer, there are an infinite number of possibilities for designing individual exhibits and sequencing them to maximize visitor attention and educational outcome (Figure 13-11).

Many of these principles, as well as human factors principles we discussed earlier in this chapter, were examined in a study by Harvey et al. (1998). They found that they could more than double the amount of time visitors spent at exhibits by adding such elements as interactive features, multisensory stimulation (e.g., sound and touch as well as vision), better lighting, and lettering that is easier to read. These features also contributed to a sense of immersion in the museum experience. Evaluating designs before and after implementation is extremely important in assuring that the exhibit accomplishes its intended purpose (Bitgood & Loomis, 1993; Klein, 1993; Miles & Clarke, 1993).

Management of Natural Lands for Leisure

Most of us look forward to (or at least appreciate) the opportunity to take a break from hard work and intensive study in order to engage in some leisure activity. Optimists once believed that leisure time would grow as a proportion of daily life. Unfortunately, the amount of available free time may be decreasing (Csikszentmihalyi & Kleiber, 1991; Zuzanek, Beckers, & Peters, 1998), whereas people's sense of being stressed and pressed for time has grown (Robinson & Godbey, 1997; Zuzanek & Smale, 1997). Perhaps we need to ensure that whatever time we get is, indeed, "quality time."

THE MEANING OF LEISURE

Before we begin our discussion it would be useful for you to reflect on your personal understanding of the term *leisure*. Of course, all of us have an intuitive idea of what leisure is, but most researchers agree that leisure will be experienced when an individual (1) is intrinsically motivated and (2) perceives freedom of choice (Neulinger, 1981; Tinsley & Tinsley, 1986). *Intrinsic motivation* is a popular term that has resisted precise definition by psychologists; basically, it is the degree to which a behavior leads to personal satisfaction and enjoyment (Smither, 1988). Perhaps the term is best understood by contrasting it with the *extrinsic motivation* prompted by some external agent, such as pay, gifts, or praise. Several researchers have examined the importance of these dimensions. Iso-Ahola (1986) found that students were more likely to perceive an activity as leisure if their participation had been voluntary, the rewards had been intrinsic, and the activity had not been work related. For example, Juniu, Tedrick, and Boyd (1996) compared amateur and professional musicians' perceptions of both practices and concerts. Amateurs viewed both the performances and practices as intrinsically mo-

tivating leisure experiences, whereas the professionals perceived both activities as work.

We may choose from many types of leisure experiences, from reading a novel to tossing a football. In Chapter 10 we briefly discussed the leisure opportunities parks and playgrounds provide for urban residents. In this chapter we will concentrate on natural settings, a more and more popular site for leisure activities such as hiking, sightseeing, or fishing. Many researchers attribute the popularity of leisure activities in natural lands to the need for restorative experiences. As we have seen repeatedly, encounters with nature balance the stresses and distractions of everyday life.

We introduced some of the most common explanations for the role of nature as a salve for life stresses in Chapter 2. Briefly, we said that North American attitudes toward nature have changed dramatically since colonial times. Nevertheless, in addition to the cultural backdrop, researchers posit a human need for contact with nature. One model emphasizes an instinctual attraction to natural elements (biophilia) and suggests that leisure contact with nature should reduce stress (Ulrich et al., 1991; Ulrich, Dimberg, & Driver, 1990). Attention restoration theory is a second, competing explanation (R. Kaplan & S. Kaplan, 1989; S. Kaplan, 1995). As Chapter 2 outlined, attention restoration theory suggests that natural landscape elements capture one's attention in an effortless, restorative fashion. Other researchers emphasize the importance of place-centered experiences (Korpela & Hartig, 1996; Roberts, 1996) and the importance of developing a sense of self-efficacy and competence (e.g., Propst & Koesler, 1998). Other motivations for spending time in the wilderness may include the need to be with our family members or to avoid them, to take risks, or to feel comforted and safe. One could add the aesthetic delight associated with many wilderness areas (e.g., breathtaking views)

Figure 13-12   Aesthetic delight and emotional experiences are often associated with natural areas.

and the emotional experience such visits can generate (see Figure 13-12).

CHARACTERISTICS OF RECREATIONISTS

Gender and race also affect recreational choices and demands (Burger, Sanchez, Gibbons, & Gochfeld, 1998; Culp, 1998; Garcia, 1996; Henderson, 1996; Philipp, 1997; Virden & Walker, 1999; Zuzanek, Robinson, & Iwasaki, 1998). For example, studies show that African Americans and Hispanic Americans desire somewhat more modern comforts in developed forest campgrounds (Floyd, Outley, Bixler, & Hammitt, 1995; Irwin, Gartner, & Phelps, 1990; Virden & Walker, 1999) than Anglo Americans. The particular types of recreational involvement visitors seek also vary at different points in the life cycle

(e.g., Larson, Gillman, & Richards, 1997). Perhaps it is unsurprising that adolescents enjoy spending leisure time with the family less than their parents do.

Even though both urban and rural residents sometimes seem similar in their attitudes toward wilderness, their definitions of what is or is not wild varies. Thus, Lutz, Simpson-Housley, and de Man (1999) report that urban residents, unlike rural people, classified photographs of areas as wilderness even though they depicted logging activity, grazing, roads, hydroelectric dams, and even villages.

THE CHALLENGE OF LAND MANAGEMENT

With some of our thoughts about the psychological effect of nature experiences in mind, let us turn to the more utilitarian issue of identifying and perhaps modifying settings to meet these needs. We will begin our discussion of land management with two questions: What does it mean to "manage" land? And, "For whom is land managed?" A person might respond to the first question with: "To take care of the land so that all of us can use it." To the second question one reply might be: "For all citizens of our nation, no matter what their religion, race, or social status." Although both answers are in some respects admirable, they reflect an anthropocentric or instrumental interpretation of management (Gee, 1994; Stokols, 1990; Thompson & Barton, 1994). In contrast, Thompson and Barton say an ecocentric view values nature for its own sake, not just its contribution to human welfare. Some (e.g., Gee, 1994) would like us to give up the term *user* as a description of people in environments because it emphasizes the place of humans as superior and treats land, vegetation, and wildlife as "resources" whose sole purpose is to serve humanity. Along these lines, the naturalist writer (and ex-forester) Aldo Leopold (1949) proposed that humans should think of themselves as part of a community that extends to soils, waters, plants, and animals. Thus, Leopold's *land ethic* removes *Homo sapiens* from the role of conqueror and makes us equal members of a natural community.

Our science, and this book, aim to understand and improve the human condition. Nevertheless, we, too, grow uncomfortable with a purely anthropocentric position. The fields of resource management are not unaware, or even necessarily unsympathetic, to the demands implied by the land ethic as a fundamental principle of management. As the research literature catches up with the popularity of this land ethic, we may find that an ecocentric view will predominate over the more traditional utilitarian, economic focus. Nevertheless, by legislative mandate the agencies charged with overseeing public lands such as national forests are required to balance a variety of demands among different recreational users and other interests such as mining and timber cutting. Done fairly, this balancing act probably demands some sort

of resource valuation (economic or psychological) and management action (Clarke, Bell, & Peterson, 1999).

Can We Love Our Lands Too Much?

In what ways do humans create management challenges? More and more of us seek untrammeled nature. Ironically, by embracing it, we challenge its existence. Especially since World War II there has been a dramatic increase in the number of citizens who desire to maintain natural areas for recreation, aesthetic appreciation, or wildlife observation not associated with hunting or fishing. You might think that finding such a place should not be hard because vast areas of North America (in some states and provinces, more than 50% of the land) are publicly owned. However, more and more people are seeking to enjoy national parks, national forests, and other natural areas, and now many facilities are stretched thin (e.g., Mitchell, 1994). The popularity of natural areas has sometimes led to conditions of relatively high population density, crowding, and overstaffing. The most common complaint by hikers in the Grand Canyon is noise—noise from tourist aircraft that give visitors a popular aerial view of the spectacular scenery. During peak periods hikers looking to "get away from it all" have to listen to the drone of aircraft noise twice every minute, and certainly the intermittent character of this noise makes adaptation to it very difficult (see the box on page 459).

As increasing numbers of visitors try to escape the stresses of city life they often bring crime, exactly the kind of urban problem many seek to avoid. Although rangers have always had law enforcement powers, traditionally they have been able to use a "low key" approach to protecting natural settings from visitors and visitors from one another (Pendleton, 1998). With increasing crime, many have advocated a more aggressive, "hard enforcement" approach (Shore, 1994). If rangers are forced to adopt urban police tactics, what will happen to the traditional ranger's role as the "camper's friend"? More generally, how can parks, forests, and wilder-

TOURIST AIRCRAFT FLIGHTS OVER NATIONAL PARKS: Noise, Noise Everywhere!

One of the most spectacular views people can experience is to tour Grand Canyon National Park or Hawaii Volcanoes National Park from the air. Many of these tours originate from well outside the park, such as flights from Las Vegas to the Grand Canyon, and the tourists on them may well never see the park from the ground. But for those visiting the parks on the ground, the aircraft overflights are a nuisance. We have noted that a major reason people visit sites such as national parks is to gain a restorative experience—to escape the stress of the everyday world and appreciate the majesty and tranquility of nature. Consider, however, that overflights at the Grand Canyon number 10,000 per month (Staples, 1996)! In parts of the Grand Canyon, aircraft noise is audible 79% of the time, with as many as 43 separate aircraft noise events within every 20-minute interval, or more than two per minute (Horonjeff et al., 1993). Consider also that such noise is intermittent and uncontrollable, two factors that make it particularly annoying and stressful.

In a laboratory study, Mace, Bell, and Loomis (1999) simulated helicopter noise at either a relatively quiet 40 dB or a fairly loud 80 dB level while participants viewed slides of national park scenes. Ratings of annoyance, solitude, tranquility, freedom, and naturalness were all negatively impacted by the noise. In addition, overall affective feelings were lower following exposure to the helicopter noise. Surprisingly, these effects were observed for both the quieter and louder noise levels. Furthermore, slides of some of the most spectacular natural scenery in the United States were rated lower on scenic beauty during exposure to helicopter noise. How can this possibly be a restorative experience?

Obviously, aircraft overflights are a serious management issue for those charged with preserving the parks and regulating visitor activities. On the one hand, tour operators are financially dependent on the flight operations, and there are plenty of paying customers willing to spend handsomely for a spectacular view from the air. Moreover, some disabled visitors could not see some areas of the parks in conventional ways, but

Figure 13-13 Tourist aircraft overflight noise is a major source of complaints among visitors to national parks. The view from the air is spectacular, but visitors on the ground complain that even low-level aircraft noise detracts from tranquility and aesthetic beauty.

could from the air. On the other hand, the overflight noise is clearly bothersome to most visitors on the ground and can be disturbing to wildlife. There is an ongoing debate about how to accommodate the competing interests involved. For some parks, overflights are strictly prohibited, and for others, aircraft cannot fly below certain altitudes. Some day, quieter engines may offer promise. For the time being, overflights remain one of the most common sources of visitor complaints.

ness areas accommodate more visitors without changing the characteristics of the natural settings that attract people to them?

Even careful hikers may cause long-term damage to fragile natural environments (Pitt &

Zube, 1987). Furthermore, the enjoyment of one recreational activity (particularly one involving advanced technology such as motorcycling or using a jet ski) may interfere with the enjoyment of other recreation activities. These are behavioral problems, but environmental managers have historically been trained in professions such as forestry, wildlife and fisheries biology, or range management. Thus, managers trained in the natural sciences to support the sustained yield of commodities such as timber now find themselves managing for the enjoyment of recreationists (Pitt & Zube, 1987).

How can they weigh the importance of different competing demands? Many of these questions have traditionally been the domain of economists; however, recreation is a resource use characterized by the behavior and experiences of people, so psychologists are increasingly involved.

What Does Recreation Management Manage?

As we began our discussion we said that leisure requires that an individual be intrinsically motivated and perceive freedom of choice (Neulinger, 1981; Tinsley & Tinsley, 1986). Simply stated, leisure means "being able to do what you want to do," but researchers emphasize that leisure is an *experience* rather than an *activity* (e.g., Driver & Tocher, 1970; Floyd & Gramann, 1997; Manfredo, Driver, & Tarrant, 1996; Tinsely & Tinsley, 1986). That is, different activities may lead to similar psychological experiences, or conversely, the same activity might yield different experiences (only some of which would be perceived as leisure) depending on the personality of the participant, the setting, or other factors. For example, the psychological experience of risk taking might be met by mountain climbing or by downhill skiing. On the other hand, the experience of skiing for one person might focus on speed, for another the accomplishment of good form, and for a third individual, an opportunity to enjoy the company of a loved one. The magnitude of the leisure experience probably varies from time to time and among individuals. For some, the feeling of freedom, increased sensitivity, and decreased awareness of the passage of time are profound. The result may be cardiovascular and health benefits (Froelicher & Froelicher, 1991; Paffenbarger, Hyde, & Dow, 1991) or more abstract benefits

like the development of self-identity (Haggard & Williams, 1991). Indeed, several researchers suggest that the leisure state can be similar to mystic or peak experiences (Csikszentmihalyi & Kleiber, 1991; Tinsley & Tinsley, 1986). One recent managerial emphasis has been to identify and manage for both first-order benefits such as relaxation or diversion and long-term benefits that impact one's quality of life (Driver, 1996).

New voices have begun to demand that managers acknowledge the importance of spiritual and other difficult-to-measure values of natural environments for both members of the dominant culture and minority groups such as Native Americans (Driver, Dustin, Baltic, Elsner, & Peterson, 1996; McDonald & Schreyer, 1991; Roberts, 1996). Although environmental psychology shares many of the views of science, the desire to measure concrete phenomena among them, adequate resource management will probably require a variety of methods—some of them quite phenomenological and foreign to experimental researchers.

Just as the same place might support different recreational experiences, very different environments can also provide quite similar kinds of leisure experiences. Thus, the leisure experience is jointly determined by environmental, social, and individual difference variables.

As illustrated in Figure 13-14, the recreation production process can be seen as one of attempting to meet human recreational demands by managing the basic resources of a given site (Driver & Brown, 1983). Consider a hypothetical piece of land in Figure 13-15A. It might be "developed" to provide water access for dozens of boaters, utility hook-ups could be installed for motor home owners, and the trail could be manicured to provide access for individuals who are unable to walk appreciable distances (Figure 13-15B). Each of these modifications—which we can think of as design alternatives—would please some segment of the public. They would be unlikely to please a fit wilderness backpacker seeking solitude. For the backcountry traveler, any signs of human modification (and certainly boat docks and parking lots) may destroy the wilderness experience. What is a resource manager to do? On purely economic

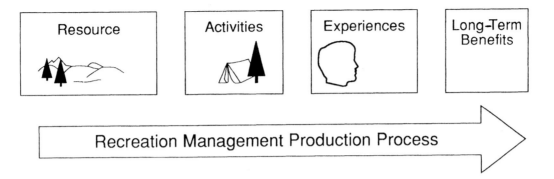

Figure 13-14 The recreation production process.

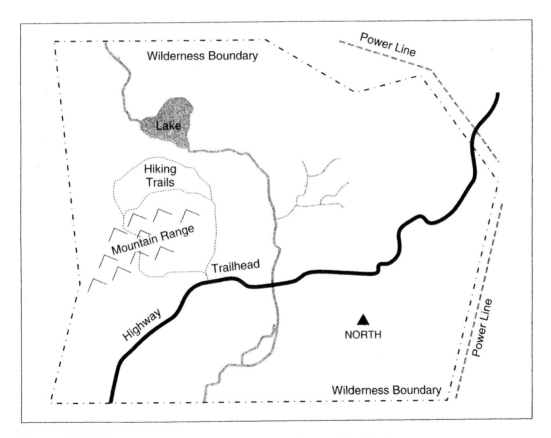

Figure 13-15A A hypothetical piece of land relatively unmodified for human use.

grounds the decision seems easy. More people can enjoy our hypothetical area (and spend more money at regional gas stations, restaurants, and convenience stores) if the area is "developed." On the other hand, every instance of development creates long-term changes in the land that affect wildlife and plant habitats, watersheds, forests, and, of course, our hypothetical backpacker.

How many people can recreate in an area, and what is the effect of their use? The ability of an area to absorb use is termed **carrying**

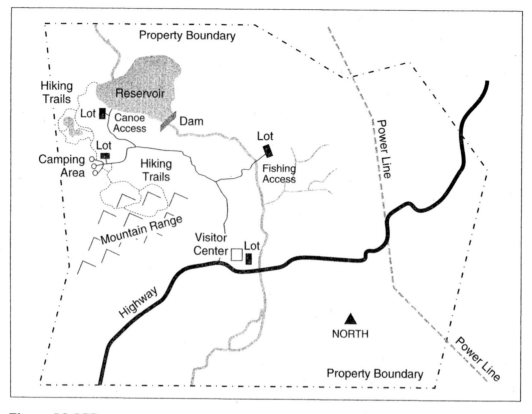

Figure 13-15B   The same piece of land developed for extensive recreational use.

capacity. In the case of recreation, carrying capacity may include the resistance of an area to ecological damage, the availability of facilities such as campsites, and *social carrying capacity,* which refers to the desired level of social interaction (Manning, Valliere, Wang, & Jacobi, 1999; Pitt & Zube, 1987; Stewart & Cole, 1999). Social carrying capacity is the most psychological of the three and essentially represents crowding. As we observed in Chapter 9, the tolerable level of density is subjective and situation specific. The tolerable level of density may vary by person, activity, or environment. Although the absolute number of encounters with other people in wilderness and other natural areas is very low, recreation resource managers are regularly concerned with user complaints of crowding. Of course, this is consistent with our discussion in Chapter 9, which emphasized that increasing density affects crowding through a

filter of an individual's expectations and preferences. Crowding is perceived when density becomes inconvenient or restricts our freedom to engage in some desired behavior.

Most groups object to loud, inconsiderate behavior (e.g., Owens, 1985; West, 1982), but if others are perceived to be similar to one's group in behavior and apparent values, the conflict and perceived crowding are minimized. In general, recreationists employing higher levels of technology are more likely to disrupt the experiences of their lower technology counterparts than the reverse. Canoeists tend to dislike meeting motorboats (Lucas, 1964; Schreyer & Nelsen, 1978); backpackers dislike encountering horses (Stankey, 1973). Thus, the low-technology recreationists are likely to be the first to be displaced. Furthermore, some activities are considered less specialized than are others. Inner-tube floating requires less skill and a less unique environment than kayaking, for instance. Crowding may be perceived at lower use levels by some user groups in particular areas. For example, in one study (Tarrant, Cordell, & Kibler, 1997), kayakers were more bothered by crowding than were rafters in areas of river rapids, but not at put-in sites. In so-called "front-country" areas such as popular sites in national parks, visitors expect to encounter a number of others. For these environments, subjective norms may allow for higher levels of density (Manning, Lime, Freimund, & Pitt, 1996).

How Much Wilderness Is Enough?

Many nations have created wilderness areas by legislative action and are considering establishing more. The acreage encompassed by national parks and monuments, national forests, and other generally natural areas far exceeds that which has been officially designated "wilderness," a term reserved for areas that are to remain undeveloped. Should more land be reserved as "true" wilderness and be lost forever for lumbering, resort development, or even campgrounds? Returning to our hypothetical example, will as many different people enjoy a given parcel of land if it is declared wilderness as would use it if it were developed with an access road, sanitary facilities, and picnic tables? The answer to the latter question is probably "no." If one measures the value of an area based only on the number of users, the wilderness seems to be an expensive luxury. On the other hand, it might be demonstrated that the wilderness provides unique experience opportunities that can be received nowhere else, and that, in the long run, the effects of recreation are beneficial not only to the individual, but to society in general.

The *Recreation Opportunity Spectrum* (ROS) is one widely used tool for managing public lands to provide various recreational experiences and benefits (Driver, Brown, Stankey, & Gregoire, 1987). Loosely, the ROS is based on a "bull's-eye" concept wherein the interior of a wild area is managed to remain pristine. More and more evidence of human modification is al-

lowed in concentric zones, with the most intensive development restricted to the peripheral zones, where the managed lands meet the privately held areas that surround them.

PRELUDE TO PRESERVATION

Our discussion of leisure experiences in natural areas highlights some of the concerns about preserving these lands for recreation and other uses, including nonhuman use. One of the factors that works against preservation is the multiple demands that people place on the resource in the first place. Often, the problem is not that people intentionally abuse the resource, but that collective demands from multiple users exceed the capacity of the resource to support all individual interests. Some critics have expressed dismay at present practices that treat landscapes and their animal inhabitants as commodities (Roberts, 1996). What are the moral rights of wildlife to their particular habitat? What if providing for their requirements forces the withdrawal from use by humans? Western attitudes traditionally seem to accept humans' exploitative use of land, vegetation, and animals. In our next and final chapter we will see how environmental psychology can be used to address this issue and help preserve the environment for all.

SUMMARY: LEISURE AND RECREATION ENVIRONMENTS

Recreation and leisure opportunities are important for a growing number of people. Management and design considerations help to determine what opportunities are available in recreation and leisure environments. We emphasized the importance of psychological experiences, rather than particular activities, in motivating recreational behavior. For instance, wilderness hiking or camping may be viewed as ways of coping with stressful aspects of everyday life, but crowding, littering, and other problems associated with camping may work against the coping function, and desire for personal growth may be as important as escape.

SUMMARY

Work environments have developed flexibility because of technological innovations allowing removal of equipment from rigid power sources and construction of expansive and high-rise spaces. The quality of the work environment also affects job satisfaction, although job security, working relationships, and other factors are usually found to have more impact on job satisfaction than does quality of the environment.

A major innovation in office design is the open-office plan, or office landscape. The advantages of the open office include reduced maintenance costs, easier communication, better workflow, and easier supervision. Disadvantages of open offices, which may in some circumstances outweigh advantages, include increased noise and other distractions, as well as reduced privacy.

The ability to treat one's workspace as a territory, as well as the right to personalize it, may serve as a form of status in the organization. Especially at high ranks in a firm, such territorial treatment and personalization may correlate with job satisfaction.

Research on classroom design has shown that windowless classrooms have no consistent effects on academic performance, but that presence of windows promotes pleasant moods. Open-plan classrooms also show mixed results,

with some research showing an increase in activity associated with open classrooms, as well as increases in noise and decreases in privacy. Apparently, an optimal level of complexity in the classroom environment promotes learning.

Libraries have a problem of periodic overuse of facilities. One proposal would separate the "normal" library functions from the study function it often serves. In addition, evidence indicates that orientation aids would help many libraries. Orientation is also a problem in museums, with some evidence suggesting that improved orientation enhances satisfaction. Exploration of a museum tends to be systematic and is heavily influenced by the attraction gradient of exhibits. Museum fatigue may be caused by overstimulation and can be alleviated by designing discontinuity into exhibits.

Perhaps as a restorative counter to work and other life stresses, recreation and leisure opportunities are increasingly important. Wilderness and camping experiences permit opportunities for personal growth and can help tame everyday stresses. However, crowding and other problems can undermine the value of leisure experiences. Managers are challenged by the task of maximizing the positive value of recreation while minimizing resource damage.

KEY TERMS

assigned workspace	fixed workspace	open-plan office	windowless
attraction gradient	human factors	scientific management	classrooms
carrying capacity	landscaped office	situational awareness	workflow
discontinuity	mapping	(SA)	
ergonomics	museum fatigue	space surround	
exit gradient	open classrooms	environment	

SUGGESTED PROJECTS

1. Take a tour of various offices around your campus and note various types of personalization. Does personalization seem to vary with status, gender of the occupant, type of job, or academic specialization?

2. In this chapter we discussed several population stereotypes regarding the human–machine system. For example, we normally expect that turning a knob in a clockwise fashion will increase the volume, temperature, or lighting level (but remember, water faucets are just the opposite!). Can you think of other examples of population stereotypes or violations of these stereotypes on your campus? Make up a short survey to give to your friends to determine the relative strength of your proposed stereotypes. There will probably be some disagreements. What are the implications of ambiguous situations?

3. Try to visit open classrooms and conventional classrooms in your area. Which seem to have the most activity? Which seem to have more noise?

4. Visit your local museum or zoo and note exploration patterns of visitors. Can you identify attraction gradients and discontinuities? Do visitors tend to use the same route through displays?

5. Explore several types of recreation environments and note the concentration of ages in them, as well as the experiences or benefits people seem to be getting from them. Are there systematic differences? What design changes would be necessary to provide other recreational opportunities?

6. Recall that recreation managers distinguish among activity opportunities, experiences, and benefits. With a few friends, try to create a list of the benefits to yourself or to society that result from recreation in outdoor environments.

CHANGING BEHAVIOR TO SAVE THE ENVIRONMENT

BEYOND TRADITIONAL BEHAVIORAL INTERVENTIONS

 # Introduction

Imagine that you are a shepherd and that you share a pasture known as "the commons" with the other shepherds of your village. Further assume that the commons cannot be enlarged—it constitutes all the land you and the others have on which to graze your animals. Although you share the pastureland, the economic benefits you gain from your herd are yours, and from time to time you are confronted with the decision of whether to purchase another sheep for your flock. The commons is becoming depleted (i.e., overgrazed), but you feel that you would enjoy the economic advantage of owning another animal. After all, the commons could support one more sheep without too much further damage. You reason that the cost (to you) of one additional sheep grazing on the commons is quite low, and you conclude that you are acting rationally by deciding to make the purchase. However, force yourself to consider what would happen if all the shepherds added one extra animal. The eventual result would be complete depletion of the commons, and all would suffer. After you have ruminated on this for a while, you become uncertain about what to do.

This story is taken from Hardin's (1968) "The Tragedy of the Commons." As you have probably realized, it offers an excellent analogy with

sources are being consumed at too high a rate, which is endangering the future availability of the resource. At a personal level we often find ourselves faced with resource-related decisions that are modern-day equivalents of whether or not to add another sheep to our herd. Should we avoid buying paper plates in order to save trees? If we use paper plates, can we avoid wasting water to wash dishes? If we drive a car today instead of taking public transportation, isn't the convenience worth any minimally additional pollution we add to the air we all breathe? In a sense, our needs are pitted against those of the larger community. We are faced with a choice between satisfying our immediate needs with the prospect of negative future consequences to society, and restricting our present consumption for the further good of the community. The way we resolve such dilemmas obviously has important implications. Hardin argues that if we want the commons to survive, each of us must give up some of our freedom. While the individual shepherd will benefit by adding to his or her flock, one must refrain for the greater good. But as logical as this seems, your experience may suggest to you that it will require more than reasoning to make people refrain from behaviors that are environmentally destructive, although personally satisfying. Unfortunately, people fre-

John Platt (1973) considers situations such as the **commons dilemma,** in which short-term personal gains conflict with long-term societal needs, to be types of **social traps,** or situations in which gains *appear* to outweigh costs. In general, Platt feels that social traps are hard to break out of, but claims it is essential for researchers to design strategies enabling us to do just that. Various methods have been suggested to help us break out of the commons dilemma. For example, researchers have tried to increase short-term costs of environmentally destructive behaviors so that they become less attractive behavioral alternatives, and have attempted to decrease the personal costs of acts that preserve the environment. Environmental psychologists have also tried to educate people (e.g., by conducting environmental seminars) to make them realize their interdependence and to make more

salient the long-term societal costs of squandering resources. Education also can encourage behaviors incompatible with those that waste precious resources. Some interventions have also supplied people with feedback about the extent to which they are depleting the commons, and others have divided up available resources into privately owned portions.

In this concluding chapter we will discuss a broad range of techniques that have been used by environmental psychologists in an attempt to study and change an array of human behaviors that are destructive of our shared environment. Some of these behaviors are easily amenable to conceptualization in terms of the "commons dilemma" and "social trap" analyses we have described, while others require a different type of conceptualization. ◘

How Can Environmental Psychology Guide Environmentally Responsible Actions?

Clearly, changing human behavior to save the environment is an extremely important topic. What unique contribution can environmental psychology make to help deal with the many environmental problems we face (e.g., insufficient and expensive fuels, air and water pollution, depletion of forests and other natural areas)? Innovative technologies, such as wind and solar energy and pollution abatement techniques, have received considerable attention. Many seem to think that solving our environmental problems requires only the right technologies. In contrast, relatively less attention has focused on strategies for preserving the environment that involve changes in people's behavior. We will argue that both technology and behavior change—sometimes involving substantial modifications in how we act on an "everyday" basis —must make significant contributions if things are to improve.

Why do we (and many other environmental psychologists) feel this way? First, in some cases

technologies have gotten us into this mess. Modern transportation has solved problems in locomotion but has caused pollution, periodic energy shortages, and unsightly commercial "strips." Modern packaging allows us to preserve all types of food but has created a tremendous litter problem. Most technologies have unfortunate "side-effects," and in this chapter we will see that psychological techniques for behavior change could help eliminate them. Second, in cases (e.g., dealing with littering) where technology cannot fix the problem, changing our behavior is the best means of coping. Even when certain technologies promise cures for environmental problems (e.g., energy-efficient appliances), particular behaviors are often necessary to ensure that people *use* the technologies. Clearly, behavior does have strong effects on the environment: We would not be exaggerating if we asserted that almost everything any of us does has either a positive or a negative impact on our environment.

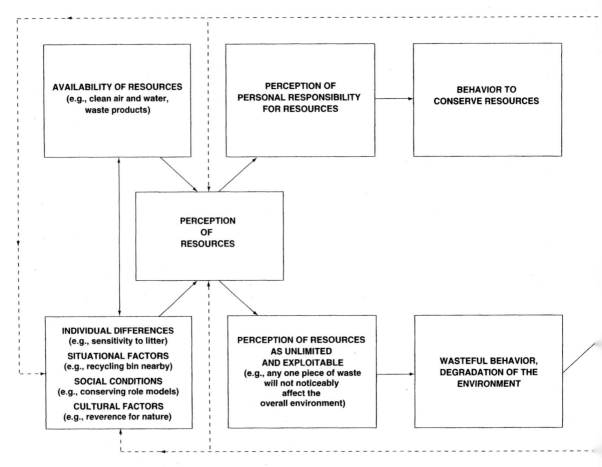

Figure 14-1   The eclectic model applied to behavior change to preserve the environment.

The perspective we take in this chapter can be viewed within the framework of our eclectic model, as depicted in Figure 14-1. Experiencing the environment as outside our ideal—too much pollution or rapidly depleting resources, for example—motivates us to seek change. Additional inputs, such as feedback about our resource consumption or incentives to change our ways, motivate us to modify our own behavior, which in turn changes our perception of the situation.

Two questions remain: (1) Will changing our behavior to save the environment require a lower quality of life, and (2) can it be done? Generally, the answer to the first is "no." If we changed our behavior so that fewer of us drove cars and more used public transportation, there would be less pollution, we would have signifi-

cantly more money to spend, we could walk or ride bicycles anywhere, we and our neighbors would be less susceptible to respiratory illnesses, and so on. In many ways, the quality of life would actually improve. We rephrase the second question: Do the behavior change techniques that we will be describing in this chapter work? We will leave that for you to decide after reading our presentation of the evidence in the coming pages. We will, however, suggest that there is lots of room for environmental psychologists to improve our environmentally relevant behaviors (e.g., Oskamp, 2000; Veitch & Gifford, 1997).

If we could influence environmentally relevant behaviors to improve the environment, what would we focus on? Encouraging environmentally protective acts (e.g., rewarding people for picking up litter) and discouraging environ-

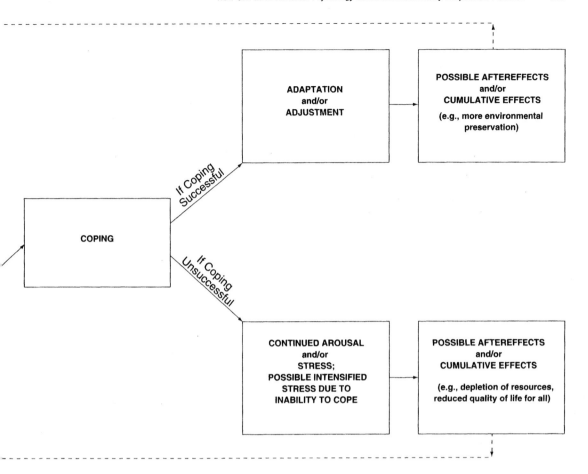

mentally destructive behavior (e.g., high fines for littering) will improve the situation. Unfortunately, programs that encourage protective behaviors do not necessarily inhibit destructive behaviors, and vice versa. Also, not all environmentally protective and destructive behaviors have the same impact on the environment. A program that stops people from littering is sure to have direct environmental impact; one that encourages people to vote for conservation-oriented legislators will probably have a more diffuse impact. We should also stress that the effects of any environmentally protective or destructive behavior are complex. Suppose we could get people to recycle all newspapers. This would save trees but might cause water pollution from the ink removal process. It would save energy since we would not need to process vir-

gin wood, but the recycling process itself uses a great deal of energy. Although most environmentally conscious people probably think that using paper cups causes less damage to the environment than using plastic ones, a case can actually be made that plastic is less harmful (Hocking, 1991)! Sometimes it is hard to figure out whether we are helping or hurting the environment (Figure 14-2). When we consider all of these preservation efforts collectively, we sometimes refer to the desired outcome as a **sustainable** future—creating a setting in which we do not destroy resources through either overconsumption or pollution, and in which we ensure that a high quality of life is available for future generations (e.g., Oskamp, 2000).

As we proceed in this final chapter, we will have more to say about the reasons people are

Figure 14-2 Actions to "save" the environment often involve trade-offs. Actions with low impact on one segment of the environment often have high impact on another segment. For example, using paper plates to save water costs trees; using washable dishes saves trees but costs water and places a burden on waste treatment facilities. A (controversial) case can actually be made that plastic cups have less impact on the environment than paper cups. The more we wash dishes and diapers, the more we place burdens on wastewater facilities such as this one.

motivated to reduce environmentally destructive behavior, but at this point we would like to make a few comments about perceived risk, because people's perceptions of high or low risk influence their willingness to engage in destructive behavior (e.g., dumping used motor oil into a storm sewer) as well as their willingness to engage in behavior that helps preserve the environment (e.g., participating in an Earth Day campaign to clean up a local stream). Slovic (e.g., 1997) points out that perception of risk is not the same as risk calculated by experts—the perceived risk may be higher or lower. Perceived risks tend to be higher if the activities associated with them are seen as uncontrollable, inequitable, catastrophic, unknown, dreadful, and likely to affect future generations; such is the case, for example, with people's perceived risks about nuclear power. Perceived risks tend to be lower if the activities associated with them are

seen as voluntary, individual, not globally catastrophic, easily reduced, and of low risk to future generations; such is the case for swimming, power lawn mowers, and food preservatives. What is the risk to the individual and to the environment for one act of littering or for one person using an automobile instead of mass transit? People's perception is that the risk is minimal, in part because using the automobile and littering are seen as voluntary and controllable and any negative consequences will probably occur long after the specific act of littering or driving. In reality, the consequences of these activities build over time and increase as more and more people engage in them. We will see what we can do about these consequences in the remainder of the chapter.

What is the range of environmental problems that we would like to improve if we could? These may be categorized as (1) problems of environmental aesthetics (e.g., prevention and control of litter, protection of natural resources, preventing urban deterioration); (2) health-related problems (e.g., pollution, radiation, high levels of noise); and (3) resource problems (e.g., overconsumption of resources such as water or energy). These categories are neither exhaustive nor mutually exclusive. Often, specific environmental problems, such as overdependence on the automobile, have an impact on all three categories. This means that specific interventions may also have an impact on more than one category.

At this point, let us examine the commons dilemma in more detail to see how we can modify the situation to improve the outlook for the environment. We will then consider a range of environmental problems and the ameliorative techniques used by environmental psychologists to help solve them. Some of these techniques may hold great promise for solving the critical problems that now confront us.

The Commons Dilemma as an Environment–Behavior Problem

Hardin's (1968, 1998) depiction of the tragedy of the commons has spawned numerous attempts to examine factors that might help us work out favorable solutions to the commons dilemma. To see how generalizable Hardin's propositions have become, it might be useful to enumerate examples of commons-like behavior beyond that of Hardin's shepherds.

EXAMPLES OF COMMONS TRAGEDIES

Hardin himself was interested in the problem of overpopulation: Seemingly self-serving motives for reproduction (e.g., having more labor to run the family farm) have a long-term negative consequence if the total population outstrips the food supply. Hardin borrowed the analogy of the commons from Lloyd (1833), who was also interested in how a selfish view could lead to disastrous overpopulation. Other examples of the commons dilemma are apparent when we consider some scenes typical on many college campuses.

Parking lots are often jammed with long lines of cars waiting for a space. The parking lot can be thought of as a commons: It is shared by all and owned by none of those who use it. Because parking spaces may be scarce, individuals acting in self-interest may arrive early to get a share of the valued resource. But as demand for parking spaces increases, you must arrive earlier and earlier to be assured of one. The result is that people who really need access later may be deprived of access because others have rushed in before them. Or consider space in campus dining areas around the noon hour. Space is limited, and many students use dining tables to socialize or to study for the next class. If they studied elsewhere, there would be room for all to eat. Scheduling of classes is also a type of commons. It appears that 10 A.M. is the most popular time for faculty to teach and for students to want to be in class. Accordingly, classroom space is scarcest at that hour, but it is underutilized at other times of the day. Budgets also have the characteristics of a commons. If the members of a group budget a fixed amount from which they all draw, such as for phone calls or photocopying, the tendency is for everyone to spend more than their share from that account. If each participant decides, then, that it is in his or her best interest to spend a fair share before someone else uses it, the resource becomes depleted very quickly—to the detriment of all. As yet another example, consider the Internet as a shared resource. Users are not charged in proportion to the volume they use, so individual overconsumption can lead to congestion at critical junctions to the detriment of all (Huberman & Lukose, 1997). Can you think of other cases of this type of dilemma (see Figure 14-3)?

Numerous examples of a commons-type tragedy show how broadly applicable the analogy is to environmental problems. Overconsumption

Figure 14-3A & 3B   Some examples of a "commons" on a college campus. What other examples can you think of?

of natural resources seems to be a repeating historical phenomenon. Cahokia was a city of 15,000 people at the confluence of the Mississippi and Missouri rivers in the 12th century. It was a technological marvel of its day, with a city wall of 20,000 logs. But overexploitation of the forest for fuel and construction robbed the landscape of adequate ground cover and led to flooding of the city and farmland in enough consecutive years to wipe out the civilization (Holden, 1996). Pringle et al. (1993) document the tragic consequences of shared use of the

Danube River. The Danube stretches 2,860 kilometers through nine different countries; about 12% of Europeans—86 million people—live within its basin. Through a canal, it links the North Sea to the Black Sea. It is a major transportation pathway but also provides drinking water, irrigation water, and hydroelectric power. However, it is also used to dispose of industrial and municipal waste. Any one nation's exploitation of the river has an environmental impact that is not terribly noticeable, but the sum of abuse by all nations is tragically detectable in the delta. Among other things, the fish harvest is down by as much as half, and some species show a 90% decline. In another resource tragedy, Stone (1999) describes how the exploitation of the Aral Sea in Central Asia has been terribly disastrous. Beginning in the 1920s, the former Soviet Union diverted water from the rivers that feed the Aral in order to boost cotton farming—and the farming was remarkably successful for several decades. But over time the Aral lost 80% of its volume of water, and with the runoff of pesticides and fertilizers, much of the formerly freshwater Aral is now saltier than the ocean. All 24 of its native fish species have vanished, and 100,000 people dependent on fishing and related industries have been displaced. The sea and surrounding soil are poisoned with sulfates, phosphates, chlorinated hydrocarbons—as much as 700 tons of salt per hectare. This shared, life-giving resource for the 35 million people in its watershed is now a source of death. Another interesting example of a commons tragedy comes from an attempt to use an environmentally friendly source of energy. Kerr (1991) and Vogel (1997) describe how geothermal energy has been exploited at The Geysers, a natural steam-generating geological formation in northern California. Production of electricity from this relatively clean energy bonanza began with one company in the 1960s, and it eventually provided 6% of California's electric power. But by 1988 the number of organizations tapping the energy source had increased to 11 users. Too many companies exploiting the limited resource soon cut total electricity output by half, and a $3.5 billion investment is threatened. The problem once again is that whereas individual use of a resource can be tolerated by the environment, the combined exploitation by a group often leads to overuse and tragic disaster. Even exploration of outer space involves a commons dilemma: The launch of a single space vehicle seems strictly an advancement of science, but the total output of debris left in orbit from multiple launches increases the danger that a future space vehicle could be destroyed by high-speed impact with "space junk" (Marshall, 1985).

CONCEPTUAL VIEWS OF COMMONS-TYPE PROBLEMS

We mentioned that Platt (1973) conceptualized the commons dilemma as a type of social trap. Platt described three such categories of social traps, each of which is relevant to environmentally destructive behavior. The commons type of trap, or **individual good–collective bad trap,** involves a group competing for a valued resource, such that destructive behavior by one participant has minor impact on the whole, but if all engage in the same individual behavior, the impact on the commons is disastrous. The **one-person trap,** or **self-trap,** involves a disastrous consequence to one person. Typical of these traps is addiction to drugs or food. The momentary pleasures of the present have disastrous consequences in the long run. The third type is the **missing hero trap.** Whereas the commons trap and self-trap involve unfortunate actions that we take, the missing hero trap involves an action that we fail to take, such as refusing to help someone in need or failing to warn others of the toxicity of a substance with which they work.

Interestingly, Platt (1973) notes that all three of these traps can be analyzed in terms of the rewards and punishments (i.e., reinforcements) associated with them. There is a positive side to the situation that we seek, and a negative side that we want to avoid. The problem is that the positive and negative have become separated in time, or the negative has been diluted across the members of a group, so that the behavior leading to the short-term positive consequence is more likely to occur. For example, in the commons problem of overharvesting whales, the immediate reward of taking one whale seems more prominent than the long-term consequence of everyone else taking more whales. In a self-trap of overconsuming food, the short-term pleasure of an extra dessert seems overwhelming relative to the long-term consequence of damage to the body and to our appearance. In a missing-hero trap, there is an unpleasant component to the behavior we should be performing: The punishment is short term but the reward is long term, so we avoid the behavior. For example, we may fail to pick up litter because the inconvenience seems to outweigh the long-term benefit of an aesthetically pleasing environment.

How, then, do we resolve social traps? Platt argues that we rearrange the positive and negative consequences of our behavior. If we engage in a destructive behavior because it has immediate rewards, such as using an automobile rather than mass transit, we can impose a system of penalties for automobile use (such as heavy freeway tolls) and rewards for mass transit use (such as free rides on high-pollution days). Or, we could increase the unit cost of a resource, such as electricity, for those who use a large quantity, and reduce the unit cost for those who use little. For example, Oskamp et al. (1994) found that financial gain—that is, monetary reward—was a major motivator for businesses engaging in office paper recycling. Similarly, Grasmick, Bursik, and Kinsey (1991) found that shame and embarrassment—that is, negative consequences—were significant motivators in an antilittering campaign. We will examine these ideas in more detail when we discuss specific use of rewards and punishments to prevent environmentally destructive behavior.

Edney (1980) points out that although this reinforcement interpretation of the commons dilemma is appealing in its simplicity, it ignores a number of human elements. For one, it ignores the long-established evidence that individuals are different from one another: They do not all respond in the same way to the same rearrangement of the circumstances for rewards and punishments. Reinforcement approaches also sidestep questions of conscience, altruism, ethics, and humanistic tendencies, and suggest that reason is dominated by questions of reward. Hopper and Nielsen (1991), for example, observed that altruistic motives influenced recycling behavior, and Axelrod and Lehman (1993) found that deeply held personal principles could guide environmentally conscious behavior. Similarly, Stern, Dietz, and Kalof (1993) observed that concern for consequences to others and to the environment, as well as concern for consequences to oneself, guide environmental consciousness.

We should mention that there are other formulations of commons-type problems, including the use of the label **resource dilemma** to describe the commons situation. To social psychologists, the problem is one of a class of **social dilemmas,** where individual interests are pitted against group interests (e.g., Komorita & Parks, 1995; Liebrand & Messick, 1996). Economists refer to a variation of the issue as the **public goods problem** (e.g., Marwell & Ames, 1979). In this situation, individuals must all contribute to a common cause, such as paying taxes for mutual self-defense or contributing to a public television station. Any one person can fail to contribute, and the public cause will survive. However, if too many people get the idea of not contributing, the common good suffers. Those who do not contribute are termed **free riders,** since they not only do not contribute, but also benefit from the public cause. For example, a person who sneaks onto a subway without paying or who poaches wildlife without a hunting license could be termed a free rider.

FINDINGS FROM LABORATORY EXPERIMENTS

In the preceding discussion we showed how Platt would rearrange rewards and punishments to solve the commons dilemma. Hardin (1968, 1998) suggests that some form of governance is necessary to manage the commons in a nondestructive manner. In his terms, we must have

SIMULATING THE COMMONS DILEMMA:
How It's Done and What Is Found

To test different techniques for helping us break out of the commons dilemma, a number of simulations have been developed that incorporate the central elements people face in such contexts. In these simulations, various interventions are manipulated to discover which ones would cause us to behave in a more constructive way. The simulations have included computer analogs (e.g., Cass & Edney, 1978; Fusco et al., 1991; Gifford & Wells, 1991), as well as noncomputerized methods involving portable (e.g., Edney, 1979) and nonportable apparatus (e.g., Edney & Harper, 1978). In addition to being useful for exploring strategies for helping us break out of the commons dilemma, simulations can be used as teaching devices to aid us in understanding the dynamics of our environmentally destructive behaviors.

To give you a feel for these simulation techniques, we will discuss Edney's (1979) **"nuts game"** simulation in some detail. Recall that commons dilemmas include (1) a limited resource that may regenerate itself somewhat but that can be endangered through overconsumption; and (2) people who have the choice between restricting current individual consumption for the good of society (and the future of the resource pool) and exploiting the resource for their own immediate good. A successful simulation would have to include these elements.

How can this be done? Edney's nuts game accomplishes it quite nicely. A small number of participants enter the lab and sit around an open bowl that originally contains 10 hexagonal nuts, obtained from a hardware store. The bowl symbolizes the pool of resources (e.g., trees or whales), and the nuts symbolize the individual resources themselves. Participants are told that their goal is to obtain as many nuts as possible. (This simulates the fact that typically we try to maximize our outcomes in life.) Players can take as many nuts as they want at any time after a trial begins. The experimenter also states that the number of nuts remaining in the bowl after every 10-second interval will be doubled. This replenishment cycle simulates natural resource regeneration rates. The above events continue until the time limit for the game is exceeded, or until the players empty the bowl.

How do participants behave during the "nuts game"? We would hope that they would take at most a few nuts out of the pool per 10-second period, which would allow the game to continue and maximize the long-term outcomes. However, in his research, Edney (1979) found that 65% of the groups depleted the pool completely before the first replenishment stage! They took out all 10 nuts (i.e., depleted the resource pool completely) during the first few seconds of the game. As in the "real world," people exploit the commons, with unfortunate results.

"mutual coercion, mutually agreed upon" in order to regulate our tendencies toward overconsumption. Laboratory investigations using commons-dilemma simulations (see the box above) have explored a number of factors, including forms of "governance," that might help us conserve the commons. It is instructive to review some of these laboratory findings when participants harvest resources from a common pool.

Structural changes to the commons are often the most effective in preserving it (Samuelson et al., 1984). This can include privatizing the shared resource, such as dividing it up into indi-

vidually owned territories instead of everyone sharing the whole (e.g., Edney & Bell, 1983; Martichuski & Bell, 1991). However, this strategy essentially eliminates the "commons" as a shared resource and is not practical for some resources, such as the air we all breathe or our national parks. Enforcing new "rules" also helps preserve the commons, a variation of Hardin's solution of "mutual coercion, mutually agreed upon." This can take the form of Platt's notions of adding rewards for conservation behavior or punishments for overharvesting (e.g., Bell, Petersen, & Hautaluoma, 1989; Birjulin, Smith, & Bell, 1993; Harvey, Bell, & Birjulin, 1993; Kline et al., 1984; Komorita, 1987; Yamagishi, 1986). Although privatization is an effective management tool when feasible, rewards and punishments are effective when privatization is not practical (Martichuski & Bell, 1991). Requiring that all participants harvest in equal amounts also leads to conservation-oriented consumption (e.g., Edney & Bell, 1983). Simply adding a mechanism to give participants feedback about the effects of their harvesting decisions can be effective (e.g., Kline et al., 1984; Seligman & Darley, 1977).

Although typically less effective than structural changes, modification to the social relationship among the participants can improve management of the commons (Table 14-1). Allowing or encouraging the participants to communicate with one another about harvesting strategies (also a form of structural change) can promote conservation (e.g., Bouas & Komorita, 1996; Dawes, McTavish, & Shaklee, 1977; Kerr & Kaufman-Gilliland, 1994; Loomis, Samuelson, & Sell, 1995). Groups who are afforded both feedback and communication are especially successful at maintaining the commons (Jorgenson & Papciak, 1981). Also, different leadership and decision-making rules have been related to commons dilemma outcomes. A study by Shippee (1978) found that personal participation in choosing a group's leadership and in implementing decisions to limit resource use led to quite successful conservation results. Identification with other group members and cohesion within the group leads to pro-ecological harvesting choices (e.g., Brewer & Kramer, 1986; Kramer & Brewer, 1984). Increasing attraction between group members is also helpful (Smith, Bell, & Fusco, 1988).

It has been found that cooperation among players improves conservation (Gifford & Hine, 1997; Loomis et al., 1995); consequently, there must be trust between participants (Edney, 1979; Moore et al., 1987; Mosler, 1993). Those who have a trusting and cooperative nature seem most able to manage the commons together (Parks, 1994). When one's individual behavior in a commons dilemma situation is subject to the scrutiny of others, he or she is less apt to overexploit the commons (Jerdee & Rosen, 1974).

Other studies have explored the effects of knowing one is interdependent with others for a resource, rather than having his or her own supply (e.g., Edney & Bell, 1984). In fact, obtaining knowledge of resource interdependence seems to increase the intensity of behaviors aimed at "getting as much as possible for oneself," which ends up depleting the commons (Cass & Edney, 1978). This suggests that rationing resources could be a useful strategy. Experiments have also studied whether educating people about the optimal strategy for using resources in commons dilemma situations leads to pro-ecological

Table 14-1	RELATIVE IMPACT OF INTERVENTIONS TO SAVE THE COMMONS	
Factor	**Impact**[°]	**Reference**
Social		
Attraction	.12	Smith, Bell, & Fusco (1988)
Group Identity	.11	Kramer & Brewer (1984)
Trust	.08	Messick et al. (1983)
Moral Suasion	.05	Edney & Bell (1983)
Structural		
Privatization	.30	Martichuski & Bell, (1991)
Communication	.15	Kerr & Kaufman-Gilliland (1994)
	.44	Jorgenson & Papciak (1981)
Reinforcement		
Feedback	.16	Jorgenson & Papciak (1981)
	.19	Seligman & Darley (1977)
Reward	.09	Birjulin, Smith, & Bell (1993)
	.17	Martichuski & Bell (1991)
	.26	Kline et al. (1984)
Punishment	.19	Grasmick, Bursik, & Kinsey (1991)
	.40	Harvey, Bell, & Birjulin (1993)

[°] *Variance accounted for or d statistic, on a scale from .00 to 1.00.*

THE WORLD AS A COMMONS:
New and Old Ecology

Aside from a number of examples of individual commons dilemmas, we might also consider the entire world as a commons. Environmental sociologist Riley Dunlap, drawing on the views of others, has suggested that a shift is occurring in how we view world resources (e.g., Dunlap & Van Liere, 1978, 1984; Dunlap, Van Liere, Mertig, & Jones, 2000). This shift takes the form of a contrast between a long-held **dominant Western world view** and a **new ecological paradigm**, or **New Environmental Paradigm** (Dunlap et al., 2000) approach to the world's resources (see also Schultz & Zelezny, 1999; Stern et al., 1995). We summarize these contrasting attitudes here.

The dominant Western world view holds that:

1. Humans are unique and have dominion over all other organisms.
2. We are masters of our own destiny—we have the intellectual and technological resources to solve any problem.
3. We have access to an infinite amount of resources.
4. Human history involves infinite progress for the better.

The new ecological paradigm holds that:

1. Humans are interdependent with other organisms, such that their preservation is to our advantage.
2. Many things we do have unintended negative consequences for the environment.
3. Some things, such as fossil fuels, are finite.
4. Ecological constraints, such as the carrying capacity of an environment, are placed upon us.

You may agree or disagree with some of the above contentions. The ones you endorse probably have much to do with how palatable you find the various strategies for managing the commons.

action. As we will see later when we discuss environmental education, often it is quite ineffective (Edney & Harper, 1978). Edney (1981) observed that the sacrifices that must be made to preserve the commons are often unequally shared among the population (i.e., some harvest less than others; see Herlocker et al., 1997; Samuelson & Messick, 1986). Perceptions of inequity seem to impair conservation efforts. Moral exhortation to be altruistic helps, but not much (Edney & Bell, 1983).

What does all of the above mean? Clearly, commons dilemma analogs can give us useful hypotheses regarding how to deal with "real-life" situations. There is certainly some evidence that these laboratory studies have implications for the real world, although the linkage is not complete. For example, we mentioned that numerous researchers (e.g., Edney & Bell, 1983)

demonstrated the efficacy of dividing the commons into territories. Acheson (1975) observed that Maine lobstermen who were highly territorial in defending their ocean claims were more successful in maintaining productivity than those who were less territorial. In addition, Thompson and Stoutemyer (1991) found that focusing on long-term consequences improved water conservation. When water shortages are severe, citizens are more willing to cooperate with authorities in conservation efforts if they trust and feel positively about the authorities (Tyler & Degoey, 1995). If group cohesion helps preserve the commons in the lab, could a similar principle hold in the real world? We noted in Chapter 12 that intentionally providing green common areas for socializing promotes community cohesion and commitment (Coley et al., 1997; Kweon et al., 1998).

A FOUNDATION FOR INTERVENTION

We conclude this section by noting that as individual action has more and more impact globally, the relevance of the commons dilemma becomes more and more apparent. As examples, (1) in connection with the greenhouse effect, individual use of fossil fuels seems harmless, but collective use dangerously warms the entire planet; (2) the individual use of chlorofluorocarbons (e.g., for air-conditioning) seems harmless, but collectively it creates an ozone hole over the planetary poles, which is why such refrig-

erants are being phased out; and (3) the locally harmless use of fossil fuels creates acid rain when the collective output of the fuels precipitates over neighboring areas. While it is easy to adopt a fatalistic attitude that we are hopelessly locked into one collective ecological disaster after another, there remains hope. Research has indeed shown that a variety of strategies can modify environmentally destructive behavior. We turn next to some strategies that have known outcomes in their applications, with the knowledge that the commons dilemma is a useful foundation for the implications of these strategies.

Overview of Strategies to Encourage Environmentally Responsible Behavior

How can psychologists use what they have learned to encourage environmentally responsible behavior? How much impact can we have if we just educate people about how selfish actions can be destructive to the environment on which we all depend? If we change attitudes, will we ensure a sustainable future? Is it effective to use rewards for pro-environmental behavior and punishments for wasting resources, or is something more drastic needed to bring about needed changes?

We will address these and related questions by describing the strategies psychologists advocate for changing behavior and the degree to which these various strategies have been shown to be effective in promoting environmentally responsible behavior. These approaches can be applied to many types of environmental behavior,

but we will emphasize primarily resource conservation (especially energy and water use), recycling, and reduction of littering. We will also take a look at some broader interventions (e.g., "greening" of industry). For further reading, we suggest you consult the March 1995 and September 1995 issues of *Environment and Behavior,* which are devoted entirely to psychologists' perspectives on litter control, recycling, and conservation behavior.

We can categorize psychological interventions as antecedent or consequent, depending on whether the intervention occurs before the target behavior or after. Antecedent interventions include attitude change, education, and prompts. Consequent interventions include rewards, punishments, and feedback.

Antecedent Strategies: Intervening Before the Behavior

Antecedent strategies precede the behavior they are attempting to change. In many cases the primary targets are attitudes, which we discussed in Chapter 2. For instance, we might try to change attitudes through persuasive or in-

formational messages about the environmental benefits of mass transit. Simply stated, the goal of these strategies is to "make people care." Other approaches assume that people have a positive attitude, but aim to show them how to behave in ways consistent with what they already "care" about. For example, we could prompt people with a "Please Recycle" sign, or we could provide information about the energy efficiency of certain appliances to improve the success of individuals who are already trying to conserve. Of course, many types of information can serve multiple functions, and one recurring hope is that those who are well informed are more likely to adopt environmentally responsible views (Newhouse, 1990). We also hope that such views translate into corresponding responsible behavior. As you might expect, those who are better informed about recyclable materials and local recycling programs are probably more likely to be recyclers themselves (Vining & Ebreo, 1990), and those who are informed about carpooling report being more willing to try it (Kearney & De Young, 1995).

ATTITUDE CHANGE AND EDUCATION

Environmental education involves making people aware of the scope and nature of environmental problems and of behavioral alternatives that might alleviate them. Often, education changes attitudes favorably toward preserving the environment. Simple persuasion has been tried with some success. In two studies of college classrooms, a letter from a fellow professor (Luyben, 1980a) or from the college president (Luyben, 1980b) made it more likely that college professors would turn off classroom lights following their lectures.

In one Michigan study, volunteers from a small village received pamphlets advocating reduced use of resources (for example, purchasing items in reusable containers, reusing aluminum foil, and avoiding overpackaged products). For some participants the arguments were primarily economic, some were given environmental reasons, and some were provided with both economic and environmental reasons to reduce waste. Subsequently, those in all three treatment groups reported more conservation behavior, and those receiving both economic and environmental rationales reported the most conservation (De Young et al., 1993). Similarly, a

California study examined a commons education group that received educational messages focusing on the long-term benefits of conservation and emphasizing the effectiveness of individual action using Hardin's tragedy of the commons metaphor. A second group received information emphasizing the short-term economic advantage of water conservation. Participants in a third group received no educational messages but were encouraged to conserve and were made aware that their water use was being monitored. For lower-middle-class residents, the appeal that focused on the long-term consequences of conservation and the necessity for individual action yielded less water consumption than did either the economic-based appeal to conserve or the control appeal (no educational message). For reasons that are unclear, the upper-middle-class residents did not show increased water conservation in response to any of the treatments (Thompson & Stoutemyer, 1991). In another case, an educational program in East Harlem, New York, increased the volume of recycled goods from 8.7 tons daily to 11.4 tons daily over a 1-year period (Margai, 1997).

Although the above examples provide success stories and reason for hope, most studies have suggested that simply educating people is not tremendously effective at changing energy-relevant behaviors, recycling efforts, or littering tendencies (e.g., Dwyer et al., 1993; Heberlein, 1975; Kempton et al., 1985; Palmer, Lloyd, & Lloyd, 1978; Winnett et al., 1978). For example, Heberlein (1975) gave people either a booklet of energy-saving tips prepared by the electric company, an informational letter educating them about the personal and social costs of not conserving energy, or an informational pamphlet actually urging people to use more energy. What were the effects of the educational strategies? None of them had any appreciable effect on behavior. In a similar vein, Geller (1981) conducted educational workshops on energy use and found that they were very effective in changing reported attitudes and intentions regarding energy use, but follow-up audits of participants' homes revealed that the changes suggested in the workshops had not been implemented. Also discouraging was a study of three Australian cities that evaluated the effect of two intensive persuasive television campaigns aimed at reducing gasoline consumption. One campaign emphasized the money-saving aspects of conservation, and the other presented gasoline conservation as a "civic duty." Despite some small reductions in gasoline consumption, the authors concluded that the television campaigns had not been cost effective.

In assessing the success of attitude change and other educational strategies in promoting pro-environmental behavior, we should reconsider an assumption we proposed earlier: Does an individual's attitude really predict the likelihood that he or she will actually behave in ways that protect the environment? Just because we change attitudes does not necessarily mean we promote behavior change (Syme et al., 1987). In North America and Europe, general concern for the environment is high (e.g., Cohen & Horm-Wingerd, 1993; Dunlap & Mertig, 1995; Wall, 1995), but the relationship between general pro-environmental attitudes and subsequent behavior is somewhat tenuous. In fact, several researchers have concluded that general attitudes toward the environment are not very predictive of eventual behavior (e.g., Geller, Winnett, & Everett, 1982; Newhouse, 1990; Olsen, 1981). This implies that convincing people an ecological disaster looms will not necessarily change behavior (Stern, 2000). Other researchers are more optimistic (e.g., Brandon & Lewis, 1999; Samuelson & Biek, 1991; Seligman, 1986; Stern & Oskamp, 1987; Vogel, 1996).

The Problem of Attitude–Behavior (Non-)Correspondence

Why might attitudes not predict behavior? What can we do to increase attitude–behavior correspondence? Most of those reading this chapter would assert that they have environmentally conscious attitudes. Yet few of these readers have probably recycled every aluminum can they have used in the past year, most have not regularly picked up litter around their neighborhood, and most have used more water than they need to when they shower. One researcher found that stated attitudes about water use, reported water use, and actual meter readings of water use corresponded very poorly in a San Antonio, Texas, sample (de Oliver, 1999). A similarly low correspondence was found among a sample of women in Mexico when asked about recycling behavior (Corral-Verdugo, 1997). A number of factors could account for low attitude–behavior correspondence, and there are things we can do to increase the correspondence (e.g., Ebreo, Hershey, & Vining, 1999; Tarrant & Cordell, 1997).

General Versus Specific Attitudes and Behaviors One problem is that general attitudes ("I'm pro-environment") may not predict specific behaviors ("I recycled every can I used in the past year"). Yet, there is reason to believe that specific attitudes ("I am conscientious about recycling newspapers") can be successful in predicting related behavior ("I put my newspapers in the recycle bin"; e.g., Tarrant & Cordell, 1997; Wall, 1995).

Salience or Accessibility Another reason that attitudes may not predict behavior well is that a specific attitude is not always salient or accessible. As we will see in a moment, prompts such as "Please Recycle" signs can remind us that we have attitudes favorable toward recycling and thus facilitate corresponding actions. If we do not think about the specific attitude, we may not be as likely to follow through with consistent behavior, so prompts and other factors can make specific attitudes more salient.

Subjective Norms and Perceived Control In Chapter 2 we also examined the Theory of Planned Behavior (or the Theory of Reasoned Action), which posits that a simple formula of "attitudes lead to behavior" is inadequate (e.g., Ajzen & Madden, 1986; Harland, Staats, & Wilke, 1999). Examine Figure 2-7 on page 33 once again. The Theory of Planned Behavior posits that subjective norms (e.g., "It is socially appropriate to act in this way and in this place") as well as perceived control (e.g., "It will be easy to engage in this action right now") and attitudes ("I support recycling") together influence behavioral intentions ("I intend to recycle this can"), and intentions in turn predict behavioral outcomes. Assume, for example, that you have not finished a can of pop, you are already a minute late getting to class, and drinks are not permitted in the classroom. Although you are a conscientious recycler, you may have to leave that aluminum can on a bench outside the classroom if there is no recycling container nearby. Thus, more than just your attitude predicts your actions in this case (e.g., Hamid & Cheng, 1995; Kaiser, Wölfing, & Fuhrer, 1999; Taylor & Todd, 1995). Formulations other than the Theory of Planned Behavior also show that including more than attitudes in the equation increases prediction of behavior (e.g., Dietz, Stern, & Guagnano, 1998; Pelletier, Dion, Tuson, & Green-Demers, 1999; Stern, Dietz, & Guagnano, 1995).

As another example, one series of studies suggests that the primary attitudinal dimensions relevant to energy concern are comfort and health, the trade-off between effort and savings, the perceived efficacy of individual conservation efforts (i.e., can one person make a difference?), and the perceived legitimacy of the energy problem (Samuelson & Biek, 1991). Our comfort and health seem to be particularly important. You will probably not be surprised to learn that messages that ask people to sacrifice comfort or health are often ineffective, even among people who express generally positive attitudes about the environment (cf. Kempton et al., 1992; Stern & Gardner, 1981). Thus, increasing the cost or the perceived ease of carrying out the behavior may decrease the correspondence between the attitude and the specific behavior (see also Guagnano, Stern, & Dietz, 1995).

Attitude Strength and Direct Experience

Another reason that an attitude may not predict behavior is that the attitude is not particularly strong. Most of us believe in saving beautiful, 100-year-old trees when possible, but few of us believe strongly enough about trees to live in one for 2 years to prevent a lumber company from harvesting it. Under what conditions do educational programs on conservation have the greatest potential for success? We think attitudes that are formed from direct behavioral experience are more predictive of later behavior than those that are more passive and abstract. For example, an educational program for high school students that included an energy audit and teaching students how to monitor home energy consumption positively affected student behaviors and those of parents (Stevens et al., 1979). This program may have been productive because it taught specific conservation behaviors, not general ones. An energy audit (especially a Type A audit, in which an auditor comes to the home, makes specific suggestions, and discusses them with the owner or occupant) also has shown promise (Geller et al., 1982). Involving homeowners actively in an energy audit (e.g., having them go up to the attic with the auditor to examine it) also makes a difference (Stern & Aronson, 1984). In addition, one study (Gonzales, Aronson, & Costanzo, 1988) examined homeowners' reactions to auditors trained to utilize certain social psychological principles (e.g., to personalize their recommendations, to induce commitment, and to frame recommendations in terms of loss rather than gain). Auditors were also trained to use "vivid" language (e.g., telling people that the cracks under their doors were equivalent to having a hole the size of a basketball in their living room, or that their attic, which had little insulation, was "naked"). Unfortunately, in comparison with a control group, although the trained auditors elicited greater compliance with their recommendations and generated more applications for finance programs to pay for home retrofitting, no differences were found in actual energy use.

Commitment and Goal Setting Another factor in attitude–behavior correspondence is the degree of commitment to an issue. The more one makes a commitment to an issue (e.g., water or energy conservation), the more likely it is that future behavior will follow. In a series of studies, Pallack, Cook, and Sullivan (1980) manipulated the degree of commitment to energy conservation and measured subsequent energy use. One group of homeowners (high commitment) was told that the list of people participating in an energy conservation study would be made public along with the experiment's results; a second group (low commitment) was assured of anonymity. Those in the high commitment condition used less energy than those in the low commitment or a third, control condition, and the effects persisted for as much as 6 months after the study had terminated.

Mass-transit systems typically consume far less fuel per passenger than private automobiles. One study (Bachman & Katzev, 1982) compared people who made a personal commitment to ride the bus, individuals who received free bus passes, and people who were both committed to riding the bus and who received free tickets. Although all three treatments increased bus use over a control group, those making a personal commitment seemed to show the most bus use, an effect that was still observed 12 weeks later (Figure 14-4).

As with energy conservation, several researchers have investigated the effects of commitment on recycling (e.g., Burn & Oskamp, 1986; Katzev & Pardini, 1987–1988; Pardini & Katzev, 1983–1984). In particular, the strength of commitment seems to be important to sustaining long-term recycling. In one study (Par-

Figure 14-4 Mass transit reduces overall fuel consumption and contributes less air pollution than the use of numerous automobiles does. Public commitment and free passes increase ridership.

dini & Katzev, 1983–1984), for example, those who made a strong (written) commitment to recycle newspapers and a group making only weaker verbal commitments both recycled more than a third group that received only information. Weeks later, however, only the strong-commitment group showed continued increases in paper recycling. In another study, Wang and Katzev (1990) asked residents of a retirement home to sign a group pledge to participate in a 4-week recycling project. For the 4 weeks of the commitment, the weight of recycled material increased 47% over baseline measures. Additional observations indicated that recycling continued at a similar high rate for at least 4 weeks following the group commitment. In a second study of college students, the same researchers compared two different commitment procedures and a reinforcement strategy. Some participants attended a 5-minute talk about paper recycling and were then asked to sign a group commitment to participate in the program. Individuals in a second treatment condition were approached individually, told of the recycling program, and asked to sign a personal pledge to participate in the program. In addition, a third (incentive) treatment group was given a flyer explaining that all residents of their dorm hallway would receive discount coupons for local businesses if 50% of the people on their hallway recycled during a given week. All three treatment groups recycled more often than a control group, but individuals who had committed to recycling showed greater recycling than those who had signed the group pledge. Perhaps the most interesting finding was that at least some of the students assigned to the individual commitment group continued to recycle for some weeks after their commitment had expired, whereas more of both the incentive and the group commitment participants reverted to about the same level as before the experimental treatment. Thus, the more committed we can make people to act in an environmentally responsible way, the more likely they will be to engage in environmentally responsible behaviors. In other studies, Cobern, Porter, Leeming, and Dwyer (1995) found that residents who made a commitment to recycle

USING A FOOT IN THE DOOR
TO ENCOURAGE RECYCLING

Another approach to encouraging conservation behavior involves the **foot-in-the-door technique,** which has been used in some classic studies in social psychology on gaining compliance. Like the salesperson who stands a better chance of making the sale by initially getting a "foot in the door," environmental psychologists may be more apt to get people to recycle (or engage in other pro-ecological behavior) after eliciting a small commitment from them. The standard "foot-in-the-door" paradigm goes like this. The experimenter first makes a small request of the target person that very few are likely to refuse (e.g., sign a petition for a highly respectable pro-environmental cause). This is followed by progressively larger requests (e.g., recycle your soft drink containers). Because people who comply with the small initial request come to view themselves as interested in preserving the environment, they are more apt to agree to the second, larger request than are participants who are never presented with the initial request. Arbuthnot et al. (1976–1977) used this strategy to increase recycling behavior. Their initial smaller requests were to have people answer survey items favoring environmental protection, to save aluminum cans for a week, and to send in a postcard urging officials to expand a local recycling program. Did being confronted with these initial requests affect long-term use of a recycling center? The answer is "yes." As long as 18 months after the initial request, participants were more apt to use the center than those not exposed to the foot-in-the-door strategy.

The above findings have some interesting implications. We might view responding to prompts, such as reminders to turn off lights, as analogous to initial requests in the foot-in-the-door technique. This would suggest that responding to the multiple "small requests" being made of us to improve the environment these days may in some way "prime" us to comply with the larger, more important environmental demands we will face in the years ahead. The phenomenon also suggests some potential problems with the use of reinforcement techniques for encouraging pro-ecological behavior. Self-perception theory in social psychology implies that often we infer our attitudes from observing our behavior and the circumstances under which it occurs (Bem, 1972). Arbuthnot et al. (1976–1977) proposed that people in their study agreed to the larger requests because after complying with the initial request, they inferred from their behavior that they had pro-environmental attitudes. They had no incentives for their initial compliance, thus they must really care about the environment. This pro-ecological self-perception may have been why long-term changes in recycling occurred in the foot-in-the-door study. On the other hand, what would people infer about themselves from their behavior after recycling because they were offered a financial reward or reinforcement? Bem would say that instead of inferring that they care about the environment—a conclusion that could be associated with continued recycling—they might conclude they merely did it for the money. Such a self-perception could lead them to stop recycling as soon as the rewards for it were removed.

grass clippings and to talk to their neighbors about recycling were two to three times more likely to recycle than a no-commitment control group, and the effect lasted at least a year. DeLeon and Fuqua (1995) found that combining a public commitment (including having one's name published in the local paper as a recycler) and feedback manipulation increased newspaper recycling volume by 40% versus a no-commitment control group.

Goal setting is a related technique. Motivation theorists tell us that specific and challenging goals are likely to result in more behavioral change than easy or general goals (e.g., Locke, 1968, 1970). Although goal setting is not a common variable in studies of conservation behavior (Dwyer et al., 1993), there is some evidence that both children (Hamad et al., 1980–1981) and college students (McCaul & Kopp, 1982) show higher compliance when they are assigned recycling goals. In fact, college students in the McCaul and Kopp study recycled 37% more.

Changing Attitudes: A Final Note

Although the tie between general attitudes and behavior is often weak, interventions aimed at changing attitudes persist. Perhaps ironically, one attraction of attitude change is its potential for generalizability. That is, behavioral change would be efficient if we could change a few global attitudes that might then promote a variety of responsible behaviors across a number of settings (e.g., Berger, 1997; Daneshvary, Daneshvary, & Schwer, 1998; Karp, 1996). If they worked, such broad programs would be more efficient than those tailored to dozens of different situations. In particular, some theorists believe that it will ultimately be more important to teach consumers to recognize social traps than to establish a number of consequent programs to promote conservation.

PROMPTS

Varieties of Prompts

Prompts (cues that convey a message) have also been used to influence conservation. Television announcers may prompt us to use energy wisely, or signs in university residence halls may remind us that "Empty rooms love darkness." **Modeling,** in which we observe another person engaging in conservation behavior, can be considered a type of prompt. Other types are **approach prompts,** which imply an incentive for engaging in a specific behavior (e.g., "Thank you for keeping the park clean"), and **avoidance prompts,** which imply a disincentive (e.g., "We frown on those who trample the grass").

These procedures are certainly cheaper than some other strategies we will discuss, but do they work? Sometimes they do indeed, especially if they are specific, well timed, well placed, and the behavior they request is easily enacted (Geller et al., 1982; Stern & Oskamp, 1987). For example, the "Empty rooms love darkness" prompt would work best if placed on the back of the door you open to leave (well

placed) and if it also said, "Turn off the lights when going out" (suggesting a specific behavioral response). An effective use of prompts to curtail unnecessary use of air-conditioners was devised by Becker and Seligman (1978), who strategically placed a light in the kitchens of homes so that the light would turn on when air-conditioning was on and outside temperatures were below 68 °F (20 °C). The prompt indicated that air-conditioning was unnecessary, and the light went off only when the air-conditioner was turned off. This achieved an energy savings of 15% (Figure 14-5).

Many studies have employed prompts and cues as antecedent strategies to prevent littering. For example, handbills with an antilitter prompt are less apt to be littered than those without a prompt (Geller et al., 1982). Generally, prompts that state the specific antilitter response desired (e.g., "Place this paper in a trash can") are more effective than less specific ones (e.g., "Keep the area clean"). Antilitter prompts are also more effective when given in close temporal proximity to an opportunity to dispose of litter, when proper litter disposal is relatively convenient, and when the prompt is phrased in polite, nondemanding language (Geller et al., 1982; Stern & Oskamp, 1987). Even under optimal circumstances, the absolute magnitude of change effected by these sorts of prompts is often relatively small (though statistically significant), and to have a meaningful effect they may have to be experienced by many people over a long time frame.

Another type of prompt is the presence of waste receptacles. Finnie (1973) reported that compared with a condition in which no trash cans were in sight, their presence reduced littering by about 15% along city streets and by nearly 30% on highways. When a greater number of trash cans were present, littering decreased still more. The value of trash cans or similar objects as antilitter prompts may depend on their attractiveness or distinctiveness. Finnie (1973) observed that colorful garbage cans reduced littering by 14.9% over baseline levels, whereas ordinary cans led to a reduction of only 3.15%. Similarly, Miller et al. (1976) reported

Figure 14-5 This decal on a door is an example of a prompt. Although relatively inexpensive, prompts have a generally small impact on overall actions.

that brightly colored cans resembling birds were much more effective than plain cans in eliciting appropriate disposal. In another study in Baton Rouge, Louisiana, it was found that implementing a recycling program served as a type of prompt against littering: After implementation of the recycling program, recyclable litter declined (though not nonrecyclable litter; Reams, Geaghan, & Gendron, 1996). Finally, an ingenious "garbage can" in the shape of a hat actually worn by students to Clemson University football games greatly reduced trash in the area of the football stadium (Miller et al., 1976; O'Neill, Blanck, & Joyner, 1980). Unfortunately, picking up someone else's litter and putting it in a receptacle is often a more costly behavior than depositing one's own litter, and it seems to be relatively unresponsive to prompts. For example, 10 experiments by Geller and associates (Geller, 1976, 1987; Geller, Mann, & Brasted, 1977) found that prompts have minimal effects on people's likelihood of picking up others' litter and disposing of it. The only exception may be in natural areas, such as campgrounds, where prompts may elicit such behavior (cf. Crump, Nunes, & Crossman, 1977). The relative ineffectiveness of prompts is especially unfortunate, since much money is spent on them by state litter control authorities and organizations such as Keep America Beautiful.

IS A LITTLE LITTER A MORE EFFECTIVE PROMPT THAN NONE AT ALL?

We suggested that more environmentally destructive behavior typically occurs when there is evidence of previous misdeeds (litter on the ground) than when there is not. Whereas this is usually true, work by Cialdini, Reno, and Kallgren (1990) brought out an interesting caveat. While they found (like previous investigators) that a perfectly clean environment produces less littering than a dirty environment, they also observed that the least littering occurs in a setting that is clean except for one piece of litter. The studies were run as follows: Participants were handed a public service-related circular as they walked down a path. Beforehand, the experimenter had positioned 0, 1, 2, 4, 8, or 16 pieces of litter in front of them. Surprisingly, 18% of the participants littered in the "no litter" condition, but only 10% littered in the "one piece of litter" condition. Beyond that, littering by participants increased proportionately to the amount of litter positioned by the experimenter. Why did Cialdini et al. observe such a "check mark" pattern for the relationship between the amount of litter in the environment and subsequent littering? They reasoned that while a perfectly clean environment makes the "no littering" norm salient, an environment clean except for one violation makes it even more salient. With increasing violations, however, the norm becomes undermined, and littering is facilitated. These findings are provocative, and if replicated in other contexts they could have practical implications for environmental education as well as environmental design. For example, do you think you would be more likely to return your shopping cart at the supermarket if all but one of the remaining carts were neatly stacked, or if there were no violations of the "return your cart" norm?

Actions of Others

Other antecedent factors that may serve as prompts include the amount of litter already in a setting, and the behavior of models. Generally, "litter begets litter"—the more littered an environment, the more littered it becomes. In fact, studies have shown up to a five-fold increase in littering in "littered" as opposed to "clean" settings (e.g., Finnie, 1973; Geller, Witmer, & Tuso, 1977; Krauss, Freedman, & Whitcup, 1978).

An exception to the "litter begets litter" finding (see also the box above) has been reported in some natural settings, where people are less apt to litter and more apt to pick up other people's trash when their picnic areas are littered than when clean. This may be because in such settings environmental cleanliness plays an especially important role for people, since they are there to appreciate natural beauty (cf. Geller et al., 1982).

Directly observing the behavior of models can serve as a prompt that reduces or produces littering. Cialdini (1977) exposed individuals to a model who littered or did not litter in a clean

or dirty environment. After seeing the model fail to litter in the clean setting, observers littered the least; after seeing the model litter in the dirty setting, they littered the most. Similarly, Jason, Zolik, and Matese (1979) found that observing a model who showed people how to pick up dog droppings with a "pooper scooper" led to a target area more free of feces.

The use of models has also been implemented as an antecedent strategy to encourage conservation. Research tells us that modeling is most effective when the model is perceived positively but is similar to the observer (cf. Bandura, 1977). Presumably, this similarity leads the observer to expect to receive rewards similar to the model if he or she performs the modeled behavior (Newhouse, 1990). In one modeling application, Winnett et al. (1981) produced a series of videotaped programs on how to adapt to cooler temperatures at home (e.g., change thermostats gradually, wear warmer clothing, use extra blankets). Models who enacted these behaviors were rewarded (i.e., the vignette ended with them being happy with each other); those who approached the situation inappropriately were punished (i.e., the vignette ended with them being angry with each other). Did the modeling intervention work? Yes, overall electricity use was down 14%, and energy used for heating decreased 26%. Other studies (Winnett et al., 1984; Winnett et al., 1985) show

that a videotaped presentation can result in energy savings for up to 9 weeks. Although these results are encouraging, other studies have found models to be less effective in preventing environmentally destructive behavior (cf. Geller et al., 1982).

Consequent Strategies: Intervening After the Behavior

So far, we have talked about antecedent strategies where the intervention occurs before the targeted behavior. **Consequent** (or **contingent**) interventions, on the other hand, occur after the target behaviors are observed. These strategies include reinforcement techniques and feedback. **Positive reinforcement** uses reward —the person gains something valuable (e.g., money) for performing environmentally constructive acts (e.g., recycling). **Negative reinforcement** offers relief from a noxious situation (e.g., high energy bills) in exchange for desirable behavior (e.g., turning down the thermostat). **Punishment,** on the other hand, means an unpleasant consequence occurs (e.g., a fine) as a result of an undesirable behavior (e.g., bypassing a catalytic converter). **Feedback** simply provides information about whether one is attaining or failing to attain an environmental goal (e.g., lower fuel consumption). Often, a specific program implements several of these strategies at once. For example, high-occupancy vehicle (HOV) lanes on freeways (lanes reserved for vehicles with several occupants) reward those who carpool, prompt those who do not carpool, punish those who do not carpool by forcing them into slower lanes and subjecting those who abuse HOV lanes to fines. Let us see how successful these different types of interventions have been (Figure 14-6).

REWARDS AND PUNISHMENTS

In the case of energy use, consumption carries built-in disincentives (the expense of natural gas, oil, or electricity). Large energy consumers such as companies and institutions have responded to

incentives (rewards) and disincentives (fines) in predictable, rational ways (Dennis et al., 1990). For individuals the price incentive is not irrelevant, and at least some energy conservation measures may become common simply because they

Figure 14-6 High-Occupancy Vehicle (HOV) lanes on freeways reward drivers who carpool. Negative reinforcement occurs because carpooling drivers are relieved of the burden of traffic jams, and positive reinforcement occurs because drivers get to their destination faster. Punishment can occur through fines for those who use the lanes without carpooling. The lanes and the signs for them also serve as prompts.

save money (Kempton et al., 1992). Unfortunately, individuals seem not to be as purely rational as companies, and this market approach does not seem to be very effective in reducing *overall* consumption. For instance, one study found that doubling the price of energy led to only a 10% decrease in use (Stern & Gardner, 1981). Individual consumers are more complex than institutions, reacting not only to economic changes, but to idiosyncratic personal factors as well (Dennis et al., 1990). Furthermore, market imperfections (for example, the landlord who buys a refrigerator may not be the person paying for the electricity it consumes if it is an inefficient model) and low energy prices reduce the effectiveness of energy cost alone in changing consumer behavior. Finally, the costs of high utility rates fall disproportionately on the poor (e.g., Francis, 1983). In sum, relying on prices to guide behavior has loopholes and is potentially unfair, so some researchers have examined the utility of adding additional consequent strategies.

Some of these reinforcement-based strategies (e.g., financial payments) have demonstrated consistent behavioral change. For example, Foxx and Hake (1977) offered people various rewards (e.g., cash, tours) to lower the number of miles they drove in private automobiles. The rewards led to a 20% reduction in miles driven, compared with a control group, where mileage increased by about 5%. Other studies have found that competitions between teams enhance mileage reductions and that giving lottery tickets (instead of cash payments) can be an effective motivator to drive less (Reichel & Geller, 1980). In housing, much of the research focuses on individually metered residences (e.g., private homes or apartments with separate meters). For example, some studies have simply paid residents of individually metered residences to lower their energy utilization. These techniques have proven quite effective in changing energy use, both alone and when combined with feedback (Cone & Hayes, 1980). Although paying people for lowering residential energy use is often effective in changing behavior, it may be difficult to implement these consequent payments in a way that is cost effective. In addition, the effect is probably reduced if the rewards seem abstract or are not immediate. For example, in one study financial incentives in the form of tax credits were offered for home retrofitting, but the credits were not sufficient to encourage a high level of this behavior (Stobaugh & Yergin, 1979).

In addition to adjusting utility rates, some studies have used reinforcers to change the pattern of our energy use. Utilities save money when they can rely on their least expensive sources of power, which is typically the case during hours of "nonpeak" demand. When demand "peaks," they must augment their supply with more expensive sources of power. Therefore, it is advantageous to utilities to shift the pattern of energy use from peak to nonpeak periods. To decrease peak demand in some places, financial incentives are provided (i.e., rates are lowered for consumers during nonpeak hours and raised during peak demand hours). These price incentives and disincentives may be effective in switching some discretionary energy-consuming activities (e.g., washing and drying clothes) from peak to nonpeak periods, with benefits to both the utility and the consumer.

Promoting energy conservation might seem to be more difficult in master-metered apartments, where people do not get information about their energy use and frequently do not directly pay for it. One energy conservation method used in such master-metered apartments is rebating all or some part of the money saved by energy-conscious residents. If $10,000 is saved through energy conservation in an apartment building, half might be divided among residents, the other half kept by management. This procedure becomes more effective for conservation as the proportion of savings given to residents increases, and when there is greater cohesion among residents (Slaven, Wodarski, & Blackburn, 1981). Walker (1979) tried another reinforcement strategy with people living in master-metered apartments. It was publicized that people with thermostats set above 74 °F (23 °C) in summer who had their windows closed when air-conditioning was on would receive a $5.00 payment. Apartments were selected randomly for inspection, and those meeting the criteria were reinforced. This technique led to a 4% to 8% savings in energy use throughout the apartment complex.

Perhaps the most persuasive demonstrations of the usefulness of contingent rewards occur in those jurisdictions that have adopted "bottle bills." For instance, the state of New York began requiring a 5-cent returnable deposit on soft drink and beer containers in 1983. Before the law went into effect, recycling rates for the state were 5% for cans, 3% for glass, and 1% for plastics. During the first year after implementation, 59% of cans, 77% of glass, and 33% of plastic containers were recycled, accounting for about a 5% reduction in solid waste by weight (Wolf & Feldman, 1991; Figure 14-7).

In addition to these widespread incentives, several researchers have demonstrated that monetary rewards or punishments can promote short-term recycling (e.g., Jacobs & Bailey, 1982). Unfortunately, behavior often quickly returns to baseline levels when the reinforcements are removed (e.g., DeYoung, 1986; Jacobs & Bailey, 1982).

Consequent strategies are also effective in encouraging people to pick up litter, and (at least in the short run) have generally been more effective than antecedent techniques (Cone & Hayes, 1980). State bottle bill legislation has been effective in reducing roadside litter by 75% (Levitt & Leventhal, 1984; Osborne & Powers, 1980). In describing antecedent strategies we noted a Clemson study in which cans in the shape of distinctive hats served as prompts for litter deposit. The hat cans were also the sources of consequent reinforcement because the Clemson hat dispensed a mechanical "Thank you" to anyone who deposited litter (Miller et al., 1976; O'Neill, Blanck, & Joyner, 1980).

Reinforcement-based techniques can be coupled with prompts. Kohlenberg and Phillips (1973) positioned a prompt that said, "Depositing Litter May Be Rewarded," and then proceeded to reward litter depositors on different reinforcement schedules. This technique precipitated a dramatic cleanup. Another study in-

Figure 14-7   Reduction of the waste stream requires that we reduce the manufacture of unnecessary products, reuse what we can, and try to recycle the rest. Implementation of "bottle bill" legislation is an example of a type of reinforcement strategy that has relatively high impact.

volved a combination of prompts, environmental education, and reinforcers (in this case, feedback). Investigators organized local newspaper coverage about the littering problem (which served as a prompt and a form of education), along with daily feedback on littering in certain target areas. These methods accounted for a decrease in litter compared with baseline conditions (Schnelle et al., 1980). Other experiments have similarly coupled reinforcers with educational techniques and prompts. In these studies, the rewards generally account for much more of the resultant improvement in the litter situation than the other methods (Cone & Hayes, 1980). However, while reinforcement methods are useful to combat littering, they sometimes require costly supervision to monitor behavior and dispense reinforcers. What alternatives are there for reinforcing litter depositors without supervision? Reinforcements can also be administered on the "honor system." A sign in a United States national forest area offered people either 25 cents (sent to them by mail) or a chance to win a larger reward if they filled a plastic trash bag with garbage and completed an information card stating their name and address. Compared with a prompt-only condition (which asked people to fill up a bag but offered no reward), the "honor system" reinforcement condition was much more successful (Powers, Osborne, & Anderson, 1973).

A clever and effective use of positive reinforcement that motivates people to pick up litter is the "litter lottery." A litter lottery offers people an opportunity to win valuable prizes just for depositing litter appropriately. In one version of the technique (Bacon-Prue et al., 1980; Hayes, Johnson, & Cone, 1975), experimenters distribute specially marked items on the ground amidst the litter that is habitually present. People are told some litter is marked in an undetectable fashion, and if an experimenter verifies that they have collected a marked item, they will be awarded a substantial prize. In another version (Kohlenberg & Phillips, 1973), the experimenter merely observes litter deposits and rewards ecologically minded people intermittently. Both techniques have achieved dramatic results toward cleaning up the environment, but both require costly human monitoring. The "marked-item" strategy can, however, be "automated" and introduced widely. Imagine the following scenario for improving litter control throughout the world. Litter with invisible identifying marks could be distributed at random, as would special trash cans (indistinguishable from ordinary ones) that would deliver valuable reinforcers automatically if a marked item were deposited. Whenever you had some spare time, you might find yourself absentmindedly picking up garbage, throwing it in a trash can, and fantasizing about future wealth! Of course, the "novelty" of such a program might wear off after a while, and it would be expensive to administer, which could detract from its over-

all usefulness. In conceiving of any technique for improving environmental conditions, one should be careful to think about ways in which it could be "subverted." One problem with the litter lottery is that it may not prevent people from throwing litter on the ground because the piece they throw away cannot possibly be "marked." Children participating in a litter lottery still disposed of litter inappropriately (La Hart & Bailey, 1975). On the other hand, the more garbage litter lottery participants deposit inappropriately, the more difficult it should be to find "marked" items. Adults may be aware of this contingency, and it may keep them from littering. Reinforcements based on the number of bags of litter collected may be problematic because people are encouraged to pick up large but not small pieces of litter. In fact, some individuals may not pick up litter from the target area at all, but may bring it from home. Even when rewards are based on some criterion for success such as a clean campsite, some people may "clean" the area by throwing trash somewhere else. However, even though these behavioral techniques can be subverted, they are still valuable tools for improving the environment for all.

FEEDBACK

Environmental feedback serves several consequent functions. At the least, it provides information about the relative effectiveness of different behaviors. It may also be reinforcing because it provides competency information; that is, it tells us when we are doing a good job. Brandon and Lewis (1999) found that feedback about energy consumption in Bath, U.K.—especially feedback through computer monitoring—substantially reduced energy consumption. Of course, feedback about energy consumption will be more effective if people know the relative importance of each of the components of total energy use (Dennis et al., 1990). For example, Costanzo et al. (1986) found that many consumers thought that turning off lights would save as much energy as using less hot water, but lighting accounts for less than 7%

of most residential electric bills. Because this percentage is small it may be swamped by other energy expenditures. This may lead consumers to conclude incorrectly that their other conservation efforts are ineffective if their utility bills do not reflect savings from their efforts to turn out lights (Kempton & Montgomery, 1982; Seligman et al., 1979). Often, energy use feedback compares our consumption this year with the same period last year. Good feedback should correct for differences between current weather and weather during the corresponding period of the previous year so that doing better or worse is not an artifact of warmer or cooler temperatures. Although the focus should probably be on individual outcomes rather than on those received by a group, combining both individual and group-based feedback can be very effective (Winnett, Neale, & Grier, 1979).

Feedback is also one approach to reducing transportation fuel use. For example, Van Houten and Nau (1981) posted signs reading "NUMBER OF PEOPLE SPEEDING LAST WEEK: _____ BEST RECORD TO DATE: _____." According to these researchers, the signs were effective in reducing traffic speed—even more effective than increasing the number of tickets issued by police (Van Houten, Nau, & Marini, 1980). In another study, Rothstein (1980) arranged for a graph of gasoline consumption to be displayed during the evening news for seven successive nights. During the feedback period, local service stations reported a consumption decrease of 31.5%.

The more frequent the energy consumption feedback, the more conservation (Seligman & Darley, 1977), though relatively infrequent feedback can sometimes be surprisingly effective (Hayes & Cone, 1981). Feedback is more effective at decreasing energy use when the cost of energy relative to peoples' incomes is high (Winkler & Winnett, 1982), when people believe that the feedback accurately reflects their energy-consuming behavior, and when the household has made a commitment to save energy (Stern & Oskamp, 1987). Some studies suggest that feedback is especially effective during periods of high energy use (e.g., hot, humid days of summer; the coldest days of winter; Cone & Hayes, 1980), and that giving customers energy reduction goals along with feedback enhances its effect (Becker, 1978).

Katzev and Mishima (1992) investigated the effects of feedback on paper recycling on a small college campus. The mailroom's recycling containers labeled "RECYCLABLE PAPER ONLY" were located near both exits in the central mailroom. During the treatment period, a large sign reading: "RECYCLABLE PAPER: _____ POUNDS COLLECTED YESTERDAY" was placed between the two exits and updated each day during the treatment. During the week that feedback was posted, the weight of recycled paper increased 76.7%. Although records were maintained only for 1 week following the feedback treatment, the amount of recycled paper remained about 46% above baseline.

Beyond Traditional Behavioral Interventions

If we examine the interventions we have discussed so far in terms of their effectiveness in promoting environmentally responsible behavior, we will find that the overall results do have a measurable impact, yet they fall short of the kind of impact we would like to believe we could really bring about (see, e.g., Huffman,

Grossnickle, Cope, & Huffman, 1995; Porter, Leeming, & Dwyer, 1995). Examine the information in Table 14-1. It is an incomplete and oversimplified summary of the relative impact of the interventions we have discussed so far. In terms of impact, we find that education and attitude change produce about a 10% improvement in outcome, prompts about a 15% improvement, and consequent (reinforcement) strategies perhaps a 10% to 20% improvement. Can we do even better?

We believe that strategies combining the above interventions with others have the potential to be even more effective. Among these are removal of barriers to individual action, sacrifice and commitment to ecocentric perspectives at the individual level, policy and technological innovations influencing individuals, and "green" adaptations by industry and other large-scale entities.

REMOVAL OF BARRIERS

We noted previously that individuals who believe it will be difficult to carry out an environmentally responsible behavior are unlikely to engage in that action. Margai (1997) reported that progress in a major recycling program was impeded by such inconveniences as building design (e.g., needing to use long hallways and dangerous stairwells to carry bags of recyclables), distance to drop-off sites (sometimes two blocks, with the goods hand-carried), and inadequacy of containers (sometimes attractive to rats). If we can remove perceived barriers to environmentally responsible action, we should be able to make favorable behavior more likely. For example, placing recycling bins in convenient locations should serve not only as a prompt, but also as a mechanism enabling committed individuals to recycle their trash. Providing mixed-use recycling bins removes the barrier of asking consumers to sort their bottles and cans and newspapers into different bins. Oskamp et al. (1996) examined recycling among residents of communities that required four types of recyclable materials to be separated and others that allowed commingled or mixed-use recycle bins. The "commingled use" communities had a participation rate of 90% versus 77% for the "separate use" communities. Mixed-used communities generated 32.1 gallons of recycled material per household per week versus 5.5 gallons for communities requiring separation. In other words, removing the barrier of separation pro-

duced six times more recycled trash volume (Figure 14-8). Potter, Dwyer, and Leeming (1995) describe how inspectors felt helpless in enforcing environmental laws in Memphis, Tennessee, until the court system was reformed (i.e., a barrier removed) to streamline and implement effective procedures. McKenzie-Mohr (2000) describes two other examples. In one, residents near Toronto needed to curtail summer water use substantially. Researchers identified barriers to compliance and had student cyclists visit the residents to discuss water conservation. They also issued lawn watering gauges and a prompt to place above outdoor faucets, and asked residents to sign a commitment to stick to conservation practices. These residents reduced watering by 54%, compared with a 15% increase for a control group. In the other example, McKenzie-Mohr describes a 27% increase in purchasing products with recycled content after researchers worked to removed barriers consumers had reported (e.g., unawareness of products with recycled content and difficulty locating them).

INDIVIDUAL SACRIFICE AND COMMITMENT TO CONSERVATION

A look at Table 14-1 suggests that appeals to pro-environmental action through moral suasion and altering social relations are relatively ineffective compared with interventions such as rewards and punishments that directly benefit or punish the individual. We can think of these alternatives in part as altruistic or group-interest motives versus individual or self-interest motives (e.g., Baron, 1997; Miller, 1999; Oskamp, 2000). While there is every reason to believe that appealing to self-interest motives is effective (e.g., pocketing savings from energy conservation, free tokens for using mass transit, payments for recycling aluminum cans), there is a growing body of evidence that people's dedication to saving the environment can lead at least some to forgo selfish benefits, assume personal responsibility, and endure personal sacrifice in order to promote conservation of the environment (e.g., Geller, 1995; Kaiser & Shimoda,

 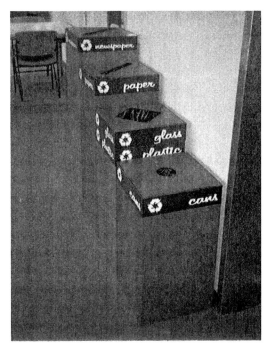

Figure 14-8A & 8B   Removal of barriers, such as allowing commingled recycling rather than having the consumer separate recyclable goods, increases environmentally responsible behavior. Note that these containers also serve as prompts.

1999; Werner & Makela, 1998). Those high in self-transcendent, biosphere-oriented or eco-centric values and low in self-interest values do tend to have much more favorable attitudes toward preserving the environment and do report more environmentally friendly behavior (e.g., De Young, 1996; Karp, 1996; Thøgersen, 1996). To be sure, there is considerable debate about whether consumption of resources is the real culprit in environmental abuse (e.g., Myers, 1997; Stern, 2000; Vincent & Panayotou, 1997). Nevertheless, we believe there is merit in individuals assuming responsibility for behaving in an environmentally friendly manner, and that developing and practicing pro-environmental values can facilitate movement in this direction. For example, Howard (1997, 2000) calls for conscious efforts to modify lifestyles in order to reduce consumption. Yet, as we will see, others believe we can achieve change without the resistance that occurs in response to calls for substantial modification of lifestyles.

POLICY AND TECHNOLOGICAL INNOVATIONS: INFLUENCING INDIVIDUALS

As desirable as changes in resource consumption behavior are, direct consumptive behavior may not be the most effective target for intervention. Some social traps may be so potent that only some form of mutual coercion or managerial decision can achieve effective change. For instance, our use of private automobiles corresponds in many ways to the social trap analysis we described earlier. The short-term benefits of the private passenger car (e.g., convenience, privacy, and prestige) accrue to the driver, while some of its negative consequences (e.g., pollution, energy consumption) are longer range, and the costs are shared by the driver and others.

Some researchers believe that the greatest savings in energy and other resource use by individuals comes from changes in technology, regulations, or building codes (Kempton, Darley, &

PAYING MORE FOR ENERGY TO KEEP THE ENVIRONMENT CLEANER

We have noted that some individuals seem willing to make sacrifices to promote the well-being of the environment. Would you be willing to increase your electric bill if it would reduce pollution? Many citizens are volunteering to do just that. In our search for more environmentally friendly means of producing electricity, one problem we continually run up against is cost. Solar power is cleaner than burning coal, but its cost currently makes it substantially cheaper to generate electricity with coal or natural gas. In some parts of the country, there is enough wind to power generators on a fairly reliable basis, but the cost is relatively high. In an interesting program, a utility in northeast Colorado agreed to build wind generators in a very windy area near Medicine Bow, Wyoming (Figure 14-9). Customers who voluntarily participate agree to increase their electric bills by 2.5 cents per kWh (on top of a base of 5.5 cents per kWh). Each wind turbine reduces the amount of coal burned by 980 tons per year, eliminating more than 4 million pounds of carbon dioxide per year. Each turbine supplies enough power for 12 businesses and 500 residents. The project was so successful with an initial two turbines that after the first year, five more were added to meet customer demand (see http://www.light-power.org/windpower.htm).

Figure 14-9   An example of a wind generator. The tower for the turbines described above is 164 feet high, and the rotor is 154 feet in diameter. This does provide a definite contrast with the nearby landscape (see Chapter 3), and residents near wind generators do sometimes complain about the impact on their view.

Stern, 1992; Stern, 1992a, 1992b, 2000; Stern & Gardner, 1981). In general, these require a one-time legislative or purchasing decision with savings that accrue automatically to each subsequent instance of energy-consuming behavior. For instance, more was probably accomplished by forcing automobile manufacturers to adopt more stringent fuel efficiency standards than could have been achieved by any conceivable effort to change the behavior of individual drivers (Stern, 1992b).

In private residences, the 1980s saw the introduction of condensing pulse combustion furnaces that have efficiencies in excess of 90% compared with the 63% efficiency ratings of typical standard natural gas furnaces sold in the 1970s. Because space heating is the most important single energy expense in the home, the benefits of more efficient furnaces and better insulation can be substantial (Hirst et al., 1986). According to Stern and Gardner (1981), replacing six major home appliances with the most efficient substitute yields an energy savings of 33.2%, compared with a 12.5% savings from the most successful behavioral interventions aimed at encouraging consumers to use energy in the most efficient way. As Stern (1992b) reminds us, builders and manufacturers decide which consumer products to produce, but they do not themselves pay for the energy used by their products, so they may lack sufficient incentive to manufacture efficient products or to build efficient homes. Of course, consumers can demand more efficient designs, but their ability to change the efficiency of buildings or appliances through their purchasing patterns is indirect. Legislated standards are also sometimes flawed. For example, because light trucks have not always been covered by the same fuel economy standards as cars, some manufacturers market certain light trucks as carlike recreational vehicles, lowering the fuel economy of United States vehicles overall (Kempton et al., 1992).

In spite of some successes, there are difficulties in establishing energy-saving technological innovations in homes and workplaces. A survey of potential computer purchasers, for example, showed enthusiasm for computers that are designed to use less electricity, but only if they were no more expensive than standard models (Nadel, 1994). In extensions to an incandescent light replacement program at Notre Dame that we describe in the next section, Howard and his colleagues (1993) attempted to sell new energy-saving fluorescent bulbs to private households. Despite an explanation of the environmental benefits and the promise of long-term savings, relatively few residential customers agreed to buy even one replacement bulb, and these purchases were primarily by individuals with higher incomes. Free-trial offers raised the participation by middle- and lower-income homes somewhat, but participation remained low (Howard et al., 1993). Individuals are cautious, often looking to their peers to separate genuine technological innovations from those that are scams or gimmicks (Dennis et al., 1990). Simply demonstrating technical superiority is not likely to be as successful as an approach that makes use of persuasive techniques from social psychology and marketing. Of course, persuasive communication may not have to be directed at all consumers. As we said, building supervisors and administrators were initially reluctant to install new energy-saving fluorescent bulbs, but once just a few individuals became convinced, large-scale changes were instituted. Thus, one effective approach seems to be a concentrated effort aimed at individuals who have the time and resources to evaluate new technologies and who make decisions resulting in high-volume purchases.

Although Stern and others (e.g., Kempton et al., 1992) may be correct in singling out technological innovations as an excellent source of energy savings, an approach that relies on improved technology also carries liabilities. Geller et al. (1982) suggest that we may underestimate the savings from behavioral interventions to promote conservation, and that some researchers have not considered all the energy waste involved in replacing old equipment with new (e.g., costs of disposal, energy used in the manufacturing process). More generally, we fear that misapplied programs focused on technological innovation might decrease the likelihood that individual consumers will accept personal responsibility for conserving energy, undermining other conservation efforts. Regardless of these points of contention, there is much to be said for developing psychological and other techniques for encouraging our purchase of the most energy-efficient equipment available.

TECHNOLOGY CHANGE IN INDUSTRY: GOING "GREEN" ON A LARGER SCALE

Stern and colleagues (e.g., Stern, 2000; Gardner & Stern, 1996) have pointed out that individual behavior accounts for about 51% of carbon monoxide production. For other pollutants and resource consumption that burden our environment, individual behavior accounts for a much smaller percentage (e.g., 5% of solid waste, one-third of energy use). The great bulk of consumption of resources and generation of pollutants and waste products is through manufacturing or other industrial processes. It seems logical, then,

Figure 14-10 Even trash that is disposed of properly becomes part of the crisis of solid waste. Source reduction through recycling and reclamation can help mitigate the problem.

that pro-environmental actions in larger entities such as corporations would have great impact on the environment.

Why should anyone—individual or corporation—engage in recycling under any circumstances? One answer is to reduce the amount of waste that ends up in landfills, since many believe that we face a "landfill crisis" (Oskamp et al., 1994) in which we are using up landfill space so rapidly that we will run out of it, at least in some locations (Figure 14-10). One way of mitigating this problem preemptively is **source reduction,** or not producing trash in the first place. We could do this by purchasing items in reusable containers, reusing aluminum foil, and avoiding products with excessive packaging. Another reason to recycle is *reclamation,* or recovering materials that can be profitably turned into useful products. Gold, for example, has been used and reused by humans for centuries. There is an intriguing and real possibility that some of the gold in the jewelry you or your friends own

was once part of Spanish, Roman, or Egyptian treasure. Recycling aluminum cans instead of making them from raw materials saves money and energy (recycling aluminum cans uses 95% less energy than refining raw materials) and conserves the world's reserves of aluminum. A bidding war broke out for used plastic soda bottles in the United States in 1995 such that their price rose from $40 to $460 per ton (Miller, 1995). Polyethylene terephthalate (PET) soft drink bottles can be made into insulating fiberfill or yarns, and high-density polyethylene (HDPE) like that in plastic milk jugs can be reused as containers for products such as detergents and shampoo. Some plastics can be recycled into durable picnic tables. In these cases, doing the right thing for the environment (source reduction, recycling) saves money in the manufacturing process.

It is often believed that it is cheaper for industry to pollute the environment than to adopt cleaner manufacturing processes. Yet, would it not make sense that if an environmentally friendly process (sometimes called a **"green" innovation**) would actually boost corporate profits, the incentive for the industry to enhance the "bottom line" economically as well as its public image should encourage "green" practices? Examples of just such innovations are growing in number, and they combine elements of antecedent interventions (e.g., attitude change about obligations to our common environment) and consequent interventions (e.g., increased corporate profits for adopting "green" policies, threat of fines from government regulatory agencies for spoiling the environment).

In 1995, Bigness reported that 10 million automobiles are junked in the United States each year, and 95% of them get recycled. It makes economic sense to do so: In 1995 it cost a firm $33 to procure and process a junked vehicle, but recycling the metals in it could gross $250. In 1999 manufacturers introduced a fiber from the kenaf plant (a plant related to hibiscus, hollyhock, and okra) that can be mixed with polypropylene to create a safe and recyclable product to replace fiberglass in car panels, seats, and trim. Kenaf also can substitute for wood pulp in paper products. It is now economically advantageous, and less damaging to the environment, to use this product (Holden, 1999b).

The value of even simple technological improvements was demonstrated by Howard et al. (1993), who devised a plan to replace incandescent bulbs in the University of Notre Dame dormitories with compact fluorescent bulbs that fit in the existing fixtures. Although the fluorescent bulbs are initially more expensive, they last longer and are more efficient, so over the lifetime of a fluorescent bulb, the energy savings far exceed the initial higher cost. In spite of considerable resistance from university administrators and building supervisors, more than 2,000 bulbs were initially placed in five experimental dormitories, each of which was paired with a similar untreated control-condition dorm. According to the researchers, each fluorescent bulb saved an average of $2.30 in energy costs each month, quickly paying back its higher cost. Following this demonstration, Notre Dame instituted a plan to replace 8,000 incandescent bulbs, designed to save more than $190,000 between August 1994 and February 1997.

The Rocky Mountain Institute, a "green" think tank in Aspen, Colorado, released a report in 1994 describing similar "green" technologies that helped industry profits. For example, similar to the Notre Dame example above, the U.S. Post Office in Reno, Nevada, spent $300,000 in lighting improvements and saved $50,000 per year in energy costs—enough to recover the investment in 6 years. A major discount retail chain built an environmentally friendly store in Lawrence, Kansas, which used natural lighting through skylights on one side of the store; cash register sales were twice as high on the naturally lighted side of the store. A major manufacturer "greened" up a new building to drastically cut energy use and found that worker absenteeism fell 15% and productivity rose 15%, more than making up for the increased cost of the building (RMI, 1994).

A similar story is told by McDonough and Braungart (1998). They describe development of a textile that is biodegradable. Manufacture of the textile requires only 38 chemicals (more than 8,000 chemicals are used in combined textile industries) and is so clean that wastewater leaving the manufacturing plant is as clean as water entering it; inspectors initially thought their measuring instruments were malfunctioning, but they actually were affirming that the effluent water was clean!

Yet another example of this "green" industrial revolution is described by Kaiser (1999b). She reviews a number of new, cleaner manufacturing processes, including an industrial park in Denmark in which waste products from one industry (e.g., gypsum from a power plant, fermentation waste from a biotech firm) become raw materials for another (e.g., gypsum becomes drywall, and fermentation waste becomes fertilizer).

In still other applications of this "green capitalism" concept, Anderson and Leal (1997) report how a paper company increased profits by converting forests to recreational use, a real-estate developer turned a flood channel into an attractive trout stream, and landowners in South Africa attracted tourists by preserving wildlife. A variation on the theme occurs when an organization such as the Nature Conservancy—a not-for-profit conservation group—uses charitable contributions to buy land and keep it in an undeveloped, wildlife-preserving state of existence; the previous owner becomes wealthier by selling the land, and the ecology of the area also benefits.

INTEGRATION OF PERSPECTIVES: WHAT YOU CAN DO

We began this chapter with Hardin's tragedy of the commons, in which individual, short-term self-interest by multiple individuals leads to ecological disaster for all. We then observed that changing individual attitudes does not always lead to substantial pro-environmental behavior, and that even individual incentives do not typically have as much impact on the environment as policy and technological change. Finally, we saw that individual commitment can make a difference and that individual and corporate adoption of "green" practices can have larger-scale impact. We would conclude by noting that the effect of any of these interventions begins with individual action. Your personal commitment to recycling, to conserving limited resources, and

to practices that do not pollute our common environment will make a difference. Yet you can also make a difference by encouraging your friends and your employer to adopt similar practices. Cobern et al. (1995) found that asking targeted residents to commit to recycling *and* to agree to encourage their neighbors to recycle led to a substantial increase in recycling by both the target residents and their neighbors. One of our former students with a degree in economics and psychology and who occupies an executive position with a recycling firm has observed that it takes psychology to persuade other industries it is in their own best interest, and the best interest of the environment, to sell their waste paper to the recycling firm rather than pay to have a trash hauler dump it in the local landfill. This is a case of an individual persuading others to adopt "green" practices, and yet the individual effort pays big dividends for multiple corporations and for our common environment. That should serve as a message of encouragement and optimism. Your individual action may appear minuscule in the commons of the world, but you can also serve as a model (prompt) for others, as an advocate for larger policy change, and as an innovator for encouraging and enabling larger entities to adopt environmentally friendly practices that benefit all humans as well as other species.

SUMMARY

Much environmentally destructive behavior can be conceptualized in terms of social traps. These are situations in which personal interests with a short-term focus conflict with societal needs that have a long-term focus. For example, littering and purchasing products that are not fuel-efficient are instances in which short-term individual convenience conflicts with the long-term needs of society. How can we escape from such traps? Research on the commons dilemma suggests that dividing some common resources into territories helps, although this is not always practical. Increasing communication and trust and fostering attraction toward and group iden-

tification with those who share the commons also are valuable strategies. Altering reinforcements or consequences is also an exceptionally effective approach; adding positive consequences for conservation behavior or punishments for exploitative behavior helps preserve the commons for all. In this chapter, environmental education and attitude change, use of environmentally relevant prompts, reinforcement-related techniques, and other methods are considered as potential means of altering environmentally destructive behavior. Each of these techniques has unique costs and benefits. Environmental education seems to be a relatively ineffective method, but may be less costly than some other methods. The use of environmentally related prompts is somewhat effective and relatively inexpensive.

Reinforcement techniques (positive reinforcement, negative reinforcement, punishment, and feedback) appear to be very effective in creating short-term change, but they have several drawbacks. For example, reinforcement techniques may be quite expensive, and evidence shows that the effects often disappear when the reinforcement contingency is withdrawn. Alternative strategies include removing barriers to conservation, individual sacrifice and commitment, and "green" policies and technological advances. These approaches seem to hold great promise at the individual and group level.

KEY TERMS

antecedent strategies
approach prompts
avoidance prompts
commons dilemma
consequent
  interventions
contingent
  interventions
dominant Western
  world view

environmental
  education
feedback
foot-in-the-door
  technique
free rider
green innovation
individual good–
  collective bad trap
missing hero trap

modeling
negative
  reinforcement
new ecological
  paradigm
New Environmental
  Paradigm
nuts game
one-person trap
positive reinforcement

prompts
public goods problem
punishment
resource dilemma
self-trap
social dilemmas
social traps
source reduction
sustainable

SUGGESTED PROJECTS

1. Design an environmental education program that you feel would have an optimal chance of effectively changing behavior in an environmentally constructive direction. Use whatever media you like, focus on whatever population you desire, and choose a target behavior that corresponds to your area of major interest in environmental psychology.

2. Select an environmentally destructive target behavior and design a prompt that would help alleviate the problem. If you can get the necessary permission from authorities, attempt to test the effectiveness of your technique.

3. Test Cialdini et al.'s hypothesis that a setting with a single violation of a norm serves as a more effective prompt than one with no evidence of norm violation. Some possible subjects for your investigation are the appropriate return of shopping carts at a local supermarket, graffiti in a restroom, or vandalism in a university building.

4. If you have lived in a college residence hall, design a program (using the techniques described in this chapter) to help alleviate noise. Keep in mind the relative effectiveness of education, prompts, reinforcements, and policy changes and innovations.

acclimation—adaptation to one specific environmental stressor, such as temperature.

acclimatization—adaptation to multiple stressors in an environment, such as humidity, temperature, and wind.

accretion measures—unobtrusive indications of behavior, involving traces of additions to the environment, such as litter or fingerprints.

adaptation—weakening of a reaction (especially psychological) to a stimulus; becoming accustomed to a particular degree of a given type of stimulation; in Wohlwill's ideas, a shift in optimal stimulation.

adaptation level (AL)—an ideal level of stimulation that leads to maximum performance or satisfaction.

adjustment—Sonnenfeld's idea of technological change of a stimulus, as opposed to adaptation, which refers to change in the response to the stimulus.

aesthetics—in architecture, an attempt to identify, to understand, and, eventually, to create those features of an environment that lead to pleasurable responses.

affect—feelings or emotional states.

affordances—in Gibson's theory of ecological perception, the properties of an object or place that give it constant and automatically detectable functions; the possibilities of use an environment provides.

aftereffects—consequences of a stimulus that occur after the stimulus has stopped; effects of a stressor on mood or behavior, often measured by task performance, that occur after termination of the stressor.

air ionization—condition in which molecules of air partially "split" into positively and negatively charged particles, or positive and negative ions.

alarm reaction—a startle response to a stressor; the first stage of Selye's GAS.

Alzheimer's disease—the most common progressive dementia, occurring primarily in the elderly and accounting for half of all nursing home admissions.

ambient stressors—chronic, global stressors such as pollution, noise, or traffic congestion.

ambient temperature—surrounding or atmospheric temperature.

amplitude—the amount of energy in a sound, as represented by the height of the sound wave, perceived psychologically as loudness.

analog representation—the theoretical case in which a cognitive map is stored in memory in a picture form that corresponds point for point to the physical environment.

annoyance—in noise, the irritating or bothersome aspect.

antecedent strategies—behavioral change techniques that occur before the target behavior and that are designed to increase the likelihood of favorable acts.

anthropocentric, anthropocentrism—viewing the natural environment from the perspective of how it meets human needs; *see also* homocentric, ecocentric.

applicability gap—a two-way communications breakdown that occurs when scientists fail to ask questions with direct application to design problems or when designers neglect to employ those principles that have empirical support.

applicants—in ecological psychology, those who meet the membership requirements of a behavior setting and who are trying to become a part of it.

appraisal—cognitive assessment of a stressor along the dimensions of harm or loss, threat, and challenge.

approach prompts—prompts that supply an incentive for engaging in a particular behavior.

archival data—data that researchers may find in others' historical records, such as police crime reports, weather records, or hospital records.

arousal—a continuum of physiological or psychological activation ranging from sleep to excitement; crowding, personal space intrusions, or other stressors can lead to overarousal.

assigned workspace—designating a particular machine or area to a particular worker in order to create a feeling of ownership.

Attention Restoration Theory (ART)—Kaplan and Kaplan's notion that extensive mental labor can lead to directed attention fatigue and that restorative experiences such as hiking a nature trail or visiting a museum can help recover the ability to concentrate.

attitude—a relatively stable tendency to evaluate a

person, object, or idea in a positive or negative way; many definitions stress the interrelationship of feelings, cognitions, and behaviors.

attraction gradient—the likelihood that the design of an exhibit will attract museum visitors to view it.

avoidance prompts—prompts that supply a disincentive for enacting a particular behavior.

background stressors—persistent, repetitive stressors whose impact is relatively gradual, including daily hassles.

barometric pressure—atmospheric pressure, as read by a barometer.

Beaufort Scale—a scale of wind force developed by Admiral Sir Francis Beaufort in 1806.

behavioral interference—the notion that under high density conditions, many negative effects are due to "getting in people's way" and similar mechanisms for blocking goals.

behavioral sink—area in which the negative effects of high density are intensified.

behavioral toxicology—a field of study that examines whether behavioral anomalies, such as language deficits, may be due to toxic exposure, such as lead poisoning.

behavior constraint—a model that emphasizes how the environment (e.g., urban life, personal restrictions) may limit or interfere with activities, leading to loss of perceived control.

behavior mapping—a structured observational technique in which behaviors are observed, recorded, and located on a map of the setting being observed.

behavior setting—the basic unit of environment–behavior relationships; in Barker's ecological psychology, an entity that encompasses the location of a large volume of behavior; consists of the interdependency between the standing patterns of behavior and the physical milieu.

biophilia—the proposition that humans have evolved a biological affinity for natural environments.

biophobia—the proposition that humans have a genetic predisposition to learn to fear certain potentially dangerous elements of nature, such as spiders and snakes.

block organizations—neighborhood organizations formed to add social cohesiveness and overcome urban ills and that work for such improvements as better lighting, police protection, or street repairs.

capacity—in ecological psychology, the maximum number of inhabitants a behavior setting can hold; in overload notions, the limited capability for processing information.

carrying capacity—the amount of use a resource can support; with respect to recreation, the carrying capacity could be variously defined as the number of people who can fit in a given area, the number who can be accommodated without resource damage, or the number who can receive a satisfactory experience such as solitude.

cataclysmic events—sudden, powerful events that require a great deal of adaptation in order for people to recover, avoid, or cope with their effects, such as natural disasters.

catecholamines—epinephrine (adrenaline), norepinephrine, and dopamine—secretions that energize various systems in the body.

challenge appraisal—a cognitive appraisal component of the stress model that focuses on the possibility of overcoming the stressor.

chill factor—*see* wind-chill index.

climate—average weather conditions or prevailing weather over a long period of time.

cognitive map—the brain's representation of the spatial environment.

coherence—in landscape evaluations, the degree to which the elements in a scene are organized and seem to fit together.

cohousing—a type of housing in which families have their own private residences but share larger facilities, such as a master kitchen and recreation area.

commodity—the functional goal (intended use) of a design.

commons dilemma—Hardin's notion that depletion of scarce resources can happen because people sharing a resource harvest it with short-term self-interest in mind rather than long-term group interest.

compensatory behaviors—in personal space, behaviors such as increased eye contact that make up for inappropriately far distances, such as leaning away from a person for inappropriately close distances.

complexity—with regard to landscape or architectural aesthetics, the variety and salience of elements in a scene.

complexity of spatial layout—the amount and difficulty of information that must be processed in order to move through the environment.

confounds—variables other than the ones being studied that also vary across different conditions, and thus that can account for systematic effects in the dependent variable.

congruence—the "fit" between user needs or preferences and the physical features of a setting.

consequent interventions—reinforcement, feedback, and other interventions that occur after the target behavior (e.g., littering, failure to recycle) occurs; *see also* contingent interventions.

conservation of resources (COR) theory—Hobfoll's proposal that the extent to which people lose important resources or are able to minimize this loss will determine how much stress is experienced from a natural disaster.

constructivism—a perspective that holds that perception is an active process in which sensory inputs are analyzed, compared with past experience, and manipulated in order to construct perceptual interpretations of the outside world.

contingent interventions—reinforcement, feedback, and other interventions that occur contingent on the target behavior's (e.g., littering, recycling) occurring or not occurring; *see also* consequent interventions.

control models—environment–behavior models that emphasize consequences of loss of perceived control; the behavior constraint model is an example.

coping—handling stressors; efforts to restore equilibrium after stressful events.

core temperature—the temperature inside the body, also called deep body temperature.

correlational research—research that does not manipulate environmental occurrences or prescribe who should be involved as participants; events occurring prior to the research and concurrent actions of other factors in the setting may interfere with the conclusions drawn.

corticosteroids—steroid compounds produced by the adrenal cortex; increased secretion is characteristic of alarm reactions.

crisis effect—the phenomenon wherein disaster events attract a great deal of attention while they are occurring (or shortly thereafter), but concern for future disasters decreases after that.

crowding—experiential state when the constraints of high density are salient to an individual.

cue utilization—the individual weights assigned by a person in making perceptual judgments based on past experiences, personality, or other characteristics.

curvilinear relationship—a relationship between two variables that is not a straight line, such as a U-shaped or inverted-U function.

daily hassles—stable, low-intensity problems encountered as part of one's routine, such as commuting.

decibels (dB)—units of measure of loudness of sound, in the form of logarithmic representations of sound pressure.

deep body temperature—*see* core temperature.

deep ecology—a form of ecocentrism that is critical of modern technology, science, and political structures, believing that these endanger nature.

defensible space—clearly bounded or semiprivate areas that appear to belong to someone.

degree of visual access—the extent to which different parts of a setting can be seen from a number of vantage points; access facilitates the learning of a new environment.

deindividuation—according to Zimbardo, feeling of anonymity, or loss of individual identity, that releases otherwise inhibited antisocial behavior.

delight—the aesthetic goal of design.

dementia—a medical term meaning long-term loss of memory and confusion.

denial—ignoring or suppressing awareness of stressors and other problems.

density-intensity—Freedman's conceptualization that high density increases the intensity of behaviors and feelings that would have occurred anyway under lower density conditions.

dependent variable—in the experimental method, the behavior of the participant that is measured by the experimenter.

descriptive approach (to landscape assessment)—landscape assessment based on the judgments of professionals trained to detect patterns, primarily based on artistic judgment.

descriptive research—research that reports behavior, emotions, or other characteristics that occur in a given setting or in response to a specific event.

design alternatives—the list of potential solutions to a design problem.

design cycle—a continuous cycle of information gathering, planning, and evaluation; the cyclical nature allows information gathered from one project to add to the knowledge for subsequent designs.

design review—a case-by-case examination of proposed new building projects that attempts to ensure that they will remain harmonious with both the existing architecture and the ongoing social fabric of a district.

determinism—in a strict sense, a philosophical

notion that circumstances have absolute causal relationships to events.

differentiation—distinctiveness; buildings or environments that are different or distinctive are more easily remembered.

diffusion of responsibility—an explanation for decreased helping behavior which posits that as the number of potential helpers increases, each one assumes less individual responsibility for helping.

directed attention fatigue (DAF)—a state of mental exhaustion similar to overload; restorative experiences are thought to alleviate DAF.

disaster events—powerful events that cause substantial disorganization, disruption, or destruction to an area, community, or series of communities.

discontinuity—the idea of changing a steady pattern in museum exhibits; breaking up the pattern helps relieve museum fatigue.

disruption—disturbance of individual, group, or organizational functioning and routine.

distortions—errors in cognitive maps based on inaccurate retrieval that leads us to put some things too close together, some too far apart, and misalign others.

districts—large geographical areas that are identified in cognitive maps; typically, the places within a district have a common character and often are given names such as the French Quarter, Chinatown, or the East End.

dominant Western world view—belief that human domination over infinite natural resources leads to inevitable progress.

ecocentric, ecocentrism—valuing nature for its own sake instead of for how it supports humans.

ecological niche—according to Gibson's ecological perception, a set of affordances that are utilized.

ecological perception—the approach that emphasizes that perception is holistic and direct; according to this view, patterns of stimulation give the perceiver immediate information about the environment—including its affordances—with little effort or cognitive activity.

ecological psychology—Barker's behavior setting approach to studying the interaction between humans and their environment.

ecological validity—the objective usefulness of various environmental stimuli in making accurate perceptual judgments.

edges—elements in cognitive maps that limit or divide features, such as paths or districts; edges may be elements such as walls, railyards, or water features.

effective temperature—an adjustment in perceived temperature to account for humidity, similar to the Temperature-Humidity Index.

ELF-EMF—extremely low frequency electromagnetic fields associated with weather disturbances or power lines.

El Niño—a weather oscillation marked by warming of equatorial waters off the Pacific coast of South America around Christmas time, resulting in warmer and drier weather in the northern United States and colder and wetter weather in the southern United States.

empirical—publicly observable.

empirical laws—statements of simple observable relationships between phenomena (often expressed in mathematical terms) that can be demonstrated time and time again.

empiricism—a position that holds that externally observable events are the only legitimate source of data.

en masse behavior pattern—in Barker's ecological psychology, the behavior of a group.

environment—one's surroundings; the word is frequently used to refer to a specific part of one's surroundings, as in social environment (referring to the people and groups among whom one lives), physical environment (all of the nonanimal elements of one's surroundings, such as cities, wilderness, or farmland), natural (nonhuman) environment, or built environment (referring specifically to that part of the environment built by humans).

environmental assessment—describing and evaluating environments, such as through EQI or PEQI methods or landscape preference methods.

environmental cognition—the ability or propensity to imagine and think about the spatial world.

environmental competence—*see* environmental press.

environmental education—making people aware of the scope and nature of environmental problems and of behavioral alternatives to alleviate them.

Environmental Emotional Reaction Index (EERI)—an assessment of the emotional reactions of humans to some component of environmental quality.

environmental justice—a movement that aims to correct prior and prevent future abuses of the environment by framing pro-environmental issues in terms of "justice" due.

environmental load—a theoretical position based on overload of information from the environment.

environmental press—a model which posits that the demands or press an environment places on its occupants as well as the competence of the occupants determine the consequences of interacting with the environment.

environmental psychology—the study of the interrelationship between behavior and experience and the built and natural environment.

Environmental Quality Index (EQI)—objective measures of environmental quality—the chemical or physical properties of water or air, for example.

environmental racism—framing environmental justice issues, such as the disproportionate location of toxic wastes in areas occupied by disadvantaged residents, in terms of discrimination against and justice for disadvantaged groups.

environmental spoiling hypothesis—the notion that perceived quality of the living environment is determined largely by the number of unpleasant contacts with others.

environmental stress model—a theoretical perspective that emphasizes how the environment can elicit stress and coping reactions when it is evaluated to be threatening.

equilibrium—a state of balance; the steady state to which stress reactions try to restore the organism.

ergonomics—the discipline that concerns itself with the design and modification of equipment and workplaces to make them better adapted to the needs of humans; also called human factors.

erosion measures—trace measures that signify something taken away or worn down (e.g., wear patterns on carpet).

ethic—a system of morals or standards held by a person, culture, or religion, such as a land ethic.

ethological models—in personal space, formulations that assume that when space is inadequate, fear and discomfort are experienced due to feelings of aggression or threat; such models are based on observations of nonhuman animals.

event duration—how long an event lasts.

exit gradient—how much a gallery exit attracts a museum visitor to leave the gallery.

experiential realism—the extent to which the experimental manipulation has impact on the participant and is representative of events that occur in the real world.

experimental method—a way of conducting research that allows inferences about what might cause a given effect; by varying two factors and studying effects of these factors under controlled conditions, one can observe specific causes for observed effects; *see also* random assignment.

external validity—the degree to which a research study's findings generalize to other contexts.

extra-individual behavior pattern—in Barker's ecological psychology, the behavior of large numbers of people.

familiar stranger—someone you have observed on many occasions but never interact with.

feedback—a technique that provides information about whether one is attaining or failing to attain an environmental goal; as such, it is a means for changing environmentally destructive behavior.

festival marketplaces—busy retail and tourist centers that combine retail space, leisure, and theater; the prototype is Boston's Quincy Market, designed by James Rouse.

firmness—the design goal of permanence and structural integrity.

fixed workspace—assignment of a specific workspace or machine to a worker on a more or less permanent basis, as opposed to sharing the machine or space with others; also called assigned workspace.

folk design tradition—architecture based on the day-to-day needs of people as they live, shop, and work; design familiar to the common person in everyday life.

foot-in-the-door technique—a technique that increases compliance with a standard request by first asking for a small favor that the respondent is likely to agree to.

forward-up equivalence—in map design, having what is forward on the ground being at the top ("up") on the map.

free rider—in the public goods problem, a person who fails to contribute to the common good but reaps the benefits of the contributions of the other participants.

frequency—the number of cycles per second in a sound wave, perceived psychologically as pitch.

frostbite—formation of ice crystals in the skin.

functionalism—a tradition within psychology that views behavior as a way of adapting to or surviving the demands of the environment.

Gaia hypothesis—the idea that heating and cooling of the earth (oceans, land, and atmosphere) as well as associated operations of living things are part of a self-regulating system.

gaps—lapses in communication among designers, users, behavioral scientists, or clients that result in design errors or oversights.

general adaptation syndrome (GAS)—Selye's stress model, which consists of the alarm reaction, the stage of resistance, and the stage of exhaustion.

generalizability—a measurement of how well a finding, relationship, or theory applies from one setting to another.

gentrification—the emergence of middle- and upper-class areas in parts of the inner city that were formerly deteriorated.

Gestalt perception, Gestalt principles, or Gestalt theory—perceptual principles developed by the Gestalt school of psychology that have heavily influenced architectural thinking; Gestalt principles are based on a holistic assumption that we read meaning—such as shape or melody—into perceptions beyond the mere sum of individual sensations.

global warming—warming of average global temperature by several degrees per century, thought by many to be due to atmospheric pollution and associated greenhouse effects, and to altered land use patterns.

grand design tradition—architecture, such as monuments or impressive facades, built to impress the populace, client, or other architects.

greenhouse effect—the excess heating of the planet due in part to carbon dioxide and other pollutants trapping too much heat close to Earth's surface.

green innovation—new practices that help sustain environmental resources, such as light bulbs that save energy or manufacturing processes that do not produce toxic wastes.

green justice—another term for environmental justice, a movement that aims to correct prior and prevent future abuses of the environment by framing the issues in terms of "justice" due.

habituation—the process (especially physiological) by which a person's responses to a particular stimulus become weaker over time.

harm or loss appraisal—a cognitive appraisal component of the stress model that focuses on damage already done.

hearing loss—permanent or temporary decrease in one's ability to hear caused by damage to the eardrum or to the tiny hair cells in the inner ear.

heat exhaustion—moderate condition of faintness, nausea, headache, and restlessness due to heat stress.

heatstroke—severe and life-threatening condition of heat stress in which the sweating mechanism breaks down.

hertz (Hz)—cycles per second of a sound wave.

heuristics—simple principles that facilitate decision making.

high density—situations characterized by high social or spatial density; a large number of people in an area.

homelessness—when a person does not have a fixed, regular, and adequate nighttime residence.

homeostatic—descriptive of automatic mechanisms that serve to maintain a state of balance, such as the sweating reaction to heat stress.

homocentric, homocentrism—viewing the natural environment from the perspective of how it meets human needs; *see also* anthropocentric, ecocentric.

human factors—the discipline that concerns itself with the design and modification of equipment and workplaces to make them better adapted to the needs of humans; also called ergonomics.

hypothalamus—a primitive part of the brain responsible in part for regulating temperature, hunger, thirst, aggression, and sex drive.

hypothermia—a life-threatening decline in core temperature.

hypothesis—a scientific hunch or formal statement of an anticipated relationship between events.

hypoxia—reduced oxygen intake associated with low air pressure conditions, such as high altitudes, or with carbon monoxide poisoning.

independent variable—in the experimental method, the circumstances the experimenter manipulates in order to determine the impact on the dependent variable.

individual difference variables—variables that reflect differences in people in terms of background, personality, or other factors.

individual good–collective bad trap—a type of social trap in which a resource is depleted because short-term positive consequences of usage are experienced by the individual but long-term negative consequences are dispersed through the group.

informed consent—written permission from a person agreeing to participate in a study after being informed of procedures, risks, and any circumstances that might alter the decision to participate.

inside density—population density indices using "inside" measures, such as number of people per residence or per room.

internal validity—the rigor with which a research study is constructed so that one knows whether observed effects are due to variables of interest as opposed to such methodological artifacts as confounds or failure to control extraneous variables.

internality–externality—personality variable that taps whether people believe they, or outside forces, control their outcomes.

interpersonal distance—the distance between people.

intervening construct—an inferred phenomenon that mediates the relationship between other events or concepts.

invasion of privacy—in research, access by the researcher into nonpublic activities of the research participant without the person's permission.

land ethic—Aldo Leopold's belief that humans share nature with a community of equal elements, including other species, soils, and water.

landmarks—structures or geographical entities that are distinctive; landmarks are usually visible from some distance and include features such as tall buildings or monuments.

landscaped office—an open office; a large office area with few walls that is designed to be flexible and to facilitate the organizational processes that take place within it.

learned helplessness—Seligman's idea that once we believe we have lost control over the things that happen to us, we cease trying to change the situation; experienced state when people "learn"

there is no contingency between their inputs and their outcomes.

legibility—the degree to which a scene is distinctive or memorable; in cognitive maps, the degree to which an area is easily learned or remembered.

lens model—Brunswik's model of perception that emphasizes the active process by which humans make judgments based on probabilistic weighting of the variety of stimuli in the environment.

levee effect—the observation that once protective precautions are taken against a potential disaster, people tend to settle and live around these precautions, even though threats are still present.

linear perspective—the principle that parallel lines will appear to converge in the distance.

linear relationship—straight-line (monotonic or rectilinear) relationship between two variables.

long, hot summer effect—the belief that heat wave conditions precipitate violence.

loudness—the physical perception of amplitude in a sound or noise.

low point—the point in a disaster at which victims perceive that the worst threat, harm, or adaptive demand has been reached; following this point, things gradually improve.

maintenance minimum—in ecological psychology, the minimum number of inhabitants needed to maintain a behavior setting.

mapping—in human factors, the relationship between the actions of an operator and those of a machine.

masking—covering up or eliminating the distinct perception of a sound or noise by adding another sound or noise of similar frequency and similar or higher amplitude.

mediating variable—a variable that operates in a sequence between other variables, such as arousal mediating the relationship between noise and aggression.

missing hero trap—a type of social trap in which individuals fail to act for the benefit of the group because the penalty to the "hero" who does act seems inordinately large.

model—a relationship between concepts that is often based on analogies or metaphors.

modeling—an antecedent strategy in which others display or model the desired pro-environmental behavior.

moderator variable—a variable that interacts with

another, such that the effects of a variable are different depending on the level of the moderator, as in the effects of noise on task performance differ depending on (or as moderated by) the level of task difficulty.

modern design—the design tradition in industrialized societies; critics suggest that a premium is placed on originality and aesthetics at the expense of evolutionary development.

multidimensional scaling (MDS)—a class of statistical procedures for displaying the relationships between concepts or places based on their similarity on several dimensions of interest.

museum fatigue—Robinson's notion that museum visitors tire because they must pay attention to so much information in exhibits.

mystery—in landscape assessment, the degree to which hidden information creates intrigue and leads a viewer to further investigate a scene.

narrow band—a sound or noise with relatively few frequencies in it.

nativism—the view that perceptual processes or other phenomena come to us automatically (i.e., are inborn) as opposed to having to be learned through experience.

natural disaster—a disaster that is caused by natural (nonhuman) factors; *see also* disaster events.

negative affect-escape model—the position that aggression increases with discomfort (such as from heat or pollution) up to a point but then declines with further discomfort as escape motives become stronger than aggression motives.

negative ions—*see* air ionization.

negative reinforcement—removing a noxious stimulus; increases desirable environmental behavior because people are motivated to avoid unpleasant stimuli (e.g., fines) or to escape an ongoing noxious stimulus (e.g., high electric bills).

new ecological paradigm, or New Environmental Paradigm—belief that humans are interdependent with a fragile natural ecology that contains limited resources; the New Environmental Paradigm is a scale for measuring this concept.

NIMBY—"Not in my back yard," or opposition to placing potential nuisance facilities (e.g., toxic waste dumps, freeways, or prisons) in or near one's neighborhood.

nodes—in cognitive maps, points where behavior is concentrated, such as at a place where major paths cross one another or intersect a landmark.

noise—sound that is undesirable or unwanted.

noise-induced permanent threshold shift (NIPTS)—hearing loss that is typically present a month or more after noise exposure ceases, characterized by increases in threshold below which sounds are inaudible.

nonperformers—in ecological psychology, those who carry out secondary roles in a behavior setting.

normative theory—approaches to design based on values and opinions rather than empirical facts.

nuts game—a simulation of the commons dilemma developed by Edney.

observation—watching people behave and recording what is seen.

one atmosphere—atmospheric pressure at sea level, or 14.7 pounds per square inch.

one-person trap—another term for self-trap.

open classrooms—schools designed with few interior walls so that students and teachers are free to move about.

open-plan office—a large office area with few walls, designed to be flexible and to facilitate the organizational processes that take place within it.

outside density—population density indices using "outside" measures, such as number of persons, dwellings, or structures per acre.

overload—a condition in which stimulation from the environment exceeds the capacity of the person to process the inputs, resulting in ignoring of some information; *see also* environmental load.

overstaffed or overstaffing—in ecological psychology, a condition of having too many participants, where the number of "participants" exceeds the capacity of the system.

overstimulation—*see* overload.

ozone—a form of oxygen in which three atoms are molecularly combined (O_3).

ozone hole—atmospheric reduction in ozone around the polar regions due in part to chlorofluorocarbons.

palliative—emotion-focused (as opposed to direct-action focused) coping processes, such as denial, using drugs, or appraising the situation as nonthreatening.

participation—intentionally soliciting the input of users in the design process, such as occurs in design review.

paths—shared travel corridors identified in cognitive maps, such as streets, walkways, or riverways.

pattern language—prescriptions for design problems presented in such a way that they allow user participation in design.

perceived control—belief that we can influence the things that are happening to us.

Perceived Environmental Quality Index (PEQI)—a subjective assessment of some characteristic of environmental quality as perceived by a human observer.

perception—the process of extracting meaning from the complex stimuli we encounter in everyday life; higher-order processing of sensory inputs.

performers—in ecological psychology, those who carry out the primary tasks in a behavior setting.

peripheral vasoconstriction—constriction (narrowing) of blood vessels in the arms and legs, as in response to cold stress.

peripheral vasodilation—dilation (widening) of blood vessels in the arms and legs, as in response to heat stress.

personal space—a body buffer zone that people maintain between themselves and others.

personal stressors—stressful events that affect one person or only a few people at a time, such as loss of a loved one or loss of a job.

person–environment congruence—the notion that the setting promotes the behavior and goals within it; the degree of "fit" between people and their environment.

phenomenological—pertaining to the "reality" of subjective personal experience as opposed to external, objective reality.

physical milieu—in Barker's ecological psychology, the physical component of the behavior setting.

physical-perceptual approach (to landscape assessment)—assessment strategies that emphasize the characteristics of the physical environment that can be statistically related to judgments of aesthetics.

piloerection—"goose bumps" or the stiffening of hairs on the skin.

pitch—the psychological perception of the frequency of a sound.

place attachment—psychological bonding to an environment.

positive ions—*see* air ionization.

positive reinforcement—when people are given positively valued stimuli for performing environmentally constructive acts.

positive theory—procedures or design components that are based on empirical observations rather than opinions or values.

possibilism—the notion that the environmental context makes possible some activities but does not force them to occur, as in climatological, geographical, or architectural probabilism.

post-occupancy evaluation (POE)—a retrospective evaluation used to suggest modifications of the present structure and to improve the available knowledge for future projects.

post-traumatic stress disorder (PTSD)—an anxiety disorder characterized by having experienced a traumatic event (like a disaster), frequent, unwanted, and uncontrollable thoughts about the event, heightened motivation to avoid reminders of the event, social withdrawal, and heightened arousal.

preindustrial vernacular architecture—the design of common buildings in nonindustrialized cultures; there is relatively little individual variation in style, but construction techniques often have evolved as good solutions to environmental demands.

preservationism—a holistic view of nature that assumes that an intact ecosystem is greater than the sum of its parts, with an emphasis on preserving nature because of this extra value above and beyond the value of the individual parts.

primary appraisal—cognitive assessment of threat.

primitive architecture—the architecture of so-called primitive societies; shelter is constructed according to standard techniques that have evolved over time; these are often quite sophisticated solutions to environmental demands.

privacy—an interpersonal boundary process by which people regulate interactions with others.

privacy gradient—different levels of privacy, as evidenced in different areas of the home.

privacy regulation model—Altman's model in which personal space, territoriality, and crowding are conceptualized as involving processes regulating privacy, or the desired degree of interaction with others.

probabilism—the notion that the environmental context makes some activities more probable than others but does not absolutely determine which will occur, as in climatological, geographical, or architectural probabilism.

procedural theory—methods of gathering data and making decisions.

prompts—cues that convey a message.

propinquity—how close people are to each other in the places they occupy, in terms of objective physical, or functional, distance.

propositional storage—the theoretical position that the environment is stored not as a series of pictures, but as interconnected concepts or ideas connected to one another by a network of associations.

prosocial behavior—helping behavior or other activity to promote the welfare of others.

psychological reactance—Brehm's notion that any time we feel our freedom of action being threatened we act to restore that freedom.

psychological stress—the behavioral and emotional components of the stress model.

public goods problem—a situation where all participants need to contribute to a common asset; if too many people fail to contribute, the common good suffers.

punishment—administering a noxious or painful stimulus to those who engage in environmentally destructive behavior.

random assignment—a critical element of the experimental method involving use of randomizing procedures (e.g., toss of a coin, use of a random number table) to determine which participants will be in which conditions of the study.

reactance—*see* psychological reactance.

reciprocal response—in personal space, a similar response to that shown by the other person, such as moving closer when the other moves closer, or turning farther away when the other turns away.

refractory period—a time interval during which an organism recovers from a stressful event.

reliable—a criterion indicating that a particular method of measuring a phenomenon is good; a reliable measure is one that gives the same reading each time, at least under the same conditions each time.

repression-sensitization—a continuum of ways of thinking about stressors; repressors tend to avoid and deny the existence of the stressor, while sensitizers tend to approach and respond emotionally to the stressor.

resource dilemma—a type of commons dilemma

in which individual greed for a resource conflicts with the needs of others who use it, such that use of the resource must be managed in order to avoid depleting it faster than it can be replenished.

resourcism—a perspective that natural resources should be conserved because of their value to humans.

Restricted Environmental Stimulation Technique (or Therapy) (REST)—a technique in which a level of sensory deprivation is used to modify behavior or relieve anxiety.

restorative environment—a setting such as a natural area or museum that can help people recover from directed attention fatigue.

reticular formation—a part of the brain that regulates arousal.

scientific management—a management philosophy that proposed that workers are primarily motivated by economic benefit, and that encouraged the analysis of work activities in order to increase efficiency.

screening—ignoring extraneous stimuli, or prioritizing demands.

Seasonal Affective Disorder (SAD)—a psychiatric diagnosis in which a depressive state seems to come and go with seasonal changes.

secondary appraisal—cognitive appraisal of coping strategies.

self-report measures—measures of some feeling, belief, attitude, recollection, etc. that rely on a person's report of how he or she feels, thinks, and so on; these may be oral reports, as in an interview, or they may be written, as with questionnaires.

self-trap—a type of social trap in which the momentary pleasures of the present have disastrous consequences in the long run, such as in addiction.

sensation—the relatively straightforward activity of human sensory systems in reacting to simple stimuli such as an individual flash of light.

sensory deprivation—a condition in which environmental stimulation is kept to a minimum.

sequential maps—maps that are predominantly listings or drawings of the ordered places a person might come upon in traveling from one place to another.

sick building syndrome—discomfort and symptoms of illness with no clear disease as a result of being in a specific building; the condition is thought to be due to a combination of individual sensitivity, actual toxic exposure, and worry over potential toxic exposure.

simulation or simulation methods—introduction of components of a real environment into an artificial laboratory setting in which control over assignment to conditions and experimental procedures can be maintained.

situational awareness (SA)—in human factors research, attending to the task at hand as opposed to being distracted or to trusting an automated system to perform the task properly.

situational conditions—the conditions that exist within a particular circumstance and that influence our actions.

sketch maps—paper-and-pencil drawings of the layout of an area from memory; use of sketch maps is a major method of studying cognitive maps.

social comparison—comparing one's assets, values, or actions with those of another.

social density—manipulations that vary group size while keeping area constant.

social dilemmas—situations where individual interests are pitted against group interests, such as in the commons dilemma.

social support—the feeling that one is cared about and valued by other people; the feeling that one belongs to a group.

social traps—Platt's notion that we get "trapped" into destructive behavior because we are motivated by short-term self-interest rather than long-term rewards (which may be shared by a group).

sociofugal—furniture arrangements, architectural designs, or social factors that discourage social interaction.

sociopetal—furniture arrangements, architectural designs, or social factors that encourage social interaction.

source reduction—prevention of pollution and waste disposal problems by avoiding production of the waste (e.g., by not overpackaging products for retail sale) or by using the waste for productive purposes (e.g., turning waste products into construction materials).

space surround environment—a museum exhibit design in which the visitor is entirely surrounded by elements of the exhibit, as opposed to the elements being in separate display cases.

spatial density—manipulations that vary area while keeping group size constant.

spatial maps—cognitive maps in which the environment has become coded in an organized and flexible representation that resembles a "bird's-eye" view.

staffing theory—in Barker's ecological psychology, the idea that performance and satisfaction are in part a function of the degree to which a behavior setting has too many or too few occupants.

stage of exhaustion—depletion of coping reserves when coping skills are pushed to the limit; the third stage of Selye's GAS.

stage of resistance—stress reaction seeking to restore balance; the second stage of Selye's GAS.

standing patterns of behavior—in Barker's ecological psychology, the social component of the behavior setting, which is interdependent with the physical milieu.

stress—a formulation that predicts that certain environmental conditions lead to a stress reaction, which may have emotional, behavioral, and physiological components; *see* psychological stress or environmental stress model.

structure matching—in you-are-here maps, the process of determining one's location by matching elements of the environment to those depicted in the map.

substantive theory—useful facts about the relationship between environmental variables such as construction techniques, color, or privacy and design goals.

survey knowledge—organization of a cognitive map spatially, or through a "bird's-eye" view.

sustainable—practices that do not overly deplete resources, such as farming techniques that do not deplete soil or water resources.

symbolic barriers—physical features that are not actual barriers to other people but which signify that an area is someone's territory, such as name and address signs, hedges, or rock borders.

synomorphic—in Barker's ecological psychology, the similar structure of the physical milieu and the standing patterns of behavior.

systemic stress—from Selye, the physiological component of the stress model.

Temperature-Humidity Index (THI)—an index that attempts to allow for the contribution of both temperature and humidity in accounting for discomfort, similar to effective temperature.

temporary threshold shifts (TTS)—hearing loss that lasts for a day or less, characterized by increases in threshold below which sounds are inaudible.

teratology—the study of the effects of chemical exposure (e.g., lead, alcohol) on the developing organism.

territoriality—a set of behaviors and cognitions an organism or group exhibits, based on perceived ownership of physical space; claim to and defense of a geographic area.

territories—relatively stationary areas, often with visible boundaries, that are at least temporarily "owned" by someone.

theory—a set of concepts and a set of relationships between the concepts.

theory of planned behavior—a theory derived from the work of Fishbein and Ajzen which proposes that attitudes plus subjective norms plus perceived control predict behavioral intentions, which in turn predict behavior; an earlier term for it was the theory of reasoned action.

theory of reasoned action—*see* theory of planned behavior.

thermoreceptors—nervous system sensory receptors that can detect changes in temperature.

threat appraisal—a cognitive appraisal component of the stress model that focuses on future dangers.

timbre—tonal quality; the purity of a sound, as determined by the number of frequencies in it.

tonal quality—timbre; the purity of a sound, as determined by the number of frequencies in it.

transactional approach—a method of inquiry that concentrates on patterns of relationships rather than on specific causes; views the environment as part of an event in time whose components are so intermeshed that no part is understandable without the simultaneous inclusion of all of the other parts.

transition—the degree to which there is direct access from a building to a street; transition facilitates spatial learning of a new environment.

understaffed—in ecological psychology, a condition of having too few participants.

understimulation—a condition of too few incoming stimuli, which leads to boredom.

unobtrusive measures—measures that can be collected without a person's awareness.

urban homesteading—a program in which (usually poor) individuals are given abandoned urban property if they agree to rehabilitate it to meet existing housing codes and occupy it for a prescribed period of time.

urban renewal—a series of integrated steps taken to maintain and upgrade the environmental, economic, or social health of an urban area.

urban villages—relatively homogenous, small social structures of neighborhoods or businesses.

valid—a criterion indicating that a particular method of measuring a phenomenon is good; a valid measure is one that actually measures what it claims to measure.

values—important beliefs and feelings that are broader than attitudes and not tied to a specific object; examples include preservationism and ecocentrism.

wayfinding—the process of using stored spatial information to plan and carry out movement in the environment.

weather—relatively short-term changes in atmospheric conditions.

Weber-Fechner function—a psychophysical principle that says the higher the magnitude of a stimulus (e.g., loudness of a sound), the greater the difference in magnitude the next higher stimulus needs to be in order for it to be detected as different.

white noise—a very wide range of unpatterned sound frequencies, such as would be found when tuning a television to a channel with no station on it.

wide band—a sound or noise with many frequencies in it.

windchill index—the effect of wind intensifying consequences of cold temperature.

windowless classrooms—classroom learning environments without windows, designed to reduce distraction and conserve energy.

workflow—the movement of information or work materials between workstations.

Yerkes–Dodson law—states that performance is maximal at intermediate levels of arousal and declines as arousal increases or decreases from this point.

you-are-here maps—maps displayed in public locations as orientation devices.

REFERENCES

Aakko, E., Remington, P., Dixon, J., & Ford L. (1999). Assessing smoke-free workplaces in Wisconsin municipal and county government buildings, 1997. *WMJ, 98(1)*, 38–41.

Abelson, P. H. (1993). Power from wind turbines. *Science, 261*, 1255.

Abraini, J. H., Martinez, E., Lemaire, C., Bisson, T., Juan de Mendoza, J.-L., & Therme, P. (1997). Anxiety, sensorimotor and cognitive performance during a hydrogen-oxygen dive and long-term confinement in a pressure chamber. *Journal of Environmental Psychology, 17*, 157–164.

Abramson, L. Y., Seligman, M. E. P., & Teasdale, J. D. (1978). Learned helplessness in humans: Critique and reformulation. *Journal of Abnormal Psychology, 87*, 49–74.

Abu-Ghazzeh, T. M. (1996). Movement and wayfinding in the King Sadu University built environment: A look at freshman orientation and environmental information. *Journal of Environmental Psychology, 16*, 303–318.

Abu-Ghazzeh, T. M. (1998). Housing layout, social interaction, and the place of contact in Abu-Nuseir, Jordan. *Journal of Environmental Psychology, 19*, 41–73.

Abu-Obeid, N. (1998). Abstract and scenographic imagery: The effect of environmental form on wayfinding. *Journal of Environmental Psychology, 18*, 159–173.

Acheson, J. M. (1975). The lobster fiefs: Economic and ecological effects of territoriality in the Maine lobster industry. *Human Ecology, 3*, 183–207.

Acking, D. A., & Kuller, R. (1972). The perception of an interior as a function of its color. *Ergonomics, 15*, 645–654.

Acton, W. I. (1970). Speech intelligibility in a background noise and noise-induced hearing loss. *Ergonomics, 13*, 546–554.

Adam, J. M. (1967). Military problems of air transport and tropical service. In C. N. Davies, P. R. Davis, & F. H. Tyrer (Eds.), *The effects of abnormal physical conditions at work* (pp. 74–80). London: E & S Livingstone.

Adams, J. R. (1973). Review of *Defensible space. Man-Environment Systems, 3*, 267–268.

Adams, L., & Zuckerman, D. (1991). The effect of lighting conditions on personal space requirements. *Journal of General Psychology, 118*, 335–340.

Adams, P. R., & Adams, G. R. (1984). Mount Saint Helens's ashfall: Evidence for a disaster stress reaction. *American Psychologist, 39*, 252–260.

Adler, A. (1943). Neuropsychiatric complications in victims of Boston's Coconut Grove disaster. *Journal of the American Medical Association, 17*, 1098–1101.

Ahrentzen, S. (1990). Managing conflict by managing boundaries: How professional homeworkers cope with multiple roles at home. *Environment and Behavior, 22*, 723–752.

Ahrentzen, S., & Evans, G. (1984). Distraction, privacy, and classroom design. *Environment and Behavior, 16*, 437–454.

Ahrentzen, S., Jue, G. M., Skorpanich, M. A., & Evans, G. W. (1982). School environments and stress. In G. W. Evans (Ed.), *Environmental stress* (pp. 224–255). New York: Cambridge University Press.

Ahrentzen, S., Levine, D. W., & Michelson, W. (1989). Space, time, and activity in the home: A gender analysis. *Journal of Environmental Psychology, 9*, 89–101.

Aiello, J. R. (1977). A further look at equilibrium theory: Visual interaction as a function of interpersonal distance. *Environmental Psychology and Nonverbal Behavior, 1*, 122–140.

Aiello, J. R. (1987). Human spatial behavior. In D. Stokols & I. Altman (Eds.), *Handbook of Environmental Psychology* (Vol. 1, pp. 505–531). New York: Wiley Interscience.

Aiello, J. R., & Aiello, T. (1974). The development of personal space: Proxemic behavior of children 6 through 16. *Human Ecology, 2*, 177–189.

Aiello, J. R., Baum, A., & Gormley, F. B. (1981a). Social determinants of residential crowding stress. *Personality and Social Psychology Bulletin, 7*, 643–649.

Aiello, J. R., DeRisi, D., Epstein, Y., & Karlin, R. (1977). Crowding and the role of interpersonal distance preference. *Sociometry, 40,* 271–282.

Aiello, J. R., Epstein, Y. M., & Karlin, R. A. (1975a). Effects of crowding on electrodermal activity. *Sociological Symposium, 14,* 43–57.

Aiello, J. R., Epstein, Y. M., & Karlin, R. A. (1975b). *Field experimental research in human crowding.* Paper presented at the meeting of the Eastern Psychological Association.

Aiello, J. R., Nicosia, G. J., & Thompson, D. E. (1979). Physiological, social, and behavioral consequences of crowding on children and adolescents. *Child Development, 50,* 195–202.

Aiello, J. R., & Thompson, D. E. (1980a). When compensation fails: Mediating effects of sex and locus of control at extended interaction distances. *Basic and Applied Social Psychology, 1,* 65–82.

Aiello, J. R., & Thompson, D. E. (1980b). Personal space, crowding, and spatial behavior in a cultural context. In I. Altman, J. F. Wohlwill, & A. Rapoport (Eds.), *Human behavior and environment* (Vol. 4, pp. 107–178). New York: Plenum.

Aiello, J. R., Thompson, D. E., & Baum, A. (1981b). The symbiotic relationship between social psychology and environmental psychology: Implications from crowding, personal space, and intimacy regulation research. In J. H. Harvey (Ed.), *Cognition and social behavior, and the environment* (pp. 423–438). Hillsdale, NJ: Erlbaum.

Aiello, J. R., Thompson, D. E., & Brodzinsky, D. M. (1983a). How funny is crowding anyway? Effects of room size, and the introduction of humor. *Basic and Applied Social Psychology, 4,* 193–207.

Aiello, J. R., Vautier, J. S., & Bernstein, M. D. (1983b, August). *Crowding stress: Impact of social support, group formation, and control.* Paper presented at annual meeting of the American Psychological Association, Anaheim, CA.

Ainsworth, R. A., Simpson, L., & Cassell, D. (1993). Effects of three colors in an office interior on mood and performance. *Perceptual and Motor Skills, 76,* 235–241.

Ajzen, I., & Fishbein, M. (1980). *Understanding attitudes and predicting social behavior.* Englewood Cliffs, NJ: Prentice-Hall.

Ajzen, I., & Madden, T. J. (1986). Prediction of goal-directed behavior: The role of intention, perceived. *Journal of Experimental Social Psychology, 22,* 453–474.

Akhtar, S. (1990). Concept of interpersonal distance in borderline personality disorder. *American Journal of Psychiatry, 147,* 260–261.

Albas, C. A. (1991). Proxemic behavior: A study of extrusion. *Journal of Social Psychology, 131,* 697–702.

Albas, D. C., & Albas, C. A. (1989). Meaning in context: The impact of eye contact and perception of threat on proximity. *Journal of Social Psychology, 129,* 525–531.

Albert, S., & Dabbs, J. M., Jr. (1970). Physical distance and persuasion. *Journal of Personality and Social Psychology, 15,* 265–270.

Aldwin, C., & Stokols, D. (1988). The effects of environmental change on individuals and groups: Some neglected issues in stress research. *Journal of Environmental Psychology, 8,* 57–75.

Aldwin, C. A., & Revenson, T. A. (1987). Does coping help? A reexamination of the relation between coping and mental health. *Journal of Personality and Social Psychology, 53,* 337–348.

Alexander, C. (1969). Major changes in environmental form required by social and psychological demands. *Ekistics, 28,* 78–85.

Alexander, C. (1972). The city as a mechanism for sustaining human contact. In R. Gutman (Ed.), *People and buildings* (pp. 406–434). New York: Basic Books.

Alexander, C. (1979). *The timeless way of building.* New York: Oxford University Press.

Alexander, C., Ishikawa, S., & Silverstein, M. (1977). *A pattern language.* New York: Oxford University Press.

Alexander, C., Silverstein, M., Angel, S., Ishikawa, S., & Abrams, D. (1975). *The Oregon experiment.* New York: Oxford University Press.

Allen, G. L. (1981). A developmental perspective on the effects of "subdividing" macrospatial experience. *Journal of Experimental Psychology: Human Learning and Memory, 7,* 120–132.

Allen, G. L., & Kirasic, K. C. (1985). Effects of the cognitive organization of knowledge on judgments of macrospatial distance. *Memory and Cognition, 13,* 218–227.

Allen, I. L. (1980). The ideology of dense neighborhood redevelopment: Cultural diversity and transcendent community experience. *Urban Affairs Quarterly, 15*, 409–428.

Allen, J. B., & Ferrand, J. L. (1999). Environmental locus of control, sympathy, and proenvironmental behavior: A test of Geller's actively caring hypothesis. *Environment and Behavior, 31*, 338–353.

Allen, V. L., & Greenberger, D. B. (1980). Destruction and perceived control. In A. Baum & J. E. Singer (Eds.), *Advances in environmental psychology* (Vol. 2, pp. 85–110). Hillsdale, NJ: Erlbaum.

Allgeier, A. R., & Byrne, D. (1973). Attraction toward the opposite sex as a determinant of physical proximity. *Journal of Social Psychology, 90*, 231–219.

Alloy, L. B., Peterson, C., Abramson, L.Y., & Seligman, M. E. P. (1984). Attributional style and the generality of learned helplessness. *Journal of Personality and Social Psychology, 46*, 681–687.

Altman, I. (1973). Some perspectives on the study of man–environment phenomena. *Representative Research in Social Psychology, 4*, 109–126.

Altman, I. (1975). *The environment and social behavior.* Monterey, CA: Brooks/Cole.

Altman, I., & Chemers, M. (1980). *Culture and environment.* Monterey, CA: Brooks/Cole.

Altman, I., Lawton, M. P., & Wohlwill, J. F. (Eds.) (1984). *Elderly people and the environment.* New York: Plenum.

Altman, I., & Low, S. M. (Eds.) (1992). *Place attachment. Human Behavior and Environment: Advances in Theory and Research* (Vol. 12). New York: Plenum.

Altman, I., Nelson, P. A., & Lett, E. E. (1972, Spring). The ecology of home environments. *Catalog of Selected Documents in Psychology* (No. 150).

Altman, I., & Rogoff, B. (1987). World views in psychology: Trait, interactional, organismic, and transactional perspectives. In I. Altman & D. Stokols (Eds.), *Handbook of environmental psychology* (Vol. I, pp. 7–40). New York: Wiley-Interscience.

Altman, I., & Vinsel, A. M. (1977). Personal space: An analysis of E. T. Hall's proxemics framework. In I. Altman & J. F. Wohlwill (Eds.), *Human behavior and environment: Ad-*

vances in theory and research (Vol. 1, pp. 181–259). New York: Plenum.

Altman, I., & Wandersman, A. (Eds.). (1987). *Neighborhood and community environments.* New York: Plenum.

Altman, I., & Werner, C. M. (Eds.). (1985). *Home environments.* New York: Plenum.

Amato, P. R. (1981). The impact of the built environment on prosocial and affiliative behaviour: A field study of the Townsville city mall. *Australian Journal of Psychology, 33*, 297–303.

Amato, P. R. (1983). Helping behavior in urban and rural environments: Field studies based on taxonomic organization of helping episodes. *Journal of Personality and Social Psychology, 45*, 571–586.

Amerigo, M., & Aragones, J. I. (1997). A theoretical and methodological approach to the study of residential satisfaction. *Journal of Environmental Psychology, 17*, 47–57.

Andersen, P. A., & Andersen, J. F. (1984). The exchange of nonverbal intimacy: A critical review of dyadic models. *Journal of Nonverbal Behavior, 8*, 327–349.

Anderson, B., Erwin, N., Flynn, D., Lewis, L., & Erwin, J. (1977). Effects of short-term crowding on aggression in captive groups of pigtail monkey. *Aggressive Behavior, 3*, 33–46.

Anderson, C. A. (1987). Temperature and aggression: Effects on quarterly, yearly, and city rates of violent and nonviolent crime. *Journal of Personality and Social Psychology, 52*, 1161–1173.

Anderson, C. A. (1989). Temperature and aggression: Ubiquitous effects of heat on occurrence of human violence. *Psychological Bulletin, 106*, 74–96.

Anderson, C. A., & Anderson, D. C. (1984). Ambient temperature and violent crime: Tests of the linear and curvilinear hypotheses. *Journal of Personality and Social Psychology, 46*, 91–97.

Anderson, C. A., & Anderson, K. B. (1996). Violent crime rate studies in philosophical context: A destructive testing approach to heat and Southern culture of violence effects. *Journal of Personality and Social Psychology, 70*, 740–756.

Anderson, C. A., & Anderson, K. B. (1998). Temperature and aggression: Paradox, controversy, and a (fairly) clear picture. In R. Geen & E. Donnerstein (Eds.), *Human aggression: Theories, research, and implications for social policy* (pp. 240–298). San Diego, CA: Academic Press.

Anderson, C. A., Anderson, K. B., & Deuser, W. E. (1996). Examining an affective aggression framework: Weapon and temperature effects on aggressive thoughts, affect, and attitudes. *Personality and Social Psychology Bulletin, 22,* 366–376.

Anderson, C. A., Bushman, B. J., & Groom, R. W. (1997). Hot years and serious and deadly assault: Empirical tests of the heat hypothesis. *Journal of Personality and Social Psychology, 73,* 1213–1223.

Anderson, C. A., & DeNeve, K. M. (1992). Temperature, aggression, and negative affect escape model. *Psychological Bulletin, 111,* 347–351.

Anderson, C. A., Deuser, W. E., & DeNeve, K. M. (1995). Hot temperatures, hostile affect, hostile cognition, and arousal: Tests of a general model of affective aggression. *Personality and Social Psychology Bulletin, 21,* 434–448.

Anderson, T. L., & Leal, D. R. (1997). *Envirocapitalists: Doing good while doing well.* Lanham, MD: Rowman & Littlefield.

Ando, Y. (1987). Effects of daily dose on the fetus and cerebral hemispheric specialization of children. *Inta-Noise '87, 2,* 941–944. Beijing, China.

Ando, Y., & Hattori, H. (1973). Statistical studies in the effects of intense noise during human fetal life. *Journal of Sound and Vibration, 27,* 101–110.

Anthony, J. L., Lonigan, C. J., & Hecht, S. A. (1999). Dimensionality of posttraumatic stress disorder symptoms in children exposed to disaster: Results from confirmatory factor analyses. *Journal of Abnormal Psychology, 108,* 326–226.

Anthony, K. H., Weidemann, S., & Chin, Y. (1990). Housing perceptions of low-income single parents. *Environment and Behavior, 22,* 147–182.

Appleton, J. (1975). *The experience of landscape.* London: Wiley.

Appleyard, D. (1970). Styles and methods of structuring a city. *Environment and Behavior, 2,* 101–117.

Appleyard, D. (1976). *Planning a pluralistic city.* Cambridge, MA: M.I.T. Press.

Appleyard, D., & Lintell, M. (1972). The environmental quality of city streets: The residents' viewpoint. *Journal of the American Institute of Planners, 38,* 84–101.

Aptekar, L., & Boore, J. A. (1990). The emotional effects of disaster on children: A review of the literature. *International Journal of Mental Health, 19,* 77–90.

Aragones, J. I., & Arredondo, J. M. (1985). Structure of urban cognitive maps. *Journal of Environmental Psychology, 5,* 197–212.

Arbuckle, T. Y., Cooney, R., Milne, J., & Melchior, A. (1994). Memory for spatial layouts in relation to age and schema typicality. *Psychology and Aging, 9,* 467–480.

Arbuthnot, J., Tedeschi, R., Wayner, M., Turner, J., Kressel, S., & Rush, R. (1976–1977). The induction of sustained recycle behavior through the foot-in-the-door technique. *Journal of Environmental Systems, 6,* 355–358.

Archer, J. (1970). Effects of population density on behavior in rodents. In J. H. Crook (Ed.), *Social behavior in birds and mammals* (pp. 169–210). New York: Academic Press.

Ardrey, R. (1966). *The territorial imperative.* New York: Atheneum.

Argyle, M., & Dean, J. (1965). Eye-contact, distance and affiliation. *Sociometry, 28,* 289–304.

Arkkelin, D. (1978). *Effects of density, sex, and acquaintance level on reported pleasure, arousal, and dominance.* Doctoral dissertation, Bowling Green State University.

Arnstein, S. R. (1969). A ladder of citizen participation. *Journal of American Institute of Planners, 35,* 217.

Arnsten, A. F. T. (1998). The biology of being frazzled. *Science, 280,* 1711–1712.

Arp III, W., & Kenny, C. (1996). Black environmentalism in the local community context. *Environment and Behavior, 28,* 267–282.

Arreola, D. D. (1981). Fences as landscape taste: Tucson's barrios. *Journal of Cultural Geography, 2,* 96–105.

Arvey, R. D., Bouchard, T. J., Jr., Segal, N., & Abraham, L. M. (1989). Job satisfaction: Genetic and environmental components. *Journal of Applied Psychology, 74,* 187–192.

Ashley, M. J., Cohen, J., Ferrence, R., Bull, S., Bondy, S., Poland, B., & Pederson, L. (1998). Smoking in the home: Changing attitudes and current practices. *American Journal of Public Health, 88,* 797–800.

Ashley, M. J., & Ferrence, R. (1998). Reducing children's exposure to environmental tobacco smoke in homes: Issues and strategies. *Tobacco Control, 7(1),* 61–65.

Asmus, C. L. (1998). Sources of apathy and predictors of annoyance to livestock odor in a rural community. *Dissertation Abstracts International, 60-03B,* 1333.

Asmus, C. L., & Bell, P. A. (1999). Effects of environmental odor and coping style on negative affect, anger, arousal, and escape. *Journal of Applied Social Psychology, 29,* 245–260.

Atlas, R., & LeBlanc, W. G. (1994, October). Environmental barriers to crime. *Ergonomics in Design,* 9–16.

Averill, J. R. (1973). Personal control over aversive stimuli and its relationship to stress. *Psychological Bulletin, 80,* 286–303.

Axelrod, L. J., & Lehman, D. R. (1993). Responding to environmental concerns: What factors guide individual action? *Journal of Environmental Psychology, 13,* 149–159.

Azuma, H. (1984). Secondary control as a heterogeneous category. *American Psychologist, 39,* 970–971.

Bachman, W., & Katzev, R. (1982). The effects of non-contingent fee bus tickets and personal commitment on urban bus ridership. *Transportation Research, 16A,* 103–108.

Bacon-Prue, A., Blount, R., Pickering, D., & Drabman, R. (1980). An evaluation of three litter control procedures — Trash receptacles, paid workers, and the marked item technique. *Journal of Applied Behavior Analysis, 13,* 165–170.

Bagley, C. (1989). Urban crowding and the murder rate in Bombay, India. *Perceptual and Motor Skills, 69,* 1241–1242.

Bahrick, L. E., Parker, J. F., Fivush, R., & Levitt, M. (1998). The effects of stress on young children's memory for a natural disaster. *Journal of Experimental Psychology: Applied 4,* 308–331.

Bailenson, J. N., Shum, M. S., & Uttal, D. H. (1998). Road climbing: Principles governing asymmetric route choices on maps. *Journal of Environmental Psychology, 18,* 251–264.

Baird, C. L., & Bell, P. A. (1995). Place attachment, isolation, and the power of a window in a hospital environment. *Psychological Reports, 76,* 847–850.

Baird, L. L. (1969). Big school, small school: A critical examination of the hypothesis. *Journal of Educational Psychology, 60,* 253–260.

Baker, M. A., & Holding, D. H. (1993). The effects of noise and speech on cognitive task performance. *Journal of General Psychology, 120,* 339–355.

Balling, J. D., & Falk, J. H. (1982). Development of visual preference for natural environments. *Environment and Behavior, 14,* 5–28.

Balling, R. C., Jr., & Cerveny, R. S. (1995). Influence of lunar phase on daily global temperatures. *Science, 267,* 1481–1483.

Balogun, S. K. (1991). Personal space as affected by religions of the approaching and the approached people. *Indian Journal of Behaviour, 15,* 45–50.

Baltes, M. M., Kindermann, T., Reisenzein, R., & Schmid, U. (1987). Further observational data on the behavioral and social world of institutions for the aged. *Psychology and Aging, 2,* 390–403.

Banbury, S., & Berry, D. C. (1998). Disruption of office-related tasks by speech and office noise. *British Journal of Psychology, 89,* 499–517.

Bandura, A. (1977). *Social learning theory.* Englewood Cliffs, NJ: Prentice-Hall.

Banzinger, G., & Owens, K. (1978). Geophysical variables and behavior: II. Weather factors as predictors of local social indicators of maladaptation in two non-urban areas. *Psychological Reports, 43,* 427–434.

Barabasz, A., & Barabasz, M. (1985). Effects of restricted environmental stimulation: Skin conductance, EEG alpha, and temperature responses. *Environment and Behavior, 17,* 239–253.

Barabasz, A., & Barabasz, M. (1986). Antarctic isolation and inversion perception: Regression phenomena. *Environment and Behavior, 18,* 285–292.

Barabasz, A. F., & Barabasz, M. (1993). *Clinical and experimental restricted environmental stimulation: New developments and perspectives.* New York: Springer.

Barash, D. P. (1973). Human ethology: Personal space reiterated. *Environment and Behavior, 5,* 67–73.

Barber, N. (1990). Home color as a territorial

marker. *Perceptual and Motor Skills, 71,* 1107–1110.

Barefoot, J. C., Hoople, H., & McClay, D. (1972). Avoidance of an act which would violate personal space. *Psychonomic Science, 28,* 205–206.

Barker, R. G. (1960). Ecology and motivation. In M. R. Jones (Ed.), *Nebraska Symposium on Motivation* (Vol. 8, pp. 1–50). Lincoln: University of Nebraska Press.

Barker, R. G. (1968). *Ecological psychology: Concepts and methods for studying the environment of human behavior.* Stanford, CA: Stanford University Press.

Barker, R. G. (1979). Settings of a professional lifetime. *Journal of Personality and Social Psychology, 37,* 2137–2157.

Barker, R. G. (1987). Prospecting in environmental psychology: Oskaloosa revisited. In D. Stokols & I. Altman (Eds.), *Handbook of environmental psychology* (Vol. II, pp. 1413–1432). New York: Wiley-Interscience.

Barker, R. G. (1990). Recollections of the Midwest Psychological Field Station. *Environment and Behavior, 22,* 503–513.

Barker, R. G., & Gump, P. V. (1964). *Big school, small school.* Stanford, CA: Stanford University Press.

Barker, R. G., & Schoggen, P. (1973). *Qualities of community life.* San Francisco: Jossey-Bass.

Barker, R. G., & Wright, H. F. (1951). *One boy's day.* New York: Row, Peterson.

Barker, R. G., & Wright, H. F. (1955). *Midwest and its children.* New York: Row, Peterson.

Barnard, W. A., & Bell, P. A. (1982). An unobtrusive apparatus for measuring interpersonal distance. *Journal of General Psychology, 107,* 85–90.

Barnett, M. A., Quackenbush, S. W., & Pierce, L. K. (1997). Perception of and reactions to the homeless: A survey of fourth-grade, high school, and college students in a small Midwestern community. *Journal of Social Distress & the Homeless, 6,* 283–302.

Baron, J. (1997). The illusion of morality as self-interest: A reason to cooperate in social dilemmas. *Psychological Science, 8,* 330–335.

Baron, R. A. (1976). The reduction of human aggression: A field study of the influence of incompatible reactions. *Journal of Applied Social Psychology, 6,* 260–274.

Baron, R. A. (1987a). Effects of negative ions on cognitive performance. *Journal of Applied Psychology, 72,* 131–137.

Baron, R. A. (1987b). Effects of negative ions on interpersonal attraction: Evidence for intensification. *Journal of Personality and Social Psychology, 52,* 547–553.

Baron, R. A., & Bell, P. A. (1975). Aggression and heat: Mediating effects of prior provocation and exposure to an aggressive model. *Journal of Personality and Social Psychology, 31,* 825–832.

Baron, R. A., & Bell, P. A. (1976). Physical distance and helping: Some unexpected benefits of "crowding in" on others. *Journal of Applied Social Psychology, 6,* 95–104.

Baron, R. A., & Byrne, D. (2000). *Social psychology* (9th ed.). Boston: Allyn and Bacon.

Baron, R. A., Russell, G. W., & Arms, R. L. (1985). Negative ions and behavior: Impact on mood, memory, and aggression among Type A and Type B persons. *Journal of Personality and Social Psychology, 48,* 746–754.

Baron, R. M., & Fisher, J. D. (1984). The equity-control model of vandalism: A refinement. In C. Levy-Leboyer (Ed.), *Vandalism: Behavior and motivations* (pp. 63–75). Amsterdam: North Holland.

Baron, R. M., & Kenny, D. A. (1986). The moderator-mediator variable distinction in social psychological research: Conceptual, strategic, and statistical considerations. *Journal of Personality and Social Psychology, 51,* 1173–1182.

Baron, R. M., Mandel, D. R., Adams, C. A., & Griffen, L. M. (1976). Effects of social density in university residential environments. *Journal of Personality and Social Psychology, 34,* 434–446.

Baron, R. M., & Rodin, J. (1978). Personal Control as a mediator of crowding. In A. Baum, J. E. Singer, & S. Valins (Eds.), *Advances in environmental psychology* (Vol. 1, pp. 145–181). Hillsdale, NJ: Erlbaum.

Barrios, B. A., Corbitt, L. C., Estes, J. P., & Topping, J. S. (1976). Effect of social stigma on interpersonal distance. *The Psychological Record, 26,* 342–348.

Bassuk, E., Rubin, L., & Lauriat, A. (1986). Characteristics of sheltered homeless families. *American Journal of Public Health, 76,* 1097–1101.

Bauer, R. M., Greve, K. W., Besch, E. L., Schramke,

C. J., Crouch, J., Hicks, A., Ware, M. R., & Lyles, W. B. (1992). The role of psychological factors in the report of building-related symptoms in Sick Building Syndrome. *Journal of Consulting and Clinical Psychology, 60*, 213–219.

Baum, A. (1987). Toxins, technology, and natural disaster. In G. R. VandenBos & B. K. Bryant (Eds.), *Cataclysms, crises, and catastrophes: Psychology in action* (pp. 9–53). Washington, DC: American Psychological Association.

Baum, A., Aiello, J., & Calesnick, L. E. (1978). Crowding and personal control: Social density and the development of learned helplessness. *Journal of Personality and Social Psychology, 36*, 1000–1011.

Baum, A., Calesnick, L. E., Davis, G. E., & Gatchel, R. J. (1982). Individual differences in coping with crowding: Stimulus screening and social overload. *Journal of Personality and Social Psychology, 43*, 821–830.

Baum, A., Cohen, L., & Hall, M. (1993). Control and intrusive memories as possible determinants of chronic stress. *Psychosomatic Medicine, 55*, 274–286.

Baum, A., & Davis, G. E. (1976). Spatial and social aspects of crowding perception. *Environment and Behavior, 8*, 527–545.

Baum, A., & Davis, G. E. (1980). Reducing the stress of high-density living: An architectural intervention. *Journal of Personality and Social Psychology, 38*, 471–481.

Baum, A., & Fisher, J. D. (1977). *Situation-related information as a mediator of responses to crowding.* Unpublished manuscript, Trinity College.

Baum, A., Fisher, J. D., & Solomon, S. (1981). Type of information, familiarity, and the reduction of crowding stress. *Journal of Personality and Social Psychology, 40*, 11–23.

Baum, A., & Fleming, I. (1993). Implications of psychological research on stress and technological accidents. *American Psychologist, 48*, 665–672.

Baum, A., Fleming, I., Israel, A., & O'Keeffe, M. K. (1992). Symptoms of chronic stress following a natural disaster and discovery of a human-made hazard. *Environment and Behavior, 24*, 347–365.

Baum, A., Fleming, R., & Davidson, L. M. (1983). Natural disaster and technological catastrophe. *Environment and Behavior, 15*, 333–354.

Baum, A., Garofalo, J. P., & Yali, A. M. (1999). Socioeconomic status and chronic stress: Does stress account for SES effects on health? *Annals of the New York Academy of Sciences, 1999*, 1–13.

Baum, A., & Gatchel, R. J. (1981). Cognitive determinants of response to uncontrollable events: Development of reactance and learned helplessness. *Journal of Personality and Social Psychology, 40*, 1078–1089.

Baum, A., & Greenberg, C. I. (1975). Waiting for a crowd: The behavioral and perceptual effects of anticipated crowding. *Journal of Personality and Social Psychology, 32*, 667–671.

Baum, A., Grunberg, N. E., & Singer, J. E. (1982). The use of physiological and neuroendocrinological measurements in the study of stress. *Health Psychology, 1*, 217–236.

Baum, A., & Koman, S. (1976). Differential response to anticipated crowding: Psychological effects of social and spatial density. *Journal of Personality and Social Psychology, 34*, 526–536.

Baum, A., & Paulus, P. B. (1987). Crowding. In D. Stokols & I. Altman (Eds.), *Handbook of environmental psychology* (Vol. I, pp. 533–570). New York: Wiley-Interscience.

Baum, A., Reiss, M., & O'Hara, J. (1974). Architectural variants of reaction to spatial invasion. *Environment and Behavior, 6*, 91–100.

Baum, A., Shapiro, A., Murray, D., & Wideman, M. (1979). Mediation of perceived crowding and control in residential dyads and triads. *Journal of Applied Social Psychology, 9*, 491–507.

Baum, A., Singer, J. E., & Baum, C. S. (1981). Stress and the environment. *Journal of Social Issues, 37*, 4–35.

Baum, A., & Valins, S. (1977). *Architecture and social behavior: Psychological studies of social density.* Hillsdale, NJ: Erlbaum.

Baum, A., & Valins, S. (1979). Architectural mediation of residential density and control: Crowding and the regulation of social contact. In L. Berkowitz (Ed.), *Advances in experimental social psychology* (Vol. 12, pp. 131–175). New York: Academic Press.

Baum, A. S., & Burnes, D. W. (1993, Spring). Facing the facts about homelessness. *Public Welfare, 20-27*, 46.

Baumeister, R. F. (1985). The championship choke. *Psychology Today, 19(4)*, 48–52.

Baumeister, R. F., & Steinhilber, A. (1984). Paradoxical effects of supportive audiences on performance under pressure: The home field disadvantage in sports championships. *Journal of Personality and Social Psychology, 47,* 85–93.

Baylis, G. C., & Driver, J. (1995). One-sided edge assignment in vision: 1. Figure-ground segmentation and attention to objects. *Current Directions in Psychological Science, 4,* 140–146.

Beal, J. B. (1974). Electrostatic fields, electromagnetic fields and ions—Mind/body/environment interrelationships. In J. G. Llaurado, A. Sances, & J. H. Battocletti (Eds.), *Biologic and clinical effects of low-frequency magnetic and electric fields* (pp. 5–20). Springfield, IL: Thomas.

Beard, R. R., & Wertheim, G.A. (1967). Behavioral impairment associated with small doses of carbon monoxide. *American Journal of Public Health, 57,* 2012–2022.

Bechtel, R. B. (1977). *Enclosing behavior.* Stroudsberg, PA: Dowden, Hutchinson, & Ross.

Bechtel, R. B. (1997). *Environment & behavior: An introduction.* Thousand Oaks, CA: Sage.

Beck, R. J., & Wood, D. (1976). Cognitive transformation of information from urban geographic fields to mental maps. *Environment and Behavior, 8,* 199–238.

Becker, F. D. (1973). Study of spatial markers. *Journal of Personality and Social Psychology, 26,* 439–445.

Becker, F. D. (1981). *Workspace: Creating environments in organizations.* New York: CBS Educational and Professional Publishing.

Becker, F. D., & Coniglio, C. (1975). Environmental messages: Personalization and territory. *Humanities, 11,* 55–74.

Becker, F. D., Gield, B., Gaylin, K., & Sayer, S. (1983). Office design in a community college: Effect of work and communication patterns. *Environment and Behavior, 15,* 699–726.

Becker, F. D., & Mayo, C. (1971). Delineating personal space and territoriality. *Environment and Behavior, 3,* 375–381.

Becker, F. D., Sommer, R., Bee, J., & Oxley, B. (1973). College classroom ecology. *Sociometry, 36,* 514–525.

Becker, L. J. (1978). The joint effect of feedback and goal setting on performance: A field study of residential energy conservation. *Journal of Applied Psychology, 63,* 228–233.

Becker, L. J., & Seligman, C. (1978). Reducing air-conditioning waste by signaling it is cool outside. *Personality and Social Psychology Bulletin, 4,* 412–415.

Beighton, P. (1971). Fluid balance in the Saraha. *Nature, 233,* 275–277.

Belk, R. W. (1992). Attachment to possessions. In I. Altman & S. M. Low (Eds.), *Place Attachment* (pp. 37–62). New York: Plenum.

Bell, B., Kara, G., & Batterson, C. (1978). Service utilization and adjustment patterns of elderly tornado victims in an American disaster. *Mass Emergencies, 3,* 71–81.

Bell, I. R., Baldwin, C. M., Russek, L. G., Schwartz, G. E., & Hardin, E. E. (1998). Early life stress, negative paternal relationships, and chemical intolerance in middle-aged women: support for a neural sensitization model. *Journal of Women's Health, 7,* 1135–1147.

Bell, P. A. (1978). Effects of heat and noise stress on primary and subsidiary task performance. *Human Factors, 20,* 749–752.

Bell, P. A. (1992). In defense of the negative affect escape model of heat and aggression. *Psychological Bulletin, 111,* 342–346.

Bell, P. A., & Baron, R. A. (1974). Environmental influences on attraction: Effects of heat, attitude similarity, and personal evaluations. *Bulletin of the Psychonomic Society, 4,* 479–481.

Bell, P. A., & Baron, R. A. (1976). Aggression and heat: The mediating role of negative affect. *Journal of Applied Social Psychology, 6,* 18–30.

Bell, P. A., & Baron, R. A. (1977). Aggression and ambient temperature: The facilitating and inhibiting effects of hot and cold environments. *Bulletin of the Psychonomic Society, 9,* 443–445.

Bell, P. A., & Byrne, D. (1978). Repression-sensitization. In H. London & J. Exner (Eds.), *Dimensions of personality* (pp. 449–485). New York: Wiley.

Bell, P. A., & Doyle, D. P. (1983). Effects of heat and noise on helping behavior. *Psychological Reports, 53,* 955–959.

Bell, P. A., & Fusco, M. E. (1986). Linear and curvilinear relationships between temperature, affect, and violence: Reply to Cotton. *Jour-*

nal of Applied Social Psychology, 16, 802–807.

Bell, P. A., & Fusco, M. E. (1989). Heat and violence in the Dallas field data: Linearity, curvilinearity, and heteroscedasticity. *Journal of Applied Social Psychology, 19,* 1479–1482.

Bell, P. A., Garnard, D. B., & Heath, D. (1984). Effects of ambient temperature and seating arrangement on personal and evironmental evaluations. *Journal of General Psychology, 110,* 197–200.

Bell, P. A., & Greene, T. C. (1982). Thermal stress: Physiological comfort, performance, and social effects of hot and cold environments. In G. W. Evans (Ed.), *Environmental stress* (pp. 75–105). London: Cambridge University Press.

Bell, P. A., Hess, S., Hill, E., Kukas, S., Richards, R. W., & Sargent, D. (1984). Noise and context-dependent memory. *Bulletin of the Psychonomic Society, 22,* 99–100.

Bell, P. A., Kline, L. M., & Barnard, W. A. (1988). Friendship and freedom of movement as moderators of sex differences in interpersonal distancing. *Journal of Social Psychology, 128,* 305–310.

Bell, P. A., Malm, W., Loomis, R. J., & McGlothin, G. E. (1985). The impact of impaired visibility on visitor enjoyment of the Grand Canyon: A test of an ordered logit utility model. *Environment and Behavior, 17,* 459–474.

Bell, P. A., Petersen, T. R., & Hautaluoma, J. E. (1989). The effect of punishment probability on overconsumption and stealing in a simulated commons. *Journal of Applied Social Psychology, 19,* 1483–1495.

Bell, P. A., & Smith, J. M. (1997). A behavior mapping method for assessing efficacy of change on special care units. *American Journal of Alzheimer's Disease, 12,* 184–189.

Belter, R. W., Foster, K. Y., Imm, P. S., & Finch, A. J., Jr. (1991, April). Parent vs. child reports of PTSD symptoms related to a catastrophic natural disaster. In J. M. Vogel (Chair), *Children's responses to natural disasters: The aftermath of Hurricane Hugo and the 1989 Bay Area earthquake.* Symposium conducted at the meeting of the Society for Research in Child Development, Seattle.

Bem, D. J. (1971). *Beliefs, attitudes, and human affairs.* Belmont, CA: Brooks/Cole.

Bem, D. J. (1972). Self perception theory. In L. Berkowitz (Ed.), *Advances in experimental social psychology* (Vol. 6, pp. 1–62). New York: Academic Press.

Bennett, N., Andreae, J., Hegarty, P., & Wade, B. (1980). *Open plan schools.* Atlantic Highlands, NJ: Humanities.

Bennett, R., Rafferty, J. M., Canivez, G. L., & Smith, J. M. (1983, May). *The effects of cold temperature on altruism and aggression.* Paper presented at the meeting of the Midwestern Psychological Association, Chicago, IL.

Benson, G. P., & Zieman, G. L. (1981). *The relationship of weather to children's behavior problems.* Unpublished manuscript, Colorado State University, Fort Collins.

Berck, J. (1992). No place to be: Voices of homeless children. *Public Welfare,* 28–33.

Berger, I. E. (1997). The demographics of recycling and the structure of environmental behavior. *Environment and Behavior, 29,* 515–531.

Berglund, B., Berglund, U., & Lindvall, T. (1987). Measurement and control of annoyance. In H.S. Koelega (Ed.), *Environmental annoyance: Characterization, measurement, and control* (pp. 29–44). New York: Elsevier.

Berk, L. E., & Goebel, B. L. (1987). High school size and extracurricular participation: A study of a small college environment. *Environment and Behavior, 19,* 53–76.

Berkowitz, L. (1970). The contagion of violence: An S-R mediational analysis of some effects of observed aggression. In W. J. Arnold & M. M. Page (Eds.), *Nebraska Symposium on Motivation* (Vol. 18, pp. 95–135). Lincoln: University of Nebraska Press.

Berkowitz, L. (1993). *Aggression: Its causes, consequences, and control.* New York: McGraw Hill.

Berlyne, D. E. (1960). *Conflict, arousal, and curiosity.* New York: McGraw-Hill.

Berlyne, D. E. (1974). *Studies in the new experimental aesthetics: Steps toward an objective psychology of aesthetic appreciation.* New York: Halsted Press.

Bernaldez, F. G., Gallardo, D., & Abello, R. P. (1987). Children's landscape preferences: From rejection to attraction. *Journal of Environmental Psychology, 7,* 169–176.

Bernard, L. C., & Krupat, E. (1994). *Health psychology.* Orlando: Harcourt Brace.

Berry, P. C. (1961). Effects of colored illumination

upon perceived temperature. *Journal of Applied Psychology, 45,* 248–250.

Berscheid, E., & Reis, H. (1998). Attraction and close relationships. In D. T. Gilbert, S. Fiske, & G. Lindzey (Eds.), *The handbook of social psychology* (Vol. I, 4th ed., pp. 193–281). New York: McGraw Hill.

Best, J. B. (1998). *Cognitive psychology* (5th ed.). Belmont: Wadsworth.

Bickman, L., Teger, A., Gabiele, T., McLaughin, C., Berger, M., & Sunaday, E. (1973). Dormitory density and helping behavior. *Environment and Behavior, 5,* 465–490.

Bigness, J. (1995, July 10). As auto companies put more plastic in their cars, recyclers can recycle less. *Wall Street Journal,* pp. B1, B4.

Bih, H. (1992). The meaning of objects in environmental transitions: Experiences of Chinese students in the United States. *Journal of Environmental Psychology, 12,* 135–147.

Bing-shuang, H., Yue-lin, Y., Ren-yi, W., & Zhu-bao, C. (1997). Evaluation of depressive symptoms in workers exposed to industrial noise. *Homeostasis in Health & Disease, 38,* 123–125.

Binney, S. E., Mason, R., Martsolf, S. W., & Detweiler, J. H. (1996). Credibility, public trust, and the transport of radioactive waste through local communities. *Environment and Behavior, 28,* 283–301.

Birjulin, A. A., Smith, J. M., & Bell, P. A. (1993). Monetary reward, verbal reinforcement, and harvest strategy of others in the commons dilemma. *Journal of Social Psychology, 133,* 207–214.

Birren, J. E., & Schaie, K. W. (1996). *Handbook of the psychology of aging* (4th ed.). San Diego: Academic Press.

Bitgood, S. C., & Loomis, R. J. (1993). Introduction: Environmental design and evaluation in museums. *Environment and Behavior, 25,* 683–697.

Bitgood, S. C., & Patterson, D. D. (1993). The effects of gallery changes on visitor reading and object viewing time. *Environment and Behavior, 25,* 761–781.

Bixler, R. D., & Floyd, M. F. (1997). Nature is scary, disgusting, and uncomfortable. *Environment and Behavior, 29,* 443–467.

Bjornsson, E., Janson, C., Norback, D., & Boman, G. (1998). Symptoms related to the sick building syndrome in a general population sample: Associations with atopy, bronchial hyper-responsiveness and anxiety. *International Journal of Tuberculosis & Lung Disease, 2,* 1023–1028.

Blackman, S., & Catalina, D. (1973). The moon and the emergency room. *Perceptual and Motor Skills, 37,* 624–626.

Blades, M. (1990). The reliability of data collected from sketch maps. *Journal of Environmental Psychology, 10,* 327–339.

Blades, M., & Spencer, C. (1987). Young children's strategies when using maps with landmarks. *Journal of Environmental Psychology, 7,* 201–217.

Blakely, K., & Snyder, M. G. (1995). *Fortress America: Gated and walled communities in the United States.* Cambridge, MA: Lincoln Institute of Land Policy.

Bleda, P. R., & Bleda, S. (1978). Effects of sex and smoking on reactions to spatial invasion at a shopping mall. *Journal of Social Psychology, 104,* 311–312.

Bleda, P. R., & Sandman, P. H. (1977). In smoke's way: Socioemotional reactions to another's smoking. *Journal of Applied Psychology, 62,* 452–458.

Block, L. K., & Stokes, G. S. (1989). Performance and satisfaction in private versus nonprivate work settings. *Environment and Behavior, 21,* 277–297.

Bolin, R. (1985). Disaster characteristics and psychosocial impacts. In B. J. Sowder (Ed.), *Disasters and mental health: Selected contemporary perspectives* (pp. 3–28). Rockville, MD: U.S. Department of Health and Human Services.

Bolin, R., & Klenow, J. D. (1982). Response of the elderly to disaster: An age-stratified analysis. *International Journal of Aging and Human Development, 16,* 283–296.

Bolt, Beranek, & Newman, Inc. (1982). Occupational noise: The subtle pollutant. In J. Ralof, *Science News, 121(21),* 347–350.

Bonio, S., Fonzi, A., & Saglione, G. (1978). Personal space and variations in the behavior of ten-year-olds. *Italian Journal of Psychology, 10,* 15–25.

Bonnes, M., Bonaiuto, M., & Ercolani, A. P. (1991). Crowding and residential satisfaction in the urban environment: A contextual approach. *Environment and Behavior, 23,* 531–552.

Bonnes, M., Giuliani, M. V., Amoni, F., & Bernard, Y.

(1987). Cross-cultural rules for the optimization of the living room. *Environment and Behavior, 19,* 204–227.

Bonnes, M., Mannetti, L., Tanucci, G., & Secchiaroli, G. (1990). The city as a multi-place system: An analysis of people-urban environment transactions. *Journal of Environmental Psychology, 10,* 37–66.

Bonta, J. (1986). Prison crowding: Searching for the functional correlates. *American Psychologist, 41,* 99–101.

Booth, A., & Edwards, J. N. (1976). Crowding and family relations. *American Sociological Review, 41,* 308–321.

Booth, W. (1988). Johnny Appleseed and the greenhouse. *Science, 242,* 19–20.

Borg, E. (1981). Noise, hearing, and hypertension (editorial). *Scandinavian Audiology, 10(2),* 125–126.

Bornstein, M. H. (1979). The pace of life revisited. *International Journal of Psychology, 14,* 83–90.

Bostrom, A., Fischhoff, B., & Morgan, M. G. (1992). Characterizing mental models of hazardous processes: A methodology and an application to radon. *Journal of Social Issues, 48,* 85–100.

Bouas, K. S., & Komorita, S. S. (1996). Group discussion and cooperation in social dilemmas. *Personality and Social Psychology Bulletin, 22,* 1144–1150.

Boucher, M. L. (1972). Effect of seating distance on interpersonal attraction in an interview situation. *Journal of Consulting and Clinical Psychology, 38,* 15–19.

Bourassa, S. C. (1990). A paradigm for landscape aesthetics. *Environment and Behavior, 22,* 787–812.

Bouska, M. L., & Beatty, P. A. (1978). Clothing as a symbol of status: Its effect on control of interaction territory. *Bulletin of the Psychonomic Society, 4,* 235–238.

Bowman, U. (1964). Alaska earthquake. *American Journal of Psychology, 121,* 313–317.

Boyce, P. R., & Rea, M. S. (1994). A field evaluation of full-spectrum, polarized lighting. *Journal of the Illuminating Engineering Society, 23,* 86–107.

Bradley, J. S. (1992). Disturbance caused by residential air conditioner noise. *Journal of the Acoustical Society of America, 93,* 1978–1986.

Brandon, G., & Lewis, A. (1999). Reducing household energy consumption: A qualitative and quantitative field study. *Journal of Environmental Psychology, 19,* 75–85.

Brauer, M., & t' Mannetje, A. (1998). Restaurant smoking restrictions and environmental tobacco smoke exposure. *American Journal of Public Health, 88,* 1834–1836.

Brehm, J. W. (1966). *A theory of psychological reactance.* New York: Academic Press.

Brehm, S. S., & Brehm, J. W. (1981). *Psychological reactance: A theory of freedom and control.* New York: Academic Press.

Brenner, H., Born, J., Novak, P., & Wanek, V. (1997). Smoking behavior and attitude toward smoking regulations and passive smoking in the workplace: A study among 974 employees in the German metal industry. *Preventive Medicine, 26,* 138–143.

Brewer, M. B., & Kramer, R. M. (1986). Choice behavior in social dilemmas: Effects of social density, group size, and decision framing. *Journal of Personality and Social Psychology, 50,* 543–549.

Briere, J., Downes, A., & Spensley, J. (1983). Summer in the city: Urban weather conditions and psychiatric emergency room visits. *Journal of Abnormal Psychology, 92,* 77–80.

Broadbent, D. E. (1958). *Perception and communication.* Oxford: Pergamon.

Broadbent, D. E. (1963). Differences and interactions between stresses. *Quarterly Journal of Experimental Psychology, 15,* 205–211.

Broadbent, D. E. (1971). *Decision and stress.* New York: Academic Press.

Brodeur, P. (1993). *The great power-line cover-up: How the utilities and the government are trying to hide the cancer hazard posed by electromagnetic fields.* Boston: Little, Brown.

Brodsky, C. M. (1983). "Allergic to everything": A medical subculture. *Psychosomatics, 24,* 731–742.

Brokemann, N. C., & Moller, A. T. (1973). Preferred seating position and distance in various situations. *Journal of Counseling Psychology, 20,* 504–508.

Bromet, E. (1980). *Preliminary report on the mental health of Three Mile Island residents.* Pittsburgh, PA: Western Psychiatric Institute, University of Pittsburgh.

Bromet, E., Hough, L., & Connell, M. (1984). Mental

health of children near the Three Mile Island reactor. *Journal of Preventive Psychiatry, 2,* 275–301.

Bronzaft, A. L., Ahern, K. D., McGinn, R., O'Connor, J., & Savino, B. (1998). Aircraft noise: A potential health hazard. *Environment and Behavior, 30,* 101–113.

Bronzaft, A. L. (1981). The effect of a noise abatement program on reading ability. *Journal of Environmental Psychology, 1,* 215–222.

Bronzaft, A. L. (1985–1986). Combating the unsilent enemy—Noise. *Prevention in Human Services, 4 (1–2),* 179–192.

Bronzaft, A. L., & McCarthy, D. P. (1975). The effects of elevated train noise on reading ability. *Environment and Behavior, 7,* 517–527.

Brooks, M. J., & Kaplan, A. (1972). The office environment: Space planning and affective behavior. *Human Factors, 14,* 373–391.

Brower, S., Dockett, K., & Taylor, R. (1983). Residents' perceptions of territorial features and perceived local threat. *Environment and Behavior, 15,* 419–437.

Brown, B. (1979, Aug.). *Territoriality and residential burglary.* Paper presented at the meeting of the American Psychological Association, New York, NY.

Brown, B. (1987). Territoriality. In D. Stokols & I. Altman (Eds.), *Handbook of environmental psychology* (Vol. 1, pp. 505–531). New York: Wiley-Interscience.

Brown, B. B. (1992). The ecology of privacy and mood in a shared living group. *Journal of Environmental Psychology, 12,* 5–20.

Brown, B. B., & Bentley, D. L. (1993). Residential burglars judge risk: The role of territoriality. Special Issue: Crime and the environment. *Journal of Environmental Psychology, 13,* 51–61.

Brown, B. B., Burton, J. R., & Sweaney, A. L. (1998). Neighbors, households, and front porches: New urbanist community tool or mere nostalgia. *Environment and Behavior, 30,* 579–600.

Brown, B. B., & Perkins, D. D. (1992). Disruptions in place attachment. In I. Altman & S. M. Low (Eds.), *Place attachment* (pp. 279–304). New York: Plenum.

Brown, B. B., & Werner, C. M. (1985). Social cohesiveness, territoriality, and holiday decorations: The influence of cul-de-sacs. *Environment and Behavior, 17,* 539–565.

Brown, C. E. (1981). Shared space invasion and race. *Personality and Social Psychology Bulletin, 7,* 103–108.

Brown, G. C., & Kirk, R. E. (1987). Geophysical variables and behavior: XXXVIII. Effects of ionized air on the performance of a vigilance task. *Perceptual & Motor Skills, 64,* 951–962.

Brown, I. D., & Poulton, E. C. (1961). Measuring the spare "mental capacity" of car drivers by a subsidiary task. *Ergonomics, 4,* 35–40.

Brown, J. G., & Burger, C. (1984). Playground design and preschool children's behaviors. *Environment and Behavior, 16,* 599–626.

Brown, K. S. (1999). Taking global warming to the people. *Science, 283,* 1440–1441.

Brown, P. (1992). Popular epidemiology and toxic waste contamination: Lay and professional ways of knowing. *Journal of Health and Social Behavior, 33,* 267–281.

Bruck, L. (1997). Welcome to Eden. *Nursing Homes, 46(1),* 28ff.

Bruin, M. J., & Cook, C. C. (1997). Understanding constraints and residential satisfaction among low-income single-parent families. *Environment and Behavior, 29,* 532–553.

Brunswik, E. (1956). *Perception and the representative design of psychological experiments.* Berkeley: University of California Press.

Brunswik, E. (1959). The conceptual framework of psychology. In O. Neurath, R. Camp, & C. Morris (Eds.), *Foundation of the unity of science: Toward an international encyclopedia of unified science.* Chicago: University of Chicago Press.

Bryant, K. J. (1982). Personality correlates of sense of direction and geographical orientation. *Journal of Personality and Social Psychology, 43,* 1318–1324.

Budd, G. M. (1973). Australian physiological research in the Antarctic and Subarctic, with special reference to thermal stress and acclimatization. In O. E. Edholm & E. K. E. Gunderson (Eds.), *Polar human biology.* London: Heineman.

Bullard, R. D., & Wright, B. H. (1993). Environmental justice for all: Community perspectives on health and research needs. *Toxicology & Industrial Health, 9,* 831–841.

Bullinger, M. (1989). Psychological effects of air pollution on healthy residents: A time-series approach. *Journal of Environmental Psychology, 9,* 103–118.

Bullinger, M., Morfeld, M., von Mackensen, S., & Brasche, S. (1999). The sick-building syndrome, do women suffer more? *Zentralblatt fur Hygiene und Umweltmedizin, 202(2-4),* 235–241.

Bunston, T., & Breton, M. (1992). Homes and homeless women. *Journal of Environmental Psychology, 12,* 149–162.

Burge, S., Hedge, A., Wilson, S., Bass, J. H., & Robertson, A. (1987). Sick building syndrome: A study of 4,373 office workers. *Annals of Occupational Hygiene, 31,* 493–504.

Burger, J., Sanchez, J., Gibbons, J. W., & Gochfeld, M. (1998). Gender differences in recreational use, environmental attitudes, and perceptions of future land use at the Savannah River site. *Environment and Behavior, 30,* 472–486.

Burgio, L., Scilley, K., Hardin, J. M., Hsu, C., & Yancey, J. (1996). Environmental "white noise": An intervention for verbally agitated nursing home residents. *Journal of Gerontology: Psychological Sciences, 51B,* P364–P373.

Burke, E. (1757). *Philosophical inquiry into the origin of our ideas of the sublime and beautiful.* Republished 1899 in *The Works of Edmund Burke* (Vol.1). Boston: Little, Brown.

Burn, S. M., & Oskamp, S. (1986). Increasing community recycling with persuasive communication and public commitment. *Journal of Applied Social Psychology, 16,* 9–41.

Burt, C. D. B. (1993). Concentration and academic ability following transition to university: An investigation of the effects of homesickness. *Journal of Environmental Psychology, 13,* 333–342.

Burton, I., Kates, R. W., & White, G.F. (1993). *The environment as hazard* (2nd ed.). New York: Guilford.

Butler, D. L., Acquino, A. L., Hissong, A. A., & Scott, P. A. (1993). Wayfinding by newcomers in a complex building. *Human Factors, 35,* 159–173.

Butler, D. L., & Biner, P. M. (1987). Preferred lighting levels: Variability among settings, behaviors, and individuals. *Environment and Behavior, 19,* 695–721.

Butler, D. L., & Biner, P. M. (1989). Effects of setting on window preferences and factors associated with those preferences. *Environment and Behavior, 21,* 17–32.

Byrne, D. (1971). *The attraction paradigm.* New York: Academic Press.

Byrne, D., Baskett, G. D., & Hodges, L. (1971). Behavioral indicators of interpersonal attraction. *Journal of Applied Social Psychology, 1,* 137–149.

Byrne, D., Ervin, C. R., & Lamberth, J. (1970). Continuity between the experimental study of attraction and real life computer dating. *Journal of Personality and Social Psychology, 16,* 157–165.

Byrne, R.W. (1979). Memory for urban geography. *Quarterly Journal of Experimental Psychology, 31,* 147–154.

Byrnes, G., & Kelly, I. W. (1992). Crisis calls and lunar cycles: A twenty-year review. *Psychological Reports, 71,* 779–785.

Cacioppo, J. T., & Petty, R. E. (1983). Foundations of social psychophysiology. In J. T. Cacioppo & R. E. Petty (Eds.), *Social psychophysiology: A sourcebook* (pp. 3–36). New York: Guilford.

Cahoon, R. L. (1972). Simple decision making at high altitude. *Ergonomics, 15,* 157–163.

Calhoun, J. B. (1962). Population density and social pathology. *Scientific American, 206,* 139–148.

Calhoun, J. B. (1964). The social use of space. In W. Mayer & R. Van Gelder (Eds.), *Physiological mammalogy* (pp. 2–187). New York: Academic Press.

Calhoun, J. B. (1967). Ecological factors in the development of behavioral anomalies. In J. Zubin & H. F. Hunt (Eds.), *Comparative psychopathology* (pp. 1–51). New York: Grune & Stratton.

Calhoun, J. B. (1970). Space and the strategy of life. *Ekistics, 29,* 425–437.

Calhoun, J. B. (1971). Space and the strategy of life. In A. H. Esser (Ed.), *Behavior and environment: The use of space by animals and men* (pp. 329–387). Bloomington: University of Indiana Press.

Calkins, M. P. (1987). *Designing for dementia.* Owings Mills, MD: National Health Publishing.

Cameron, L. D., Brown, P. M., & Chapman, J. G. (1998). Social value orientations and decisions to take proenvironmental action. *Journal of Applied Social Psychology, 28,* 675–697.

Campbell, D. E. (1979). Interior office design and visitor response. *Journal of Applied Psychology, 64,* 648–653.

Campbell, D. E. (1982). Lunar-lunacy research: When enough is enough. *Environment and Behavior, 14,* 418–424.

Campbell, D. E., & Beets, J. L. (1977). Meteorological variables and behavior: An annotated bibliography. *JSAS Catalog of Selected Documents in Psychology, 7,* 1 (Ms. No. 1403).

Campbell, D. E., & Beets, J. L. (1981). *Human response to naturally occurring weather phenomena: Effects of wind speed and direction.* Unpublished manuscript, Humboldt State University.

Campbell, D. E., & Herren, K. (1978). Interior arrangement of the faculty office. *Psychological Reports, 43,* 234.

Campbell, J. (1983). Ambient stressors. *Environment and Behavior, 15,* 355–380.

Campbell, J. B. (1992). Extraversion and noise sensitivity: A replication of Dornic and Ekehammar's study. *Personality and Individual Differences, 13,* 935–955.

Canino, G. J., Bravo, M., Rubio-Stipec, M., & Woodbury, M. (1990). The impact of disaster on mental health: Prospective and retrospective analyses. *International Journal of Mental Health, 19,* 51–69.

Cannon, W. B. (1929). *Bodily changes in pain, hunger, fear, and rage.* Boston: Branford.

Cannon, W. B. (1931). Studies on the conditions of activity in the endocrine organs, XXVII. Evidence that the medulli-adrenal secretion is not continuous. *American Journal of Physiology, 98,* 447–452.

Carlisle, S. G. (1982). French homes and French character. *Landscape, 26,* 13–23.

Carlopio, J. R., & Gardner, D. (1992). Direct and interactive effects of the physical work environment on attitudes. *Environment and Behavior, 24,* 579–601.

Carp, F. (1987). Environment and aging. In D. Stokols & I. Altman (Eds.), *Handbook of environmental psychology* (Vol. 1, pp. 329–360). New York: Wiley-Interscience.

Carpman, J. R., & Grant, M. A. (1993). *Design that cares: Planning health facilities for patients and visitors* (2nd ed.). Chicago: American Hospital Publishing, Inc.

Carpman, J. R., Grant, M. A., & Simmons, D. A. (1983–84). Wayfinding in the hospital environment: The impact of various floor numbering alternatives. *Journal of Environmental Systems, 13,* 353–364.

Carvalho, M., George, R. V., & Anthony, K. H. (1997). Residential satisfaction in *condomínios exclusivos* (gate-guarded neighborhoods) in Brazil. *Environment and Behavior, 29,* 734–768.

Cass, R., & Edney, J. J. (1978). The commons dilemma: A simulation testing the effects of resource visibility and territorial division. *Human Ecology, 6,* 371–386.

Cavatorta, A., Falzoi, M., Romamelli, A., Cigala, F., Ricco, M., Bruschi, G., Franchini, L., & Borghetti, A. (1987). Adrenal response in the pathogenesis of arterial hypertension in workers exposed to high noise levels. *Hypertension, 5,* 463–466.

Cermak, J. E., Davenport, A. G., Plate, E. J., & Viegas, D. X. (1995). *Wind climate in cities.* Dordrecht, The Netherlands: Kluwer.

Chaiken, S., & Stangor, C. (1987). Attitudes and attitude change. *Annual Review of Psychology, 38,* 575–630.

Chang, V. (1999). Lead abatement and prevention of developmental disabilities. *Journal of Intellectual & Developmental Disability, 24,* 161–168.

Chaouloff, F., & Zamfir, O. (1993). Psychoneuroendocrine outcomes of short-term crowding stress. *Physiology and Behavior, 54,* 767–770.

Chapko, M. K., & Solomon, M. (1976). Air pollution and recreation behavior. *Journal of Social Psychology, 100,* 149–150.

Chapman, J. C., Christian, J. J., Pawlikowski, M. A., & Michael, S. D. (1998). Analysis of steroid hormone levels in female mice at high population density. *Physiology & Behavior, 64,* 529–533.

Chapman, S., Borland, R., Scollo, M., Brownson, R. C., Dominello, A., & Woodward, S. (1999). The impact of smoke-free workplaces on declining cigarette consumption in Australia and the United States. *American Journal of Public Health, 89,* 1018–1023.

Charry, J. M., & Hawkinshire, F. B. W. (1981). Effects of atmospheric electricity on some substrates of disordered social behavior. *Journal of Personality and Social Psychology, 41*, 185–197.

Chawla, L. (1992). Childhood place attachments. In I. Altman & S. M. Low (Eds.), *Place attachment* (pp. 63–86). New York: Plenum.

Cherek, D. R. (1985). Effects of acute exposure to increased levels of background industrial noise on cigarette smoking behavior. International *Archives of Occupational and Environmental Health, 56*, 23–30.

Cherulnik, P. D. (1993). *Applications of environment-behavior research: Case studies and analysis.* New York: Cambridge University Press.

Cherulnik, P. D., & Wilderman, S. K. (1986). Symbols of status in urban neighborhoods: Contemporary perceptions of nineteenth-century Boston. *Environment and Behavior, 18*, 604–622.

Cheyne, J. A., & Efran, M. G. (1972). The effect of spatial and interpersonal variables on the invasion of group-controlled territories. *Sociometry, 35*, 477–489.

Christensen, R. (1982). Alaskan winters: A mental health hazard? *Alaskan Medicine, 24*, 89.

Christensen, R. (1984). Cabin fever: A folk belief and the misdiagnosis of complaints. *Journal of Mental Health Administration, 11*, 2–3.

Christian, J. J. (1955). Effects of population size on the adrenal glands and reproductive organs of male mice in populations of fixed size. *American Journal of Physiology, 182*, 292–300.

Churchman, A., & Mitrani, M. (1997). The role of the physical environment in culture shock. *Environment and Behavior, 29*, 64–86.

Cialdini, R. (1977). *Littering as a function of extant litter.* Unpublished manuscript, Arizona State University.

Cialdini, R. B., & Kenrick, D. T. (1976). Altruism as hedonism: A social development perspective on the relationship of negative mood state and helping. *Journal of Personality and Social Psychology, 54*, 907–914.

Cialdini, R. B., Reno, R. R., & Kallgren, C. A. (1990). A focus theory of normative conduct: Recycling the concept of norms to reduce littering in public places. *Journal of Personality and Social Psychology, 58*, 1015–1026.

Cini, M. A., Moreland, R. L., & Levine, J. M. (1993). Group staffing levels and responses to prospective and new group members. *Journal of Personality and Social Psychology, 65*, 723–734.

Citterio A., Sinforiani, E., Verri, A., Cristina, S., Gerosa, E., & Nappi, G. (1998). Neurological symptoms of the sick building syndrome: Analysis of a questionnaire. *Functional Neurology, 13*, 225–230.

Clarke, A., Bell, P. A., & Peterson, G. L. (1999). The influence of attitude priming and social responsibility on the valuation of environmental public goods using paired comparisons. *Environment and Behavior, 31*, 838–857.

Clemente, F., & Kleiman, M. B. (1977). Fear of crime in the United States. *Social Forces, 51*, 176–531.

Clitheroe, H. C., Stokols, D., & Zmuidzinas, M. (1998). Conceptualizing the context of environment and behavior. *Journal of Environmental Psychology, 18*, 103–112.

Coakley, J. (Ed.) (1993). The territorial management of ethnic conflict. *Regional Politics & Policy, 3* (1). London: Frank Cass.

Cobern, M. K., Porter, B. E., Leeming, F. C., & Dwyer, W. O. (1995). The effect of commitment on adoption and diffusion of grass cycling. *Environment and Behavior, 27*, 213–232.

Cochran, C., & Hale, W. (1984). Personal space requirements in indoor versus outdoor locations. *Journal of Psychology, 117*, 121–123.

Cohen, D. (1996). Law, social policy, and violence: The impact of regional cultures. *Journal of Personality and Social Psychology, 70*, 961–978.

Cohen, D., Nisbett, R. E., Bowdle, B. F., & Schwarz, N. (1996). Insult, aggression, and the Southern culture of honor: An "experimental ethnography." *Journal of Personality and Social Psychology, 70*, 945–960.

Cohen, H., Moss, S., & Zube, E. (1979). Pedestrians and wind in the urban environment. In A. D. Seidel & S. Danford (Eds.), *Environmental design: Research, theory, and application* (pp. 71–82). Washington, DC: Environmental Design Research Association.

Cohen, J. L., Sladen, B., & Bennett, B. (1975). The effects of situational variables on judgments of crowding. *Sociometry, 38*, 273–281.

Cohen, M. R. (1973). Environmental information versus environmental attitudes. *Journal of Environmental Education, 5*, 5–8.

Cohen, R., Goodnight, J. A., Poag, C. K., Cohen, S., Nichol, G. T., & Worley, P. (1986). Easing the transition of kindergarten: The affective and cognitive effects of different spatial familiarization experiences. *Environment and Behavior, 18*, 330–345.

Cohen, S. (1978). Environmental load and the allocation of attention. In A. Baum, J. E. Singer, & S. Valins (Eds.), *Advances in environmental psychology* (Vol. 1, pp. 1–29). Hillsdale, NJ: Erlbaum.

Cohen, S. (1980). Aftereffects of stress on human performance and social behavior: A review of research and theory. *Psychological Bulletin, 87*, 578–604.

Cohen, S., Evans, G. W., Krantz, D. S., & Stokols, D. (1980). Physiological, motivational, and cognitive effects of aircraft noise on children: Moving from the laboratory to the field. *American Psychologist, 35*, 231–243.

Cohen, S., Evans, G. W., Krantz, D. S., Stokols, D., & Kelly, S. (1981). Aircraft noise and children: Longitudinal and cross-sectional evidence on adaptation to noise and the effectiveness of noise abatement. *Journal of Personality and Social Psychology, 40*, 331–345.

Cohen, S., Evans, G. W., Stokols, D., & Krantz, D. S. (1986). *Behavior, health and environmental stress.* New York: Plenum.

Cohen, S., Glass, D. C., & Singer, J. E. (1973). Apartment noise, auditory discrimination, and reading ability in children. *Journal of Experimental Social Psychology, 9*, 407–422.

Cohen, S., & Hoberman, H. M. (1983). Positive events and social supports as buffers of life change stress. *Journal of Applied Social Psychology, 13*, 99–125.

Cohen, S., & Horm-Wingerd, D. (1993). Children and the environment: Ecological awareness among preschool children. *Environment and Behavior, 25*, 103–120.

Cohen, S., Kamarck, T., & Mermelstein, R. (1983). A global measure of perceived stress. *Journal of Health and Social Behavior, 24*, 385–396.

Cohen, S., & Lezak, A. (1977). Noise and inattentiveness to social cues. *Environment and Behavior, 9*, 559–572.

Cohen, S., & Spacapan, S. (1978). The aftereffects of stress: An attentional interpretation. *Environmental Psychology and Nonverbal Behavior, 3*, 43–57.

Cohen, S., & Spacapan, S. (1984). The social psychology of noise. In D. M. Jones & A. J. Chapman (Eds.), *Noise and society* (pp. 221–245). Chichester: Wiley.

Cohen, S., & Weinstein, N. D. (1982). Nonauditory effects of noise on behavior and health. In G. W. Evans (Ed.), *Environmental stress* (pp. 45–74). Cambridge: Cambridge University Press.

Cohen, S., & Wills, T. A. (1985). Stress, social support, and the buffering hypothesis. *Psychological Bulletin, 8*, 310–357.

Cohn, E. G. (1993). The prediction of police calls for service: The influence of weather and temporal variables on rape and domestic violence. *Journal of Environmental Psychology, 13*, 71–83.

Cohn, E. G. (1996). The effect of weather and temporal variations on calls for police service. *American Journal of Police, 15*, 23–43.

Cohn, E. G., & Rotton, J. (1997). Assault as a function of time and temperature: A moderator-variable time-series analysis. *Journal of Personality and Social Psychology, 72*, 1322–1334.

Coley, R. L., Kuo, F. E., & Sullivan, W. C. (1997). Where does community grow? The social context created by nature in urban public housing. *Environment and Behavior, 29*, 468–494.

Colligan, M. J., & Murphy, L. R. (1979). Mass psychogenic illness in organizations: An overview. *Journal of Occupational Psychology, 52*, 77–90.

Collins, A. M., & Quillian, M. R. (1969). Retrieval time from semantic memory. *Journal of Verbal Learning and Verbal Behavior, 8*, 240–247.

Collins, D. L., Baum, A., & Singer, J. (1983). Coping with chronic stress at Three Mile Island: Psychological and biochemical evidence. *Health Psychology, 2*, 149–166.

Committee for Health Care for Homeless People. (1988). *Homelessness, health and human needs.* Washington, DC: National Academy Press.

Commoner, B. (1963). *Science and survival.* New York: Viking.

Cone, J. D., & Hayes, S. C. (1980). *Environmental problems/behavioral solutions.* Monterey, CA: Brooks/Cole.

Conroy, J., III, & Sundstrom, E. (1977). Territorial dominance in a dyadic conversation as a function of similarity of opinion. *Journal of Personality and Social Psychology, 35, 570–576.*

Cook, C. C. (1988). Components of neighborhood satisfaction: Responses from urban and suburban single-parent women. *Environment and Behavior, 20, 115–149.*

Cook, M. (1970). Experiments on orientation and proxemics. *Human Relations, 23, 61–76.*

Cook, M., & Mineka, S. (1989). Observational conditioning of fear to fear-relevant versus fear-irrelevant stimuli in Rhesus monkeys. *Journal of Abnormal Psychology, 95, 195–207.*

Cook, M., & Mineka, S. (1990). Selective observations in the observational conditioning of fear in Rhesus monkeys. *Journal of Experimental Psychology: Animal Behavior Processes, 16, 372–389.*

Cook-Deegan, R. M. (1987). *Losing a million minds: Confronting the tragedy of Alzheimer's disease and other dementias.* Washington, DC: U.S. Government Printing Office.

Cooley, J. D., Wong, W. C., Jumper, C. A., & Straus, D. C. (1998). Correlation between the prevalence of certain fungi and sick building syndrome. *Occupational & Environmental Medicine, 55, 579–584.*

Cooper, M., & Rodman, M. C. (1994). Accessibility and quality of life in housing cooperatives. *Environment and Behavior, 26, 49–70.*

Cooper Marcus, C. (1996). *House as a mirror of self: Exploring the deeper meaning of home.* Berkeley, CA: Conari Press.

Cooper Marcus, C., & Sarkissian, W. (1985). *Housing as if people mattered.* Berkeley, CA: University of California Press.

Cornell, E. H., & Hay, D. H. (1984). Children's acquisition of a route via different media. *Environment and Behavior, 16, 627–642.*

Cornell, E. H., Heth, C. D., & Alberts, D. M. (1994). Place recognition and wayfinding by children and adults. *Memory and Cognition, 22, 633–643.*

Cornell, E. H., Heth, C. D., & Skoczylas, M. J. (1999). The nature and use of route expectancies following incidental learning. *Journal of Environmental Psychology, 19, 209–229.*

Cornoldi, C., & McDaniel, M. A. (1991). *Imagery and cognition.* New York: Springer.

Corral-Verdugo, V. (1997). Dual 'realities' of conservation behavior: Self-report vs observations of re-use and recycling behavior. *Journal of Environmental Psychology, 17, 135–145.*

Costanzo, M., Archer, D., Aronson, E., & Pettgrew, T. (1986). Energy conservation behavior: The difficult path from information to action. *American Psychologist, 41, 521–528.*

Cotton, J. L. (1986). Ambient temperature and violent crime. *Journal of Applied Social Psychology, 16, 786–801.*

Couclelis, H., Golledge, R. G., Gale, N., & Tobler, W. (1987). Exploring the anchor-point hypothesis of spatial cognition. *Journal of Environmental Psychology, 7, 99–122.*

Covington, J., & Taylor, R. (1989). Gentrification and crime: Robbery and larceny changes in appreciating Baltimore neighborhoods during the 1970s. *Urban Affairs Quarterly, 25, 140–170.*

Cox, V. C., Paulus, P. B., & McCain, G. (1984). Prison crowding research: The relevance for prison housing standards and a general approach regarding crowding phenomena. *American Psychologist, 39, 1148–1160.*

Cox, V. C., Paulus, P., McCain, G., & Karlovac, M. (1982). The relationship between crowding and health. In A. Baum & J. E. Singer (Eds.), *Advances in environmental psychology* (Vol. 4, pp. 271–294). Hillsdale, NJ: Erlbaum.

Craig, K. J., Cruikshank, M., Gabelnick, D. A., & Baum, A. (1992, August). *Age, coping, and chronic stress following traumatic events.* Symposium at the annual meeting of the American Psychological Association, Washington, DC.

Craik, K. H. (1983). The psychology of the large-scale environment. In N. R. Feimer & E. S. Geller (Eds.), *Environmental psychology: Directions and perspectives* (pp. 67–105). New York: Praeger.

Craik, K. H., & Feimer, N. R. (1987). Environmental

assessment. In D. Stokols & I. Altman (Eds.), *Handbook of environmental psychology* (pp. 891–918). New York: Wiley.

Craik, K. H., & Zube, E. H. (1976). *Perceiving environmental quality: Research and applications.* New York: Plenum.

Cranz, G. (1982). *The politics of park design: A history of urban parks in America.* Cambridge, MA: M.I.T. Press.

Crawford, J. O., & Bolas, S. M. (1996). Sick building syndrome, work factors and occupational stress. *Scandinavian Journal of Work, Environment & Health, 22,* 243–250.

Crawford, M., & Unger, R. (2000). *Women and gender: A feminist psychology* (3rd ed.). New York: McGraw-Hill.

Crook, M. A., & Langdon, F. J. (1974). The effects of aircraft noise on schools in the vicinity of the London Airport. *Journal of Sound and Vibration, 34,* 241–248.

Crouch, A., & Nimran, U. (1989). Perceived facilitators and inhibitors of work performance in an office environment. *Environment and Behavior, 21,* 206–226.

Crowe, M. J. (1968). Toward a "definitional model" of public perceptions of air pollution. *Journal of the Air Pollution Control Association, 18,* 154–157.

Crowe, T. (1991). *Crime prevention through environmental design: Applications of architectural design and space management concepts.* Boston: National Crime Prevention Institute/Butterworth-Heinemann.

Crozier, W. R., & Burgess, G. (1992). The influence of type of accommodation upon the spontaneous self-concept of the elderly. *Journal of Environmental Psychology, 12,* 129–133.

Crump, S. L., Nunes, D. L., & Crossman, E. K. (1977). The effects of litter on littering behavior in a forest environment. *Environment and Behavior, 9,* 137–146.

Csikszentmihalyi, M., & Kleiber, D. A. (1991). Leisure and self-actualization. In B. L. Driver, P. J. Brown, & G. L. Peterson (Eds.), *Benefits of leisure* (pp. 91–102). State College, PA: Venture Publishing, Inc.

Culotta, E. (1995). Will plants profit from high CO_2? *Science, 268,* 654–656.

Culp, R. H. (1998). Adolescent girls and outdoor recreation: A case study examining constraints and effective programming. *Journal of Leisure Research, 30,* 356–379.

Culver, R., Rotton, J., & Kelly, I. W. (1988). Geophysical variables and behavior: XLIX. Moon myths and mechanisms: A critical examination of purported explanations of lunar-lunacy relations. *Psychological Reports, 62,* 683–710.

Cunningham, M. R. (1979). Weather, mood, and helping behavior: Quasi experiments with the sunshine Samaritan. *Journal of Personality and Social Psychology, 37,* 1947–1956.

Cunningham, M. R., Steinberg, J., & Grev, R. (1980). Wanting to and having to help: Separate motivations for positive mood and guilt-induced helping. *Journal of Personality and Social Psychology, 38,* 181–192.

Cuthbertson, B. H., & Nigg, J. M. (1987). Technological disaster and the nontherapeutic community: A question of true victimization. *Environment and Behavior, 19,* 462–483.

D' Atri, D. A., Fitzgerald, E. F., Kasl, S. V., & Ostfeld, A. M. (1981). Crowding in prison: The relationship between changes in housing mode and blood pressure. *Psychosomatic Medicine, 43,* 95–105.

Da Costa Nunez, R. (1994). *Hopes, dreams and promise: The future of homeless children in America.* New York: Homes for the Homeless, Inc.

Dabbs, J. M. (1971). Physical closeness and negative feelings. *Psychonomic Science, 23,* 141–143.

Dalholm, E. H., & Rydberg-Mitchell, B. (1992). Communicating with lay people: Architecture & comportment/Architecture & behavior. *Architecture & Behavior, 8,* 241–262.

Damon, A. (1977). The residential environment, health, and behavior. Simple research opportunities, strategies, and some findings in the Solomon Islands and Boston, Massachusetts. In L. E. Hinckle, Jr., & W. C. Loring (Eds.), *The effect of the man-made environment on health and behavior.* Atlanta: Centers for Disease Control, Public Health Service.

Daneshvary, N., Daneshvary, R., & Schwer, R. K. (1998). Solid-waste recycling behavior and support for curbside textile recycling. *Environment and Behavior, 30,* 144–161.

Daniel, T. C., & Vining, J. (1983). Methodological issues in the assessment of landscape quality. In I. Altman & J. F. Wohlwill (Eds.), *Behav-*

ior and the natural environment (pp. 39–84). New York: Plenum.

Daves, W. F., & Swaffer, P. W. (1971). Effect of room size on critical interpersonal distance. *Perceptual and Motor Skills, 33,* 926.

Davidson, L. M., & Baum, A. (1986). Chronic stress and post traumatic stress disorders. *Journal of Consulting and Clinical Psychology, 54,* 303–308.

Davidson, L. M., Baum, A., & Collins, D. L. (1982). Stress and control-related problems at Three Mile Island. *Journal of Applied Social Psychology, 12,* 349–359.

Davidson, L. M., Fleming, I., & Baum, A. (1986). Post-traumatic stress as a function of chronic stress and toxic exposures. In C. Figley (Ed.), *Trauma and its wake* (pp. 55–77). New York: Brunner Mazel.

Davidson, L. M., Fleming, R., & Baum, A. (1987). Chronic stress, catecholamines, and sleep disturbance at Three Mile Island. *Journal of Human Stress, 13,* 75–83.

Dawes, R. M., McTavish, J., & Shaklee, H. (1977). Behavior, communication and assumptions about other people's behaviors in a commons dilemma situation. *Journal of Personality and Social Psychology, 35,* 1–11.

Day, K. (1994). Conceptualizing women's fear of sexual assault on campus: A review of causes and recommendations for change. *Environment and Behavior, 26,* 742–765.

Day, L. L. (1992). Placemaking by design: Fitting a large new building into a historic district. *Environment and Behavior, 24,* 326–346.

de Oliver, M. (1999). Attitudes and inaction: A case study of the manifest demographics of urban water conservation. *Environment and Behavior, 31,* 372–394.

De Young, R. (1986). Some psychological aspects of recycling: The structure of conservation satisfactions. *Environment and Behavior, 18,* 435–449.

De Young, R. (1996). Some psychological aspects of reduced consumption behavior: The role of instrinsic satisfaction and competence motivation. *Environment and Behavior, 28,* 358–409.

De Young, R., Duncan, A., Frank, J., Gill, N., Rothman, S., Shenot, J., Shotkin, A., & Zweizig, M. (1993). Promoting source reduction behavior: The role of motivational behavior. *Environment and Behavior, 25,* 70–85.

Dean, L., Pugh, W., & Gunderson, E. (1975). Spatial and perceptual components of crowding: Effects on health and satisfaction. *Environment and Behavior, 7,* 225–236.

Dean, L., Pugh, W., & Gunderson, E. (1978). The behavioral effects of crowding. *Environment and Behavior, 10,* 419–431.

Dean, L. M., Willis, F. N., & Hewitt, J. (1975). Initial interaction distance among individuals equal and unequal in military rank. *Journal of Personality and Social Psychology, 32,* 294–299.

Deaux, K. K., & LaFrance, M. (1998). Gender. In D. T. Gilbert, S. Fiske, & G. Lindzey (Eds.), *The handbook of social psychology* (Vol. I, 4th ed., pp. 788–827). New York: McGraw Hill.

Decker, J. (1994). The validation of computer simulations for design guideline dispute resolution. *Environment and Behavior, 26,* 421–443.

DeFronzo, J. (1984). Climate and crime: Tests of an FBI assumption. *Environment and Behavior, 16,* 185–210.

DeGiovanni, F. F., & Paulson, N. A. (1984). Household diversity in revitalizing neighborhoods. *Urban Affairs Quarterly, 20*(2), 211–232.

DeGroot, I. (1967). Trends in public attitudes toward air pollution. *Journal of the Air Pollution Control Association, 17,* 679–681.

Delahanty, D., Dougall, A. L., Craig, K. L., Jenkins F. J., & Baum, A. (1997). Chronic stress and natural killer cell activity following exposure to traumatic death. *Psychosomatic Medicine, 59,* 467–476.

DeLeon, I. G., & Fuqua, R. W. (1995). The effects of commitment and group feedback on curbside recycling. *Environment and Behavior, 27,* 233–250.

DeLoache, J. S. (1987). Rapid change in the functioning of very young children. *Science, 238,* 1556–1557.

Dennis, M. L., Soderstrom, E. J., Koncinski, W. S., & Cavanaugh, B. (1990). Effective dissemination of energy-related information: Applying social psychology and evaluation research. *American Psychologist, 45,* 1109–1117.

DePaulo, B., & Friedman, H. (1998). Nonverbal communication. In D. T. Gilbert, S. Fiske, & G. Lindzey (Eds.), *The handbook of social psychology* (Vol. II, 4th ed., pp. 3–39). New York: McGraw Hill.

Derogatis, L. R. (1977). *The SCL-90 Manual 1: Scoring, administration, and procedures for the SCL-90*. Baltimore: Johns Hopkins University School of Medicine, Clinical Psychometrics Unit.

DeSanctis, M., Halcomb, C. G., & Fedoravicius, A. S. (1981). *Meteorological determinants of human behavior: A holistic environmental perspective with special reference to air ionization and electrical field effects*. Unpublished manuscript, Texas Tech University, Lubbock, TX.

Desjarlais, R. (1997). *Shelter blues*. Philadelphia: University of Pennsylvania Press.

Desor, J. A. (1972). Toward a psychological theory of crowding. *Journal of Personality and Social Psychology, 21,* 79–83.

Devlin, A. S. (1992). Psychiatric ward renovation: Staff perception and patient behavior. *Environment and Behavior, 24,* 66–84.

Devlin, A. S., & Bernstein, J. (1995). Interactive wayfinding: Use of cues by men and women. *Journal of Environmental Psychology, 15,* 23–38.

Devlin, K. (1990). An examination of architectural interpretation: Architects versus nonarchitects. *Journal of Architectural and Planning Research, 7,* 235–244.

Devlin, K., & Nasar, J. L. (1989). The beauty and the beast: Some preliminary comparisons of "high" versus popular residential architecture and public versus architect judgments of same. *Journal of Environmental Psychology, 9,* 333–344.

Dew, M. A., Bromet, E. J., & Schulberg, H. C. (1987). Application of a temporal persistence model to community residents' long-term beliefs about the Three Mile Island nuclear accident. *Journal of Applied Social Psychology, 17,* 1071–1091.

Dexter, E. (1904). School deportment and weather. *Educational Review, 19,* 160–168.

Diamond, J. (1998). *Guns, germs, and steel: The fates of human societies*. New York: W.W. Norton.

Dickason, J. D. (1983). The origin of the playground: The role of the Boston women's clubs, 1885–1890. *Leisure Sciences, 6,* 83–98.

Dickson, D. (1987). Adjusting to an aging population. *Science, 236,* 772–773.

Dienstbier, R. A. (1989). Arousal and physiological toughness: Implications for mental and physical health. *Psychological Review, 96,* 84–100.

Dietz, T., Stern, P. C., & Guagnano, G. A. (1998). Social structural and social psychological bases of environmental concern. *Environment and Behavior, 30,* 450–471.

Dill, C. A., Gilden, E. R., Hill, P. C., & Hanselka, L. L. (1982). Federal human subjects regulations: A methodological artifact? *Personality and Social Psychology Bulletin, 8,* 417–425.

Dillman, D., & Tremblay, K., Jr. (1977). The quality of life in rural America. *Annals of the American Academy of Political and Social Sciences, 429,* 115–129.

Dober, R. P. (1992). *Campus design*. New York: Wiley.

Dober, R. P. (1996). *Campus architecture: Building in the groves of the academe*. New York: McGraw-Hill.

Dohrenwend, B. P., Dohrenwend, B. S., Kasl, S. V., & Warheit, G. J. (1979). *Report of the Task Group on Behavioral Effects to the President's Commission on the Accident at Three Mile Island* (pp. 18–21). Washington, DC.

Dollinger, S. J., & Cramer, P. (1990). Children's defensive responses and emotional upset following a disaster: A projective assessment. *Journal of Personality Assessment, 541,* 116–127.

Donnerstein, E., & Wilson, D. W. (1976). Effects of noise and perceived control on ongoing and subsequent aggressive behavior. *Journal of Personality and Social Psychology, 34,* 774–781.

Dooley, D., Rook, K., & Catalano, R. (1987). Job and non-job stressors and their moderators. *Journal of Occupational Psychology, 60,* 115–132.

Doring, H. J., Hauf, G., & Seiberling, M. (1980). Effects of high intensity sound on the contractile function of the isolated ileum of guinea pigs and rabbits. In *Noise as a public health problem, Proceedings of the Third International Congress* (ASHA Report No. 10). Rockville, MD: American Speech and Hearing Association.

Dornic, S., & Ekehammar, B. (1990). Extraversion, neuroticism, and noise sensitivity. *Personality and Individual Differences, 11,* 989–992.

Dougall, A. L., Craig, K. J., & Baum, A. (1999).

Assessment of characteristics of intrusive thoughts and their impact on distress among victims of traumatic events. *Psychosomatic Medicine, 61,* 38–48.

Downs, R., & Stea, D. (1973). *Image and the environment: Cognitive mapping and spatial behavior.* Chicago: Aldine.

Downs, R. M., & Stea, D. (1977). *Maps in minds: Reflections on cognitive mapping.* New York: Harper & Row.

Drabek, T. E. (1986). *Human system responses to disaster: An inventory of sociological findings.* New York: Springer-Verlag.

Drabek, T. E., & Stephenson, J. S. (1971). When disaster strikes. *Journal of Applied Social Psychology, 1,* 187–203.

Driver, B. L. (1996). Benefits-driven management of natural areas. *Natural Areas Journal, 16,* 94–99.

Driver, B. L., & Brown, P. J. (1983). Contributions of behavioral scientists to recreation resource management. In I. Altman & J. F. Wohlwill (Eds.), *Behavior and the environment* (pp. 307–339). New York: Plenum.

Driver, B. L., Brown, P., Stankey, G., & Gregoire, T. (1987). The ROS planning system: Evolution, basic concepts, and research needed. *Leisure Sciences, 9,* 201–212.

Driver, B. L., Dustin, D., Baltic, T., Elsner, G., & Peterson, G. (1996). Nature and the human spirit: Overview. In B. L. Driver, D. Dustin, T. Baltic, G. Elsner, & G. Peterson (Eds.), *Nature and the human spirit: Toward an expanded land management ethic* (pp. 3–8). State College, PA: Venture.

Driver, B. L., & Tocher, R. (1970). Toward a behavioral interpretation of recreational engagements, with implications for planning. In B.L. Driver (Ed.), *Elements of outdoor recreation planning* (pp. 9–31). Ann Arbor: University Microfilms.

Dubos, R. (1965). *Man adapting.* New Haven, CT: Yale University Press.

Duffy, M., Bailey, S., Beck, B., & Barker, D.G. (1986). Preferences in nursing home design: A comparison of residents, administrators, and designers. *Environment and Behavior, 18,* 246–257.

Duke, M. P., & Nowicki, S. (1972). A new measure and social learning model for interpersonal distance. *Journal of Experimental Research in Personality, 6,* 119–132.

Duke, M. P., & Wilson, J. (1973). The measurement of interpersonal distance in pre-school children. *Journal of Genetic Psychology, 123,* 361–362.

Duncan, J. S. (1985). The house as a symbol of social structure: Notes on the language of objects among collectivistic groups. In I. Altman & C. M. Werner (Eds.), *Home environments* (pp. 133–151). New York: Plenum.

Dunlap, R. E., & Mertig, A. G. (1995). Global concern for the environment: Is affluence a prerequisite? *Journal of Social Issues, 51,* 121–137.

Dunlap, R. E., & Van Liere, K. D. (1978, Summer). The "New Environmental Paradigm": A proposed measuring instrument and preliminary results. *Journal of Environmental Education, 9,* 10–19.

Dunlap, R. E., & Van Liere, K. D. (1984). Commitment to the dominant social paradigm and concern for environmental quality. *Social Science Quarterly, 65,* 1013–1028.

Dunlap, R. E., Van Liere, K., Mertig, A., & Howell, R. (1992, August). *Measuring endorsement of an ecological world view: A revised NEP scale.* Paper presented in the Annual Meeting of the Rural Sociology Society, Philadelphia, PA.

Dunlap, R. E, Van Liere, K. D., Mertig, A. G., & Jones, R. E. (2000). Measuring endorsement of the New Ecological Paradigm: A revised NEP scale. *Journal of Social Issues, 56,* in press.

Durham, T. W., McCammon, S. L., & Allison, E. J. (1985). The psychological impact of disaster on rescue personnel. *Annals of Emergency Medicine, 14,* 664–668.

Duvall, D., & Booth, A. (1978). The housing environment and women's health. *Journal of Health and Social Behavior, 19,* 410–417.

Dwyer, W. O., Leeming, F. C., Cobern, M. K., Porter, B. E., & Jackson, J. M. (1993). Critical review of behavioral interventions to preserve the environment: Research since 1980. *Environment and Behavior, 25,* 275–321.

Dyson, M. L. & Passmore, N. I. (1992). Inter-male spacing and aggression in African painted reed frogs, Hyperolius marmoratus. *Ethology, 91,* 237–247.

Eagly, A. H., & Chaiken, S. (1993). *The psychology of attitudes.* Fort Worth, TX: Harcourt Brace Jovanovich.

Easterbrook, J. A. (1959). The effects of emotion on cue-utilization and the organization of behavior. *Psychological Review, 66,* 183–201.

Eaton, S. B., Fuchs, L. S., & Snook-Hill, M. (1998). Personal-space preference among male elementary and high school students with and without visual impairments. *Journal of Visual Impairment & Blindness, 92,* 769–782.

Ebbesen, E. B., Kjos, G. L., & Konecni, V. J. (1976). Spatial ecology: Its effects on the choice of friends and enemies. *Journal of Experimental Social Psychology, 12,* 505–518.

Eberts, E. H., & Lepper, M. R. (1975). Individual consistency in the proxemic behavior of pre-school children. *Journal of Personality and Social Psychology, 32,* 481–489.

Ebreo, A., Hershey, J., & Vining, J. (1999). Reducing solid waste: Linking recycling to environmentally responsible consumerism. *Environment and Behavior, 31,* 107–135.

Eckenrode, J., & Gore, S. (1981). Stressful events and social supports: The significance of context. In B. H. Gottlieb (Ed.), *Social networks and social support* (pp. 43–68). Beverly Hills, CA: Sage.

Edmonds, E. M., & Smith, L. R. (1985). Students' performance as a function of sex, noise, and intelligence. *Psychological Reports, 56,* 727–730.

Edney, J. J. (1972). Property, possession and permanence: A field study in human territoriality. *Journal of Applied Psychology, 2,* 275–282.

Edney, J. J. (1974). Human territoriality. *Psychological Bulletin, 81,* 959–975.

Edney, J. J. (1975). Territoriality and control: A field experiment. *Journal of Personality and Social Psychology, 31,* 1108–1115.

Edney, J. J. (1976). Human territories: Comment on functional properties. *Environment and Behavior, 8,* 31–48.

Edney, J. J. (1979). The nuts game: A concise commons dilemma analogue. *Environmental Psychology and Nonverbal Behavior, 3,* 252–254.

Edney, J. J. (1980). The commons problem: Alternative perspectives. *American Psychologist, 35,* 131–150.

Edney, J. J. (1981). Paradoxes on the commons: Scarcity and the problem of equality. *Journal of Community Psychology, 9,* 3–34.

Edney, J. J., & Bell, P. A. (1983). The commons dilemma: Comparing altruism, the Golden Rule, perfect equality of outcomes, and territoriality. *The Social Science Journal, 20,* 23–33.

Edney, J. J., & Bell, P. A. (1984). Sharing scarce resources: Group-outcome orientation, external disaster, and stealing in a simulated commons. *Small Group Behavior, 15,* 87–108.

Edney, J. J., & Harper, C. S. (1978). The effects of information in a resource management problem: A social trap analog. *Human Ecology, 6,* 387–395.

Edney, J. J., & Uhlig, S. R. (1977). Individual and small group territories. *Small Group Behavior, 8,* 457–468.

Edwards, D. J. A. (1972). Approaching the unfamiliar: A study of human interaction distances. *Journal of Behavioral Sciences, 1,* 249–250.

Efran, M. G., & Cheyne, J. A. (1974). Affective concomitants of the invasion of shared space: Behavioral physiological, and verbal indicators. *Journal of Personality and Social Psychology, 29,* 219–226.

Eggertsen, R., Svensson, A., Magnusson, M., & Andren, L. (1987). Hemodynamic effects of loud noise before and after central sympathetic nervous stimulation. *Acta Medica Scandinavica, 221,* 159–164.

Ehrlich, P. (1968). *The population boom.* New York: Ballantine.

Eiser, J. R., Podpadec, T. J., Reicher, S. D., & Stevenage, S. V. (1998). Muddy waters and heavy metal: Time and attitudes guide judgements of pollution. *Journal of Environmental Psychology, 18,* 199–208.

Ekblad, S. (1996). Ecological psychology in Chinese societies. In M. H. Bond (Ed.), *The handbook of Chinese psychology* (pp. 379–392). Hong Kong: Oxford University Press.

Ellermeier, W., & Hellbrueck, J. (1998). Is level irrelevant in "irrelevant speech?" Effects of loudness, signal-to-noise ratio, and binaural unmasking. *Journal of Experimental Psychology: Human Perception & Performance, 24,* 1406–1414.

Elliott, S. J., Taylor, S. M., Hampson, C., Dunn, J., Eyles, J., Walter, S., & Streiner, D. (1997). 'It's not because you like it better . . .': Residents' reappraisal of a landfill site. *Journal of Environmental Psychology, 17,* 229–241.

Ellison-Potter, P. A., Bell, P. A., & Deffenbacher, J. L. (2000). The effects of anonymity, aggressive stimuli, and trait anger on aggressive driving behavior: A laboratory simulation. *Journal of Applied Social Psychology,* in press.

Elm, D., Warren, S., & Madill, H. (1998). The effects of auditory stimuli on functional performance among cognitively impaired elderly. *Canadian Journal of Occupational Therapy, 65,* 30–36.

Engelhart, S., Burghardt, H., Neumann, R., Ewers, U., Exner, M., & Kramer, M. H. (1999). Sick building syndrome in an office building formerly used by a pharmaceutical company: A case study. *Indoor Air-International Journal of Indoor Air Quality & Climate, 9,* 139–143.

Environmental Protection Agency (2000). Report cited in *Fort Collins Coloradoan,* Gannett News Service, February 6, p. A-1.

Epel, E. S., Bandura, A., & Zimbardo, P. G. (1999). Escaping homelessness: The influences of self-efficacy and time perspective on coping with homelessness. Journal of Applied Social Psychology, 29, 575–596.

Epstein, Y. M., & Karlin, R. A. (1975). Effects of acute experimental crowding. *Journal of Applied Social Psychology, 5,* 34–53.

Erikson, K. T. (1976). Loss of communality at Buffalo Creek. *American Journal of Psychiatry, 133,* 302–305.

Ersland, S., Weisaeth, L., & Sund, A. (1989). The stress upon rescuers involved in an oil rig disaster. "Alexander L. Fain Kielland" 1980. *Acta Psychiatrica Scandinavia, Supplementum, 355,* 38–49.

Essa, E. L., Hilton, J. M., & Murray, C. I. (1992). The relationship between weather and preschoolers' behavior. *Children's Environments Quarterly, 7,* 32–36.

Evans, G. W. (1978). Human spatial behavior: The arousal model. In A. Baum & Y. Epstein (Eds.), *Human response to crowding* (pp. 283–302). Hillsdale, NJ: Erlbaum.

Evans, G. W. (1979a). Behavioral and physiological consequences of crowding in humans. *Journal of Applied Social Psychology, 9,* 27–46.

Evans, G. W. (1979b). Design implications of spatial research. In J. Aeillo & A. Baum (Eds.), *Residential crowding and design* (pp. 197–215). New York: Plenum.

Evans, G. W. (1980). Environmental cognition. *Psychological Bulletin, 88,* 259–287.

Evans, G. W., Brennan, P. L., Skorpanich, M. A., & Held, D. (1984). Cognitive mapping and elderly adults: Verbal and location memory for urban landmarks. *Journal of Gerontology, 39,* 452–457.

Evans, G. W., Bullinger, M., & Hygge, S. (1998). Chronic noise exposure and physiological response: A prospective study of children living under environmental stress. *Psychological Science, 9,* 75–77.

Evans, G. W., & Cohen, S. (1987). Environmental stress. In D. Stokols & I. Altman (Eds.), *Handbook of environmental psychology* (Vol. 1, pp. 571–610). New York: Wiley-Interscience.

Evans, G. W., & Howard, H. R. B. (1972). A methodological investigation of personal space. In W. J. Mitchell (Ed.), *Environmental design: Research and practice,* Proceedings of EDRA3/AR8 Conference. Los Angeles: University of California.

Evans, G. W., & Howard, R. B. (1973). Personal space. *Psychological Bulletin, 80,* 334–344.

Evans, G. W., Hygge, S., & Bullinger, M. (1995). Chronic noise and psychological stress. *Psychological Science, 6,* 333–338.

Evans, G. W., & Jacobs, S. V. (1981). Air pollution and human behavior. *Journal of Social Issues, 37,* 95–125.

Evans, G. W., & Jacobs, S. V. (1982). Air pollution and human behavior. In G. W. Evans (Ed.), *Environmental stress* (pp. 105–132). Cambridge: Cambridge University Press.

Evans, G. W., Jacobs, S. V., Dooley, D., & Catalano, R. (1987). The interaction of stressful life events and chronic strains on community mental health. *American Journal of Community Psychology, 15,* 23–34.

Evans, G. W., Jacobs, S. V., & Frager, N. B. (1982). Behavioral responses to air pollution. In A. Baum & J. Singer (Eds.), *Advances in environmental psychology* (Vol. 4, pp. 237–270). Hillsdale, NJ: Erlbaum.

Evans, G. W., & Lepore, S. J. (1992). Conceptual and analytic issues in crowding research. *Journal of Environmental Psychology, 12,* 163–173.

Evans, G. W., & Lepore, S. J. (1993). Household crowding and social support: A quasi-experimental analysis. *Journal of Personality and Social Psychology, 65,* 308–316.

Evans, G. W., & Lepore, S. J. (1997). Moderating and mediating processes in environment-behavior research. In G. T. Moore & R. W. Marans (Eds.), *Advances in environment, behavior, and design* (Vol. 4, pp. 255–285). New York: Plenum.

Evans, G. W., Lepore, S. J., & Schroeder, A. (1996). The role of interior design elements in human responses to crowding. *Journal of Personality and Social Psychology, 70,* 41–46.

Evans, G. W., Lepore, S. J., Shejwal, B. R., & Palsane, M. N. (1998). Chronic residential crowding and children's well-being: An ecological perspective. *Child Development, 69,* 1514–1523.

Evans, G. W., & Lovell, B. (1979). Design modification in an open-plan school. *Journal of Educational Psychology, 71,* 41–49.

Evans, G. W., Marrero, D. G., & Butler, P. A. (1981). Environmental learning and cognitive mapping. *Environment and Behavior, 13,* 83–104.

Evans, G. W., & Maxwell, L. (1997). Chronic noise exposure and reading deficits: The mediating effects of language acquisition. *Environment and Behavior, 29,* 638–656.

Evans, G. W., & McCoy, J. M. (1998). When buildings don't work: The role of architecture in human health. *Journal of Environmental Psychology, 18,* 85–94.

Evans, G. W., Palsane, M. N., Lepore, S. J., & Martin, J. (1989). Residential density and psychological health: The mediating effects of social support. *Journal of Personality and Social Psychology, 57,* 994–999.

Evans, G. W., & Pezdek, K. (1980). Cognitive mapping: Knowledge of real-world distance and location. *Journal of Experimental Psychology: Human Learning & Memory, 6,* 13–24.

Evans, G. W., Smith, C., & Pezdek, K. (1982). Cognitive maps and urban form. *American Planning Association Journal, 48,* 232–244.

Ewert, A. W. (1994). Playing the edge: Motivation and risk taking in a high-altitude wildernesslike environment. *Environment and Behavior, 26,* 3–24.

Eyles, J., Taylor, S. M., Johnson, N., & Baxter, J. (1993). Worrying about waste: Living close to solid waste disposal facilities in Southern Ontario. *Social Science and Medicine, 37,* 805–812.

Eysenck, M. W. (1975). Interactive effects of noise, activation level and dominance in memory latencies. *Journal of Experimental Psychology, 104,* 143–148.

Fanger, P. O., Breum, N. O., & Jerking, E. (1977). Can colour and noise influence man's thermal comfort? *Ergonomics, 20,* 11–18.

Faupel, C. E., & Styles, S. P. (1993). Disaster education, household preparedness, and stress responses following Hurricane Hugo. *Environment and Behavior, 25,* 228–249.

Fay, T. H. (Ed.). (1991). *Noise and health.* New York: New York Academy of Medicine.

Fazio, R. H. (1990). Multiple processes by which attitudes guide behavior: The MODE model as an integrative framework. In M. P. Zanna (Ed.), *Advances in experimental social psychology* (pp. 75–109). San Diego, CA: Academic Press.

Fazio, R. H., Sanbonmatsu, D. M., Powell, M. C., & Kardes, F. R. (1986). On the automatic activation of attitudes. *Journal of Personality and Social Psychology, 50,* 229–238.

Fazio, R. H., & Zanna, M. P. (1981). Direct experience and attitude–behavior consistency. *Advances in Experimental Social Psychology, 14,* 161–202.

Feather, N. T. (1961). The relationship of persistence at a task to expectation of success and achievement-related motives. *Journal of Abnormal and Social Psychology, 63,* 552–561.

Fein, G. G., Schwartz, P. M., Jacobson, S. W., & Jacobson, J. L. (1983). Environmental toxins and behavioral development: A new role for psychological research. *American Psychologist, 38,* 118–119.

Feldman, R. M. (1990). Settlement-identity: Psychological bonds with home place in a mobile society. *Environment and Behavior, 22,* 183–229.

Feldman, R. M. (1996). Constancy and change in attachments to types of settlements. *Environment and Behavior, 28,* 419–445.

Felipe, N. J., & Sommer, R. (1966). Invasions of personal space. *Social Problems, 14,* 206–214.

Feller, R. A. (1968). Effect of varying corridor illumination on noise level in a residence hall. *The Journal of College Personnel, 9,* 150–152.

Ferguson, E. L., & Hegarty, M. (1994). Properties of cognitive maps constructed from texts. *Memory and Cognition, 22,* 455–473.

Festinger, L. A. (1954). A theory of social comparison processes. *Human Relations, 7,* 117–140.

Festinger, L. A. (1957). *A theory of cognitive dissonance.* Stanford, CA: Stanford University Press.

Festinger, L. A., Schachter, S., & Back, K. (1950). *Social pressures in informal groups*. New York: Harper & Row.

Fidell, S., Barber, D. S., & Schultz, T. J. (1991). Updating a dosage-effect relationship for the prevalence of annoyance due to general transportation noise. *Journal of the Acoustical Society of America, 89,* 221–233.

Fidell, S., & Silvati, L. (1991). An assessment of the effect of residential acoustic insulation on prevalence of annoyance in an airport community. *Journal of the Acoustical Society of America, 89,* 244–247.

Finnie, W. C. (1973). Field experiments in litter control. *Environment and Behavior, 5,* 123–144.

Firestone, I. J. (1977). Reconciling verbal and nonverbal models of dyadic communication. *Environmental Psychology and Nonverbal Behavior, 2,* 30–44.

Fischer, C. S. (1973). Urban malaise. *Social Forces, 52,* 221–235.

Fischer, C. S. (1976). *The urban experience*. New York: Harcourt Brace Jovanovich.

Fischman, J. (1996). California social climbers: Low water prompts high status. *Science, 272,* 811–812.

Fishbein, M., & Ajzen, I. (1975). *Belief, attitude, intention, and behavior: An introduction to theory and research*. Reading, PA: Addison-Wesley.

Fisher, B., & Nasar, J. L. (1992). Fear of crime in relation to three exterior site features: Prospect, refuge, and escape. *Environment and Behavior, 24,* 35–56.

Fisher, J. D. (1974). Situation-specific variables as determinants of perceived environmental aesthetic quality and perceived crowdedness. *Journal of Research in Personality, 8,* 177–188.

Fisher, J. D., & Baron, R. M. (1982). An equity-based model of vandalism. *Population and Environment, 5,* 182–200.

Fisher, J. D., & Baum, A. (1980). Situational and arousal-based messages and the reduction of crowding stress. *Journal of Applied Social Psychology, 10,* 191–201.

Fisher, J. D., & Byrne, D. (1975). Too close for comfort: Sex differences in response to invasion of personal space. *Journal of Personality and Social Psychology, 32,* 15–21.

Fishman, R. (1982). *Urban utopias in the twentieth century: Ebenezer Howard, Frank Lloyd Wright, and Le Corbusier*. Cambridge, MA: MIT Press.

Fleming, I. (1985). *The stress reducing functions of specific types of social support for victims of a technological catastrophe*. Unpublished doctoral dissertation, University of Maryland, College Park.

Fleming, I., Baum, A., & Weiss, L. (1987). Social density and perceived control as mediators of crowding stress in high-density residential neighborhoods. *Journal of Personality and Social Psychology, 52,* 899–906.

Fleming, R., Baum, A., Gisriel, M. M., & Gatchel, R. J. (1982). Mediation of stress at Three Mile Island by social support. *Journal of Human Stress, 8,* 14–22.

Florin, P., & Wandersman, A. (1990). An introduction to citizen participation, voluntary organizations, and community development: Insights for empowerment through research. *American Journal of Community Psychology, 18,* 41–54.

Floyd, M. F., & Gramann, J. H. (1997). Experience-based setting management: Implications for market segmentation of hunters. *Leisure Sciences, 19,* 113–127.

Floyd, M. F., Outley, C. W., Bixler, R. D., & Hammitt, W. E. (1995). Effects of race, environmental preference and negative affect on recreation preferences. In *Abstracts from the 1995 National Recreation and Park Association symposium on leisure research* (p. 88). Arlington, VA: National Recreation and Park Association.

Flynn, C. B. (1979). *Three Mile Island telephone survey*. U.S. Nuclear Regulatory Commission (NUREF/CR-1093).

Flynn, R. (1995). America's cities: Centers of culture, commerce, and community or collapsing hope? *Urban Affairs Review, 30,* 635–640.

Fogarty, S. J., & Hemsley, D. R. (1983, March). Depression and the accessibility of memories: A longitudinal study. *British Journal of Psychiatry, 142,* 232–237.

Folkman, S. (1984). Personal control and stress and coping processes: A theoretical analysis. *Journal of Personality and Social Psychology, 46,* 839–852.

Forbes, G., & Gromoll, H. (1971). The lost letter technique as a measure of social variables:

Some exploratory findings. *Social Forces, 50,* 113–115.

Forgays, D. G., & Forgays, D. K. (1992). Creativity enhancement through flotation isolation. *Journal of Environmental Psychology, 12,* 329–335.

Forgays, D. G., Forgays, D. K., Pudvah, M., & Wright, D. (1991). A direct comparison of the "wet" and "dry" flotation environments. *Journal of Environmental Psychology, 11,* 179–187.

Fortenberry, J. H., Maclean, J., Morris, P., & O'Connell, M. (1978). Mode of dress a perceptual cue to deference. *Journal of Social Psychology, 104,* 139–140.

Foxx, R. M., & Hake, D. F. (1977). Gasoline conservation: A procedure for measuring and reducing the driving of college students. *Journal of Applied Behavior Analysis, 10,* 61–74.

Francis, R. S. (1983). Attitudes toward industrial pollution, strategies for protecting the environment, and environmental–economic trade-offs. *Journal of Applied Social Psychology, 13,* 310–327.

Franck, K. A. (1984). Exorcising the ghost of physical determinism. *Environment and Behavior, 16,* 441–435.

Franck, K. D., Unseld, C. T., & Wentworth, W. E. (1974). *Adaptation of the newcomer: A process of construction.* Unpublished manuscript, City University of New York.

Franke, R. H., & Kaul, J. D. (1978). The Hawthorne experiments: First statistical interpretation. *American Sociological Review, 43,* 623–643.

Frankel, A. S., & Barrett, J. (1971). Variations in personal space as a function of authoritarianism, self-esteem, and racial characteristics of a stimulus situation. *Journal of Consulting and Clinical Psychology, 37,* 95–98.

Fraser, T. M. (1989). *The worker at work.* Bristol, PA: Taylor & Francis.

Frederickson, L. M., & Anderson, D. H. (1999). A qualitative exploration of the wilderness experience as a source of spiritual inspiration. *Journal of Environmental Psychology, 19,* 21–39.

Freedman, J. L. (1975). *Crowding and behavior.* San Francisco: Freeman.

Freedman, J. L., Birsky, J., & Cavoukian, A. (1980).

Environmental determinants of behavioral contagion: Density and number. *Basic and Applied Social Psychology, 1,* 155–161.

Freedman, J. L., Klevansky, S., & Ehrlich, P. I. (1971). The effect of crowding on human task performance. *Journal of Applied Social Psychology, 1,* 7–26.

Freedman, J. L., Levy, A. S., Buchanan, R. W., & Price, J. (1972). Crowding and human aggressiveness. *Journal of Experimental Social Psychology, 8,* 528–548.

Freedman, J. L., & Perlick, D. (1979). Crowding, contagion, and laughter. *Journal of Experimental Social Psychology, 15,* 295–303.

Freedy, J. R., Shaw, D. L., Jarrell, M. P., & Masters, C. R. (1992). Towards an understanding of the psychological impact of natural disasters: An application of the conservation of resources stress model. *Journal of Traumatic Stress, 5,* 441–454.

Frese, M. (1985). Stress at work and psychosomatic complaints: A causal interpretation. *Journal of Applied Psychology, 70,* 314–328.

Fried, M. (1963). Grieving for a lost home. In L. J. Duhl (Ed.), *The urban condition* (pp. 151–171). New York: Basic Books.

Fried, M., & Gleicher, P. (1961). Some sources of residential satisfaction in an urban slum. *Journal of the American Institute of Planners, 27,* 305–315.

Frisancho, A. R. (1993). *Human adaptation and accommodation.* Ann Arbor, MI: University of Michigan Press.

Fritz, C. E., & Marks, E. S. (1954). The NORC studies of human behavior in disaster. *Journal of Social Issues, 10,* 26–41.

Froelicher, V. F., & Froelicher, E. S. (1991). Cardiovascular benefits of physical activity. In B. L. Driver, P. J. Brown, & G. L. Peterson (Eds.), *Benefits of leisure* (pp. 59–72). State College, PA: Venture Publishing.

Frug, G. E. (1999). *City making.* Princeton, NJ: Princeton University Press.

Fruhstorfer, B., Pritsch, M. G., Pritsch, M. B., Clement, H. W., & Wesemann, W. (1988). Effects of daytime noise load on the sleep–wake cycle and endocrine patterns in man. III. 24 hours secretion of free and sulfate conjugated catecholamines. *International Journal of Neuroscience, 43,* 53–62.

Fry, A. M., & Willis, F. N. (1971). Invasion of personal space as a function of the age of the invader. *Psychological Record, 2,* 385–389.

Fuhrer, U., Kaiser, F. G., & Hartig, T. (1993). Place attachment and mobility during leisure time. *Journal of Environmental Psychology, 13,* 309–321.

Fuller, T. D., Edwards, J. N., Sermsri, S., & Vorakitphokatorn, S. (1993). Housing, stress, and physical well-being: Evidence from Thailand. *Social Science and Medicine, 36,* 1417–1428.

Fuller, T. D., Edwards, J. N., Vorakitphokatorn, S., & Sermsri, S. (1993). Household crowding and family relations in Bangkok. *Social Problems, 40,* 410–430.

Fullerton, C. S., McCarroll, J. E., Ursano, R. S., & Wright, K. M. (1992). Psychological responses of rescue workers: Fire fighters and trauma. *American Journal of Orthopsychiatry, 62,* 371–378.

Fusco, M. E., Bell, P. A., Jorgensen, M. D., & Smith, J. M. (1991). Using a computer to study the commons dilemma. *Simulation & Gaming, 22,* 67–74.

Gaines, T. A. (1991). *The campus as a work of art.* New York: Praeger.

Gale, N., Golledge, R. G., Pelligrino, J. W., & Doherty, S. (1990). The aquisition and integration of route knowledge in an unfamiliar neighborhood. *Journal of Environmental Psychology, 10,* 3–26.

Galea, L. A. M., & Kimura, D. (1993). Sex differences in route-learning. *Personality and Individual Differences, 14,* 53–65.

Galer, I. A. R. (Ed.). (1987). *Applied ergonomics handbook.* London: Butterworths.

Gallant, S. J., Hamilton, J. A., Popiel, D. A., & Morokoff, P. J. (1991). Daily moods and symptoms: Effects of awareness of study focus, gender, menstrual cycle phase, and day of the week. *Health Psychology, 10,* 180–189.

Galle, O. R., Gove, W. R., & McPherson, J. M. (1972). Population density and pathology: What are the relationships for man? *Science, 176,* 23–30.

Galster, G., & Hesser, G. (1981). Residential satisfaction: Compositional and contextual correlates. *Environment and Behavior, 13,* 735–759.

Ganellen, R. J., & Blaney, P. H. (1984). Hardiness and social support as mediators of the effects of life stress. *Journal of Personality and Social Psychology, 47,* 156–163.

Gans, H. J. (1962). *The Organvillagers.* New York: Free Press.

Garber, J., & Seligman, M. E. P. (Eds.). (1981). *Human helplessness: Theory and applications.* New York: Academic Press.

Garcia, K. D., & Wierwille, W. W. (1985). Effects of glare on performance of VDT reading-comprehension task. *Human Factors, 27,* 163–173.

Garcia, M. T. (1996). Hispanic perspectives and values. In B.L. Driver, D. Dustin, T. Baltic, G. Elsner, & G. Peterson (Eds.), *Nature and the human spirit: Toward an expanded land management ethic* (pp. 145–151). State College, PA: Venture.

Gardner, G. T. (1978). Effects of federal human subjects regulations on data obtained in environmental stress research. *Journal of Personality and Social Psychology, 36,* 628–634.

Gardner, G. T., & Stern, P. C. (1996). Environmental problems and human behavior. Needham Heights, MA: Allyn & Bacon.

Gärling, T., Biel, A., & Gustafsson, M. (1998). Different kinds and roles of environmental uncertainty. *Journal of Environmental Psychology, 18,* 75–83.

Gärling, T., Böök, A., & Ergezen, N. (1982). Memory for the spatial layout of the everyday physical environment. *Scandinavian Journal of Psychology, 23,* 23–35.

Gärling, T., Böök, A., & Lindberg, E. (1984). Cognitive mapping of large-scale environments: The interrelationship between action plans, acquisition, and orientation. *Environment and Behavior, 16,* 3–34.

Gärling, T., Böök, A., & Lindberg, E. (1986). Spatial orientation and wayfinding in the designed environment: A conceptual analysis and some suggestions for postoccupany evaluation. *Journal of Architectural Planning Research, 3,* 55–64.

Gärling, T., & Evans, G. W. (1991). *Environment, cognition, and action.* New York: Oxford University Press.

Gärling, T., & Golledge, R. (Eds.). (1993). *Behaviour and environment: Psychological and geographical approaches.* Amsterdam: North Holland, Elsevier.

Gärling, T., Lindberg, E., Carreiras, M., & Böök, A. (1986). Reference systems in cognitive maps. *Journal of Environmental Psychology, 6,* 1–18.

Garmezy, N., & Rutter, M. (1985). Acute reactions to stress. In M. Rutter & L. Hersov (Eds.), *Child psychiatry: Modern approaches* (2nd ed., pp. 152–176). Oxford: Blackwell Scientific.

Garreau, J. (1991). *Edge city.* New York: Doubleday.

Garzino, S. J. (1982). Lunar effects on mental behavior: A defense of the empirical research. *Environment and Behavior, 4,* 395–417.

Gaster, S. (1992). Historic changes in children's access to U.S. cities. *Children's Environments, 9,* 23–36.

Gatchel, R. J., & Newberry, B. (1991). Psychophysiological effects of toxic chemical contamination exposure: A community field study. *Journal of Applied Social Psychology, 21,* 1961–1976.

Gatchel, R. J., Schaeffer, M. A., & Baum, A. (1985). A psychological field study of stress at Three Mile Island. *Psychophysiology, 22,* 175–181.

Gaydos, H. F. (1958). Effect on complex manual performance of cooling the body while maintaining the hands at normal temperatures. *Journal of Applied Physiology, 12,* 373–376.

Gaydos, H. F., & Dusek, E. R. (1958). Effects of localized hand cooling versus total body cooling on manual performance. *Journal of Applied Physiology, 12,* 377–380.

Gee, M. (1994). Questioning the concept of the "user." *Journal of Environmental Psychology, 14,* 113–124.

Geen, R. G. (1984). Preferred stimulation levels in introverts and extroverts: Effects on arousal and performance. *Journal of Personality and Social Psychology, 46,* 1303–1312.

Geen, R. G., & McCown, E. J. (1984). Effects of noise and attack on aggression and physiological arousal. *Motivation and Emotion, 8,* 231–241.

Geen, R. G., & O'Neal, E. C. (1969). Activation of cue-elicited aggression by general arousal. *Journal of Personality and Social Psychology, 11,* 289–292.

Geller, E. S. (1976). *Behavioral approaches to environmental problem solving: Littering and recycling.* Symposium presentation at the meeting of the Association for the Advancement of Behavior Therapy, New York.

Geller, E. S. (1981). Evaluating energy conservation programs: Is verbal report enough? *Journal of Consumer Research, 8,* 331–335.

Geller, E. S. (1987). Environmental psychology and applied behavior analysis: From strange bedfellows to a productive marriage. In D. Stokols & I. Altman (Eds.), *Handbook of environmental psychology* (pp. 361–388). New York: Wiley-Interscience.

Geller, E. S. (1995). Actively caring for the environment: An integration of behaviorism and humanism. *Environment and Behavior, 27,* 184–195.

Geller, E. S., Mann, M., & Brasted, W. (1977, August). *Trash can design: A determinant of litter-related behavior.* Paper presented at the meeting of the American Psychological Association, San Francisco, CA.

Geller, E. S., Winnett, R. A., & Everett, P. B. (1982). *Preserving the environment: New strategies for behavior change.* New York: Pergamon.

Geller, E. S., Witmer, J. F., & Tuso, M. E. (1977). Environmental interventions for litter control. *Journal of Applied Psychology, 62,* 344–351.

Gergen, K. J., Gergen, M. K., & Barton, W. H. (1973). Deviance in the dark. *Psychology Today, 7,* 129–130.

Gibbs, L. (1982). *Love Canal: My story.* Albany, NY: SUNY Press.

Gibbs, M. S. (1986). Psychopathological consequences of exposure to toxins in the water supply. In A. H. Lebovits, A. Baum, & J. Singer (Eds.), *Advances in environmental psychology* (pp. 47–70). Hillsdale, NJ: Erlbaum.

Gibson, B., Harris, P., & Werner, C. (1993). Intimacy and personal space: A classroom demonstration. *Teaching of Psychology, 20,* 180–181.

Gibson, B., & Werner, C. (1994). Airport waiting areas as behavior settings: The role of legibility cues in communicating the setting program. *Journal of Personality and Social Psychology, 66,* 1049–1060.

Gibson, J. J. (1979). *An ecological approach to visual perception.* Boston: Houghton Mifflin.

Giel, R., & Ormel, J. (1977). Crowding and subjective health in the Netherlands. *Social Psychiatry, 12,* 37–42.

Gifford, R. (1997). *Environmental Psychology: Principles and practice* (2nd ed.). Boston: Allyn & Bacon.

Gifford, R., Hine, D. W., Muller-Clemm, W., Reynolds, D. J., Jr., & Shaw, K. T. (2000). Decoding modern architecture: A lens model approach for understanding the aesthetic differences of architects and laypersons. *Environment and Behavior, 32,* 163–187.

Gifford, R., & Hine, D. W. (1997). Toward cooperation in commons dilemmas. *Canadian Journal of Behavioural Science, 29,* 167–179.

Gifford, R., & Peacock, J. (1979). Crowding: More fearsome than crime-provoking? *Psychologia, 22,* 79–83.

Gifford, R., & Sacilotto, P. A. (1993). Social isolation and personal space: A field study. *Canadian Journal of Behavioural Science, 25,* 165–174.

Gifford, R., & Wells, J. (1991). FISH: A commons dilemma simulation. *Behavior Research Methods, Instrumentation, and Computers, 23,* 437–441.

Gilderbloom, J. I., & Markham, J. P. (1996). Housing modification needs of the disabled elderly: What really matters? *Environment and Behavior, 28,* 512–535.

Gilman, B. I. (1916). Museum fatigue. *Scientific Monthly, 12,* 62–64.

Gilson, R. D. (1995). Special issue preface. *Human Factors, 37,* 3–4.

Giuliani, M. V., & Feldman, R. (1993). Place attachment in a developmental and cultural context. *Journal of Environmental Psychology, 13,* 267–274.

Glacken, C. J. (1967). *Traces on the Rhodian shore.* Berkeley, CA: University of California Press.

Glass, D. C., & Singer, J. E. (1972). *Urban stress.* New York: Academic Press.

Glass, D. C., Singer, J. E., & Friedman, L.W. (1969). Psychic cost of adaptation to an environmental stressor. *Journal of Personality and Social Psychology, 12,* 200–210.

Glass, D. C., Singer, J. E., Leonard, H. S., Krantz, D., Cohen, S., & Cummings, H. (1973). Perceived control of aversive stimulation and the reduction of stress responses. *Journal of Personality, 41,* 577–595.

Gleser, G., Green, B., & Winget, C. (1981). *Prolonged psychosocial effects of disaster: A study of Buffalo Creek.* New York: Academic Press.

Goehring, J. B. (1999, January). *Aggressive driving: Background and overview report.* Paper presented at the Aggressive Driving and the Law Symposium, National Highway Traffic Safety Administration, Washington, DC.

Gold, J. R. (1982). Territoriality and human spatial behavior. *Progress in Human Geography, 6,* 44–67.

Golden, D. (1999, December 16). The last full moon of the millennium must be momentous. *The Wall Street Journal,* pp. A1, A8.

Goldman, L. K., & Glantz, S. A. (1998). Evaluation of antismoking advertising campaigns. *JAMA, 279,* 323–324.

Goldsmith, J. R. (1968). Effects of air pollution on human health. In A. C. Stern (Ed.), *Air pollution* (2nd ed., Vol. 1, pp. 335–386). New York: Academic Press.

Goldstein, E. B. (1999). *Sensation and perception* (5th ed.). Pacific Grove, CA: Brooks/Cole.

Golledge, R. G. (1987). Environmental cognition. In D. Stokols & I. Altman (Eds.), *Handbook of environmental psychology* (pp. 131–174). New York: Wiley.

Golledge, R. G., Smith, T. R., Pellegrino, J. W., Doherty, S., & Marshall, S. P. (1985). A conceptual model and empirical analysis of children's acquisition of spatial knowledge. *Journal of Environmental Psychology, 5,* 125–152.

Golledge, R. G., & Stimson, R. J. (1997). *Spatial behavior.* New York: Guilford.

Gomes, L. M. P., Martinho-Pimenta, A. J. F., & Castelo, B. (1999). Effects of occupational exposure to low frequency noise on cognition. *Aviation Space & Environmental Medicine, 70*(3), Suppl, A115–A118.

Gonzales, M. H., Aronson, E., & Costanzo, M. A. (1988). Using social cognition to promote energy conservation: A quasi-experiment. *Journal of Applied Social Psychology, 18,* 1049–1066.

Goodman, G. H., & McAndrew, F. T. (1993). Domes and Astroturf: A note on the relationship between the physical environment and the performance of Major League baseball players. *Environment and Behavior, 25,* 121–125.

Goranson, R. E., & King, D. (1970). *Rioting and daily temperature: Analysis of the U.S. riots in 1967.* Toronto: York University.

Gordon, M. S., Riger, S., Lebailly, R., & Heath, L.

(1980). Crime, women, and the quality of urban life. *Signs, 5*, 144–160.

Gormley, F. P., & Aiello, J. R. (1982). Social density, interpersonal relationships, and residential crowding stress. *Journal of Applied Social Psychology, 12*, 222–336.

Gottman, J. (1966). The growing city as a social and political process. *Transactions of the Bartlett Society, 5*, 9–46.

Gould, P., & White, R. (1982). *Mental maps* (2nd ed.). Boston: Allen & Unwin.

Gove, W. R., & Hughes, M. (1983). *Crowding in the household.* New York: Academic Press.

Gramann, J. H., & Burdge, R. J. (1984). Crowding perception determinants at intensively developed outdoor recreation sites. *Leisure Sciences, 6*, 167–186.

Grandjean, E., Hunting, W., & Pidermann, M. (1983). VDT workstation design: Preferred settings and their effects. *Human Factors, 25*, 161–175.

Grasmick, H. G., Bursik, R. J., Jr., & Kinsey, K. A. (1991). Shame and embarrassment as deterrents to noncompliance with the law: The case of an antilittering campaign. *Environment and Behavior, 23*, 233–251.

Gratz, R. B. (1989). *The living city.* New York: Simon & Schuster.

Green, B. L., Grace, M. C., Lindy, J. D., Gleser, G. C., Leonard, A. C., Korol, M., & Winget, C. (1990a). Buffalo Creek survivors in the second decade: Stability of stress symptoms. *American Journal of Orthopsychiatry, 60*, 43–54.

Green, B. L., Grace, M. C., Vary, M.G., Krammer, T. L., Gleser, G. C., & Leonard, A. C. (1994). Children of disaster in the second decade: A 17-year follow-up of Buffalo Creek survivors. *Journal of the American Academy of Child and Adolescent Psychiatry, 33*, 71–79.

Green, B. L., Lindy, J. D., Grace, M. C., Gleser, G. C., & Leonard, A. C. (1990b). Buffalo Creek survivors in the second decade: Comparison with unexposed and non-litigant groups. *Journal of Applied Social Psychology, 20*, 1033–1050.

Green, D. M., & Fidell, S. (1991). Variability in the criterion for reporting annoyance in community noise surveys. *Journal of the Acoustical Society of America, 89*, 234–243.

Greenbaum, P. E., & Greenbaum, S. D. (1981).

Territorial personalization: Group identity and social interaction in a Slavic-American neighborhood. *Environment and Behavior, 13*, 574–589.

Greenberg, C. I. (1979). Toward an integration of ecological psychology and industrial psychology: Undermanning theory, organization size, and job enrichment. *Environmental Psychology and Nonverbal Behavior, 3*, 228–242.

Greenberg, C. I., & Baum, A. (1979). Compensatory responses to anticipated densities. *Journal of Applied Social Psychology, 9*, 1–12.

Greenberg, C. I., & Firestone, I. (1977). Compensatory responses to crowding: Effects of personal space intrusion and privacy reduction. *Journal of Personality and Social Psychology, 35*, 637–644.

Greenberg, M. S., & Ruback, R. B. (1984). Criminal victimization: Introduction and overview. *Journal of Social Issues, 40*, 1–8.

Greenberger, D. B., & Allen, V. C. (1980). Destruction and complexity: An application of aesthetic theory. *Personality and Social Psychology Bulletin, 6*, 479–483.

Greenbie, B. B. (1982). The landscape of social symbols. *Landscape Research, 7*, 2–6.

Greene, L. R. (1977). Effects of verbal evaluation feedback and interpersonal distance on behavioral compliance. *Journal of Consulting Psychology, 24*, 10–14.

Greene, T. C., & Bell, P. A. (1980). Additional considerations concerning the effects of "warm" and "cool" wall colors on energy conservation. *Ergonomics, 23*, 949–954.

Greene, T. C., & Bell, P. A. (1986). Environmental stress. In M. A. Baker (Ed.), *Sex differences in human performance* (pp. 81–106). London: Wiley.

Greene, T. C., & Connelly, C. M. (1988). Computer analysis of aesthetic districts. In D. Lawrence, R. Habe, A. Hacker, & D. Sherrods (Eds.), *People's needs/planet management: Paths to co-existence* (pp. 333–335). Washington, DC: Environmental Design Research Association.

Greene, T. C., & Warden, A. G. (2000). *Mapping pleasantness in familiar campus landscapes.* Unpublished manuscript, St. Lawrence University.

Grieshop, J. I., & Stiles, M. C. (1989). Risk and home-pesticide users. *Environment and Behavior, 21*, 699–716.

Griffiths, I. D. (1975). The thermal environment. In D. C. Canter (Ed.), *Environmental interaction: Psychological approaches to our physical surroundings* (pp. 21–52). New York: International Universities Press.

Griffiths, I. D., & Raw, G. J. (1987). Community and individual response to changes in traffic noise exposure. In H. S. Koelega (Ed.), *Environmental annoyance: Characterization, measurement, and control* (pp. 333–343). Amsterdam: Elsevier Science Publishers.

Griffitt, W. (1970). Environmental effects on interpersonal affective behavior: Ambient effective temperature and attraction. *Journal of Personality and Social Pyschology, 15,* 240–244.

Griffitt, W., & Veitch, R. (1971). Hot and crowded: Influences of population density and temperature on interpersonal affective behavior. *Journal of Personality and Social Psychology, 17,* 92–98.

Groat, L. (1982). Meaning in post-modern architecture: An examination using the multiple sorting task. *Journal of Environmental Psychology, 2,* 3–22.

Grob, A. (1995). A structural model of environmental attitudes and behaviour. *Journal of Environmental Psychology, 15,* 209–220.

Grossman, L. M. (1987, June 17). City pedestrian malls fail to fulfill promise of revitalizing downtown. *The Wall Street Journal,* p. 27.

Grunberg, J., & Eagle, P. F. (1990). Shelterization: How the homeless adapt to shelter living. *Hospital and Community Psychiatry, 41,* 521–525.

Guagnano, G. A., Stern, P. C., & Dietz, T. (1995). Influences on attitude–behavior relationships: A natural experiment with curbside recycling. *Environment and Behavior, 27,* 699–718.

Gubrium, J. F. (1974). Victimization in old age: Available evidence and three hypotheses. *Crime and Delinquency, 20,* 245–250.

Guenther, R. (1982, Aug. 4). Ways are found to minimize pollutants in airtight houses. *The Wall Street Journal,* p. 25.

Gulian, E., & Thomas, J. R. (1986). The effects of noise, cognitive set, and gender on mental arithmetic performance. *British Journal of Psychology, 77,* 503–511.

Gulliver, F. P. (1908). Orientation of maps. *Journal of Geography, 7,* 55–58.

Gump, P. V. (1974, August). Operating environments in schools of open and traditional design. *School Review, 84,* 574–593.

Gump, P. V. (1984). School environments. In I. Altman & J. F. Wohlwill (Eds.), *Children and the environment* (pp. 131–174). New York: Plenum.

Gump, P. V. (1987). School and classroom environments. In D. Stokols & I. Altman (Eds.), *Handbook of environmental psychology* (Vol. 1, pp. 691–732). New York: Wiley-Interscience.

Gunderson, E. K. E. (1968). Mental health problems in Antarctica. *Archives of Environmental Health, 17,* 558–564.

Guterbock, T. M. (1990). The effect of snow on urban density patterns in the United States. *Environment and Behavior, 22,* 358–386.

Haase, R. S., & Pepper, D. T. (1972). Nonverbal components of empathic communication. *Journal of Counseling Psychology, 19,* 417–424.

Haber, G. M. (1980). Territorial invasion in the classroom: Invadee response. *Environment and Behavior, 12,* 17–31.

Haggard, L. M., & Werner, C. M. (1990). Situational support, privacy regulation, and stress. *Basic and Applied Social Psychology, 11,* 313–337.

Haggard, L. M., & Williams, D. R. (1991). Self-identity benefits of leisure activities. In B. L. Driver, P. J. Brown, & G. L. Peterson (Eds.), *Benefits of leisure* (pp. 103–119). State College, PA: Venture.

Hall, E. T. (1959). *The silent language.* New York: Doubleday.

Hall, E. T. (1963). A system for the notation of proxemic behavior. *American Anthropologist, 65,* 1003–1026.

Hall, E. T. (1966). *The hidden dimension.* New York: Doubleday.

Hall, E. T. (1968). Proxemics. *Current Anthropology, 9,* 83–107.

Hall, M., & Baum, A. (1995). Intrusive thoughts as determinants of distress in parents of children with cancer. *Journal of Applied Social Psychology, 25,* 1215–1230.

Hall, P. (1988). *Cities of tomorrow: An intellectual history of urban planning and design.* Oxford: Basil Blackwell.

Halpern, D. (1995). *Mental health and the built environment: More than bricks and mortar?* London: Taylor & Francis.

Hamad, C. D., Bettinger, R., Cooper, D., & Semb, G. (1980–1981). Using behavioral procedures to establish an elementary school paper recycling program. *Journal of Environmental Systems, 10,* 149–156.

Hambrick-Dixon, P. J. (1986). Effects of experimentally imposed noise on task performance of Black children attending day care centers near elevated subway trains. *Developmental Psychology, 22,* 259–264.

Hamid, P. N., & Cheng, S-T. (1995). Predicting antipollution behavior: The role of molar behavioral intentions, past behavior, and locus of control. *Environment and Behavior, 27,* 679–698.

Ham-Rowbottom, K. A., Gifford, R., & Shaw, K. T. (1999). Defensible space theory and the police: Assessing the vulnerability of residences to burglary. *Journal of Environmental Psychology, 19,* 117–129.

Hancock, P. A. (1986). Sustained attention under thermal stress. *Psychological Bulletin, 99,* 263–281.

Handford, H. A., Dickerson Mayes, S., Mattison, R. E., Humphrey, J., Bagnato, S., Bixler, E., & Kales, J. (1986). Child and parent reaction to the Three Mile Island nuclear accident. *Journal of the American Academy of Child Psychiatry, 25,* 346–356.

Haney, C., & Zimbardo, P. (1998). The past and future of U.S. prison policy. *American Psychologist, 53,* 709–727.

Hansson, R. O., Noulles, D., & Bellovich, S. J. (1982). Social comparison and urban-environmental stress. *Personality and Social Psychology Bulletin, 8,* 68–73.

Hanyu, K., & Itsukushima, Y. (1995). Cognitive distance of stairways: Distance, traversal time, and mental walking time estimations. *Environment and Behavior, 27,* 579–591.

Harada, M. (1977). Cogenital alkyl mercury poisoning (congenital Minamata disease). *Pediatrician, 6,* 58–68.

Hardin, G. (1968). The tragedy of the commons. *Science, 162,* 1243–1248.

Hardin, G. (1998). Extensions of "the tragedy of the commons." *Science, 280,* 682–683.

Hardoff, D., Pamarthi, M. F., Feldman, J., & Jacobson, M. S. (1997). Altered lipid profiles in passive smoking urban adolescents as indicated by urinary cotinine. *International Journal of Adolescent Medicine & Health, 9,* 181–186.

Harland, P., Staats, H., & Wilke, H. A. M. (1999). Explaining proenvironmental intention and behavior by personal norms and the Theory of Planned Behavior. *Journal of Applied Psychology, 29,* 2505–2528.

Harries, K. D., & Stadler, S. J. (1988). Heat and violence: New findings from Dallas field data, 1980–1981. *Journal of Applied Social Psychology, 18,* 129–138.

Harris, P. B., & Brown, B. B. (1996). The home and identity display: Interpreting resident territoriality from home exteriors. *Journal of Environmental Psychology, 16,* 187–203.

Harrison, A. A., Clearwater, Y. A., & McKay, C. P. (Eds.). (1991). *From Antarctica to outer space: Life in isolation and confinement.* New York: Springer.

Hart, R. A. (1987). Children's participation in planning and design. In C. Simon & T. G. David (Eds.), *Spaces for children* (pp. 217–239). New York: Plenum.

Hartig, T., & Evans, G. W. (1993). Psychological foundations of nature experience. In T. Gärling & R. G. Golledge (Eds.), *Behavior and environment: Psychological and geographical approaches* (pp. 427–457). Amsterdam: North Holland.

Hartig, T., Mang, M., & Evans, G. W. (1991). Restorative effects of natural environment experience. *Environment and Behavior, 23,* 3–26.

Hartshorn, G., & Bynum, N. (1999). Tropical forest synergies. *Science, 286,* 2093–2094.

Hartsough, D. M., & Savitsky, J. C. (1984). Three Mile Island: Psychology and environmental policy at a crossroads. *American Psychologist, 39,* 1113–1122.

Harvey, M.L., & Bell, P.A. (1995). The moderating effect of threat on the relationship between population concern and environmental concern. *Population and Environment, 17,* 121–133.

Harvey, M. L., Bell, P. A., & Birjulin, A. A. (1993). Punishment and type of feedback in a simulated commons dilemma. *Psychological Reports, 73,* 447–450.

Harvey, M. L., Loomis, R. J., Bell, P. A., & Marino, M. (1998). The influence of museum exhibit design on immersion and psychological flow. *Environment and Behavior, 30,* 601–627.

Hasell, M. J., & Peatross, F. D. (1990). Exploring connections between women's changing roles and house forms. *Environment and Behavior, 22,* 3–26.

Hay, R. (1998). Sense of place in developmental context. *Journal of Environmental Psychology, 18*, 5–29.

Hayduk, L. A. (1978). Personal space: An evaluative and orienting overview. *Psychological Bulletin, 85*, 117–134.

Hayduk, L. A. (1981). The permeability of personal space. *Canadian Journal of Behavioral Science, 17*, 140–149.

Hayduk, L. A. (1983). Personal space: Where we now stand. *Psychological Bulletin, 94*, 293–335.

Hayduk, L. A. (1985). Personal space: The conceptual and measurement implications of structural equation models. *Canadian Journal of Behavioral Science, 17*, 140–149.

Hayduk, L. A. (1994). Personal space: Understanding the simplex model. *Journal of Nonverbal Behavior, 18*, 245–260.

Hayes, S. C., & Cone, J. D. (1981). Reduction of residential consumption of electricity through simple monthly feedback. *Journal of Applied Behavior Analysis, 14*, 81–88.

Hayes, S. C., Johnson, V. S., & Cone, J. D. (1975). The marked item technique: A practical procedure for litter control. *Journal of Applied Behavior Analysis, 8*, 381–386.

Hayward, D. G., Rothenberg, M., & Beasley, R. R. (1974). Children's play and urban playground environments: A comparison of traditional, contemporary, and adventure playground types. *Environment and Behavior, 6*, 131–169.

Hayward, J. (1989). Urban parks: Research, planning, and social change. In I. Altman & E. H. Zube (Eds.), *Public places and spaces* (pp. 193–216). New York: Plenum.

Heaton, A. W., & Sigall, H. (1989). The "championship choke" revisited: The role of fear of acquiring a negative identity. *Journal of Applied Social Psychology, 19*, 1019–1033.

Hebb, D. O. (1972). *Textbook of psychology* (3rd ed.). Philadelphia: Saunders.

Heberlein, T. A. (1975). Conservation information: The energy crisis and electricity consumption in an apartment complex. *Energy Systems and Policy, 1*, 105–117.

Hedge, A. (1984). Evidence of a relationship between office design and self-reports of ill health among office workers in the United Kingdom. *Journal of Architectural and Planning Research, 1*, 163–174.

Hedge, A., Burge, P. S., Robertson, A. S., Wilson, S., & Harris-Bass, J. (1989). Work-related illness in offices: A proposed model of the sick building syndrome. *Environment International, 15*, 143–158.

Hedge, A., Erickson, W. A., & Rubin, G. (1994). The effects of alternative smoking policies on indoor air quality in 27 office buildings. *Annals of Occupational Hygiene, 38*, 265–278.

Hedge, A., Mitchell, G. E., & McCarthy, J. (1993). Effects of furniture integrated breathing zone filtration system, indoor air quality, Sick Building Syndrome, productivity, and absenteeism. *Indoor Air, 3*, 328–336.

Hediger, H. (1950). *Wild animals in captivity.* London: Butterworth.

Hedman, R., & Jaszewski, A. (1984). *Fundamentals of urban design.* Washington, DC: Planners Press.

Heerwagen, J. H., & Heerwagan, D. R. (1986). Lighting and psychological comfort. *Lighting Design and Applications, 16*, 47–51.

Heerwagen, J. H., & Orians, G. H. (1986). Adaptions to windowlessness: A study of the use of visual decor in windowed and windowless offices. *Environment and Behavior, 18*, 623–639.

Heerwagen, J. H., & Orians, G. H. (1993). Humans, habitats, and aesthetics. In S. R. Kellert & E. O. Wilson (Eds.), *The biophilia hypothesis* (pp. 138–172). Washington, DC: Island Press.

Heft, H. (1979a). The role of environmental features in route-learning: Two exploratory studies of wayfinding. *Environmental Psychology and Nonverbal Behavior, 3*, 172–185.

Heft, H. (1979b). Background and focal environmental conditions of the home and attention in young children. *Journal of Applied Social Psychology, 9*, 47–69.

Heft, H. (1981). An examination of constructivist and Gibsonian approaches to environmental psychology. *Population and Environment, 4*, 227–245.

Heft, H. (1983). Wayfinding as the perception of information over time. *Population and Environment, 6*, 133–150.

Heft, H. (1989). Affordances and the body: An intentional analysis of Gibson's ecological approach to visual perception. *Journal for the Theory of Social Behavior, 19*, 1–30.

Heft, H., & Wohlwill, J. F. (1987). Environmental cognition in children. In D. Stokols & I. Altman (Eds.), *Handbook of environmental psychology* (pp. 175–203). New York: Wiley.

Heller, J., Groff, B., & Solomon, S. (1977). Toward an

understanding of crowding: The role of physical interaction. *Journal of Personality and Social Psychology, 35,* 183–190.

Helson, H. (1964). *Adaptation level theory.* New York: Harper & Row.

Hempel, L. C. (1997). Population in context: A typology of environmental driving forces. *Population & Environment: A Journal of Interdisciplinary Studies, 18,* 439–461.

Henderson, K. (1996). Feminist perspectives, female ways of being, and nature. In B. L. Driver, D. Dustin, T. Baltic, G. Elsner, & G. Peterson (Eds.), *Nature and the human spirit: Toward an expanded land management ethic* (pp. 153–162). State College, PA: Venture

Hendrick, C., Wells, K. S., & Faletti, M. V. (1982). Social and emotional effects of geographical relocation on elderly retirees. *Journal of Personality and Social Psychology, 42,* 951–962.

Henig, J. R. (1982). Neighborhood response to gentrification: Conditions to mobilization. *Urban Affairs Quarterly, 17,* 343–358.

Henry, D. O. (1994). Prehistoric cultural ecology in southern Jordan. *Science, 265,* 336–341.

Hensley, W. E. (1982). Professor proxemics: Personality and job demands as factors of faculty office arrangement. *Environment and Behavior, 14,* 581–591.

Herlocker, C. E., Allison, S. T., Foubert, J. D., & Beggan, J. K. (1997). Intended and unintended overconsumption of physical, spatial, and temporal resources. *Journal of Personality and Social Psychology, 73,* 992–1004.

Hern, W. M. (1991). Proxemics: The application of theory to conflict arising from antiabortion demonstrations. *Population and Environment: A Journal of Interdisciplinary Studies, 12,* 379–388.

Herzberg, F. (1966). *Work and the nature of man.* Cleveland, OH: World Publishing.

Herzberg, F., Mausner, B., & Snyderman, B. (1959). *The motivation to work.* New York: Wiley.

Herzog, T. R. (1989). A cognitive analysis for preference of urban nature. *Journal of Environmental Psychology, 9,* 27–43.

Herzog, T. R. (1992). A cognitive analysis of preference for urban spaces. *Journal of Environmental Psychology, 12,* 237–248.

Herzog, T. R., & Barnes, G. J. (1999). Tranquility and preference revisited. *Journal of Environmental Psychology, 19,* 171–181.

Herzog, T. R., Black, A. M., Fountaine, K. A., &

Knotts, D. (1997). Reflection and attentional recovery as distinctive benefits of restorative environments. *Journal of Environmental Psychology, 17,* 165–170.

Herzog, T. R., & Bosley, P. J. (1992). Tranquility and preference as affective qualities of natural environments. *Journal of Environmental Psychology, 12,* 115–127.

Herzog, T. R., & Gale, T. A. (1996). Preferences for urban buildings as a function of age and nature context. *Environment and Behavior, 28,* 44–72.

Herzog, T. R., & Miller, E. J. (1998). The role of mystery in perceived danger and environmental preference. *Environment and Behavior, 30,* 429–449.

Herzog, T. R., & Smith, G. A. (1988). Danger, mystery, and environmental preference. *Enviroment and Behavior, 20,* 320–344.

Heshka, S., & Nelson, Y. (1972). Interpersonal speaking distance as a function of age, sex, and relationship. *Sociometry, 35,* 491–498.

Heshka, S., & Pylypuk, A. (1975, June). *Human crowding and adrenocortical activity.* Paper presented at the meeting of the Canadian Psychological Association, Quebec.

Heth, C. D., & Cornell, E. H. (1998). Characteristics of travel by persons lost in Albertan wilderness areas. *Journal of Environmental Psychology, 18,* 223–235.

Heth, C. D., Cornell, E. H., & Alberts, D. M. (1997). Differential use of landmarks by 8- and 12-year-old children during route reversal navigation. *Journal of Environmental Psychology, 17,* 199–213.

Hillier, B., & Hanson, J. (1984). *Social logic of space.* London: Cambridge University Press.

Hillmann, R. B., Brooks, C. I., & O'Brien, J. P. (1991). Differences in self-esteem of college freshmen as a function of classroom seating-row preference. *Psychological Record, 41,* 315–320.

Hiroto, D. S. (1974). Locus of control and learned helplessness. *Journal of Experimental Psychology, 102,* 187–193.

Hirst, E., Clinton, J., Geller, H., & Kroner, W. (1986). *Energy efficiency in buildings: Progress and promise.* Washington, DC: American Council for an Energy-Efficient Economy.

Hirtle, S. C., & Jonides, J. (1985). Evidence of hierarchies in cognitive maps. *Memory and Cognition, 13,* 208–217.

Hirtle, S. C., & Sorrows, M. E. (1998). Designing a

multimodal tool for allocating buildings on a college campus. *Journal of Environmental Psychology, 18,* 265–276.

Hiss, T. (1990). *The experience of place.* New York: Alfred A. Knopf.

Hobbs, P. V., & Radke, L. F. (1992). Airborne studies of the smoke from the Kuwait oil fires. *Science, 256,* 987–991.

Hobfoll, S. E. (1989). Conservation of resources: A new attempt at conceptualizing stress. *American Psychologist, 44,* 513–524.

Hobfoll, S. E. (1991). Traumatic stress: A theory based on rapid loss of resources. *Anxiety Research, 4,* 187–197.

Hocking, M. B. (1991). Paper versus polystyrene: A complex choice. *Science, 251,* 504–505.

Hodgson, M. J., & Morey, P. R. (1989). Allergic and infectious agents in the outdoor air. *Immunology and Allergy Clinics of North America, 9,* 399–412.

Holahan, C. J. (1972). Seating patterns and patient behavior in an experimental dayroom. *Journal of Abnormal Psychology, 80,* 115–124.

Holahan, C. J. (1976). Environmental change in a psychiatric setting: A social systems analysis. *Human Relations, 29,* 153–166.

Holahan, C. J., & Saegert, S. (1973). Behavioral and attitudinal effects of large-scale variation in the physical environment of psychiatric wards. *Journal of Abnormal Psychology, 83,* 454–462.

Holden, C. (1995). EMF good for trees? *Science, 267,* 451.

Holden, C. (1996). The last of the Cahokians. *Science, 272,* 351.

Holden, C. (1999a). Healthy pit stop. *Science, 284,* 1115.

Holden, C. (1999b). Stuff of car seats? *Science, 284,* 583.

Holding, C. S. (1992). Clusters and reference points in cognitive representations of the environment. *Journal of Environmental Psychology, 12,* 45–56.

Holgate, S. T., Samet, J. M., Maynard, R. L., & Koren, H. S. (Eds.) (1999). *Air pollution and health.* San Diego: Academic Press.

Holman, E. A., & Silver, R. C. (1994, August). *The relationship between place attachment and social relationships in coping with the southern California firestorms.* Paper presented at the meeting of the American Psychological Association, Los Angeles, CA.

Holmes, R. M. (1992). Children's artwork and non-verbal communication. *Child Study Journal, 22,* 157–166.

Hong, S., Candelone, J.-P., Patterson, C. C., & Boutron, C. F. (1994). Greenland ice evidence of hemispheric lead pollution two millennia ago by Greek and Roman civilizations. *Science, 265,* 1841–1843.

Hoppe, P., & Martinac, I. (1998) Indoor climate and air quality. Review of current and future topics in the field of ISB study group 10. *International Journal of Biometerorology, 42,* 1–7.

Hopper, J. R., & Nielsen, J. M. (1991). Recycling as altruistic behavior: Normative and behavioral strategies to expand participation in a community recycling program. *Environment and Behavior, 23,* 195–220.

Horonjeff, R. D., Kimura, Y., Miller, N. P., Robert, W. E., Rossano, C. F., & Sanchez, G. (1993). *Acoustic data collected at Grand Canyon, Haleakala, and Hawaii Volcanoes National Parks.* National Park Service, USDI, Report No. 290940.18, Denver, CO.

Hourihan, K. (1984). Context-dependent models of residential satisfaction: An analysis of housing groups in Cork, Ireland. *Environment and Behavior, 16,* 369–393.

Houts, P. S., Miller, R. W., Tokuhata, G. K., & Ham, K. S. (1980, April 8). *Health-related behavioral impact of the Three Mile Island nuclear incident.* Report submitted to the TMI Advisory Panel on Health Research Studies of the Pennsylvania Department of Health, Part I.

Howard, G. S. (1997). *Ecological psychology: Creating a more Earth-friendly human nature.* Notre Dame, IN: University of Notre Dame Press.

Howard, G. S. (2000). Adapting human lifestyles for the twenty-first century. *American Psychologist, 55,* 509–514.

Howard, G. S., Delgado, E., Miller, D., & Gubbins, S. (1993). Transforming values into actions: Ecological preservation. *The Counseling Psychologist, 21,* 582–596.

Hubbard, P. (1996). Conflicting interpretations of architecture: An empirical investigation. *Journal of Environmental Psychology, 16,* 75–92.

Hubel, D. H., & Wiesel, T. N. (1979). Brain mechanisms of behavior. *Scientific American, 241,* 150–162.

Huberman, B. A., & Lukose, R. M. (1997). Social

dilemmas and Internet congestion. *Science, 277*, 535–537.

Huerta, F., & Horton, R. (1978). Coping behavior of elderly flood victims. *The Gerontologist, 18*, 541–546.

Huffman, K. T., Grossnickle, W. F., Cope, J. G., & Huffman, K. P. (1995). Litter reduction: A review and integration of the literature. *Environment and Behavior, 27*, 153–183.

Hughes, J., & Goldman, M. (1978). Eye contact, facial expression, sex, and the violation of personal space. *Perceptual and Motor Skills, 46*, 579–584.

Hummel, C. F., Levitt, L., & Loomis, R. J. (1973). *Research strategies for measuring attitudes toward pollution.* Unpublished manuscript, Colorado State University.

Hummel, C. F., Loomis, R. J., & Hebert, J. A. (1975). *Effects of city labels and cue utilization on air pollution judgments* (Working Papers in Environmental-Social Psychology, No. 1). Unpublished manuscript, Colorado State University.

Hunt, M. E. (1984). Environmental learning without being there. *Environment and Behavior, 16*, 307–334.

Hunter, A. (1978). Persistence of local sentiments in mass society. In D. Street (Ed.), *Handbook of contemporary urban life* (pp. 133–162). San Francisco: Jossey-Bass.

Huntington, E. (1915). *Civilization and climate.* New Haven, CT: Yale University Press.

Huntington, E. (1945). *Mainsprings of civilization.* New York: Wiley.

Hygge, S. (1992). Heat and performance. In D. M. Jones & A. P. Smith (Eds.), *Handbook of human performance* (pp. 79–104). New York: Wiley.

Hygge, S. (1993). Classroom experiments on the effects of aircraft, traffic, train, and verbal noise on long-term recall and recognition and memory. In M. Vallet (Ed.), *Noise as a public health problem: Proceedings of the Sixth International Congress* (pp. 531–538). Paris: INGRETS.

Hytten, K., & Hasle, A. (1989). Fire fighters: A study of stress and coping. *Acta Psychiatrica Scandinavia, 80*, 50–55.

Ickes, W., Patterson, M. L., Rajecki, D. W., & Tanford, S. (1982). Behavioral and cognitive consequences of reciprocal versus compensatory responses to preinteraction expectancies. *Social Cognition, 1*, 160–190.

Iltus, S., & Hart, R. (1994). Participatory planning and design of recreational spaces with children. *Architecture & Comportment / Architecture & Behavior, 10*, 361–370.

Im, S. (1984). Visual preferences in enclosed urban spaces: An exploration of a scientific approach to environmental design. *Environment and Behavior, 16*, 235–262.

Imamoğlu, E. O., & Kiliç, N. (1999). A social psychological comparison of the Turkish elderly residing at high or low quality institutions. *Journal of Environmental Psychology, 19*, 231–242.

Intergovernmental Panel on Climate Change (1991). *Climate change.* Washington, DC: Island Press.

Ironson, G., Wynings, C., Schneierman, N., Baum, A., Rodriguez, M., Greenwood, D., Benight, C., Antoni, M., LaPerriere, A., Huang, H., Klimas, N., & Fletcher, M. A. (1997). Posttraumatic stress symptoms, intrusive thoughts, loss, and immune function after Hurricane Andrew. *Psychosomatic Medicine, 59*, 128–141.

Irwin, P. N., Gartner, W. C., & Phelps, C. C. (1990). Mexican American/Anglo cultural differences as recreation style determinants. *Leisure Sciences, 12*, 335–348.

Ising, H., & Melchert, H. U. (1980). Endocrine and cardiovascular effects of noise. In *Noise as a public health problem: Proceedings of the Third International Congress* (pp. 194–203). (ASHA Report No. 10). Rockville, MD: American Speech and Hearing Association.

Ising, H., Rebebtisch, E., Poustka, F., & Curio, I. (1990). Annoyance and health risk caused by military low-altitude flight noise. *International Archives of Occupational and Environmental Health, 62*, 357–363.

Iso-Ahola, S. E. (1986). A theory of substitutability of leisure behavior. *Leisure Science, 8*, 367–389.

Issar, A. S. (1995). Climate change and the history of the Middle East. *American Scientist, 83*, 350–355.

Ittelson, W. H. (1970). Perception of the large-scale environment. *Transactions of the New York Academy of Sciences, 32*, 807–815.

Ittelson, W. H. (1973). Environmental perception and contemporary perceptual theory. In W. H. Ittelson (Ed.), *Environment and cognition* (pp. 1–19). New York: Seminar Press.

Ittelson, W. H. (1978). Environmental perception and urban experience. *Environment and Behavior, 10,* 193–213.

Ittelson, W. H., Proshansky, H. M., & Rivlin, L. G. (1970). A study of bedroom use on two psychiatric wards. *Hospital and Community Psychiatry, 21,* 25–28.

Ittelson, W. H., Proshansky, H. M., & Rivlin, L. G. (1972). Bedroom size and social interaction of the psychiatric ward. In J. Wohlwill & D. Carson (Eds.), *Environment and the social sciences* (pp. 95–104). Washington, DC: American Psychological Association.

Ittelson, W. H., Proshansky, H. M., Rivlin, L. G., & Winkel, G. H. (1974). *An introduction to environmental psychology.* New York: Holt, Rinehart and Winston.

Ittelson, W. H., Rivlin, L. G., & Proshansky, H. M. (1976). The use of behavioral maps in environmental psychology. In H. M. Proshansky, W. H. Ittelson, & L. G. Rivlin (Eds.), *Environmental psychology: People and their physical settings* (pp. 658–668). New York: Holt, Rinehart and Winston.

Iwata, O. (1992). Crowding and behavior in Japanese public spaces: Some observations and speculations. 10th International Congress of the International Association for Cross-Cultural Psychology: Symposium on cross-cultural perspectives on crowding and behavior (1990, Nara, Japan). *Social Behavior and Personality, 20,* 57–70.

Jackson, E. L. (1981). Responses to earthquake hazard: The west coast of America. *Environment and Behavior, 3,* 387–416.

Jackson, K. (1985). *Crabgrass frontier.* London: Oxford University Press.

Jackson, S. W. (1986). *Melancholia and depression from Hippocratic times to modern times.* New Haven, CT: Yale University Press.

Jacobs, H. E., & Bailey, J. S. (1982). Evaluating participation in a residential program. *Journal of Environmental Systems, 13,* 245–254.

Jacobs, K. W., & Blandino, S. E. (1992). Effects of paper on which the profile of mood states is printed on the psychological states it measures. *Perceptual and Motor Skills, 75,* 267–271.

Jacobs, S. V., Evans, G. W., Catalano, R., & Dooley, D. (1984). Air pollution and depressive symptomatology: Exploratory analyses of intervening psychosocial factors. *Population and Environment, 7,* 260–272.

Jain, U. (1993). Concomitants of population density in India. *Journal of Social Psychology, 133,* 331–336.

James, B. (1984). A few words about the home field advantage. In B. James (Ed.), *The Bill James baseball abstract 1984.* New York: Ballantine.

James, W. (1979). The dilemma of determinism. In F. H. Burkhardt, F. Bowers, & I. K. Skrupkelis (Eds.), *Will to believe* (pp. 114–140). Cambridge, MA: Harvard University Press.

Jamison, R. N., Anderson, K. O., & Slater, M. A. (1996). Weather changes and pain: Perceived influence of local climate on pain complaint in chronic pain patients. *Pain, 61,* 309–315.

Janney, J., Minoru, M., & Holmes, T. (1977). Impact of natural catastrophe on life events. *Journal of Human Stress, 3,* 22–34.

Jansen, G. N. (1973). Non-auditory effects of noise—Physiological and psychological reactions in man. *Proceedings of the International Congress on Noise as a Public Health Problem. Dubrovnik, Yugoslavia, May 13–18.* Washington, DC: U.S. Environmental Protection Agency.

Jason, L. A., Zolik, E. S., & Matese, F. (1979). Prompting dog owners to pick up dog droppings. *American Journal of Community Psychology, 7,* 339–351.

Jencks, C. (1994). *The homeless.* Cambridge, MA: Harvard University Press.

Jentsch, F., Barnett, J., Bowers, C. A., & Salas, E. (1999). Who is flying this plane anyway? What mishaps tell us about crew members' role assignment and air crew situation awareness. *Human Factors, 41,* 1–14.

Jerdee, T. H., & Rosen, B. (1974). The effects of opportunity to communicate and visibility of individual decisions on behavior in the common interest. *Journal of Applied Psychology, 59,* 712–716.

Jessen, G., Steffensen, P., & Jensen, B. (1998). Seasons and meteorological factors in suicidal behaviour: Findings and methodological considerations from a Danish study. *Archives of Suicide Research, 4,* 263–280.

Job, R. F. S. (1988). Community response to noise: A review of factors influencing the relationship between noise exposure and reaction. *Journal of the Acoustical Society of America, 83,* 991–1001.

Johannes, L. (1996, March 6). More people warm to simulated sunlight but scientists doubt it cures winter blues. *The Wall Street Journal*, p. B1.

Johnson, J. E., & Leventhal, H. (1974). Effects of accurate expectations and behavioral instructions on reactions during a noxious medical examination. *Journal of Personality and Social Psychology, 29*, 710–718.

Johnson, R., & Richards, B. (1988, November 2). The buyer of Sears Tower will face rather shattering problem: Windows. *The Wall Street Journal*, p. A8.

Johnson-Laird, P. N. (1996). Images, models, and propositional representations. In M. De Vega, M. J. Intons-Peterson, P. N. Johnson-Laird, M. Denis, & M. Marschark (Eds.), *Models of visuospatial cognition* (pp. 90–127). New York: Oxford University Press.

Joiner, D. (1971). Office territory. *New Society, 7*, 660–663.

Jones, C. J., Nesselroade, J. R., & Birkel, R. C. (1991). Examination of staffing level effects in the family household: An application of P-technique factor analysis. *Journal of Environmental Psychology, 11*, 59–73.

Jones, D. M., Smith, A. P., & Broadbent, D. E. (1979). Effects of moderate intensity noise on the Bakan Vigilance Task. *Journal of Applied Psychology, 64*, 627–634.

Jones, F. N., & Tauscher, J. (1978). Residence under an airport landing pattern as a factor in teratism. *Archives of Environmental Health, 33*, 10–12.

Jones, J. W. (1978). Adverse emotional reactions of nonsmokers to secondary cigarette smoke. *Environmental Psychology and Nonverbal Behavior, 3*, 125–127.

Jones, J. W., & Bogat, G. A. (1978). Air pollution and human aggression. *Psychological Reports, 43*, 721–722.

Jones, R. T., Ribbe, D. P., & Cunningham, P. (1994). Psychosocial correlates of fire disaster among children and adolescents. *Journal of Traumatic Stress, 7*, 117–122.

Jorgenson, D. O., & Dukes, F. O. (1976). Deindividuation as a function of density and group membership. *Journal of Personality and Social Psychology, 34*, 24–39.

Jorgenson, D. O., & Papciak, A. S. (1981). The effects of communication, resource feedback and identifiability on behavior in a simulated commons. *Journal of Experimental Social Psychology, 17*, 373–385.

Judge, P. G., & deWaal, F. B. (1993). Conflict avoidance among rhesus monkeys: Coping with short-term crowding. *Animal Behaviour, 46*, 221–232.

Jung, J. (1984). Social support and its relation to health: A critical examination. *Basic and Applied Social Psychology, 5*, 143–149.

Juniu, S., Tedrick, T., & Boyd, R. (1996). Leisure or work?: Amateur and professional musicians' perception of rehearsal and performance. *Journal of Leisure Research, 28*, 44–56.

Kahneman, D. (1973). *Attention and effort*. Englewood Cliffs, NJ: Prentice-Hall.

Kaiser, F. G., & Shimoda, T. A. (1999). Responsibility as a predictor of ecological behavior. *Journal of Environmental Psychology, 19*, 243–253.

Kaiser, F. G., Wölfing, S., & Fuhrer, U. (1999). Environmental attitude and ecological behaviour. *Journal of Environmental Psychology, 19*, 1–19.

Kaiser, J. (1999a). Getting to the roots of carbon loss, chili's gain. *Science, 285*, 1198–1199.

Kaiser, J. (1999b). Turning engineers into resource accountants. *Science, 285*, 685–686.

Kaitilla, S. (1993). Satisfaction with public housing in Papua New Guinea: The case of West Taraka housing scheme. *Environment and Behavior, 25*, 514–545.

Kalkstein, L. S., & Davis, R. E. (1989). Weather and human mortality: An evaluation of demographic and interregional responses in the United States. *Annals of the Association of American Geographers, 79*, 44–64.

Kammann, R., Thompson, R., & Irwin, R. (1979). Unhelpful behavior in the street: City size or immediate pedestrian density? *Environment and Behavior, 11*, 245–250.

Kaniasty, K., & Norris, F. (1991, June). *In Search of "altruistic community": Social support following Hurricane Hugo*. Paper presented at the Third Biennial Conference on Community Research and Action, Tempe, AZ.

Kaniasty, K., & Norris, F. (1993). A test of the social support deterioration model in the context of natural disaster. *Journal of Personality and Social Psychology, 64*, 395–408.

Kaniasty, K., Norris, F., & Murrell, S. A. (1990). Received and perceived social support following natural disaster. *Journal of Applied Social Psychology, 20*, 85–144.

Kant, I. (1790). *The critique of pure reason* (republished in 1929). New York: Macmillan.

Kaplan, R. (1975). Some methods and strategies in the prediction of preference. In E. H. Zube, R. O. Brush, & J. G. Fabos (Eds.), *Landscape assessment* (pp. 118–129). Stroudsburg, PA: Dowden, Hutchinson, & Ross.

Kaplan, R. (1984). The impact of urban nature: A theoretical analysis. *Urban Ecology, 8,* 189–197.

Kaplan, R. (1985). Nature at the doorstep: Residential satisfaction and the nearby environment. *Journal of Architectural Planning Research, 2,* 115–127.

Kaplan, R. (1987). Validity in environment/behavior research. *Environment and Behavior, 19,* 495–500.

Kaplan, R., & Kaplan, S. (1987). The garden as a restorative experience. In M. Francis & R. T. Hester, Jr. (Eds.), *Meanings of the garden* (pp. 334–341). Davis, CA: University of California, Davis.

Kaplan, R., & Kaplan, S. (1989). *The experience of nature: A psychological perspective.* New York: Cambridge University Press.

Kaplan, R., Kaplan, S., & Ryan, R. L. (1998). *With people in mind: Design and management of everyday nature.* Washington DC: Island Press.

Kaplan, S. (1975). An informal model for the prediction of preference. In E. H. Zube, R. O. Brush, & J. G. Fabos (Eds.), *Landscape assessment* (pp. 92–101). Stroudsburg, PA: Dowden, Hutchinson, & Ross.

Kaplan, S. (1987). Aesthetics, affect, and cognition: Environmental preference from an evolutionary perspective. *Environment and Behavior, 19,* 3–32.

Kaplan, S. (1995). The restorative benefits of nature: Toward an integrative framework. *Journal of Environmental Psychology, 15,* 169–182.

Kaplan, S., Bardwell, L. V., & Slakter, D. B. (1993). The museum as a restorative environment. *Environment and Behavior, 25,* 725–742.

Kaplan, S., & Kaplan, R. (1978). *Humanscape: Environments for people.* North Scituate, MA: Duxbury Press.

Kaplan, S., & Kaplan, R. (1982). *Cognition and the environment: Functioning in an uncertain world.* New York: Praeger.

Kaplan, S., & Kaplan, R. (1989). The visual environment: Public participation in design and planning. *Journal of Social Issues, 45,* 59–86.

Karabenick, S. A., & Meisels, M. (1972). Effects of performance evaluation on interpersonal distance. *Journal of Personality, 40,* 275–286.

Karlin, R. A., Epstein, Y., & Aiello, J. (1978). Strategies for the investigation of crowding. In A. Esser & B. Greenbie (Eds.), *Design for communality and privacy* (pp. 71–88). New York: Plenum.

Karlin, R. A., McFarland, D., Aiello, J. R., & Epstein, Y. M. (1976). Normative mediation of reactions to crowding. *Environmental Psychology and Nonverbal Behavior, 1,* 30–40.

Karlin, R. A., Rosen L., & Epstein, Y. (1979). Three into two doesn't go: A follow-up of the effects of overcrowded dormitory rooms. *Personality and Social Psychology Bulletin, 5,* 391–395.

Karp, D. G. (1996). Values and their effect on pro-environmental behavior. *Environment and Behavior, 28,* 111–133.

Kastka, J. (1980). *Noise annoyance reduction in residential areas by traffic control technics.* 10th International Congress on Acoustics, Sydney.

Kates, R. W., Haas, J. E., Amaral, D. J., Olson, R. A., Ramos, R., & Olson, R. (1973). Human impact of the Managua earthquake: Transitional societies are peculiarly vulnerable to natural disasters. *Science, 182,* 981–989.

Katovich, M. (1986). Ceremonial openings in bureaucratic encounters: From shuffling feet to shuffling papers. In N. K. Denzin (Ed.), *Studies in symbolic interaction* (Vol. 6, pp. 307–333). Greenwich, CT: JAI Press.

Katz, P. (1937). *Animals and men.* New York: Longmans, Green.

Katzev, R., & Mishima, H. R. (1992). The use of posted feedback to promote recycling. *Psychological Reports, 71,* 259–264.

Katzev, R. D., & Pardini, A.U. (1987–1988). The comparative effectiveness of reward and commitment approaches in motivating community recycling. *Journal of Environmental Systems, 17,* 93–113.

Kaya, N., & Erkip, F. E. (1999). Invasion of personal space under the condition of short-term crowding: A case study on an automatic teller machine. *Journal of Environmental Psychology, 19,* 183–189.

Keane, C. (1991). Socioenvironmental determinants of community formation. *Environment and Behavior, 23,* 27–46.

Keane, C. (1998). Evaluating the influence of fear of crime as an environmental mobility restrictor on women's routine activities. *Environment and Behavior, 30,* 60–74.

Kearney, A. R., & De Young, R. (1995). A knowledge-based intervention for promoting carpooling. *Environment and Behavior, 27,* 650–678.

Keating, J., & Snowball, H. (1977). Effects of crowding and depersonalization on perception of group atmosphere. *Perceptual and Motor Skills, 44,* 431–435.

Keller, L. M., Bouchard, T. J., Jr., Arvey, R. D., Segal, N. L., & Davis, R. V. (1992). Work values: Genetic and environmental influences. *Journal of Applied Psychology, 77,* 79–88.

Kelley, H., & Arrowwood, A. (1960). Coalitions in the triad: Critique and experiment. *Sociometry, 23,* 231–244.

Kelly, I. W., Rotton, J., & Culver, R. (1985–86). The moon was full and nothing happened. *Skeptical Inquirer, 10,* 129–143.

Kelly, J. T. (1985). Trauma: With the example of San Francisco's shelter programs. In P. W. Brickner, L. K. Scharer, B. Conanan, A. Elvy, & M. Savarese (Eds.), *Health care of homeless people* (pp. 77–91). New York: Springer.

Kempton, W., Darley, J. M., & Stern, P. C. (1992). Psychological research for the new energy problems: Strategies and opportunities. *American Psychologist, 47,* 1213–1223.

Kempton, W., Harris, C. K., Keith, J. G., & Weil, J. S. (1985). Do consumers know what works in energy conservation? *Marriage and Family Review, 9,* 115–133.

Kempton, W., & Montgomery, L. (1982). Folk quantification of energy. *Energy—The International Journal, 7,* 817–827.

Kendler, K. S., Neale, M. C., Kessler, R. C., Heath, A. C., & Eaves, L. J. (1992). The genetic epidemiology of phobias in women. *Archives of General Psychiatry, 49,* 273–281.

Kenrick, D. T., & Johnson, G. A. (1979). Interpersonal attraction in aversive environments. A problem for the classical conditioning paradigm. *Journal of Personality and Social Psychology, 87,* 572–579.

Kenrick, D. T., & MacFarlane, S. W. (1986). Ambient temperature and horn honking: A field study of the heat/aggression relationship. *Environment and Behavior, 18,* 179–191.

Kent, S. (1991). Partitioning space: Cross-cultural factors influencing domestic spatial segmentation. *Environment and Behavior, 23,* 438–473.

Kent, S. J., von Gierke, H. E., & Tolan, G. D. (1986). Analysis of the potential association between noise-induced hearing loss and cardiovascular disease in USAF aircrew members. *Aviation Space Environmental Medicine, 4,* 348–361.

Kerr, J. H., & Tacon, P. (1999). Psychological responses to different types of locations and activities. *Journal of Environmental Psychology, 19,* 287–294.

Kerr, N. L., & Kaufman-Gilliland, C. M. (1994). Communication, commitment, and cooperation in social dilemmas. *Journal of Personality and Social Psychology, 66,* 513–529.

Kerr, R. (1997). Greenhouse forecasting still cloudy. *Science, 276,* 1040–1042.

Kerr, R. A. (1991). Geothermal tragedy of the commons. *Science, 253,* 134–135.

Kerr, R. A. (1992a). New assaults seen on Earth's ozone shield. *Science, 255,* 797–798.

Kerr, R. A. (1992b). When climate twitches, evolution takes great leaps. *Science, 257,* 1622–1624.

Kerr, R. A. (1993). Ozone takes a nosedive after the eruption of Mt. Pinatubo. *Science, 260,* 490–491.

Kerr, R. A. (1995). It's official: First glimmer of greenhouse warming seen. *Science, 270,* 1565–1567.

Kerr, R. A. (1996). New dawn for sun–climate links? *Science, 271,* 1360–1361.

Kerr, R. A. (1998). Warming's unpleasant surprise: Shivering in the greenhouse? *Science, 28,* 156–158.

Kerr, R. A. (1999a). Big El Niños ride the back of slower climate change. *Science, 283,* 1108–1109.

Kerr, R. A. (1999b). Has a great river in the sea slowed down? *Science, 286,* 1061–1062.

Kerr, R. A. (1999c). Will the Arctic Ocean lose all its ice? *Science, 286,* 1828.

Kevan, S. M. (1980). Perspectives on season of suicide: A review. *Social Science and Medicine, 14,* 369–378.

Kiehl, J. T. (1999). Solving the aerosol puzzle. *Science, 283,* 1273–1275.

Kildeso, J., Wyon, D., Skov, T., & Schneider, T. (1999). Visual analogue scales for detecting changes in symptoms of the sick building syndrome in an intervention study. *Scandinavian Journal of Work, Environment, & Health, 25,* 361–367.

Kingston, S. G., & Hoffman-Goetz, L. (1996). Effect of environmental enrichment and housing density on immune system reactivity to acute exercise stress. *Physiology & Behavior, 60,* 145–150.

Kira, A. (1976). *The bathroom.* New York: Viking.

Kirk, N. L. (1988). Factors affecting perceptions of safety in a campus environment. In D. Lawrence, R. Habe, A. Hacker, & D. Sherrod (Eds.), *People's needs/planet management: Paths to co-existence* (pp. 215–221). Washington, DC: Environmental Design Research Association.

Kitchin, R. M. (1994). Cognitive maps: What are they and why study them? *Journal of Environmental Psychology, 14,* 1–19.

Kitchen, R. M. (1996). Methodological convergence in cognitive mapping research: Investigating configurational knowledge. *Journal of Environmental Psychology, 16,* 163–185.

Kitchen, R. M., Blades, M., & Golledge, R. G. (1997). Relations between psychology and geography. *Environment and Behavior, 29,* 554–573.

Kleeman, W. B. (1988). The politics of office design. *Environment and Behavior, 20,* 537–549.

Klein, H.-J., (1993). Tracking visitor circulation in museum settings. *Environment and Behavior, 25,* 782–800.

Klein, K., & Beith, B. (1985). Re-examination of residual arousal as an explanation of aftereffects: Frustration tolerance versus response speed. *Journal of Applied Psychology, 70,* 642–650.

Klein, K., & Harris, B. (1979). Disruptive effects of disconfirmed expectancies about crowding. *Journal of Personality and Social Psychology, 37,* 769–777.

Kline, L. M., & Bell, P. A. (1983). Privacy preference and interpersonal distancing. *Psychological Reports, 53,* 1214.

Kline, L. M., Bell, P. A., & Babcock, A. M. (1984). Field dependence and interpersonal distance. *Bulletin of the Psychonomic Society, 22,* 421–422.

Kline, L. M., Harrison, A., Bell, P. A., Edney, J. J., &

Hill, E. (1984). Verbal reinforcement and feedback as solutions to a simulated commons dilemma. *Psychological Documents, 14,* 24 (ms. No. 2648).

Kloor, K. (1999). A surprising tale of life in the city. *Science, 286,* 663.

Kmiecik, C., Mausar, P., & Banziger, G. (1979). Attractiveness and interpersonal space. *Journal of Social Psychology, 108,* 227–278.

Knave, B. G., Wibom, R., Vas, M., Hedstrong, L. D., & Bergqvist, U. O. V. (1985). Work with video display terminals among office employees: I. Subjective symptoms and discomfort. *Scandinavian Journal of Work, Environment, & Health, 11,* 457–466.

Knez, I. (1995). Effects of indoor lighting on mood and cognition. *Journal of Environmental Psychology, 15,* 39–51.

Knowles, E. S. (1972). Boundaries around social space: Dyadic responses to an invader. *Environment and Behavior, 4,* 437–447.

Knowles, E. S. (1973). Boundaries around group interaction: The effect of group and member status on boundary permeability. *Journal of Personality and Social Psychology, 26,* 327–331.

Knowles, E. S. (1978). The gravity of crowding: Application of social physics to the effects of others. In A. Baum & Y. Epstein (Eds.), *Human response to crowding* (pp. 183–218). Hillsdale, NJ: Erlbaum.

Knowles, E. S. (1980a). An affiliative conflict theory of personal and group spatial behavior. In P. B. Paulus (Ed.), *Psychology of group influence* (pp. 133–188). Hillsdale, NJ: Erlbaum.

Knowles, E. S. (1980b). Convergent validity of personal space measures: Consistent results with low intercorrelations. *Journal of Nonverbal Behavior, 4,* 240–248.

Knowles, E. S. (1983). Social physics and the effects of others: Tests of the effects of audience size and distance on social judgments and behavior. *Journal of Personality and Social Psychology, 45,* 1263–1279.

Knowles, E. S., & Bassett, R. I. (1976). Groups and crowds as social entities: Effects of activity, size and member similarity on nonmembers. *Journal of Personality and Social Psychology, 34,* 837–845.

Knowles, E. S., & Brickner, M. A. (1981). Social cohesion effects on spatial cohesion. *Person-*

ality and Social Psychology Bulletin, 7, 309–313.

Knowles, E. S., & Johnson, P. K. (1974). Intrapersonal consistence and interpersonal distance. *JSAS Catalog of Selected Documents in Psychology, 4,* 124.

Knowles, E. S., Kreuser, B., Haas, S., Hyde, M., & Schuchart, G. E. (1976). Group size and the extension of social space boundaries. *Journal of Personality and Social Psychology, 33,* 647–654.

Koelega, H. S., & Brinkman, J. A. (1986). Noise and vigilance: An evaluative review. *Human Factors, 28,* 465–481.

Kohlenberg, R., & Phillips, T. (1973). Reinforcement and rate of litter depositing. *Journal of Applied Behavior Analysis, 6,* 391–396.

Kohler, W. (1970). *Gestalt psychology: An introduction to new concepts in modern psychology.* New York: Liveright.

Kojima, H. (1984). A significant stride toward the comparative study of control. *American Psychologist, 39,* 972–973.

Komorita, S. S. (1987). Cooperative choice in decomposed social dilemmas. *Personality and Social Psychology Bulletin, 13,* 53–63.

Komorita, S. S., & Parks, C. D. (1995). *Social dilemmas.* Boulder, CO: Westview Press.

Konar, E., Sundstrom, E., Brady, C., Mandel, D., & Rice, R. (1982). Status markers in the office. *Environment and Behavior, 14,* 561–580.

Konarski, E. A., Riddle, J. I., & Walker, J. (1994). Case study of the relation between census reduction and injuries to residents in an ICF/MR. *Mental Retardation, 32,* 132–136.

Konecni, V. J. (1975). The mediation of aggressive behavior: Arousal level versus anger and cognitive labeling. *Journal of Personality and Social Psychology, 32,* 706–712.

Konecni, V. J., Libuser, L., Morton, H., & Ebbesen, E. B. (1975). Effects of a violation of personal space on escape and helping responses. *Journal of Experimental Social Psychology, 11,* 288–299.

Koneya, M. (1976). Location and interaction in row-and-column seating arrangements. *Environment and Behavior, 8,* 265–283.

Koop, C. E. (1986). *The health consequences of involuntary smoking: A report of the Surgeon General.* Rockville, MD: U.S. Department of Health and Human Services, Public Health Service, Centers for Disease Control, Center for Health Promotion and Education, Office on Smoking and Health.

Korpela, K., & Hartig, T. (1996). Restorative qualities of favorite places. *Journal of Environmental Psychology, 16,* 221–233.

Korte, C. (1980). Urban–nonurban differences in social behavior and social psychological models of urban impact. *Journal of Social Issues, 36,* 29–51.

Korte, C., Ypma, I., & Toppen, A. (1975). Helpfulness in Dutch society as a function of urbanization and environmental input level. *Journal of Personality and Social Psychology, 32,* 996–1003.

Koscheyev, V. S., Martens, V. K., Kosenkov, A. A., Lartzev, M. A., & Leon, G. R. (1993). Psychological status of Chernobyl nuclear power plant operators after the natural disaster. *Journal of Traumatic Stress, 6,* 561–568.

Koss, M. P., Gudycz, C. A., & Wisiniewski, N. (1987). The scope of rape: Incidence and prevalence of sexual aggression and victimization in a national sample in higher education. *Journal of Counseling and Clinical Psychology, 55,* 162–170.

Kosslyn, S. M. (1980). *Image and mind.* Cambridge, MA: Harvard University Press.

Kosslyn, S. M. (1983). *Ghosts in the mind's machine: Creating images in the brain.* New York: W. W. Norton.

Kosslyn, S. M., Ball, T. M., & Reiser, B. J. (1978). Visual images preserve metric spatial information: Evidence from studies of image scanning. *Journal of Experimental Psychology: Human Perception and Performance, 4,* 47–60.

Kostof, S. (1987). *America by design.* New York: Oxford University Press.

Kovach, E. J., Jr., Surrette, M. A., & Aamodt, M. G. (1988). Following informal street maps: Effects of map design. *Environment and Behavior, 20,* 683–699.

Kramer, R. M., & Brewer, M. B. (1984). Effects of group identity on resource use in a simulated commons dilemma. *Journal of Personality and Social Psychology, 46,* 1044–1057.

Krauss, R. M., Freedman, J. L., & Whitcup, M. (1978). Field and laboratory studies of littering. *Journal of Experimental Social Psychology, 14,* 109–122.

Kristal-Boneh, E., Harari, G., & Green, M. S. (1997). Heart rate response to industrial work at different outdoor temperatures with or without temperature control system at the plant. *Ergonomics, 40,* 729–736.

Kroling, P. (1985). Natural and artificial air ion—a biologically relevant climatic factor? *International Journal of Biometerology, 29,* 233–242.

Kryter, K. D. (1990). Aircraft noise and social factors in psychiatric hospital admission rates: A re-examination of some data. *Psychological Medicine, 20,* 395–411.

Kryter, K. D. (1994). *The handbook of learning and effects of noise.* San Diego: Academic Press.

Kuipers, B. (1982). The "map in the head" metaphor. *Environment and Behavior, 14,* 202–220.

Küller, R., & Lindsten, C. (1992). Health and behavior of children in classrooms with and without windows. *Journal of Environmental Psychology, 12,* 305–317.

Kuo, F. E., Bacaicoa, M., & Sullivan, W. C. (1998). Transforming inner-city landscapes: Trees, sense of safety, and preference. *Environment and Behavior, 30,* 28–59.

Kupritz, V. W. (1998). Privacy in the work place: The impact of building design. *Journal of Environmental Psychology, 18,* 341–356.

Kweon, B.-S., Sullivan, W. C., & Wiley, A. R. (1998). Green common spaces and the social integration of inner-city older adults. *Environment and Behavior, 30,* 504–519.

La Hart, D., & Bailey, J. S. (1975). Reducing children's littering on a nature trail. *Journal of Environmental Education, 7,* 37–45.

Lalli, M. (1992). Urban-related identity: Theory, measurement, and empirical findings. *Journal of Environmental Psychology, 12,* 285–303.

Landesberger, H. A. (1958). *Hawthorne revisited.* Ithaca, NY: Cornell University Press.

Landy, F. J. (1989). *Psychology of work behavior.* Pacific Grove, CA: Brooks/Cole.

Lang, J. (1987). *Creating architectural theory: The role of the behavioral sciences in environmental design.* New York: Van Nostrand Reinhold.

Lang, J. (1988). Understanding normative theories of architecture. *Environment and Behavior, 20,* 601–632.

Langer, E. J., & Rodin, J. (1976). The effects of choice and enhanced personal responsibility for the aged: A field experiment in an institutional setting. *Journal of Personality and Social Psychology, 34,* 191–198.

Langer, E. J., & Saegert, S. (1977). Crowding and cognitive control. *Journal of Personality and Social Psychology, 35,* 175–182.

Larsen, L., Adams, J., Deal, B., Kweon, B., & Tyler, E. (1998). Plants in the workplace: The effects of plant density on productivity, attitudes, and perceptions. *Environment and Behavior, 30,* 261–281.

Larson, J. H., & Lowe, W. (1990). Family cohesion and personal space in families with adolescents. *Journal of Family Issues, 11,* 101–108.

Larson, R. W., Gillman, S. A., & Richards, M. H. (1997). Divergent experiences of family leisure: Fathers, mothers, and young adolescents. *Journal of Leisure Research, 29,* 78–97.

Laska, S. B. (1990). Homeowner adaptation to flooding: An application of the general hazards coping theory. *Environment and Behavior, 22,* 320–357.

Lassen, C. L. (1973). Effects of proximity on anxiety and communication in the initial psychiatric interview. *Journal of Abnormal Psychology, 81,* 226–232.

Latané, B., & Darley, J. M. (1970). *The unresponsive bystander: Why doesn't he help?* New York: Appleton-Century-Crofts.

Latané, B., Liu, J., Nowak, A., Bonevento, M., & Zheng, L. (1995). Distance matters: Physical space and social impact. *Personality and Social Psychology Bulletin, 21,* 795–805.

Latta, R. M. (1978). Relation of status incongruous to personal space. *Personality and Social Psychology Bulletin, 4,* 143–146.

Lavelle, M., & Coyle, M. (Eds.) (1992, September). The racial divide in environmental law. Unequal protection. *National Law Journal, Supplement,* 21.

Lavrakas, P. J. (1982). Fear of crime and behavior restriction in urban and suburban neighborhoods. *Population and Environment, 5,* 242–264.

Lawson, B. R., & Walters, D. (1974). The effects of a new motorway on an established residential area. In D. Canter & T. Lee (Eds.), *Psychology and the built environment* (pp. 132–138). New York: Wiley.

Lawton, C. A. (1996). Strategies for indoor wayfinding: The role of orientation. *Journal of Environmental Psychology, 16,* 137–145.

Lawton, M. P. (1975). Competence, environmental press, and the adaptation of older people. In P. G. Windley & G. Ernst (Eds.), *Theory development in environment and aging* (pp. 13–83). Washington, DC: Gerontological Society.

Lawton, M. P., & Nahemow, L. (1973). Ecology and the aging process. In C. Eisdorfer & M. P. Lawton (Eds.), *The psychology of adult development and aging* (pp. 619–674). Washington, DC: American Psychological Association.

Lazarus, R. (1966). *Psychological stress and the coping process.* New York: McGraw-Hill.

Lazarus, R. S. (1998). *Fifty years of research and theory by R.S. Lazarus: An analysis of historical and perennial issues.* Malwah, NJ: Erlbaum.

Lazarus, R. S., & Cohen, J. B. (1977). Environmental stress. In I. Altman & J. F. Wohlwill (Eds.), *Human behavior and the environment: Current theory and research* (Vol. 2, pp. 89–127). New York: Plenum.

Lazarus, R. S., DeLongis, A., Folkman, S., & Gruen, R. (1985). Stress and adaptational outcomes: The problem of confounded measures. *American Psychologist, 40,* 770–779.

Lazarus, R. S., & Folkman, S. (1984). *Stress, appraisal, and coping.* New York: Springer.

Lazarus, R. S., & Launier, R. (1978). Stress-related transactions between person and environment. In L. A. Pervin & M. Lewis (Eds.), *Perspectives in interactional psychology* (pp. 287–327). New York: Plenum.

Leather, P., Pyrgas, M., Beale, D., & Lawrence, C. (1998). Windows in the workplace: Sunlight, view, and occupational stress. *Environment and Behavior, 30,* 739–762.

Leavitt, J., & Saegert, S. (1989). *From abandonment to hope: Community-households in Harlem.* New York: Columbia University Press.

LeBlanc, J. (1956). Impairment of manual dexterity in the cold. *Journal of Applied Physiology, 9,* 62–64.

LeBlanc, J. (1962). Local adaptation to cold of Gaspé fisherman. *Journal of Applied Physiology, 17,* 950–952.

LeBlanc, J. (1975). *Man in the cold.* Springfield, IL: Thomas.

LeBlanc, S. A. (1999). *Prehistoric warfare in the American Southwest.* Salt Lake City, UT: University of Utah Press.

Lebo, C. P., & Oliphant, K. P. (1968). Music as a source of acoustical trauma. *Laryngoscope, 78,* 1211–1218.

Lebovits, A., Bryne, M., & Strain, J. (1986). The case of asbestos-exposed workers: A psychological evaluation. In A. Lebovits, A. Baum, & J. Singer (Eds.), *Advances in environmental psychology* (Vol. 6, pp. 3–17). Hillsdale, NJ: Erlbaum.

Lee, D. H. K. (1964). Terrestrial animals in dry heat: Man in the desert. In D. B. Dill, E. G. Adolph, & C. G. Wilbur (Eds.), *Handbook of physiology* (pp. 551–582). Washington, DC: The American Physiological Society.

Lee, J., & Moray, N. (1992). Trust, control strategies and allocation of function in human–machine systems. *Ergonomics, 35,* 1243–1270.

Lehman, D. R., Wortman, C. B., & Williams, A. F. (1987). Long-term effects of losing a spouse or child in a motor vehicle crash. *Journal of Personality and Social Psychology, 52,* 218–231.

Leithead, C. S., & Lind, A. R. (1964). *Heat stress and heat disorders.* London: Cassell.

Lemke, S., & Moos, R. H. (1986). Quality of residential settings for elderly adults. *Journal of Gerontology, 41,* 268–276.

Leopold, A. (1949). *A Sand County almanac: And sketches here and there.* New York: Oxford University Press.

Lepore, S. J., Evans, G. W., & Palsane, M. N. (1991). Social hassles and psychological health in the context of chronic crowding. *Journal of Health and Social Behavior, 32,* 357–367.

Lepore, S. J., Evans, G. W., & Schneider, M. L. (1991). Dynamic role of social support in the link between chronic stress and psychological distress. *Journal of Personality and Social Psychology, 61,* 899–909.

Lepore, S. J., Evans, G. W., & Schneider, M. L. (1992). Role of control and social support in explaining the stress of hassles and crowding. *Environment and Behavior, 24,* 795–811.

Lercher, P. (1996). Environmental noise and health: An integrated research perspective. *Environment International, 22,* 117–129.

Lercher, P., Hortnagel, J., & Kofler, W. W. (1993). Work noise annoyance and blood pressure:

Combined effects with stressful working conditions. *International Archives of Occupational and Environmental Health, 65,* 23–28.

Lerner, R. N., Iwawaki, S., & Chihara, T. (1976). Development of interpersonal space schemata among Japanese children. *Developmental Psychology, 12,* 466–467.

Lester, D. (1995). An extension of the association between population density and mental illness to suicidal behavior. *Journal of Social Psychology, 135,* 657–658.

Lester, D. (1996). A hazardous environment and city suicide rates. *Perceptual and Motor Skills, 82,* 1330.

Levi, D., & Kocher, S. (1999). Virtual nature: The future effects of information technology on our relationship to nature. *Environment and Behavior, 31,* 203–226.

Levine, A. G. (1982). *Love Canal: Science, politics and people.* Lexington, MA: Lexington Books, D.C. Heath.

Levine, A., & Stone, R. (1986). Threats to people and what they value. Residents' perceptions of the hazards of Love Canal. In A. H. Lebovits, A. Baum, & J. Singer (Eds.), *Advances in environmental psychology* (Vol. 6, pp. 109–130). Hillsdale, NJ: Erlbaum.

Levine, B. D., & Stray-Gundersen, J. (1997). "Living high-training low": Effect of moderate-altitude acclimatization with low-altitude training on performance. *Journal of Applied Physiology, 83,* 102–112.

Levine, M. (1982). You-are-here maps: Psychological considerations. *Environment and Behavior, 14,* 221–237.

Levine, M., Marchon, I., & Hanley, G. (1984). The placement and misplacement of you-are-here maps. *Environment and Behavior, 16,* 139–157.

Levine, R. (1988, November). City stress index: 25 best, 25 worst. *Psychology Today, 22,* 53–58.

Levine, R. V., Martinez, T. S., Brase, G., & Sorenson, K. (1994). Helping in 36 U.S. cities. *Journal of Personality and Social Psychology, 67,* 69–82.

Levitt, L., & Leventhal, G. (1984, August). *Litter reduction: How effective is the New York State bottle bill?* Paper presented at the meeting of the American Psychological Association, Toronto, Canada.

Levy-Leboyer, C., & Naturel, V. (1991). Neighborhood noise annoyance. *Journal of Environmental Psychology, 11,* 75–86.

Lewin, K. (1951). Formalization and progress in psychology. In D. Cartwright (Ed.), *Field theory in social science.* New York: Harper.

Lewis, C. A. (1973). People–plant interaction: A new horticultural perspective. *American Horticulturist, 52,* 18–25.

Lewis, D. A., & Maxfield, M. G. (1980). Fear in the neighborhoods: An investigation of the impact of crime. *Journal of Research in Crime and Delinquency, 17,* 160–169.

Lewis, J., Baddeley, A. D., Bonham, K. G., & Lovett, D. (1970). Traffic pollution and mental efficiency. *Nature, 225,* 95–97.

Ley, D., & Cybriwsky, R. (1974a). The spatial ecology of stripped cars. *Environment and Behavior, 6,* 53–68.

Ley, D., & Cybriwsky, R. (1974b). Urban graffiti as territorial markers. *Annals of the Association of American Geographers, 64,* 491–505.

Li, S. (1994). Users' behaviour of small urban spaces in winter and marginal seasons. *Architecture & Comportement, 10,* 95–109.

Lieber, A. L., & Sherin, C. R. (1972). Homicides and the lunar cycle: Toward a theory of lunar influence on human emotional disturbance. *American Journal of Psychiatry, 129,* 101–106.

Liebrand, W. B. G., & Messick, D. M. (Eds.) (1996). *Frontiers in social dilemmas research.* New York: Springer Verlag.

Lifton, R. J., & Olson, E. (1976). The human meaning of total disaster. The Buffalo Creek experience. *Psychiatry, 39,* 1–18.

Lima, B. R., Pai, S., Cavis, L., Haro, J. M., Lima, A. M., Toledo, V., Lozano, J., & Santacruz, H. (1991). Psychiatric disorders in primary health care clinics one year after a major Latin American disaster. *Stress Medicine, 7,* 25–32.

Lindberg, E., Hartig, T., Garvill, J., & Gärling, T. (1992). Residential-location preferences across the lifespan. *Journal of Environmental Psychology, 12,* 187–198.

Linet, M. S., Hatch, E. E., Kleinerman, R. A., Robison, L. L., Kaune, W. T., Friedman, D. R., Severson, R. K., Haines, C. M., Hartsock, C. T., Niwa, S., Wacholder, S., & Tarone, R. E. (1997). Residential exposure to magnetic fields and acute lymphoblastic leukemia in children. *New England Journal of Medicine, 337,* 1–7.

Lipman, A. (1967). Chairs as territory. *New Society, 20,* 564–566.

Lipsey, M. W. (1977). Attitudes toward the environment and pollution. In S. Oskamp (Ed.), *Attitudes and opinions.* Englewood Cliffs, NJ: Prentice-Hall.

Lipton, S. G. (1977). Evidence of central city revival. *American Planning Association Journal, 45,* 136–147.

Lloyd, A. J., & Shurley, J. T. (1976). The effects of sensory perceptual isolation on single motor unit conditioning. *Psychophysiology, 13,* 340–361.

Lloyd, R., & Steinke, T. (1986). The identification of regional boundaries in cognitive maps. *Professional Geographer, 38,* 149–159.

Lloyd, W. F. (1833). *Two lectures on the checks to population.* New York: Augustus M. Kelley (facsimile edition, 1968).

Locke, E. A. (1968). Toward a theory of task motivation and incentives. *Organizational Behavior and Human Performance, 3,* 157–189.

Locke, E. A. (1970). Job satisfaction and job performance: A theoretical analysis. *Organizational Behavior and Human Performance, 5,* 484–500.

Lockhard, J. S., McVittie, R. I., & Isaac, L. M. (1977). Functional significance of the affiliative smile. *Bulletin of the Psychonomic Society, 9,* 367–370.

Loewen, L. J., & Suedfeld, P. (1992). Cognitive and arousal effects of masking office noise. *Environment and Behavior, 24,* 381–395.

Lomranz, J., Shapira, A., Choresh, N., & Gilat, Y. (1975). Children's personal space as a function of age and sex. *Developmental Psychology, 11,* 541–545.

London, B., Lee, B., & Lipton, S. G. (1986). The determinants of gentrification in the United States: A city level analysis. *Urban Affairs Quarterly, 21,* 369–387.

Long, G. T., Selby, J. W., & Calhoun, L. G. (1980). Effects of situational stress and sex on interpersonal distance preference. *Journal of Psychology, 105,* 231–237.

Lonigan, C. J., Anthony, J. L., & Shannon, M. P. (1998). Diagnostic efficacy of posttraumatic symptoms in children exposed to disaster. *Journal of Clinical Child Psychology, 27,* 255–267.

Lonigan, C. J., Shannon, M. P., Taylor, C. M., Finch, A. J., Jr., & Sallee, F. R. (1994). Children exposed to disaster: II. Risk factors for the development of post-traumatic symptomatology. *Journal of the American Academy of Child & Adolescent Psychiatry, 33,* 94–105.

Loo, C. (1972). The effects of spatial density on the social behavior of children. *Journal of Applied Social Psychology, 4,* 219–226.

Loo, C. (1978). Density, crowding, and preschool children. In A. Baum & Y. Epstein (Eds.), *Human response to crowding* (pp. 371–388). Hillsdale, NJ: Erlbaum.

Loo, C., & Kennelly, D. (1979). Social density: Its effects on behaviors and perceptions of preschoolers. *Environmental Psychology and Nonverbal Behavior, 3,* 131–146.

Loo, C. M., & Ong, P. (1984). Crowding perceptions, attitudes, and consequences of crowding among the Chinese. *Environment and Behavior, 16,* 55–67.

Loo, C., & Smetana, J. (1978). The effects of crowding on the behavior and perception of 10-year-old boys. *Environmental Psychology and Nonverbal Behavior, 2,* 226–249.

Loomis, D. K., Samuelson, C. D., & Sell, J. A. (1995). Effects of information and motivational orientation on harvest of a declining renewable resource. *Society and Natural Resources, 8,* 1–18.

Loomis, R. J. (1987). *Museum visitor evaluation: New tool for management.* Nashville, TN: American Association for State and Local History.

Lorenz, K. (1966). *On aggression.* New York: Harcourt Brace Jovanovich.

Lott, B. S., & Sommer, R. (1967). Seating arrangements and status. *Journal of Personality and Social Psychology, 7,* 90–95.

Love, K. D., & Aiello, J. R. (1980). Using projective techniques to measure interaction distance: A methodological note. *Personality and Social Psychology Bulletin, 6,* 102–104.

Lovelock, J. (1988). *The ages of Gaia.* New York: W. W. Norton.

Lovelock, J. (1998). A book for all seasons. *Science, 280,* 832–833.

Low, S. M. (1997). Urban spaces as representations of culture: The plaza in Costa Rica. *Environment and Behavior, 29,* 3–33.

Low, S. M., & Altman, I. (1992). Place attachment: A conceptual inquiry. In I. Altman & S. M. Low (Eds.), *Place attachment* (pp. 1–12). New York: Plenum.

Lublin, J. S. (1985, June 20). The suburban life: Trees, grass plus noise, traffic and pollution. *The Wall Street Journal*, p. 29.

Lucas, R. C. (1964). *The recreational capacity of the Quetico-Superior area* (Research Paper No. LS-15). St. Paul, MN: U.S. Department of Agriculture, Lake States Forest Experiment Station.

Lundberg, U. (1976). Urban commuting: Crowdedness and catecholamine excretion. *Journal of Human Stress, 2,* 26–32.

Luquette, A. J., Landiss, C. W., & Merki, D. J. (1970). Some immediate effects of a smoking environment on children of elementary school age. *Journal of School Health, 40,* 533–536.

Lutz, A. R., Simpson-Housley, P., & de Man, A. F. (1999). Wilderness: Rural and urban attitudes and perceptions. *Environment and Behavior, 31,* 259–266.

Luyben, P. D. (1980a). Effects of informational prompts on energy conservation in college classrooms. *Journal of Applied Behavior Analysis, 13,* 611–617.

Luyben, P. D. (1980b). Effects of a presidential prompt on energy conservation in college classrooms. *Journal of Environmental Systems, 10,* 17–25.

Lyles, W. B., Greve, K. W., Bauer, R. M., Ware, M. R., Schramke, C. J., Crouch, J., & Hicks, A. (1991). Sick building syndrome. *Southern Medical Journal, 84,* 65–72.

Lynch, K. (1960). *The image of the city.* Cambridge, MA: M.I.T. Press.

Lynch, R. M., & Kipen, H. (1998). Building-related illness and employee lost time following application of hot asphalt roof: A call for prevention. *Toxicology & Industrial Health, 14,* 857–968.

Lyons, E. (1983). Demographic correlates of landscape preference. *Environment and Behavior, 15,* 487–511.

Maccoby, E. E. (1990). Gender and relationships: A developmental account. *American Psychologist, 45,* 513–520.

Maccoby, E. E., & Jacklin, C. (1974). *The psychology of sex.* Stanford, CA: Stanford University Press.

MacDonald, J. E., & Gifford, R. (1989). Territorial cues and defensible space theory: The burglar's point of view. *Journal of Environmental Psychology, 9,* 193–205.

MacDougall, J. M., Dembroski, T. M., Slaats, S., Herd, J. A., & Eliot, R. S. (1983, September).

Selective cardiovascular effects of stress and cigarette smoking. *Journal of Human Stress, 9,* 13–21.

Mace, B. L., Bell, P. A., & Loomis, R. J. (1999). Aesthetic, affective, and cognitive effects of noise on natural landscape assessment. *Society & Natural Resources, 12,* 225–242.

Mace, B. L., & Greene, T. C. (1997, April). *Campus landscapes: Preference, familiarity, and districts of spatial cognition.* Presented at the meeting of the Rocky Mountain Psychological Association, Reno, NV.

MacEachren, A. M. (1992). Learning spatial information from maps: Can orientation-specificity be overcome? *Professional Geographer, 44,* 431–443.

Madsen, G. E., Dawson, S. E., & Spykerman, B. R. (1996). Perceived occupational and environmental exposures: A case study of former uranium millworkers. *Environment and Behavior, 28,* 571–590.

Maher, C. R., & Lott, D. F. (1995). Definitions of territoriality used in the study of variation on vertebrate spacing systems. *The Association for the Study of Animal Behaviour, 49,* 1581–1597.

Maida, C. A., Gordon, N. S., Steinberg, A., & Gordon, G. (1989). Psychological impact of disasters: Victims of the Baldwin Hills fire. *Journal of Traumatic Stress, 2,* 37–47.

Maier, S. F., Watkins, L. R., & Fleshner, M. (1994). Psychoneuroimmunology: The interface between behavior, brain, and immunity. *American Psychologist, 49,* 1004–1017.

Main, T. (1998). How to think about homelessness: Balancing structural and individual causes. *Journal of Social Distress and the Homeless, 7,* 41–54.

Malkin, J. (1992). *Hospital interior architecture: Creating healing environments for special patient populations.* New York: Van Nostrand Reinhold.

Mandel, D. R., Baron, R. M., & Fisher, J. D. (1980). Room utilization and dimension of density. *Environment and Behavior, 12,* 308–319.

Manfredo, M. J., Driver, B. L., & Tarrant, M. A. (1996). Measuring leisure motivation: A meta-analysis of the recreation experience preference scales. *Journal of Leisure Research, 28,* 188–213.

Manning, R. E., Lime, D. W., Freimund, W. A., & Pitt, D. G. (1996). Crowding norms at frontcountry sites: A visual standard approach to setting standards of quality. *Leisure Sciences, 18,* 39–59.

Manning, R. E., Valliere, W. A., Wang, B., & Jacobi, C. (1999). Crowding norms: Alternative measurement approaches. *Leisure Sciences, 21,* 97–115.

Marans, R. W., & Rodgers, W. (1975). Toward an understanding of community satisfaction. In A. Hawley & V. Rock (Eds.), *Metropolitan America in contemporary perspective* (pp. 229–352). New York: Halsted Press.

Margai, F. L. (1997). Analyzing changes in waste reduction behavior in a low-income urban community following a public outreach program. *Environment and Behavior, 29,* 769–792.

Markham, S. (1947). *Climate and the energy of nations.* New York: Oxford University Press.

Markowitz, J. S., & Gutterman, E. M. (1986). Predictors of psychological distress in the community following two toxic chemical incidents. In A. H. Lebovits, A. Baum, & J. E. Singer (Eds.), *Advances in environmental psychology* (Vol. 6, pp. 89–107). Hillsdale, NJ: Erlbaum.

Markowitz, J. S., Gutterman, E. M., Link, B., & Rivera, M. (1987). Psychological response of firefighters to a chemical fire. *Journal of Human Stress, 13,* 84–93.

Marling, K. A. (Ed.). (1997). *Designing Disney's theme parks: The architecture of reassurance.* Paris: Flammarion.

Marshall, E. (1985). Space junk grows with weapons tests. *Science, 230,* 424–425.

Martichuski, D. K., & Bell, P. A. (1991). Reward, punishment, privatization, and moral suasion in a commons dilemma. *Journal of Applied Social Psychology, 21,* 1356–1369.

Martichuski, D. K., & Bell, P. A. (1993). Treating excess disabilities in special care units: A review of interventions. *American Journal of Alzheimer's Care & Research, 8*(5), 8–13.

Martin, R. A., Kuiper, N. A., Olinger, L. J., & Dobbin, J. (1987). Is stress always bad? Telic versus paratelic dominance as a stress-moderating variable. *Journal of Personality and Social Psychology, 53,* 970–982.

Martin, R. A., & Lefcourt, H. M. (1983). The sense of humor as a moderator of the relationship between stressors and moods. *Journal of Personality and Social Psychology, 45,* 1313–1324.

Martindale, D. A. (1971). Territorial dominance behavior in dyadic verbal interactions. *Proceedings of the Annual Convention of the American Psychological Association, 6,* 305–306.

Martinelli, A. M. (1999). A theoretical model for the study of active and passive smoking in military women: An at-risk population. *Military Medicine, 164,* 475–480.

Marwell, G., & Ames, R. E. (1979). Experiments on the provisions of public goods. *American Journal of Sociology, 84,* 1335–1360.

Maschke, C., Ising, H., & Arndt, D. (1995). Nachtlicher verkehrslarm und gesundheit. *Bundesgesund Heitsldatt, 38,* 130–136.

Maslow, A. H., & Mintz, N. C. (1956). Effects of esthetic surrounding: I. Initial effects of three esthetic conditions upon perceiving "energy" and "well-being" in faces. *Journal of Psychology, 41,* 247–254.

Mathews, K. E., & Canon, L. K. (1975). Environmental noise level as a determinant of helping behavior. *Journal of Personality and Social Psychology, 32,* 571–577.

Mathews, K. E., Canon, L. K., & Alexander, K. (1974). The influence of level of empathy and ambient noise on the body buffer zone. *Proceedings of the American Psychological Association Division of Personality and Social Psychology, 1,* 367–370.

Matthews, H. M. (1992). *Making sense of place: Children's understanding of large-scale environments.* London: Harvester Wheatsheav.

Matus, V. (1988). *Design for northern climates: Cold-climate planning and environmental design.* New York: Van Nostrand Reinhold.

Mausner, C. (1999). A kaleidoscope model: Defining natural environments. *Journal of Environmental Psychology, 16,* 335–348.

Maxfield, M. G. (1984). The limits of vulnerability in explaining fear of crime: A comparative neighborhood analysis. *Research in Crime and Delinquency, 21,* 233–249.

Maxwell, L. E. (1996). Multiple effects of home and day care crowding. *Environment and Behavior, 28,* 494–511.

Mazumdar, S., & Mazumdar, S. (1993). Sacred space and place attachment. *Journal of Environmental Psychology, 13,* 231–242.

Mazumdar, S., & Mazumdar, S. (1999). 'Women's significant spaces': Religion, space, and community. *Journal of Environmental Psychology, 19,* 159–170.

McCain, G., Cox, V. C., & Paulus, P. B. (1976). The

relationship between illness complaints and degree of crowding in a prison environment. *Environment and Behavior, 8,* 283–290.

McCallum, R., Rasbult, C., Hong, G., Walden, T., & Schopler, J. (1979). Effect of resource availability and importance of behavior on the experience of crowding. *Journal of Personality and Social Psychology, 37,* 1304–1313.

McCammon, S., Durham, T. W., Allison, E. J., & Williamson, J. E. (1988). Emergency workers' cognitive appraisal and coping with traumatic events. *Journal of Traumatic Stress, 1,* 353–372.

McCarthy, D. O., Ouimet, M. E., & Dunn, J. M. (1992). The effects of noise stress on leukocyte function in rats. *Research in Nursing and Health, 15,* 131–137.

McCarthy, D. P., & Saegert, S. (1979). Residential density, social overload, and social withdrawal. In J. R. Aiello & A. Baum (Eds.), *Residential crowding and design* (55–75). New York: Plenum.

McCaul, K. D., & Kopp, J. T. (1982). Effects of goal setting and commitment on increasing metal recycling. *Journal of Applied Psychology, 67,* 377–379.

McCauley, C., Coleman, G., & DeFusco, P. (1977). Commuters' eye contact with strangers in city and suburban train stations: Evidence of short-term adaptation to interpersonal overload in the city. *Environmental Psychology and Nonverbal Behavior, 2,* 215–225.

McCauley, C., & Taylor, J. (1976). Is there overload of acquaintances in the city? *Environmental Psychology and Nonverbal Behavior, 1,* 41–55.

McDonald, B. L., & Schreyer, R. (1991). Spiritual benefits of leisure participation and leisure settings. In B. L. Driver, P. J. Brown, & G. L. Peterson (Eds.), *Benefits of leisure* (pp. 179–194). State College, PA: Venture.

McDonald, T. P., & Pellegrino, J. W. (1993). Psychological perspectives on spatial cognition. In T. Gärling & R. G. Golledge (Eds.), *Behavior and environment: Psychological and geographical approaches* (pp. 47–82). Amsterdam: Elsevier Science Publishers B.V.

McDonough, W., & Braungart, M. (1998, October). The NEXT Industrial Revolution. *Atlantic Monthly,* 82–92.

McElroy, J. C., Morrow, P. C., & Wall, L. C. (1983).

Generalizing impact of object language to other audiences: Peer response to office design. *Psychological Reports, 53,* 315–322.

McFarland, R. A. (1972). Psychophysiological implications of life at high altitude and including the role of oxygen in the process of aging. In M. K. Yousef, S. M. Horvath, & R. W. Bullard (Eds.), *Physiological adaptations: Desert and mountain* (pp. 157–181). New York: Academic Press.

McFarlane, A. C., Policansky, S. K., & Irwin, C. (1987). A longitudinal study of the psychological morbidity in children due to a natural disaster. *Psychological Medicine, 17,* 727–738.

McGuinness, D., & Sparks, J. (1979). Cognitive style and cognitive maps: Sex differences in representations. *Journal of Mental Imagery, 7,* 101–118.

McGuire, W. J., & Gaes, G. G. (1982). *The effects of crowding versus age composition in aggregated prison assault rates.* Unpublished manuscript, Office of Research, Federal Prison System, Washington, DC.

McKenzie-Mohr, D. (2000). Fostering sustainable behavior through community-based social marketing. *American Psychologist, 55,* 531–537.

McKinney, K. D. (1998). Parental perceptions of children's privacy needs: Conceptions of Privacy. *Journal of Family Issues, 19,* 75–100.

McKinnon, W., Weisse, C. S., Reynolds, C. R., Bowles, C. A., & Baum, A. (1989). Chronic stress, leukocyte subpopulations, and humoral response to latent viruses. *Health Psychology, 8,* 389–402.

McLean, E. K., & Tarnopolsky, A. (1977). Noise, distress, and mental health. *Psychological Medicine, 7,* 19–62.

McLure, J., Walkey, F., & Allen, M. (1999). When earthquake damage is seen as preventable: Attributions, locus of control and attitudes to risk. *Applied Psychology: An International Review, 48,* 239–256.

McNamara, T. P. (1986). Mental representations of spatial relations. *Cognitive Psychology, 18,* 87–121.

McNamara, T. P., & Diwadkar, V. A. (1997). Symmetry and asymmetry in human spatial memory. *Cognitive Psychology, 34,* 160–190.

McNamara, T. P., Hardey, J. K., & Hirtle, S. C. (1989). Subjective hierarchies in spatial memory. *Journal of Experimental Psychology: Learning, Memory, and Cognition, 15,* 211–227.

Medalia, N. Z. (1964). Air pollution as a socio-environmental health problem: A survey report. *Journal of Health and Human Behavior, 5,* 154–165.

Meer, J. (1986, May). The strife of bath. *Psychology Today, 20*(5), 6.

Mehrabian, A. (1976–77). A questionnaire measure of individual differences in stimulus screening and associated differences in arousability. *Environmental Psychology and Nonverbal Behavior, 1,* 89–103.

Mehrabian, A., & Diamond, S. G. (1971a). Effects of furniture arrangement, props, and personality on social interaction. *Journal of Personality and Social Psychology, 20,* 18–30.

Mehrabian, A., & Diamond, S. G. (1971b). Seating arrangement and conversation. *Sociometry, 34,* 281–289.

Meisels, M., & Guardo, C. J. (1969). Development of personal space schemata. *Child Development, 49,* 1167–1178.

Melick, M. E. (1978). Life change and illness: Illness behavior of males in the recovery period of a natural disaster. *Journal of Health and Social Behavior, 19,* 335–342.

Melton, A. W. (1933). Studies of installation at the Pennsylvania Museum of Art. *Museum News, 10,* 5–8.

Melton, A. W. (1936). Distribution of attention in galleries in a museum of science and industry. *Museum News, 14,* 5–8.

Melton, A. W. (1972). Visitor behavior in museums: Some early research in environmental design. *Human Factors, 14,* 393–403.

Mendell, M. J., & Smith, A. H. (1990). Consistent patterns of elevated symptoms in air-conditioned office buildings: A re-analysis of epidemiologic studies. *American Journal of Public Health, 80,* 1193–1199.

Mendelsohn, R., & Orcutt, G. (1979). An empirical analysis of air pollution dose-response curves. *Journal of Environmental Economics and Management, 6,* 85–106.

Menozzi, M., Von Buol, A., Waldmann, H., Kuendig, S., Krueger, H., & Spieler, W. (1999). Training in ergonomics at VDU workplaces. *Ergonomics, 42,* 835–845.

Mercer, G. W., & Benjamin, M. L. (1980). Spatial behavior of university undergraduates in double-occupancy residence rooms: An inventory of effects. *Journal of Applied Social Psychology, 10,* 32–44.

Merchant, C. (1992). *Radical ecology: The search for a livable world.* New York: Routledge.

Merry, S. E. (1981). Defensible space undefended: Social factors in crime control through environmental design. *Urban Affairs Quarterly, 16,* 397–422.

Merry, S. E. (1987). Crowding, conflict, and neighborhood regulation. In I. Altman & A. Wandersman (Eds.), *Neighborhood and community environments* (pp. 35–68). New York: Plenum.

Mesch, G. S., & Manor, O. (1998). Social ties, environmental perception, and local attachment. *Environment and Behavior, 30,* 504–519.

Messick, D. M., Wilke, H., Brewer, M. B., Kramer, R. M., Zemke, P. E., & Lui, L. (1983). Individual adaptations and structural change as solutions to social dilemmas. *Journal of Personality and Social Psychology, 44,* 294–309.

Meyers, D. G. (1999). *Social psychology,* (6th ed.). Boston: McGraw-Hill.

Michelini, R. L., Passalacqua, R., & Cusimano, J. (1976). Effects of seating arrangement on group participation. *Journal of Social Psychology, 99,* 179–186.

Michelson, W. (1977). *Environmental choice, human behavior, and residential satisfaction.* New York: Oxford University Press.

Middlemist, R. D., Knowles, E. S., & Matter, C. F. (1976). Personal space invasions in the lavatory: Suggestive evidence for arousal. *Journal of Personality and Social Psychology, 33,* 541–546.

Miedema, H. M. E., & Vos, H. (1999). Demographic and attitudinal factors that modify annoyance from transportation noise. *Journal of the Acoustical Society of America, 105,* 3336–3344.

Miles, R., & Clarke, G. (1993). Setting off on the right foot: Front-end evaluation. *Environment and Behavior, 25,* 698–709.

Mileti, D. S. (1999). *Disasters by design: A reassessment of natural hazards in the United States.* Washington, DC: National Academy of Sciences, Joseph Henry Press.

Mileti, D. S., & Fitzpatrick, C. (1993). *The great earthqauke experiment: Risk communication and public action.* Boulder, CO: Westview Press.

Mileti, D. S., & Sorensen, J. S. (1990). *Communication of emergency public warnings: A social science perspective and state-of-the-art assessment.* Oak Ridge, TN: Oak Ridge National Laboratory.

Milgram, S. (1970). The experience of living in cities. *Science, 167,* 1461–1468.

Milgram, S. (1977). *The individual in a social world.* Reading, MA: Addison-Wesley.

Milgram, S., & Jodelet, D. (1976). Psychological maps of Paris. In H. Proshansky, W. Ittelson, & L. Rivlin (Eds.), *Environmental psychology* (pp. 104–124). New York: Holt, Rinehart and Winston.

Millar, K., & Steels, M. J. (1990). Sustained peripheral vasoconstriction while working in continuous intense noise. *Aviation, Space, and Environmental Medicine, 61,* 695–698.

Miller, D. T. (1999). The norm of self-interest. *American Psychologist, 54,* 1053–1060.

Miller, I. W., III, & Norman, W. H. (1979). Learned helplessness in humans: A review and attribution theory model. *Psychological Bulletin, 86,* 93–118.

Miller, J. D. (1974). Effects of noise on people. *Journal of the Acoustical Society of America, 56,* 729–764.

Miller, J. P. (1995, July 26). Bidding war breaks out for used plastic soda bottles. *The Wall Street Journal,* p. B4.

Miller, M. (1982). Cited in J. Raloff, Occupational noise—the subtle pollutant. *Science News, 121,* 347–350.

Miller, M., Albert, M., Bostick, D., & Geller, E. S. (1976, March). *Can the design of a trash can influence litter-related behavior?* Paper presented at the meeting of the Southeastern Psychological Association, New Orleans, LA.

Miller, S., & Nardini, R. M. (1977). Individual differences in the perception of crowding. *Environmental Psychology and Nonverbal Behavior, 2,* 3–13.

Miller, S., Rossbach, J., & Munson, R. (1981). Social density and affiliative tendency as determinants of dormitory and residential outcomes. *Journal of Applied Social Psychology, 11,* 356–365.

Miransky, J., & Langer, E. J. (1978). Burglary (non)-prevention: An instance of relinquishing control. *Personality and Social Psychology Bulletin, 4,* 399–405.

Mitchell, J. G. (1994, October). Our national parks. *National Geographic, 186,* 2–55.

Mitchell, M. Y., Force, J. E., Carroll, M. S., & McLaughlin, W. J. (1991). Forest places of the heart: Incorporating special places into public management. *Journal of Forestry, 4,* 32–37.

Mocellin, J. S., Suedfeld, P., Bernadelz, J. P., & Barbarito, M. E. (1991). Levels of anxiety in polar environments. *Journal of Environmental Psychology, 11,* 265–275.

Moeser, S. D. (1988). Cognitive mapping in a complex building. *Environment and Behavior, 20,* 21–49.

Moffatt, A. S. (1997). Resurgent forests can be greenhouse gas sponges. *Science, 277,* 315–316.

Montano, D., & Adamopoulous, J. (1984). The perception of crowding in interpersonal situations: Affective and behavioral responses. *Environment and Behavior, 16,* 643–667.

Montello, D. (1988). Classroom seating location and its effect on course achievement. *Journal of Environmental Psychology, 8,* 149–157.

Mooney, K. M., Cohn, E. S., & Swift, M. B. (1992). Physical distance and AIDS: Too close for comfort? *Journal of Applied Social Psychology, 22,* 1442–1452.

Moore, D. P., & Moore, J. W. (1996). Posthurricane burnout: An island township's experience. *Environment and Behavior, 28,* 134–155.

Moore, G. (1987). Environment and behavior research in North America: History, developments, and unresolved issues. In D. Stokols & I. Altman (Eds.), *Handbook of environmental psychology* (Vol. 2, pp. 1359–1410).

Moore, R. C. (1989). Playgrounds at the crossroads: Policy and action research needed to ensure a viable future for public playgrounds in the United States. In I. Altman & E. H. Zube (Eds.), *Public places and spaces* (pp. 83–120). New York: Plenum.

Moore, S. F., Shaffer, L. S., Pollak, E. L., & Taylor-Lemke, P. (1987). The effects of interpersonal trust and prior common problem experience on commons management. *Journal of Social Psychology, 127,* 19–29.

Moos, R. H. (1976). *The human context: Environmental determinants of behavior.* New York: Wiley.

Moos, R. H., & Gerst, M. S. (1974). *University Residence Environment Scale.* Palo Alto, CA: Consulting Psychologists Press.

Moos, R. H., & Lemke, S. (1985). Specialized living environments for older people. In J. E. Birren & K. W. Schaie (Eds.), *Handbook of the psychology of aging* (2nd ed., pp. 864–899). New York: Van Nostrand Reinhold.

Moos, W. S. (1964). The effects of "Föhn" weather on accident rates in the city of Zurich (Switzerland). *Aerospace Medicine, 35,* 643–645.

Moran, R., Anderson, R., & Paoli, P. (Eds.) (1990). *Building for people in hospitals: Workers and consumers.* Shankill, Dublin: European Foundation for the Improvement of Living and Working Conditions.

Morasch, B., Groner, N., & Keating, J. (1979). Type of activity and failure as mediators of perceived crowding. *Personality and Social Psychology Bulletin, 5,* 223–226.

Morata, T. C. (1998). Assessing occupational hearing loss: Beyond noise exposures. *Scandinavian Audiology, Supplement, 27(Suppl 48),* 111–116.

Moreland, R. L., & Zajonc, R. B. (1982). Exposure effects in person perception: Familiarity, similarity, and attraction. *Journal of Experimental Social Psychology, 18,* 395–415.

Morgan, D. G., & Stewart, N. J. (1998a). High versus low density special care units: Impact on the behaviour of elderly residents with dementia. *Canadian Journal on Aging, 17,* 143–165.

Morgan, D. G., & Stewart, N. J. (1998b). Multiple occupancy versus private rooms on dementia care units. *Environment and Behavior, 30,* 487–503.

Morris, E. W. (1987). Comment on "Castles in the Sky." *Environment and Behavior, 19,* 115–119.

Morrow, L. M., & Weinstein, C. S. (1982). Increasing children's use of literature through program and physical design changes. *Elementary School Journal, 83,* 131–137.

Morrow, P. C., & McElroy, J. C. (1981). Interior office design and visitor response. *Journal of Applied Psychology, 66,* 646–630.

Moser, G., & Levy-Leboyer, C. (1985). Inadequate environment and situation control: Is a malfunctioning phone always an occasion for aggresion? *Environment and Behavior, 17,* 520–533.

Mosler, H. J. (1993). Self-dissemination of environmentally responsible behavior: The influence of trust in a commons dilemma game.

Journal of Environmental Psychology, 13, 111–123.

Mudur, G. (1995). Monsoon shrinks with aerosol models. *Science, 270,* 1922.

Muecher, H., & Ungeheuer, H. (1961). Meteorological influences on reaction time, flicker fusion frequency, job accidents, and use of medical treatment. *Perceptual and Motor Skills, 12,* 163–168.

Munroe, R. L., Munroe, R. H., & Vutpakdi, K. (1999). Some psychological correlates of population density in East Africa. *Journal of Social Psychology, 139,* 392–393.

Murphy-Berman, V., & Berman, J. (1978). Importance of choice and sex invasions of personal space. *Personality and Social Psychology Bulletin, 4,* 424–428.

Muzi, G., Abbritti, G., Accattoli, M. P., & dell'Omo, M. (1998). Prevalence of irritative symptoms in a nonproblem air-conditioned office building. *International Archives of Occupational & Environmental Health, 71,* 372–378.

Myers, N. (1997). Consumption: Challenge to sustainable development . . . *Science, 276,* 53–57.

Myers, P. (1978). *Neighborhood conservation and the elderly.* Washington, DC: Conservation Foundation.

Nadel, B. (1994, April). Energy star PCs: Power to the PC. *PC Magazine: The Independent Guide to Personal Computing, 13,* 114ff.

Nagy, E., Yasunaga, S., & Kose, S. (1995). Japanese office employees' psychological reactions to their underground and above-ground offices. *Journal of Environmental Psychology, 15,* 123–134.

Nahemow, L., & Lawton, M. P. (1973). Toward an ecological theory of adaptation and aging. In W. F. E. Preisser (Ed.), *Environmental design research* (Vol. 1, pp. 24–32). Stroudsberg, PA: Dowden, Hutchinson, & Ross.

Nasar, J. L. (1994). Urban design aesthetics: The evaluative qualities of building exteriors. *Environment and Behavior, 26,* 377–401.

Nasar, J. L., & Fisher, B. S. (1992). Fear of crime in relation to three exterior site features: Prospect, refuge, and escape. *Environment and Behavior, 24,* 35–65.

Nasar, J. L, & Fisher, B. S. (1993). Hot spots of fear and crime: A multi-method investigation.

Journal of Environmental Psychology, 13, 187–206.

Nasar, J. L., & Jones, K. M. (1997). Landscapes of fear and stress. *Environment and Behavior, 29,* 291–323.

Nasar, J. L., & Min, M. S. (1984, August). *Modifiers of perceived spaciousness and crowding: A cross cultural study.* Paper presented at the annual meeting of the American Psychological Association, Toronto, Canada.

Nasar, J. L., & Upton, K. (1997). Landscapes of fear and stress. *Environment and Behavior, 29,* 291–323.

Nash, B. C. (1981). The effects of classroom spatial organization on four- and five-year-old children's learning. *British Journal of Educational Psychology, 51,* 144–155.

Nash, R. (1982). *Wilderness and the American mind* (3rd ed.). New Haven, CT: Yale University Press.

National Institute of Environmental Health Sciences (1999). *Health effects from exposure to power-line frequency electric and magnetic fields.* Washington, DC. National Institutes of Health Publication No. 99-4493.

National Research Council (U.S.) (1997). *Possible health effects of exposure to residential electric and magnetic fields.* Washington, DC: National Academy Press.

Navarro, P. L., Simpson-Housley, P., & DeMan, A. F. (1987). Anxiety, locus of control, and appraisal of air pollution. *Perceptual and Motor Skills, 64,* 811–814.

Needleman, H. L., Schell, A. S., Bellinger, D., Leviton, A., & Alldred, E. N. (1990). The long-term effects of exposure to low doses of lead in childhood: An 11-year follow-up report. *New England Journal of Medicine, 322,* 83–88.

Neill, S. R. St. J. (1982). Experimental alterations in playroom layout and their effect on staff and child behavior. *Educational Psychology, 2,* 103–119.

Neiman, L. (1988). A critical review of resiliency literature and its relevance to homeless children. *Children's Environment Quarterly, 5,* 17–25.

Neulinger, J. (1981). *To leisure: An introduction.* Boston: Allyn & Bacon.

Newcombe, N. (1985). Method for the study of spatial cognition. In R. Cohen (Ed.), *The de-velopment of spatial cognition* (pp. 1–12). Hillsdale, NJ: Erlbaum.

Newcombe, N., Huttenlocher, J., Sandberg, E., Lie, E., & Johnson, S. (1999). What do misestimations and asymmetries in spatial judgment indicate about spatial representation? *Journal of Experimental Psychology, 25,* 986–996.

Newell, P. B. (1997). A cross-cultural examination of favorite places. *Environment and Behavior, 29,* 495–514.

Newell, P. B. (1998). A cross-cultural comparison of privacy definitions and functions: A systems approach. *Journal of Environmental Psychology, 18,* 357–371.

Newhouse, N. (1990). Implications of attitude and behavior research for environmental conservation. *The Journal of Environmental Education, 22,* 1.

Newman, C. J. (1976). Children of disaster. Clinical observations at Buffalo Creek. *American Journal of Psychiatry, 133,* 306–309.

Newman, J., & McCauley, C. (1977). Eye contact with strangers in city, suburb, and small town. *Environment and Behavior, 9,* 547–558.

Newman, O. (1972). *Defensible space.* New York: Macmillan.

Newman, O. (1975). Reactions to the defensible space study and some further findings. *International Journal of Mental Health, 4,* 48–70.

Newman, O. (1995). Defensible space: A new physical planning tool for urban revitalization. *Journal of the American Planning Association, 61,* 149–155.

Newman, O., & Franck, K. (1982). The effects of building size on personal crime and fear of crime. *Population and Environment, 5,* 203–220.

Newsweek (1994, July 11). Drawing up safer cities, *124*(2), p. 57.

Nicosia, G. J., Hymen, D., Karlin, R. A., Epstein, Y. M., & Aiello, J. R. (1979). Effects of bodily contact on reactions to crowding. *Journal of Applied Social Psychology, 9,* 508–523.

Nivison, M. E., & Endresen, I. M. (1993). An analysis of relationships among environmental noise, annoyance and sensitivity to noise, and the consequences for health and sleep. *Journal of Behavioral Medicine, 16,* 257–276.

Nolen-Hoeksema, S., & Morrow, J. (1991). A prospective study of depression and posttraumatic stress symptoms after a natural disaster: The 1989 Loma Prieta earthquake. *Journal of Personality and Social Psychology, 61,* 115–121.

Norberg-Schulz, C. (1979). *Genius loci: Towards a phenomenology of architecture.* New York: Rizzoli.

Norlander, T., Bergman, H., & Archer, T. (1998). Effects of flotation REST on creative problem solving and originality. *Journal of Environmental Psychology, 18,* 399–408.

Norman, D. A. (1988). *The psychology of everyday things.* New York: Basic Books.

Normoyle, J., & Lavrakas, P. J. (1984). Fear of crime in elderly women: Perceptions of control, predictability, and territoriality. *Personality and Social Psychology Bulletin, 10,* 191–202.

Norris, F. H., & Kaniasty, K. (1996). Received and perceived social support in times of stress: A test of the social support deterioration deterrence model. *Journal of Personality and Social Psychology, 71,* 498–511.

Norris, F. H., & Kaniasty, K. (1992). Reliability of delayed self-reports in research. *Journal of Traumatic Stress, 5,* 575–588.

Norris, F. H., & Murrell, S. (1984). Protective functions of resources related to life events, global stress, and depression in older adults. *Journal of Health and Social Behavior, 23,* 145–159.

Norris, F. H., Smith, T., & Kaniasty, K. (1999). Revisiting the experience-behavior hypothesis: The effects of Hurricane Hugo on hazard preparedness and other self-protective acts. *Basic & Applied Social Psychology, 21,* 37–47.

Norris, F. H., & Uhl, G. A. (1993). Chronic stress as a mediator of acute stress: The case of Hurricane Hugo. *Journal of Applied Social Psychology, 23,* 1263–1284.

Norris-Baker, C. (1999). Aging on the old frontier and the new: A behavior setting approach to the declining small towns of the midwest. *Environment and Behavior, 31,* 240–258.

North, A. C., & Hargreaves, D. J. (1999). Can music move people? The effects of musical complexity and silence on waiting time. *Environment and Behavior, 31,* 136–149.

North, C. S., Smith, E. M., & Spitznagel, E. L. (1994). Posttraumatic stress disorder in survivors of a mass shooting. *American Journal of Psychiatry, 151,* 82–88.

Novaco, R. W., Stokols, D., Campbell, J., & Stokols, J. (1979). Transportation stress and community psychology. *American Journal of Community Psychology, 4,* 361–380.

Nowak, R. (1994). Chronobiologists out of sync over light therapy patients. *Science, 263,* 1217–1218.

Nurminen, T., & Kurppa, K. (1989). Occupational noise exposure and course of pregnancy. *Scandinavian Journal of Work and Environmental Health, 15,* 117–124.

Oak Ridge Associated Universities (1992, June). *Health effects of low-frequency electric and magnetic fields (ORAU 92/F8).* Oak Ridge, TN: Oak Ridge Associated Universities.

Oak Ridge Associated Universities Panel (1993). EMF and cancer. *Science, 260,* 13–14.

Oelschlaeger, M. (1991). *The idea of wilderness: From prehistory to the age of ecology.* New Haven, CT: Yale University Press.

Ohta, R. J., & Ohta, B. M. (1988). Special units for Alzheimer's disease patients: A critical look. *The Gerontologist, 28,* 803–808.

Oldham, G. (1988). Effects of changes in workspace partitions and spatial density on employee reactions: A quasi-experiment. *Journal of Applied Psychology, 73,* 253–258.

Oldham, G. R., & Brass, D. J. (1979). Employee reactions to an open-plan office: A naturally occurring quasi-experiment. *Administrative Science Quarterly, 24,* 267–284.

Oliver, L. C., & Shackleton, B. W. (1998). The indoor air we breath. *Public Health Reports, 113,* 398–409.

Ollendick, D. G., & Hoffman, M. (1982). Assessment of psychological reactions in disaster victims. *Journal of Comparative Psychology, 10,* 157–167.

Olsen, M. E. (1981). Consumers' attitudes toward energy conservation. *Journal of Social Issues, 37,* 108–131.

Olsen, R. (1978). *The effect of the hospital environment.* Unpublished doctoral dissertation, City University of New York.

Olson, J., & Zanna, M. (1993). Attitudes and attitude change. *Annual Review of Psychology, 44,* 117–154.

Omata, K. (1992). Spatial organization of activities of Japanese families. *Journal of Environmental Psychology, 12,* 259–267.

Omata, K. (1995). Territoriality in the house and its relationship to the use of rooms and the psychological well-being of Japanese married women. *Journal of Environmental Psychology, 15*, 147–154.

O'Neal, E. C., Brunault, M. A., Carifio, M. S., Troutwine, R., & Epstein, J. (1980). Effects of insult on personal space preferences. *Journal of Nonverbal Behavior, 5*, 56–62.

O'Neil, M. J. (1994). Workspace adjustability, storage, and enclosure as predictors of employee reactions and performance. *Environment and Behavior, 26*, 504–526.

O'Neill, G. W., Blanck, L. S., & Joyner, M. A. (1980). The use of stimulus control over littering in a natural setting. *Journal of Applied Behavior Analysis, 13*, 379–381.

O'Neill, S. M., & Paluck, B. J. (1973). Altering territoriality through reinforcement. *Proceedings of the 81st Annual Convention of the American Psychological Association, Montreal, Canada, 8*, 901–902.

Oppewal, H., & Timmermans, H. (1999). Modeling consumer perception of public space in shopping centers. *Environment and Behavior, 31*, 45–65.

O'Riordan, T. (1976). Attitudes, behavior, and environmental policy issues. In I. Altman & J. F. Wohlwill (Eds.), *Human behavior and environment: Advances in theory and research* (Vol. 1, pp. 1–36). New York: Plenum.

Ornstein, S. (1992). First impressions of the symbolic meanings connoted by reception area design. *Environment and Behavior, 24*, 85–110.

Osborne, J. G., & Powers, R. B. (1980). Controlling the litter problem. In G. L. Martin & J. G. Osborne (Eds.), *Helping the community: Behavioral applications* (pp. 103–168). New York: Plenum.

Oseland, N., & Donald, I. (1993). The evaluation of space in homes: A facet study. *Journal of Environmental Psychology, 13*, 251–261.

OSHA (1981). Occupational noise exposure. "Hearing Conservation Amendment" (20 CFR Part 1910). *Federal Register, 45*, 11 (January 16).

Oskamp, S. (2000). Psychology for a sustainable society. *American Psychologist, 55*, 496–508.

Oskamp, S., Williams, R., Unipan, J., Steers, N., Mainieri, T., & Kurland, G. (1994). Psychological factors affecting paper recycling by businesses. *Environment and Behavior, 26*, 477–503.

Oskamp, S., Zelezny, L., Schultz, P. W., Hurin, S., & Burkhardt, R. (1996). Commingled versus separated curbside recycling: Does sorting matter? *Environment and Behavior, 28*, 73–91.

Osmond, H. (1957). Function as the basis of psychiatric ward design. *Mental Hospitals (Architectural Supplement), 8*, 23–29.

Ostfeld, R. S., Canham, C. D., & Pugh, S. R. (1993). Intrinsic density-dependent regulation of vole populations. *Nature, 366*, 259–261.

Owens, P. L. (1985). Conflict as a social interaction process in environment and behavior research: The example of leisure and recreation research. *Journal of Environmental Psychology, 5*, 243–259.

Oxley, D., & Barrera, M., Jr. (1984). Undermanning theory and the workplace: Implications of setting size for job satisfaction and social support. *Environment and Behavior, 16*, 211–234.

Oxley, D., Haggard, L. M., Werner, C. M., & Altman, I. (1986). Transactional qualities of neighborhood social networks: A case study of "Christmas Street." *Environment and Behavior, 18*, 640–677.

Oxman, R., & Carmon, N. (1986). Responsive public housing: An alternative for low-income families. *Environment and Behavior, 18*, 258–284.

Pablant, P., & Baxter, J. C. (1975, July). Environmental correlates of school vandalism. *Journal of the American Institute of Planners*, 270–279.

Pacelle, M. (1996, August 7). Some urban planners say downtowns need a lot more congestion. *The Wall Street Journal*, pp. A1, A16.

Paffenbarger, R. S., Jr., Hyde, R. T., & Dow, A. (1991). Health benefits of physical activity. In B. L. Driver, P. J. Brown, & G. L. Peterson (Eds.), *Benefits of leisure* (pp. 49–57). State College, PA: Venture.

Page, R. A. (1977). Noise and helping behavior. *Environment and Behavior, 9*, 559–572.

Page, R. A. (1978, May). *Environmental influences on prosocial behavior: The effect of temperature.* Paper presented at the meeting of the Midwestern Psychological Association, Chicago, IL.

Palamarek, D. L., & Rule, B. G. (1979). The effects

of temperature and insult on the motivation to retaliate or escape. *Motivation and Emotion, 3,* 83–92.

Pallack, M. S., Cook, D. A., & Sullivan, J. J. (1980). Commitment and energy conservation. In L. Bickman (Ed.), *Applied Social Psychology Annual, 1,* 235–253.

Palmer, M. H., Lloyd, M. E., & Lloyd, K. D. (1978). An experimental analysis of electricity conservation procedures. *Journal of Applied Behavior Analysis, 10,* 665–672.

Palmstierna, T., Huitfeldt, B., & Wistedt, B. (1991). The relationship of crowding and aggressive behavior on a psychiatric intensive care unit. *Hospital and Community Psychiatry, 42,* 1237–1240.

Parasuraman, R. (1997). Humans and automation: Use, misuse, disuse, abuse. *Human Factors, 39,* 230–253.

Pardini, A. U., & Katzev, R. D. (1983–1984). The effects of strength of commitment on newspaper recycling. *Journal of Environmental Systems, 13,* 245–254.

Parker, G. (1977). Cyclone Tracy and Darwin evacuees. On the restoration of the species. *British Journal of Psychiatry, 130,* 548–555.

Parks, C. D. (1994). The predictive ability of social values in resource dilemmas and public goods games. *Personality and Social Psychology Bulletin, 20,* 431–438.

Parmelee, P., & Lawton, M. P. (1990). The design of special environments for the aged. In J. Birren & K. W. Schaie (Eds.), *Handbook of the psychology of aging* (3rd ed., pp. 464–488). New York: Academic Press.

Parr, A. E. (1966). Psychological aspects of urbanology. *Journal of Social Issues, 22,* 39–45.

Parsons, H. M. (1978). What caused the Hawthorne effect? A scientific detective story. *Administration & Society, 10,* 259–283.

Parsons, R. (1991). The potential influences of environmental perception on human health. *Journal of Environmental Psychology, 11,* 1–23.

Parsons, R., Tassinary, L. G., Ulrich, R. S., Hebl, M. R., & Grossman-Alexander, M. (1998). The view from the road: Implications for stress recovery and immunization. *Journal of Environmental Psychology, 18,* 113–140.

Passchier-Vermeer, W. (1993). *Noise & health.* The Hague: Health Council of the Netherlands.

Passini, R. (1984). Spatial representations, a wayfinding perspective. *Journal of Environmental Psychology, 4,* 153–164.

Patterson, A. H. (1978). Territorial behavior and fear of crime in the elderly. *Environmental Psychology and Nonverbal Behavior, 3,* 131–144.

Patterson, M. L. (1976). An arousal model of interpersonal intimacy. *Psychological Review, 83,* 235–245.

Patterson, M. L. (1977). Interpersonal distance, affect, and equilibrium theory. *Journal of Social Psychology, 101,* 205–214.

Patterson, M. L. (1978). Arousal change and the cognitive labeling: Pursuing the mediators of intimacy exchange. *Environmental Psychology and Nonverbal Behavior, 3,* 17–22.

Patterson, M. L., Kelly, C. E., Kondracki, B. A., & Wulf, L. J. (1979). Effects of seating arrangement on small-group behavior. *Social Psychology Quarterly, 42,* 181–185.

Patterson, M. L., Mullens, S., & Romano, J. (1971). Compensatory reactions to spatial intrusion. *Sociometry, 34,* 114–121.

Patterson, M. L., & Sechrest, L. B. (1970). Interpersonal distance and impression formation. *Journal of Personality, 38,* 161–166.

Paulhus, D. (1983). Sphere-specific measures of perceived control. *Journal of Personality and Social Psychology, 44,* 1253–1265.

Paulus, P. B. (1980). Crowding. In P. B. Paulus (Ed.), *Psychology of group influence* (pp. 245–290). Hillsdale, NJ: Erlbaum.

Paulus, P. B. (1988). *Prison crowding: A psychological perspective.* NY: Springer-Verlag.

Paulus, P. B., Annis, A. B., Seta, J. J., Schkade, J. K., & Matthews, R. W. (1976). Crowding does affect task performance. *Journal of Personality and Social Psychology, 34,* 248–253.

Paulus, P. B., Cox, V., McCain, G., & Chandler, J. (1975). Some effects of crowding in a prison environment. *Journal of Applied Social Psychology, 5,* 86–91.

Paulus, P. B., & Matthews, R. (1980). Crowding, attribution, and task performance. *Basic and Applied Social Psychology, 1,* 3–13.

Paulus, P. B., McCain, G., & Cox, V. (1978). Death rates, psychiatric commitments, blood pressure and perceived crowding as a function of institutional crowding. *Environmental Psychology and Nonverbal Behavior, 3,* 107–116.

Paulus, P. B., McCain, G., & Cox, V. (1981). Prison standards: Some pertinent data on crowding. *Federal Probation, 15,* 48–54.

Paulus, P. B., Nagar, D., & Camacho, L. M. (1991). Environmental and psychological factors in reactions to apartments and mobile homes. *Journal of Environmental Psychology, 11,* 143–161.

Pearce, G. P., & Patterson, A. M. (1993). The effect of space restriction and provision of toys during rearing on the behaviour, productivity and physiology of male pigs. *Applied Animal Behaviour Science, 36,* 11–28.

Pearce, P. L. (1977). Mental souvenirs: A study of tourists and their city maps. *Australian Journal of Psychology, 29,* 203–210.

Pedersen, D. M. (1997). Psychological functions of privacy. *Journal of Environmental Psychology, 17,* 147–156.

Pedersen, D. M. (1999). Dimensions of environmental competence. *Journal of Environmental Psychology, 19,* 303–308.

Pelletier, L. G., Dion, S., Tuson, K., & Green-Demers, I. (1999). Why do people fail to adopt environmental protective behaviors? Toward a taxonomy of environmental motivation. *Journal of Applied Social Psychology, 29,* 2481–2504.

Pendleton, M. R. (1998). Policing the park: Understanding soft enforcement. *Journal of Leisure Research, 30,* 552–571.

Pennebaker, J. W., & Newtson, D. (1983). Observation of a unique event: The psychological impact of the Mount Saint Helens volcano. In H. T. Reiss (Ed.), *Naturalistic approaches to studying social interaction. New directions for methodology of social and behavioral science* (No. 15, pp. 93–109). San Francisco: Jossey-Bass.

Penwarden, A. D. (1973). Acceptable wind speeds in towns. *Building Science, 8,* 259–267.

Pepler, R. D. (1972). The thermal comfort of students in climate controlled and non-climate controlled schools. *ASHRAE Transactions, 78,* 97–109.

Perkins, D. D., Meeks, J. W., & Taylor, R. B. (1992). The physical environment of street blocks and resident perceptions of crime and disorder: Implications for theory and measurement. *Journal of Environmental Psychology, 12,* 21–34.

Perry, J. D., & Simpson, M. E. (1987). Violent crimes in a city: Environmental determinants. *Environment and Behavior, 19,* 77–90.

Persinger, M. A., Ludwig, H. W., & Ossenkopf, K. P. (1973). Psychophysiological effects of extremely low frequency electromagnetic fields: A review. *Perceptual and Motor Skills, 26,* 1131–1159.

Persinger, M. A., Tiller, S. G., & Koren, S. A. (1999). Background sound pressure fluctuations (5dB) from overhead ventilation systems increase subjective fatigue of university students during three-hour lectures. *Perceptual & Motor Skills, 88,* 451–456.

Peterson, C., & Seligman, M. E. P. (1984). Causal explanations as a risk factor for depression: Theory and evidence. *Psychological Review, 91,* 347–374.

Peterson, G. L., Brown, T. C., McCollum, D. W., Bell, P. A., Birjulin, A. A., & Clarke, A. (1996). Chapter Eight: Moral responsibility effects in valuation of WTA for public and private goods by the method of paired comparison. In W. L. Adamowicz, P. C. Boxall, M. K. Luckert, W. E. Phillips, & W. A. White (Eds.), *Forestry economics and the environment* (pp. 134–159). Wallingford, United Kingdom: CAB International.

Peterson, R. L. (1975, August). *Air pollution and attendance in recreation behavior settings in the Los Angeles basin.* Paper presented at the meeting of the American Psychological Association, Chicago, IL.

Phelan, J., Link, B. G., Moore, R. E., & Stueve, A. (1997). The stigma of homelessness: The impact of the label "homeless" on attitudes toward poor persons. *Social Psychology Quarterly, 60,* 323–337.

Philip, D. (1996). The practical failure of architectural psychology. *Journal of Environmental Psychology, 16,* 277–284.

Philipp, S. T. (1997). Race, gender, and leisure benefits. *Leisure Studies, 19,* 191–207.

Philp, R. B., Fields, G. N., & Roberts, W. A. (1989). Memory deficit caused by compressed air equivalent to 36 meters of seawater. *Journal of Applied Psychology, 74,* 443–446.

Piaget, J., & Inhelder, B. (1967). *The child's conception of space.* New York: Norton.

Pinheiro, J. Q. (1998). Determinants of cognitive maps of the world as expressed in sketch maps. *Journal of Environmental Psychology, 18,* 321–339.

Pitt, D., & Zube, E. (1987). Management of natural environments. In D. Stokols & I. Altman (Eds.), *Handbook of environmental psychology* (pp. 1009–1042). New York: Wiley-Interscience.

Platt, J. (1973). Social traps. *American Psychologist, 28,* 641–651.

Plotkin, W. B. (1978). Long-term eyes-closed alpha-enhancement training: Effects on alpha amplitudes and on experimental state. *Psychophysiology, 15,* 40–52.

Pollack, L. M., & Patterson, A. H. (1980). Territoriality and fear of crime in elderly and non-elderly homeowners. *Journal of Social Psychology, 111,* 119–129.

Pollet, D., & Haskell, P. C. (1979). *Sign systems for libraries.* New York: Bowker.

Pontell, H. N., & Welsh, W. N. (1994). Incarceration as a deviant form of social control: Jail overcrowding in California. *Crime and Delinquency, 40,* 18–36.

Porter, B. E., Leeming, F. C., & Dwyer, W. O. (1995). Solid waste recovery: A review of behavioral programs to increase recycling. *Environment and Behavior, 27,* 122–152.

Porteus, J. (1977). *Environment and behavior.* Reading, MA: Addison-Wesley.

Potter, L. E., Dwyer, W. O., & Leeming, F. C. (1995). Encouraging proenvironmental behavior: The environmental court as contingency manager. *Environment and Behavior, 27,* 196–212.

Poulton, E. C. (1970). *The environment and human efficiency.* Springfield, IL: Thomas.

Poulton, E. C. (1977). Continuous noise masks auditory feedback and inner speech. *Psychological Bulletin, 88,* 3–32.

Poulton, E. C. (1979). *The environment at work.* Springfield, IL: Thomas.

Poulton, E. C., Hunt, J. C. R., Mumford, J. C., & Poulton, J. (1975). Mechanical disturbance produced by steady and gusty winds of moderate strength: Skilled performance and semantic assessments. *Ergonomics, 18,* 651–673.

Powers, R. B., Osborne, J. G., & Anderson, E. G. (1973). Positive reinforcement of litter removal in the natural environment. *Journal of Applied Behavior Analysis, 6,* 579–586.

Preiser, W. F. E. (1972). Application of unobtrusive observation techniques in building performance appraisal. In B. E. Foster (Ed.), *Performance concept in buildings.* (Special Publication No. 361, Vol. 1). Washington, DC: National Bureau of Standards.

Preiser, W. F. E. (1973). An analysis of unobtrusive observations of pedestrian movement and stationary behavior in a shopping mall. In R. Kuller (Ed.), *Architectural psychology* (pp. 287–300). Stroudsburg, PA: Dowden, Hutchinson, & Ross.

Prerost, F. J. (1982). The development of the mood inhibiting effects of crowding during adolescence. *Journal of Psychology, 110,* 197–202.

Prerost, F. J., & Brewer, R. K. (1980). The appreciation of humor by males and females during conditions of crowding experimentally induced. *Psychology, A Journal of Human Behavior, 17,* 15–17.

Pressman, N. (1994). Climatic factors in play areas and public space. *Architechture & Comportement/Architecture & Behavior, 10,* 417–427.

Pringle, C., Vellidis, G., Heliotis, F., Bandacu, D., & Cristofor, S. (1993). Environmental problems of the Danube delta. *American Scientist, 81,* 350–361.

Pringle, H. (1997). Death in Norse Greenland. *Science, 275,* 924–926.

Propst, D. B., & Koesler, R. A. (1998). Bandura goes outdoors: Role of self-efficacy in the outdoor leadership development process. *Leisure Sciences, 20,* 319–344.

Proshansky, H. M. (1972). Methodology in environmental psychology: Problems and issues. *Human Factors, 14,* 451–460.

Proshansky, H. M. (1973). Theoretical issues in "environmental psychology." *Representative Research in Social Psychology, 4,* 93–107.

Proshansky, H. M. (1976). Environmental psychology and the real world. *American Psychologist, 31,* 303–310.

Proshansky, H. M., Ittelson, W. H., & Rivlin, L. G. (Eds.) (1970). *Environmental Psychology: Man and his physical setting.* New York: Holt, Rinehart and Winston.

Proulx, G. (1993). A stress model for people facing a fire. *Journal of Environmental Psychology, 13,* 137–147.

Provins, K. A. (1958). Environmental conditions and driving efficiency: A review. *Ergonomics, 2,* 63–88.

Pruchno, R. A., Dempsey, N. P., Carder, P., &

Koropeckyj-Cox, T. (1993). Multigenerational households of caregiving families: Negotiating shared space. *Environment and Behavior, 25,* 349–366.

Pylyshyn, Z. W. (1973). What the mind's eye tells the mind's brain: A critique of mental imagery. *Psychological Bulletin, 80,* 1–24.

Pylyshyn, Z. W. (1981). The imagery debate: Analogue media versus tacit knowledge. *Psychological Review, 88,* 16–45.

Quarantelli, E. L. (Ed.) (1978). *Disasters: Theory and research.* Beverly Hills, CA: Sage.

Quarantelli, E. L. (1998). *What is a disaster?* New York: Routledge.

Ragneskog, H., Gerdner, L., Josefsson, K., & Kihlgren, M. (1998). Probable reasons for expressed agitation in persons with dementia. *Clinical Nursing Research, 7,* 189–206.

Raloff, J. (1982). Occupational noise—The subtle pollutant. *Science News, 121,* 347–350.

Ramadier, T., & Moser, G. (1998). Social legibility, the cognitive map and urban behavior. *Journal of Environmental Psychology, 18,* 307–319.

Rankin, R. E. (1969). Air pollution control and public apathy. *Journal of the Air Pollution Control Association, 19,* 565–569.

Raphael, B., Singh, B., Bradbury, L., & Lambert, F. (1983). Who helps the helpers? The effects of a disaster on the rescue workers. *Omega—Journal of Death & Dying, 14,* 9–20.

Rapoport, A. (1969). *House form and culture.* Englewood Cliffs, NJ: Prentice-Hall.

Rapoport, A. (1975). Toward a redefinition of density. *Environment and Behavior, 7,* 133–158.

Read, M. A., Sugawara, A. I., & Brandt, J. A. (1999). Impact of space and color in the physical environment on preschool children's cooperative behavior. *Environment and Behavior, 31,* 413–428.

Reams, M. A., Geaghan, J. P., & Gendron, R. C. (1996). The link between recycling and litter: A field study. *Environment and Behavior, 28,* 92–110.

Reddy, D. M., Baum, A., Fleming, R., & Aiello, J. R. (1981). Mediation of social density by coalition formation. *Journal of Applied Social Psychology, 11,* 529–537.

Reichel, D. A., & Geller, E. S. (1980, March). *Group versus individual contingencies to conserve transportation energy.* Paper presented at the meeting of the Southeastern Psychological Association, Washington, DC.

Reichner, R. (1979). Differential responses to being ignored: The effects of architectural design and social density on interpersonal behavior. *Journal of Applied Social Psychology, 9,* 13–26.

Reifman, A. S., Larrick, R. P., & Fein, S. (1991). Temper and temperature on the diamond: The heat–aggression relationship in major league baseball. *Personality and Social Psychology Bulletin, 17,* 580–585.

Reisenzein, R. (1983). The Schachter theory of emotion: Two decades later. *Psychological Bulletin, 94,* 239–264.

Reiter, R. (1985). Frequency distribution of positive and negative small ions, based on many years' recordings at two mountain stations located at 740 and 1780 m ASL. *International Journal of Biometeorology, 29,* 223–231.

Reizenstein, J. E. (1982). Hospital design and human behavior: A review of the recent literature. In A. Baum & J. E. Singer (Eds.), *Advances in environmental psychology* (Vol. 4, pp. 137–170). Hillsdale, NJ: Erlbaum.

Relph, E. (1976). *Place and placeness.* London: Pion.

Remland, M. S., Jones, T. S., & Brinkman, H. (1991). Proxemic and haptic behavior in three European countries. *Journal of Nonverbal Behavior, 15,* 215–232.

Remland, M. S., Jones, T. S., & Brinkman, H. (1995). Interpersonal distance, body orientation, and touch: Effects of culture, gender, and age. *Journal of Social Psychology, 135,* 281–297.

Rent, G. S., & Rent, C. S. (1978). Low income housing factors related to residential satisfaction. *Environment and Behavior, 10,* 459–488.

Reusch, J., & Kees, W. (1956). *Nonverbal communication: Notes on the visual perception of human relations.* Berkeley, CA: University of California Press.

Riad, J. K., & Norris, F. H. (1996). The influence of relocation on the environmental, social, and psychological stress experienced by disaster victims. *Environment and Behavior, 28,* 163–182.

Riad, J. K., Norris, F. H., & Ruback, R. B. (1999). Predicting evacuation in two major disasters: Risk perception, social influences, and access to resources. *Journal of Applied Social Psychology, 29,* 918–934.

Richards, P. (1979). Middle class vandalism and the age–status conflict. *Social Problems, 26,* 482–497.

Riggio, R. (2000) *Introduction to industrial/organizational psychology* (3rd ed.). Upper Saddle River, NJ: Prentice Hall.

Riley, R. B. (1992). Attachment to the ordinary landscape. In I. Altman & S. M. Low (Eds.), *Human behavior and environment: Advances in theory and research* (pp. 13–35). New York: Plenum.

Rim, Y. (1975). Psychological test performance during climatic heat stress from desert winds. *International Journal of Biometeorology, 19,* 37–40.

Rind, B. (1996). Effect of beliefs about weather conditions on tipping. *Journal of Applied Social Psychology, 26,* 137–147.

Rivlin, L. G. (1990). The significance of home and homelessness. *Marriage and Family Review, 15,* 39–56.

Rivlin, L. G., & Wolfe, M. (1985). *Institutional settings in children's lives.* New York: Wiley-Interscience.

Rivlin, L. G., Wolfe, M., & Beyda, M. (1973). Age-related differences in the use of space. In W. F. E. Preiser (Ed.), *Environmental design research* (Vol.1, pp. 191–203). Stroudsberg, PA: Dowden, Hutchinson, & Ross.

Robbins, L. N., Fischbach, R. L., Smith, E. M., Cottler, L. B., Solomon, S. D., & Goldring, E. (1986). Impact of disaster on previously assessed mental health. In J. H. Shore (Ed.), *Disaster stress studies: New methods and findings* (pp. 22–48). Washington, DC: American Psychiatric Association.

Roberts, E. (1996). Place and spirit in public land management. In B. L. Driver, D. Dustin, T. Baltic, G. Elsner, & G. Peterson (Eds.), *Nature and the human spirit: Toward an expanded land management ethic* (pp. 61–80). State College, PA: Venture Publishing.

Roberts, S. M., & Schein, R. H. (1993). The entrepreneurial city: Fabricating urban development in Syracuse, New York. *Professional Geographer, 45,* 21–33.

Robinson, E. S. (1928). *The behavior of the museum visitor.* (No. 5 in Publications of the American Association of Museums New Series.) Washington, DC: American Association of Museums.

Robinson, J. P., & Godbey, G. (1997). *Time for life: The surprising ways Americans use their time.* University Park, PA: The Pennsylvania State University Press.

Robinson, M. B., & Robinson, C. E. (1997). Environmental characteristics associated with residential burglaries of student apartment complexes. *Environment and Behavior, 29,* 657–675.

Rochford, E. B., Jr., & Blocker, T. J. (1991). Coping with "natural" hazards as stressors: The predictors of activism in a flood disaster. *Environment and Behavior, 23,* 171–194.

Rock, I., & Palmer, S. (1990, December). The legacy of Gestalt psychology. *Scientific American,* 84–90.

Rocky Mountain Institute (1994). *Greening the building and the bottom line.* Aspen, CO: Rocky Mountain Institute.

Rodin, J. (1976). Crowding, perceived choice and response to controllable and uncontrollable outcomes. *Journal of Experimental Social Psychology, 12,* 564–578.

Rodin, J. (1986). Aging and health: Effects of the sense of control. *Science, 233,* 1271–1276.

Rodin, J., & Baum, A. (1978). Crowding and helplessness: Potential consequences of density and loss of control. In A. Baum & Y. Epstein (Eds.), *Human response to crowding* (pp. 389–401). Hillsdale, NJ: Erlbaum.

Rodin, J., & Langer, E. J. (1977). Long-term effects of a control-relevant intervention with the institutionalized aged. *Journal of Personality and Social Psychology, 35,* 897–902.

Rodin, J., Solomon, S., & Metcalf, J. (1978). Role of control in mediating perceptions of density. *Journal of Personality and Social Psychology, 36,* 989–999.

Roethlisberger, F. J., & Dickson, W. J. (1939). *Management and the worker.* Cambridge, MA: Harvard University Press.

Rohe, W. M. (1982). The response to density in residential settings: The mediating effects of social and personal variables. *Journal of Applied Social Psychology, 12,* 292–303.

Rohe, W. M., & Basolo, V. (1997). Long-term effects of homeownership on the self-perceptions and social interaction of low-income persons. *Environment and Behavior, 29,* 793–819.

Rook, K. S., & Dooley, D. (1985). Applying social support research: Theoretical problems and future directions. *Journal of Social Issues, 41,* 5–28.

Rose, L. (1987). Workplace video display terminals and visual fatigue. *Journal of Occupational Medicine, 29,* 321–324.

Rosen, S., Bergman, M., Plestor, M., Plestor, D., El-Mofty, A., & Satti, M. (1962). Presbycosis study of a relatively noise-free population in the Sudan. *Annals of Otology, Rhinology, and Laryngology, 71,* 727–743.

Rosenfeld, D. (2000). Suppression of rain and snow by urban and industrial air pollution. *Science, 287,* 1793–1796.

Rosenfeld, H. M., Breck, R. E., Smith, S. E., & Kehoe, S. (1984). Intimacy-mediators of the proximity-gaze compensation effect: Movement, controversial role, acquaintance, and gender. *Journal of Nonverbal Behavior, 8,* 235–249.

Rosenthal, N. E. (1993). *Winter blues: Seasonal Affective Disorder: What it is and how to overcome it.* New York: The Guilford Press.

Rosenthal, N. E., Sack, D. A., Gillen, J. C., Lewy, A. J., Goodwin, F. K., Davenport, Y., Mueller, P. S., Newssome, D. A., & Wehr, T. A. (1984). Seasonal Affective Disorder: A description of the syndrome and preliminary findings with light therapy. *Archives of General Psychiatry, 41,* 72–80.

Ross, D. M., Haas, G. E., Loomis, R. J., & Malm, W. C. (1984). *Visibility impairment and visitor enjoyment.* Cooperative Institute for Research in the Atmosphere, Colorado State University, Report No. 5.

Ross, H. (1987). *Just for living: Aboriginal perceptions of housing in northwest Australia.* Canberra: Aboriginal Studies Press.

Rossano, M. J., West, S. O., Robertson, T. J., Wayne, M. C., & Chase, R. B. (1999). The acquisition of route and survey knowledge from computer models. *Journal of Environmental Psychology, 19,* 101–115.

Roth, S., & Cohen, L. J. (1986). Approach, avoidance, and coping with stress. *American Psychologist, 41,* 813–819.

Rothbaum, F., Weisz, J. R., & Snyder, S. S. (1982). Changing the world and changing the self: A two-process model of perceived control. *Journal of Personality and Social Psychology, 42,* 5–37.

Rothenberg, M., & Rivlin, L. (1975). *An ecological approach to the study of open classrooms.* Paper presented at a conference on Ecological Factors in Human Development, University of Surrey, England.

Rothstein, R. N. (1980). Television feedback used to modify gasoline consumption. *Behavior Therapy, 11,* 683–688.

Rotter, J. (1966). Generalized expectancies for internal versus external control of reinforcement. *Psychological Monographs, 80* (Whole No. 609).

Rotton, J. (1983). Affective and cognitive consequences of malodorous pollution. *Basic and Applied Social Psychology, 4,* 171–191.

Rotton, J. (1986). Determinism redux: Climate and cultural correlates of violence. *Environment and Behavior, 18,* 346–368.

Rotton, J. (1987a). Clearing the air about ions. *Skeptical Inquirer, 11,* 305–306.

Rotton, J. (1987b). Hemmed in and hating it: Effects of shape of a room on tolerance for crowding. *Perceptual and Motor Skills, 64,* 285–286.

Rotton, J. (1990). Individuals under stress. In C. E. Kimble (Ed.), *Social psychology: Living with people* (pp. 505–543). New York: W. C. Brown.

Rotton, J. (1993a). Atmospheric and temporal correlates of sex crimes: Endogenous factors do not explain seasonal differences in rape. *Environment and Behavior, 25,* 625–642.

Rotton, J. (1993b). Geophysical variables and behavior: LXXIII. Ubiquitous errors: A reanalysis of Anderson's (1987) "temperature and aggression." *Psychological Reports, 73,* 259–271.

Rotton, J., Barry, T., Frey, J., & Soler, E. (1978). Air pollution and interpersonal attraction. *Journal of Applied Social Psychology, 8,* 57–71.

Rotton, J., Barry, T., & Kimble, C. A. (1985, August). *Climate and crime: Coping with multicolinearity.* Paper presented at the meeting of the American Psychological Association, Los Angeles, CA.

Rotton, J., & Cohn, E. G. (1999). Errors of commission and omission: Comment on Anderson and Anderson's (1998) "temperature and aggression." *Psychological Reports, 85,* 611–620.

Rotton, J., & Cohn, E. G. (2000). Violence is a curvilinear function of temperature in Dallas: Replication and reconciliation. *Journal of Personality and Social Psychology,* in press.

Rotton, J., Dubitsky, S. S., Milov, A., White, S. M., & Clark, M. C. (1997). Distress, elevated cortisol, cognitive deficits, and illness following a natural disaster. *Journal of Environmental Psychology, 17,* 85–98.

Rotton, J., & Frey, J. (1984). Psychological costs of air pollution: Atmospheric conditions, seasonal trends, and psychiatric emergencies. *Population and Environment, 7,* 3–16.

Rotton, J., & Frey, J. (1985). Air pollution, weather, and violent crimes: Concomitant time-series analysis of archival data. *Journal of Personality and Social Psychology, 49,* 1207–1220.

Rotton, J., Frey, J., Barry, T., Milligan, M., & Fitzpatrick, M. (1979). The air pollution experience and interpersonal aggression. *Journal of Applied Social Psychology, 9,* 397–412.

Rotton, J., & Kelly, I. W. (1985a). A scale for assessing belief in lunar effects: Reliability and concurrent validity. *Psychological Reports, 57,* 239–245.

Rotton, J., & Kelly, I. W. (1985b). Much ado about the full moon: A meta-analysis of lunar-lunacy research. *Psychological Bulletin, 97,* 286–306.

Rotton, J., & Kelly, I. W. (1987). Comment on "The lunar-lunacy relationship": More ado about the full moon. *Psychological Reports, 61,* 733–734.

Rotton, J., Kelly, I. W., & Elortegui, P. (1986). Assessing belief in lunar effects: Known-groups validation. *Psychological Reports, 59,* 171–174.

Rotton, J., Oszewski, D., Charleston, M., & Soler, E. (1978). Loud speech, conglomerate noise, and behavior aftereffects. *Journal of Applied Psychology, 63,* 360–365.

Rotton, J., Shats, M., & Standers, R. (1990). Temperature and pedestrian tempo: Walking without awareness. *Environment and Behavior, 22,* 650–674.

Rotton, J., & White, S. M. (1996). Air pollution, the sick building syndrome, and social behavior. *Environment International, 22,* 53–60.

Rowe, J. W., & Kahn, R. L. (1987). Human aging: Usual and successful. *Science, 237,* 143–149.

Ruback, R. B., & Carr, T. S. (1984). Crowding in a women's prison: Attitudinal and behavioral effects. *Journal of Applied Social Psychology, 14,* 57–68.

Ruback, R. B., & Juieng, D. (1997). Territorial defense in parking lots: Retaliation against waiting drivers. *Journal of Applied Social Psychology, 27,* 821–834.

Ruback, R. B., & Pandey, J. (1991). Crowding, perceived control, and relative power: An analysis of households in India. *Journal of Applied Social Psychology, 21,* 315–344.

Ruback, R. B., & Pandey, J. (1992). Very hot and really crowded: Quasi-experimental investigations of Indian "Tempos." *Environment and Behavior, 24,* 527–554.

Ruback, R. B., & Pandey, J. (1996). Gender differences in perceptions of household crowding: Stress, affiliation, and role obligations in rural India. *Journal of Applied Social Psychology, 26,* 417–436.

Ruback, R. B., Pape, K. D., & Doriot, P. (1989). Waiting for a phone: Intrusion on callers leads to territorial defense. *Social Psychology Quarterly, 52,* 232–241.

Ruback, R. B., & Riad, J. K. (1994). The more (men), the less merry: Social density, social burden, and social support. *Sex Roles, 30,* 743–763.

Ruback, R. B., & Snow, J. N. (1993). Territoriality and nonconscious racism at waterfountains: Intruders and drinkers (Blacks and Whites) are affected by race. *Environment and Behavior, 25,* 250–267.

Rubin, E. S., Cooper, R. N., Frosch, R. A., Lee, T. H., Marland, G., Rosenfeld, A. H., & Stine, D. D. (1992). Realistic mitigation options for global warming. *Science, 257,* 148–266.

Rubinstein, R. L., & Parmelee, P. A. (1992). Attachment to place and the representation of the life course by the elderly. In I. Altman & S. M. Low (Eds.), *Place attachment* (pp. 139–163). New York: Plenum.

Rubonis, A. V., & Bickman, L. (1991). Psychological impairment in the wake of disaster: The disaster–psychopathology relationship. *Psychological Bulletin, 109,* 384–399.

Rullo, G. (1987). People and home interiors: A bibliography of recent psychological research. *Environment and Behavior, 19,* 250–259.

Rumsey, N., Bull, R., & Gahagan, D. (1982). The effects of facial disfigurement on the proxemic behavior of the general public. *Journal of Applied Social Psychology, 12,* 137–150.

Russell, J. A., & Lanius, U. F. (1984). Adaptation level and the affective appraisal of environments. *Journal of Environmental Psychology, 4,* 119–135.

Russell, J. A., & Mehrabian, A. (1978). Environmental, task, and temperamental effects on work performance. *Humanitas, 14,* 75–95.

Russell, J. A., & Pratt, G. (1980). A description of

the affective quality attributed to environments. *Journal of Personality and Social Psychology, 38,* 311–322.

Russell, J. A., & Snodgrass, J. (1987). Emotion and the environment. In D. Stokols & I. Altman (Eds.), *Handbook of environmental psychology* (Vol. 1, pp. 245–280). New York: Wiley-Interscience.

Russell, J. A., & Ward, L. M. (1982). Environmental psychology. *Annual Review of Psychology, 33,* 651–688.

Russell, J. A., Ward, L. M., & Pratt, G. (1981). Affective quality attributed to environments: A factor analytic study. *Environment and Behavior, 13,* 259–288.

Russell, M., Cole, P., & Brown, E. (1973). Absorption by non-smokers of carbon monoxide from room air polluted by tobacco smoke. *Lancet, 1,* 576–579.

Russell, M. B., & Bernal, M. E. (1977). Temporal and climatic variables in naturalistic observation. *Journal of Applied Behavior Analysis, 10,* 399–405.

Rutland, A., Custance, D., & Campbell, R. N. (1993). The ability of three- to four-year-old children to use a map in a large-scale environment. *Journal of Environmental Psychology, 13,* 365–372.

Ryden, M. B., Bossenmaier, M., & McLachlan, C. (1991). Aggressive behavior in cognitively impaired nursing home residents. *Research in Nursing and Health, 14,* 87–95.

Saal, F. E., & Knight, P. A. (1988). *Industrial/organizational psychology: Science and practice.* Pacific Grove, CA: Brooks/Cole.

Sabatino, D. A., Meald, J. E., Rothman, S. G., & Miller, T. L. (1978). Destructive norm-violating social behavior among adolescents. A review of protective efforts. *Adolescence, 13,* 675–680.

Sadalla, E. K., Burroughs, J., & Quaid, M. (1980). House form and social identity. In R. Stough (Ed.), *Proceedings of the 11th International Meeting of the Environmental Design Research Association, 11,* 201–206.

Sadalla, E. K., Burroughs, W. J., & Staplin, L. J. (1980). Reference points in spatial cognition. *Journal of Experimental Psychology: Human Learning and Memory, 6,* 516–528.

Sadalla, E. K., & Magel, S. G. (1980). The perception of traversed distance. *Environment and Behavior, 12,* 65–79.

Sadalla, E. K., & Sheets, V. S. (1993). Symbolism in building materials: Self-preservation and cognitive components. *Environment and Behavior, 25,* 155–180.

Sadalla, E. K., Sheets, V., & McCreath, H. (1990). The cognition of urban tempo. *Environment and Behavior, 22,* 230–254.

Sadalla, E. K., & Staplin, L. J. (1980a). An information storage model for distance cognition. *Environment and Behavior, 12,* 183–193.

Sadalla, E. K., & Staplin, L. J. (1980b). The perception of traversed distance: Intersections. *Environment and Behavior, 12,* 167–182.

Saegert, S. (1975). Effects of spatial and social density on arousal, mood, and social orientation. *Dissertation Abstracts International, Jan., 35(7-B):* 3649.

Saegert, S. (1978). High density environments: Their personal and social consequences. In A. Baum & Y. M. Epstein (Eds.), *Human response to crowding* (pp. 259–276). Hillsdale, NJ: Erlbaum.

Saegert, S. (1982). Environment and children's mental health: Residential density and low income children. In A. Baum & J. E. Singer (Eds.), *Handbook of psychology and health* (Vol. 2, pp. 247–281). Hillsdale, NJ: Erlbaum.

Saegert, S., MacIntosh, E., & West, S. (1975). Two studies of crowding in urban public spaces. *Environment and Behavior, 1,* 159–184.

Salib, E., & Gray, N. (1997). Weather conditions and fatal self-harm in North Cheshire. *British Journal of Psychiatry, 171,* 473–477.

Samdahl, D. M., & Christensen, H. H. (1985). Environmental cues and vandalism: An exploratory study of picnic table carving. *Environment and Behavior, 17,* 445–458.

Sampson, R. J., Raudenbush, S. W., & Earls, F. (1997). Neighborhoods and violent crime: A multilevel study of collective efficacy. *Science, 277,* 918–924.

Samuelson, C. D., & Biek, M. (1991). Attitudes toward energy conservation: A confirmatory factor analysis. *Journal of Applied Social Psychology, 21,* 549–568.

Samuelson, C. D., & Messick, D. M. (1986). Inequities in access to and use of shared resources in social dilemmas. *Journal of Personality and Social Psychology, 51,* 960–967.

Samuelson, C. D., Messick, D. M., Rutte, C. G., & Wilke, H. (1984). Individual and structural solutions to resource dilemmas in two cultures. *Journal of Personality and Social Psychology, 47,* 94–104.

Sancar, F., & Eyikan, B. (1998). Studio instructors talk about skills, knowledge, and professional roles in architecture and landscape architecture. *Environment and Behavior, 30,* 378–397.

Sanders, M. S., & McCormick, E. J. (1993). *Human factors in engineering and design* (7th ed.). New York: McGraw-Hill.

Sandman, P. M., Weinstein, N. D., & Klotz, M. L. (1987). Public response to the risk from geological radon. *Journal of Communication, 37,* 93–108.

Sapolsky, R. M. (1997). The importance of a well-groomed child. *Science, 277,* 1620–1621.

Sarter, N. B., & Woods, D. D. (1995). How in the world did we ever get into that mode? Mode awareness in supervisionary control. *Human Factors, 37,* 5–19.

Saunders, M., Gustanski, J., & Lawton, M. (1974). Effect of ambient illumination on noise levels of groups. *Journal of Applied Psychology, 59,* 527–528.

Savinar, J. (1975). The effect of ceiling height on personal space. *Man–Environment Systems, 5,* 321–324.

Savitz, D. A., & Calle, E. E. (1987). Leukemia and occupational exposure to electromagnetic fields: Review of epidemiologic surveys. *Journal of Occupational Medicine, 29,* 47–51.

Savitz, D. A., Wachtel, H., Barnes, F. A., John, E. M., & Tvrdik, J. G. (1988). Case-control study of childhood cancer and exposure to 60-Hz magnetic fields. *American Journal of Epidemiology, 128,* 21–38.

Schachter, S., & Singer, J. E. (1962). Cognitive, social, and physiological determinants of emotional states. *Psychological Review, 69,* 379–399.

Schaeffer, G. H., & Patterson, M. L. (1980). Intimacy, arousal, and small group crowding. *Journal of Personality and Social Psychology, 38,* 283–290.

Schaeffer, M., Baum, A., Paulus, P., & Gaes, G. (1988). Architecturally mediated effects of social density in prison. *Environment and Behavior, 20,* 3–19.

Scheflen, A. E. (1971). Living space in an urban ghetto. *Family Process, 10,* 429–450.

Scheflen, A. E. (1976). *Human territories: How we behave in space–time.* Englewood Cliffs, NJ: Prentice-Hall.

Scheier, M. F., Carver, C. S., & Gibbons, F. X. (1979). Self-directed attention, awareness of bodily states, and suggestibility. *Journal of Personality and Social Psychology, 37,* 1576–1588.

Schettino, A. P., & Borden, R. J. (1976). Group size versus group density: Where is the affect? *Personality and Social Psychology Bulletin, 2,* 67–70.

Schiavo, R. S., Kobashi, K., Quinn, C., Sefscik, A., & Synn, K. M. (1995). Territorial influences on the permeability of group spatial boundaries. *The Journal of Social Psychology, 135,* 27–29.

Schiffenbauer, A. I. (1979). Designing for high-density living. In J. R. Aiello & A. Baum (Eds.), *Residential crowding and design* (pp. 229–240). New York: Plenum.

Schkade, J. (1978). The effects of expectancy set and crowding on task performance. *Dissertation Abstracts International, Jan., 38(7-B):* 3474.

Schmidt, D. E., & Keating, J. P. (1979). Human crowding and personal control: An integration of the research. *Psychological Bulletin, 86,* 680–700.

Schmidt, K. (1998). Coming to grips with the world's greenhouse gases. *Science, 281,* 504–506.

Schmitz, S. (1997). Gender related strategies in environmental development: Effects of anxiety on wayfinding in and representation of a three-dimensional maze. *Journal of Environmental Psychology, 17,* 215–228.

Schneider, F. W., Lesko, W. A., & Garrett, W. A. (1980). Helping behavior in hot, comfortable, and cold temperatures. *Environment and Behavior, 12,* 231–240.

Schneiderman, N. (1982). Animal behavior models of coronary heart disease. In D. S. Krantz, A. Baum, & J. E. Singer (Eds.), *Handbook of psychology and health* (Vol. 3, pp. 19–56). Hillsdale, NJ: Erlbaum.

Schnelle, J. F., Gendrich, J. G., Beegle, G. P., Thomas, M. M., & McNess, M. P. (1980). Mass media techniques for prompting behavior change in the community. *Environment and Behavior, 12,* 157–166.

Schooler, T. Y., Dougall, A. L., & Baum, A. (1999). Cues, frequency, and the disturbing nature of intrusive thoughts: Patterns seen in rescue workers after the crash of Flight 427. *Journal of Traumatic Stress, 12,* 571–584.

Schopler, J., & Stockdale, J. (1977). An interference

analysis of crowding. *Environmental Psychology and Nonverbal Behavior, 1,* 81–88.

Schopler, J., & Walton, M. (1974). *The effects of structure, expected enjoyment, and participant's internality–externality upon feelings of being crowded.* Unpublished manuscript, University of North Carolina.

Schreyer, R., & Nelson, M. L. (1978). *Westwater and desolation canyons: Whitewater river recreation.* Logan, UT: Institute for the Study of Outdoor Recreation and Tourism, Utah State University.

Schultz, D. B. (1965). *Sensory restriction.* New York: Academic Press.

Schultz, P. W., & Zelezny, L. (1999). Values as predictors of environmental attitudes: Evidence for consistency across 14 countries. *Journal of Environmental Psychology, 19,* 255–265.

Schulz, R. (1976). Effects of control and predictability on the physical and psychological well-being of the institutionalized aged. *Journal of Personality and Social Psychology, 33,* 563–573.

Schulz, R., & Hanusa, B. H. (1978). Long-term effects of control and predictability-enhancing interventions: Findings and ethical issues. *Journal of Personality and Social Psychology, 36,* 1194–1201.

Schulz, R., & Heckhausen, J. (1996). A life span model of successful aging. *American Psychologist, 51,* 702–714.

Schutte, N., Malouff, J., Lawrence, E., Glazer, K., & Cabrales, E. (1992). Creation and validation of a scale measuring perceived control over the institutional environment. *Environment and Behavior, 24,* 366–380.

Schuyler, D. (1986). *The new urban landscape: The redefinition of city form in nineteenth-century America.* Baltimore: Johns Hopkins University Press.

Schwartz, B., & Barsky, S. P. (1977). The home advantage. *Social Forces, 55,* 641–661.

Schwartz, D. C. (1968, July–August). On the ecology of political violence: "The long hot summer" as a hypothesis. *American Behavioral Scientist,* 24–28.

Schwartz, S. E., & Andreae, M. O. (1996). Uncertainty in climate change caused by aerosols. *Science, 272,* 1121–1122.

Scott, A. L. (1993). A beginning theory of personal space boundaries. *Perspectives in Psychiatric Care, 29,* 12–21.

Scott, M. J., & Canter, D. V. (1997). Picture or place? A multiple sorting study of landscape. *Journal of Environmental Psychology, 17,* 263–281.

Searleman, A., & Herrmann, D. (1994). *Memory from a broader perspective.* New York: McGraw-Hill.

Sebba, R., & Churchman, A. (1983). Territories and territoriality in the home. *Environment and Behavior, 15,* 191–210.

Segal, M. W. (1974). Alphabet and attraction: An unobtrusive measure of the effect of propinquity in a field setting. *Journal of Personality and Social Psychology, 30,* 655–657.

Segerstrom, S. C., Solomon, G. F., Kemeny, M. E., & Fahey, J. L. (1998). Relationship of worry to immune sequelae of the Northridge earthquake. *Journal of Behavioral Medicine, 21,* 433–450.

Seibert, P. S., & Anooshian, L. J. (1993). Indirect expression of preference in sketch maps. *Environment and Behavior, 25,* 607–624.

Seidel, A. D. (1985). What is success in E & B research utilization? *Environment and Behavior, 17,* 47–70.

Seligman, C. (1986). Energy consumption, attitudes, and behavior. In M. J. Saks & L. Saxe (Eds.), *Advances in applied social psychology* (Vol. 3, pp. 153–180). Hillsdale, NJ: Erlbaum.

Seligman, C., & Darley, J. M. (1977). Feedback as a means of decreasing residential energy consumption. *Journal of Applied Psychology, 62,* 363–368.

Seligman, C., Kriss, M., Darley, J. M., Fazio, R. H., Becker, L. J., & Pryor, J. B. (1979). Predicting summer energy consumption from homeowners' attitudes. *Journal of Applied Social Psychology, 9,* 70–90.

Seligman, M. E. P. (1970). On the generality of the laws of learning. *Psychological Review, 77,* 406–418.

Seligman, M. E. P. (1975). *Helplessness.* San Francisco: Freeman.

Sell, R. (1976). *Cooperation and competition as a function of residential environment, consequences of game strategy choices, and perceived control.* Doctoral dissertation, State University of New York–Stony Brook.

Selten, J. P., van der Graaf, Y., van Duursen, R., Gispen-de Wied, C., & Kahn, R. S. (1999). Psychotic illness after prenatal exposure to the 1953 Dutch Flood Disaster. *Schizophrenia Research, 35*, 243–245.

Selye, H. (1956). *The stress of life.* New York: McGraw-Hill.

Shafer, E., Jr., Hamilton, J. F., & Schmidt, E. A. (1969). Natural landscape preferences: A predictive model. *Journal of Leisure Research, 1*, 1–19.

Shaffer, D. R., & Sadowski, C. (1975). This table is mine: Resect for marked barroom tables as a function of gender of spatial marker and desirability of locale. *Sociometry, 38*, 408–419.

Sharpe, G. W. (1976). *Interpreting the environment.* New York: Wiley.

Sheets, V. L., & Manzer, C. D. (1991). Affect, cognition, and urban vegetation. *Environment and Behavior, 3*, 285–304.

Sherrod, D. R. (1974). Crowding, perceived control and behavioral aftereffects. *Journal of Applied Social Psychology, 4*, 171–186.

Sherrod, D. R., Armstrong, D., Hewitt, J., Madonia, B., Speno, S., & Fenyd, D. (1977). Environmental attention, affect and altruism. *Journal of Applied Social Pscyhology, 7*, 359–371.

Sherrod, D. R., & Downs, R. (1974). Environmental determinants of altruism: The effects of stimulus overload and perceived control on helping. *Journal of Experimental Social Psychology, 10*, 468–479.

Shinn, M. (1997). Family homelessness: State or trait? *American Journal of Community Psychology, 25*, 755–766.

Shippee, G. E. (1978). *Leadership, group participation, and avoiding the tragedy of the commons.* Unpublished doctoral dissertation, Arizona State University, Tempe, AZ.

Shippee, G. E., Burroughs, J., & Wakefield, S. (1980). Dissonance theory revisited: Perception of environmental hazards in residential areas. *Environment and Behavior, 12*, 35–51.

Shlay, A. B. (1985). Castles in the sky: Measuring housing and neighborhood ideology. *Environment and Behavior, 17*, 593–626.

Shlay, A. B. (1987). Who governs housing preferences: Comment on Morris. *Environment and Behavior, 19*, 121–136.

Shoen, K. (1991). *Districts of pleasantness and danger.* Unpublished manuscript, St. Lawrence University.

Shore, D. (1994). Bad lands. *Outside, 19* (7), 56–71.

Shore, J. H., Tatum, E. L., & Vollmer, W. M. (1986). Psychiatric reactions to disaster: The Mount St. Helens experience. *American Journal of Psychiatry, 143*, 590–595.

Shusterman, D. (1992). Critical review: The health significance of environmental odor pollution. *Archives of Environmental Health, 47*, 76–87.

Sieber, J. E. (1998). Planning ethically responsible research. In L. Bickman & D. J. Rog (Eds.), *Handbook of applied social research methods* (pp. 127–156). Thousand Oaks, CA: Sage.

Sieber, W. J., Rodin, J., Larson, L., Ortega, S., Cummings, N., Levy, S., Whiteside, T., & Herberman, R. (1992). Modulation of human natural killer cell activity by exposure to uncontrollable stress. *Brain, Behavior, and Immunity, 6*, 141–156.

Siegel, J. M., & Steele, C. M. (1980). Environmental distraction and interpersonal judgements. *British Journal of Social and Clinical Psychology, 19*, 23–32.

Sigelman, C. K., & Adams, R. M. (1990). Family interactions in public: Parent–child distance and touching. *Journal of Nonverbal Behavior, 14*, 63–75.

Sikorska, E. (1999). Organizational determinants of resident satisfaction with assisted living. *Gerontologist, 39*, 450–456.

Sime, J. D. (1999). What is environmental psychology? Texts, content and context. *Journal of Environmental Psychology, 19*, 191–206.

Simmel, G. (1957). The metropolis and mental life. In K. H. Wolff (Ed. & Trans.), *The sociology of Georg Simmel* (pp. 409–424). London: The Free Press of Glencoe.

Simon, H. A. (1960). *The new science of management decisions.* New York: Harper & Row.

Sims, J. H., & Baumann, D. D. (1972). The tornado threat: Coping styles of the North and South. *Science, 176*, 1386–1391.

Singer, J. E., Lundberg, U., & Frankenhaeuser, M. (1978). Stress on the train: A study of urban commuting. In A. Baum, J. E. Singer, & S. Valins (Eds.), *Advances in environmental psychology* (Vol. 1, pp. 41–56). Hillsdale, NJ: Erlbaum.

Sinha, S. P., & Mukherjee, N. (1996). The effect of perceived cooperation on personal space requirements. *Journal of Social Psychology, 136*, 655–657.

Sinha, S. P., Nayyar, P., & Mukherjee, N. (1995). Perception of crowding among children and adolescents. *Journal of Social Psychology, 135,* 263–268.

Skeen, D. R. (1976). Influence of interpersonal distance in serial learning. *Psychological Reports, 39,* 579–582.

Skjaeveland, O., & Gärling, T. (1997). Effects of interactional space on neighbouring. *Journal of Environmental Psychology, 17,* 181–198.

Skogan, W., & Maxfield, M. (1981). *Coping with crime.* Beverly Hills, CA: Sage.

Skolnick, P., Frasier, L., & Hadar, I. (1977). Do you speak to strangers? A study of invasions of personal space. *European Journal of Social Psychology, 7,* 375–381.

Skorjanc, A. D. (1991). Differences in interpersonal distance among nonoffenders as a function of perceived violence of offenders. *Perceptual and Motor Skills, 73,* 659–662.

Skotko, V. P., & Langmeyer, D. (1977). The effects of interaction distance and gender on self-disclosure in the dyad. *Sociometry, 40,* 178–182.

Skov, T., Cordtz, T., Jensen, L. K., Saugman, P., Schmidt, K., & Theilade, P. (1991). Modifications of health behavior in response to air pollution notifications in Copenhagen. *Social Science and Medicine, 33,* 621–626.

Slangen-de Kort, Y. A. W., Midden, C. J. H., & van Wagenberg, A. F. (1998). Predictors of the adaptive problem-solving of older persons in their homes. *Journal of Environmental Psychology, 18,* 187–197.

Slaven, R. E., Wodarksi, J. S., & Blackburn, B. L. (1981). A group contingency for electricity conservation in master-metered apartments. *Journal of Applied Behavior Analysis, 14,* 357–363.

Sloan, A. W. (1979). *Man in extreme environments.* Springfield, IL: Thomas.

Sloane, P. D., Lindeman, D. A., Phillips, C., Moritz, D. J., & Koch, G. (1995). Evaluating Alzheimer's special care units: Reviewing the evidence and identifying potential sources of study bias. *The Gerontologist, 35,* 103–111.

Sloane, P. D., & Mathew, L. J. (Eds.). (1991). *Dementia units in long-term care.* Baltimore, MD: Johns Hopkins University Press.

Slovic, P. (1987). Perception of risk. *Science, 236,* 280–285.

Slovic, P. (1997). Trust, emotion, sex, politics, and science: Surveying the risk-assessment battlefield. In M. H. Bazerman, D. M. Messick,

A. E. Tenbrunsel, & K. A. Wade-Benzoni (Eds.), *Environment, ethics, and behavior* (pp. 277–313). San Francisco: New Lexington Press.

Slovic, P., Fischhoff, B., & Lichtenstein, S. (1981). Perceived risk: Psychological factors and social implications. *Proceedings of the Royal Society of London, A 376,* 17–34.

Smith, A. P., & Jones, D. M. (1992). Noise and performance. In D. M. Jones & A. P. Smith (Eds.), *Handbook of human performance* (Vol. 1, pp. 1–18). San Diego, CA: Academic Press.

Smith, A. P., & Stansfield, S. (1986). Aircraft noise exposure, noise sensitivity, and everyday errors. *Environment and Behavior, 18,* 214–226.

Smith, E. M., North, C. S., McCool, R. E., & Shea, J. M. (1990). Acute postdisaster psychiatric disorders: Identification of persons at risk. *American Journal of Psychiatry, 147,* 202–206.

Smith, G. C. (1991). Grocery shopping patterns of the ambulatory elderly. *Environment and Behavior, 23,* 86–114.

Smith, J. M., Bell, P. A., & Fusco, M. E. (1988). The influence of attraction on a simulated commons dilemma. *Journal of General Psychology, 115,* 277–283.

Smith, M. J., Cohen, B. G. F., & Stammerjohn, L. W. (1981). An investigation of health complaints and job stress in video display operations. *Human Factors, 28,* 387–400.

Smith, P., & Connolly, K. (1977). Social and aggressive behavior in preschool children as a function of crowding. *Social Science Information, 16,* 601–620.

Smith, R. J., & Knowles, E. S. (1979). Affective and cognitive mediators of reactions to spatial invasions. *Journal of Experimental Social Psychology, 15,* 437–452.

Smither, R. D. (1988). *The psychology of work and human performance.* New York: Harper & Row.

Soleri, P. (1970). *Arcology: City in the image of man.* Cambridge, MA: MIT Press.

Soleri, P. (1993). *Arcosante: An urban laboratory?* Mayer, AZ: The Cosanti Press.

Solomon, S. D., & Canino, G. J. (1990). Appropriateness of DSM-III-R criteria for Post-traumatic Stress Disorder. *Comprehensive Psychiatry, 31,* 227–237.

Sommer, R. (1959). Studies in personal space. *Sociometry, 22,* 247–260.

Sommer, R. (1965). Further studies of small group ecology. *Sociometry, 28,* 337–348.

Sommer, R. (1967). Classroom ecology. *Journal of Applied Behavioral Science, 3,* 500.

Sommer, R. (1969). *Personal space.* Englewood Cliffs, NJ: Prentice Hall.

Sommer, R. (1972). *Design awareness.* San Francisco: Rinehart Press.

Sommer, R. (1989). Farmers' markets as community events. In I. Altman & E. H. Zube (Eds.), *Public places and spaces* (pp. 57–82). New York: Plenum.

Sommer, R. (1998). Shopping at the co-op. *Journal of Environmental Psychology, 18,* 45–53.

Sommer, R., & Olsen, H. (1980). The soft classroom. *Environment and Behavior, 12,* 3–16.

Sommer, R., & Ross, H. (1958). Social interaction on a geriatrics ward. *International Journal of Social Psychiatry, 4,* 128–133.

Sommer, R., & Summit, J. (1995). An exploratory study of preferred tree form. *Environment and Behavior, 27,* 540–557.

Sommer, R., & Wicker, A. W. (1991). Gas station psychology: The case for specialization in ecological psychology. *Environment and Behavior, 23,* 131–149.

Sommers, P., & Moos, R. (1976). The weather and human behavior. In R. H. Moos (Ed.), *The human context: Environmental determinants of behavior* (pp. 73–107). New York: Wiley.

Sonnenfeld, J. (1966). Variable values in space and landscape: An inquiry into the nature of environmental necessity. *Journal of Social Issues, 22,* 71–82.

Southworth, M., & Owens, P. M. (1993). The evolving metropolis: Studies of community, neighborhood, and street form at the urban edge. *Journal of the American Planning Association, 59,* 271–287.

Spreckelmeyer, K. F. (1993). Office relocation and environmental change: A case study. *Environment and Behavior, 25,* 181–204.

Spyker, J. M. (1975). Assessing the impact of low level chemicals on development: Behavioral and latent effects. *Federation Processings, 34,* 1835–1844.

Srivastava, P., & Mandal, M. K. (1990). Proximal spacing to facial affect expressions in schizophrenia. *Comprehensive Psychiatry, 31,* 119–124.

Srivastava, R. K. (1974). Undermanning theory in the context of mental health care environments. In D. H. Carson (Ed.), *Man–environment interactions* (Part 2, pp. 245–258). Stroudsberg, PA: Dowden, Hutchinson, & Ross.

Stahl, S. M., & Lebedun, M. (1974). Mystery gas: An analysis of mass hysteria. *Journal of Health and Social Behavior, 15,* 44–50.

Stankey, G. H. (1973). *Visitor perception of wilderness recreation carrying capacity* (Research Paper NO. INT-142, p. 62). Ogden, UT: U.S. Department of Agriculture, Intermountain Forest and Range Experiment Station.

Stansfield, S. (1992). Noise, noise sensitivity, and psychiatric disorder: Epidemiological and psychophysiological studies. *Psychological Medicine, monograph supplement, 22,* 44.

Stansfield, S. A., Sharp, D. S., Gallacher, J., & Babisch, W. (1993). Road traffic noise, noise sensitivity, and psychological disorder. *Psychological Medicine, 23,* 977–985.

Staples, S. L. (1996). Human response to environmental noise. *American Psychologist, 51,* 143–150.

Staples, S. L. (1997). Public policy and environmental noise: Modeling exposure or understanding. *American Journal of Public Health, 87,* 2063–2067.

Staples, S. L., Cornelius, R. R., & Gibbs, M. S. (1999). Noise disturbance from a developing airport: Perceived risk or general annoyance? *Environment and Behavior, 31,* 692–710.

Stea, D., & Blaut, J. M. (1973). Some preliminary observations on spatial learning in school children. In R. Downs & D. Stea (Eds.), *Image and the environment* (pp. 226–234). Chicago: Aldine.

Steblay, N. M. (1987). Helping behavior in rural and urban environments: A meta analysis. *Psychological Bulletin, 102,* 346–356.

Steele, F. (1981). *The sense of place.* Boston: CBI.

Steinglass, P., & Gerrity, E. (1990). Natural disasters and Post-Traumatic Stress Disorder: Short-term versus long-term recovery in two disaster-affected communities. *Journal of Applied Social Psychology, 20,* 1746–1765.

Steinitz, C. (1968). Meaning and congruence of urban form and activity. *Journal of the American Institute of Plannners, 34,* 233–248.

Stellman, J. M., Klitzman, S., Gordon, G. C., & Snow, B. R. (1987). Work environments and the well-being of clerical and VDT workers. *Journal of Occupational Behaviour, 8,* 95–114.

Stermer, M. (1998). Notes from an Eden Alternative pioneer. *Nursing Homes, 47*(1), pp. 5ff.

Stern, P. C. (1992a). Psychological dimensions of global environmental change. *Annual Review of Psychology, 43,* 269–302.

Stern, P. C. (1992b). What psychology knows about energy conservation. *American Psychologist, 47,* 1224–1232.

Stern, P. C. (2000). Psychology, sustainability, and the science of human–environment interactions. *American Psychologist, 55,* 523–530.

Stern, P. C., & Aronson, E. (Eds.) (1984). *Energy use: The human dimension.* New York: Freeman.

Stern, P. C., & Dietz, T. (1994). The value basis of environmental concern. *Journal of Social Issues, 56,* 121–145.

Stern, P. C., Dietz, T., & Guagnano, G. A. (1995). The New Ecological Paradigm in social-psychological context. *Environment and Behavior, 27,* 723–743.

Stern, P. C., Dietz, T., & Kalof, L. (1993). Value orientations, gender, and environmental concern. *Environment and Behavior, 25,* 322–348.

Stern, P. C., & Gardner, G. T. (1981). Psychological research and energy policy. *American Psychologist, 4,* 329–342.

Stern, P. C., & Oskamp, S. (1987). Managing scarce environmental resources. In D. Stokols & I. Altman (Eds.), *Handbook of environmental psychology* (Vol. 2, pp. 1043–1088). New York: Wiley-Interscience.

Stevens, A., & Coupe, P. (1978). Distortions in judged spatial relations. *Cognitive Psychology, 10,* 422–437.

Stevens, J. C. (1991). Thermal sensibility. In M. A. Heller & W. Schiff (Eds.), *The psychology of touch* (pp. 61–90). Hillsdale, NJ: Erlbaum.

Stevens, W., Kushler, M., Jeppesen, J., & Leedom, N. (1979). *Youth energy education strategies: A statistical evaluation.* Lansing, MI: Energy Extension Service, Department of Commerce.

Stewart, T. R. (1987). Developing an observer-based measure of environmental annoyance. In H. S. Koelega (Ed.), *Environmental annoyance: Characterization, measurement, and control* (pp. 213–222). New York: Elsevier.

Stewart, T. R., Middleton, P., & Ely, D. (1983). Urban visual air quality judgments: Reliability and validity. *Journal of Environmental Psychology, 3,* 129–145.

Stewart, W. P., & Cole, D. N. (1999). In search of situational effects in outdoor recreation: Different methods, different results. *Leisure Sciences, 21,* 269–286.

Stires, L. (1980). Classroom seating location, student grades and attitudes: Environment or selection? *Environment and Behavior, 12,* 241–254.

Stobaugh, R., & Yergin, D. (1979). *Energy future: Report of the energy project of the Harvard Business School.* New York: Random House.

Stohlgren, T. J., Chase, T. N., Pielke, R. A., Kittel, T. G. F., & Brown, J. S. (1998). Evidence that local land use practices influence regional climate, vegetation, and stream flow patterns in adjacent natural areas. *Global Change Biology, 4,* 495–504.

Stokols, D. (1972). On the distinction between density and crowding: Some implications for future research. *Psychological Review, 79,* 275–278.

Stokols, D. (1976). The experience of crowding in primary and secondary environments. *Environment and Behavior, 8,* 49–86.

Stokols, D. (1978). A typology of crowding experiences. In A. Baum & Y. Epstein (Eds.), *Human response to crowding* (pp. 219–255). Hillsdale, NJ: Erlbaum.

Stokols, D. (1979). A congruence analysis of human stress. In I. G. Sarason & C. D. Spielberger (Eds.), *Stress and anxiety* (Vol. 6, pp. 27–53). New York: Wiley.

Stokols, D. (1983). Editor's introduction: Theoretical directions of environment and behavior research. *Environment and Behavior, 15,* 259–272.

Stokols, D. (1990). Instrumental and spiritual views of people–environment relations. *American Psychologist, 45,* 641–646.

Stokols, D., & Altman, I. (Eds.) (1987). *Handbook of environmental psychology* (Vol. 1, pp. xi–xii). New York: Wiley.

Stokols, D., & Novaco, R. W. (1981). Transportation and well-being: An ecological perspective. In I. Altman, J. F. Wohlwill, & P. B. Everett (Eds.), *Transportation and behavior* (pp. 85–130). New York: Plenum.

Stokols, D., Rall, M., Pinner, B., & Schopler, J. (1973). Physical, social and personal determinants of the perception of crowding. *Environment and Behavior, 5*, 87–117.

Stone, G. L., & Morden, C. J. (1976). Effect of distance on verbal productivity. *Journal of Counseling Psychology, 23*, 486–488.

Stone, N. J., & English, A. J. (1998). Task type, posters, and workspace color on mood, satisfaction, and performance. *Journal of Environmental Psychology, 18*, 175–185.

Stone, R. (1999). Coming to grips with the Aral Sea's grim legacy. *Science, 284*, 30–33.

Storms, M. D., & Thomas, G. C. (1977). Reactions to physical closeness. *Journal of Personality and Social Psychology, 35*, 412–418.

Strahilevitz, N., Strahilevitz, A., & Miller, J. E. (1979). Air pollution and the admission rate of psychiatric patients. *American Journal of Psychiatry, 136*, 206–207.

Strayer, J., & Roberts, W. (1997). Children's personal distance and their empathy: Indices of interpersonal closeness. *International Journal of Behavioral Development, 20*(3), 385–403.

Strodtbeck, F., & Hook, H. (1961). The social dimension of a 12-man jury table. *Sociometry, 24*, 397–415.

Strube, M. J., & Werner, C. (1984). Psychological reactance and the relinquishment of control. *Personality and Social Psychology Bulletin, 10*, 225–234.

Suedfeld, P. (1975). The benefits of boredom: Sensory deprivation reconsidered. *American Scientist, 63*, 60–69.

Suedfeld, P. (1980). *Restricted environmental stimulation: Research and clinical applications.* New York: Wiley.

Suedfeld, P. (1991). Polar psychology: An overview. *Environment and Behavior, 23*, 653–665.

Suedfeld, P. (1998). What can abnormal environments tell us about normal people? Polar stations as natural psychology laboratories. *Journal of Environmental Psychology, 18*, 95–102.

Suedfeld, P., & Baker-Brown, G. (1986). Restricted environmental therapy and aversive conditioning in smoking cessation: Active and placebo effects. *Behavior Research and Therapy, 24*, 421–428.

Suedfeld, P., & Mocellin, J. S. P. (1987). The "sensed presence" in unusual environments. *Environment and Behavior, 19*, 33–52.

Suedfeld, P., Roy, C., & Landon, P. B. (1982). Restricted environmental stimulation therapy in the treatment of essential hypertension. *Behavior Research and Therapy, 20*, 553–559.

Suedfeld, P., Schwartz, G., & Arnold, W. (1980). Study of restricted environmental stimulation therapy (REST) as a treatment for autistic children. *Journal of Autism and Developmental Disorders, 10*, 337–378.

Suedfeld, P., Turner, J. W., Jr., & Fine, T. H. (Eds.) (1990). *Restricted environmental stimulation: Theoretical and empirical developments in flotation REST.* New York: Springer.

Sullivan, M. A., Saylor, C., & Foster, S. C. (1991). Post-hurricane adjustment of preschoolers and their families. *Advances in Behavior, Research, and Therapy, 13*, 163–172.

Sulman, F. G., Danon, A., Pfeifer, Y., Tal, E., & Weller, C. P. (1970). Urinalysis of patients suffering from climatic heat stress (Sharav). *International Journal of Biometeorology, 14*, 45–53.

Summit, J., & Sommer, R. (1999). Further studies of preferred tree shapes. *Environment and Behavior, 31*, 550–576.

Sundeen, R. A., & Mathieu, J. T. (1976). Fear of crime and its consequences among elderly in three urban communities. *The Gerontologist, 16*, 211–219.

Sundstrom, E. (1975). An experimental study of crowding: Effects of room size, intrusion, and goal-blocking on nonverbal behaviors, self-disclosure, and self-reported stress. *Journal of Personality and Social Psychology, 32*, 645–654.

Sundstrom, E. (1978). Crowding as a sequential process: Review of research on the effects of population density on humans. In A. Baum & Y. M. Epstein (Eds.), *Human response to crowding* (pp. 31–116). Hillsdale, NJ: Erlbaum.

Sundstrom, E. (1986). *Work places: The psychology of the physical environment in offices and factories.* New York: Cambridge University Press.

Sundstrom, E. (1987). Work environments: Offices and factories. In D. Stokols & I. Altman (Eds.), *Handbook of environmental psychology* (pp. 733–782). New York: Wiley-Interscience.

Sundstrom, E., & Altman, I. (1976). Personal space

and interpersonal relationships: Research review and theoretical model. *Human Ecology, 4,* 47–67.

Sundstrom, E., & Sundstrom, M. G. (1977). Personal space invasions: What happens when the invader asks permission? *Environmental Psychology and Nonverbal Behavior, 2,* 76–82.

Sundstrom, E., Town, J. P., Rice, R. W., Osborn, D. P., & Brill, M. (1994). Office noise, satisfaction, and performance. *Environment and Behavior, 26,* 195–222.

Susa, A. M., & Benedict, J. O. (1994). The effects of playground design on pretend play and divergent thinking. *Environment and Behavior, 26,* 560–579.

Suter, T. W., Buzzi, R., Woodson, P. P., & Battig, K. (1983, December). Psychophysiological correlates of conflict solving and cigarette smoking. *Activitas Nervosa Superior, 25,* 261–272.

Suttles, G. D. (1968). *The social order of the slum: Ethnicity and territory in the inner city.* Chicago: University of Chicago Press.

Sweeney, P. D., Anderson, K., & Bailey, S. (1986). Attributional style in depression: A meta-analytic review. *Journal of Personality and Social Psychology, 50,* 974–991.

Syme, G. L., Seligman, C., Kantola, S. J., & MacPherson, D. K. (1987). Evaluating a television campaign to promote petrol conservation. *Environment and Behavior, 19,* 444–461.

Szarek, M. J., Bell, I. R., & Schwartz, G. E. (1997). Validation of a brief screening measure of environmental chemical sensitivity: The Chemical Odor Intolerance Index. *Journal of Environmental Psychology, 17,* 345–351.

Szilagyi, A. D., & Holland, W. E. (1980). Changes in social density: Relationships with functional interaction and perceptions of job characteristics, role stress, and work satisfaction. *Journal of Applied Psychology, 65,* 28–33.

Talbot, J. F., Kaplan, R., Kuo, F. E., & Kaplan, S. (1993). Factors that enhance effectiveness of visitor maps. *Environment and Behavior, 25,* 743–760.

Talbot, J. F., & Kaplan, S. (1986). Perspectives on wilderness: Re-examining the value of extended wilderness experiences. *Journal of Environmental Psychology, 6,* 177–188.

Talbott, E. O., Findlay, R. C., Kuller, L. H., Lenkner, L. A., Matthews, K. A., Day, R. D., & Ishii, E. K. (1990). Noise-induced hearing loss and blood pressure. *Journal of Occupational Medicine, 32,* 690–697.

Tanner, C. (1999). Constraints on environmental behaviour. *Journal of Environmental Psychology, 19,* 145–157.

Tarrant, M. A., & Cordell, H. K. (1997). The effect of respondent characteristics on general environmental attitude–behavior correspondence. *Environment and Behavior, 29,* 618–637.

Tarrant, M. A., Cordell, H. K., & Kibler, T. L. (1997). Measuring perceived crowding for high-density river recreation: The effects of situational conditions and personal factors. *Leisure Sciences, 19,* 97–112.

Tasso, J., & Miller, E. (1976). The effects of the full moon on human behavior. *Journal of Psychology, 93,* 81–83.

Taubes, G. (1993). The ozone backlash. *Science, 260,* 1580–1583.

Taubes, G. (1997). Apocalypse not. *Science, 278,* 1004–1006.

Taylor, A. F., Wiley, A., Kuo, F. E., & Sullivan, W. C. (1998). Growing up in the inner city: Green spaces as places to grow. *Environment and Behavior, 30,* 2–27.

Taylor, A. J. W., & Frazier, A. G. (1982). The stress of postdisaster body handling and victim identification work. *Journal of Human Stress, 8,* 4–12.

Taylor, D. E. (1989). Blacks and the environment: Toward an explanation of the concern and action gap between Blacks and Whites. *Environment and Behavior, 21,* 175–205.

Taylor, D. E. (1999). Mobilizing for environmental justice in communities of color. In Aley, J., Burch, W. R., Conover, B., & Field, D. (Eds.), *Ecosystem management: Adaptive strategies for natural resources organizations in the 21st century* (pp. 33–67). Philadelphia: Taylor and Francis.

Taylor, D. E. (2000). The rise of the environmental justice paradigm—Injustice framing and the social construction of environmental discourses. *American Behavioral Scientist, 43,* 508–580.

Taylor, F. W. (1911). *The principles of scientific management.* New York: Harper and Brothers.

Taylor, H. A., & Tversky, B. (1996). Perspectives in spatial descriptions. *Journal of Memory and Language, 35,* 371–391.

Taylor, H. R., West, S. K., Rosenthal, F. S., Munoz, B., Newland, H. S., Abbey, H., & Emmett, E. A.

(1988). Effect of ultraviolet radiation on cataract formation. *New England Journal of Medicine, 319,* 1429–1433.

Taylor, K. (1999). Rapid climate change. *American Scientist, 87,* 320–327.

Taylor, R. (1990, April). Heavy metal, heavy toll: The lasting legacy of low-level lead in children. *Journal of NIH Research, 2,* 57–60.

Taylor, R. B. (1988). *Human territorial functioning: An empirical, evolutionary perspective on individual and small group territorial cognitions, behaviors, and consequences.* Cambridge: Cambridge University Press.

Taylor, R. B., & Brooks, D. K. (1980). Temporary territories: Responses to intrusions in a public setting. *Population and Environment, 3,* 135–145.

Taylor, R. B., & Brower, S. (1985). Home and near-home territories. In I. Altman & C. Werner (Eds.), *Human behavior and environment: Current theory and research, Vol. 8: Home environments* (pp. 183–212). New York: Plenum.

Taylor, R. B., & Covington, J. (1988). Neighborhood changes. *Ecology and Violent Criminology, 26,* 553–591.

Taylor, R. B., Gottfredson, S. D., & Brower, S. (1980). The defensibility of defensible space: A critical review and a synthetic framework for future research. In T. Hirshi & M. Gottfredson (Eds.), *Understanding crime* (pp. 53–71). Newbury Park, CA: Sage.

Taylor, R. B., Gottfredson, S. D., & Brower, S. (1981). Territorial cognitions and social climate in urban neighborhoods. *Basic and Applied Social Psychology, 2,* 289–303.

Taylor, R. B., Gottfredson, S., & Brower, S. (1984). Understanding block crime and fear. *Journal of Research in Crime and Delinquency, 21,* 303–331.

Taylor, R. B., & Hale, M. (1986). Testing alternative models of fear of crime. *Journal of Law and Criminology, 77,* 151–189.

Taylor, R. B., & Lanni, J. C. (1981). Territorial dominance: The influence of the resident advantage in triadic decision making. *Journal of Personality and Social Psychology, 41,* 909–915.

Taylor, R. B., & Stough, R. R. (1978). Territorial cognition: Assessing Altman's typology. *Journal of Personality and Social Psychology, 36,* 418–423.

Taylor, S., & Todd, P. (1995). An integrated model of waste management behavior: A test of household recycling and composting intentions. *Environment and Behavior, 27,* 603–630.

Taylor, S. E, Peplau, L. A., & Sears, D. O. (2000). *Social psychology* (10th ed.). Upper Saddle River, NJ: Prentice Hall.

Taylor, S. M. (1984). A path model of aircaft noise annoyance. *Journal of Sound and Vibration, 96,* 243–260.

Teare, J. F., Smith, G. L., Osgood, D. W., Peterson, R. W., Authier, K., & Daly, D. L. (1995). Ecological influences in youth crisis shelters: Effects of social density and length of stay on youth problem behaviors. *Journal of Child and Family Studies, 4,* 89–101.

Teculescu, D. B., Sauleau, E. A., Massin, N., Bohadana, A. B., Buhler, O., Benamghar, L., & Mur, J. M. (1998). Sick-building symptoms in office workers in northeastern France: A pilot study. *International Archives of Occupational & Environmental Health, 71,* 353–356.

Teicher, M. H., Clod, C. A., Oren, D. A., Schwartz, P. J., Luetke, C., Brown, C., & Rosenthal, N. E. (1995). The phototherapy light visor: More to it than meets the eye. *American Journal of Psychiatry, 153,* 1110–1111.

Tennen, H., & Eller, S. J. (1977). Attributional components of learned helplessness and facilitation. *Journal of Personality and Social Psychology, 35,* 265–271.

Tennis, G. H., & Dabbs, J. M. (1975). Sex, setting and personal space: First grade through college. *Sociometry, 38,* 385–394.

Terman, M., Amira, L., Terman, J. S., & Ross, D. C. (1996). Predictors of response and nonresponse to light treatment for winter depression. *American Journal of Psychiatry, 153,* 1423–1429.

Terr, L. C. (1979). Children of Chowchilla. *Psychoanalytical Study of Child, 34,* 547–643.

Terr, L. C. (1983). Chowchilla revisited: The effects of psychic trauma four years after a schoolbus kidnapping. *American Journal of Psychiatry, 140,* 1543–1550.

Terry, R. L., & Lower, M. (1979). Perceptual withdrawal from an invasion of personal space. *Personality and Social Psychology Bulletin, 5,* 396–397.

Thalhofer, N. N. (1980). Violation of a spacing norm

in high social density. *Journal of Applied Social Psychology, 10,* 175–183.

Theil, P. (1994). Beyond design review: Implications for design practice, education, and research. *Environment and Behavior, 26,* 363–376.

Theorell, T. (1990). Family history of hypertension—An individual trait interacting with spontaneously occurring job stressors. *Scandinavian Journal of Work and Environmental Health, 16(Suppl. 1),* 74–79.

Thøgersen, J. (1996). Recycling and morality: A critical review of the literature. *Environment and Behavior, 28,* 536–558.

Thoits, P. A. (1982). Conceptual, methodological, and theoretical problems in studying social support as a buffer against life stress. *Journal of Health and Social Behavior, 23,* 145–159.

Thomas, W. H. (1996). *Life worth living: How someone you love can still enjoy life in a nursing home: The Eden alternative in action.* Acton, MA: VanderWyk & Burnham.

Thompson, D. R. (1993). *Considering the museum visitor: An interactional approach to environmental design.* Unpublished doctoral dissertation, University of Wisconsin–Milwaukee.

Thompson, J., Chung, M. C., & Rosser, R. (1994). The Marchioness disaster: Preliminary report on psychological effects. *British Journal of Clinical Psychology, 33,* 75–77.

Thompson, M. P., Norris, F. H., & Hanacek, B. (1993). Age differences in the psychological consequences of Hurricane Hugo. *Psychology and Aging, 8,* 606–616.

Thompson, S. C. G. (1981). Will it hurt less if I can control it? A complex answer to a simple question. *Psychological Bulletin, 90,* 89–101.

Thompson, S. C. G., & Barton, M. (1994). Ecocentric and anthropocentric attitudes toward the environment. *Journal of Environmental Psychology, 14,* 149–157.

Thompson, S. C. G., & Stoutemyer, K. (1991). Water use as a commons dilemma: The effects of education that focuses on long-term consequences and individual action. *Environment and Behavior, 23,* 314–333.

Thorn, A. (1998). Building-related health problems: Reflections on different symptom prevalence among pupils and teachers. *International Journal of Circumpolar Health, 57,* 249–256.

Thorndyke, P. W. (1981). Distance estimation from cognitive maps. *Cognitive Psychology, 13,* 526–550.

Thorndyke, P. W., & Hayes-Roth, B. (1982). Differences in spatial knowledge acquired from maps and navigation. *Cognitive Psychology, 14,* 560–589.

Thorne, R., Hall, R., & Munro-Clark, M. (1982). Attitudes toward detached houses, terraces and apartments: Some current pressures towards less preferred but more accessible alternatives. In P. Bart, A. Chen, & G. Francescato (Eds.), *Knowledge for design: Proceedings of the 13th Environmental Design Research Association Conference* (pp. 435–448). Washington, DC: Environmental Design Research Association.

Tichener, J., & Kapp, F. I. (1976). Family and character change at Buffalo Creek. *American Journal of Psychiatry, 133,* 295–299.

Timasheff, N. S. (1967). *Sociological theory: Its nature and growth* (3rd ed.). New York: Random House.

Tinsley, H. E., & Tinsley, D. J. (1986). A theory of the attributes, benefits, and causes of leisure experience. *Leisure Sciences, 8,* 1–45.

Tobin, G. A., & Ollenburger, J. C. (1996). Predicting levels of post-disaster distress in adults following the 1993 floods in the upper midwest. *Environment and Behavior, 28,* 340–357.

Toffler, A. (1980). *The third wave.* New York: Bantam Books.

Tognoli, J. (1980). Differences in women's and men's responses to domestic space. *Sex Roles, 6,* 833–842.

Tognoli, J. (1987). Residential environments. In D. Stokols & I. Altman (Eds.), *Handbook of environmental psychology* (Vol. 1, pp. 655–690). New York: Wiley-Interscience.

Tolman, E. C. (1948). Cognitive maps in rats and men. *Psychological Review, 55,* 189–208.

Tolman, E. C., Ritchie, B. F., & Kalish, D. (1946). Studies in spatial learning I. Orientation and the short-cut. *Journal of Experimental Psychology, 36,* 13–24.

Tomaka, J., Blascovich, J., Kibler, J., & Ernst, J. M. (1997). Cognitive and physiological antecedents of threat and challenge appraisal. *Journal of Personality and Social Psychology, 73,* 63–72.

Topf, M. (1992a). Stress effects of personal control

over hospital noise. *Behavioral Medicine, 18*, 84–94.

Topf, M. (1992b). Stress effects of personal control over hospital noise on sleep. *Research in Nursing and Health, 15*, 19–28.

Topf, M. (1994). Theoretical considerations for research on environmental stress and health. *IMAGE: Journal of Nursing Scholarship, 26*, 289–293.

Trites, D., Galbraith, F. D., Sturdavent, M., & Leckwart, J. F. (1970). Influence of nursing unit design on the activities and subjective feelings of nursing personnel. *Environment and Behavior, 2*, 303–334.

Tromp, S. W. (1980). *Biometeorology: The impact of weather and climate on humans and their environment.* Philadelphia: Heyden.

Trowbridge, C. C. (1913). On fundamental methods of orientation and "imaginary maps." *Science, 88*, 888–896.

Truscott, J. C., Parmalee, P., & Werner, C. (1977). Plate touching in restaurants: Preliminary observations of a food-related marking behavior in humans. *Journal of Personality and Social Psychology, 3*, 425–428.

Tuan, Y. (1974). *Topophilia: A study of environmental perception, attitude, and values.* Englewood Cliffs, NJ: Prentice-Hall.

Turk, A., Turk, J., Wittes, J. T., & Wittes, R. (1974). *Environmental science.* Philadelphia: Saunders.

Turnage, J. J. (1990). The challenge of new workplace technology for psychology. *American Psychologist, 45*, 171–178.

Turner, P. V. (1984). *Campus: An American planning tradition.* Cambridge, MA: M.I.T. Press.

Tversky, B. (1981). Distortions in memory for maps. *Cognitive Psychology, 13*, 407–433.

Tversky, B. (1993). Cognitive maps, cognitive collages, and spatial mental models. In A. U. Frank & I. Campari (Eds.), *Spatial information theory: A theoretical basis for GIS.* Berlin: Springer-Verlag, pp. 14–24.

Tye, M. (1991). *The imagery debate.* Cambridge, MA: M.I.T. Press.

Tyler, T. R. (1981). Perceived control and behavioral reactions to crime. *Personality and Social Psychology Bulletin, 7*, 212–217.

Tyler, T. R., & Degoey, P. (1995). Collective restraint in social dilemmas: Procedural justice and social identification effects on support for authorities. *Journal of Personality and Social Psychology, 69*, 482–497.

U.S. Bureau of the Census (1990). *Statistical abstracts of the United States: 1990.* Washington, DC: U.S. Government Printing Office.

U.S. Bureau of the Census (1998). http://www.census.gov/population/www/estimates/popest.html.

U.S. Bureau of the Census (1998). http://www.census.gov/population/www/projections/popproj.html.

U.S. Riot Commission (1968). *Report of the National Advisory Commission on Civil Disorders.* New York: Bantam Books.

Ugwuegbu, D. C., & Anuseim, A. U. (1982). Effects of stress on interpersonal distance in a simulated interview situation. *Journal of Social Psychology, 116*, 3–7.

Ulrich, R. S. (1977). Visual landscape preference: A model and applications. *Man–Environment Systems, 7*, 279–292.

Ulrich, R. S. (1979). Visual landscapes and psychological well-being. *Landscape Research, 4*, 17–23.

Ulrich, R. S. (1984). View through a window may influence recovery from surgery. *Science, 224*, 420–421.

Ulrich, R. S. (1986). Human responses to vegetation and landscapes. *Landscape and Urban Planning, 13*, 29–44.

Ulrich, R. S. (1993). Biophilia and the conservation ethic. In S. R. Kellert & E. O. Wilson (Eds.), *The biophilia hypothesis* (pp. 73–137). Washington, DC: Island Press.

Ulrich, R. S. (1997). Improving medical outcomes with environmental design. *Journal of Healthcare Design, lx*, 3–7.

Ulrich, R. S., Dimberg, U., & Driver, B. L. (1991). Psychophysiological indicators of leisure benefits. In B. L. Driver, P. J. Brown, & G. L. Peterson (Eds.), *Benefits of leisure* (pp. 73–89). State College, PA: Venture.

Ulrich, R. S., & Parsons, R. (1992). The influences of passive experiences with plants on human well-being and health. In D. Relf (Ed.), *The role of horticulture in human well-being and social development* (pp. 93–105). Portland, OR: Timber Press.

Ulrich, R. S., Simons, R. F., Losito, B. D., Fiorito, E., Miles, M. A., & Zelson, M. (1991). Stress recovery during exposure to natural and urban environments. *Journal of Environmental Psychology, 11*, 201–230.

Unger, D., & Wandersman, A. (1983). Neighboring and its role in block organizations: An exploratory report. *American Journal of Community Psychology, 11,* 291–300.

United Nations (1998). *Revision of the World Population Estimates and Projections.* Population Division, Department of Economic and Social Affairs. Retrieved February 1, 2000 from the World Wide Web: http://www.popin.org/pop1998/1.htm.

Ursano, R. J., McCaughey, B. G., & Fullerton, C. S. (1994). *Individual and community responses to trauma and disaster: The structure of human chaos.* Cambridge: Cambridge University Press.

Uzzell, D. L. (1995). The myth of the indoor city. *Journal of Environmental Psychology, 15,* 299–310.

Vallet, M. (1987). The effects of non-acoustic factors on annoyance due to traffic noise. In H. S. Koelega (Ed.), *Environmental annoyance: Characterization, measurement, and control* (pp. 371–382). Amsterdam: Elsevier Science Publishers.

Van de Vliert, E., Schwartz, S. H., Huismans, S. E., Hofstede, G., & Daan, S. (1999). Temperature, cultural masculinity, and domestic political violence: A cross-national study. *Journal of Cross-Cultural Psychology, 30,* 291–314.

Van Houten, R., & Nau, P. A. (1981). A comparison of the effects of posted feedback and increased police surveilance on highway speeding. *Journal of Applied Behavior Analysis, 14,* 261–271.

Van Houten, R., Nau, P. A., & Marini, Z. (1980). An analysis of public posting in reducing speeding behavior on an urban highway. *Journal of Applied Behavioral Analysis, 13,* 283–295.

van Vliet, W. (1981). Neighborhood evaluations by city and suburban children. *Journal of the American Planning Association, 47,* 458–466.

van Vliet, W. (1983). Families in apartment buildings: Sad stories for children? *Environment and Behavior, 156,* 211–234.

van Vliet, W. (1998). *The encyclopedia of housing.* Thousand Oaks, CA: Sage.

Vanetti, E. J., & Allen, G. L. (1988). Communicating environmental knowledge: The impact of verbal and spatial abilities on the production and comprehension of route directions. *Environment and Behavior, 20,* 667–682.

Varady, D. P., & Walker, C. C. (1999). Vouchering out distressed subsidized developments: Does moving lead to enhanced feelings of safety? *Environment and Behavior, 31,* 3–27.

Vaske, J. J., Donnelly, M. P., & Petruzzi, J. P. (1996). Country of origin, encounter norms, and crowding in a frontcountry setting. *Leisure Sciences, 18,* 161–176.

Vaughan, E. (1993). Individual and cultural differences in adaptation to environmental risks. *American Psychologist, 48,* 673–680.

Veitch, J. A. (1997). Revisiting the performance and mood effects of information about lighting and fluorescent lamp type. *Journal of Environmental Psychology, 17,* 253–262.

Veitch, J. A., & Gifford, R. (1996). Assessing beliefs about lighting effects on health, performance, mood, and social behavior. *Environment and Behavior, 28,* 446–470.

Veitch, J. A., & Gifford, R. (1997). Editors' introduction to the special issue: Behavioural origins and solutions of environmental problems. *Canadian Journal of Behavioural Science, 29,* 138–143.

Venturi, R. (1966). *Complexity and contradiction in architecture.* New York: Museum of Modern Art.

Verderber, S. (1986). Dimensions of person–window transactions in the hospital environment. *Environment and Behavior, 18,* 450–466.

Vergano, D. (1999). EMF researcher made up data, ORI says. *Science, 285,* 23–25.

Vincent, J. R., & Panayotou, T. (1997). . . . Or distraction? *Science, 276,* 53–55.

Vining, J., Daniel, T. C., & Schroeder, H. W. (1984). Predicting scenic values in forested residential landscapes. *Journal of Leisure Research, 16,* 124–135.

Vining, J., & Ebreo, A. (1990). What makes a recycler? A comparison of recyclers and non-recyclers. *Environment and Behavior, 22,* 55–73.

Vinsel, A., Brown, B., Altman, I., & Foss, C. (1980). Privacy regulation, territorial displays, and effectiveness of individual functioning. *Journal of Personality and Social Psychology, 39,* 1104–1115.

Virden, R. J., & Walker, G. J. (1999). Ethnic/racial and gender variations among meanings given to, and preferences for, the natural environment. *Leisure Sciences, 21,* 219–239.

Vogel, G. (1997). How to avoid running out of steam. *Science, 275,* 761.

Vogel, J. M., & Vernberg, E. M. (1993). Part 1: Children's psychological responses to disasters. *Journal of Clinical Child Psychology, 22,* 464–484.

Vogel, S. (1996). Farmers' environmental attitudes and behavior: A case study for Austria. *Environment and Behavior, 28,* 591–613.

von Mackensen, S., Bullinger, M., & Morfeld, M. (1999). The sick building syndrome as a subjective perception—Theoretical approach and assessment methods. *Zentralblatt fur Hygiene und Umweltmedizin, 202(2-4),* 243–248.

Wahlgren, D. R., Hovell, M. F., Meltzer, S. B., Hofstetter, C. R., & Zakarian, J. M. (1997). Reduction of environmental tobacco smoke exposure in asthmatic children. A 2-year follow-up. *Chest, 111,* 81–88.

Walden, T. A., Nelson, P. A., & Smith, D. E. (1981). Crowding, privacy, and coping. *Environment and Behavior, 13,* 205–224.

Walker, J. M. (1979). Energy demand behavior in a master-meter apartment complex: An experimental analysis. *Journal of Applied Psychology, 64,* 190–196.

Wall, G. (1995). General versus specific environmental concern: A Western Canadian case. *Environment and Behavior, 27,* 294–316.

Walmsley, D. J., & Lewis, G. J. (1989). The pace of pedestrian flows in cities. *Environment and Behavior, 21,* 123–150.

Wan, G. H., & Li, C. S. (1999). Dampness and airway inflammation and systemic symptoms in office building workers. *Archives of Environmental Health, 54,* 58–63.

Wandersman, A. (1981). A framework of participation in community organizations. *Journal of Applied Behavioral Science, 17,* 27–58.

Wandersman, A., & Hess, R. (Eds.) (1985). *Beyond the individual: Environmental approaches and prevention.* New York: Haworth.

Wang, T. H., & Katzev, R. D. (1990). Group commitment and resource conservation: Two field experiments on promoting recycling. *Journal of Applied Social Psychology, 20,* 265–275.

Wang, Y. T., & Wang, X. Y. (1997). Suicide and meteorological factors in Huhhot, Inner Mongolia. *Crisis, 18,* 115–117.

Wann, D. L., & Weaver, K. A. (1993). The relationship between interaction levels and impression formation. *Bulletin of the Psychonomic Society, 31,* 548–550.

Ward, L. M., & Russell, J. A. (1981). Cognitive set and the perception of place. *Environment and Behavior, 13,* 610–632.

Ward, L. M., & Suedfeld, P. (1973). Human responses to highway noise. *Environmental Research, 6,* 306–326.

Ward, S. L., Newcombe, N., & Overton, W. F. (1986). Turn left at the church or three miles north: A study of direction giving and sex differences. *Environment and Behavior, 18,* 192–213.

Wargocki, P., Wyon, D. P., Baik, Y. K., Clausen, G., & Fanger, P. O. (1999). Perceived air quality, sick building syndrome (SBS) symptoms and productivity in an office with two different pollution loads. *Indoor Air—International Journal of Indoor Air Quality & Climate, 9,* 165–179.

Warren, D. H. (1994). Self-localization on plan and oblique maps. *Environment and Behavior, 26,* 71–98.

Warren, D. H., & Scott, T. E. (1993). Map alignment in traveling multisegment routes. *Environment and Behavior, 25,* 643–666.

Warzecha, S., Fisher, J. D., & Baron, R. M.(1988). The equity-control model as a predictor of vandalism among college students. *Journal of Applied Social Psychology, 18,* 80–91.

Waterson, R. (1991). *The living house: An anthropology of architecture in South East Asia.* Singapore: Oxford University Press.

Webb, E. J., Campbell, D. T., Schwartz, R. D., Sechrest, L., & Grove, J. B. (1981). *Nonreactive measures in social sciences* (2nd ed.). Dallas: Houghton Mifflin.

Webb, W. M., & Worchel, S. (1993). Prior experience and expectation in the context of crowding. *Journal of Personality and Social Psychology, 65,* 512–521.

Weenig, M. W. H., Schmidt, T., & Midden, C. J. H. (1990). Social dimensions of neighborhoods and the effectiveness of information programs. *Environment and Behavior, 22,* 27–54.

Wehr, T. A., Jacobsen, F. M., Sack, D. A., Arendt, J., Tamarkin, L., & Rosenthal, N. E. (1986). Phototherapy of seasonal affective disorder. *Archives of General Psychiatry, 43,* 870–875.

Wehr, T. A., Sack, D. A., & Rosenthal, N. E. (1987). Seasonal affective disorder with summer depression and winter hypomania. *American Journal of Psychiatry, 144,* 1602–1603.

Weidemann, S., & Anderson, J. R. (1982). Residents' perceptions of satisfaction and safety: A basis for change in multifamily housing. *Environment and Behavior, 14,* 695–724.

Weil, R. J., & Dunsworth, F. A. (1958). Psychiatric aspects of disaster—a case history. Some experiences during the Springhill, Nova Scotia, mining disaster. *Canadian Psychiatric Association Journal, 3,* 11–17.

Weiner, F. H. (1976). Altruism, ambiance, and action: The effects of rural and urban rearing on helping behavior. *Journal of Personality and Social Psychology, 76,* 457–461.

Weinstein, C. S. (1979). The physical environment of the school: A review of the research. *Review of Educational Research, 49,* 577–610.

Weinstein, C. S. (1981). Classroom design as an external condition for learning. *Educational Technology, 21,* 12–19.

Weinstein, N. (1978). Individual differences in reactions to noise: A longitudinal study in a college dormitory. *Journal of Applied Psychology, 63,* 458–466.

Weinstein, N. D. (1976). The statistical prediction of environmental preferences. *Environment and Behavior, 8,* 611–626.

Weisman, J. (1981). Wayfinding and the built environment. *Environment and Behavior, 13,* 189–204.

Weisner, T., & Weibel, J. (1981). Home environments and family lifestyles in California. *Environment and Behavior, 13,* 417–460.

Weiss, B. (1983). Behavioral toxicology and environmental health science. *American Psychologist, 38,* 1174–1187.

Weiss, P. (1996). Industry group assails climate chapter. *Science, 272,* 1734.

Weisse, C. S., Pato, C. N., McAllister, C. G., Littman, R., Breier, A., Paul, S. M., & Baum, A. (1990). Differential effects of controllable and uncontrollable acute stress on lymphocyte proliferation and leukocyte percentages in humans. *Brain, Behavior, and Immunity, 4,* 339–351.

Weisz, J. R., Rothbaum, F. M., & Blackburn, T. C. (1984). Standing out and standing in: The psychology of control in America and Japan. *American Psychologist, 39,* 955–969.

Wellens, A. R., & Goldberg, M. L. (1978). The effects of interpersonal distance and orientation upon the perception of social relationships. *Journal of Psychology, 99,* 39–47.

Wellman, J. D., & Buhyoff, G. J. (1980). Effects of regional familiarity on landscape preferences. *Journal of Environmental Management, 11,* 105–110.

Wener, R. (1977). Non-density factors in the perception of crowding. *Dissertation Abstracts International, 37D,* 3569–3570.

Wener, R., Frazier, F. W., & Farbstein, J. (1985). Three generations of evaluation and design of correctional facilities. *Environment and Behavior, 17,* 71–95.

Wener, R., Frazier, F. W., & Farbstein, J. (1987, June). Building better jails. *Psychology Today, 21,* 40–44, 48–49.

Wener, R., & Kaminoff, R. D. (1983). Improving environmental information: Effects of signs on perceived crowding and behavior. *Environment and Behavior, 15,* 3–20.

Wener, R., & Keys, C. (1988). The effects of changes in jail population densities on crowding, sick call, and spatial behavior. *Journal of Applied Social Psychology, 18,* 852–866.

Werner, C. M. (1987). Home interiors: A time and place for interpersonal relationships. *Environment and Behavior, 19,* 169–179.

Werner, C. M., Altman, I., & Oxley, D. (1985). Temporal aspects of homes: A transactional perspective. In I. Altman & C. M. Werner (Eds.), *Home environments* (pp. 1–32). New York: Plenum.

Werner, C. M., Brown, B. B., & Damron, G. (1981). Territorial marking in the game arcade. *Journal of Personality and Social Psychology, 41,* 1094–1104.

Werner, C. M., & Makela, E. (1998). Motivations and behaviors that support recycling. *Journal of Environmental Psychology, 18,* 373–386.

Werner, C. M., Peterson-Lewis, S., & Brown, B. B. (1989). Inferences about homeowners' sociability: Impact of Christmas decorations and other cues. *Journal of Environmental Psychology, 9,* 279–296.

West, P. C. (1982). Effects of user behavior on the perception of crowding in backcountry forest recreation. *Forest Science, 28,* 95–105.

Weyant, J. M. (1978). Effects of mood states, costs, and benefits on helping. *Journal of Personality and Social Psychology, 36,* 1169–1176.

White, M. (1975). Interpersonal distance as affected by room size, status, and sex. *Journal of Social Psychology, 95,* 241–249.

White, M., Kasl, S. V., Zahner, G. E. P., & Will, J. C. (1987). Perceived crime in the neighborhood and mental health of women and children. *Environment and Behavior, 19,* 588–613.

White, S. M., & Rotton, J. M. (1998). Type of commute, behavioral aftereffects, and cardiovascular activity: A field experiment. *Environment and Behavior, 30,* 763–780.

Whyte, W. H. (1980). *The social life of small urban spaces.* New York: The Conservation Foundation.

Wicker, A. W. (1969). Size of church membership and members' support of church behavior settings. *Journal of Personality and Social Psychology, 13,* 278–288.

Wicker, A. W. (1973). Undermanning theory and research: Implications for the study of psychological and behavorial effects of excess populations. *Representative Research in Social Psychology, 4,* 185–206.

Wicker, A. W. (1979). *An introduction to ecological psychology.* Monterey, CA: Brooks/Cole.

Wicker, A. W. (1987). Behavior settings reconsidered: Temporal stages, resources, internal dynamics, context. In D. Stokols & I. Altman (Eds.), *Handbook of environmental psychology* (Vol. II, pp. 613–653). New York: Wiley-Interscience.

Wicker, A. W., & Kauma, C. (1974). Effects of a merger of a small and a large organization on members' behaviors and experiences. *Journal of Applied Psychology, 59,* 24–30.

Wicker, A. W., & Kirmeyer, S. (1976). From church to laboratory to national park: A program of research on excess and insufficient populations in behavior settings. In S. Wapner, S. B. Cohen, & B. Kaplan (Eds.), *Experiencing the environment* (pp. 157–185). New York: Plenum.

Wicker, A. W., Kirmeyer, S. L., Hanson, L., & Alexander, D. (1976). Effects of manning levels on subjective experiences, performance, and verbal interaction in groups. *Organizational Behavior and Human Performance, 17,* 251–274.

Wicker, A. W., McGrath, J. E., & Armstrong, G. E. (1972). Organization size and behavior setting capacity as determinants of member participation. *Behavioral Science, 17,* 499–513.

Wicker, A. W., & Mehler, A. (1971). Assimilation of new members in a large and a small church. *Journal of Applied Psychology, 55,* 151–156.

Wiesenfeld, E. (1997a). Construction of the meaning of a barrio house: The case of a Caracas barrio. *Environment and Behavior, 29,* 34–63.

Wiesenfeld, E. (1997b). From individual need to community consciousness: The dialectics between land appropriation and eviction threat (A case study of a Venezuelan "barrio"). *Environment and Behavior, 29,* 198–212.

Wieslander, G., Norback, D., Nordstrom, K., Walinder, R., & Venge, P. (1999). Nasal and ocular symptoms, tear film stability and biomarkers in nasal lavage, in relation to building-dampness and building design in hospital. *International Archives of Occupational & Environmental Health, 72,* 451–61.

Wilding, J., & Mohindra, N. (1980). Effects of subvocal suppression, articulating aloud and noise on sequence recall. *British Journal of Psychology, 71,* 247–261.

Williams, B. L., Brown, S., & Greenberg, M. (1999). Determinants of trust perceptions among residents surrounding the Savannah River Nuclear Weapons Site. *Environment and Behavior, 31,* 354–371.

Williams, D. R., Patterson, M. E., Roggenbuck, J. W., & Watson, A. E. (1992). Beyond the commodity metaphor: Examining emotional and symbolic attachment to place. *Leisure Sciences, 14,* 29–46.

Williams, G. W., McGinnis, M. Y., & Lumia, A. R. (1992). The effects of olfactory bulbectomy and chronic psychosocial stress on serum glucocorticoids and sexual behavior in female rats. *Physiology and Behavior, 52,* 755–760.

Williams, J. L., Rodeheaver, D. G., & Huggins, D. W. (1999). A comparative evaluation of a new generation jail. *American Journal of Criminal Justice, 23,* 223–246.

Willis, F. N. (1966). Initial speaking distance as the function of the speakers' relationship. *Psychonomic Science, 5,* 221–222.

Willner, P., & Neiva, J. (1986). Brief exposure to uncontrollable but not to controllable noise biases the retrieval of information from memory. *British Journal of Clinical Psychology, 25,* 93–100.

Wills, T. A. (1981). Downward comparison principles in social psychology. *Psychological Bulletin, 90,* 245–271.

Wilson, E. O. (1975). *Sociobiology.* Cambridge, MA: Harvard University Press.

Wilson, E. O. (1993). Biophilia and the conservation ethic. In S. R. Kellert & E. O. Wilson (Eds.), *The biophilia hypothesis* (pp. 31–41). Washington, DC: Island Press.

Wilson, G., & Baldassare, M. (1996). Overall "sense of community" in a suburban region: The effects of localism, privacy, and urbanization. *Environment and Behavior, 28*, 27–43.

Wilson, M. (1999). *Arcosanti archetype: The rebirth of cities by Renaissance thinker Paolo Soleri.* Fountain Hills, AZ: Marie Wilson Enterprises, Inc.

Wilson, M. A. (1996). The socialization of architectural preference. *Journal of Environmental Psychology, 16*, 33–44.

Wineman, J. D. (1982). Office design and evaluation: An overview. *Environment and Behavior, 14*, 271–298.

Wineman, J. D. (1986). *Behavioral issues in office design.* New York: Van Nostrand Reinhold.

Winkel, G. H. (1987). Implications of environmental context for validity assessments. In D. Stokols & I. Altman (Eds.), *Handbook of environmental psychology* (Vol. I, pp. 71–98). New York: Wiley-Interscience.

Winkler, R. C., & Winnett, R. A. (1982). Behavioral interventions in resource management. *American Psychologist, 37*, 421–435.

Winneke, G., & Kastka, J. (1987). Comparison of odour-annoyance data from different industrial sources: Problems and implications. In H. S. Koelega (Ed.), *Environmental annoyance: Characterization, measurement, and control* (pp. 129–138). Amsterdam: Elsevier Science Publishers.

Winnett, R. A., Hatcher, J., Leckliter, I., Ford, T. R., Fishback, J. F., Riley, A. W., & Love, S. (1981). *The effects of videotape modeling and feedback on residential comfort, the thermal environment and electricity consumption: Winter and summer studies.* Unpublished manuscript. Department of Psychology, Virginia Polytechnic Institute and State University.

Winnett, R. A., Kagel, J. H., Battalio, R. C., & Winkler, R. C. (1978). Effects of monetary rebates, feedback and information on residential energy conservation. *Journal of Applied Psychology, 63*, 73–78.

Winnett, R. A., Leckliter, I. N., Chinn, D. E., & Stahl, B. (1984). Reducing energy consumption: The long-term effects of a single

TV program. *Journal of Communication, 34*, 37–51.

Winett, R. A., Leckliter, I. N., Chinn, D. E., Stahl, B. N., & Love, S. Q. (1985). The effects of videotape modeling via cable television on residential energy conservation. *Journal of Applied Behavior Analysis, 18*, 33–44.

Winnett, R. A., Neale, M. S., & Grier, H. C. (1979). The effects of self-monitoring and feedback on residential electricity consumption. *Journal of Applied Behavior Analysis, 12*, 173–184.

Wirth, I. (1938). Urbanism as a way of life. *American Journal of Sociology, 44*, 1–24.

Wohlwill, J. F. (1966). The physical environment: A problem for a psychology of stimulation. *Journal of Social Issues, 22*, 29–38.

Wohlwill, J. F. (1970). The emerging discipline of environmental psychology. *American Psychologist, 25*, 303–312.

Wohlwill, J. F. (1974). Human response to levels of environmental stimulation. *Human Ecology, 2*, 127–147.

Wolf, N., & Feldman, E. (1991). *Plastics: America's packaging dilemma.* Washington, DC: Island Press.

Womble, P., & Studebaker, S. (1981). Crowding in a national park campground: Katmai National Monument in Alaska. *Environment and Behavior, 13*, 557–573.

Wong, C. Y., Sommer, R., & Cook, E. J. (1992). The soft classroom 17 years later. *Journal of Environmental Psychology, 12*, 337–343.

Wood, J. M., Bootzin, R. R., Rosenhan, D., & Nolen-Hoeksema, S. (1992). Effects of the 1989 San Francisco earthquake on frequency and content of nightmares. *Journal of Abnormal Psychology, 101*, 219–224.

Wood, J., Lugg, D. J., Hysong, S. J., & Harm, D. L. (1999). Psychological changes in hundred-day remote Antarctic field groups. *Environment and Behavior, 31*, 299–337.

Woods, J. E. (1988). Recent developments for heating, cooling, and ventilating buildings. *State of the Art Reviews* (Stockholm, Sweden, Swedish Council for Building Research, Healthy Buildings), *1*, 99–197.

Woodson, P. P., Buzzi, R., Nil, R., & Battig, K. (1986). Effects of smoking on vegetative reactivity to noise in women. *Psychophysiology, 23*, 272–282.

Woodward, N. J., & Wallston, B. S. (1987). Age and

health care beliefs: Self-efficacy as a mediator of low desire for control. *Psychology and Aging, 2,* 3–8.

Wooldredge, J. D., & Winfree, L. T. (1992). An aggregate-level study of inmate suicides and deaths due to natural causes in U.S. jails. *Journal of Research in Crime and Delinquency, 29,* 466–479.

Worchel, S., & Brown, E. H. (1984). The role of plausibility in influencing environmental attributions. *Journal of Experimental Social Psychology, 20,* 86–96.

Worchel, S., & Teddlie, C. (1976). The experience of crowding: A two-factor theory. *Journal of Personality and Social Psychology, 34,* 36–40.

Wortman, C. B., & Brehm, J. W. (1975). Responses to uncontrollable outcomes: An integration of reactance theory and the learned helplessness model. In L. Berkowitz (Ed.), *Advances in experimental social psychology* (Vol. 8, pp. 277–336). New York: Academic Press.

Wright, K. M., Ursano, R. J., Bartone, P. T., & Ingraham, L. H. (1990). The shared experience of catastrophe: An expanded classification of the disaster community. *American Journal of Orthopsychiatry, 60,* 35–42.

Yamagishi, T. (1986). The provision of a sanctioning system as a public good. *Journal of Personality and Social Psychology, 51,* 110–116.

Yamamoto, T., Sawada, H., Minami, H., Ishii, S., & Inoue, W. (1992). Transition from the university to the workplace. *Environment and Behavior, 24,* 189–205.

Yancey, W. L. (1972). Architecture, interaction, and social control: The case of a large scale housing project. In J. F. Wohlwill & D. H. Carson (Eds.), *Environment and the social sciences: Perspectives and applications* (pp. 126–136). Washington, DC: American Psychological Association.

Yinon, Y., & Bizman, A. (1980). Noise, success, and failure as determinants of helping behavior. *Personality and Social Psychology Bulletin, 6,* 125–130.

Young, M., & Willmott, P. (1957). *Family and kinship in East London.* Baltimore: Penguin.

Yule, W., & Williams, R. (1990). Post-traumatic stress reactions in children. *Journal of Traumatic Stress, 3,* 279–295.

Zajonc, R. B. (1984). On the primacy of affect. *American Psychologist, 39,* 117–123.

Zajonc, R. B., Murphy, S. T., & Inglehart, M. (1989). Feeling and facial efference: Implications of the vascular theory of emotion. *Psychological Review, 96,* 395–416.

Zakay, D., Hayduk, L. A., & Tsal, Y. (1992). Personal space and distance misperception: Implications of a novel observation. *Bulletin of the Psychonomic Society, 30,* 33–35.

Zeisel, J. (1975). *Sociology and architectural design* (Russell Sage Social Science Frontiers Series, No. 6). New York: Free Press.

Zeisel, J. (1981). *Inquiry by design: Tools for environment–behavior research.* Monterey, CA: Brooks/Cole.

Zhou, J., Oldham, G. R., & Cummings, A. (1998). Employee reactions to the physical work environment: The role of childhood residential attributes. *Journal of Applied Social Psychology, 28,* 2213–2238.

Zika, S., & Chamberlain, K. (1987). Relation of hassles and personality to subjective well-being. *Journal of Personality and Social Psychology, 53,* 155–162.

Zillmann, D. (1979). *Hostility and aggression.* Hillsdale, NJ: Erlbaum.

Zillmann, D. (1983). Arousal and aggression. In R. G. Geen & E. Donnerstein (Eds.), *Aggression: Theoretical and empirical reviews* (Vol. 1, pp. 75–101). New York: Academic Press.

Zimbardo, P. G. (1969). The human choices: Individuation, reason, and order versus deindividuation, impulse, and chaos. In W. J. Arnold & D. Levine (Eds.), *Nebraska Symposium on Motivation* (pp. 237–307). Lincoln: University of Nebraska Press.

Zimmer, K., & Ellermeier, W. (1998). Short form of a questionnaire measuring individual noise sensitivity. *Umweltpsychologic, 2,* 54–63.

Zimmer, K., & Ellermeier, W. (1999). Psychometric properties of four measures of noise sensitivity: A comparison. *Journal of Environmental Psychology, 19,* 295–302.

Zimring, C., & Reizenstein, J. (1980). Post occupancy evaluation: An overview. *Environment and Behavior, 12,* 429–450.

Zimring, C., Carpman, J. R., & Michelson, W. (1987). Design for special populations: Mentally retarded persons, children, hospital visitors. In D. Stokols & I. Altman (Eds.), *Handbook of environmental psychology* (Vol. 2, pp. 919–949). New York: Wiley-Interscience.

Zlutnick, S., & Altman, I. (1972). Crowding and

human behavior. In J. Wohlwill & D. Carson (Eds.), *Environment and the social sciences: Perspectives and applications* (pp. 44–58). Washington, DC: American Psychological Association.

Zube, E. H., Pitt, D. G., & Evans, G. W. (1983). A lifespan developmental study of landscape assessment. *Journal of Environmental Psychology, 3,* 115–128.

Zube, E. H., Sell, J. L., & Taylor, J. G. (1982). Landscape perception: Research, application and theory. *Landscape Planning, 9,* 1–33.

Zube, E. H., Vining, J., Law, C. S., & Bechtel, R. B. (1985). Perceived urban residential quality: A cross-cultural bimodal study. *Environment and Behavior, 17,* 327–350.

Zubek, J. P. (Ed.). (1969). *Sensory deprivation: Fifteen years of research.* New York: Appleton-Century-Crofts.

Zuckerman, M. (1979). *Sensation seeking: Beyond the optimal level of arousal.* Hillsdale, NJ: Erlbaum.

Zuzanek, J., Beckers, T., & Peters, P. (1998). The "hurried leisure class" revisited: Dutch and Canadian trends in the use of time from the 1970s to the 1990s. *Leisure Studies, 17,* 1–19.

Zuzanek, J., Robinson, J. P., & Iwasaki, Y. (1998). The relationships between stress, health, and physically active leisure as a function of lifecycle. *Leisure Sciences, 20,* 253–275.

Zuzanek, J., & Smale, B. J. A. (1997). More work–less leisure? Changing allocations of time in Canada, 1981 to 1992. *Society and Leisure, 20,* 73–106.

Zweigenhaft, R. (1976). Personal space in the faculty office: Desk placement and the student. *Journal of Applied Psychology, 61,* 529–532.

COPYRIGHT ACKNOWLEDGMENTS

NAME INDEX

M

SUBJECT INDEX